STRUCTURAL ANALYSIS

VOLUME I

STRUCTURAL ANALYSIS

VOLUME I

By

Dr. R. Vaidyanathan
B.E. (Hons), M.Sc. (Engg.), Ph.D., F.I.E., M.I.S.T.E.
Formerly Principal, Government College
of Engineering, Tirunelveli
Tamil Nadu

Dr. P. Perumal
B.E., (Distn.), M.E. (Struct.), Ph.D., F.I.E., M.I.S.T.E.
(Professor and Head of Civil Engineering (Retd.)
Government College of Engineering Salem)
Professor, Dr. Mahalingam College of
Engineering and Technology, Pollachi
Tamil Nadu

LAXMI PUBLICATIONS (P) LTD
(An ISO 9001:2015 Company)
BENGALURU • CHENNAI • GUWAHATI • HYDERABAD • JALANDHAR
KOCHI • KOLKATA • LUCKNOW • MUMBAI • RANCHI
NEW DELHI

STRUCTURAL ANALYSIS-I

Printed and bound in India
Typeset at : Goswami Associates, Delhi
First Edition : 2004, *Fourth Edition* : 2016, 2017, 2018, *Edition* : 2019
ISBN : 978-81-318-0781-1

Branches

Bengaluru	080-26 75 69 30	
Chennai	044-24 34 47 26,	24 35 95 07
Guwahati	0361-254 36 69,	251 38 81
Hyderabad	040-27 55 53 83,	27 55 53 93
Jalandhar	0181-222 12 72	
Kochi	0484-237 70 04,	405 13 03
Kolkata	033-22 27 43 84	
Lucknow	0522-220 99 16	
Mumbai	022-24 93 12 61	
Ranchi	0651-220 44 64	

PUBLISHED IN INDIA BY

Laxmi Publications (P) Ltd.
(An ISO 9001:2015 Company)
113, GOLDEN HOUSE, DARYAGANJ,
NEW DELHI - 110002, INDIA
Telephone : 91-11-4353 2500, 4353 2501
Fax : 91-11-2325 2572, 4353 2528
www.laxmipublications.com info@laxmipublications.com

C—00198/019/05
Printed at: Ajit Printing Press, Delhi.

CONTENTS

PREFACE TO THE FOURTH EDITION

PREFACE TO THE FOURTH EDITION

We feel happy to note the continuous increase in the reception of the third edition of Structural Analysis, Volume I, the thoroughly revised book, now in its fourth edition. Improvements in figures, simplifying calculations, addition of many problems on slope deflection method, moment distribution method, matrix flexibility method and matrix stiffness method have been done. We have made the contents appropriate for a wide range of readers so that any student or professional would find the desired contents.

M/s Laxmi Publications have been very tolerant and patient with the Authors. Our thanks are due to M/s Laxmi Publications whose professional comments and suggestions have made value addition to this book. The authors owe the publishers a lot for their flexibility and accommodation.

The continued reinforcing feed backs and encouraging words from Professors and students all over, in various forms, acts as catalysts in our effort.

—Authors

PREFACE TO THE FIRST EDITION

The authors have great pleasure in presenting to the student community the first volume of their book on Structural Analysis. This book meets the syllabus for Structural Analysis I of Anna University, Tamil Nadu. However, this also will serve the Students and Teachers of all other Universities, AMIE students and practicing Engineers. The objective of this book is to present the material in a simple way. The authors have taken care to answer all the doubts that they themselves had when they were students of Civil Engineering.

This volume contains five chapters namely, Deflection of determinate structures, Slope Deflection method, Moment Distribution method (including Naylor's Method), Matrix Flexibility method and Matrix Stiffness method. All the chapters are presented in a simple and lucid style with a lot of numerical examples and illustrative sketches for the benefit of the teachers and students.

A large number of problems are also available at the end of each chapter to enable the student to test his/her understanding of the subject. The authors have freely consulted many good books on the subject and the help is gratefully acknowledged.

The authors are aware that the book is not perfect. They would appreciate if errors discovered, if any, are brought to their attention. Suggestions from the readers for improvement of this book will be gratefully acknowledged. The next edition is sure to take into account all such suggestions and would be more free from errors and omissions.

The authors are very confident that the staff and student community who are in need of a single, simple book covering the entire syllabus with simple presentation will get the fullest benefits.

—Authors

# ACKNOWLEDGEMENTS

We are not sure we saw farther than others, but we had a lot of giants on whose slippery shoulders we could stand to peek into the intricacies of Structural Analysis. We thank them all.

First and foremost, we acknowledge the following authors whose books enlightened us in several aspects.

Punmia, Jain & Jain, Gupta Pandit & Gupta, Prakash Rao, Ramamrutham, Reddy, Khurmi, Manicka selvam, Rubinstein, Negi and Jangid, Vazirani & Ratwani and Kinney.

We acknowledge help from the following students and research scholars in doing numerical verifications, compilation of problems and fair copying.

Ramalakshmi, Shanmugaraj, Kolanjinathan, Balamurugan, Prabhu, Manimegalai, Saravanan, Udaya, Shirly, Mohanraj, Thiruvannamalai and Shaji.

The following were patient enough to unscramble our maze of numbers and sketches and gave them font and figures. We are really indebted to *M.S. Shahul, M.S. Computers, Melappalayam, Tirunelveli, M.S. Sudharson, Kural Computers, Samathanamuram, Tirunelveli & L. Murugesan, DTP Designer, Mepco Schlenk Engineering College, Sivakasi.*

M/s Laxmi Publications have been really very quick, prompt, understanding and considerate when we were regularly passing deadlines and overhauling our own texts. We are thankful to them for their accommodative partnership and hope to deserve the same in future.

We both are greatly beholden to Seetha and Saraswathi, our wives and our family for putting up with our night owl habits and irregular mealtimes. They have suffered a lot and we are trying to prove they have not suffered in vain.

A lot of people gave us words of encouragement which sustained us in our grind. We say, "thank you folks ; the words may not have cost you much, but they were invaluable to us".

Several sets of students have served as sounding boards for us, the brilliant, the slow, the inattentive and the indifferent. But all of them have helped us decide what kind of presentation would be effective.

We therefore dedicate this book to the student community.

—Authors

डॉ एन लक्ष्मणन
Dr. N. Lakshmanan
निदेशक
Director

स्ट्रक्चरल इंजीनियरिंग रिसर्च सेंटर-मद्रास
STRUCTURAL ENGINEERING RESEARCH CENTRE -MADRAS
वैज्ञानिक तथा औद्योगिक अनुसंधान परिषद्
(Council of Scientific & Industrial Research)
डा.सं./P.B. No. 8287, तरमणि, Taramani,
चेन्नै /Chennai - 600 113, भारत/INDIA

FOREWORD

The two volume publication on Structural Analysis by Dr. R.Vaidyanathan and his disciple Dr. P. Perumal is totally geared to meet the syllabus of Anna University.

The first volume deals comprehensively with analytical methods while volume two deals with moving loads and analysis of different structural configurations.

The speciality of the book is that it addresses the students as if a teacher would address them in a classroom.

Thanks to several decades of teaching Structural Engineering subjects in almost all the Government colleges in the state, the authors have had access to the students' minds and most of their doubts and difficulties and have used this book to answer them and clarify many a mystery relating to the behaviour of various structural shapes.

They have rendered the mathematical parts to painfree simplicity and concentrated on all practical applications of every structural shape like arches, suspension cable, rigid frames and curved beams.

I am sure students of Civil Engineering all over the world find these volumes immensely helpful in aiding their comprehension of structural behaviour and analysis.

(N. Lakshmanan)

RINA QUACER
ISO 9001

Phone : (O) 2254 2139 (R) 2550 7550
Fax : 91-44-2254 1508 / 2254 1973
Grams : SERCENTER
E-mail : nlaxman@sercm.csir.res.in
director@sercm.csir.res.in
director@cssercm.ren.nic.in
http://www.sercm.org

CHAPTER 0

INTRODUCTION

0.1 HISTORICAL

The much maligned and misinterpreted ***vasthu sasthra*** is really the mother of structural analysis. Inherited without the benefit of basic principles, the procedure oriented sasthra has fallen into the hands of the ignorant and the mercenary and has ceased to make scientific sense.

But the fact remains that several centuries ago massive temples, palaces and stupas have been built in India and these could not have been achieved without scientific analysis and pragmatic application.

The ***pyramids*** of Egypt, Qutab minar and the ***Colossus*** are standing proofs that engineers of bygone ages have developed diverse methods of reckoning loads, transfer of forces, concept of stresses and techniques of spanning distances and areas.

0.2 KEY ISSUES

In structures the key question is "How much ***stress*** is induced and where, due to given ***loads***"? The structure can be a column, a beam, a floor slab, an arch, a truss or whatever (Fig. 0.1).

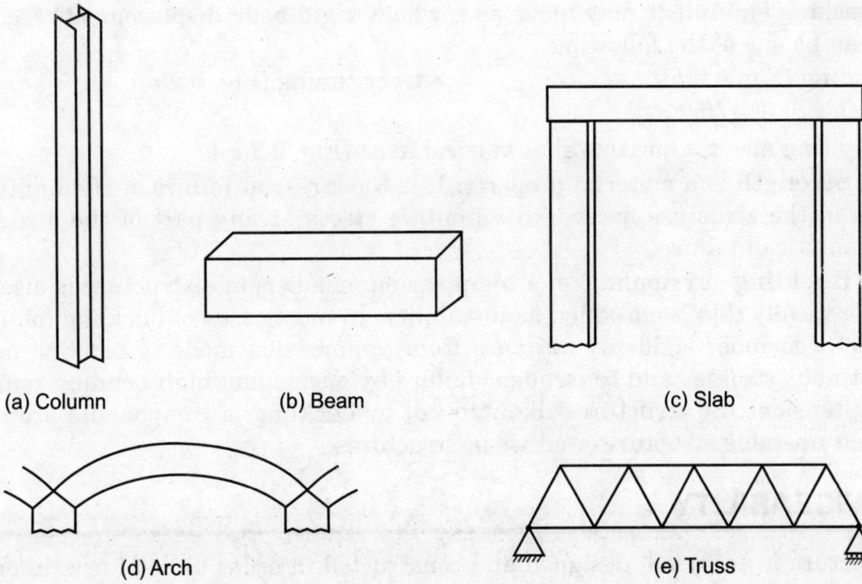

Fig. 0.1

So when we analyse a structure we apply the basic principles of mechanics and for any structure geometry and applied forces we proceed to determine ***reactive forces, internal forces, deformations, deflections*** etc. with the aim of determining ***if the structure is safe*** or not.

0.3 STRENGTH AND STABILITY

A structure must be both ***strong*** and ***stable***. If it is not strong it will break (or snap or crack or crush) and collapse. This can happen when the material is not strong enough to take the stress coming on it.

If a structure is not stable it will move or fall or slide; even if the material of the component parts has not broken (or failed).

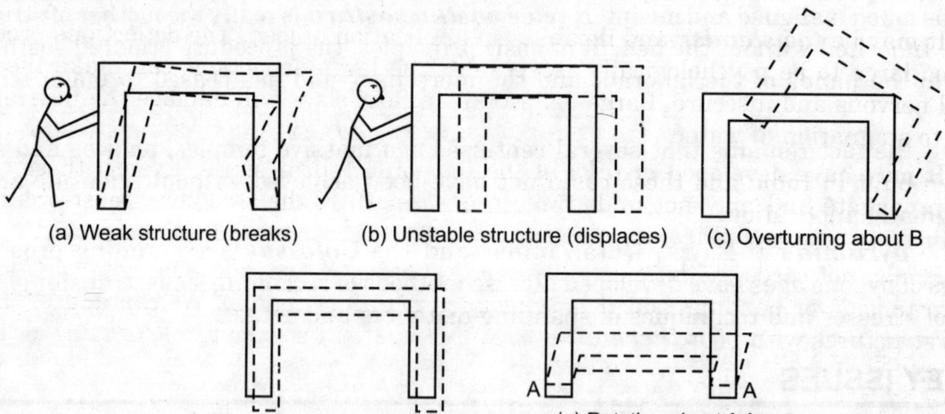

(a) Weak structure (breaks) (b) Unstable structure (displaces) (c) Overturning about B

(d) Sinking (e) Rotating about A A

Fig. 0.2

0.3.1. The body of an ***unstable structure*** may remain intact without any yielding, crushing, cracking etc. But it may move as a whole (rigid body displacement) Fig. 0.2. This movement can be one of the following:

- Sliding [Fig. 0.2(*b*)]
- Over tuning [Fig. 0.2(*c*)]
- Sinking [Fig. 0.2(*d*)]
- Rotating about a horizontal or vertical axis [Fig. 0.2(*e*)].

0.3.2. Strength is a material property. It is the stress at failure or the limiting stress. If the stress in the structure exceeds the limiting stress at any part of the structure, this condition is a state of failure.

0.3.3. Buckling or crippling (of a compression member) in a structure is also a form of failure. Theoretically this is classified as instability. In most cases of buckling (of a compression member) a member suddenly switches from compression mode to bending mode. This leads to enormous stresses and consequent failure by developing high bending compressions and bending tensions. So structures likely to fail by buckling of components are borderline cases between unstable structures and weak structures.

0.4 SERVICEABILITY

A structure is analysed, designed and constructed in order to be of use (in order to be ***serviceable***). There can be cases in which a stable and strong structure can become unserviceable (Fig. 0.3).

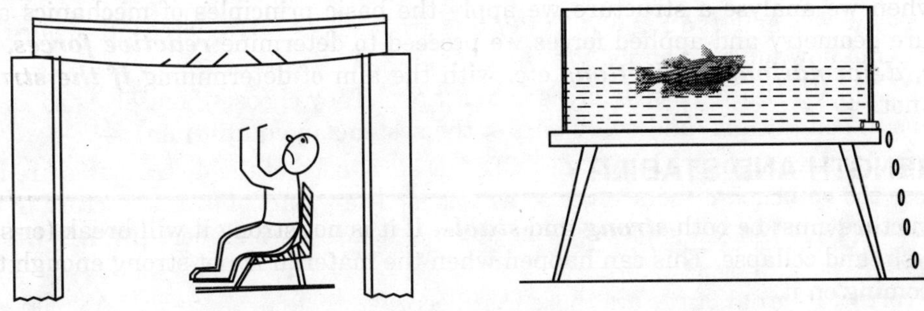

(a) Sags, cracks (b) Leaks

Fig. 0.3. Unserviceable structures.

(*i*) It may be ***too slender*** and flexible. On application of loads the deflections produced may be too large to be psychologically acceptable. The large deflections can make the occupants feel nervous and insecure. Large deflections in floor slabs will damage the flooring and also lead to stagnation of water.

(*ii*) It may have developed ***cracks*** in the tension or shear zones. Even if these cracks may not propagate and may not undermine the strength of the structure, most occupants/ users of such structures will feel it unsafe to use them. Cracks may shelterpests and vermin.

(*iii*) It may not serve the intended purpose. A leaking water tank may not overturn, nor collapse in tension or bending. But it is unusable (or unserviceable). We can add to the list buildings/structures with poor ***thermal/acoustic*** insulation, poor natural ***lighting*** or excessive ***glare.***

0.5 DETERMINACY

0.5.1. Forces in members of some structures can be computed simply by applying the conditions of statics, namely, $\Sigma H = 0$, $\Sigma V = 0$ and $\Sigma M = 0$ (for a 2D case). External and internal forces that can keep the structure in equilibrium under applied loads are unique. They are the only possible forces developed. Such structures are called ***statically determinate*** or simply ***determinate*** (Fig. 0.4).

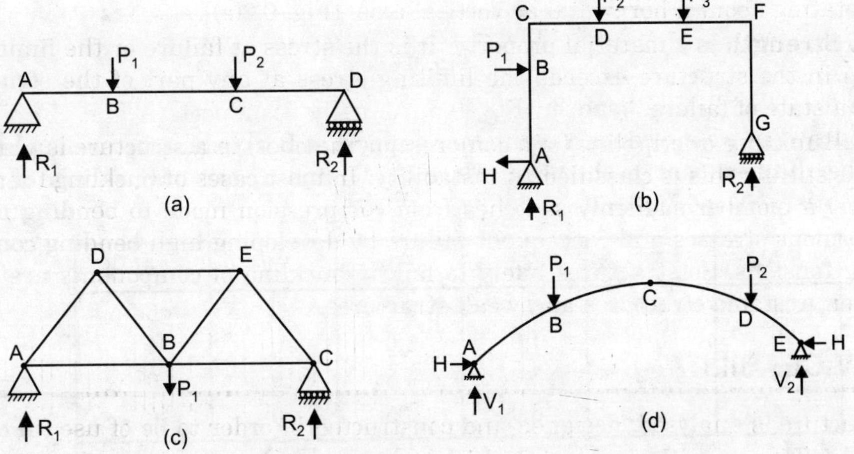

Fig. 0.4. Determinate structures.

In cases *a*, *b*, *c* and *d* in Fig. 0.4, P_1, P_2 etc. are loads or active forces. It is our choice to apply them, and how much to apply. We can *choose* P_1, P_2, P_3, ...

R_1, R_2, V_1, V_2, H are passive forces or reactions. They choose themselves so as to keep the system/s in equilibrium (and every part of the systems in equilibrium).

Forces, moments, shears in AB, BC, CD, DE etc of each of the 4 structures are internal forces. These too will choose themselves so as to keep every part of each structure in equilibrium. For example, in Fig. 0.4 (*c*), joint B has internal forces in 4 members BA, BD, BE and BC plus the load P.

The correct internal forces are those that keep junction B in equilibrium (Fig. 0.5).

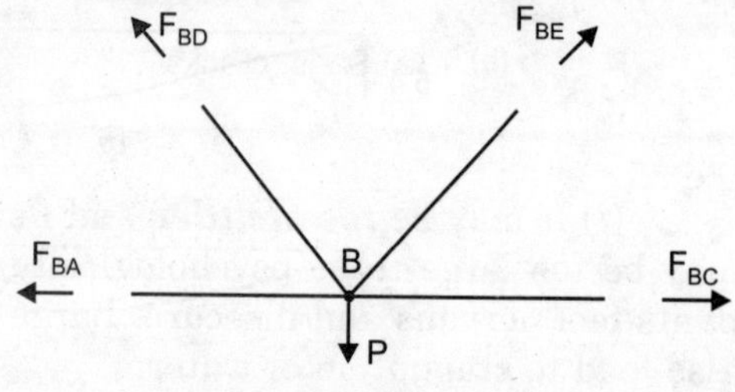

Fig. 0.5

Determinate structures are good for the analyst (Often, only to the analyst). The calculations are simple. When there is change of temperature, the structure may change its shape a little, but no change occurs in the reactions or internal forces. When one or more of the supports move or sink a little, no change occurs in the internal/reactive forces.

0.5.2. Indeterminate structures are not strictly indeterminate, but only ***statically indeterminate.*** This happens to structures when we provide more supports than the minimum required for stability. Learned people prefer to call these ***hyperstatic structures.*** These are also called ***redundant structures,*** since there is a redundancy (excess) of supports or members.

In pin jointed frames, redundancy caused by too many members is called ***internal redundancy***. Then there is ***external redundancy*** caused by too many supports. When we introduce additional supports/members, we generally ensure more safety and more work (in analysis).

Redundant beams cannot be solved by conditions of equilibrium alone.

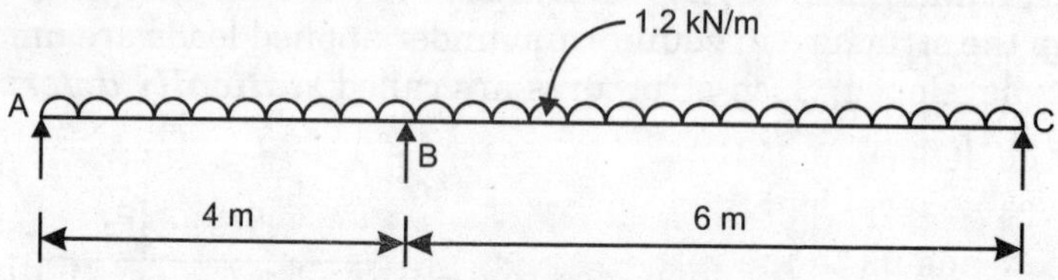

Fig. 0.6

Take for example the beam in Fig. 0.6. Actually 2 supports A and C are enough to ensure stability. But we introduce a redundant support at B to reduce forces on A and C and to reduce the bending moment on the beam. Here R_A, R_B and R_C can have several sets of values that would satisfy $\Sigma V = 0$, $\Sigma M = 0$.

S.No.	R_A	R_B	R_C	ΣV	ΣM
1.	6.0	0	6.0	0	0
2.	5.4	1.0	5.6	0	0
3.	4.2	3.0	4.8	0	0
4.	2.4	6.0	3.6	0	0

There can be several such combinations. Which of them is the right set? This is like looking for a man with blue jeans and yellow tee shirt in Chennai central. You find dozens of them fitting the description. So, the conditions of statics are not enough for us to find R_A, R_B and R_C the passive forces. We have to use ***deformation conditions*** in addition.

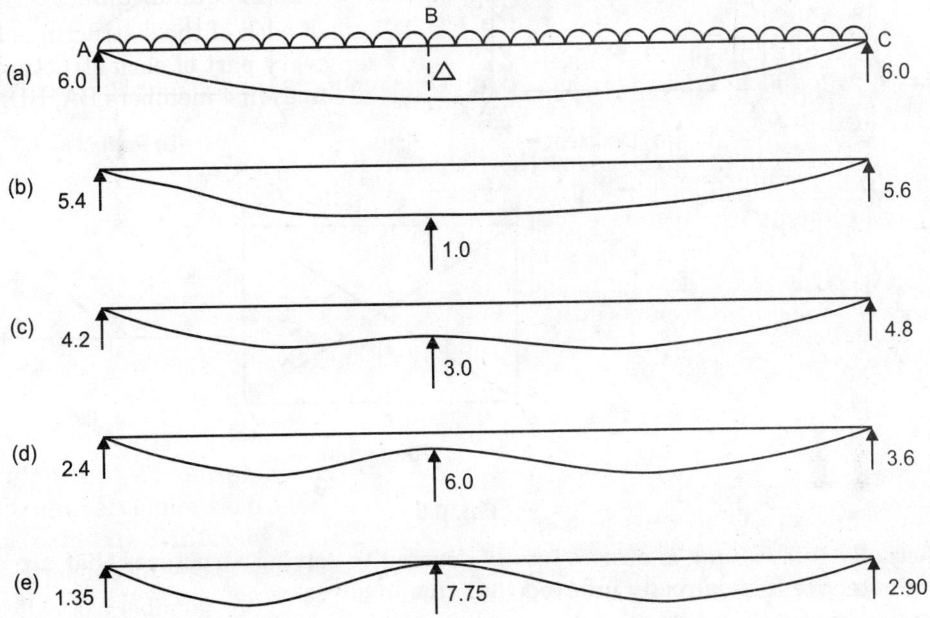

Fig. 0.7

Let us have a close look at Fig. 0.7.

Figure 0.7 (*a*) shows the beam with 2 supports only. It develops a deflection Δ as shown. Let us say we want to reduce the deflection by supporting the beam at B. We take a prop and push the beam up at B. And up, and up, till the level of B is the same as those of A and C. This is how we get several answer sets for R_A, R_B and R_C. The correct answer is one where R_B is able to arrest the deflection at B completely. This is what would happen if the supports were initially at the same level and the load transferred to the beam.

This also leads us to the inference that if the supports are (initially) not at the same level, or if the supports should sink, the reactions (and hence the bending moments in the beam) would alter.

Internal redundancy occurs in pin jointed frames. This is, when we have more members than the minimum required for the frame to keep its shape (Fig. 0.8).

Figure 0.8 (*a*) does not have enough members. It is deficient. It cannot act as a truss. It cannot keep its shape. It is a mechanism.

Figure 0.8 (*b*) is perfect. It has just the required number of members. It can take loads at the nodes. The member forces can be computed using equilibrium conditions ($\Sigma V = 0$, $\Sigma H = 0$) at each joint.

Figure 0.8 (*c*) has more members than the minimum needed. There can be several sets of member forces satisfying conditions of equilibrium. Of these, only one set is right, the set that satisfies displacement criteria or the set that corresponds to minimum strain energy stored in the truss.

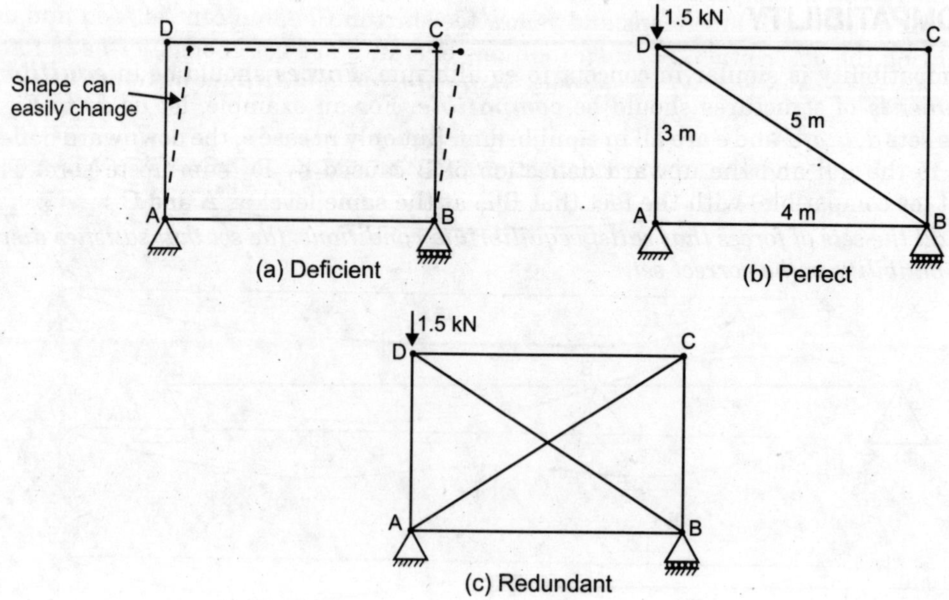

Fig. 0.8

Actually this volume is almost totally devoted to solving structures that are statically indeterminate. We have already inferred that this involves:

(*i*) Computation of deflection

(*ii*) Minimization of strain energy and other techniques.

0.5.3. Determinate Versus Redundant Structures: The essential differences between statically determinate structures and redundant structures are summarized in Table 0.1.

Table 0.1

S.No.	*Determinate Structures*	*Indeterminate Structures*
1.	It can be analysed from conditions of statics alone. So, easy to analyse.	It cannot be. We would need in addition compatibility of displacement/minimum energy condition. Analysis requires more effort.
2.	Reactive forces are independent of member properties like E, A, I, etc.	Dependent on member properties.
3.	Temperature changes do not create additional forces/stresses.	They do.
4.	There would be no forces/stresses due to lack of fit.	Lack of fit would produce internal stresses.
5.	They do not have any second line of defence. If one of the members fail, the structure would collapse.	For every additional redundancy there is some additional safety against collapse.

0.6 COMPATIBILITY

Compatibility is similar in concept to equilibrium. ***Forces*** should be in ***equilibrium. Displacements*** of structures should be ***compatible***. For an example, let us go to Fig. 0.7. Here force sets *a*, *b*, *c*, *d* and *e* are all in equilibrium. But only in case *e*, the downward deflection at B due to the udl and the upward deflection of B caused by R_B sum up to zero. This is consistent (or compatible) with the fact that B is at the same level as A and C.

Of all the sets of forces that satisfy equilibrium conditions, the set that satisfies displacement compatibility is the correct set.

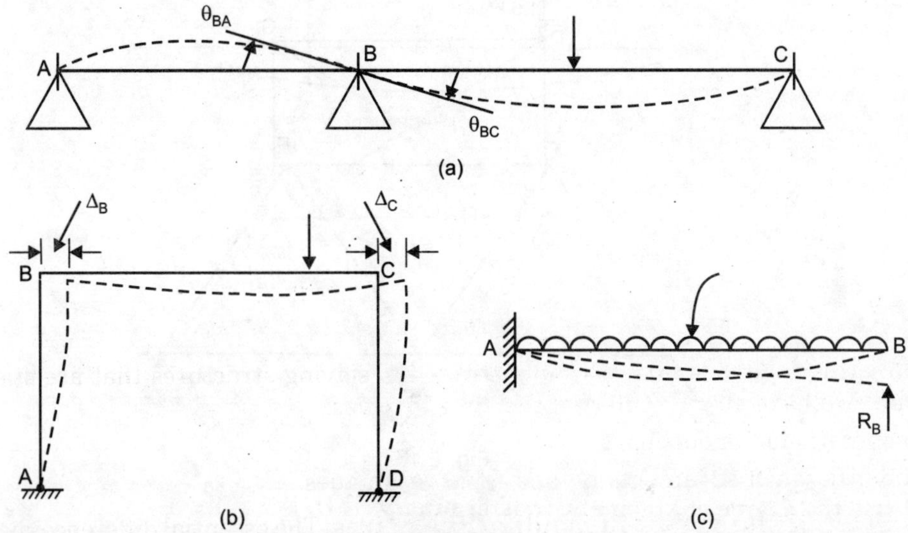

Fig. 0.9

In Fig. 0.9 (*a*) we have a 2 span continuous beam. Let us suppose it deflects as shown. BA will have a slope θ_{BA} ; BC will have a slope θ_{BC}. But the beam ABC is continuous and there is no hinge at B. So ABC can have only one slope at B. Hence $\theta_{BA} = \theta_{BC}$. This is a compatibility condition.

In Fig. 0.9 (*b*) let the portal sway to the right as shown. B moves by Δ_B. C moves by Δ_C. If we take only bending displacements and ignore axial displacements (which are usually negligible), then B and C shall have the same displacement *i.e.*, $\Delta_B = \Delta_C$. This again is a compatibility condition.

In Fig. 0.9 (*c*), we deal with a propped cantilever. When we consider that the downward deflection of B (of the cantilever) is equal and opposite to the upward deflection due to R_B, we are again using a compatibility condition.

0.7 DEGREES OF FREEDOM (dof)

The freedom for movement in the X,Y and rotational directions of every significant point in a structure is an important parameter in a structure. In Fig. 0.10 (*a*), A and B can move axially. That is 2 degrees of freedom (2 dof). But actually if AB cannot deform (rigid body), we only have one condition (compatibility).

$$\Delta_A = \Delta_B$$

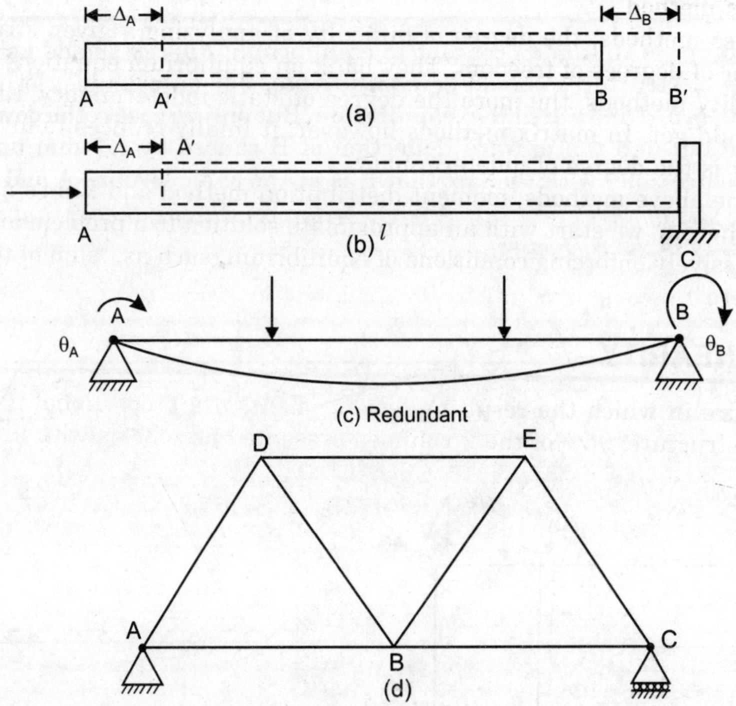

Fig. 0.10

Hence the degree of kinematic indeterminacy (K.I.) is 2 – 1 = 1.

In Fig. 0.10 (*b*), only A has the freedom to move. There is no compatibility condition to be satisfied. Hence KI = 1.

To find Δ_A we need the equilibrium equation alone

$$P = k\Delta = \frac{AE}{l} . \Delta$$

Hence,
$$\Delta_A = \frac{Pl}{AE}$$

Beam AB in Fig. 0.10 (*c*) has 2 dof, θ_A and θ_B. There is no compatibility condition relating them. Hence this is 2 degrees kinematically indeterminate. To find θ_A and θ_B we need 2 equations of equilibrium. We would use these in chapter 5. For the truss in Fig. 0.10 (*d*), A cannot move; B, D and E have 2 dof each and C has only horizontal dof. Hence there are 7 dof for the truss. To find their displacements, we need 7 conditions of equilibrium like in the matrix equation.

$$\underset{7}{\{F\}} = \underset{7\times7}{(k)} \underset{7}{\{\delta\}} \quad \text{and} \quad \{\delta\} = [k]^{-1}\{F\}$$

0.8 TYPES OF ANALYSIS

Analysis types are broadly 2 in number

1. ***Stiffness methods***
   Moment distribution
   Slope deflection
2. ***Flexibility methods***
   Energy methods
   Matrix flexibility method

Matrix stiffness — Column analogy method

Kani's method

In stiffness methods, the degree of difficulty of analyzing a given structure increases with the number of degrees of freedom. They bank on equilibrium equations.

In flexibility methods, the more the degree of static indeterminacy, the more involved the problem would get. In matrix methods however, it finally reduces to inverting an $\mathbf{n \times n}$ matrix, where n is the degree of indeterminacy.

Among the above methods, moment distribution method and Kani's method are iterative methods. Therein, we start with an approximate solution to a problem and go on refining the same, successively enforcing conditions of equilibrium (such as, "sum of the moments at a joint shall be zero").

0.9 NON-LINEARITY

A structure in which the resulting displacements are proportional to applied loads is called a linear structure. 95% of the 2 volumes presented here deals with linear structures.

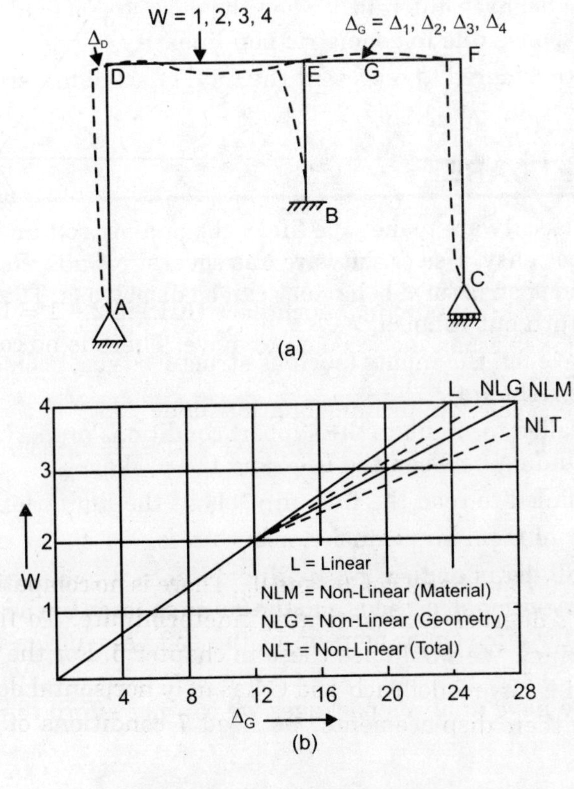

Fig. 0.11. Non-linear behaviour.

Linear structures permit us to add or subtract the effects of loads. If $W = W_1 + W_2$, then Δ, the deflection due to W is equal to the sum of Δ_1 and Δ_2, which are deflections due to W_1 and W_2 respectively. So, in a linear structure, $\Delta = \Delta_1 + \Delta_2$.

A structure would behave linearly when

(*i*) The material is linear, elastic all over.

(*ii*) The resulting displacements are small and do not alter the geometry of the structure.

If anyone of the above conditions gets violated, the structure behaves non-linearly.

Usually when loads are large, stresses induced are large. The material in the highly stressed zones becomes elastoplastic or plastic. Hence the behaviour of the structure alters. Its response, namely, deflection due to a unit increase in load changes. This behaviour is called ***material non-linearity***.

In Fig. 0.11, Δ_G is plotted against W. The curve L shows the linear response. The curve NLM represents material non-linear behaviour (in the absence of geometric non-linearity) when the deflection at G due to W = 2 is 12 mm, then the deflection due to W = 4 will not be 24 mm but more, (in this case 28 mm). [Only very rarely does a material get stiffer at higher stresses].

Figure 0.11 (*a*) shows the displaced configuration of the original frame ABCDEF. The displaced frame is not the same as the original frame (especially if the displacements are large). The altered frame behaves differently when the next step of load comes on it. Line NLG in Fig. 0.11 (*b*) is the response due to geometric non-linearity.

Line NLT can be considered to represent the total effect of material and geometric non-linearities.

0.10 ANALYSIS SOFTWARE

Structural analysis software makes the life of the analyst soft and easy. But one should take care not to take it too easy. Use of software has several pitfalls. Resort to software without a good comprehension of structural behaviour can be disastrous. There are several reasons why a package would churn out rubbish.

1. We may not have fed the inputs (such as structure type, geometry, material properties, loads) the right way.
2. We may have failed to simulate the support conditions properly. This is a major area of misunderstanding between the user and the packages.
3. We may have failed to read the fine print about the limitations.
4. There is plenty of room for confusion about units.

This is a only small list of endless possibilities. Hence it is absolutely necessary that a structure be solved by conventional methods and the output offered by the software be validated. The validation is more for our comprehension of the software usage than for the package itself.

So just because we have analysis packages we can not afford to neglect our structural analysis.

CHAPTER 1

DEFLECTION OF DETERMINATE STRUCTURES

1.1 GENERAL

A structure has to be stable, strong and sufficiently stiff. A stable structure will not topple, slide or roll. A strong structure will have all its parts under safe stress levels. A stiff structure will not deflect or elongate or shorten too much. Excessive deflections would make a structure functionally unsuitable.

So it is necessary to analyse structures to find their deflections as well as their stresses. Besides checking the functional acceptability, computation of deflection also serves as a first step in solving indeterminate structures.

Deflections of trusses are caused by the lengthening and shortening of its members due to tensions and compressions respectively. The displaced configurations of a truss can be determined by plotting the assemblage taking into account the deformation of its members. It can also be arrived at by considering the strain energy of the truss and the work done by the loads as they suffer displacements.

1.2 PRINCIPLE OF VIRTUAL WORK

The principle of virtual work was developed by Johann Bernoulli in 1717, and this is the most versatile of the methods available for computing deflections of structures. This method is sometimes referred to as the unit-load method. The term virtual work means the work done by a real force acting through a virtual displacement or a virtual force acting through a real displacement. The virtual work is not a real quantity but an imaginary one.

The principle of virtual work is based on the conservation of energy for a structure, which implies that work done on a structure by external loads is equal to the internal energy stored in the structure. ($U_e = U_i$).

Take a deformable structure of any shape or size and apply a series of external loads {P} to it. It will cause internal forces {F} at points throughout the structure. It is necessary that the external and internal loads be related by equations of equilibrium. As a consequence of the above loadings, external displacements Δ will occur at the P loads and internal displacements {δ} will occur at each point of internal forces {F}. These displacements do not have to be elastic, and they may not be related to the loads. However, the external and internal displacements must be related by the compatibility of the displacements. In other words, if the external displacements are known, the corresponding internal displacements are uniquely defined.

In general, the principle of virtual work and energy states that

$$\underbrace{\Sigma P \Delta}_{\text{Work of external loads}} = \underbrace{\Sigma F \delta}_{\text{Work of internal forces}} \qquad \text{...(1.1)}$$

1.3 METHOD OF VIRTUAL FORCES

Forces and displacements are mutually interrelated. A force on a member stretching a member does work since the force moves. The member itself, because it is kept stretched by the force, stores internal energy. Obviously the internal energy stored will be equal to the work done by the force on the member.

Thus in a structure made up of several members and several external forces, there are two basic energy quantities.

(i) 'W_e'—Work done by external forces as they move through some displacements (most usually in the gravity direction)

(ii) 'W_i'—Energy stored by all the strained members due to the internal forces induced in them.

The displacement caused in simple pin jointed determinate frames can easily be determined by using the method of virtual forces, as described below in 3 steps.

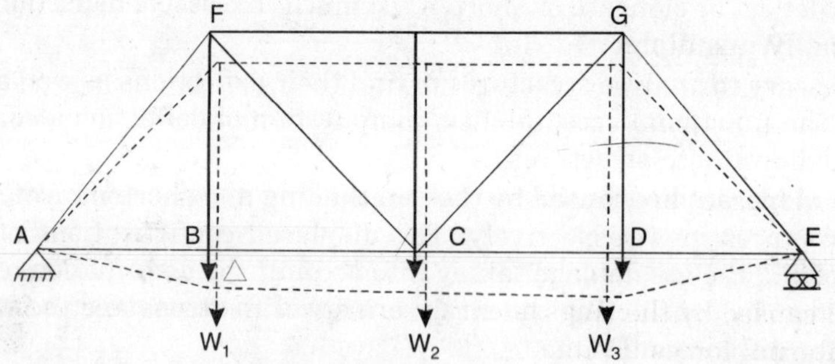

Fig. 1.1

Step 1. Suppose we want to determine the displacement at B of the truss in Fig. 1.1 due to the load system W_1 W_2 W_3 = {W}. We have to first solve for the internal forces F_1 F_2 F_3.....F_n = {F} in the members due to {W}. This is a question of statics. We know that every member will have an additional elongation (or shortening) of F_1l/AE, F_2l/AE, F_3l/AE......due to F_1 F_2 F_3........... These member displacements will increase from 0 to {Fl/AE} as the internal forces increase from zero to {F}.

Step 2. Now apply only a unit load (virtual force) at B in the direction of the desired deflection Δ. Find the internal forces due to the unit load (k_1 k_2 k_3 .... = {k}). We will forget about the external work done and internal energy stored due to the unit load and proceed to the next step.

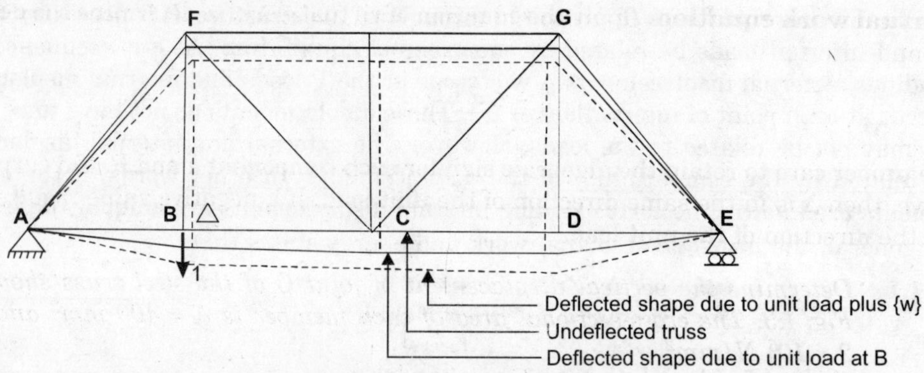

Fig. 1.2

Step 3. Now apply the {W} system in addition to the unit load, which has been applied first.

Energy Equation :

We will now have the deflected truss. The deflections are due to the unit load plus {W} system.

The external work done W_e due to the imposition of {W} is in 2 parts.

W_{e1} due to the unit load displacing through Δ at B.

W_{e2} due to the {W} system displacing progressively from zero to Δ_1, Δ_2, Δ_3, ...

And the internal work done is also due to two causes.

W_{i1} due to the pre-existing internal force system $k_1\ k_2\ k_3$ ......displacing through member displacements $F_1 l/AE$, $F_2 l/AE$ ...... caused by {W} system

W_{i2} due to the current force system $F_1\ F_2\ F_3$ ...... displacing progressively from 0 to $F_1 l/AE$, $F_2 l/AE$, $F_3 l/AE$, ... etc.

$$W_{e1} + W_{e2} = W_e = 1.\Delta + \frac{W_1\Delta_1}{2} + \frac{W_2\Delta_2}{2} + \ldots\ldots$$

$$W_{i1} + W_{i2} = W_i = \Sigma\frac{kFL}{AE} + \Sigma\frac{F^2L}{2AE}$$

If {W} system were alone applied on the structure without the unit load preceeding it we would have got the relation $W_{e2} = W_{i2}$. Therefore, we can infer that $1.\Delta = \Sigma\dfrac{kFL}{AE}$. Thus, we get the deflection Δ at B caused by the {W} system in terms of the internal forces {k} due to a unit load at B and the internal forces {F} due to the {W} system.

1.4 DEFLECTIONS OF PIN JOINTED PLANE FRAMES—PROCEDURE FOR ANALYSIS

Sign Convention : Assume that tensile forces are positive and compressive forces are negative.

1. Virtual forces k. Remove all the real loads from the truss. Place a unit load on the truss at the joint and in the direction of the desired displacement. Use the method of joints or the method of sections and calculate the internal forces k in each member of the truss.

2. Real forces F. These forces are caused only by the real loads acting on the truss. Use the method of sections or the method of joints to determine the forces F in each member.

3. Virtual work equation. Apply the equation of virtual work, to determine the desired displacement.

i.e.,
$$1\,.\,\Delta = \Sigma\frac{kFL}{AE}.$$

Take proper care to retain the algebraic sign for each component k and F. If Δ turns out to be positive, then Δ is in the same direction of the unit load. If a negative value results, Δ is opposite to the direction of the unit load.

Example 1.1 *Determine the vertical displacement of joint C of the steel truss shown in Fig. 1.3. The cross-sectional area of each member is A = 400 mm² and E = 2 × 10⁵ N/mm².*

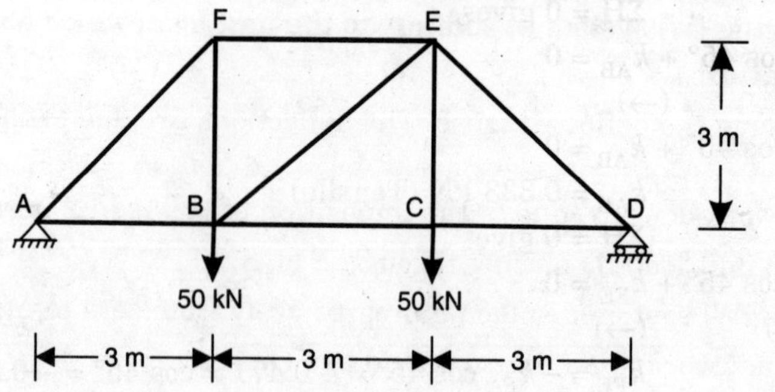

Fig. 1.3

Solution We know that $1 \cdot \Delta = \Sigma \dfrac{kFL}{AE}$

Virtual forces k: Remove all the (external) loads and apply a unit vertical force at joint C of the truss. Analyse the truss using the method of joints.

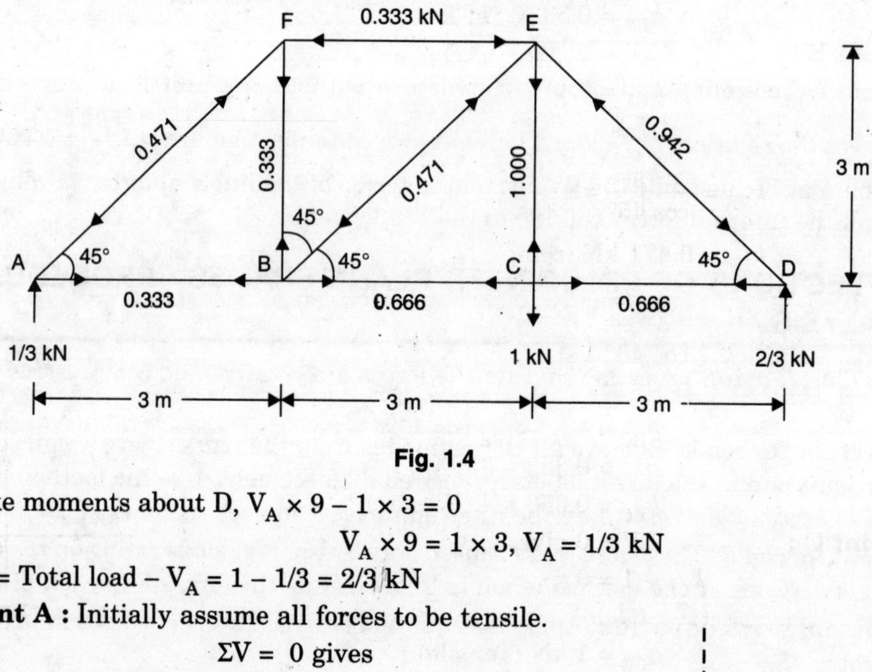

Fig. 1.4

Take moments about D, $V_A \times 9 - 1 \times 3 = 0$

$$V_A \times 9 = 1 \times 3,\ V_A = 1/3 \text{ kN}$$

V_D = Total load $- V_A = 1 - 1/3 = 2/3$ kN

Joint A : Initially assume all forces to be tensile.

$$\Sigma V = 0 \text{ gives}$$

$$k_{AF} \cos 45° + 1/3 = 0$$
$$(\uparrow) \qquad (\uparrow)$$

$$k_{AF} \cos 45° = -1/3$$

$$k_{AF} = -\frac{1}{3 \cos 45°} = -0.471 \text{ kN}$$

i.e., $k_{AF} = 0.471$ kN (comp.)

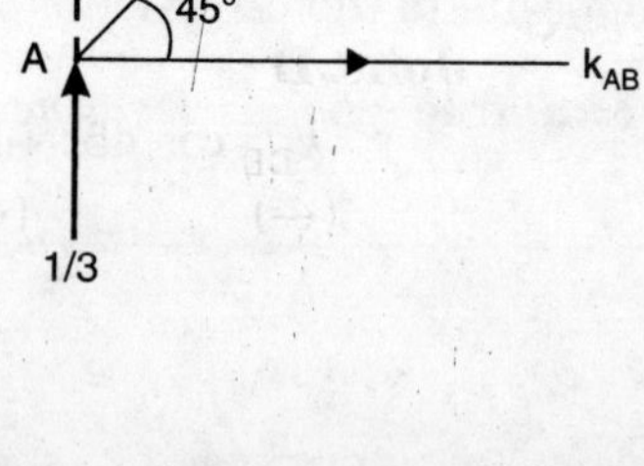

$\Sigma H = 0$ gives

$k_{AF} \cdot \cos 45° + k_{AB} = 0$

(→) (→)

$(-0.471) \cos 45° + k_{AB} = 0$

$k_{AB} = 0.333$ kN (Tensile)

Joint F : $\Sigma H = 0$ gives

$k_{FA} \cos 45° + k_{FE} = 0$

(→) (→)

$k_{FE} = -k_{FA} \cos 45° = -0.471 \times \cos 45° = -0.333$ kN

$k_{FE} = 0.333$ kN (comp.)

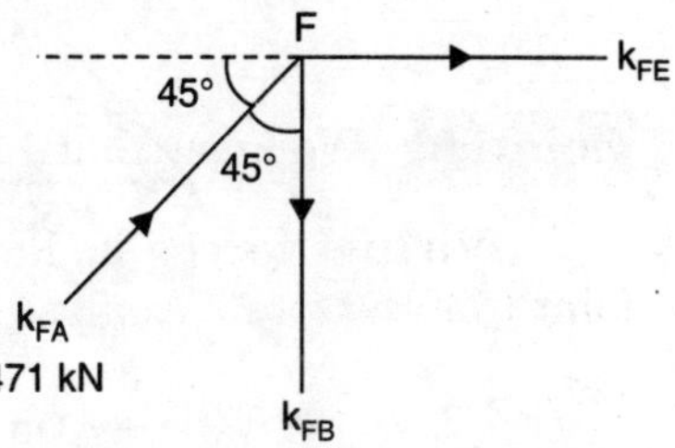

$\Sigma V = 0$ gives

$k_{FA} \cos 45° - k_{FB} = 0$

(↑) (↓)

$k_{FA} \cos 45° = k_{FB}$

$0.471 \cos 45° = k_{FB}$

∴ $k_{FB} = 0.333$ kN (Tensile)

Joint B : $\Sigma V = 0$ gives

$k_{BE} \cos 45° + k_{BF} = 0$

(↑) (↑)

$$k_{BE} = \frac{-k_{BF}}{\cos 45°} = -\frac{0.333}{\cos 45°} = -0.471$$

$k_{BE} = 0.471$ kN (comp.)

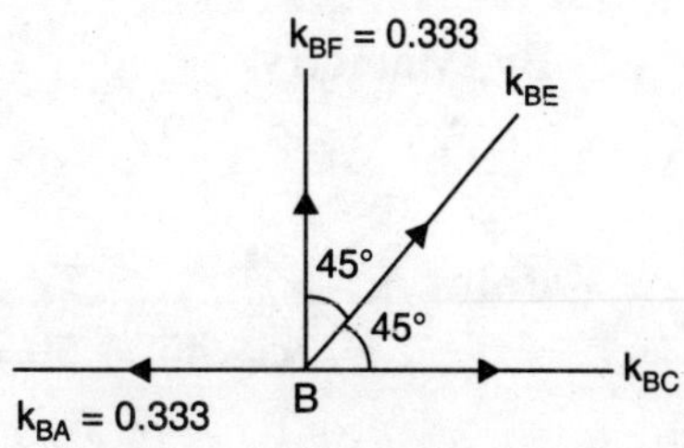

$\Sigma H = 0$ gives

$k_{BC} - k_{BA} + k_{BE} \cos 45° = 0$

(→) (←) (→)

$k_{BC} = k_{BA} - k_{BE} \cos 45°$

$= 0.333 - (-0.471) \cos 45°$

$k_{BC} = 0.666$ kN (Tensile)

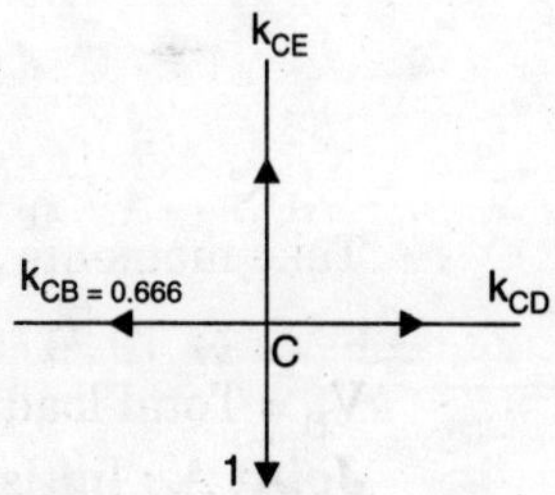

Joint C : $\Sigma V = 0$ gives

$k_{CE} - 1 = 0$

(↑) (↓)

∴ $k_{CE} = 1$ kN (Tensile)

$\Sigma H = 0$ gives

$k_{CD} - k_{CB} = 0$

(→) (←)

$k_{CD} = k_{CB} = 0.666$

∴ $k_{CD} = 0.666$ kN (Tensile)

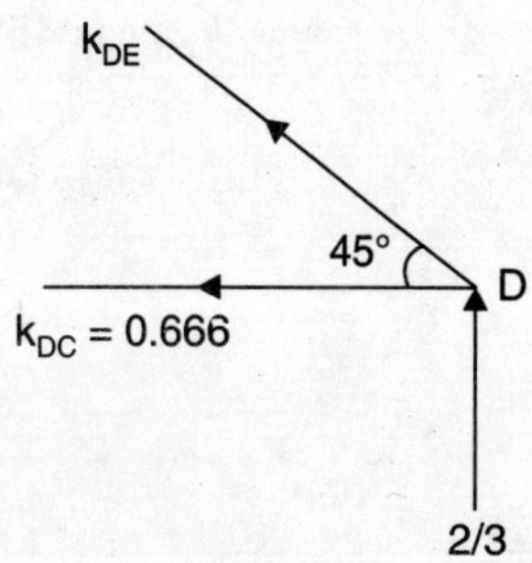

Joint D : $\Sigma H = 0$ gives

$k_{DE} \cos 45° + k_{DC} = 0$

(←) (←)

$$k_{DE} = -k_{DC}/\cos 45°$$
$$= -0.666/\cos 45°$$
$$= -0.942.$$
$$k_{DE} = 0.942 \text{ kN (comp.)}$$

Real forces, F: The real forces in the members due to the given system of external loads are calculated using the method of joints.

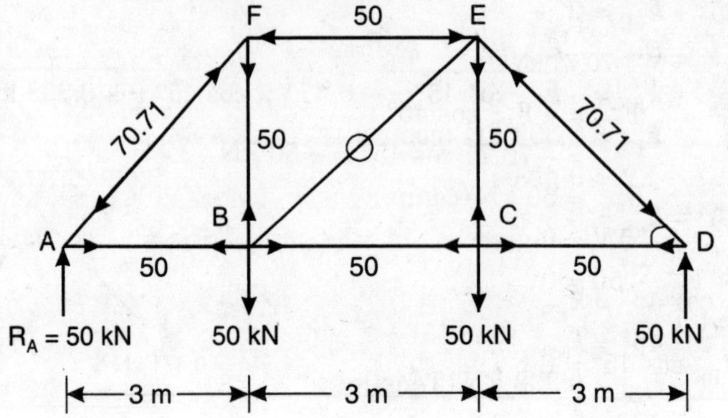

Fig. 1.5

By symmetry

$$R_A = R_D = \frac{\text{Total load}}{2} = 50 \text{ kN}$$

Joint A : $\Sigma V = 0$ gives

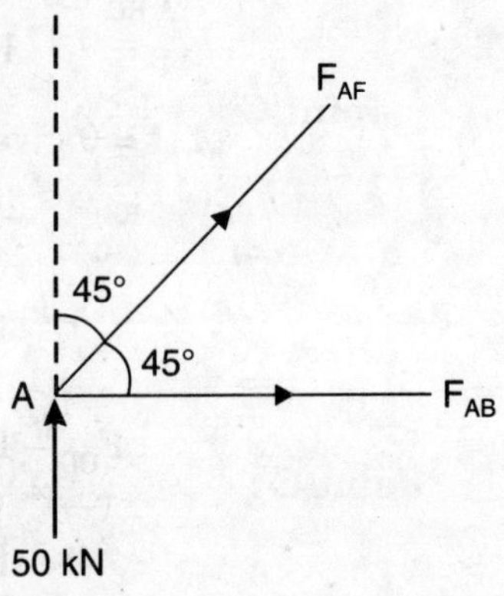

$$\underset{(\uparrow)}{F_{AF} \cos 45°} + \underset{(\uparrow)}{50} = 0$$

$$F_{AF} = -50/\cos 45° = -70.71 \text{ kN}$$
$$F_{AF} = 70.71 \text{ kN (comp.)}$$
$$\Sigma H = 0 \text{ gives}$$
$$\underset{(\rightarrow)}{F_{AB}} + \underset{(\rightarrow)}{F_{AF} \cos 45°} = 0$$
$$F_{AB} + (-70.71) \cos 45° = 0$$
$$F_{AB} = 70.71 \cos 45°$$
$$F_{AB} = 50 \text{ kN (Tensile)}$$

Joint F : $\Sigma V = 0$ gives

$$\underset{(\uparrow)}{F_{FA} \cos 45°} - \underset{(\downarrow)}{F_{FB}} = 0$$
$$-F_{FB} = -F_{FA} \cos 45°$$
$$F_{FB} = 70.71 \cos 45°$$
$$F_{FB} = 50 \text{ kN (Tensile)}$$
$$\Sigma H = 0 \text{ gives}$$
$$\underset{(\rightarrow)}{F_{FA} \cos 45°} + \underset{(\rightarrow)}{F_{FE}} = 0$$

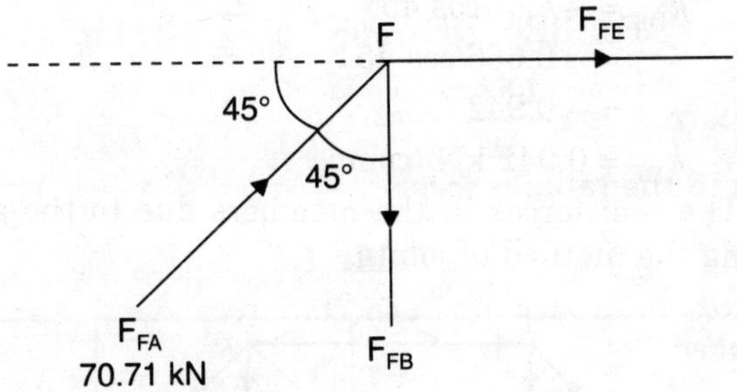

$$F_{FE} = -F_{FA}\cos 45^\circ$$
$$= -70.71\cos 45^\circ = -50 \text{ kN}$$
$$F_{FE} = 50 \text{ kN (comp.)}$$

Joint B : $\Sigma V = 0$ gives

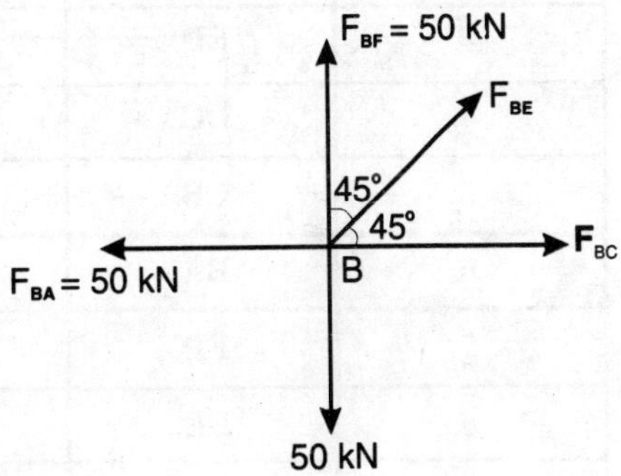

$$\underset{(\uparrow)}{F_{BE}\cos 45^\circ} + \underset{(\uparrow)}{F_{BF}} - \underset{(\downarrow)}{50} = 0$$
$$F_{BE}\cos 45^\circ = -F_{BF} + 50$$
$$= -50 + 50 = 0$$
$$\therefore \quad F_{BE} = 0$$

$\Sigma H = 0$ gives

$$\underset{(\rightarrow)}{F_{BC}} + \underset{(\rightarrow)}{F_{BE}\cos 45^\circ} - \underset{(\leftarrow)}{F_{BA}} = 0$$
$$F_{BC} + 0 - 50 = 0 \qquad (\because \; F_{BE} = 0)$$
$$F_{BC} = 50 \text{ kN (Tensile)}$$

Joint C :

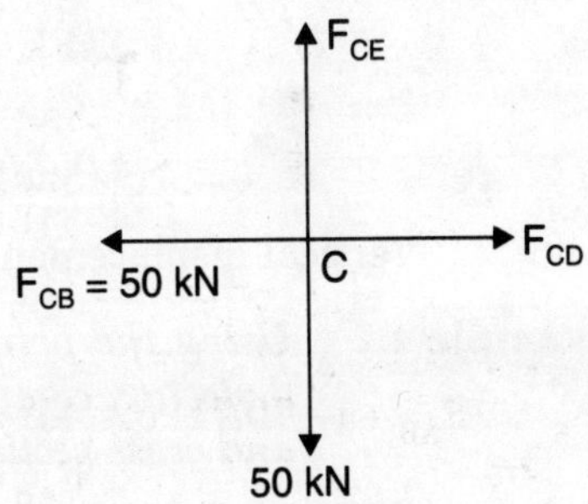

$\Sigma V = 0$ gives

$$\underset{(\uparrow)}{F_{CE}} - \underset{(\downarrow)}{50} = 0$$
$$F_{CE} = 50 \text{ kN(Tensile)}$$

$\Sigma H = 0$ gives

$$\underset{(\rightarrow)}{F_{CD}} - \underset{(\leftarrow)}{F_{CB}} = 0$$
$$F_{CD} = F_{CB}$$
$$F_{CD} = 50 \text{ kN (Tensile)}$$

Joint D : $\Sigma H = 0$ gives

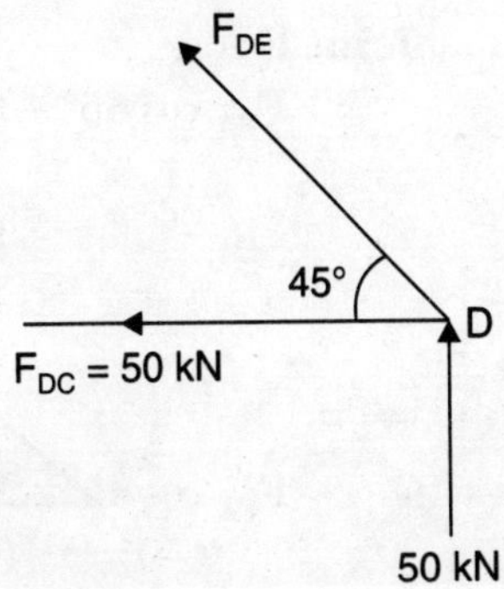

$$\underset{(\leftarrow)}{F_{DE}\cos 45^\circ} + \underset{(\leftarrow)}{F_{DC}} = 0$$
$$F_{DE} = -F_{DC} / \cos 45^\circ$$
$$= -50 / \cos 45^\circ$$
$$= -70.71 \text{ kN}$$
$$F_{DE} = 70.71 \text{ kN (comp.)}$$

Virtual work equation :

$$1.\Delta = \Sigma \frac{kFL}{AE}$$

$$AF = BE = DE = \sqrt{3^2 + 3^2} = 4.243 \text{ m}$$

i.e., $$1. \text{ kN} \times (\Delta_C)_v = \Sigma \frac{kFL}{AE}$$

Arrange the data in the tabular form.

Table 1.1

S.No.	*Member*	*k*	*F (kN)*	*L (m)*	*kFL (kN m)*
1.	AF	– 0.471	– 70.71	4.243	141.311
2.	FE	– 0.333	– 50.00	3.00	49.950
3.	ED	– 0.942	– 70.71	4.243	282.621
4.	DC	0.666	50.00	3.00	99.900
5.	CB	0.666	50.00	3.00	99.900
6.	BA	0.333	50.00	3.00	49.950
7.	FB	0.333	50.00	3.00	49.950
8.	BE	– 0.471	0	4.243	0
9.	EC	1.000	50.00	3.00	150.00
				Σ *k*FL =	923.582

$$\Sigma kFL = 923.582 \times (1000)^2 \text{ N mm}$$

$$\therefore \quad (\Delta_C)_v = \Sigma \frac{kFL}{AE} = \frac{923.582 \times (1000)^2}{400 \times 2 \times 10^5} = 11.54 \text{ mm.}$$

∴ Vertical displacement of joint C = 11.54 mm (downward).

Example 1.2 *Using the principle of virtual work, determine the vertical and horizontal deflection components of joint C of the truss in Fig. 1.6. E = 200 × 10^6 kN/m^2 and cross-sectional area of each bar = 150 × 10^{-6} m^2.*

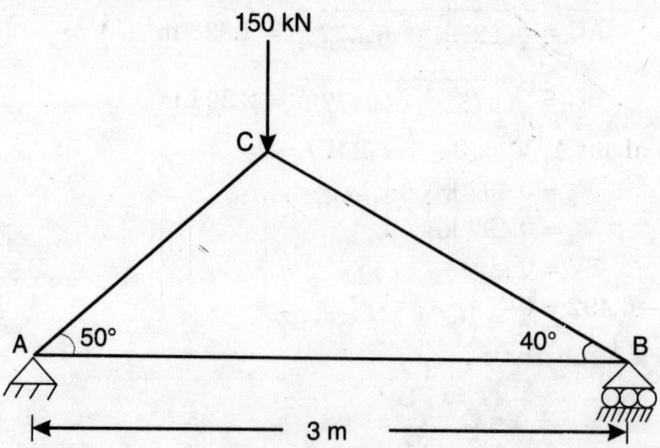

Fig. 1.6

Solution We know that $1.\Delta = \Sigma \dfrac{kFL}{AE}$

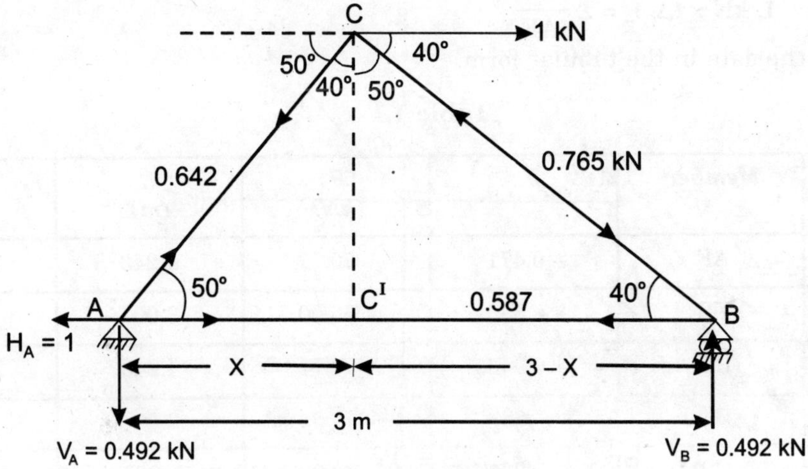

Fig. 1.7. k_h values.

Virtual forces, k_h : (for horizontal displacement at C)

Remove the real (external) load and apply a unit horizontal force at C of the truss.

$$\Sigma H = 0 \text{ gives } H_A = 1 \text{ kN } (\leftarrow)$$

$$\frac{CC^I}{x} = \tan 50°$$

$$CC^I = x \tan 50° = (3 - x) \tan 40°$$

$$x = (3 - x) \tan 40°/ \tan 50°$$

$$x = (3 - x) \times 0.704$$

$$x = 2.112 - 0.704x$$

$$x + 0.704x = 2.112$$

$$x = 2.112/1.704 = 1.239 \text{ m}$$

$$(3 - x) = 3.0 - 1.239 = 1.761 \text{ m}$$

$$\therefore \quad CC^I = 1.239 \times \tan 50° = 1.477 \text{ m}$$

$$AC = \sqrt{(1.239)^2 + (1.477)^2} = 1.928 \text{ m}$$

$$BC = \sqrt{(1.761)^2 + (1.477)^2} = 2.298 \text{ m}$$

Taking moments about A, $V_B \times 3 - 1 \times 1.477 = 0$

$$V_B = 0.492 \text{ kN } (\uparrow)$$

$$V_A = 0.492 \text{ kN } (\downarrow)$$

Joint A : $\Sigma V = 0$ gives

$k_{AC} \cdot \cos 40° - 0.492 = 0$

$(\uparrow) \qquad (\downarrow)$

$k_{AC} = 0.492/\cos 40° = 0.642$ kN (Tensile)

$\Sigma H = 0$ gives

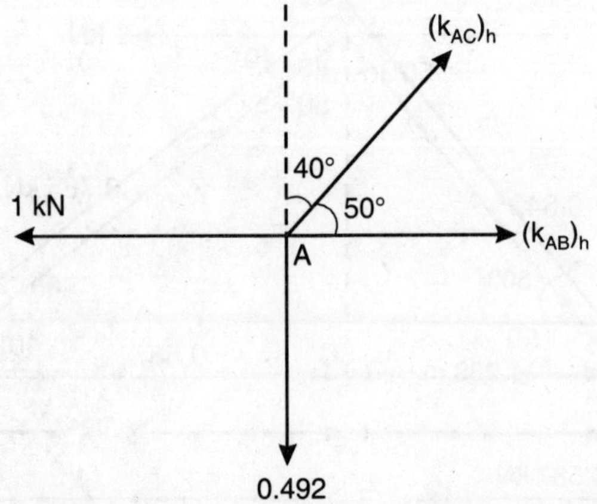

$$k_{AC} \cdot \cos 50° + k_{AB} - 1 = 0$$
$$(\rightarrow) \quad (\rightarrow)\ (\leftarrow)$$

$$0.642 \cos 50° + k_{AB} - 1 = 0$$

$$k_{AB} = 0.587 \text{ kN (Tensile)}$$

Joint C : $\Sigma V = 0$ gives

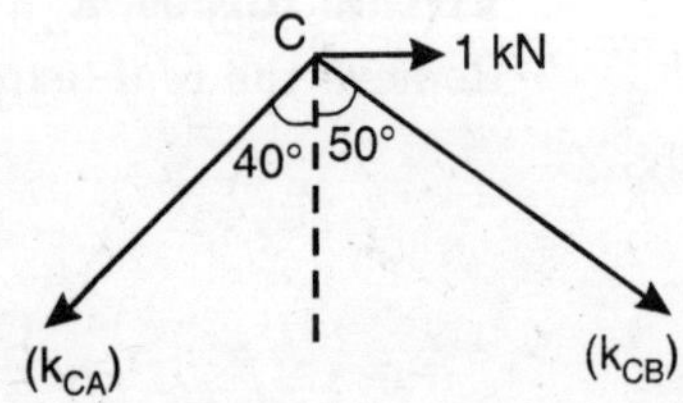

$$k_{CA} \cos 40° + k_{CB} \cos 50° = 0$$
$$(\downarrow) \quad (\downarrow)$$

$$k_{CB} = -k_{CA} \cos 40°/\cos 50°$$
$$= -0.642 \times 1.192$$
$$= -0.765 \text{ kN}$$
$$k_{CB} = 0.765 \text{ kN (comp.)}$$

The k values are tabulated as below (col. 3)

Table 1.2

S.No.	*Member*	$(k)_h$	$(k)_V$	*F (kN)*	*L (m)*	$Fk_h.L$	$Fk_V.L$
1.	AB	0.587	0.492	73.8	3.0	129.962	108.929
2.	AC	0.642	– 0.766	– 114.9	1.928	– 142.22	169.693
3.	CB	– 0.765	– 0.643	– 96.45	2.298	169.556	142.516
					Σ	157.298	421.135

Virtual forces, k_V : (for vertical displacement at C)

Remove the external load and apply a unit vertical force at C.

Taking moments about B, $V'_A \times 3 - 1 \times 1.761 = 0$

$$V'_A = 1.761/3 = 0.587 \text{ kN}$$
$$V'_B = 1 - 0.587 = 0.413 \text{ kN}$$

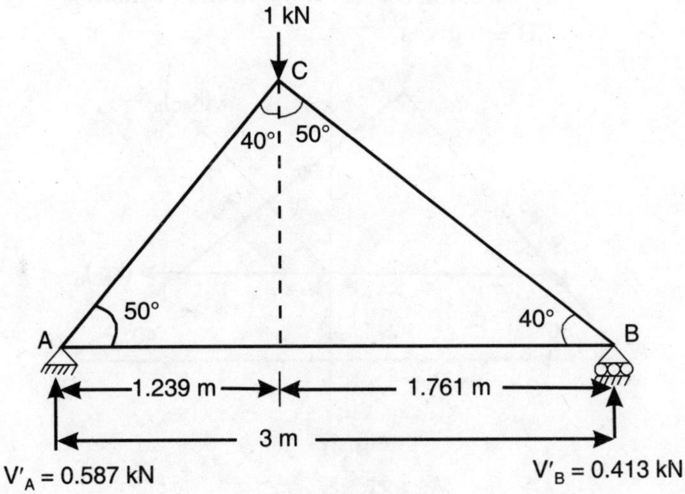

Fig. 1.8

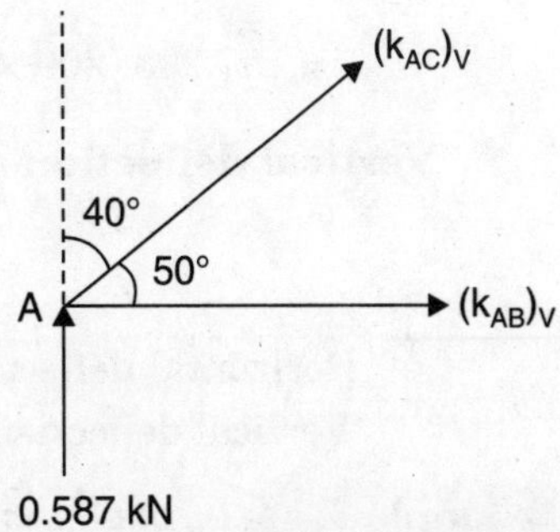

Joint A : $\Sigma V = 0$ gives

$$k_{AC.}\cos 40° + 0.587 = 0$$

$$k_{AC} = -0.587/\cos 40°$$

$$= -0.766 \text{ kN}$$

$$k_{AC} = 0.766 \text{ kN (comp.)}$$

$$\Sigma H = 0 \text{ gives}$$

$$\underset{(\rightarrow)}{k_{AC}} . \cos 50° + \underset{(\rightarrow)}{k_{AB}} = 0$$

$$k_{AB} = -k_{AC}. \cos 50°$$

$$= -(-0.766) \cos 50°$$

$\therefore$ $$k_{AB} = 0.492 \text{ (Tensile)}$$

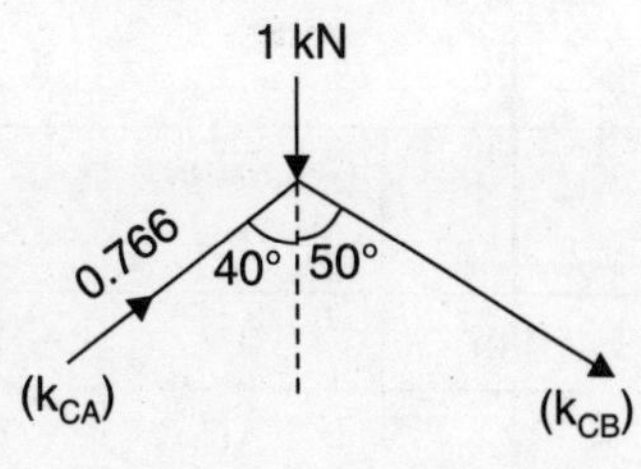

Joint C : $\Sigma V = 0$ gives

$$\underset{(\uparrow)}{0.766 \cos 40°} - \underset{(\downarrow)}{1} - \underset{(\downarrow)}{k_{CB}} . \cos 50° = 0$$

$$0.587 - 1 = k_{CB} . \cos 50°$$

$$-0.643 = k_{CB}$$

$$k_{CB} = 0.643 \text{ kN (comp.)}$$

The values are shown in column 4 of the table.

Real forces, F: The real forces in the members due to the given external load are calculated as below.

The only given force is of magnitude 150 kN and applied at C vertically. Therefore the forces in the members will be 150 times the k_V values.

$\therefore$ $$F_{AB} = 0.492 \times 150 = 73.8 \text{ kN (Tensile)}$$

$$F_{AC} = 0.766 \times 150 = 114.9 \text{ kN (Comp.)}$$

$$F_{CB} = 0.643 \times 150 = 96.45 \text{ kN (Comp.)}$$

These values are shown in column 5 of the table.

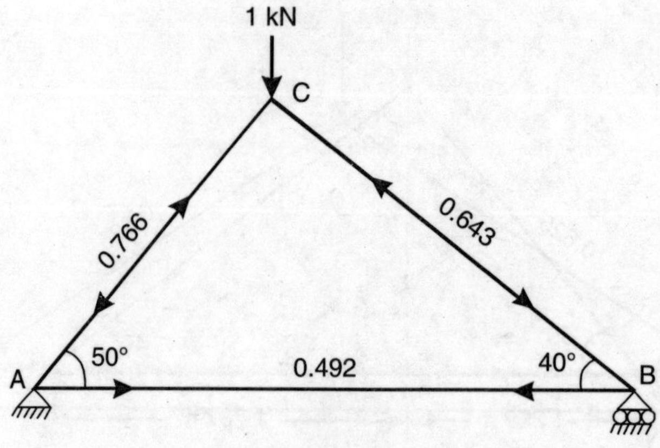

Fig. 1.9

Virtual work equation : $1.\Delta = \Sigma \dfrac{k\text{FL}}{\text{AE}}$

Horizontal deflection of point C : $1.(\Delta_C)_h = \Sigma k_h \dfrac{\text{FL}}{\text{AE}}$

$= 157.298/(200 \times 10^6 \times 150 \times 10^{-6}) = 0.00524 \text{ m} = 5.24 \text{ mm } (\rightarrow)$

Vertical deflection of point C : $1.(\Delta_C)_V = \Sigma k_V \dfrac{\text{FL}}{\text{AE}}$

$= 421.135/(200 \times 10^6 \times 150 \times 10^{-6})$

$= 0.014 \text{ m} = 14 \text{ mm } (\downarrow)$

$\therefore$ Horizontal deflection of point C = 5.24 mm ($\rightarrow$)

Vertical deflection of point C = 14 mm ($\downarrow$).

Example 1.3 *In the truss shown in Fig. 1.10 no load acts. The member AB gets 4 mm too short. The cross-sectional area of each member is A = 300 mm² and E = 200 GPa. Determine the vertical displacement of joint C.*

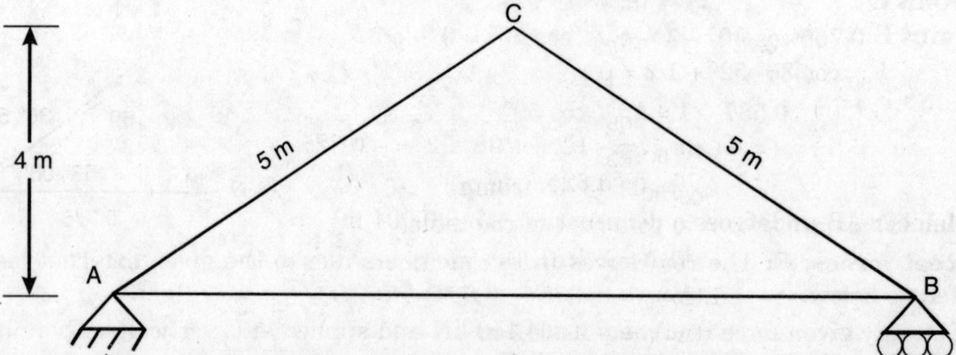

Fig. 1.10

Solution **Virtual forces, k:** Since the vertical displacement of joint C is required, a vertical force of 1 kN is applied at C. The force k in each member is determined as follows:

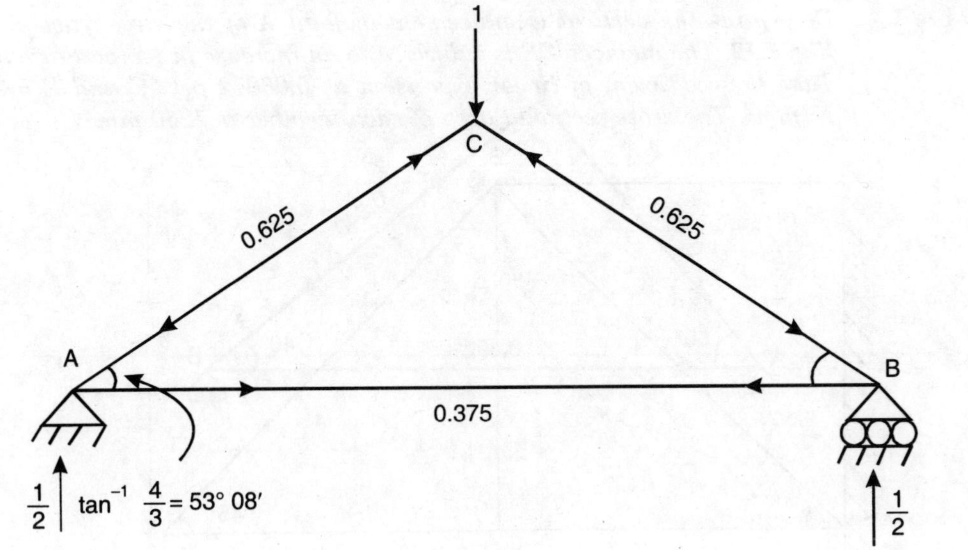

Fig. 1.11

By symmetry, $R_A = R_B = 1/2\ (\uparrow)$

Joint A : $\Sigma V = 0$ gives

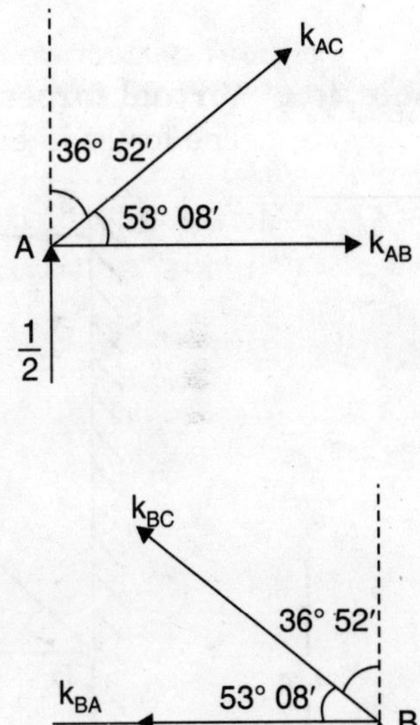

$$\underset{(\uparrow)}{k_{AC}.\cos 36° 52'} + \underset{(\uparrow)}{1/2} = 0$$

$$k_{AC} = -1/2 \cos 36° 52'$$

$$= -0.625 = 0.625 \text{ (comp.)}$$

$\Sigma H = 0$ gives

$$\underset{(\rightarrow)}{k_{AB}} + \underset{(\rightarrow)}{k_{AC} \cos 53° 08'} = 0$$

$$k_{AB} + (-0.625) \cos 53° 08' = 0$$

$$k_{AB} = 0.375 \text{ kN (Tensile)}$$

Joint B : $\Sigma V = 0$ gives

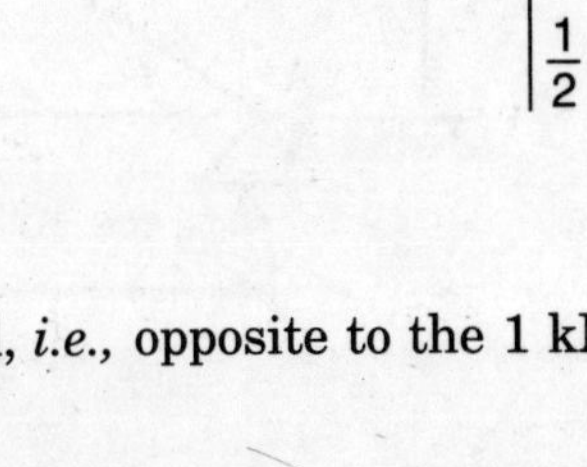

$$\underset{(\uparrow)}{k_{BC} \cos 36° 52'} + \underset{(\uparrow)}{1/2} = 0$$

$$k_{BC} = -1/2 \cos 36° 52' = -0.625$$

$$\therefore \quad k_{BC} = 0.625 \text{ (comp.)}$$

Member AB undergoes a deformation, $\Delta L = 0.004$ m

$$\therefore \quad 1.\Delta = \Sigma(k.\Delta L)$$

$$1.(\Delta_C)_V = (0.375)(-0.004) + 0 + 0$$

$$(\Delta_C)_V = -0.0015 \text{ m}$$

$$(\Delta_C)_V = -1.5 \text{ mm}$$

The negative sign indicates that joint C is displaced upward, *i.e.,* opposite to the 1 kN vertical load.

$$\therefore \quad (\Delta_C)_V = 1.5 \text{ mm } (\uparrow).$$

Example 1.4 *Derermine the vertical displacement of joint A of the steel truss shown in Fig. 1.12. The member BE is subjected to an increase in temperature of 70°C. Take the coefficient of linear expansion as 0.00012 per °C and E = 2 × 10^5 N/mm^2. The cross-sectional area of each member is 1250 mm^2.*

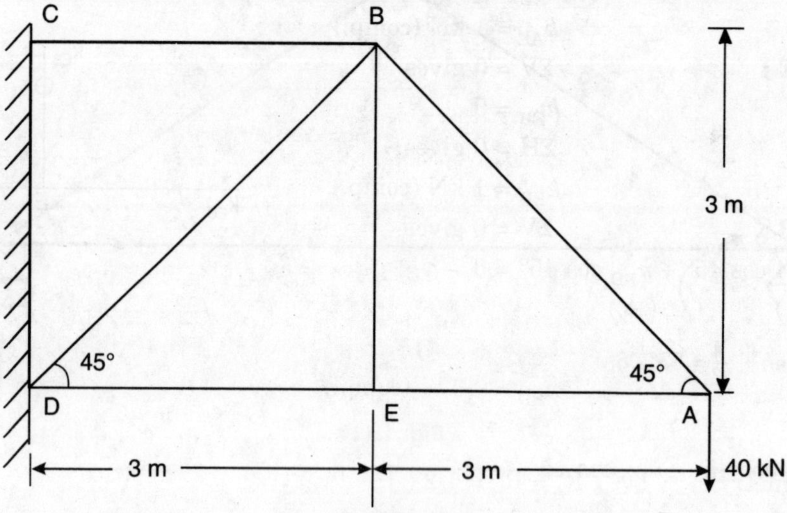

Fig. 1.12

Solution **Virtual forces, k:** Remove the external load and apply a unit vertical load at A. The forces in each member are calculated as below :

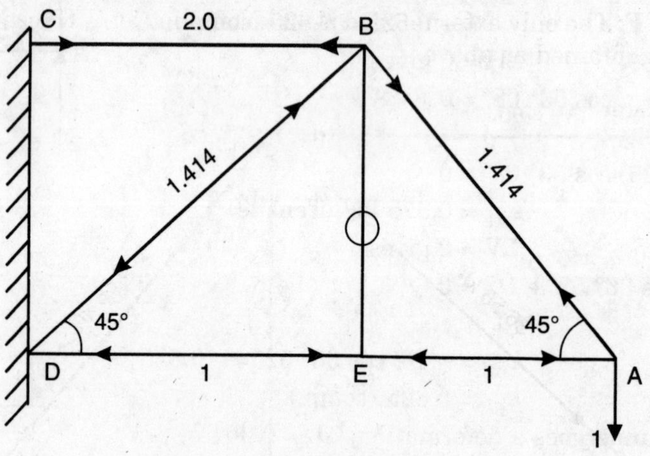

Fig. 1.13

Joint A : $\Sigma V = 0$ gives

$$k_{AB} \cos 45° - 1 = 0$$
$$(\uparrow) \qquad (\downarrow)$$

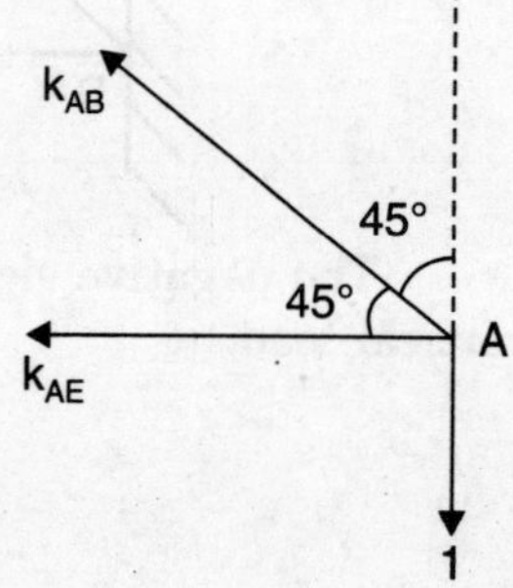

$$k_{AB} = \frac{1}{\cos 45°}$$

$$k_{AB} = 1.414 \text{ kN} \quad \text{(Tensile)}$$

$$\Sigma H = 0 \text{ gives}$$

$$k_{AE} + k_{AB} \cos 45° = 0$$
$$(\leftarrow) \quad (\leftarrow)$$

$$k_{AE} = -k_{AB} \cos 45° = -1.414 \cos 45° = -1 \text{ kN}$$

$$k_{AE} = 1 \text{ kN (comp.)}$$

Joint E : $\Sigma V = 0$ gives

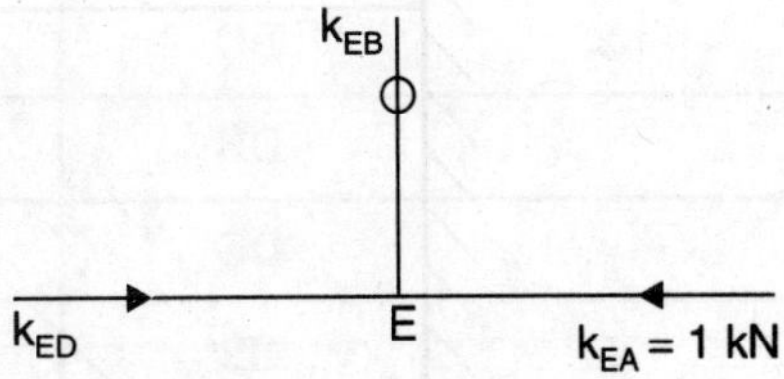

$$k_{EB} = 0$$

$$\Sigma H = 0 \text{ gives}$$

$$k_{ED} = 1 \text{ kN (comp.)}$$

Joint B : $\Sigma V = 0$ gives

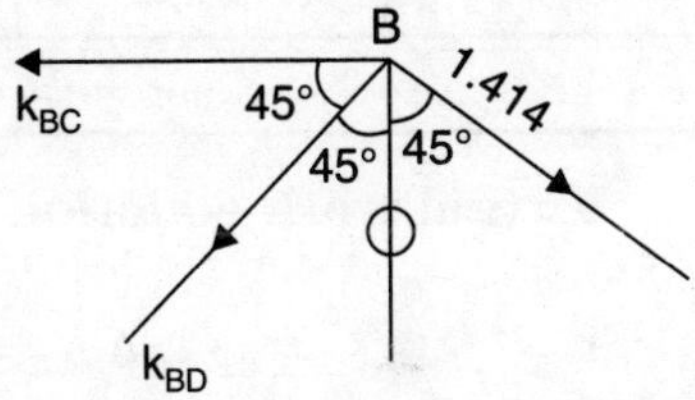

$$1.414 \cos 45° + k_{BD} \cos 45° = 0$$
$$(\downarrow) \quad (\downarrow)$$

$$k_{BD} = -1.414$$

i.e., $k_{BD} = 1.414$ (comp.)

$$\Sigma H = 0 \text{ gives}$$

$$1.414 \cos 45° - k_{BD} \cos 45° - k_{BC} = 0$$
$$(\rightarrow) \quad (\leftarrow) \quad (\leftarrow)$$

$$1.414 \cos 45° - (-1.414) \cos 45° - k_{BC} = 0$$

$$2 \times 1.414 \cos 45° = k_{BC}$$

$$k_{BC} = 2 \text{ kN (Tensile)}$$

Real forces, F: The only external load is 40 kN applied at A. Hence the F values will be 40 times the values obtained as above.

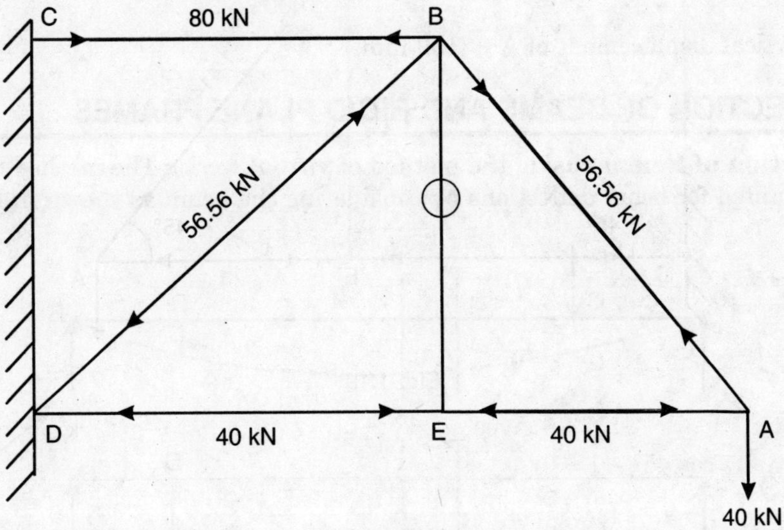

Fig. 1.14

Table 1.3

S.No.	*Member*	*k*	*F (kN)*	*L (m)*	*Σ k FL*
1.	AB	1.414	56.56	4.243	339.34
2.	BC	2.0	80.00	3.0	480.00
3.	DB	– 1.414	– 56.56	4.243	339.34
4.	DE	– 1.0	– 40.0	3.0	120.00
5.	EB	0	0	3.0	0
6.	EA	– 1.0	– 40.0	3.0	120.00
		0			**Σ = 1398.68**

Virtual work equation :

$$1.(\Delta_A)_V = \Sigma\frac{kFL}{AE} + \Sigma(k\ \alpha\ DT.L)$$

$$= \frac{1398.68}{1250 \times 10^{-6} \times 2 \times 10^{8}} + 0 \qquad (\because\ k_{BE} = 0)$$

$$= 5.59 \times 10^{-3} \text{ m} = 5.59 \text{ mm}$$

$$A = 1250 \text{ mm}^2 = 1250 \times 10^{-6} \text{ m}^2$$

$$E = 2 \times 10^5 \text{ N/mm}^2 = \frac{2 \times 10^5}{1000} \times (1000)^2 = 2 \times 10^8 \text{ kN/m}^2.$$

∴ Vertical displacement of A = 5.59 mm.

1.5 DEFLECTION OF BEAMS AND RIGID PLANE FRAMES

Deflection of Beams (using the method of virtual work). The method of virtual work may be formulated for beam deflections by considering the beam as shown in Fig. 1.15.

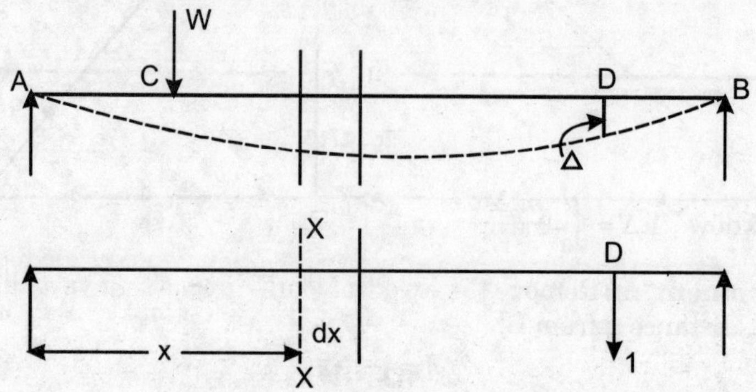

Fig. 1.15

Here the vertical (say) displacement Δ of point D is to be determined. To compute Δ a virtual unit load acting in the direction of Δ (vertical) is placed on the beam at D, and the internal virtual moment m is determined by the method of sections at an arbitrary location x from the left support. When the real loads act on the beam, the point D is displaced downward by Δ. Provided these loads cause linear elastic material response, then the element dx rotates by $d\theta$ [= (M/EI)dx]. Here M is the internal moment at x caused by the real loads. Consequently, the external virtual work done by the unit load is 1.Δ, and the internal virtual work done by the moment m is $m.d\theta = m(\text{M/EI})dx$. Summing the effects on all the elements dx along the beam requires an integration and hence

$$1.\Delta = \int_0^l m \frac{\text{M}}{\text{EI}} dx$$

where 1 = *external virtual unit load acting on the beam in the desired direction of* Δ

m = *internal virtual moment in the beam, expressed as a function of x caused by the above load.*

Δ = *external displacement of the point (at which the displacement is required) caused by the real loads acting on the beam.*

M = *internal moment in the beam expressed as a function of x and caused by the real loads.*

E = *modulus of elasticity of the material, and*

I = *moment of inertia of cross-sectional area, about the neutral axis.*

In a similar manner, the slope dy/dx or θ at a point on the beam's elastic curve can be determined by applying a unit couple or moment at the point, and the corresponding internal moments m_θ have to be determined. The work of the unit couple is 1.θ.

Then
$$1.\theta = \int_0^l m_\theta \frac{\text{M}}{\text{EI}} dx.$$

Example 1.5 *Using the method of virtual work, determine the vertical displacement of point B of the beam shown in Fig. 1.16. Take E = 2 × 10^5 MPa, I = 825 × 10^7 mm^4.*

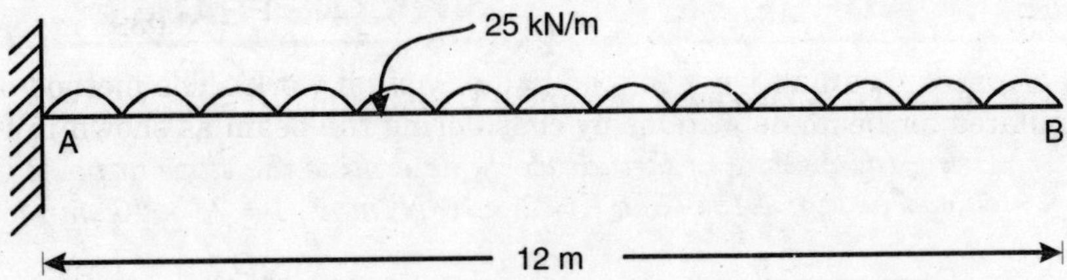

Fig. 1.16

Solution We know , $1.\Delta = \int_0^l \frac{m.\text{M}dx}{\text{EI}}$

Virtual moment, m: Remove the external load. Apply a unit vertical load at B. Consider a section x-x at a distance x from B.

$$m = -1.x \qquad (\because \text{ Hogging moment})$$

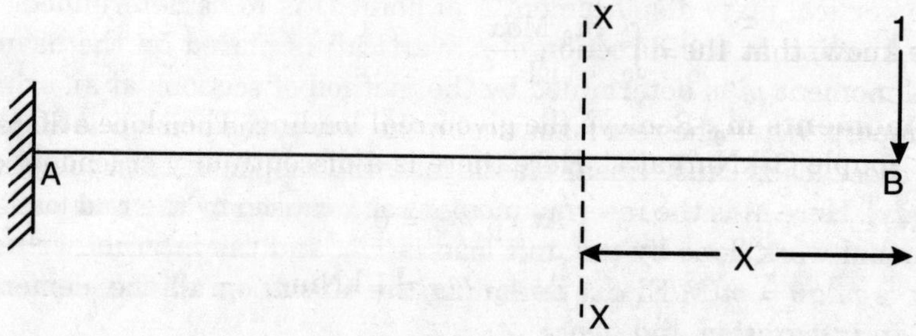

Fig. 1.17

Real moment, M: Using the same x co-ordinate, the internal moment (due to the given loading) M is formulated as,

$$M = -25\ .x.x/2 \qquad \text{(Hogging)}$$

$$M = -12.5x^2$$

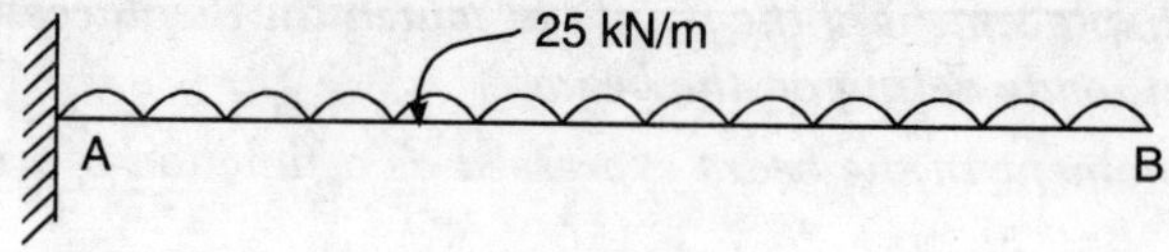

Fig. 1.18

Virtual work equation: $1.(\Delta_B)_V = \int_0^l \frac{m.Mdx}{EI}$

$$= \int_0^{12} \frac{(-1.x)(-12.5x^2)\,dx}{EI} = \int_0^{12} \frac{12.5x^3\,dx}{EI}$$

$$= \left[\frac{12.5x^4}{EI \times 4}\right]_0^{12} = \frac{12.5 \times 12^4}{EI \times 4} = \frac{64800}{EI}$$

$$= \frac{64800}{2 \times 10^8 \times 825 \times 10^{-5}} = 0.0393 \text{ m (or) } 39.3 \text{ mm.}$$

Hence the vertical displacement of point B = 39.3 mm.

Example 1.6 *Using the method of virtual work, determine the slope at point C of the beam shown in Fig. 1.19. Take $E = 2 \times 10^5 N/mm^2$, $I = 75 \times 10^6\ mm^4$.*

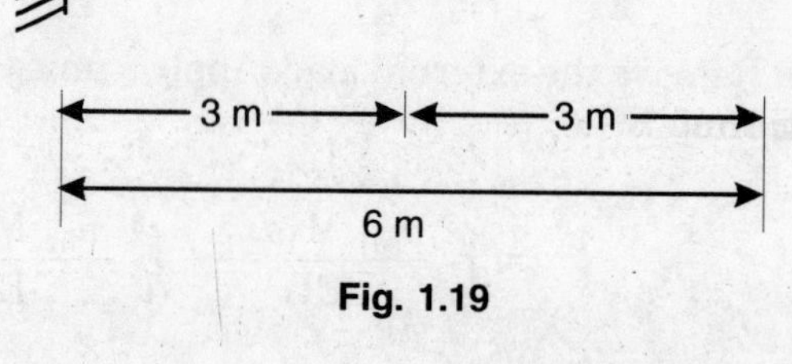

Fig. 1.19

Solution We know, that $1.\theta = \int_0^l \frac{m_\theta \, Mdx}{EI}$

Virtual moments m_θ: Remove the given real loading. The slope at C is determined by applying a unit couple (1 kNm) at C. Here there is a discontinuity of loading on the beam.

$$\text{At } x_1,\ m_{\theta_1} = 0$$

$$\text{At } x_2,\ m_{\theta_2} = 1 \text{ kNm}$$

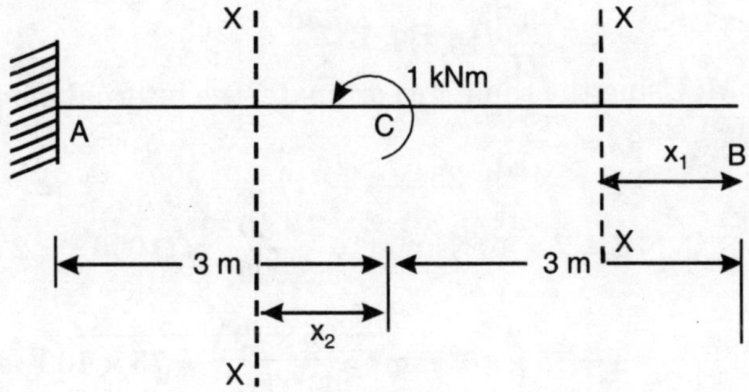

Fig. 1.20

Two x coordinates as shown must be selected in order to determine the total virtual strain energy in the beam.

$$m_{\theta_1} = 0, \quad m_{\theta_2} = 1 \text{ kNm}$$

Real moments M: Using the same coordinates x_1 and x_2, the internal moments M, due to the given external loading, are computed as below.

$$M_1 = -40.x_1 \qquad (\because \text{ Hogging})$$
$$M_2 = -40(x_2 + 3)$$

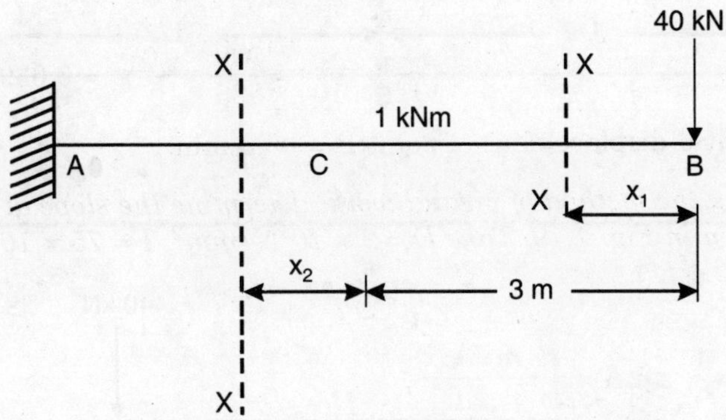

Fig. 1.21

Virtual work equation: $$1.\theta_C = \int \frac{m_\theta \, Mdx}{EI}$$

$$= \int_0^3 \frac{m_{\theta 1} \, M_1 dx_1}{EI} + \int_0^3 \frac{m_{\theta 2} \, M_2 dx_2}{EI}$$

$$= \int_0^3 \frac{(0)(-40x_1)\,dx_1}{EI} + \int_0^3 \frac{(1)[-40(x_2+3)]\,dx_2}{EI}$$

$$= 0 + \int_0^3 -\frac{40}{EI}(x_2+3)\,dx_2$$

$$= \left[-\frac{40}{EI}\left(\frac{x_2^{\,2}}{2}+3x_2\right)\right]_0^3 = -\frac{40}{EI}\left(\frac{3^2}{2}+3\times3\right)$$

$$= -\frac{40}{EI}(13.5) = -\frac{540}{EI}$$

$$\theta_C = -\frac{540}{EI}$$

$$E = 2\times10^5\ \text{N/mm}^2 = \frac{2\times10^5}{1000}\times(1000)^2 = 2\times10^8\ \text{kN/m}^2$$

$$I = 75\times10^6\ \text{mm}^4 = \frac{75\times10^6}{(1000)^4} = 75\times10^{-6}\ \text{m}^4$$

$$\therefore \quad \theta_C = \frac{540}{2\times10^8\times75\times10^{-6}} = -0.036\ \text{radians.}$$

The negative sign indicates that θ_C is opposite to the direction of the virtual couple assumed.

$$\therefore \quad \theta_C = -0.036\ \text{radians.}$$

Example 1.7 *Using the method of virtual work, determine the vertical deflection under the load point for the beam shown in Fig. 1.22 EI is constant.*

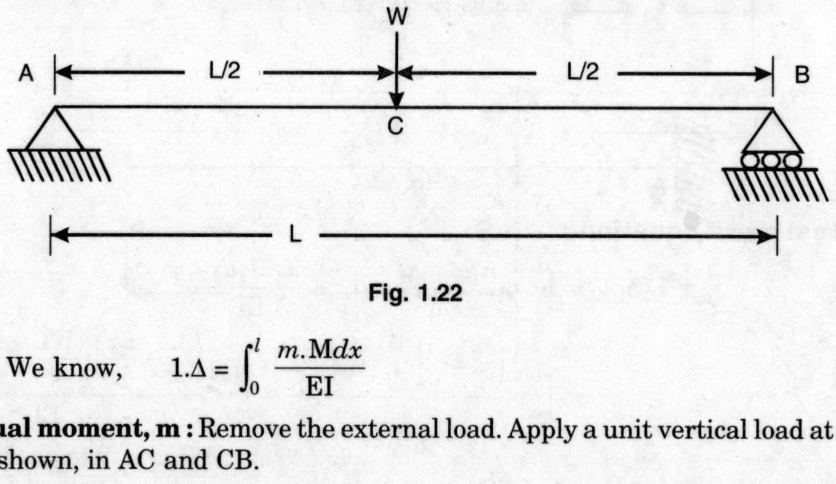

Fig. 1.22

Solution We know, $$1.\Delta = \int_0^l \frac{m.M\,dx}{EI}$$

Virtual moment, m : Remove the external load. Apply a unit vertical load at C. Consider sections as shown, in AC and CB.

By symmetry $$V_A = V_B = \frac{1}{2}$$

$$m_1 = \frac{1}{2}.x_1 \qquad \left(\text{Limits } 0 \text{ to } \frac{L}{2}\right)$$

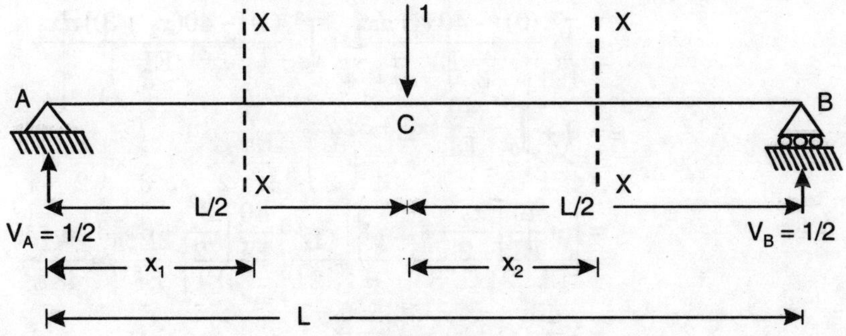

Fig. 1.23

$$m_2 = \frac{1}{2}\left(\frac{L}{2} + x_2\right) - 1.x_2 = \frac{L}{4} + \frac{x_2}{2} - x_2$$

$$m_2 = \frac{L}{4} - \frac{x_2}{2} \qquad \left(\text{Limits 0 to } \frac{L}{2}\right)$$

Real moment, M : Using the same x coordinates, the internal moment (due to the given loading) M is obtained as,

$$M_1 = \frac{W}{2} . x_1$$

$$M_2 = \frac{W}{2}\left(\frac{L}{2} + x_2\right) - W . x_2 = \frac{WL}{4} + \frac{Wx_2}{2} - Wx_2 = \frac{WL}{4} - \frac{Wx_2}{2}$$

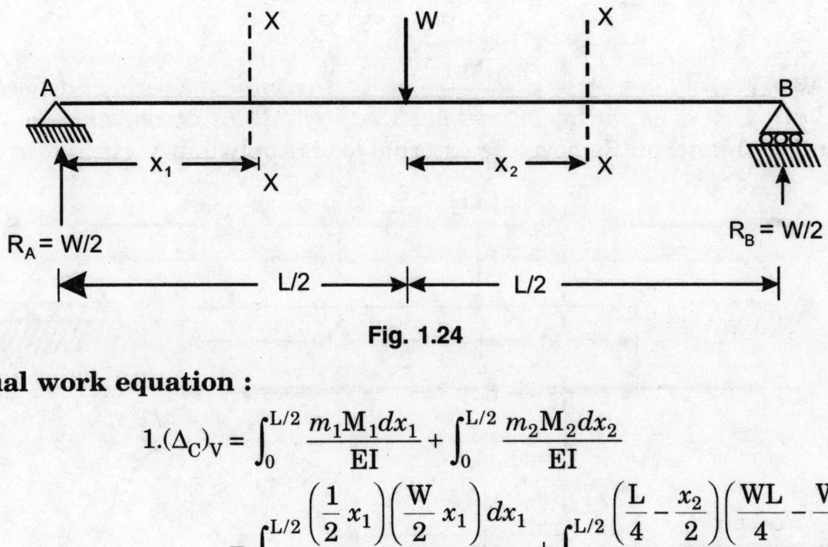

Fig. 1.24

Virtual work equation :

$$1.(\Delta_C)_V = \int_0^{L/2} \frac{m_1 M_1 dx_1}{EI} + \int_0^{L/2} \frac{m_2 M_2 dx_2}{EI}$$

$$= \int_0^{L/2} \frac{\left(\frac{1}{2} x_1\right)\left(\frac{W}{2} x_1\right) dx_1}{EI} + \int_0^{L/2} \frac{\left(\frac{L}{4} - \frac{x_2}{2}\right)\left(\frac{WL}{4} - \frac{Wx_2}{2}\right) dx_2}{EI}$$

$$= \int_0^{L/2} \frac{W{x_1}^2 dx_1}{4EI} + \int_0^{L/2} \frac{\left(\frac{WL^2}{16} - \frac{WLx_2}{8} - \frac{WLx_2}{8} + \frac{W{x_2}^2}{4}\right) dx_2}{EI}$$

$$= \frac{W}{4EI} \int_0^{L/2} {x_1}^2 dx_1 + \frac{W}{4EI} \int_0^{L/2} \left(\frac{L^2}{4} - \frac{2Lx_2}{2} + {x_2}^2\right) dx_2$$

$$= \frac{W}{4EI}\left[\left(\frac{x_1^{\,3}}{3}\right)_0^{L/2} + \left(\frac{L^2}{4}x_2 - L\frac{x_2^{\,2}}{2} + \frac{x_2^{\,3}}{3}\right)_0^{L/2}\right]$$

$$= \frac{W}{4EI}\left[\frac{1}{3}\left(\frac{L}{2}\right)^3 + \frac{L^2}{4}\left(\frac{L}{2}\right) - \frac{L}{2}\left(\frac{L}{2}\right)^2 + \frac{1}{3}\times\left(\frac{L}{2}\right)^3\right]$$

$$= \frac{W}{4EI}\left[\frac{L^3}{24} + \frac{L^3}{8} - \frac{L^3}{8} + \frac{L^3}{24}\right] = \frac{W}{4EI}\left[\frac{2L^3}{24}\right] = \frac{WL^3}{48EI}$$

i.e., $(\Delta_C)_V = \frac{WL^3}{48EI}$ = Deflection of the beam under the load point.

This problem can also be done by carrying out integration for AC alone and multiplying by 2.

Example 1.8 *A simply supported beam of span 6 m is subjected to a concentrated load of 45 kN at 2 m from the left support. Calculate the deflection under the load point. Take $E = 200 \times 10^6$ kN/m² and $I = 14.0 \times 10^{-6}$ m⁴.*

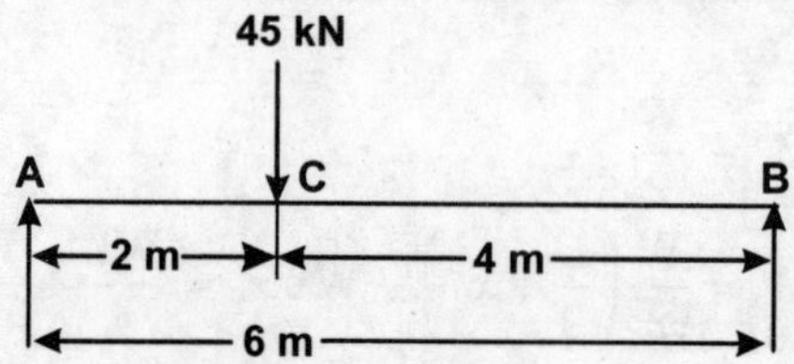

Fig. 1.25

Solution We know $1 \,.\, \Delta = \int_0^l \frac{mMdx}{EI}$

Virtual moment, m: Remove the external load. Apply unit vertical load at C.

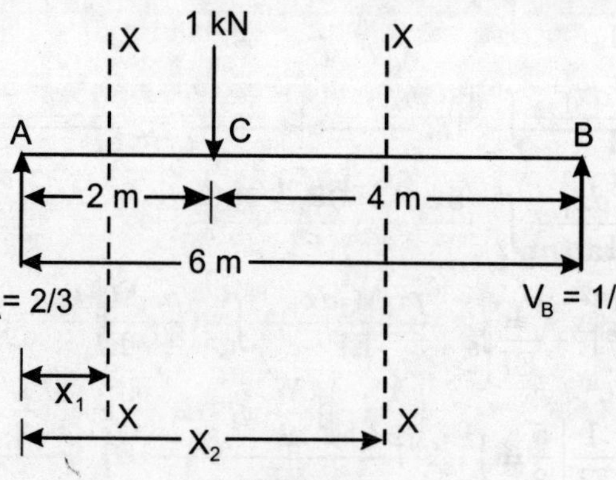

Fig. 1.26

Taking moments about B

$V_A \times 6 - 1 \times 4 = 0$ $\qquad V_A = \frac{2}{3}$ kN

$V_B = \text{Total load} - V_A$ $\qquad V_B = \frac{1}{3}$ kN

Considering sections *x-x* in AC and CB as shown

$$m_1 = \frac{2}{3} \cdot x_1 \qquad \text{(Limits 0 to 2 m)}$$

$$m_2 = \frac{2}{3} x_2 - 1(x_2 - 2) \qquad \text{(Limits 2 to 6 m)}$$

Real moment, M: Using the above *x* coordinates, the internal moment (due to the given loading), M is obtained as

$$R_A = \frac{2}{3} \times 45 = 30 \text{ kN}$$

$$R_B = \frac{1}{3} \times 45 = 15 \text{ kN} \qquad \text{(Since there is only one concentrated load of 45 kN at C)}$$

$$M_1 = 30 \times x_1$$
$$M_2 = 30x_2 - 45(x_2 - 2)$$

Virtual work equation :

$$1.(\Delta_C)_V = \int_0^l \frac{mMdx}{EI} = \int_0^2 \frac{m_1 M_1 dx_1}{EI} + \int_2^6 \frac{m_2 M_2 dx_2}{EI}$$

$$= \int_0^2 \frac{\left(\frac{2}{3} x_1\right)(30x_1)\, dx_1}{EI} + \int_2^6 \frac{\left(\frac{2}{3} x_2 - (x_2 - 2)\right)[30x_2 - 45(x_2 - 2)]\, dx_2}{EI}$$

$$= \int_0^2 \frac{20{x_1}^2 dx_1}{EI} + \int_2^6 \frac{[20{x_2}^2 - 30x_2(x_2 - 2) - 30x_2(x_2 - 2) + 45(x_2 - 2)^2]\, dx_2}{EI}$$

$$= \int_0^2 \frac{20{x_1}^2 dx_1}{EI} + \int_2^6 \frac{[20{x_2}^2 - 2 \times 30x_2(x_2 - 2) + 45({x_2}^2 - 4x_2 + 4)]\, dx_2}{EI}$$

$$= \int_0^2 \frac{20{x_1}^2 dx_1}{EI} + \int_2^6 \frac{(20{x_2}^2 - 60{x_2}^2 + 120x_2 + 45{x_2}^2 - 180x_2 + 180)\, dx_2}{EI}$$

$$= \int_0^2 \frac{20{x_1}^2 dx_1}{EI} + \int_2^6 \frac{(5{x_2}^2 - 60x_2 + 180)\, dx_2}{EI}$$

$$= \left[\frac{20}{EI} \cdot \frac{{x_1}^3}{3}\right]_0^2 + \frac{1}{EI}\left[5 \cdot \frac{{x_2}^3}{3} - 60 \cdot \frac{{x_2}^2}{2} + 180x_2\right]_2^6$$

$$= \frac{20}{EI} \cdot \frac{8}{3} + \frac{1}{EI}\left[\frac{5}{3}(6^3 - 2^3) - \frac{60}{2}(6^2 - 2^2) + 180\,(6 - 2)\right]$$

$$= \frac{1}{EI}[53.33 + 346.67 - 960 + 720] = \frac{160}{EI}$$

$$= \frac{160}{200 \times 10^6 \times 14 \times 10^{-6}} = 0.0571 \text{ m} \quad \text{(or)} \quad 57.1 \text{ mm}$$

$\therefore$ Deflection under the load point = 57.1 mm.

Alternative method:

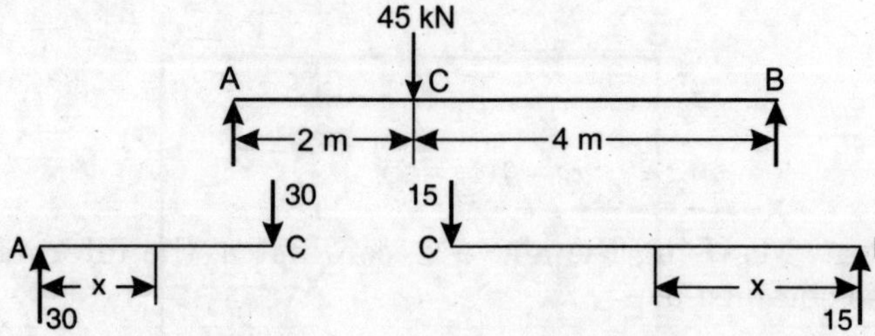

	Portion AC	Portion CB
M	$30x$	$15x$
m	$\dfrac{2}{3}x$	$\dfrac{1}{3}x$, since $R_A = \dfrac{2}{3}$ and $R_B = \dfrac{1}{3}$
$\int_0^2 mMdx$	$\left[\dfrac{60}{3}\dfrac{x^3}{3}\right]_0^2$	$\left[\dfrac{15}{3}\dfrac{x^3}{3}\right]_0^4$
	$= 20 \times \dfrac{8}{3}$	$= 5 \times \dfrac{64}{3}$

$$\Delta_C = \left(\frac{160}{3} + \frac{320}{3}\right)\text{EI} = 0.0571 \text{ m} = 57.1 \text{ mm}$$

Example 1.9 *Using the principle of virtual work, determine the horizontal displacement of support D of the frame shown in Fig. 1.27. The values of I are indicated along the members. E = 200 × 10⁶ kN/m² and I = 300 × 10⁻⁶ m⁴.*

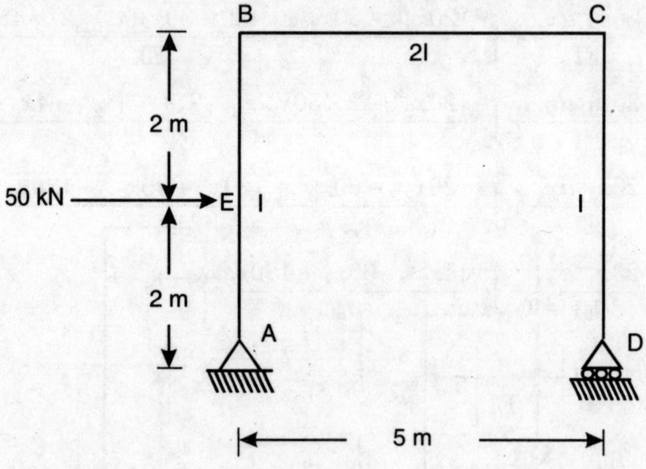

Fig. 1.27

Solution We know that, $1.\Delta = \int_0^l m \dfrac{M}{EI}\, dx$

Virtual moments, m: Remove the external load and apply a unit load in the horizontal direction (direction of the desired displacement) at D. The support reactions and internal virtual moments are computed as under.

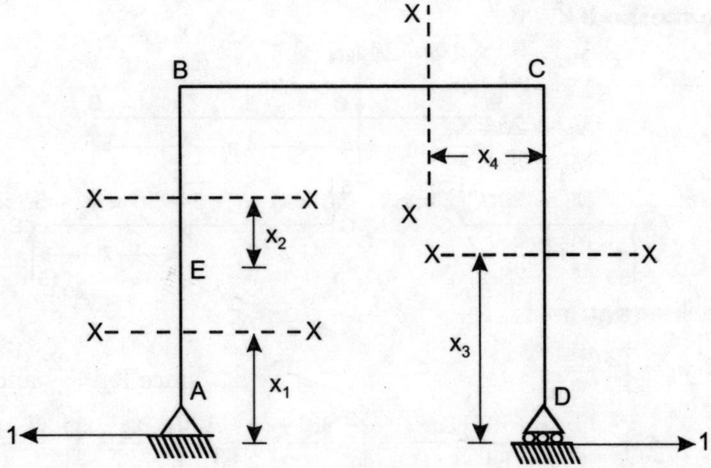

Fig. 1.28

$\Sigma H = 0$ gives $H_A = 1$ kN (←)

(Sign for moments : Left clockwise +ve ; Right anticlockwise +ve)

$m_1 = 1.x_1$ limits 0 to 2 m

$m_2 = 1.(2 + x_2)$ limits 0 to 2 m

$m_3 = 1.x_3$ limits 0 to 4 m

$m_4 = 1 \times 4$ limits 0 to 5 m

Real moments, M: Due to the given loading, the support reactions and real moments are computed as under.

$\Sigma H = 0$ gives $H_A = 50$ kN (←)

Taking moments about A,

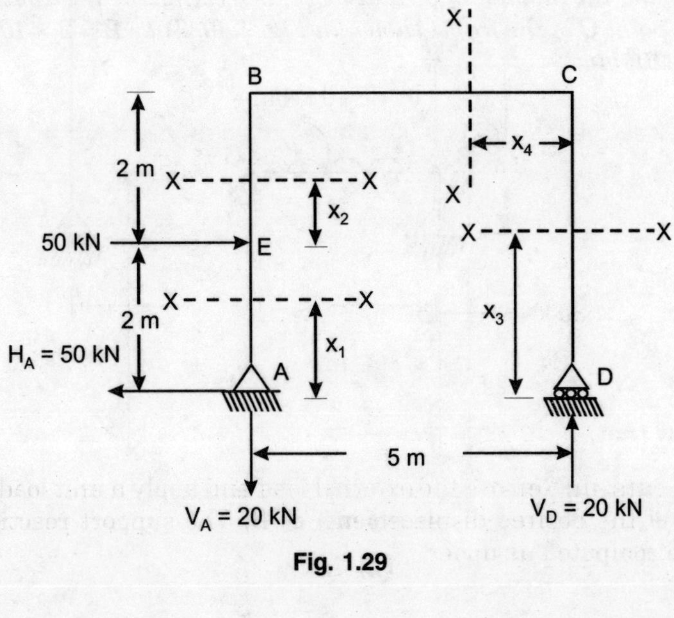

Fig. 1.29

$V_D \times 5 - 50 \times 2 = 0$

$V_D = 50 \times 2/5 = 20$ kN (↑)

$\Sigma V = 0$ gives

$V_A = 20$ kN (↓)

$M_1 = 50 \times x_1$

$M_2 = 50 \times (2 + x_2) - 50 \times x_2 = 100 + 50 \times x_2 - 50 \times x_2 = 100$ kNm

$M_3 = 0$

$M_4 = 20 \times x_4$

Virtual work equation :

$$1.(\Delta_D)_h = \int \frac{m M dx}{EI}$$

$$= \int_0^2 \frac{(1.x_1)(50x_1)\,dx_1}{EI} + \int_0^2 \frac{1(2+x_2)(100)\,dx_2}{EI} + \int_0^4 \frac{(1.x_3)\times 0\,dx_3}{EI} + \int_0^5 \frac{(1\times 4)(20\times x_4)\,dx_4}{E2I}$$

$$= \int_0^2 \frac{(50{x_1}^2)\,dx_1}{EI} + \int_0^2 \frac{(200+100x_2)\,dx_2}{EI} + 0 + \int_0^5 \frac{40x_4\,dx_4}{EI}$$

$$= \frac{1}{EI}\left\{\left[50\frac{{x_1}^3}{3}\right]_0^2 + \left[200x_2 + 100\frac{{x_2}^2}{2}\right]_0^2 + \left[\frac{40{x_4}^2}{2}\right]_0^5\right\}$$

$$= \frac{1}{EI}[133.33 + 600 + 500] = \frac{1233.33}{EI}$$

$$= \frac{1233.33}{200\times 10^6 \times 300 \times 10^{-6}} = 0.02056 \text{ m} = 20.56 \text{ mm}$$

∴ Horizontal displacement of support D, $(\Delta_D)_h = 20.56$ mm (→).

Example 1.10 *Using the method of virtual work, determine the horizontal displacement of a point C of the frame shown in Fig. 1.30. Take $E = 2 \times 10^5$ N/mm², $I = 2.27 \times 10^9$ mm⁴.*

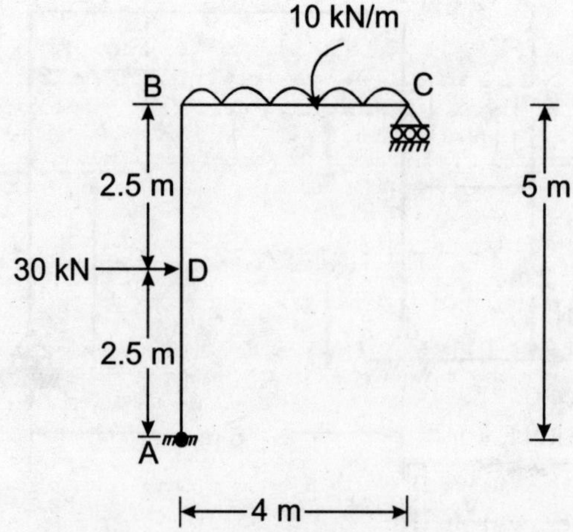

Fig. 1.30

Solution We know that $1.\Delta = \int_0^l \frac{m.Mdx}{EI}$

Virtual moments, m: Remove the external load and apply unit horizontal load at C. The support reactions and internal virtual moments are computed as shown.

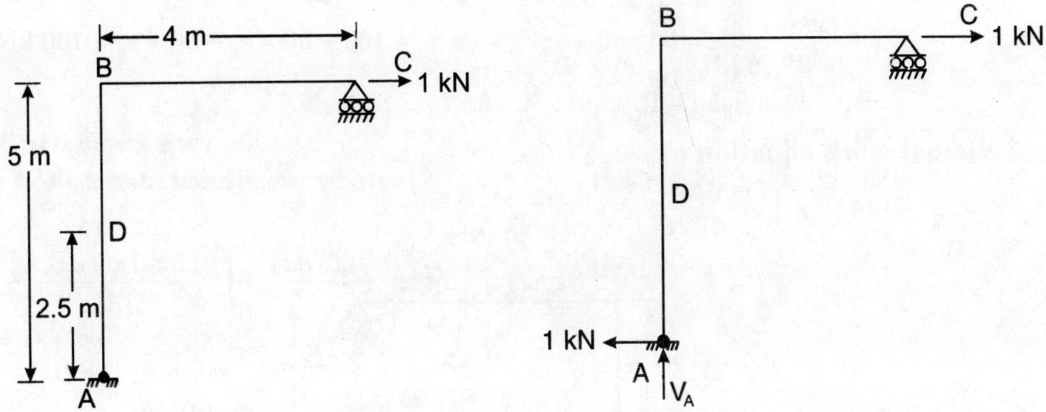

(a) (b)

Fig. 1.31

$\Sigma H = 0$ gives $\qquad H_A = 1$ kN (←)

Taking moments about C, $\quad V_A \times 4 + 1 \times 5 = 0$

(↻) (↻)

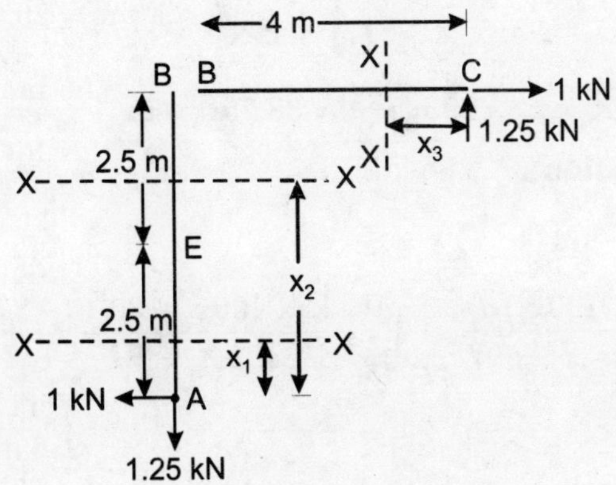

Fig. 1.32

$4V_A = -5$

$V_A = -5/4 = -1.25 \qquad$ *i.e.,* $\qquad V_A = 1.25$ kN (↓)

$\Sigma V = 0$ gives

$V_C = 1.25$ kN (↑)

$m_1 = 1.x_1$ (x_1 varies from 0 to 2.5 m)

$m_2 = 1.x_2$ (x_2 varies from 2.5 to 5.0 m)

$m_3 = 1.25x_3$ (x_3 varies from 0 to 4 m)

Real moments: Due to the given external loading, the support reactions and real moments are computed as shown below.

$\Sigma H = 0$ gives $\qquad H_A = 30$ kN (←)

Taking moments about A, $\quad R_C \times 4 - 10 \times 4 \times 4/2 - 30 \times 2.5 = 0$

$$4R_C = 80 + 75 = 155$$

$$R_C = 38.75 \text{ kN}$$

$$R_A = \text{Total load} - R_C = 10 \times 4 - 38.75 = 1.25 \text{ kN}$$

$M_1 = 30 \times x_1$ $\qquad$ (x_1 varies from 0 to 2.5)

$M_2 = 30x_2 - 30(x_2 - 2.5)$ $\qquad$ (x_2 to be substituted from 2.5 to 5 m)

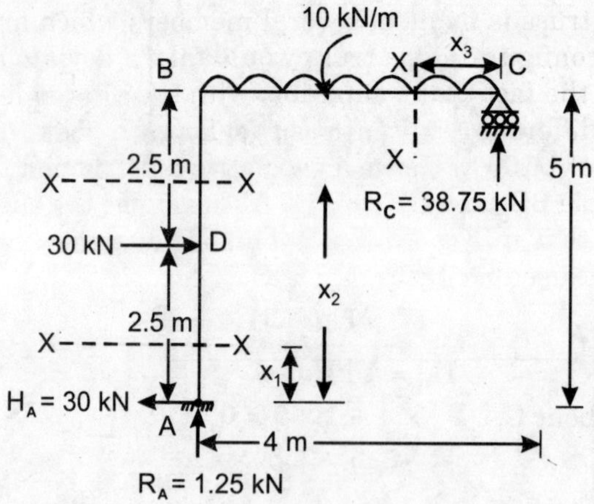

Fig. 1.33

$M_3 = 38.75 \times x_3 - 10.x_3 . \frac{x_3}{2} = 38.75x_3 - 5x_3^2$ $\qquad$ (x_3 varies from 0 to 4 m)

Virtual work equation :

$$1.(\Delta_C)_h = \int \frac{mMdx}{EI}$$

$$= \int_0^{2.5} \frac{(1.x_1)(30x_1)\,dx_1}{EI} + \int_{2.5}^{5.0} \frac{(1.x_2)[30x_2 - 30(x_2 - 2.5)]\,dx_2}{EI}$$

$$+ \int_0^4 \frac{(1.25x_3)(38.75x_3 - 5x_3^2)\,dx_3}{EI}$$

$$= \int_0^{2.5} \frac{30x_1^2 dx_1}{EI} + \int_{2.5}^{5.0} \frac{(30x_2^2 - 30x_2^2 + 75x_2)\,dx_2}{EI} + \int_0^4 \frac{(48.44x_3^2 - 6.25x_3^3)\,dx_3}{EI}$$

$$(\Delta_C)_h = \int_0^{2.5} \frac{30x_1^2 dx_1}{EI} + \int_{2.5}^{5.0} \frac{+75x_2 dx_2}{EI} + \int_0^4 \frac{(48.44x_3^2 - 6.25x_3^3)\,dx_3}{EI}$$

$$= \frac{30}{EI}\left[\frac{x_1^3}{3}\right]_0^{2.5} + \frac{75}{EI}\left[\frac{x_2^2}{2}\right]_{2.5}^{5.0} + \frac{1}{EI}\left[48.44\,\frac{x_3^3}{3} - 6.25\,\frac{x_3^4}{4}\right]_0^4$$

$$= \frac{1}{EI}\left[156.25 + \frac{75}{2}(5^2 - 2.5^2) + \frac{48.44}{3} \times 4^3 - \frac{6.25}{4} \times 4^4\right]$$

$$= \frac{1}{EI}[156.25 + 703.125 + 1033.387 - 400] = \frac{1}{EI}[1492.762]$$

$$E = 2 \times 10^8 \text{ kN/m}^2$$

$$I = 2.27 \times 10^9 \text{ mm}^4 = 2.27 \times 10^9 \times 10^{-12} = 2.27 \times 10^{-3} \text{ m}^4$$

$$\therefore \quad (\Delta_C)_h = \frac{1492.762}{2 \times 10^8 \times 2.27 \times 10^{-3}} = 0.00329 \text{ m}$$

i.e. $(\Delta_C)_h = 3.29$ mm.

Horizontal displacement of point C = 3.29 mm.

1.6 DEFLECTION OF TRUSSES

Williot Diagram. A truss is made of several members which are in compression or in tension. So from the initial configuration, a truss would take a deviated (or deflected) position when loaded. This is due to the fact that compatible with the altered lengths of members due to strains there would be a different resultant position for each node . The changes in lengths are very small (in the order of 1/1000). Hence a geometric construction taking into account the changes in length, for example BB_a and BB_c in Fig. 1.34 can give the changed position of B (B′) only approximately. Notice that in Fig. 1.35. BB_a and BB_c are shown greatly magnified and not to the same scale as that used for the truss ABC.

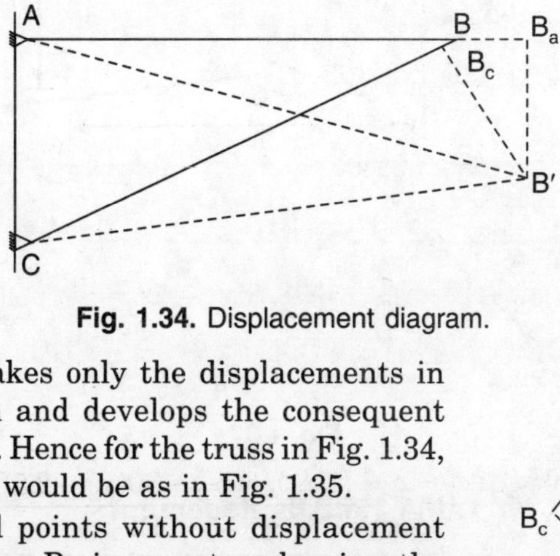

Fig. 1.34. Displacement diagram.

Williot's diagram takes only the displacements in its graphical construction and develops the consequent displacements of the nodes. Hence for the truss in Fig. 1.34, the displacement diagram would be as in Fig. 1.35.

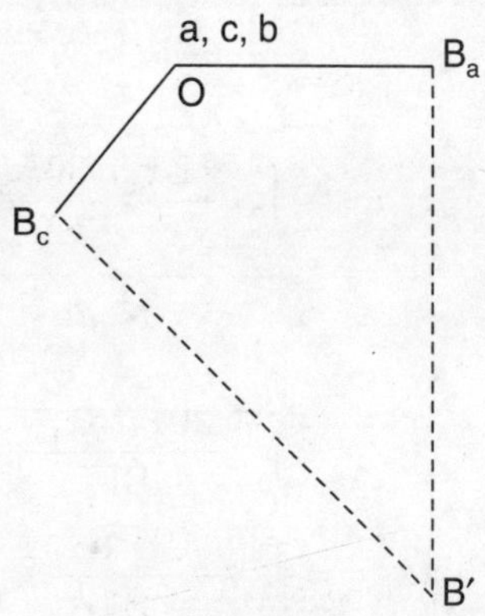

Fig. 1.35. Williot's diagram.

In this diagram all points without displacement would be at the origin. $0 - B_a$ is a vector showing the elongation of AB along AB. $0 - B_c$ is a vector showing the movement of B along BC due to shortening of BC. *a* and *c* are of course points fixed in position. Normals to $0 - B_a$ at B_a and to $0 - B_c$ at B_c intersect at B′ to give the displaced position of B from its original undisplaced position.

Figures 1.34 and 1.35 differ only in minute details.

1. Figure 1.34 relegates all undisplaced points to the origin. This would include the initial positions of all nodes in a truss.

2. The distance from the origin of any newly plotted position of a node indicates its vector displacement from its undisplaced position.

Example 1.11 will further amplify the plotting and interpretation of Williot's displacement diagram.

Example 1.11 *Due to the load P in a 5 member truss in Fig. 1.36, the elongations of the members in mm are given in Table 1.4. Draw the Williot diagram and determine the displacements of B, C and D.*

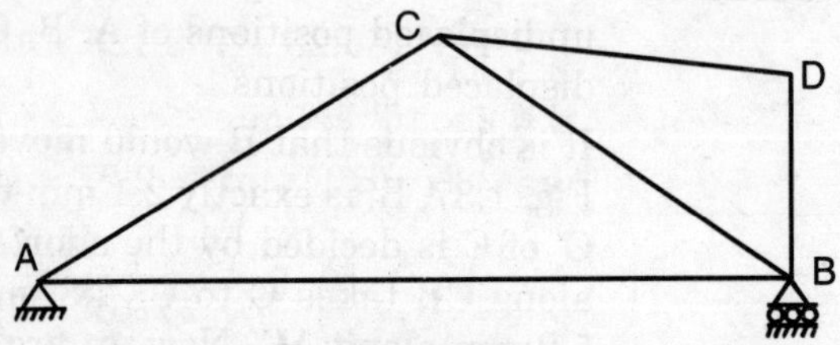

Fig. 1.36

Table 1.4

Member	*Length (mm)*	*Elongations (mm)*
AB	*6000*	*– 2.1*
AC	*4000*	*5.2*
BC	*4000*	*– 2.8*
BD	*2000*	*0*
CD	*3000*	*6.1*

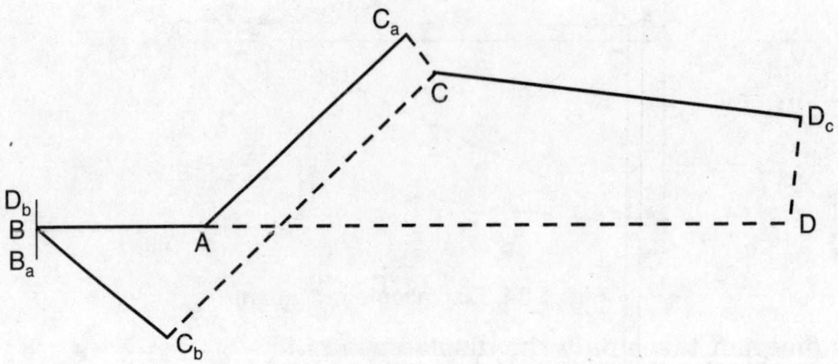

Fig. 1.37

Table 1.5. Displacements.

Node	*X-displacement (mm)*	*Y-displacement (mm)*
A	*0*	*0*
B	*– 2.10*	*0*
C	*4.10*	*3.15*
D	*10.20*	*0*

Solution The Williot's diagram would have *a, b, c* and *d* at the origin. These are the undisplaced positions of A, B, C and D. We shall look to locate B′, C′ and D′ the displaced positions.

It is obvious that B would move to its left because AB shortens by 2.1 mm. So, in Fig. 1.37, B′ is exactly 2.1 mm to the left of *a*. B′ is also B*a*. The displaced position C′ of C is decided by the shortening of BC and the elongation of AC. Shortening along CB takes C to C_b 2.8 mm along CB. Elongation along AC takes C to C_a, 5.2 mm along AC. Now an arc at C_b perpendicular to BC and another arc at C_a perpendicular to AC intersect C′. This is the displaced position of C. The arc in Williot's diagram are approximated by perpendicular lines C_b C′ and C_a C′. This is justified because taking into account our 10 : 1 scale, the arc radius would be 40,000 mm. These arcs are shown in dotted lines.

To fix D′, likewise, mark the elongations of CD from C′ and BD from B′. That would give us D_c and D_b. Draw perpendiculars (arcs) to CD at D_c, and BD at D_b. In this case, since BD has zero elongation, D_b will be at B′ itself. Table 1.5 shows the displacements of nodes measured from the Williot's diagram.

Example 1.12 *Table 1.6 shows the lengths and deformations of the members of the cantilever truss, shown in Fig. 1.38. Construct a Williot's diagram and tabulate the displacement of nodes.*

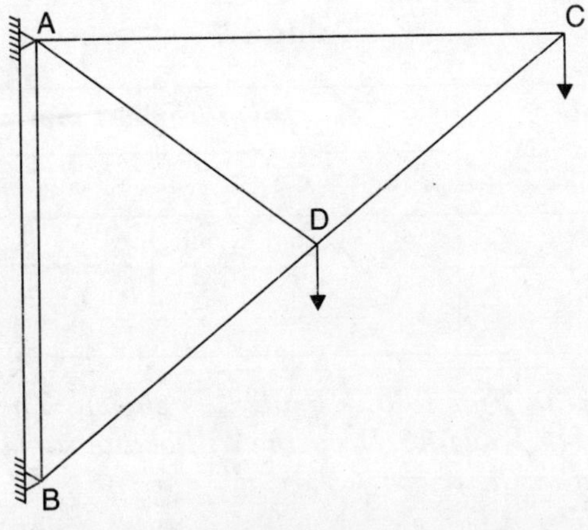

Fig. 1.38

Table 1.6

Member	*Length mm*	*Elongation mm*
AC	*6225*	*15.0*
AD	*4242*	*4.0*
BD	*4242*	*– 10.5*
DC	*4242*	*– 12.0*

Solution Figure 1.39 shows the Williot's diagram of displacements. Table 1.7 shows the displacements.

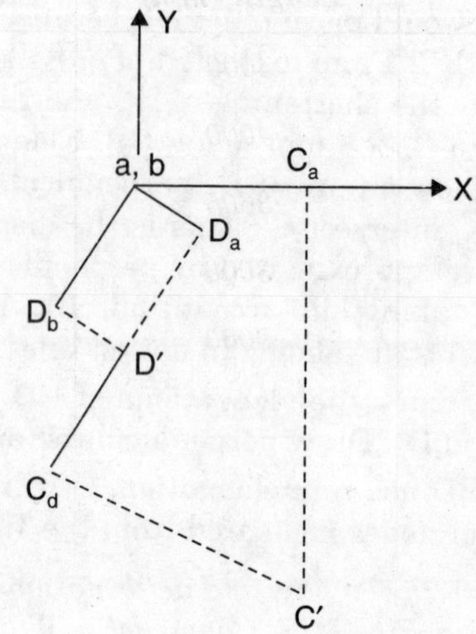

Fig. 1.39. Williot's diagram.

Table 1.7

Node	*Displacement mm*	
	x	*y*
C	+ 15.0	– 47.6
D	– 5.2	– 11.5

Example 1.13 *The truss in Fig. 1.40 is loaded as shown. The member deformations are tabulated in Table 1.8. Draw the Williot diagram and determine the maximum deflection in the truss.*

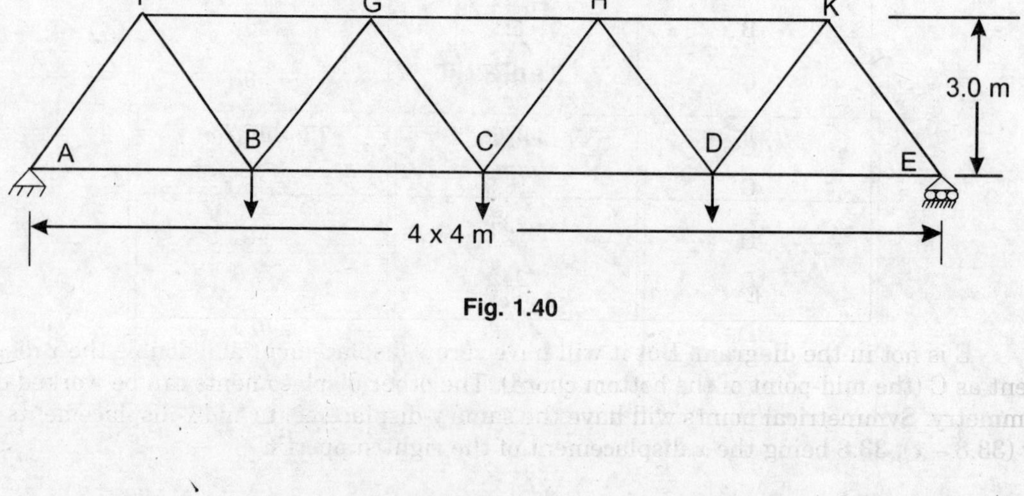

Fig. 1.40

Table 1.8

Members	*Length (mm)*	*Elongation (mm)*
AB, DE	*4000*	*5.1*
BC, CD	*4000*	*10.6*
AF, EK	*3605*	*– 2.2*
FG, HK	*3000*	*– 4.2*
GH	*3000*	*– 6.1*
BF, DK	*3605*	*4.8*
BG, DH	*3605*	*– 2.3*
CG, CH	*3605*	*1.8*

Solution To draw the Williot's diagram, we must be able to find one member (or 2 adjacent nodes) whose magnitude and direction of relative displacement are known. Luckily here we have GH which, because of symmetry (of deformations) will remain horizontal. G′ and H′ will not be at the origin. But starting from G′ and H′ we can locate the origin *a* (or *e*).

By symmentry, it is enough if we do the left half of the truss alone.

Figure 1.41 shows the Williot's diagram. We proceed from G′ as under.

G′, H′, C*g*, C*h*, C′, B*c*, B*g*, B′, F*b*, F*g*, F′, A*f*, A*b*, A′. Now we take A′ as the origin and reckon the displacements of the other nodes (see Table 1.9).

Table 1.9

Node	*Displacement (mm)*	
	x	y
A	0	0
B	5.1	– 35
C	16.9	– 50.1
F	23.1	– 16.7
G	18.9	– 46.1
H	12.8	– 46.1
E*	33.8	0

E is not in the diagram. But it will have zero y displacement and double the x displacement as C (the mid-point of the bottom chord). The other displacements can be worked out by symmetry. Symmetrical points will have the same y-displacements and x-displacements given by $(33.8 - x)$, 33.8 being the x displacement of the right support E.

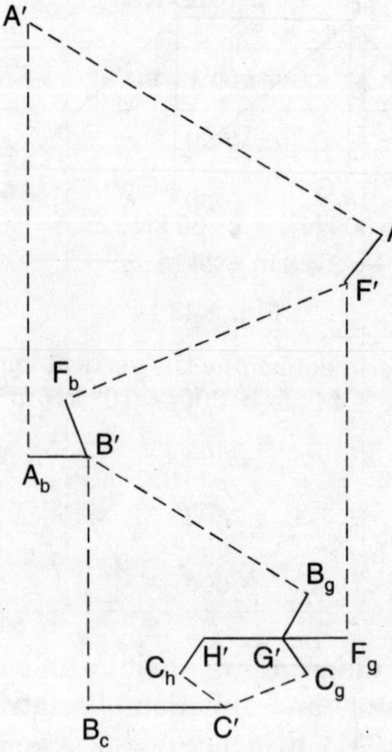

Fig. 1.41 Williot's diagram.

PROBLEMS

1. Using the method of virtual work, find the vertical deflection component of point E of the truss shown in Fig. 1.42. Cross-sectional areas of members are : AE and FD = 2500 mm^2 EF and EC = 1875 mm^2 ; AB, BC, CD, EB and FC = 1250 mm^2 ; E = 200 kN/mm^2.

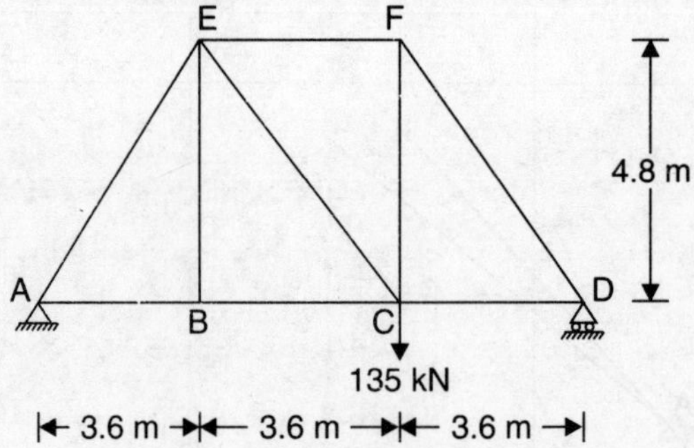

Fig. 1.42

[**Ans.** Vertical deflection of E = 2.217 mm]

2. For the truss shown in Fig. 1.43. Calculate the horizontal displacement of joint D due to the applied loading. The area of upper and lower chords = 1400 mm^2, web members = 1300 mm^2 and E = 200 × 10^3 N/mm^2 use the method of virtual work.

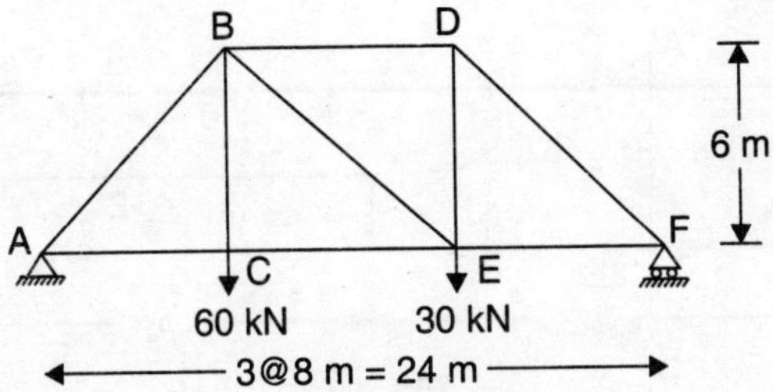

Fig. 1.43

3. Using the principle of virtual work, determine the vertical and horizontal deflection components of joint C of the truss in Fig. 1.44. $E = 200 \times 10^6$ kN/m^2 and sectional area of each bar $A = 100 \times 10^{-6}$ m^2.

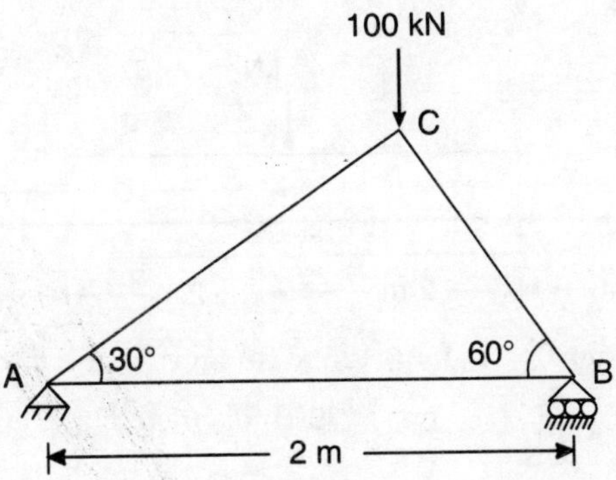

Fig. 1.44

[**Ans.** δ_{CV} = 7.75 mm (↓), δ_{CH} = 0.5 mm (←)]

4. The pin jointed steel frame shown in Fig. 1.45 is hinged to support at A and is over a roller bearing at B. If there is a uniform rise of temperature in all the members equal to 10°C above the temperature at which the frame was erected, calculate the horizontal movement of roller at B. Coefficient of liner expansion $\alpha = 1.1 \times 10^{-5}$; $E = 2 \times 10^5$ N/mm^2. Use the method of virtual work.

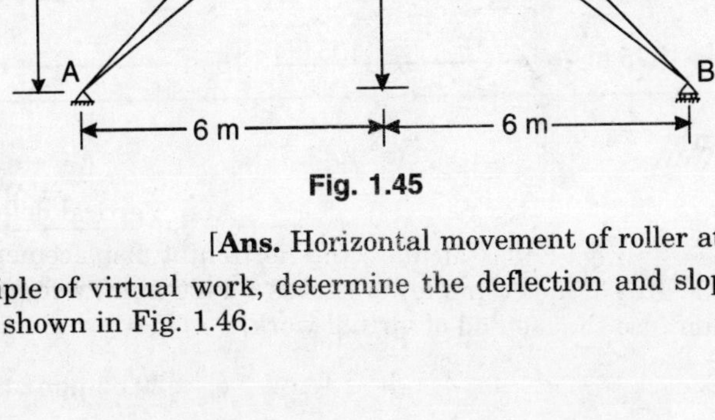

Fig. 1.45

[**Ans.** Horizontal movement of roller at B = 1.32 mm outwards]

5. Using the principle of virtual work, determine the deflection and slope at the free end of the cantilever beam shown in Fig. 1.46.

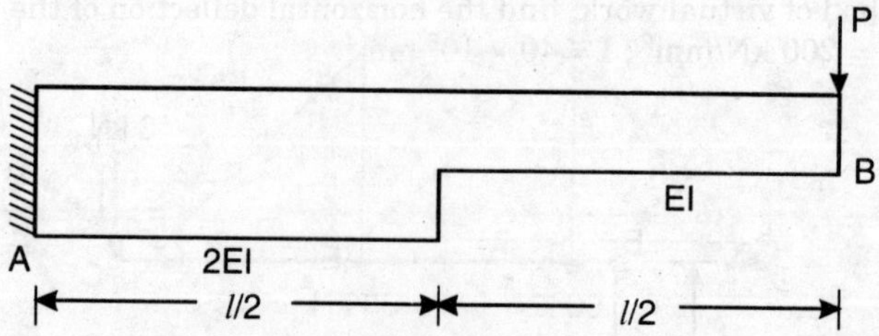

Fig. 1.46

$$\left[\textbf{Ans.}\ \delta_B = \frac{3Pl^3}{16EI},\ \theta_B = \frac{5Pl^2}{16EI}\right]$$

6. Find the vertical deflection of the centre of the beam shown in Fig. 1.47 E = 200 kN/mm^2. Use the method of virtual work.

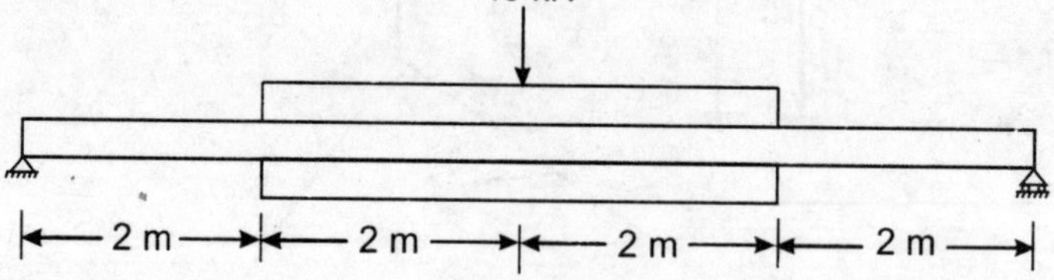

Fig. 1.47

[**Ans.** Δ = 8.22 mm]

7. Using the method of virtual work, find the rotation at A of the beam shown in Fig. 1.48
E = 200 kN/mm^2 ; I = 1.66 × 10^8 mm^4

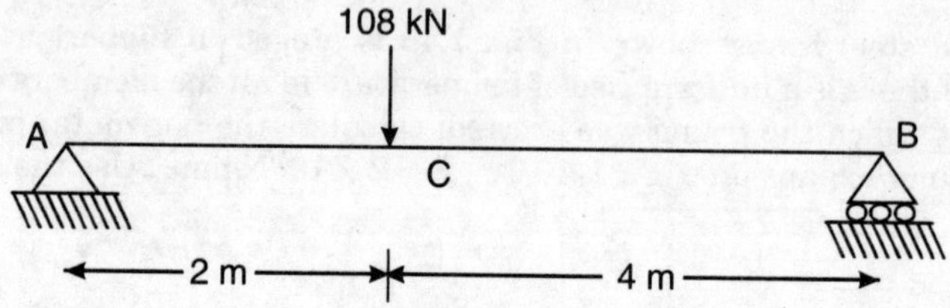

Fig. 1.48

[**Ans.** δ_A = 0.00723 radians (clockwise)]

8. Using the principle of virtual work, determine the deflection at the centre of the beam and slope at end A of the beam shown in Fig. 1.49. Take E = 200 × 10^6 kN/m^2 and I = 13 × 10^{-6} m^4.

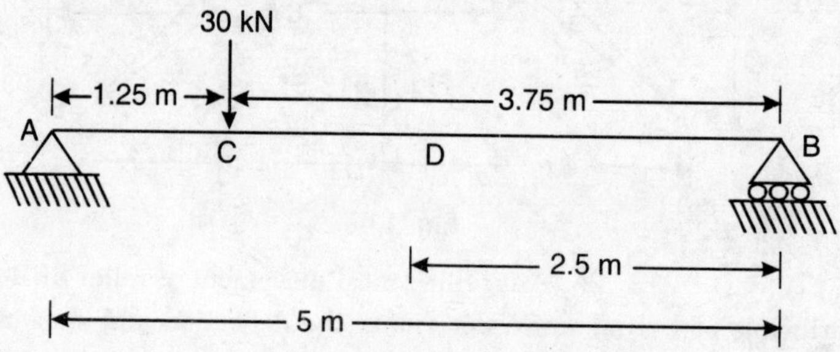

Fig. 1.49

[**Ans.** δ_D = 20.65 mm ; δ_A = 1.578 × 10^{-3} radians]

9. Using the method of virtual work, find the horizontal deflection of the free end of the post shown in Fig. 1.50. $E = 200\ kN/mm^2$; $I = 40 \times 10^6\ mm^4$.

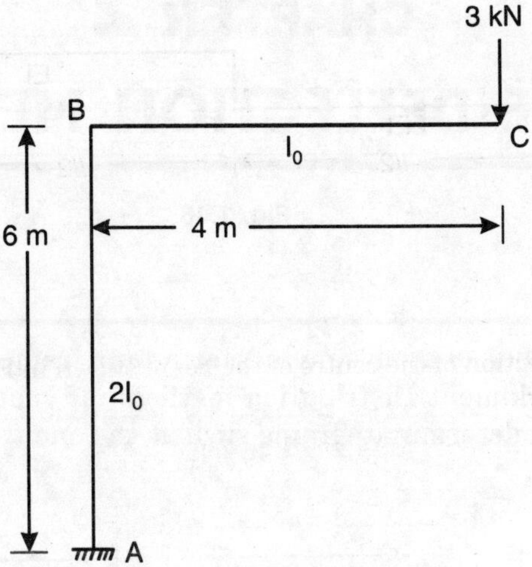

Fig. 1.50

[**Ans.** $\Delta_{CH} = 13.5$ mm]

10. Determine the horizontal displacement at support D of the frame shown in Fig. 1.51. Relative I values are indicated along the members. $E = 200 \times 106\ kN/m^2$ and $I = 300 \times 10^{-6}\ m^4$. Use the principle of virtual work.

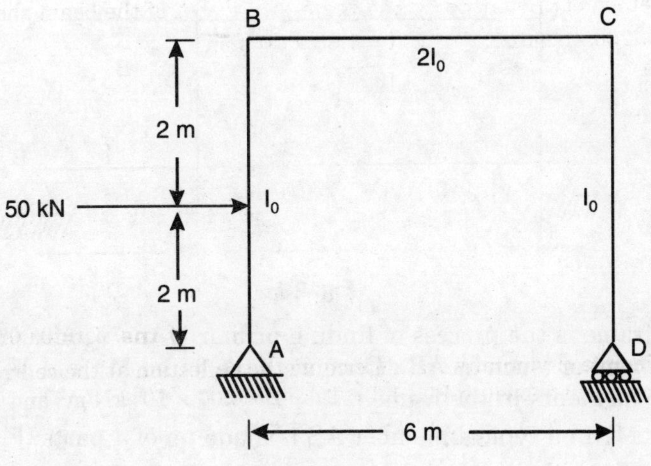

Fig. 1.51

[**Ans.** $\delta_{DH} = 22.2$ mm]

CHAPTER 2
SLOPE DEFLECTION METHOD

2.1 INTRODUCTION

This century old method of analysing indeterminate frames in flexure is the forerunner of other methods such as Moment Distribution method and matrix stiffness method. This is an algebraic method. Any indeterminate frame such as the one shown in Fig. 2.1 is made up of members in flexure.

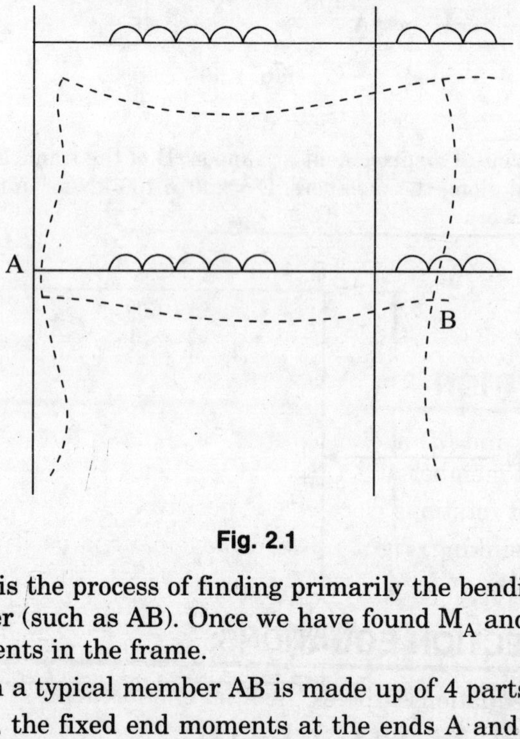

Fig. 2.1

Solving the frame is the process of finding primarily the bending moments at the ends of each and every member (such as AB). Once we have found M_A and M_B in each member we have solved for the moments in the frame.

The moment M_A in a typical member AB is made up of 4 parts (Fig. 2.2).

(*a*) M_{FAB} and M_{FBA}, the fixed end moments at the ends A and B due to the transverse loading on the member, when A and B are restrained from rotation or vertical displacement (θ_A, θ_B, Δ_A and Δ_B are assumed zero in this step) as in Fig. 2.2(*a*).

(*b*) Moments due to the rotation θ_A of A only (keeping the rotation at B and vertical displacements at A and B as zero).

(*c*) Moments due to the rotation θ_B of B only and

(*d*) Moments due to the deflections Δ_A at A and Δ_B at B (and no rotations at A or B).

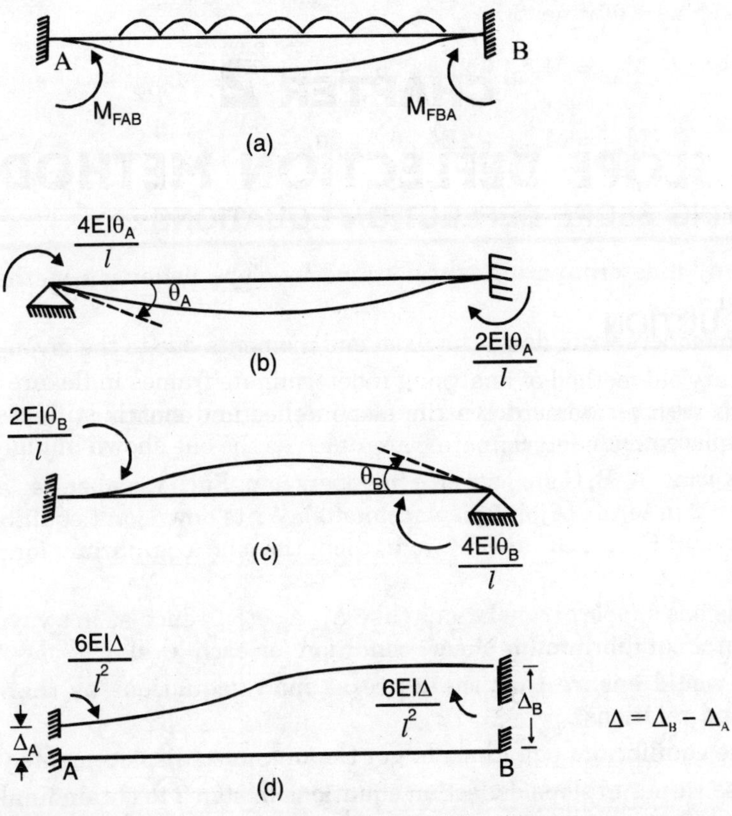

Fig. 2.2. (a) to (d)

2.2 SIGN CONVENTION

Figure 2.2(*b*), (*c*) and (*d*) show the positive signs of forces, moments, deflections and rotations of the ends of members.

For moments and rotations clockwise is positive.

For differential sinking, right upward movement is positive. For support moments, clockwise is positive.

2.3 SLOPE DEFLECTION EQUATIONS

Slope deflection equations express the final end moments of each member in a frame in terms of

(*i*) fixed end moments due to external loads.

(*ii*) moments due to rotation at A (to the final value of rotation θ_A).

(*iii*) moments due to rotation at B.

(*iv*) moments due to differential transverse displacement of B above A.

The values $4EI\theta/l$, $2EI\theta/l$ and $6EI\Delta/l^2$ obtain from earlier results (ref. 1). Hence the slope deflection equations are

$$M_{AB} = M_{FAB} + \frac{4EI\theta_A}{l} + \frac{2EI\theta_B}{l} + \frac{6EI\Delta}{l^2}$$

If I/l is taken as k (stiffness)

$$M_{AB} = M_{FAB} + 2Ek\left(2\theta_A + \theta_B + \frac{3\Delta}{l}\right) \quad ...(2.1)$$

$$M_{BA} = M_{FBA} + 2Ek\left(\theta_A + 2\theta_B + \frac{3\Delta}{l}\right) \quad ...(2.2)$$

2.4 EMPLOYING SLOPE DEFLECTION EQUATIONS

To solve any indeterminate flexural frame by slope deflection method the following steps are adopted :

1. For all members write down the fixed end moments due to the given loading (refer to Table A_1).

2. For each member write down the slope deflection equations in terms of the joint rotations and displacements. (θ_A, θ_B, ..., Δ_A, Δ_B ...).

3. At every joint (A, B, C etc.) several members join. Each member has a moment acting on it (given in step 2 in terms of joint displacements). Write down joint equilibrium equations. For example if at joint C, AC, BC and DC join, then, the joint equilibrium for moments is $M_{CA} + M_{CB} + M_{CD} = 0$.

If the frame has support translations (like Δ_A, Δ_B etc...) such as in swaying frames write down the horizontal equilibrium (or shear) condition for each level.

This step would ensure that there are as many equations as there are unknown displacements (and rotations).

4. Solve the equilibrium equations to get the unknown displacements (θ_A, θ_B etc).

5. Use these values in slope deflection equations in step 2 to obtain final joint moments.

6. Combine the joint moments of each member with the free bending moments (due to external loads) to get the final bending moment diagram in each member.

Example 2.1 will demonstrate the steps above.

2.5 ANALYSIS OF CONTINUOUS BEAMS AND FRAMES

Example 2.1 *Analyse the continuous beam loaded as shown in Fig. 2.3 by the slope deflection method and sketch the bending moment diagram, given :*

$$2I_{AB} = I_{BC} = 2I_{CD} = 2I.$$

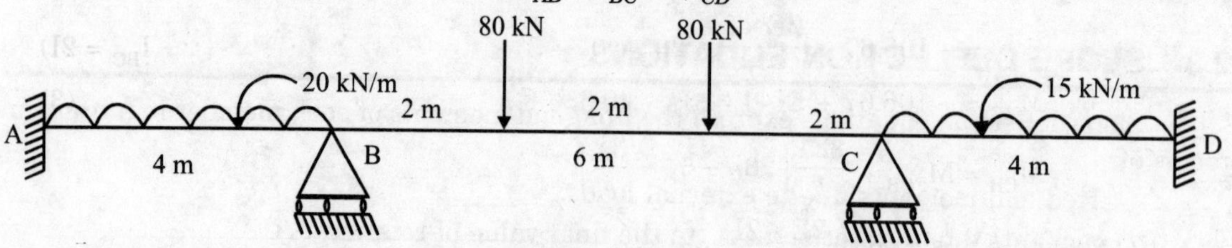

Fig. 2.3

Solution $I_{AB} = I_{CD} = I$, $I_{BC} = 2I$, $\theta_A = \theta_D = 0$ (A and D are fixed)

Unknowns θ_B, θ_C

(*a*) **Fixed end moments :**

Span AB :

$$M_{FAB} = -\frac{wl^2}{12} = -\frac{20 \times 4^2}{12} = -26.67 \text{ kNm}$$

$$M_{FBA} = +\frac{wl^2}{12} = \frac{20 \times 4^2}{12} = 26.67 \text{ kNm}$$

Span BC :

$$M_{FBC} = -\left[\frac{Wa_1b_1^2}{l^2} + \frac{Wa_2b_2^2}{l^2}\right] = -\left[\frac{80 \times 2 \times 4^2}{6^2} + \frac{80 \times 4 \times 2^2}{6^2}\right]$$

$$= -106.67 \text{ kNm}$$

$$M_{FCB} = \left[\frac{Wa_1^2b_1}{l^2} + \frac{Wa_2^2b_2}{l^2}\right] = \left[\frac{80 \times 2^2 \times 4}{6^2} + \frac{80 \times 4^2 \times 2}{6^2}\right]$$

$$= 106.67 \text{ kNm}$$

Span CD :

$$M_{FCD} = -\frac{wl^2}{12} = -\frac{15 \times 4^2}{12} = -20 \text{ kNm}$$

$$M_{FDC} = \frac{wl^2}{12} = \frac{15 \times 4^2}{12} = 20 \text{ kNm}$$

(*b*) **Slope deflection equations :**

$$M_{AB} = M_{FAB} + \frac{2EI}{l}\left[2\theta_A + \theta_B + \frac{3\delta}{l}\right] \quad (\delta = 0, \text{ since no settlement of supports})$$

$$= -26.67 + \frac{2EI}{4}[0 + \theta_B + 0]$$

$$= -26.67 + 0.5\,EI\theta_B \quad ...(1)$$

$$M_{BA} = M_{FBA} + \frac{2EI}{l}\left(2\theta_B + \theta_A + \frac{3\delta}{l}\right)$$

$$= 26.67 + \frac{2EI}{4}(2\theta_B + 0 + 0)$$

$$M_{BA} = 26.67 + EI\theta_B \quad ...(2)$$

$$M_{BC} = M_{FBC} + \frac{2EI}{l}\left(2\theta_B + \theta_C + \frac{3\delta}{l}\right)$$

$$= -106.67 + \frac{2E \times 2I}{6}(2\theta_B + \theta_C + 0) \quad (\because \ I_{BC} = 2I)$$

$$M_{BC} = -106.67 + EI\,(1.333\theta_B + 0.666\,\theta_C) \quad ...(3)$$

$$M_{CB} = M_{FCB} + \frac{2EI}{l}\left(2\theta_C + \theta_B + \frac{3\delta}{l}\right)$$

$$= 106.67 + \frac{2E \times 2I}{6}(2\theta_C + \theta_B + 0)$$

$$M_{CB} = 106.67 + EI\,(1.333\,\theta_C + 0.666\,\theta_B) \quad ...(4)$$

$$M_{CD} = M_{FCD} + \frac{2EI}{l}\left(2\theta_C + \theta_D + \frac{3\delta}{l}\right)$$

$$= -20 + \frac{2EI}{4}(2\theta_C + 0 + 0)$$

$$M_{CD} = -20 + EI\theta_C \qquad ...(5)$$

$$M_{DC} = M_{FDC} + \frac{2EI}{l}\left(2\theta_D + \theta_C + \frac{3\delta}{l}\right)$$

$$= 20 + \frac{2EI}{4}(0 + \theta_c + 0)$$

$$M_{DC} = 20 + EI\,(0.5\theta_C) \qquad ...(6)$$

(*c*) **Equilibrium equations :**

$$M_{BA} + M_{BC} = 0 \qquad ...(7)$$

$$M_{CB} + M_{CD} = 0 \qquad ...(8)$$

Adding (2) and (3), (7) becomes

$$26.67 + EI\theta_B - 106.67 + EI\,(1.333\,\theta_B + 0.666\,\theta_C) = 0$$

$$2.333\theta_B + 0.666\theta_C = \frac{80}{EI} \qquad ...(9)$$

Adding (4) and (5), (8) becomes

$$106.67 + EI\,(0.666\theta_B + 1.333\theta_C) - 20 + EI\theta_C = 0$$

$$0.666\theta_B + 2.333\theta_C = -\frac{86.67}{EI} \qquad ...(10)$$

$$\begin{bmatrix} 2.333 & 0.666 \\ 0.666 & 2.333 \end{bmatrix}\begin{Bmatrix} \theta_B \\ \theta_C \end{Bmatrix} = \frac{1}{EI}\begin{Bmatrix} 80 \\ -86.67 \end{Bmatrix}$$

$$\begin{Bmatrix} \theta_B \\ \theta_C \end{Bmatrix} = \frac{1}{EI}\begin{Bmatrix} 80 \\ -86.67 \end{Bmatrix}\begin{bmatrix} 2.333 & 0.666 \\ 0.666 & 2.333 \end{bmatrix}^{-1}$$

$$= \frac{1}{EI}\begin{Bmatrix} 16 \\ -17.34 \end{Bmatrix}\begin{bmatrix} 2.333 & -0.666 \\ -0.666 & 2.333 \end{bmatrix}$$

$$\begin{Bmatrix} \theta_B \\ \theta_C \end{Bmatrix} = \frac{1}{EI}\begin{Bmatrix} 48.88 \\ -51.11 \end{Bmatrix}$$

(*d*) **Final moments :** Substituting the values of θ_B and θ_C in equations (1) to (6)

$$M_{AB} = -26.67 + 0.5\,EI\left(\frac{48.88}{EI}\right) = -2.23 \text{ kNm}$$

$$M_{BA} = 26.67 + EI\left(\frac{48.88}{EI}\right) = 75.55 \text{ kNm}$$

$$M_{BC} = -106.67 + EI\,(1.333\theta_B + 0.666\theta_C)$$

$$= -106.67 + EI\left[1.333\left(\frac{48.88}{EI}\right) + 0.666\left(\frac{-51.11}{EI}\right)\right] = -75.55 \text{ kNm}$$

$$M_{CB} = 106.67 + EI\,(0.666\theta_B + 1.333\theta_C)$$

$$= 106.67 + EI\left[0.666\left(\frac{48.88}{EI}\right) + 1.333\left(\frac{-51.11}{EI}\right)\right] = +71.09 \text{ kNm}$$

$$M_{CD} = -20 + EI\theta_C = -20 + EI\left(\frac{-51.11}{EI}\right) = -71.1 \text{ kNm}$$

$$M_{DC} = 20 + EI\,(0.5\theta_C) = 20 + EI\left[0.5\times\left(\frac{-51.11}{EI}\right)\right]$$

$$= -5.56 \text{ kNm}$$

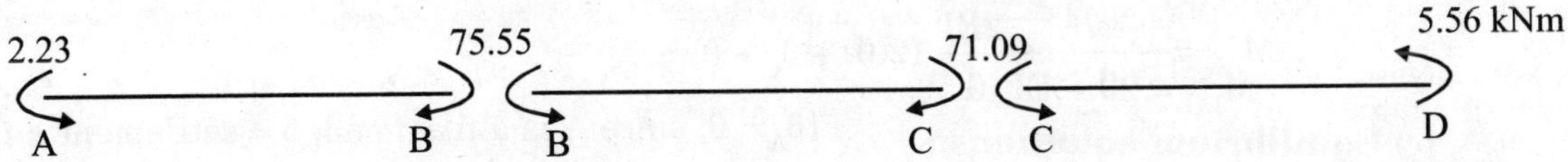

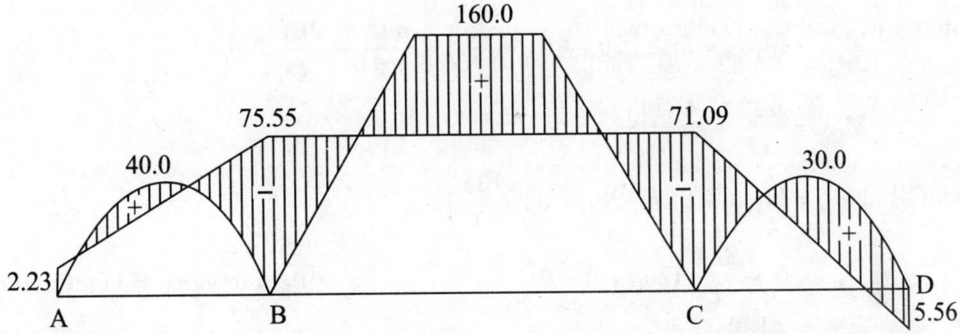

Fig. 2.4. Bending moment diagram.

Example 2.2 *Analyse the two span continuous beam loaded as shown in Fig. 2.5 by the slope deflection method and sketch the bending moment and shear force diagrams.*

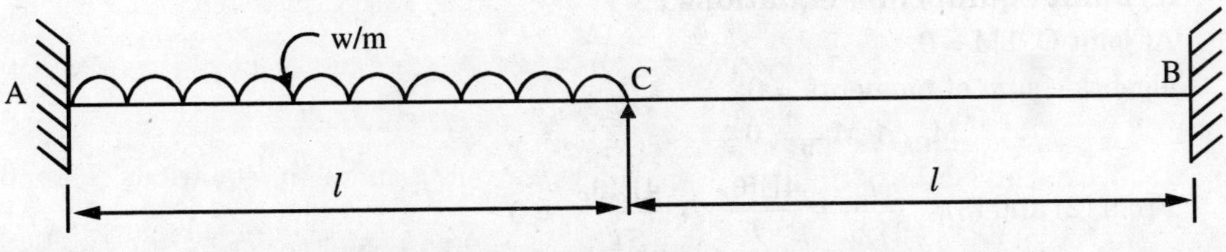

Fig. 2.5

Solution (*a*) **Fixed end moments :**

Span AC : $M_{FAC} = -\dfrac{wl^2}{12}$

$M_{FCA} = \dfrac{wl^2}{12}$

Span CB : $M_{FCB} = 0$

$M_{FBC} = 0$

(*b*) **Slope deflection equations :**

Span AC : M_{AC} = Final moment at A

$$M_{AC} = \text{Fixed moment at A} + 2Ek\left(2\,(\text{slope at A}) + \text{slope at C} + \frac{3\delta}{l}\right)$$

$$= M_{FAC} + \frac{2EI}{l}\left(2\theta_A + \theta_C + \frac{3\delta}{l}\right)$$

$$M_{AC} = -\frac{wl^2}{12} + \frac{2EI}{l}\,(2(0) + \theta_C + 0)$$

[$\theta_A = 0$, since A is a fixed end, δ = settlement = 0]

$$M_{AC} = -\frac{wl^2}{12} + \frac{2EI\theta_C}{l} \quad ...(1)$$

$$M_{CA} = M_{FCA} + 2Ek\left(2\theta_C + \theta_A + \frac{3\delta}{l}\right) = \frac{wl^2}{12} + \frac{2EI}{l}\,(2\theta_C)$$

$$M_{CA} = \frac{wl^2}{12} + \frac{4EI\theta_C}{l} \quad ...(2)$$

Span CB : $M_{CB} = M_{FCB} + 2Ek\left(2\theta_C + \theta_B + \frac{3\delta}{l}\right)$

$$= 0 + \frac{2EI}{l}\,(2\theta_C + 0 + 0)$$

[$\theta_B = 0$, since B is a fixed end]

$$M_{CB} = \frac{4EI\theta_C}{l} \quad ...(3)$$

$$M_{BC} = M_{FBC} + 2Ek\left(2\theta_B + \theta_C + \frac{3\delta}{l}\right) = 0 + \frac{2EI}{l}\,(2(0) + \theta_C + 0)$$

$$M_{BC} = \frac{2EI\theta_C}{l} \quad ...(4)$$

(*c*) **Joint equilibrium equations :**

At joint C, $\Sigma M = 0$.

i.e., algebraic sum of moments = 0

$$M_{CA} + M_{CB} = 0$$

From (2) and (3), $\frac{wl^2}{12} + \frac{4EI\theta_C}{l} + \frac{4EI\theta_C}{l} = 0$

$$\frac{wl^2}{12} + \frac{8EI\theta_C}{l} = 0, \quad \frac{8EI\theta_C}{l} = -\frac{wl^2}{12}$$

$$\theta_C = -\frac{wl^3}{12 \times 8\,EI}$$

$$\theta_C = -\frac{wl^3}{96EI}$$

(*d*) **Final moments :**

$$M_{AC} = -\frac{wl^2}{12} + \frac{2EI\theta_C}{l} = -\frac{wl^2}{12} + \frac{2EI}{l}\left(-\frac{wl^3}{96EI}\right)$$

$$= -\frac{wl^2}{12} - \frac{wl^2}{48} = -wl^2\left(\frac{1}{12} + \frac{1}{48}\right) = -wl^2\left(\frac{4+1}{48}\right)$$

$$M_{AC} = -\frac{5wl^2}{48}$$

$$M_{CA} = \frac{wl^2}{12} + \frac{4EI\theta_C}{l} = \frac{wl^2}{12} + \frac{4EI}{l}\left(-\frac{wl^3}{96EI}\right)$$

$$= \frac{wl^2}{12} - \frac{wl^2}{24} = \frac{2wl^2 - wl^2}{24} = \frac{wl^2}{24}$$

$$M_{CB} = \frac{4EI\theta_C}{l} = \frac{4EI}{l}\left(-\frac{wl^3}{96EI}\right) = -\frac{wl^2}{24}$$

$$M_{BC} = \frac{2EI\theta_C}{l} = \frac{2EI}{l}\left(-\frac{wl^3}{96EI}\right) = -\frac{wl^2}{48}$$

(*e*) **To draw S.F.D :** Span AC :

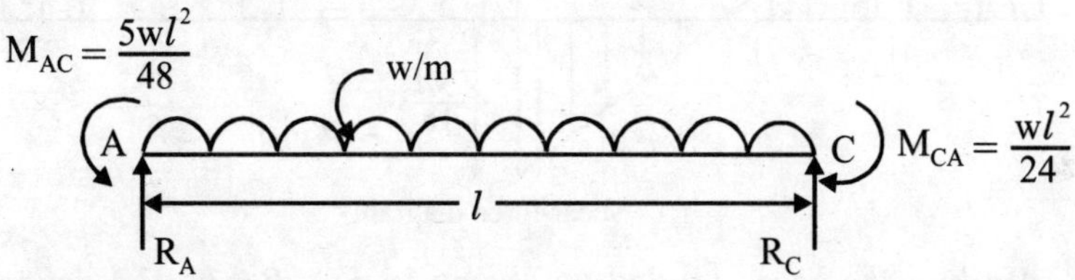

Taking moments about A,

$$R_C \times l - \frac{w.l.l}{2} + \frac{5w.l^2}{48} - \frac{wl^2}{24} = 0$$

$$R_C \times l - wl^2\left(\frac{1}{2} - \frac{5}{48} + \frac{1}{24}\right) = 0$$

$$R_C \times l = wl^2\left(\frac{24 - 5 + 2}{48}\right) = wl^2\left(\frac{21}{48}\right)$$

$$R_C = \frac{7wl}{16}$$

$$R_A = \text{Total load} - R_C = wl - \frac{7wl}{16}$$

$$R_A = \frac{9wl}{16}$$

Span CB :

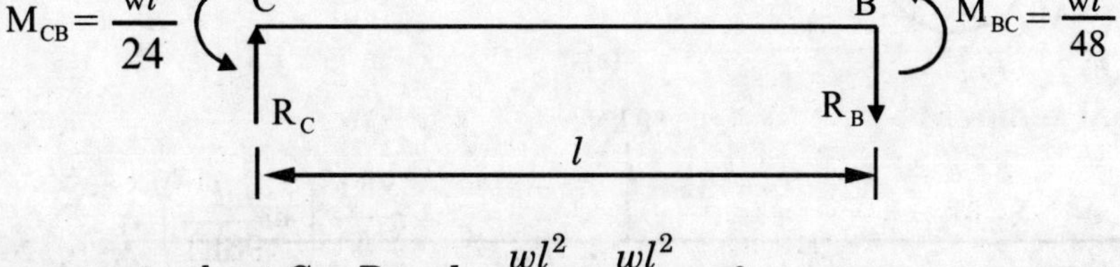

Taking moments about C, $R_B \times l - \frac{wl^2}{48} - \frac{wl^2}{24} = 0$

(↻) (↺) (↺)

$$R_B = \frac{wl}{16}\ (\downarrow), \quad R_C = \frac{wl}{16}\ (\uparrow)$$

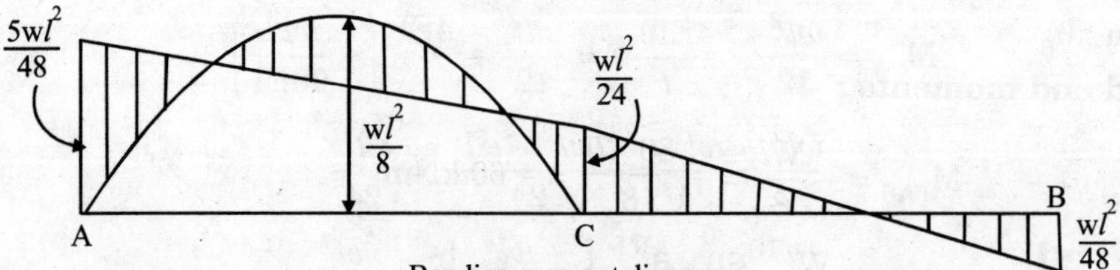

Bending moment diagram

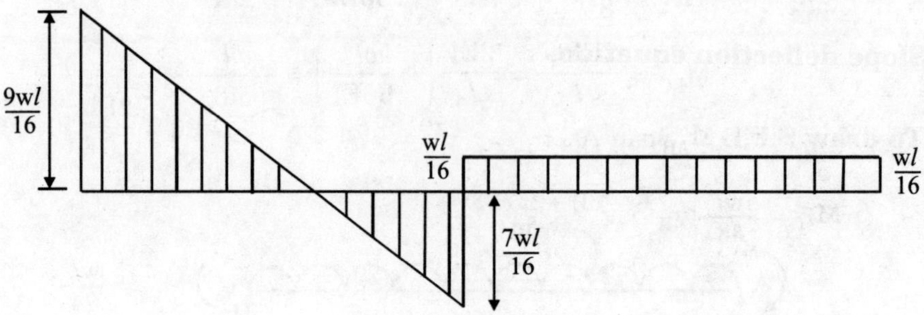

Shear force diagram

Example 2.3 *Analyse the beam loaded as shown in Fig. 2.6 by the deflection method and draw Bending moment diagram and Shear force diagram. EI is constant.*

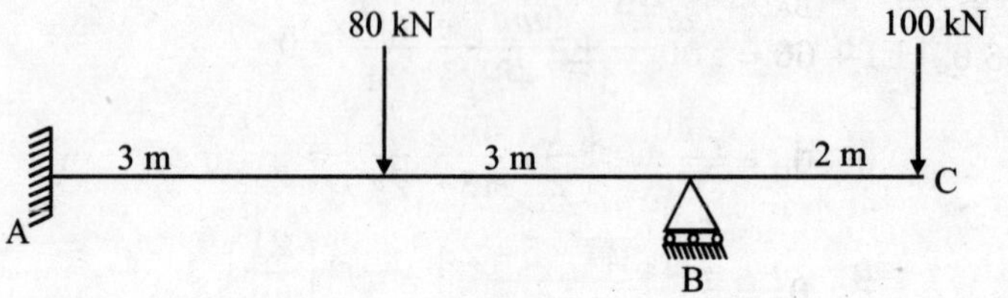

Fig. 2.6

Solution

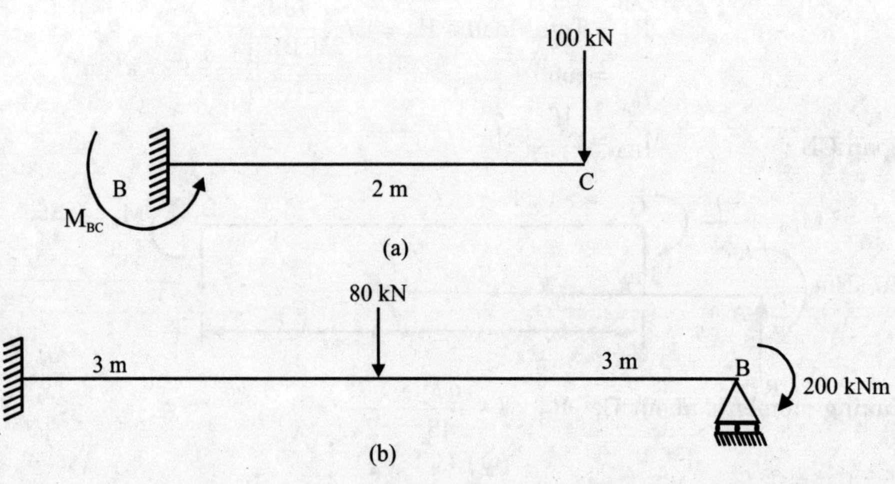

Fig. 2.7

Since $M_{BC} = -200$ kNm

$M_{BA} = 200$ kNm

Unknown : θ_B

(*a*) **Fixed end moments :**

$$M_{FAB} = -\frac{Wl}{8} = -\frac{80 \times 6}{8} = -60 \text{ kNm}$$

$$M_{FBA} = \frac{Wl}{8} = \frac{80 \times 6}{8} = 60 \text{ kNm}$$

(*b*) **Slope deflection equations :**

$$M_{AB} = M_{FAB} + \frac{2EI}{l}\left[2\theta_A + \theta_B + \frac{3\delta}{l}\right] = -60 + \frac{2EI}{6}[0 + \theta_B + 0]$$

$$= -60 + EI\,(0.333\theta_B)$$

$$M_{BA} = M_{FBA} + \frac{2EI}{l}\left[\theta_A + 2\theta_B + \frac{3\delta}{l}\right] = 60 + \frac{2EI}{6}[0 + 2\theta_B + 0]$$

$$= 60 + EI\,(0.666\theta_B)$$

(*c*) **Equilibrium equation :**

$$M_{BA} = 200 \text{ kNm}$$

$$(0.666\,\theta_B)\,EI + 60 = 200$$

$$0.666\theta_B = \frac{140}{EI}$$

$$\theta_B = \frac{210.21}{EI}$$

(*d*) **Final moments :**

$$M_{AB} = -60 + EI\,(0.333\theta_B)$$

$$= -60 + EI\left\{0.333 \times \left(\frac{210.21}{EI}\right)\right\} = 9.997 \text{ kNm}$$

$$M_{BA} = 200 \text{ kNm}$$

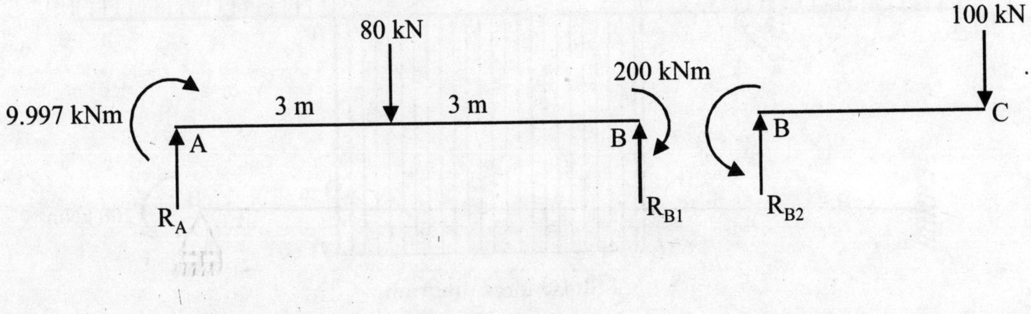

(*e*) **To draw S.F.D :**

$R_A \times 6 - 80 \times 3 + 9.997 + 200 = 0$

$6R_A = 80 \times 3 - 200 + 9.997$ $\qquad R_{B2} = 100$ kN

$R_A = 8.33$ kN

$R_A + R_{B1} = 80$

$R_{B1} = 80 - 8.33 = 71.667$ kN

$R_B = R_{B1} + R_{B2} = 71.667 + 100$

$R_B = 171.667$ kN

(*f*) **Simply supported BM :**

$M_{AB} = Wl/4 = (80 \times 6)/4 = 120$ kNm

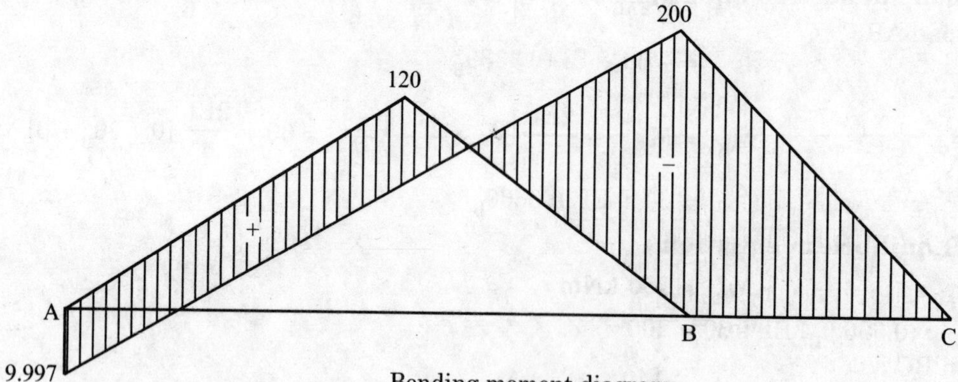

Bending moment diagram

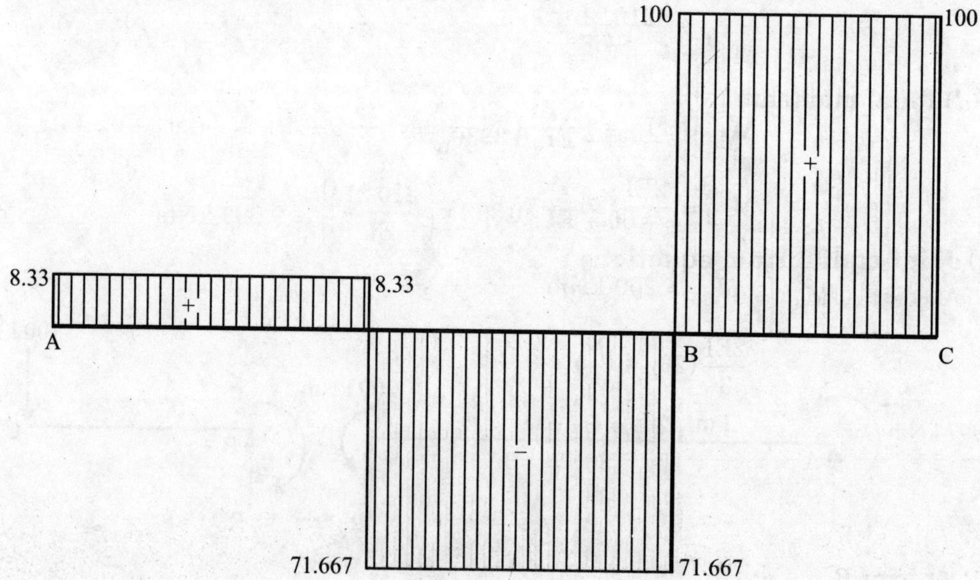

Shear force diagram

Example 2.4 *Analyse the continuous beam shown in Fig. 2.8 by the slope deflection method and sketch the bending moment diagram.*

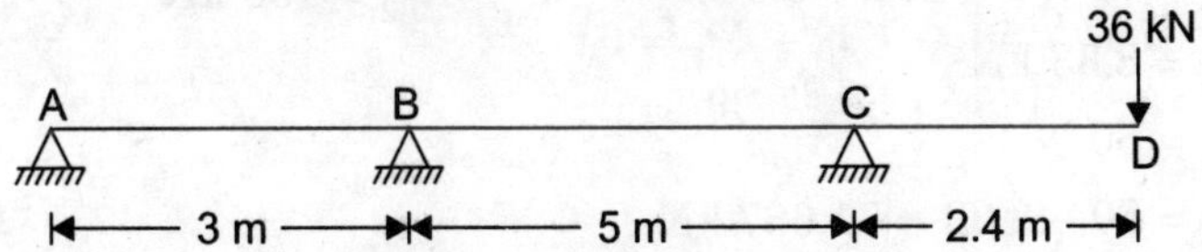

Fig. 2.8

Solution (*a*) **Fixed end moments :**

$$M_{FAB} = M_{FBA} = M_{FBC} = M_{FCB} = 0$$
$$M_{CD} = -36 \times 2.4 = -86.4 \text{ kNm}$$

(*b*) **Slope deflection equations :**

Span AB :

$$M_{AB} = M_{FAB} + \frac{2EI}{l}\left(2\theta_A + \theta_B + \frac{3\delta}{l}\right) = 0 + \frac{2EI}{3}(2\theta_A + \theta_B + 0)$$

$$M_{AB} = \frac{2EI}{3}(2\theta_A + \theta_B) \quad \text{...(1)}$$

$$M_{BA} = M_{FBA} + \frac{2EI}{l}\left(2\theta_B + \theta_A + \frac{3\delta}{l}\right) = 0 + \frac{2EI}{3}(2\theta_B + \theta_A + 0)$$

$$M_{BA} = \frac{2EI}{3}(2\theta_B + \theta_A) \quad \text{...(2)}$$

Span BC :

$$M_{BC} = M_{FBC} + \frac{2EI}{l}\left(2\theta_B + \theta_C + \frac{3\delta}{l}\right) = 0 + \frac{2EI}{5}(2\theta_B + \theta_C + 0)$$

$$M_{BC} = \frac{2EI}{5}(2\theta_B + \theta_C) \quad \text{...(3)}$$

$$M_{CB} = M_{FCB} + \frac{2EI}{l}\left(2\theta_C + \theta_B + \frac{3\delta}{l}\right) = 0 + \frac{2EI}{5}(2\theta_C + \theta_B + 0)$$

$$M_{CB} = \frac{2EI}{5}(2\theta_C + \theta_B) \quad \text{...(4)}$$

(*c*) **Joint equilibrium equations :**

At joint A, $M_{AB} = 0$

i.e.,
$$\frac{2EI}{3}(2\theta_A + \theta_B) = 0$$
$$2\theta_A + \theta_B = 0$$
$$2\theta_A = -\theta_B$$
$$\theta_A = -\frac{\theta_B}{2} \quad \text{...(5)}$$

At joint B,

$$M_{BA} + M_{BC} = 0$$

$$\frac{2EI}{3}(2\theta_B + \theta_A) + \frac{2EI}{5}(2\theta_B + \theta_C) = 0$$

$$\frac{2EI}{15}[5(2\theta_B + \theta_A) + 3(2\theta_B + \theta_C)] = 0$$

$$10\theta_B + 5\theta_A + 6\theta_B + 3\theta_C = 0$$

$$16\theta_B + 5\theta_A + 3\theta_C = 0$$

Substituting (5)

$$16\theta_B + 5 \times \left(\frac{-\theta_B}{2}\right) + 3\theta_C = 0$$

$$16\theta_B - \frac{5}{2}\theta_B + 3\theta_C = 0$$

$$13.5\theta_B + 3\theta_C = 0 \qquad \text{...(6)}$$

At joint C,

$$M_{CB} + M_{CD} = 0$$

$$\frac{2EI}{5}(2\theta_C + \theta_B) - 86.4 = 0$$

$$\frac{2EI}{5}(2\theta_C + \theta_B) = 86.4$$

$$2\theta_C + \theta_B = \frac{86.4 \times 5}{2EI}$$

$$\theta_B + 2\theta_C = \frac{216}{EI} \qquad \text{...(7)}$$

(7) × 13.5 ⇒

$$13.5\theta_B + 27\theta_C = \frac{2916}{EI} \qquad \text{...(8)}$$

$$13.5\theta_B + 3\theta_C = 0 \qquad \text{...(6)}$$

(8) – (6) gives

$$24\theta_C = \frac{2916}{EI}$$

$$\theta_C = \frac{121.5}{EI} \qquad \text{...(9)}$$

Substituting eqn. (9) in eqn. (7)

$$\theta_B = \frac{216}{EI} - 2\theta_C = \frac{216}{EI} - 2 \times \frac{121.5}{EI}$$

$$\theta_B = -\frac{27}{EI} \qquad \text{...(10)}$$

From eqn. (5)

$$\theta_A = -\frac{\theta_B}{2} = -\left(-\frac{27}{EI}\right) \times \frac{1}{2}$$

$$\theta_A = \frac{13.5}{EI} \qquad \text{...(11)}$$

Hence,

$$\theta_A = \frac{13.5}{EI}\ ;\ \theta_B = -\frac{27}{EI}\ ;\ \theta_C = \frac{121.5}{EI}$$

(*d*) **Final moments :**

Substituting θ values in eqns. (1) to (4),

$$M_{AB} = \frac{2EI}{3}(2\theta_A + \theta_B) = \frac{2EI}{3}\left(2\times\frac{13.5}{EI} - \frac{27}{EI}\right) = 0$$

$$M_{AB} = 0$$

$$M_{BA} = \frac{2EI}{3}(2\theta_B + \theta_A) = \frac{2EI}{3}\left(-2\times\frac{27}{EI} + \frac{13.5}{EI}\right)$$

$$= \frac{2EI}{3}\times\left(-\frac{40.5}{EI}\right) = -27 \text{ kNm}$$

$$M_{BC} = \frac{2EI}{5}(2\theta_B + \theta_C) = \frac{2EI}{5}\left(-2\times\frac{27}{EI} + \frac{121.5}{EI}\right)$$

$$= \frac{2EI}{5}\times\frac{67.5}{EI} = 27 \text{ kNm}$$

$$M_{CB} = \frac{2EI}{5}(2\theta_C + \theta_B) = \frac{2EI}{5}\left(2\times\frac{121.5}{EI} - \frac{27}{EI}\right)$$

$$= \frac{2EI}{5}\times\frac{216}{EI} = 86.4 \text{ kNm}$$

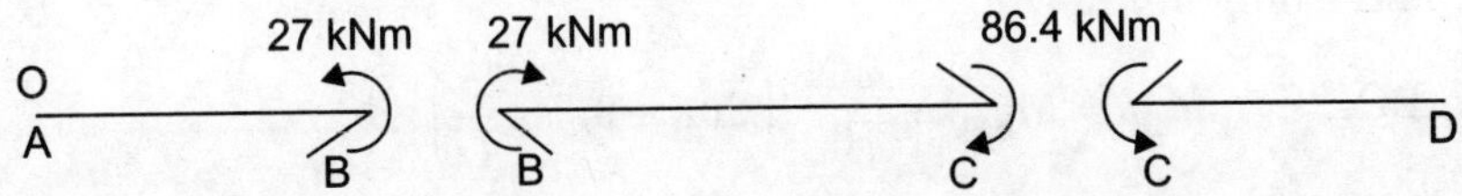

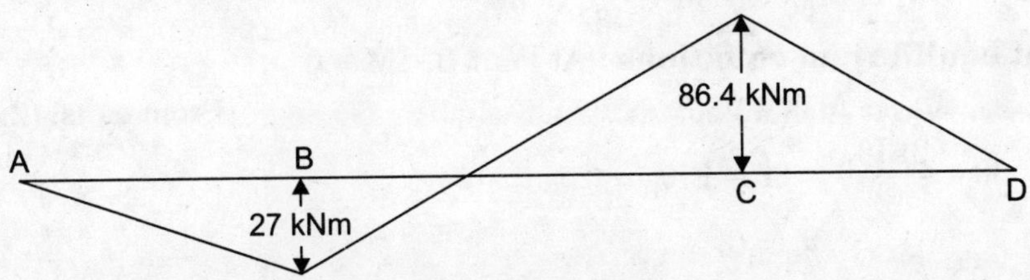

Bending moment diagram

Example 2.5 *Analyse the continuous beam loaded as shown in Fig. 2.9 by the slope deflection method. Sketch the bending moment and shear force diagrams.*

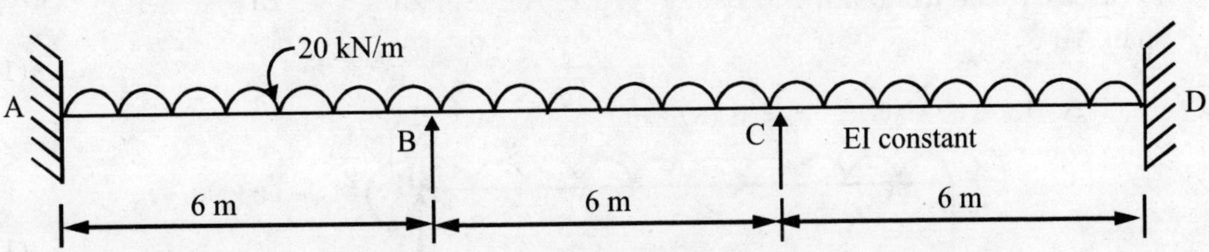

Fig. 2.9

Solution (*a*) **Fixed end moments :**

$$M_{FAB} = M_{FBC} = M_{FCD} = -\frac{wl^2}{12} = -60 \text{ kNm}$$

$$M_{FBA} = M_{FCB} = M_{FDC} = \frac{wl^2}{12} = 60 \text{ kNm.}$$

(*b*) **Slope deflection equations :** The structure is symmetrical so is the loading. There is no sinking of supports. Hence the following conditions prevail.

(*i*) $\theta_A = \theta_D = 0$ (*ii*) $\delta = 0$ for all spans

(*iii*) $\theta_B = \theta_C$.

Hence there is only one unknown displacement, namely θ_B.

For span AB, the general slope deflection equation is

$$M_{AB} = M_{FAB} + \frac{2EI}{6}\left(2\theta_A + \theta_B + \frac{3\delta}{l}\right)$$

$$M_{AB} = -60 + \frac{2EI}{6} \times \theta_B \quad \text{...(2.3)}$$

Since $\theta_A = 0$ and $\delta = 0$.

$$M_{BA} = 60 + \frac{2EI}{6} \times 2\theta_B \quad \text{...(2.4)}$$

No other slope deflection equation is needed.

Since θ_B is the only unknown.

For span BC, $$M_{BC} = M_{FBC} + \frac{2EI}{6}\left(2\theta_B + \theta_C + \frac{3\delta}{l}\right)$$

$$M_{BC} = -60 + \frac{2EI}{6} \times 3\theta_B \quad \text{...(2.5)}$$

(*c*) **Joint equilibrium equations :** At joint B, $\Sigma M = 0$

$$M_{BA} + M_{BC} = 0 \quad \text{[From eqns. (2.4) and (2.5)]}$$

$$60 + \frac{2EI\theta_B}{3} - 60 + EI\theta_B = 0$$

Hence, $\theta_B = 0$.

(*d*) **Final moments :**

$$M_{AB} = -60,\ M_{BA} = 60$$
$$M_{BC} = -60,\ M_{CB} = 60$$
$$M_{CD} = -60,\ M_{DC} = 60$$

(*e*) **Shear force diagram :**

Span AB :

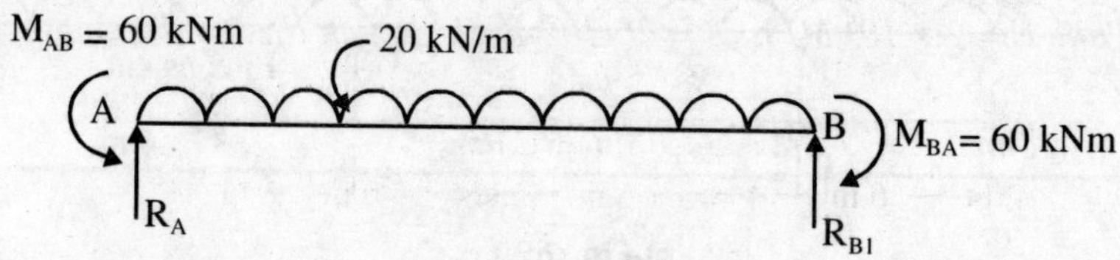

Taking moments about A, on the free body diagram of span AB,

$$-R_{B1} \times 6 - M_{AB} + M_{BA} + \frac{wl^2}{2} = 0$$

$$-R_{B1} \times 6 - 60 + 60 + \frac{20 \times 6^2}{2} = 0$$

$$R_{B1} = 60 \; ; R_A = 60$$

Similarly in span BC, $R_{B2} = R_{C1} = 60$

$$R_B = R_{B1} + R_{B2} = 120 \text{ kN.}$$

(*f*) **BMD and SFD :**

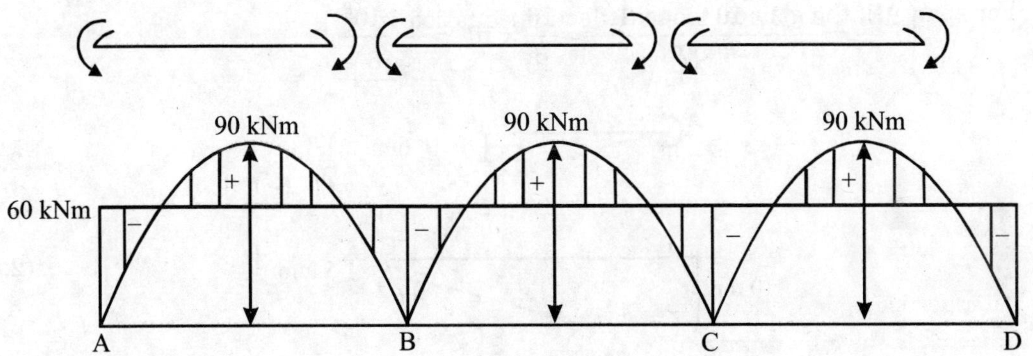

Bending moment diagram

$$\text{Simply supported span bending moment} = \frac{wl^2}{8} = \frac{20 \times 6^2}{8} = 90 \text{ kNm}$$

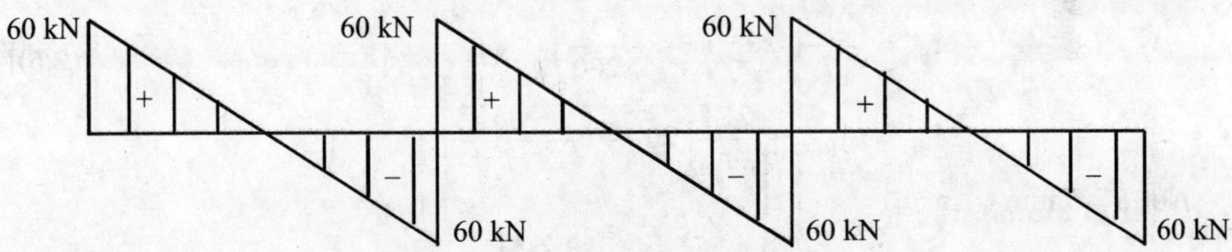

Shear force diagram

Example 2.6 *Analyse the continuous beam shown in Fig. 2.10 by the slope deflection method. The supports B and C sink 10 mm and 5 mm respectively and the support D rotates through an anti-clockwise angle of 0.1 radian. There are no loads on the beam. Values of E and I are constant throughout the length of the beam. Take* $E = 2 \times 10^5$ *MPa ;* $I = 4 \times 10^7$ mm^4*. Sketch the bending moment diagram.*

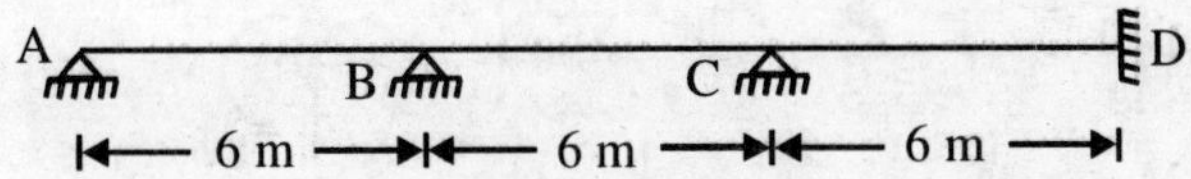

Fig. 2.10

Solution **Fixed end moments :** Since all the spans are not loaded, the fixed end moments in all the span = 0 due to loading. The fixed end moments due to settlement of supports B and C and rotation of end D will be included in the slope-deflection equations.

Slope deflection equations :

$$M_{AB} = M_{FAB} + \frac{2EI}{l}\left(2\theta_A + \theta_B + \frac{3\delta}{l}\right)$$

$$= 0 + \frac{2EI}{l}\left(2\theta_A + \theta_B - \frac{3\times 10}{6000}\right)$$

$$\left(\frac{2EI}{l} = \frac{2\times 2\times 10^5 \times 4\times 10^7}{6000} = \frac{16}{6}\times 10^9 = \frac{8\times 10^9}{3}\right)$$

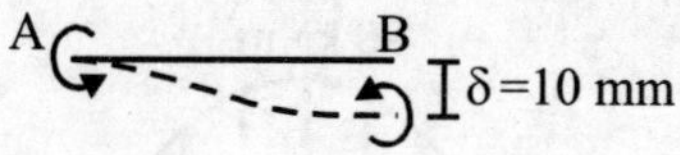

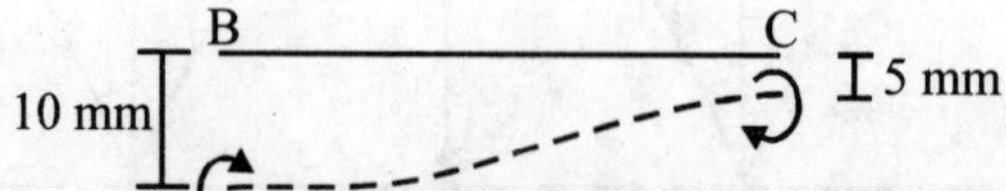

$$M_{AB} = \frac{2EI}{l}\left(2\theta_A + \theta_B - \frac{1}{200}\right) \quad ...(1)$$

$$M_{BA} = M_{FBA} + \frac{2EI}{l}\left(2\theta_B + \theta_A + \frac{3\delta}{l}\right) = 0 + \frac{2EI}{l}\left(2\theta_B + \theta_A - \frac{3\times 10}{6000}\right)$$

$$M_{BA} = \frac{2EI}{l}\left(2\theta_B + \theta_A - \frac{1}{200}\right) \quad ...(2)$$

$$M_{BC} = M_{FBC} + \frac{2EI}{l}\left(2\theta_B + \theta_C + \frac{3\delta}{l}\right)$$

Net δ = 5 mm ($\uparrow$ at B)

$$= 0 + \frac{2EI}{l}\left(2\theta_B + \theta_C + \frac{3\times (5)}{6000}\right)$$

$$M_{BC} = \frac{2EI}{l}\left(2\theta_B + \theta_C + \frac{1}{400}\right) \quad ...(3)$$

$$M_{CB} = M_{FCB} + \frac{2EI}{l}\left(2\theta_C + \theta_B + \frac{3\delta}{l}\right) = 0 + \frac{2EI}{l}\left(2\theta_C + \theta_B + \frac{3\times (+5)}{6000}\right)$$

$$M_{CB} = \frac{2EI}{l}\left(2\theta_C + \theta_B + \frac{1}{400}\right) \quad ...(4)$$

$$M_{CD} = M_{FCD} + \frac{2EI}{l}\left(2\theta_C + \theta_D + \frac{3\delta}{l}\right)$$

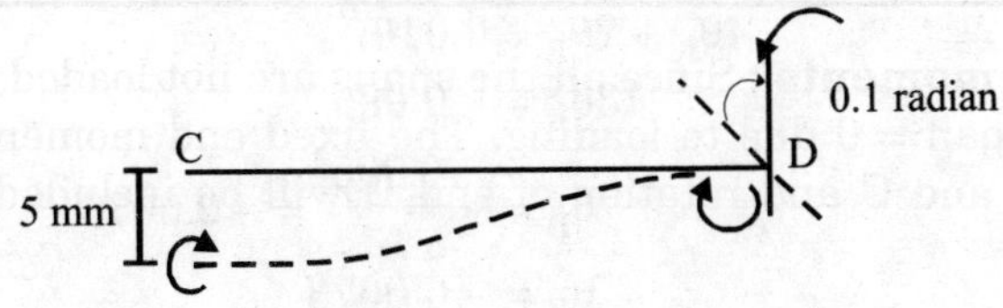

$$M_{CD} = 0 + \frac{2EI}{l}\left(2\theta_C + \theta_D + \frac{3 \times 5}{6000}\right)$$

$$M_{CD} = M_{FDC} + \frac{2EI}{l}\left(2\theta_C - 0.1 + \frac{1}{400}\right) \quad ...(5)$$

$$M_{DC} = M_{FDC} + \frac{2EI}{l}\left(2\theta_D + \theta_C + \frac{3\delta}{l}\right)$$

$$M_{DC} = 0 + \frac{2EI}{l}\left(-0.2 + \theta_C + \frac{3 \times (+5)}{6000}\right)$$

$$M_{DC} = \frac{2EI}{l}\left(-0.2 + \theta_C + \frac{1}{400}\right) \quad ...(6)$$

Equilibrium equations : There are three unknowns θ_A, θ_B and θ_C. The equilibrium equations to solve the above unknowns are :

$$M_{AB} = 0 \quad \text{(at A)} \quad ...(7)$$

$$M_{BA} + M_{BC} = 0 \quad \text{(at B)} \quad ...(8)$$

$$M_{CB} + M_{CD} = 0 \quad \text{(at C)} \quad ...(9)$$

Substituting eqn. (1) in eqn. (7), $\frac{2EI}{l}\left(2\theta_A + \theta_B - \frac{1}{200}\right) = 0$

$$2\theta_A + \theta_B - \frac{1}{200} = 0$$

$$2\theta_A + \theta_B = 0.005 \quad ...(10)$$

Substituting eqns. (2) and (3) in eqn. (8)

$$\frac{2EI}{l}\left(2\theta_B + \theta_A - \frac{1}{200}\right) + \frac{2EI}{l}\left(2\theta_B + \theta_C + \frac{1}{400}\right) = 0$$

$$\frac{2EI}{l}\left(4\theta_B + \theta_A + \theta_C - \frac{1}{400}\right) = 0$$

$$\theta_A + 4\theta_B + \theta_C - 0.0025 = 0$$

$$\theta_A + 4\theta_B + \theta_C = 0.0025 \quad ...(11)$$

Substituting eqns. (4) and (5) in eqn. (9)

$$\frac{2EI}{l}\left(2\theta_C + \theta_B + \frac{1}{400}\right) + \frac{2EI}{l}\left(2\theta_C - 0.1 + \frac{1}{400}\right) = 0$$

$$\frac{2EI}{l}(4\theta_C + \theta_B - 0.1 + 0.005) = 0$$

$$4\theta_C + \theta_B - 0.095 = 0$$

$$4\theta_C + \theta_B = 0.095 \quad ...(12)$$

(11) × 4, $\quad 4\theta_A + 16\theta_B + 4\theta_C = 0.010 \quad ...(13)$

(13) – (12), $\quad 4\theta_A + 15\theta_B = -0.085$

(10) × 2, $4\theta_A + 2\theta_B = 0.010$

(14) – (15), $13\theta_B = -0.095$

$$\theta_B = -\frac{0.095}{13} = -0.0073$$

$$\theta_B = -0.0073$$

From (11), $\theta_C = 0.0025 - \theta_A - 4\theta_B$

$$= 0.0025 - 0.0062 - 4(-0.0073)$$

$$\theta_C = 0.0255 \; ; \; \theta_A = 0.0062$$

Final moments: Substituting the values of θ_A, θ_B and θ_C in eqns. (1) to (6)

$$M_{AB} = \frac{2EI}{l}\left(2\theta_A + \theta_B - \frac{1}{200}\right) = \frac{8 \times 10^9}{3}\left(2 \times 0.0062 - 0.0073 - \frac{1}{200}\right)$$

$M_{AB} = 266667$ Nmm $= 26.67 \times 10^4$ Nmm $= 0.027 \times 10^7$ Nmm

$$M_{BA} = \frac{2EI}{l}\left(2\theta_B + \theta_A - \frac{1}{200}\right) = \frac{8 \times 10^9}{3}\left(2 \times (-0.0073) + 0.0062 - \frac{1}{200}\right)$$

$= -3.573 \times 10^7$ Nmm

$$M_{BC} = \frac{2EI}{l}\left(2\theta_B + \theta_C + \frac{1}{400}\right) = \frac{8 \times 10^9}{3}\left(2 \times (-0.0073) + 0.0255 + \frac{1}{400}\right)$$

$= 3.573 \times 10^7$ Nmm

$$M_{CB} = \frac{2EI}{l}\left(2\theta_C + \theta_B + \frac{1}{400}\right) = \frac{8 \times 10^9}{3}\left(2 \times 0.0255 - 0.0073 + \frac{1}{400}\right)$$

$= 12.32 \times 10^7$ Nmm

$$M_{CD} = \frac{2EI}{l}\left(2\theta_C - 0.1 + \frac{1}{400}\right) = \frac{8 \times 10^9}{3}\left(2 \times 0.0255 - 0.1 + \frac{1}{400}\right)$$

$= -12.40 \times 10^7$ Nmm

$$M_{DC} = \frac{2EI}{l}\left(-0.2 + \theta_C + \frac{1}{400}\right) = \frac{8 \times 10^9}{3}\left(-0.2 + 0.0255 + \frac{1}{400}\right)$$

$= -4.587 \times 10^8$ Nmm $= -45.87 \times 10^7$ Nmm

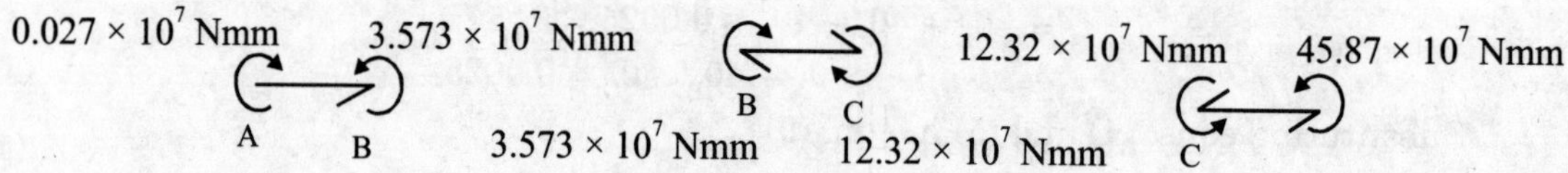

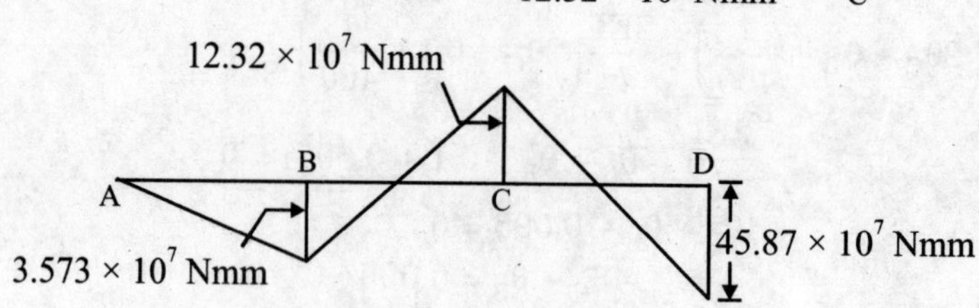

Bending moment diagram

Example 2.7 *Analyse the continuous beam loaded as shown in Fig. 2.11. by the slope deflection method. Support B sinks by 10 mm. Take E = 2 × 10^5 N/mm^2, I = 16 × 10^7 mm^4. Sketch the bending moment and shear force diagrams.*

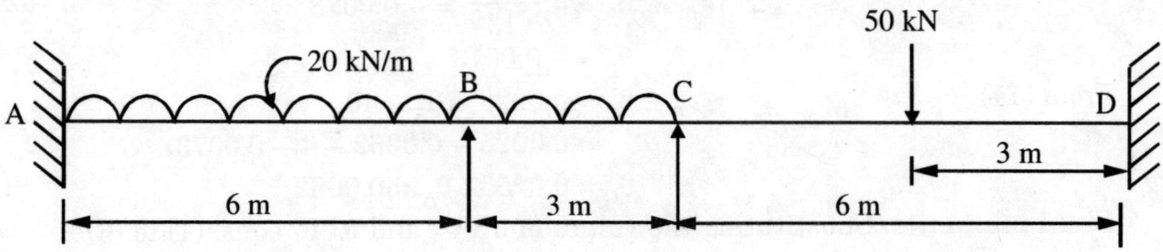

Fig. 2.11

Solution (*a*) **Fixed end moments :**

Span AB :
$$M_{FAB} = -\frac{wl^2}{12} = -\frac{20\times 6^2}{12} = -60 \text{ kNm}$$

$$M_{FBA} = \frac{wl^2}{12} = \frac{20\times 6^2}{12} = 60 \text{ kNm}$$

Span BC :
$$M_{FBC} = -\frac{wl^2}{12} = -\frac{20\times (3)^2}{12} = 15 \text{ kNm}$$

$$M_{FCB} = \frac{wl^2}{12} = \frac{20\times (3)^2}{12} = 15 \text{ kNm}$$

Span CD :
$$M_{FCD} = -\frac{wl}{8} = -\frac{50\times 6}{8} = -37.5 \text{ kNm}$$

$$M_{FDC} = \frac{wl}{8} = \frac{50\times 6}{8} = 37.5 \text{ kNm}$$

(*b*) **Slope deflection equations :**

Span AB :
$$M_{AB} = M_{FAB} + 2Ek\left(2\theta_A + \theta_B + \frac{3\delta}{l}\right) \qquad \left(k = \frac{I}{l}\right)$$

$\theta_A = 0$, Since A is fixed ; δ = 10 mm (–ve) Since the right support settles *down.*

$$= -60 + 2Ek\left(0 + \theta_B - \frac{3\times 10}{6000}\right)$$

$$M_{AB} = -60 + \frac{EI}{3}\left(0 + \theta_B - \frac{1}{200}\right) \qquad ...(1)$$

$$M_{BA} = M_{FBA} + 2Ek\left(2\theta_B + \theta_A + \frac{3\delta}{l}\right)$$

$$= 60 + \frac{2EI}{6}\left(2\theta_B - \frac{3\times 10}{6000}\right)$$

$$M_{BA} = 60 + \frac{EI}{3}\left(2\theta_B - \frac{3\times 10}{6000}\right) \qquad ...(2)$$

Span BC : $M_{BC} = M_{FBC} + 2Ek\left(2\theta_B + \theta_B + \dfrac{3\delta}{l}\right)$

$$= -15 + \frac{2EI}{3}\left(2\theta_B + \theta_C + \frac{3 \times 10}{3000}\right)$$

$$M_{BC} = -15 + \frac{2EI}{3}\left(2\theta_B + \theta_C + \frac{1}{100}\right) \quad ...(3)$$

$$M_{CB} = M_{FCB} + 2Ek\left(2\theta_C + \theta_B + \frac{3\delta}{l}\right)$$

$$M_{CB} = 15 + \frac{2EI}{3}\left(2\theta_C + \theta_B + \frac{1}{100}\right) \quad ...(4)$$

Span CD : $M_{CD} = M_{FCD} + 2Ek\left(2\theta_C + \theta_D + \dfrac{3\delta}{l}\right)$

$$= -37.5 + \frac{2EI}{6}\left(2\theta_C + \theta_D + \frac{3 \times 0}{6000}\right) \qquad (\theta_D = 0, \because \text{ D is fixed})$$

$$M_{CD} = -37.5 + \frac{EI}{3}(2\theta_C) \quad ...(5)$$

$$M_{DC} = M_{FDC} + 2Ek\left(2\theta_D + \theta_C + \frac{3\delta}{l}\right) = -37.5 + \frac{2EI}{6}(0 + \theta_C + 0)$$

$$M_{DC} = 37.5 + \frac{EI\theta_C}{3} \quad ...(6)$$

(*c*) **Joint equilibrium equations :**

Joint B : $M_{BA} + M_{BC} = 0$

From eqns. (2) and (3), $60 + \dfrac{EI}{3}\left(2\theta_B - \dfrac{1}{200}\right) + (-15) + \dfrac{2EI}{3}\left(2\theta_B + \theta_C + \dfrac{1}{100}\right) = 0$

$$45 + \frac{EI}{3}\left(2\theta_B - \frac{1}{200} + 4\theta_B + 2\theta_C + \frac{2}{100}\right) = 0$$

$$45 + \frac{EI}{3}\left(6\theta_B + 2\theta_C + \frac{3}{200}\right) = 0$$

$$EI\left(6\theta_B + 2\theta_C + \frac{3}{200}\right) = -45 \times 3 = -135 \quad ...(7)$$

Joint C : $M_{CB} + M_{CD} = 0$

From eqns. (4) and (5),

$$15 + \frac{2EI}{3}\left(2\theta_C + \theta_B + \frac{1}{100}\right) - 37.5 + \frac{2EI}{3}\theta_C = 0$$

$$15 + \frac{2EI}{3}\left(2\theta_C + \theta_B + \frac{1}{100} + \theta_C\right) - 37.5 = 0$$

$$EI\left(3\theta_C + \theta_B + \frac{1}{100}\right) = 33.75$$

$$EI\left(\theta_B + 3\theta_C + \frac{1}{100}\right) = 33.75 \qquad ...(8)$$

(8) × (6)

$$EI\left(6\theta_B + 18\theta_C + \frac{6}{100}\right) = 202.5 \qquad ...(9)$$

$$EI\left(6\theta_B + 2\theta_C + \frac{3}{200}\right) = -135 \qquad ...(7)$$

(9) – (7)

$$EI\left(16\theta_C + \frac{9}{200}\right) = 337.5 \qquad ...(10)$$

$$EI = 2 \times 10^5 \times 16 \times 10^7 = 32 \times 10^{12} \text{ Nmm}^2$$

$$EI = 32 \times 10^6 \text{ Nm}^2$$

$$32 \times 10^6 \left(16\theta_C + \frac{9}{200}\right) = 337.5 \times 10^3$$ [337.5 is in kNm.]

$$\theta_C = -\frac{1}{464}$$

$$EI\left(6\theta_B + \frac{2(-1)}{464} + \frac{3}{200}\right) = -135$$

$$\theta_B = -\frac{1}{402}$$

(*d*) **Final moments :**

From (1),

$$M_{AB} = -60 + \frac{EI}{3}\left(-\frac{1}{402} - \frac{1}{200}\right)$$

$$M_{AB} = -139.843 \text{ kNm.}$$

From (2),

$$M_{BA} = 60 + \frac{EI}{3}\left(2 \cdot \frac{(-1)}{402} - \frac{1}{200}\right)$$

$$M_{BA} = -46.354 \text{ kNm}$$

From (3),

$$M_{BC} = -15 + \frac{2 \times 32 \times 10^3}{3}\left(2\frac{(-1)}{402} + \frac{-1}{464} + \frac{1}{100}\right)$$

$$M_{BC} = 46.3 \text{ kNm}$$

From (4),

$$M_{CB} = 15 + \frac{2 \times 32 \times 10^3}{3}\left(2\frac{(-1)}{464} + \frac{-1}{402} + \frac{1}{100}\right)$$

$$M_{CB} = 83.35 \text{ kNm}$$

From (5),

$$M_{CD} = -37.5 + \frac{EI}{3}\left(2\frac{(-1)}{464}\right)$$

$$M_{CD} = -83.477 \text{ kNm}$$

From (6),

$$M_{DC} = 37.5 + \frac{EI}{3}\frac{(-1)}{464}$$

$$M_{DC} = 14.511 \text{ kNm}$$

(*e*) **To draw S.F.D. :**

Span AB :

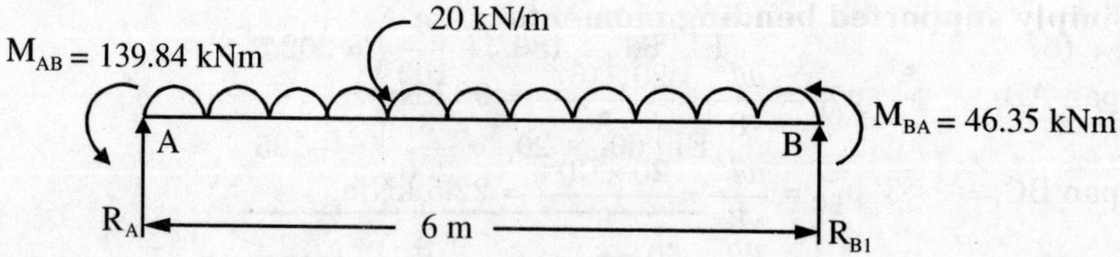

Taking moments about A.

$$\frac{wl^2}{2} - M_{BA} - M_{AB} - R_{B1}(l) = 0$$

$$\frac{20 \times (6)^2}{2} - 46.35 - 139.84 - R_{B1}(6) = 0$$

$$R_{B1} = 28.97 \text{ kN}$$

$$R_A = 20 \times 6 - 28.97 \text{ kN}$$

$$R_A = 91.03 \text{ kN}$$

Span BC :

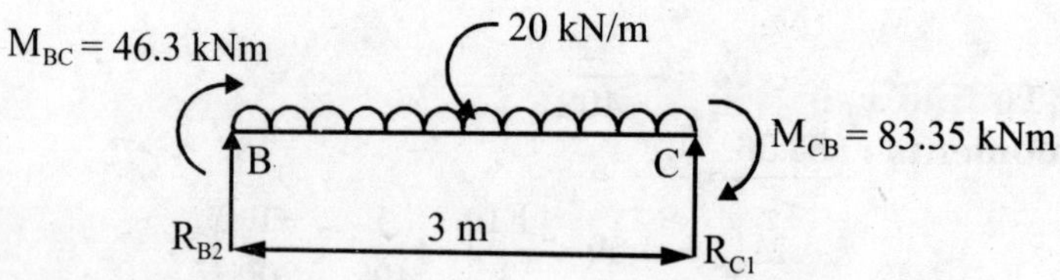

Taking moments about B

$$\frac{wl^2}{2} + M_{CB} + M_{BC} - R_{C1}(l) = 0$$

$$\frac{20 \times (3)^2}{2} + 83.35 + 46.3 - R_{C1}(3) = 0$$

$$R_{C1} = 73.22 \text{ kN}$$

$$R_{B2} = 20 \times 3 - 73.22 \text{ kN}$$

$$R_{B2} = -13.21 \text{ kN}$$

Span CD :

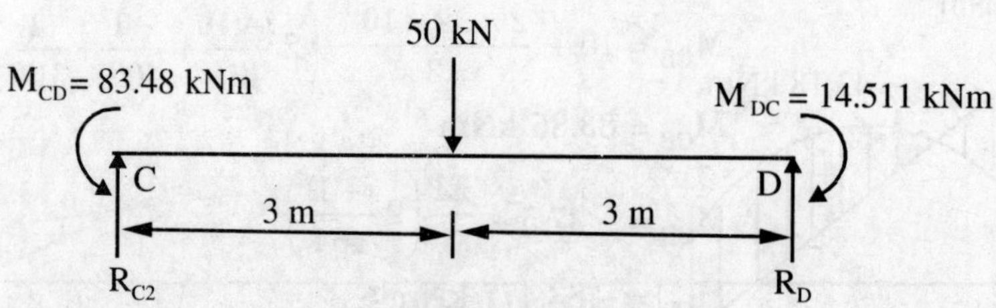

Taking moments about C

$$M_{DC} + W(a) - M_{CD} - R_D(l) = 0$$

$$14.511 + 50(3) - 83.48 - R_D(6) = 0$$

$R_D = 13.5$ kN

$R_{C2} = 36.5$ kN

(*f*) **Simply supported bending moments :**

In span AB, $\mu_{AB} = \dfrac{wl^2}{8} = \dfrac{20 \times (6)^2}{8} = 90$ kNm.

In span BC, $\mu_{BC} = \dfrac{wl^2}{8} = \dfrac{20 \times (3)^2}{8} = 22.5$ kNm.

In span CD, $\mu_{CD} = \dfrac{Wl}{4} = \dfrac{50 \times 6}{4} = 75$ kNm.

(*g*) **Net positive bending moments :**

To find x_1 : $\dfrac{139.84}{x_1} = \dfrac{46.354}{6 - x_1}$ (refer bending moment diagram below)

$$3.016 = \frac{x_1}{6 - x_1}$$

$$(6 - x_1)\ 3.016 = x_1$$

$$18.1 - 3.016x_1 = x_1$$

$$x_1 = 4.51 \text{ m}$$

In span AB, $M_{AB}\ (\text{Net}) = 90 - \left\{\dfrac{139.84}{4.51} \times (4.51 - 3)\right\} = 43.18$ kNm

In span BC, **To find x_2 :**

$$\frac{83.36}{x_2} = \frac{46.354}{3 - x_2}$$

$$1.8 = \frac{x_2}{3 - x_2}$$

$$1.8\ (3 - x_2) = x_2$$

$$5.4 - 1.8x_2 = x_2$$

$$x_2 = 1.93 \text{ m}$$

$$+\text{ve moment} = 22.5 - \left\{\frac{83.36}{1.93} \times (1.93 - 1.5)\right\} = 3.989 \text{ kNm}$$

Span CD, Net +ve moment $= 75 - \left\{14.511 + \dfrac{(83.36 - 14.511)}{6} \times 3\right\}$

$= 26.07$ kNm.

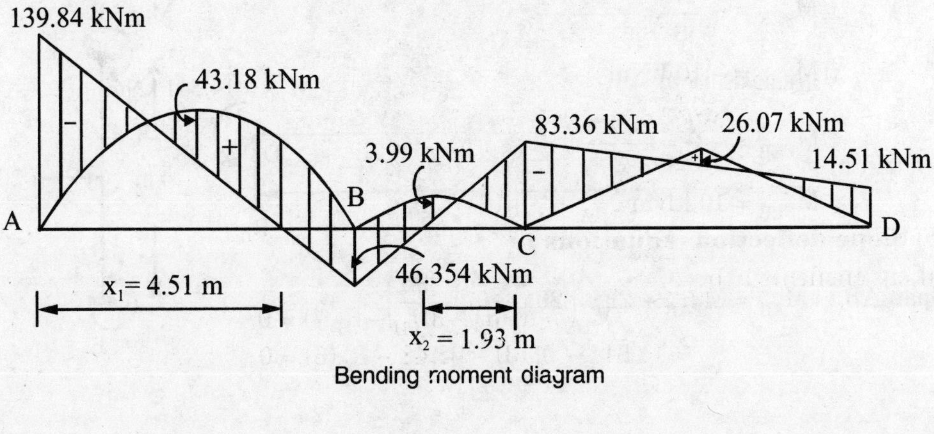

Bending moment diagram

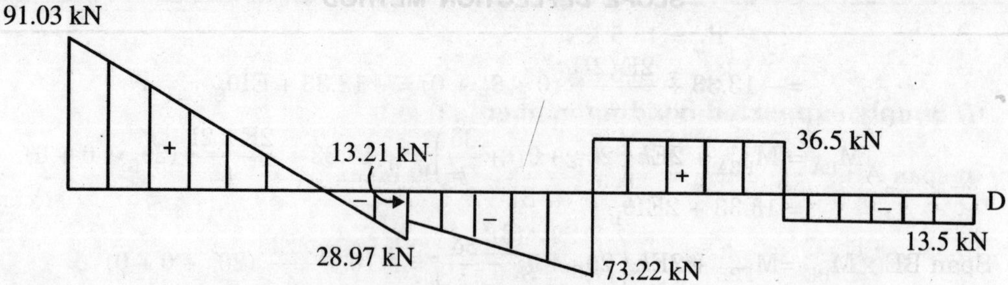

Shear force diagram

Example 2.8 *Analyse the structure shown in Fig. 2.12 by the slope deflection method. Sketch the bending moment and shear force diagrams.*

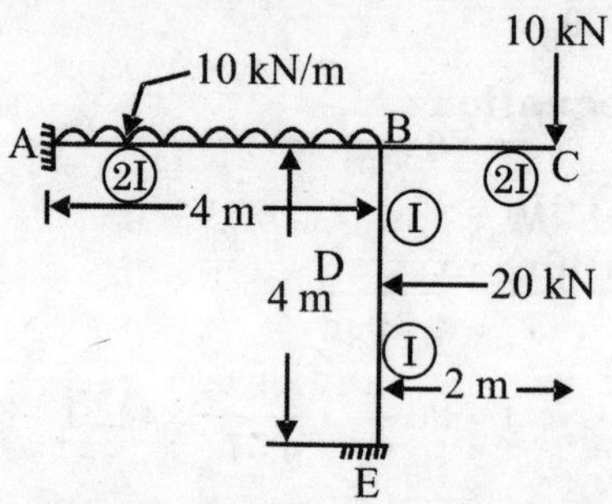

Fig. 2.12

Solution (*a*) **Fixed end moments :**

Span AB : $M_{FAB} = -\dfrac{wl^2}{12} = -\dfrac{10 \times 16}{12} = -13.33$ kNm

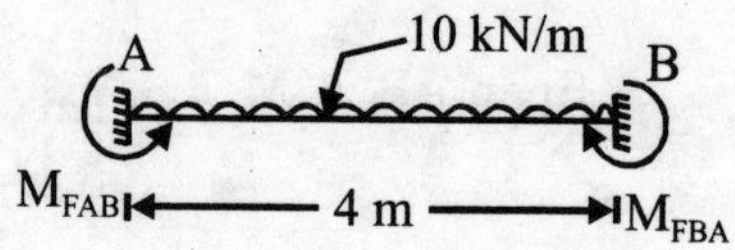

$M_{FBA} = \dfrac{wl^2}{12} = \dfrac{10 \times 16}{12}$

$M_{FBA} = 13.33$ kNm

Span BC :

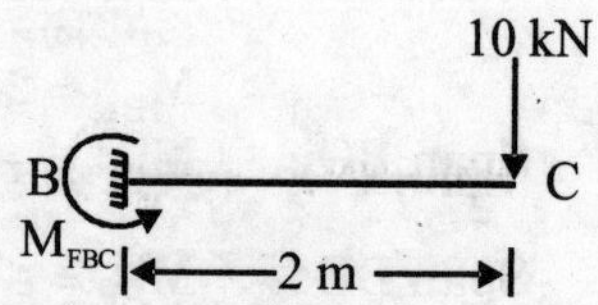

$M_{FBC} = -10 \times 2 = -20$ kNm

Span BE :

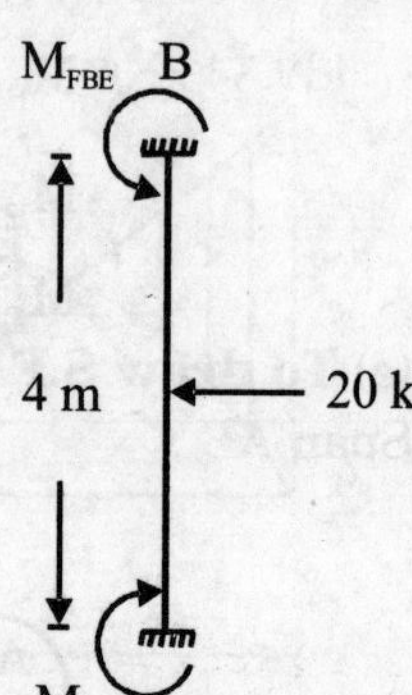

$M_{FBE} = -\dfrac{Wl}{8} = -\dfrac{20 \times 4}{8}$

$M_{FBE} = -10$ kNm

$M_{FEB} = \dfrac{Wl}{8} = \dfrac{20 \times 4}{8}$

$M_{FEB} = 10$ kNm

(*b*) **Slope deflection equations :**

Span AB : $M_{AB} = M_{FAB} + 2Ek\left(2\theta_A + \theta_B + \dfrac{3\delta}{l}\right)$

$$= -13.33 + \frac{2E(2I)}{4}(0 + \theta_B + 0) = -13.33 + EI\theta_B \quad ...(1)$$

$$M_{BA} = M_{FBA} + 2Ek\left(2\theta_B + \theta_A + \frac{3\delta}{l}\right) = +13.33 + \frac{2E \times 2I}{4}(2\theta_B + 0 + 0)$$

$$= 13.33 + 2EI\theta_B \quad ...(2)$$

Span BE : $M_{BE} = M_{FBE} + 2Ek\left(2\theta_B + \theta_E + \frac{3\delta}{l}\right) = -10 + \frac{2EI}{4}(2\theta_B + 0 + 0)$

$$M_{BE} = -10 + EI\theta_B \quad ...(3)$$

$$M_{EB} = 10 + 2Ek\left(2\theta_E + \theta_B + \frac{3\delta}{l}\right) = 10 + \frac{2EI}{4}(\theta_B + 0)$$

$$M_{EB} = 10 + \frac{EI}{2}\theta_B \quad ...(4)$$

(*c*) **Joint equilibrium equations :**

At Joint B, $\Sigma M = 0$

$$M_{BA} + M_{BC} + M_{BE} = 0$$

$$13.33 + 2EI\theta_B - 10 + EI\theta_B - 20 = 0$$

$$3EI\theta_B - 16.67 = 0$$

$$\theta_B = \frac{16.67}{3EI}$$

$$\theta_B = \frac{5.557}{EI}$$

(*d*) **Final moments :**

Substituting the value of θ_B in eqns. (1) to (4)

Span AB : $M_{AB} = -13.33 + EI\left(\frac{5.557}{EI}\right)$

$M_{AB} = -7.773$ kNm

$M_{BA} = 13.33 + 2EI\left(\frac{5.557}{EI}\right)$

$M_{BA} = 24.447$ kNm

Span BC : $M_{BC} = -20$ kNm

Span BE : $M_{BE} = -10 + EI\left(\frac{5.557}{EI}\right)$

$M_{BE} = -4.443$ kNm

$M_{EB} = 10 + \frac{EI}{2}\left(\frac{5.557}{EI}\right)$

$M_{EB} = 12.778$ kNm

(*e*) **To draw S.F.D. :**

Span AB :

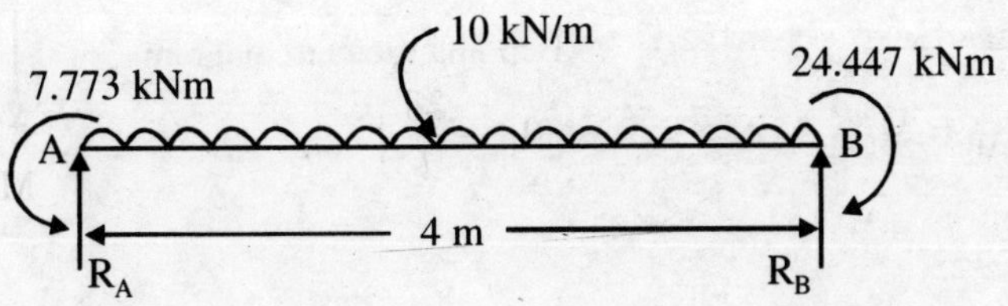

Taking moments about B,

$$R_A \times 4 - 10 \times 4(4/2) - 7.773 + 24.447 = 0$$
$$R_A = 15.832 \text{ kN}$$
$$R_{BH} = 10 \times 4 - R_A = 24.168 \text{ kN}$$

Span BE :

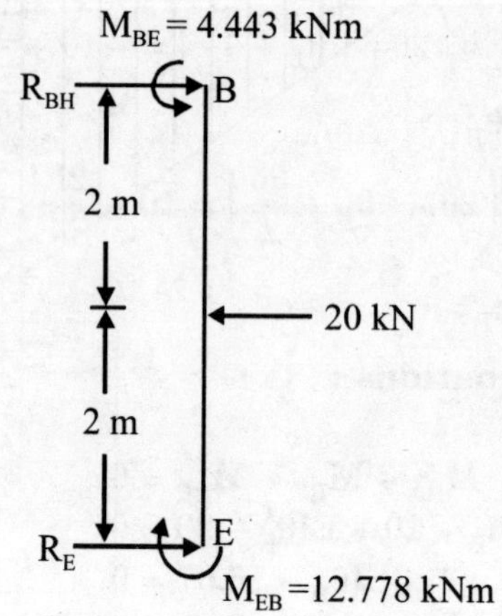

Taking moments about B,

$$-R_E(4) - 4.443 + 12.778 + 20(2) = 0$$
$$R_E = 12.084 \text{ kN}$$
$$R_{BH} = \text{Total load} - R_E = 7.916 \text{ kN}$$
$$R_{BH} = 7.916 \text{ kN}$$

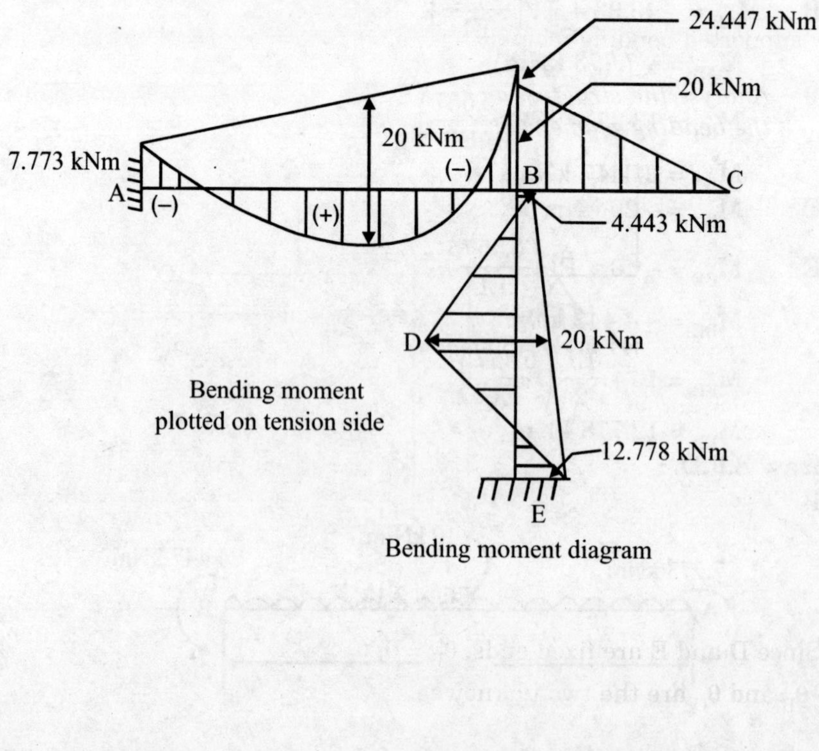

Bending moment diagram

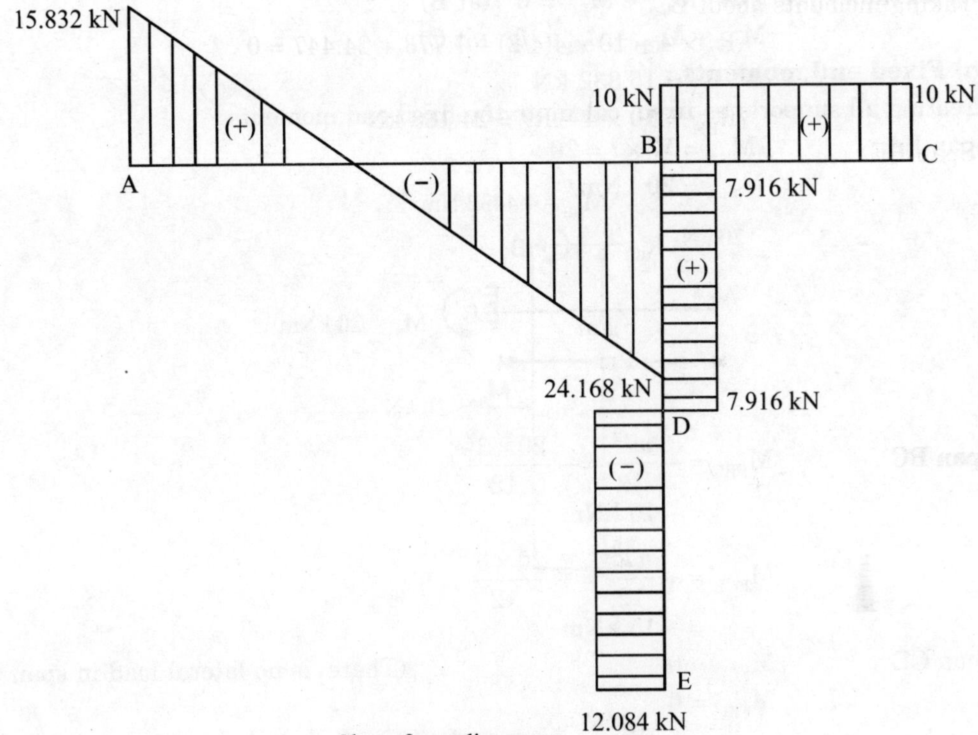

Shear force diagram

$$\text{Simply suppoted bending moment in AB} = \frac{wl^2}{8} = 10 \times \frac{16}{8} = 20 \text{ kNm}$$

$$\text{Simply supported bending moment in BE} = \frac{Wl}{4} = \frac{20 \times 4}{4} = 20 \text{ kNm.}$$

Example 2.9 *Analyse the structure in Fig. 2.13 by the slope deflection method and sketch the bending moment and shear force diagrams.*

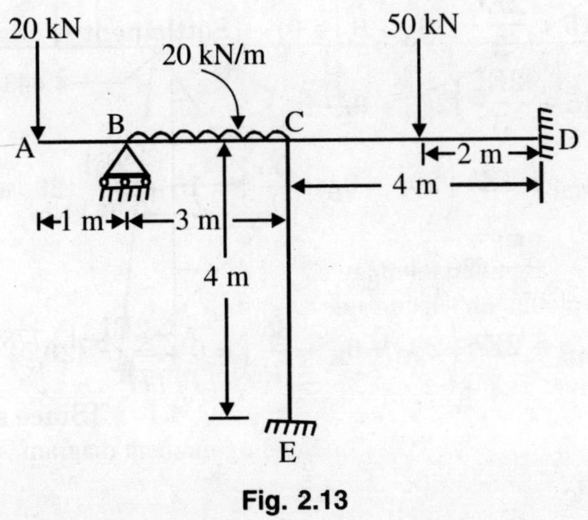

Fig. 2.13

Solution Since D and E are fixed ends, $\theta_D = 0$, $\theta_E = 0$

Hence θ_B and θ_C are the two unknowns.

Equilibrium equations : $M_{BA} + M_{BC} = 0$ (at B)

$M_{CB} + M_{CD} + M_{CE} = 0$ (at C)

(*a*) Fixed end moments :

Treating all supports as fixed, calculate the fixed end moments.

Span BA : $M_{BA} = W \times l = 20 \times 1$

$= 20$ kNm

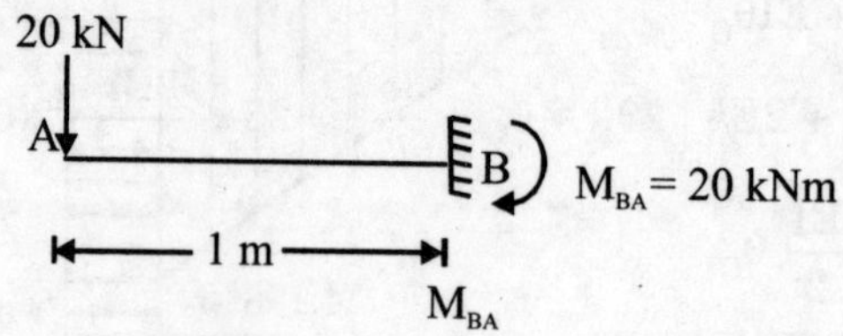

Span BC : $M_{FBC} = -\dfrac{wl^2}{12} = -\dfrac{20 \times 3^2}{12}$

$= -15$ kNm

$M_{FCB} = +\dfrac{wl^2}{12} = +\dfrac{20 \times 3^2}{12}$

$= +15$ kNm

Span CE : $M_{FCE} = 0$ (There is no lateral load in span CE)

$M_{FEC} = 0$

Span CD : $M_{FCD} = -\dfrac{Wl}{8} = -\dfrac{50 \times 4}{8} = -25$ kNm

$M_{FDC} = +\dfrac{Wl}{8} = +\dfrac{50 \times 4}{8} = +25$ kNm

(*b*) Slope deflection equations :

Span BC : $M_{BC} = M_{FBC} = 2Ek\left(2\theta_B + \theta_C + \dfrac{3\delta}{l}\right)$

$M_{BC} = -15 + \dfrac{2EI}{3}(2\theta_B + \theta_C + 0)$ [Settlement is zero in the given problem]

$M_{BC} = -15 + \dfrac{2EI}{3}(2\theta_B + \theta_C)$ ...(1)

$M_{CB} = M_{FCB} + 2Ek\left(2\theta_C + \theta_B + \dfrac{3\delta}{l}\right) = 15 + \dfrac{2EI}{3}(2\theta_C + \theta_B + 0)$

$M_{CB} = 15 + \dfrac{2EI}{3}(2\theta_C + \theta_B)$ ...(2)

Span CE : $M_{CE} = M_{FCE} + 2Ek\left(2\theta_C + \theta_E + \dfrac{3\delta}{l}\right) = 0 + \dfrac{2EI}{4}(2\theta_C + 0 + 0)$

[Since support 'E' is fixed one, slope at E, $\theta_E = 0$]

$M_{CE} = EI\theta_C$ ...(3)

$M_{EC} = M_{FEC} + 2Ek\left(2\theta_E + \theta_C - \dfrac{3\delta}{l}\right) = 0 + \dfrac{2EI}{4}(2(0) + \theta_C - 0)$

$$M_{EC} = \frac{EI}{2}\,\theta_C \qquad ...(4)$$

Span CD : $M_{CD} = M_{FCD} + 2Ek\left(2\theta_C + \theta_D + \frac{3\delta}{l}\right)$

$$= -25 + \frac{2EI}{4}(2\theta_C + 0 + 0) \qquad (\because \text{ D is fixed, } \theta_D = 0)$$

$$M_{CD} = -25 + EI\theta_C \qquad ...(5)$$

$$M_{DC} = M_{FDC} + 2Ek\left(2\theta_D + \theta_C + \frac{3\delta}{l}\right) = 25 + \frac{2EI}{4}\left(2(0) + \theta_C + \frac{3(0)}{1}\right)$$

$$M_{DC} = 25 + \frac{EI}{2}\,\theta_C \qquad ...(6)$$

(*c*) **Joint equilibrium equations :**

At Joint B, $M_{BA} + M_{BC} = 0$

$$20 + (-15) + (2/3)\,EI\,(2\theta_B + \theta_C) = 0$$

$$(2/3)EI\,(2\theta_B + \theta_C) = -5$$

$$2\theta_B + \theta_C = -\frac{7.5}{EI} \qquad ...(I)$$

At Joint C, $M_{CB} + M_{CD} + M_{CE} = 0$

$$15 + (2/3)\,EI\,(2\theta_C + \theta_B) + (-25) + EI\theta_C + EI\theta_C = 0$$

$$-10 + (2/3)\,(2\theta_C + \theta_B) + 2EI\theta_C = 0$$

$$-10 + (2/3)\,(2\theta_C + \theta_B + 3\theta_C) = 0$$

$$(2/3)\,(5\theta_C + \theta_B) = 10$$

$$5\theta_C + \theta_B = \frac{15}{EI} \qquad ...(II)$$

Solving (I) and (II), θ_C and θ_B can be found out multiplying equation (II) by 2

$$2\theta_B + 10\theta_C = \frac{30}{EI} \qquad ...(III)$$

$$2\theta_B + \theta_C = -\frac{7.5}{EI} \qquad ...(I)$$

(III) — (I) $\qquad 9\theta_C = \frac{37.5}{EI}$

$$\theta_C = \frac{1}{0.24EI} = \frac{4.167}{EI}$$

Substituting θ_C in (I) $\qquad 2\theta_B = -\frac{7.5}{EI} - \frac{1}{0.24EI}$

$$\theta_B = -\frac{5.833}{EI}$$

Substituting the valves of θ_B and θ_C in eqns. (1) to (6), final moments are calculated as below :

(*d*) **Final moments :**

Span BC : $\qquad M_{BC} = -15 + \frac{2}{3}EI\left[2\left(\frac{-5.833}{EI}\right) + \frac{4.167}{EI}\right] = -15 + (-5)$

$$M_{BC} = -20 \text{ kNm}$$

$$M_{CB} = 15 + \frac{2}{3}EI\left[2\left(\frac{4.167}{EI}\right) - \frac{5.833}{EI}\right]$$

$$M_{CB} = 16.667 \text{ kNm}$$

Span CE : $M_{CE} = EI\,\frac{(4.167)}{EI}$

$$M_{CE} = 4.167 \text{ kNm}$$

$$M_{EC} = \frac{EI}{2} \times \frac{(4.167)}{EI}$$

$$M_{EC} = 2.084 \text{ kNm}$$

Span CD : $M_{CD} = -25 + EI\theta_C = -25 + EI\,\frac{(4.167)}{EI}$

$$M_{CD} = -20.833 \text{ kNm}$$

$$M_{DC} = 25 + EI\theta_C = 25 + EI\,\frac{(4.167)}{EI}$$

$$M_{DC} = 27.084 \text{ kNm}$$

(*e*) **Shear force diagram :**

Span BA :

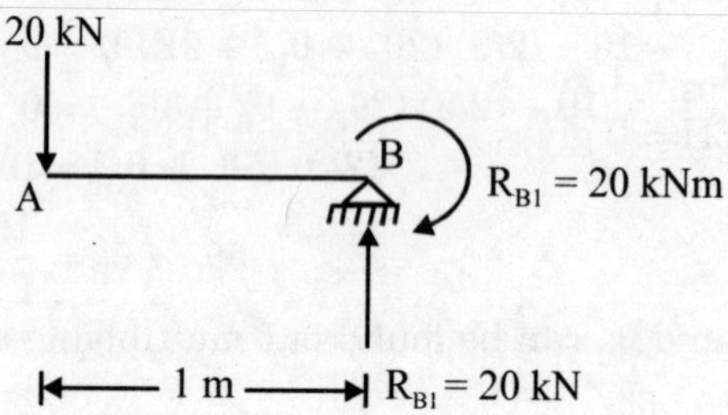

Span BC :

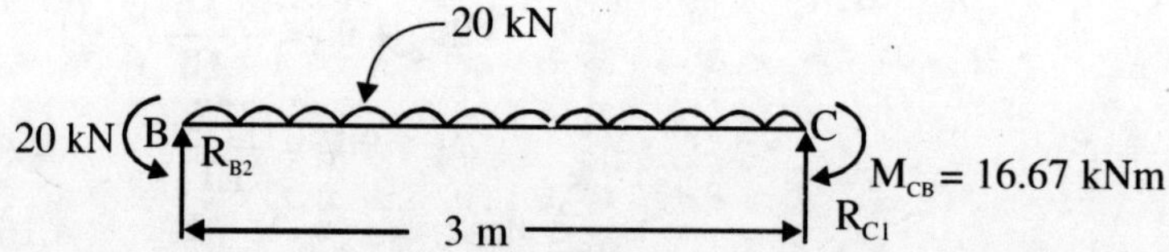

Taking moments about C, $R_{B2} \times 3 - 20 \times 3 \times 1.5 - 20 + 16.67 = 0$

$$R_{B2} = (1/3)(20 \times 3 \times 1.5 + 20 - 16.67) = 31.11 \text{ kN}$$

$$R_{C1} = 20 \times 3 - R_{B2} = 60 - 31.11 = 28.89 \text{ kN}$$

Span CD :

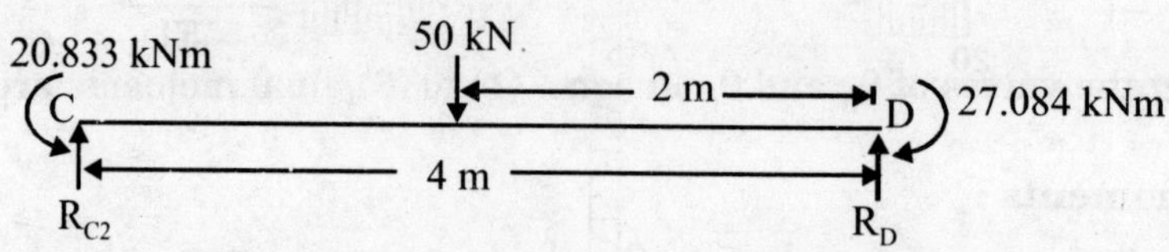

Taking moments about D,

$$R_{C2} \times 4 + 27.084 - 20.833 - 50 \times 2 = 0$$

$$R_{C2} = 23.437 \text{ kN}$$

$$R_D = 26.563 \text{ kN}$$

Span CE :

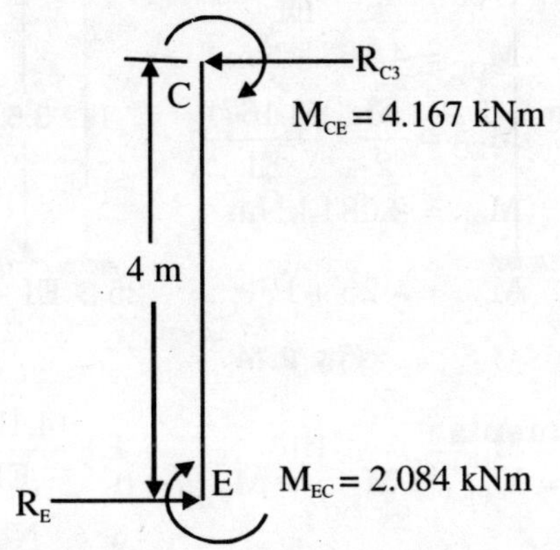

Taking moments about E, $R_{C3} \times 4 - 4.167 - 2.084 = 0$

(↺) (↻) (↻)

$$R_{C3} = 1.563 \text{ kN } (\leftarrow)$$

$\Sigma H = 0$ gives, $R_E = 1.563$ kN $(\rightarrow)$

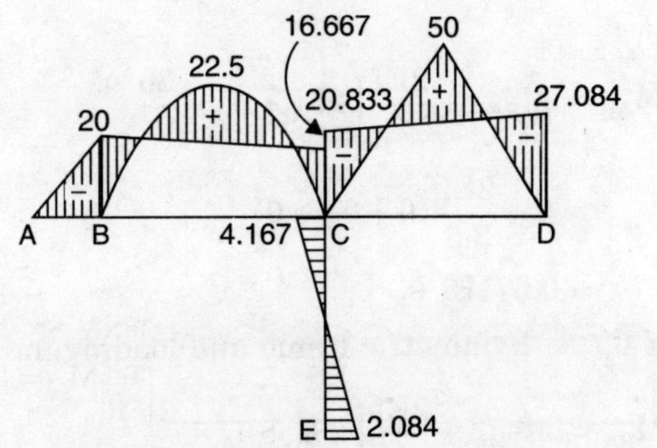

Bending moment diagram

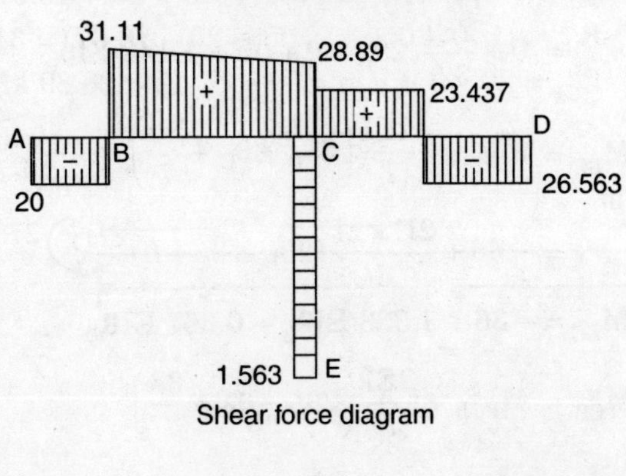

Shear force diagram

Example 2.10 *Analyse the portal frame shown in Fig. 2.14 by the slope deflection method and sketch the bending moment diagram.*

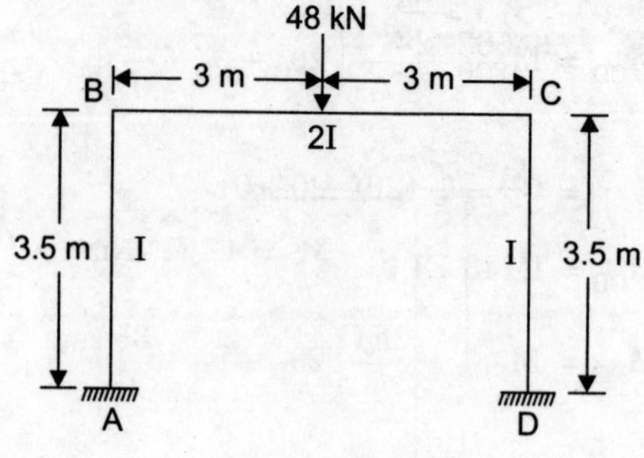

Fig. 2.14

Solution (*a*) **Fixed end moments :**

$$M_{FAB} = M_{FBA} = M_{FCD} = M_{FDC} = 0$$

[∵ No lateral load on AB and CD]

$$M_{FBC} = -\frac{Wl}{8} = -\frac{48 \times 6}{8} = -36 \text{ kNm}$$

$$M_{FCB} = \frac{Wl}{8} = \frac{48 \times 6}{8} = 36 \text{ kNm}$$

(*b*) **Slope deflection equations :**

Span AB :

$$M_{AB} = M_{FAB} + \frac{2EI}{l}\left(2\theta_A + \theta_B + \frac{3\delta}{l}\right)$$

$$= 0 + \frac{2EI}{3.5}(0 + \theta_B + 0) \qquad [\theta_A = 0, \because \text{ A is fixed}]$$

$$= 0.571EI\,\theta_B \qquad \ldots(1)$$

[$\delta = 0$, ∵ Symmetric frame and loading, no settlement in supports.]

$$M_{BA} = M_{FBA} + \frac{2EI}{l}\left(2\theta_B + \theta_A + \frac{3\delta}{l}\right)$$

$$= 0 + \frac{2EI}{3.5}(2\theta_B + 0 + 0) = 1.143\,EI\theta_B \qquad \ldots(2)$$

Span BC :

$$M_{BC} = M_{FBC} + \frac{2EI}{l}\left(2\theta_B + \theta_C + \frac{3\delta}{l}\right)$$

$$= -36 + \frac{2E \times 2I}{6}(2\theta_B + \theta_C + 0) \qquad [\because I_{BC} = 2I]$$

$$M_{BC} = -36 + 1.333\,EI\theta_B + 0.667\,EI\theta_C \qquad \ldots(3)$$

$$M_{CB} = M_{FCB} + \frac{2EI}{l}\left(2\theta_C + \theta_B + \frac{3\delta}{l}\right)$$

$$= 36 + \frac{2E \times 2I}{6}(2\theta_C + \theta_B + 0)$$

$$M_{CB} = 36 + 1.333\ EI\theta_C + 0.667\ EI\theta_B \quad ...(4)$$

Span CD : $$M_{CD} = M_{FCD} + \frac{2EI}{l}\left(2\theta_C + \theta_D + \frac{3\delta}{l}\right)$$

$$= 0 + \frac{2EI}{3.5}(2\theta_C + 0 + 0) \quad [\theta_D = 0, \because \text{ D is fixed}]$$

$$M_{CD} = 1.143\ EI\ \theta_C \quad ...(5)$$

$$M_{DC} = M_{FDC} + \frac{2EI}{l}\left(2\theta_D + \theta_C + \frac{3\delta}{l}\right)$$

$$= 0 + \frac{2EI}{3.5}(0 + \theta_C + 0) = 0.571EI\theta_C \quad ...(6)$$

(*c*) **Joint equilibrium equations :**

Joint B :

$$M_{BA} + M_{BC} = 0$$

$$1.143EI\theta_B + 1.333EI\theta_B + 0.667EI\theta_C - 36 = 0$$

$$2.476EI\theta_B + 0.667EI\theta_C = 36$$

$$2.476\theta_B + 0.667\theta_C = \frac{36}{EI} \quad ...(7)$$

Joint C :

$$M_{CB} + M_{CD} = 0$$

$$36 + 1.333EI\theta_C + 0.667EI\theta_B + 1.143EI\theta_C = 0$$

$$0.667EI\theta_B + 2.476EI\theta_C + 36 = 0$$

$$0.667\theta_B + 2.476\theta_C = -\frac{36}{EI} \quad ...(8)$$

(7) × 2.476 → $$6.13\theta_B + 1.65\theta_C = \frac{89.136}{EI} \quad ...(9)$$

(8) × 0.667 → $$0.44\theta_B + 1.65\theta_C = \frac{-24.012}{EI} \quad ...(10)$$

(9) – (10) → $$5.69\theta_B = \frac{113.148}{EI}$$

$\therefore$ $$\theta_B = \frac{19.89}{EI} \quad ...(11)$$

Substituting eqn. (11) in eqn. (7), we get $$\theta_C = -\frac{19.89}{EI} \quad ...(12)$$

(*d*) **Final moments :**

Substituting the values of θ_B and θ_C in eqns. (1) to (6)

(1) → $$M_{AB} = 0.571\,EI \times \frac{19.89}{EI} = 11.36 \text{ kNm}$$

(2) ⇒ $$M_{BA} = 1.143\,EI \times \frac{19.89}{EI} = 22.73 \text{ kNm}$$

(3) ⇒ $$M_{BC} = -36 + 1.333\,EI \times \frac{19.89}{EI} + 0.667\,EI\left(-\frac{19.89}{EI}\right)$$

$$= -22.73 \text{ kNm}$$

(4) ⇒ $$M_{CB} = 36 + 1.333\,EI\left(-\frac{19.89}{EI}\right) + 0.667\,EI\,\frac{19.89}{EI}$$

$$= 22.73 \text{ kNm}$$

(5) ⇒ $$M_{CD} = 1.143\,EI\left(-\frac{19.89}{EI}\right) = -22.73 \text{ kNm}$$

(6) ⇒ $$M_{DC} = 0.571\,EI\left(-\frac{19.89}{EI}\right)$$

$$= -11.36 \text{ kNm}$$

(*e*) **Simply supported span bending moments :**

In span AB, $\mu_{AB} = 0$

In span BC, $\mu_{BC} = \frac{Wl}{4} = \frac{48 \times 6}{4} = 72 \text{ kNm}$

In span CD, $\mu_{CD} = 0$

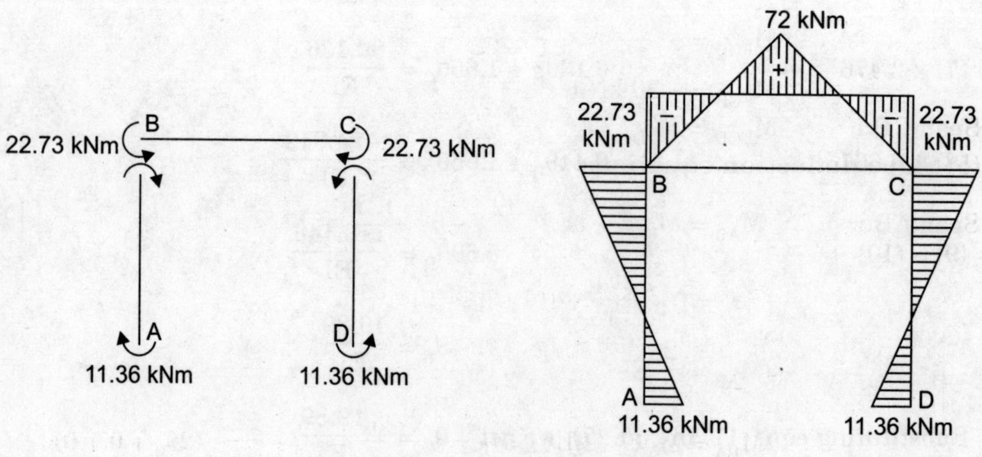

Bending moment diagram

Example 2.11 *Analyse the portal frame loaded as shown in Fig. 2.15 by the slope deflection method and sketch the bending moment and shear force diagrams.*

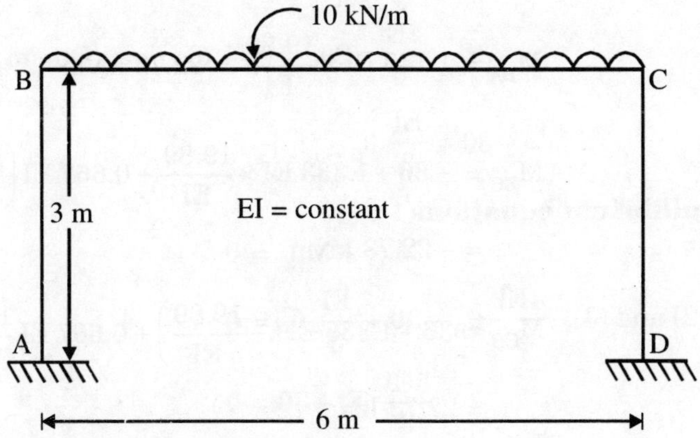

Fig. 2.15

Solution The frame, end conditions and loadings are symmetrical. Hence there will be no side sway. A and D are fixed ends ($\therefore\ \theta_A = \theta_D = 0$). The unknowns will be θ_B and θ_C. The equilibrium equations to solve for the unknowns are,

At joint B, $M_{BA} + M_{BC} = 0$

At joint C, $M_{CB} + M_{CD} = 0$

However, since the frame and loading are symmetrical, $\theta_C = -\theta_B$ and it is enough, if we find θ_B or θ_C using anyone of the equilibrium equations.

(*a*) **Fixed end moments :**

Span BA : $M_{FAB} = M_{FBA} = 0$

Span BC : $M_{FBC} = -\dfrac{wl^2}{12} = -\dfrac{10 \times 6^2}{12}$

$M_{FBC} = -30$ kNm

$M_{FCB} = +\dfrac{wl^2}{12} = +\dfrac{10 \times 6^2}{12}$

$M_{FCB} = +30$ kNm

Span CD : $M_{FCD} = M_{FDC} = 0$

(*b*) **Slope deflection equations :**

Span AB : $M_{AB} = M_{FAB} + 2Ek\left(2\theta_A + \theta_B + \dfrac{3\delta}{l}\right)$ $\left(k = \dfrac{I}{l}\right)$

$= 0 + \dfrac{2EI}{3}\,(2(0) + \theta_B + 0)$

$M_{AB} = \dfrac{2}{3}\,EI\theta_B$ ...(1)

$M_{BA} = M_{FBA} + 2Ek\left(2\theta_B + \theta_A + \dfrac{3\delta}{l}\right) = 0 + \dfrac{2EI}{3}\,(2\theta_B + 0 + 0)$

$= \dfrac{4EI}{3}\,\theta_B$ ...(2)

Span BC : $M_{BC} = M_{FBC} + 2Ek\left(2\theta_B + \theta_C + \dfrac{3\delta}{l}\right)$

$= -30 + \dfrac{2EI}{6}(2\theta_B - \theta_B + 0)$ (Due to symmetry $\theta_C = -\theta_B$)

$= -30 + \dfrac{EI}{3}\theta_B$ ...(3)

(*c*) **Joint equilibrium equations :**

$$M_{BA} + M_{BC} = 0$$

From eqns. (2) and (3), $\dfrac{4EI}{3}\theta_B - 30 + \dfrac{EI}{3}\theta_B = 0$

$$\frac{5EI}{3}\theta_B - 30 = 0$$

$$\theta_B = \frac{30 \times 3}{5EI}$$

$$\theta_B = \frac{18}{EI}$$

$$\theta_C = -\theta_B = -\frac{18}{EI}$$

(*d*) **Final moments :**

Span AB : $M_{AB} = \dfrac{2EI}{3}\left[\dfrac{18}{EI}\right] = 12 \text{ kNm}$

$M_{BA} = \dfrac{4EI}{3}\left[\dfrac{18}{EI}\right] = 24 \text{ kNm}$

Span BC : $M_{BC} = -30 + \dfrac{EI}{3}\left[\dfrac{18}{EI}\right] = -24 \text{ kNm}$

$M_{CB} = 24$ kNm (by symmetry)

Span CD : $M_{CD} = -24$ kNm (by symmetry)

$M_{DC} = -12$ kNm (by symmetry)

(*e*) **To draw S.F.D. :**

Span AB :

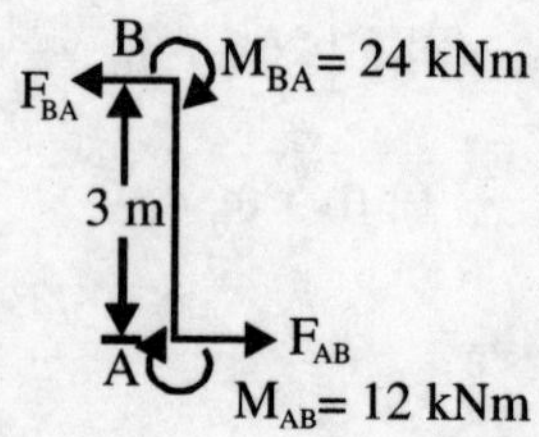

Taking moments about A, $-F_{BA}(3) + M_{BA} + M_{AB} = 0$

$$24 + 12 = F_{BA}(3)$$

$$F_{BA} = 12 \text{ kN} = F_{AB}$$

Span BC :

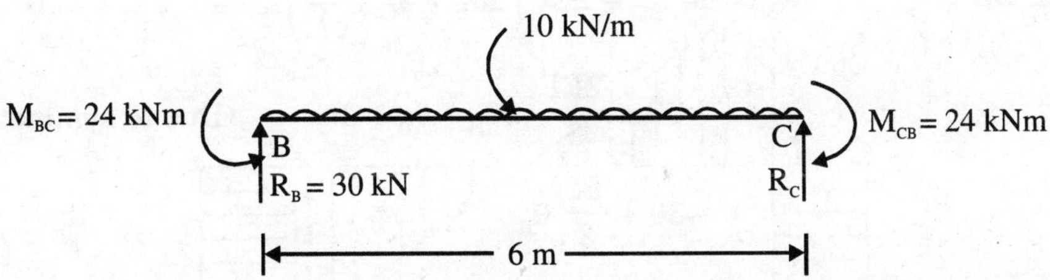

Since the moments M_{BC} end M_{CB} are equal and opposite, there will not be any change in shear due to the support moments.

$$\therefore \qquad R_B = R_C = \frac{\text{Total load}}{2} = \frac{10 \times 6}{2} = 30 \text{ kN}$$

Span CD :

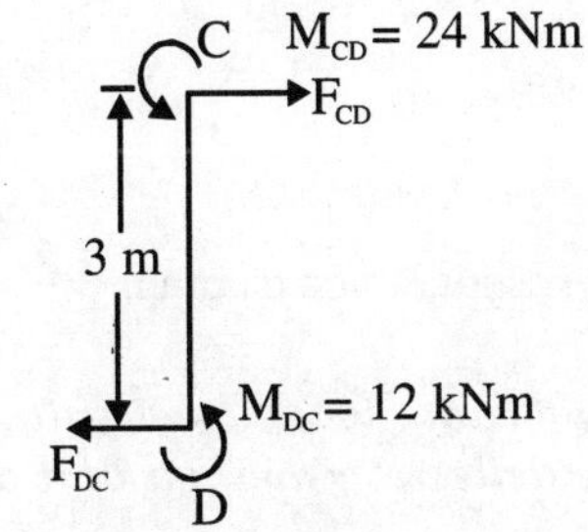

$$F_{CD} = 12 \text{ kN} = F_{DC} \qquad \text{(by symmetry)}$$

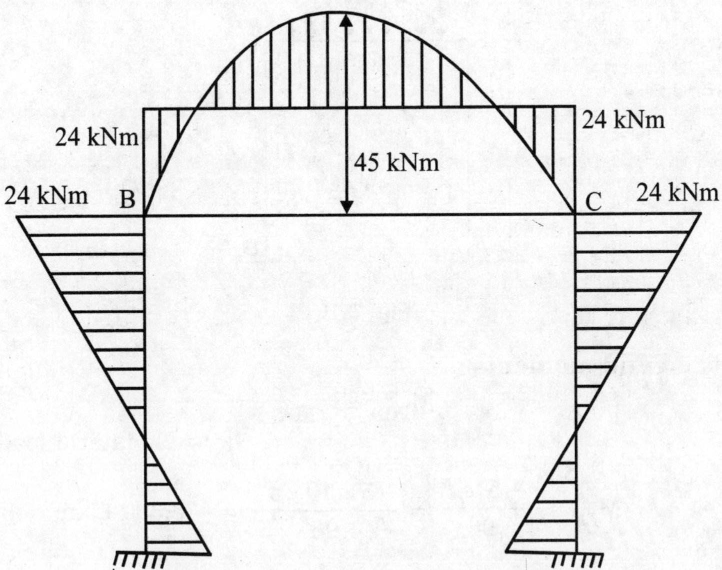

Bending moment diagram

$$\text{Simply supported bending moment in span BC} = \frac{wl^2}{8} = 10 \times \frac{6^2}{8} = 45 \text{ kNm}$$

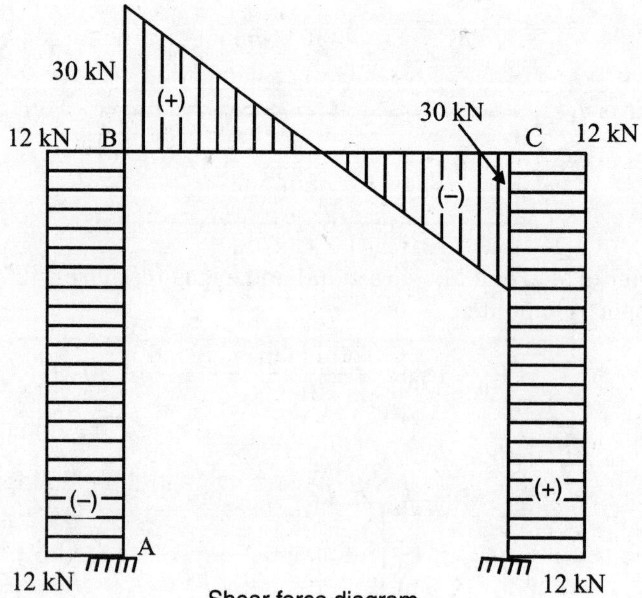

Shear force diagram

Example 2.12 *Analyse the portal frame loaded with uniformly varying load as shown in Fig. 2.16 by slope deflection method and draw BMD and SFD.*

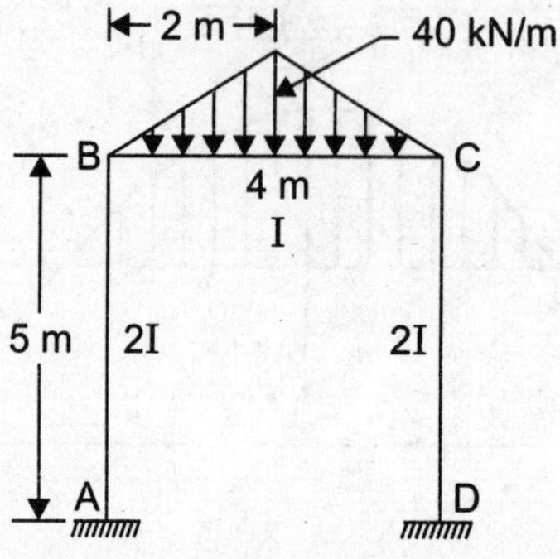

Fig. 2.16

Solution (*a*) **Fixed end moments :**

$$M_{FAB} = M_{FBA} = M_{FCD} = M_{FDC} = 0$$

[∵ No lateral load on AB and CD]

$$M_{FBC} = \frac{-5\,wl^2}{96} = \frac{-5 \times 40 \times 4^2}{96} = -33.333 \text{ kNm}$$

$$M_{FCB} = \frac{5\,wl^2}{96} = \frac{5 \times 40 \times 4^2}{96} = 33.333 \text{ kNm}$$

(*b*) **Slope deflection equations :**

Span AB : $$M_{AB} = M_{FAB} + \frac{2EI}{l}\left(2\theta_A + \theta_B + \frac{3\delta}{l}\right)$$

$$= 0 + \frac{2E \times 2I}{5}(0 + \theta_B + 0) \qquad [\because \ I_{AB} = 2I]$$

$$= \frac{4EI}{5}\theta_B \qquad ...(1)$$

[$\theta_A = 0$, since A is fixed $\delta = 0$, symmetrical loading and structure, Hence no sway]

$$M_{BA} = M_{FBA} + \frac{2EI}{l}\left(2\theta_B + \theta_A + \frac{3\delta}{l}\right)$$

$$= 0 + \frac{2E \times 2I}{5}(2\theta_B + 0 + 0) = \frac{8EI}{5}\theta_B \qquad ...(2)$$

Span BC :

$$M_{BC} = M_{FBC} + \frac{2EI}{l}\left(2\theta_B + \theta_C + \frac{3\delta}{l}\right)$$

$$= -33.333 + \frac{2EI}{4}(2\theta_B + \theta_C + 0)$$

$$= -33.333 + EI\left(\theta_B + \frac{\theta_C}{2}\right) \qquad ...(3)$$

$$M_{CB} = M_{FCB} + \frac{2EI}{l}\left(2\theta_C + \theta_B + \frac{3\delta}{l}\right)$$

$$= 33.333 + \frac{2EI}{4}(2\theta_C + \theta_B + 0)$$

$$= 33.333 + EI\left(\theta_C + \frac{\theta_B}{2}\right) \qquad ...(4)$$

Span CD :

$$M_{CD} = M_{FCD} + \frac{2EI}{l}\left(2\theta_C + \theta_D + \frac{3\delta}{l}\right)$$

$$= 0 + \frac{2E \times 2I}{5}(2\theta_C + 0 + 0) \qquad [\theta_D = 0, \text{ D is fixed and } I_{CD} = 2I]$$

$$= \frac{8EI}{5}\theta_C \qquad ...(5)$$

$$M_{DC} = M_{FDC} + \frac{2EI}{l}\left(2\theta_D + \theta_C + \frac{3\delta}{l}\right)$$

$$= 0 + \frac{2E \times 2I}{5}(0 + \theta_C + 0)$$

$$= \frac{4EI}{5}\theta_C \qquad ...(6)$$

(*c*) **Joint equilibrium equations :**

Joint B :

$$M_{BA} + M_{BC} = 0$$

$$\frac{8EI}{5}\theta_B + EI\left(\theta_B + \frac{\theta_C}{2}\right) - 33.333 = 0$$

$$\frac{13EI}{5}\theta_B + \frac{EI\theta_C}{2} - 33.333 = 0$$

$$2.6\theta_B + 0.5\theta_C = \frac{33.333}{EI} \quad ...(7)$$

Joint C : $M_{CB} + M_{CD} = 0$

$$33.333 + EI\left(\theta_C + \frac{\theta_B}{2}\right) + \frac{8EI}{5}\theta_C = 0$$

$$\frac{EI}{2}\theta_B + \frac{13EI}{5}\theta_C = -33.333$$

$$0.5\theta_B + 2.6\theta_C = -\frac{33.333}{EI} \quad ...(8)$$

$$(7) \times 2.6 \Rightarrow \quad 6.76\theta_B + 1.3\theta_C = \frac{86.67}{EI} \quad ...(9)$$

$$(8) \times 0.5 \Rightarrow \quad 0.25\theta_B + 1.3\theta_C = \frac{-16.67}{EI} \quad ...(10)$$

$$(9) - (10) \Rightarrow \quad 6.51\theta_B = \frac{103.34}{EI}$$

$$\therefore \quad \theta_B = \frac{15.87}{EI}$$

Since the frame is symmetrical, $\theta_B = -\theta_C = \dfrac{15.87}{EI}$

(*d*) **Final moments :**

Substituting θ_B and θ_C in eqns. (1) to (6),

$$(1) \Rightarrow \quad M_{AB} = \frac{4EI}{5} \times \frac{15.87}{EI} = 12.70 \text{ kNm}$$

$$(2) \Rightarrow \quad M_{BA} = \frac{8EI}{5} \times \frac{15.87}{EI} = 25.40 \text{ kNm}$$

$$(3) \Rightarrow \quad M_{BC} = -33.333 + EI\left(\frac{15.87}{EI} - \frac{15.87}{2EI}\right) = -25.40 \text{ kNm}$$

$$(4) \Rightarrow \quad M_{CB} = 33.333 + EI\left(-\frac{15.87}{EI} + \frac{15.87}{2EI}\right) = 25.40 \text{ kNm}$$

$$(5) \Rightarrow \quad M_{CD} = \frac{8EI}{5}\left(-\frac{15.87}{EI}\right) = -25.40 \text{ kNm}$$

$$(6) \Rightarrow \quad M_{DC} = \frac{4EI}{5}\left(-\frac{15.87}{EI}\right) = -12.70 \text{ kNm}$$

(*e*) **Simply supported bending moments :**

In span AB, $\mu_{AB} = 0$

In span BC

By symmetry,

$$R_B = R_C = \frac{\text{Total load}}{2}$$

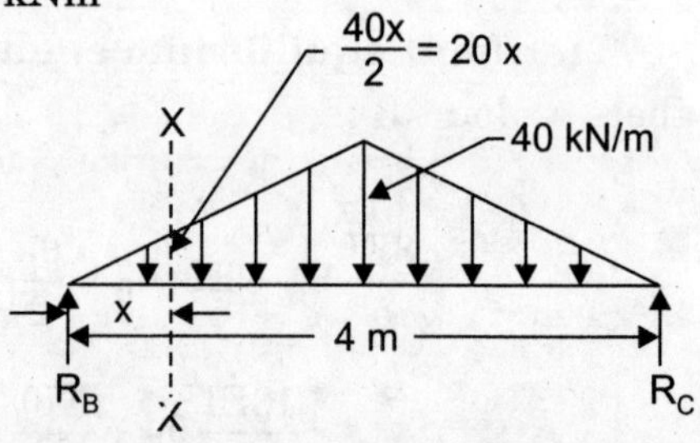

$$= \frac{1}{2} \times \frac{1}{2} wl = \frac{1}{2} \times \frac{1}{2} \times 40 \times 4$$
$$= 40 \text{ kN}$$

Bending moment at section X-X,

$$M_X = R_A x - \left(\frac{1}{2} x \times 20x\right) \times \frac{x}{3} = 40x - \frac{10}{3} x^3$$

The bending moment has cubic variation.

Here, due to symmetry, bending moment will be maximum at centre, *i.e.*, $x = 2$ m

$$\mu_{BC} = 40 \times 2 - \frac{10}{3} \times 2^3 = 53.33 \text{ kNm}$$

In span CD, $\mu_{CD} = 0$

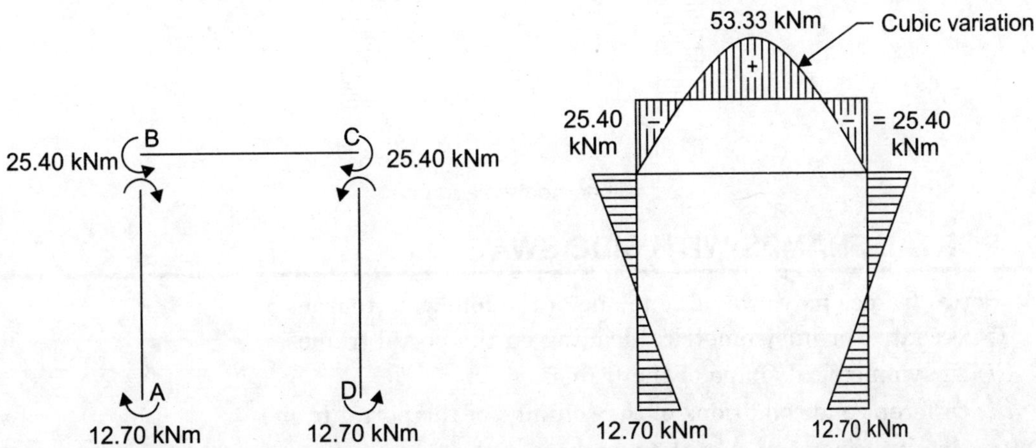

Bending moment diagram

(*f*) **Shear force diagram**

Span AB :

Taking moment about A,

$$-F_{BA} \times 5 + M_{AB} + M_{BA} = 0$$
$$5F_{BA} = 12.70 + 25.40$$
$$F_{BA} = 7.62 \text{ kN } (\leftarrow)$$
$$F_{AB} = 7.62 \text{ kN } (\rightarrow)$$

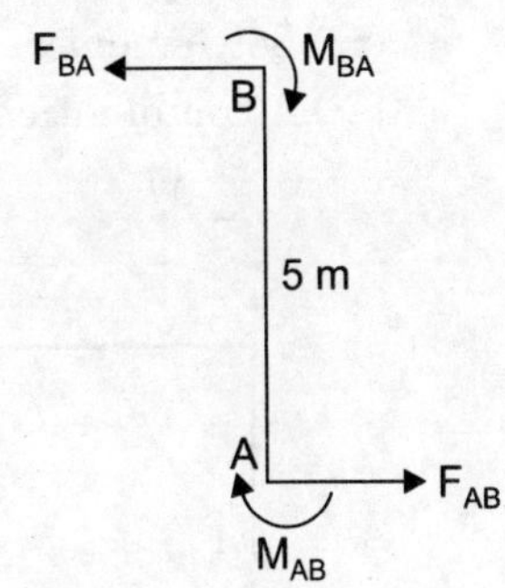

Span BC :

Since, the moments M_{BC} and M_{CB} are equal and opposite, there will be no change in shear due to the support moments.

$$\therefore \quad R_B = R_C = \frac{\text{Total load}}{2}$$
$$= \frac{1/2 \times 40 \times 4}{2}$$
$$= 40 \text{ kN}$$

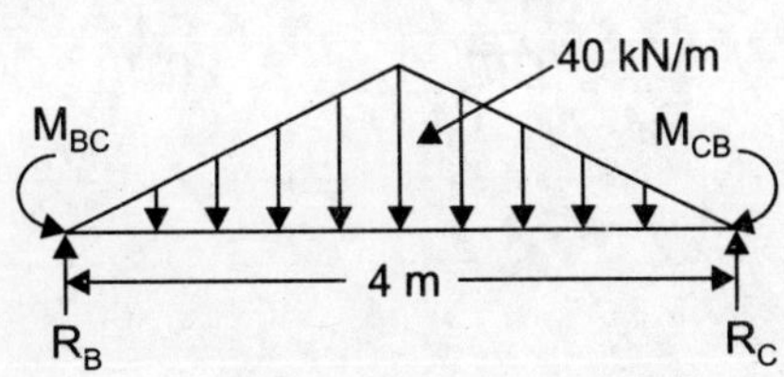

Span CD :

Taking moment about C,

$$F_{DC} \times 5 - M_{CD} - M_{DC} = 0$$

$$5F_{DC} = M_{CD} + M_{DC}$$

$$= 25.40 + 12.70$$

$$F_{DC} = 7.62 \text{ kN } (\leftarrow)$$

$$F_{CD} = 7.62 \text{ kN } (\rightarrow)$$

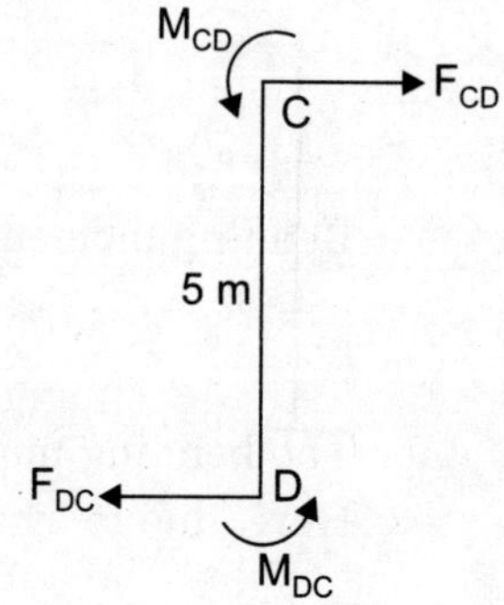

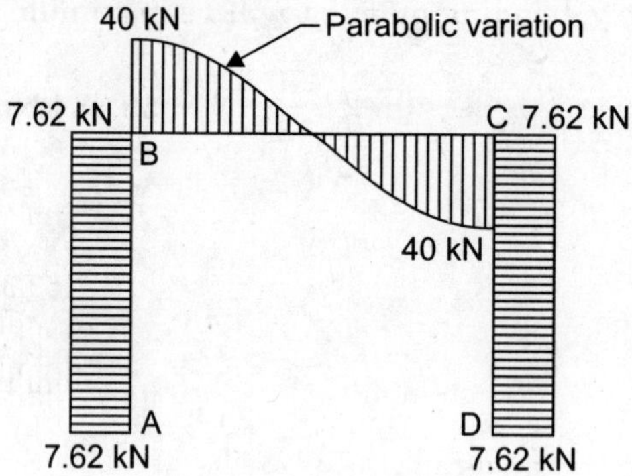

Shear force diagram

2.6 PORTAL FRAMES WITH SIDE SWAY

Portal frames may sway due to one of the following reasons :

(*i*) Eccentric or unsymmetrical loading on the portal frames.

(*ii*) Unsymmetrical shape of the frames.

(*iii*) Different end conditions of the columns of the portal frame.

(*iv*) Non-uniform section of the members of the frame.

(*v*) Horizontal loading on the columns of the frame

(*vi*) Settlement of the supports of the frame.

(*vii*) A combination of the above.

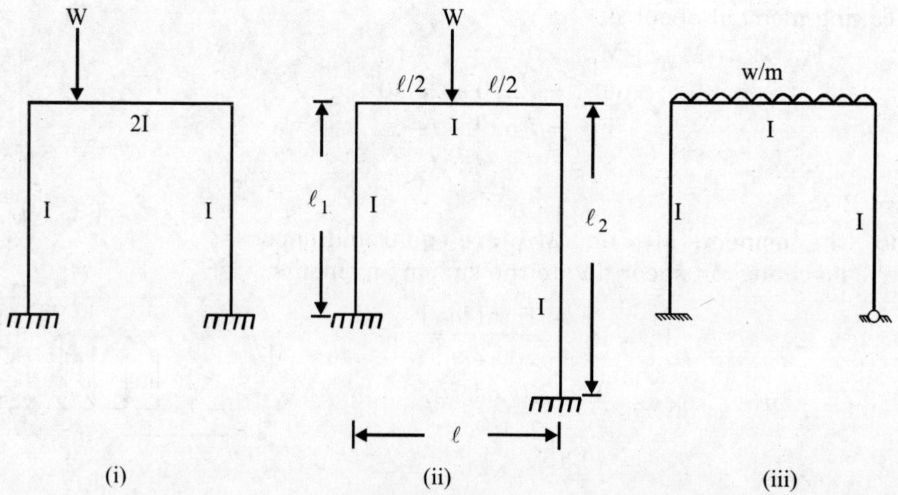

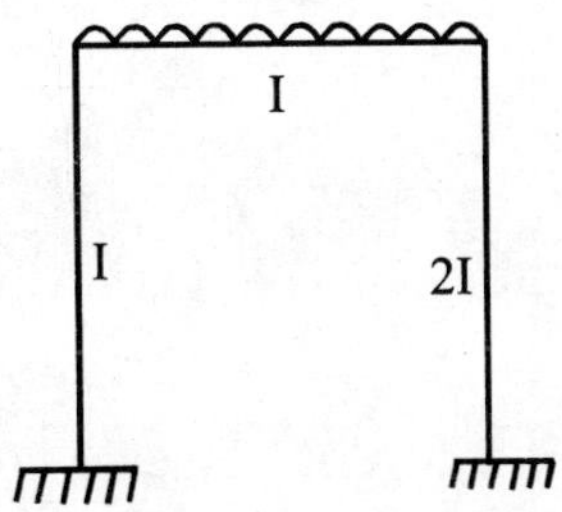

(iv)

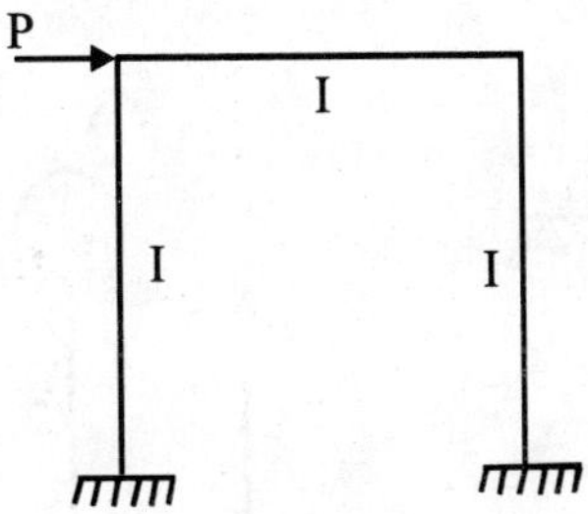

(v)

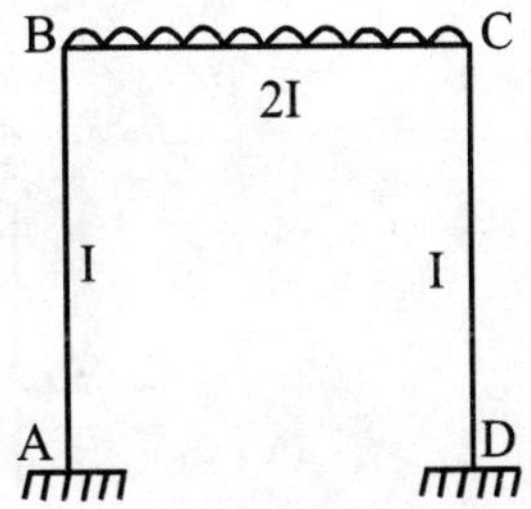

(vi)

D Settles down/Sinks by δ

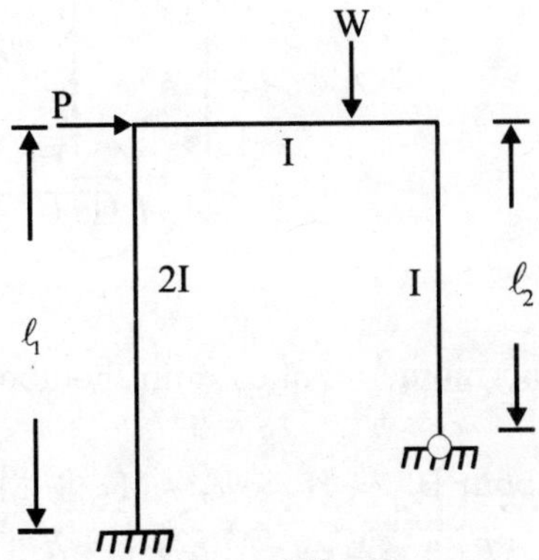

(vii)

Fig. 2.17

The amount of joint movement or sway is not known and forms an additional unknown. In slope deflection method, to evaluate the additional unknown, shear equation is made use of.

The most general equation :

Case 1 :

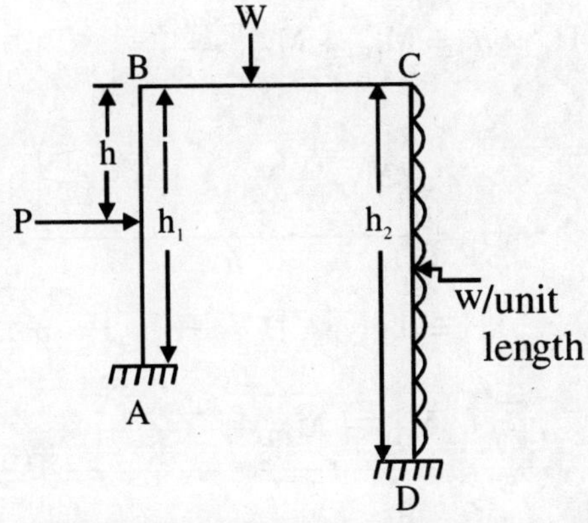

Unknowns are θ_B, θ_C and δ. For solution, equilibrium equations are

$$M_{BA} + M_{BC} = 0 \quad \text{...(1) at B}$$

$$M_{CB} + M_{CD} = 0 \quad \text{...(2) at C}$$

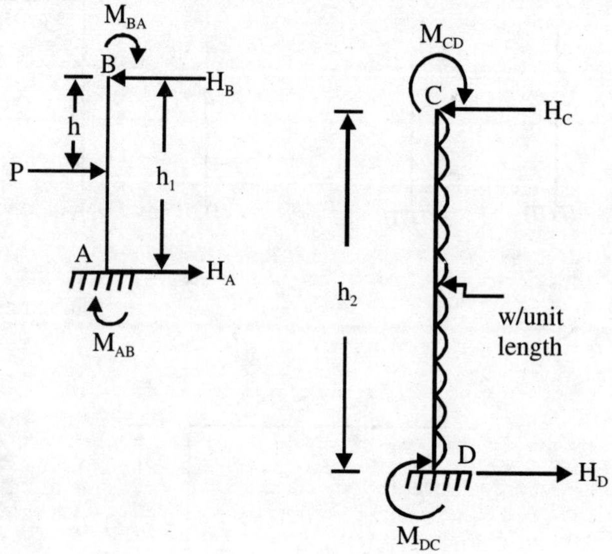

Shear equation : The end moments of column members are assumed to be clockwise, in the derivation.

In AB, taking moments about B, $-H_A \times h_1 + M_{AB} + M_{BA} - P \times h = 0$

$$-H_A \times h_1 = -M_{AB} - M_{BA} + P \times h$$

$$H_A \times h_1 = M_{AB} + M_{BA} - P \times h$$

$$H_A = \frac{(M_{AB} + M_{BA} - P \times h)}{h_1}$$

In CD, taking moments about C,

$$-H_D \times h_2 + M_{DC} + M_{CD} + w \times h_2 \times \frac{h_2}{2} = 0$$

$$H_D \times h_2 = M_{DC} + M_{CD} + \frac{wh_2^{\,2}}{2}$$

$$H_D = \frac{\left(M_{DC} + M_{CD} + \dfrac{w \times h_2^{\,2}}{2}\right)}{h_2}$$

$$\Sigma H = 0 \text{ gives, } H_A + H_D + P - wh_2 = 0$$

$$\underset{(\rightarrow)}{\frac{M_{AB} + M_{BA} - P \times h}{h_1}} + \underset{(\rightarrow)}{\frac{M_{DC} + M_{CD} + \dfrac{w \times h_2^{\,2}}{2}}{h_2}} + \underset{(\rightarrow)}{P} - \underset{(\leftarrow)}{w \times h_2} = 0$$

is the most general shear equation which will get modified, depending upon the problem, as below :

Case 2 :

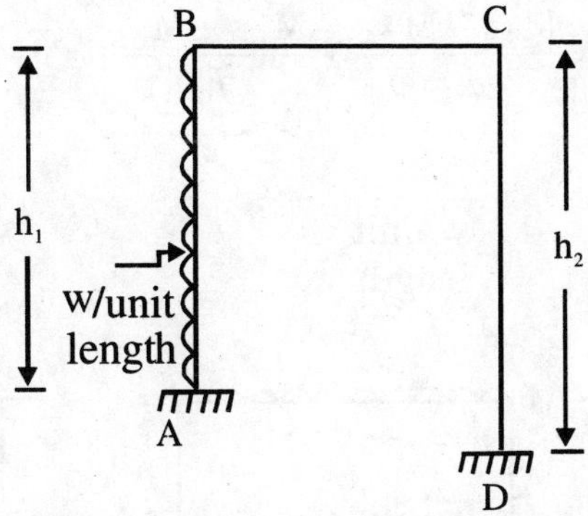

In column AB, udl is acting instead of P and column CD is free from udl. ∴ The shear equation gets modified as

$$\frac{M_{AB} + M_{BA} - \dfrac{wh_1^2}{2}}{h_1} + \frac{M_{DC} + M_{CD}}{h_2} + w \times h_1 = 0$$

Case 3 :

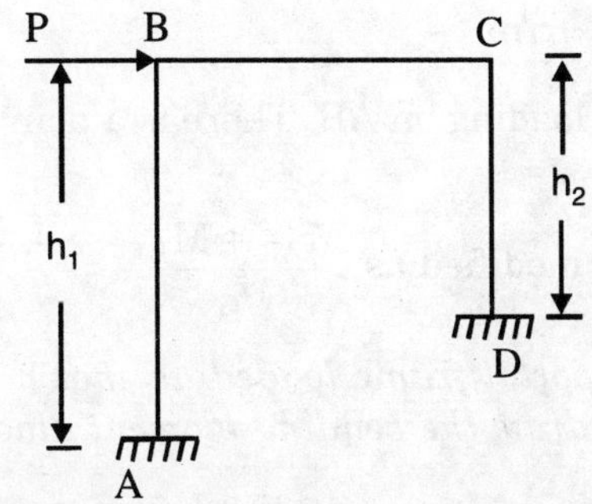

Here, Lateral load acts at the top of AB. (∴ $h = 0$) There is no udl on CD

∴ The shear equation gets modified as $\dfrac{M_{AB} + M_{BA}}{h_1} + \dfrac{M_{DC} + M_{CD}}{h_2} + P = 0$

Case 4 :

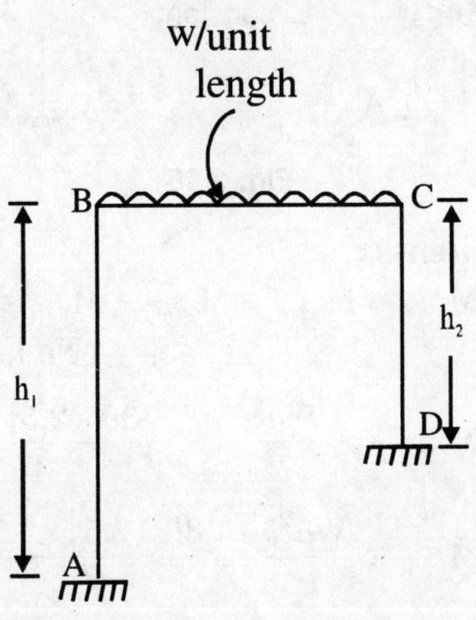

Here, there is no lateral load on AB and CD

Hence the shear equation gets modified as

$$\frac{M_{AB}+M_{BA}}{h_1}+\frac{M_{DC}+M_{CD}}{h_2}=0$$

Case 5 :

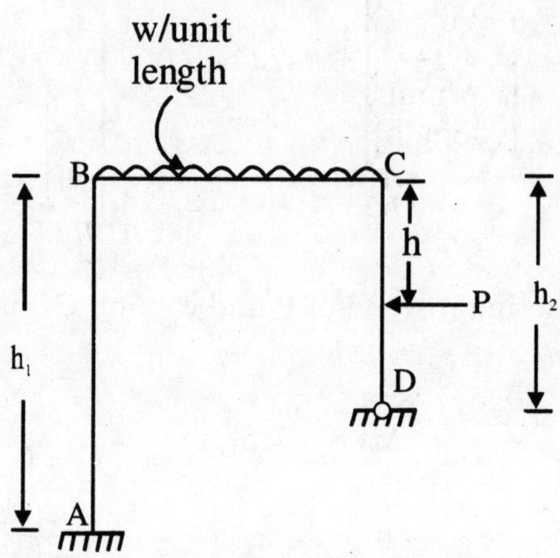

Here, there is no lateral loading on AB. There is a point load P acting at h from the top C. Support D is hinged. ∴ $M_{DC} = 0$

The shear equation gets modified as $\frac{M_{AB}+M_{BA}}{h_1}+\frac{M_{CD}+Ph}{h_2}-P=0.$

Example 2.13 *Analyse the portal frame loaded as shown in Fig. 2.18 by slope deflection method and draw the bending moment diagram.*

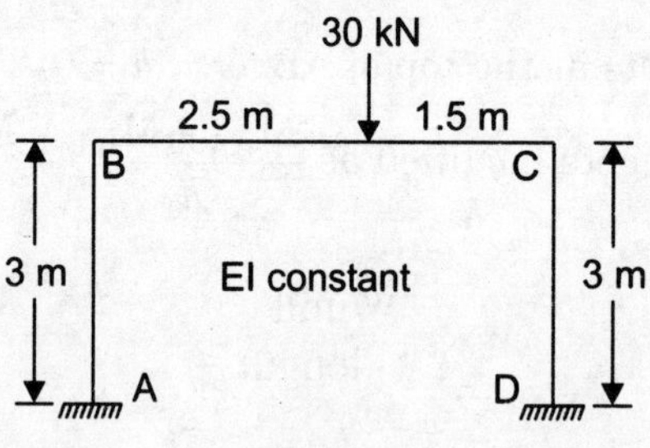

Fig. 2.18

Solution (*a*) **Fixed end moments :**

$$M_{FAB} = M_{FBA} = M_{FCD} = M_{FDC} = 0$$

[∵ No lateral load on spans AB and CD]

$$M_{FBC} = -\frac{Wab^2}{l^2} = -\frac{30\times2.5\times1.5^2}{4^2} = -10.547 \text{ kNm}$$

$$M_{FCB} = \frac{Wa^2b}{l^2} = \frac{30\times2.5^2\times1.5}{4^2} = 17.578 \text{ kNm}$$

(*b*) **Slope deflection equations :**

Eccentric or unsymmetrical loading on the portal frame. Hence the frame will sway. Let B and C move horizontally by 'δ'.

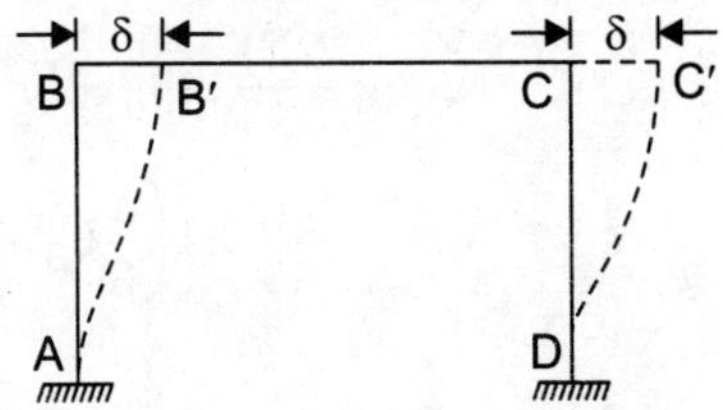

Span AB :

$$M_{AB} = M_{FAB} + \frac{2EI}{l}\left(2\theta_A + \theta_B + \frac{3\delta}{l}\right)$$

$$= 0 + \frac{2EI}{3}\left(0 + \theta_B - \frac{3\delta}{3}\right)$$

[$\theta_A = 0$, since support A is fixed]

$$= \frac{2EI}{3}(\theta_B - \delta) \quad ...(1)$$

$$M_{BA} = M_{FBA} + \frac{2EI}{l}\left(2\theta_B + \theta_A + \frac{3\delta}{l}\right)$$

$$= 0 + \frac{2EI}{3}\left(2\theta_B + 0 - \frac{3\delta}{3}\right) = \frac{2EI}{3}(2\theta_B - \delta) \quad ...(2)$$

Span BC :

$$M_{BC} = M_{FBC} + \frac{2EI}{l}\left(2\theta_B + \theta_C + \frac{3\delta}{l}\right) = -10.547 + \frac{2EI}{4}(2\theta_B + \theta_C + 0)$$

$$= -10.547 + \frac{EI}{2}(2\theta_B + \theta_C) \quad ...(3)$$

$$M_{CB} = M_{FCB} + \frac{2EI}{l}\left(2\theta_C + \theta_B + \frac{3\delta}{l}\right) = 17.578 + \frac{2EI}{4}(2\theta_C + \theta_B + 0)$$

$$= 17.578 + \frac{EI}{2}(2\theta_C + \theta_B) \quad ...(4)$$

Span CD :

$$M_{CD} = M_{FCD} + \frac{2EI}{l}\left(2\theta_C + \theta_D + \frac{3\delta}{l}\right)$$

$$= 0 + \frac{2EI}{3}\left(2\theta_C + 0 - \frac{3\delta}{3}\right)$$

[$\theta_D = 0$, D is fixed]

$$= \frac{2EI}{3}(2\theta_C - \delta) \quad ...(5)$$

$$M_{DC} = M_{FDC} + \frac{2EI}{l}\left(2\theta_D + \theta_C + \frac{3\delta}{l}\right)$$

$$= 0 + \frac{2EI}{l}\left(0 + \theta_C - \frac{3\delta}{3}\right) = \frac{2EI}{3}(\theta_C - \delta) \quad ...(6)$$

(*c*) **Joint equilibrium equations :**

Joint B :

$$M_{BA} + M_{BC} = 0$$

$$\frac{2EI}{3}(2\theta_B - \delta) - 10.547 + \frac{EI}{2}(2\theta_B + \theta_C) = 0$$

$$\frac{2EI}{3}(2\theta_B - \delta) + \frac{EI}{2}(2\theta_B + \theta_C) = 10.547$$

$$EI\left(\frac{4}{3}\theta_B - \frac{2}{3}\delta + \theta_B + \frac{\theta_C}{2}\right) = 10.547$$

$$\frac{7}{3}\theta_B + \frac{\theta_C}{2} - \frac{2\delta}{3} = \frac{10.547}{EI} \quad \text{...(7)}$$

Joint C :

$$M_{CB} + M_{CD} = 0$$

$$17.578 + \frac{EI}{2}(2\theta_C + \theta_B) + \frac{2EI}{3}(2\theta_C - \delta) = 0$$

$$17.578 + EI\left(\theta_C + \frac{\theta_B}{2}\right) + EI\left(\frac{4}{3}\theta_C - \frac{2\delta}{3}\right) = 0$$

$$EI\left(\frac{7}{3}\theta_C + \frac{\theta_B}{2} - \frac{2\delta}{3}\right) = -17.578$$

$$\frac{\theta_B}{2} + \frac{7}{3}\theta_C - \frac{2\delta}{3} = -\frac{17.578}{EI} \quad \text{...(8)}$$

(*d*) **Shear equation :**

$$\frac{M_{AB} + M_{BA}}{l_{AB}} + \frac{M_{CD} + M_{DC}}{l_{CD}} = 0$$

$$M_{AB} + M_{BA} + M_{CD} + M_{DC} = 0 \quad [\because \; l_{AB} = l_{CD}]$$

$$\frac{2EI}{3}(\theta_B - \delta) + \frac{2EI}{3}(2\theta_B - \delta) + \frac{2EI}{3}(2\theta_C - \delta) + \frac{2EI}{3}(\theta_C - \delta) = 0$$

$$\frac{2EI}{3}(\theta_B - \delta + 2\theta_B - \delta + 2\theta_C - \delta + \theta_C - \delta) = 0$$

$$3\theta_B + 3\theta_C - 4\delta = 0 \quad \text{...(9)}$$

$$(7) - (8) \Rightarrow \quad \frac{11\theta_B}{6} - \frac{11}{6}\theta_C = \frac{28.125}{EI} \quad \text{...(10)}$$

$$(7) \times (2) \Rightarrow \quad \frac{14\theta_B}{3} + \theta_C - \frac{4\delta}{3} = \frac{21.094}{EI} \quad \text{...(11)}$$

$$(9) \div (3) \Rightarrow \quad \theta_B + \theta_C - \frac{4\delta}{3} = 0 \quad \text{...(12)}$$

$$(11) - (12) \Rightarrow \quad \frac{11}{3}\theta_B = \frac{21.094}{EI}$$

$$\therefore \quad \theta_B = \frac{5.753}{EI} \quad \text{...(13)}$$

Substituting θ_B in equation (10)

$$\frac{11}{6}\left(\frac{5.753}{EI}\right)-\frac{11}{6}\theta_C=\frac{28.125}{EI}$$

$$\therefore \qquad \theta_C=-\frac{9.588}{EI} \qquad \ldots(14)$$

Substituting the values of θ_B and θ_C in eqn. (9),

$$3\left(\frac{5.753}{EI}\right)-4\delta+3\left(-\frac{9.588}{EI}\right)=0$$

$$4\delta=\frac{17.259}{EI}-\frac{28.764}{EI}$$

$$\delta=-\frac{2.876}{EI}$$

(*e*) **Final moments :**

Substituting θ_B, θ_C and δ in eqns. (1) to (6),

$$(1)\Rightarrow \qquad M_{AB}=\frac{2EI}{3}\left[\frac{5.753}{EI}-\left(-\frac{2.876}{EI}\right)\right]$$

$$=5.753 \text{ kNm}$$

$$(2)\Rightarrow \qquad M_{BA}=\frac{2EI}{3}\left[\left(2\times\frac{5.753}{EI}\right)+\frac{2.876}{EI}\right]$$

$$=9.588 \text{ kNm}$$

$$(3)\Rightarrow \qquad M_{BC}=-10.547+\frac{EI}{2}\left[\left(2\times\frac{5.753}{EI}\right)-\frac{9.588}{EI}\right]$$

$$=-9.588 \text{ kNm}$$

$$(4)\Rightarrow \qquad M_{CB}=17.578+\frac{EI}{2}\left(\frac{5.753}{EI}-2\times\frac{9.588}{EI}\right)$$

$$=10.866 \text{ kNm}$$

$$(5)\Rightarrow \qquad M_{CD}=\frac{2EI}{3}\left[\left(2\times-\frac{9.588}{EI}\right)+\frac{2.876}{EI}\right]$$

$$=-10.866 \text{ kNm}$$

$$(6)\Rightarrow \qquad M_{DC}=\frac{2EI}{3}\left(-\frac{9.588}{EI}+\frac{2.876}{EI}\right)$$

$$=-4.475 \text{ kNm}$$

(f) Simply supported bending moments :

In spans AB, CD, $\mu_{AB} = \mu_{CD} = 0$

In span BC, $\mu_{BC} = \dfrac{Wab}{l} = \dfrac{30 \times 2.5 \times 1.5}{4}$

$= 28.125$ kNm

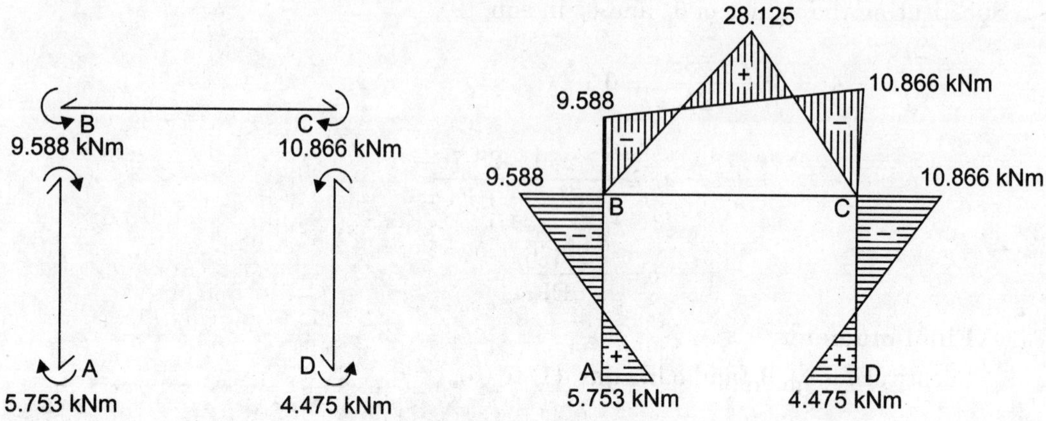

Bending moment diagram

Example 2.14 *Analyse the portal frame loaded as shown in Fig. 2.19 by the slope deflection method and sketch the bending moment and shear force diagram.*

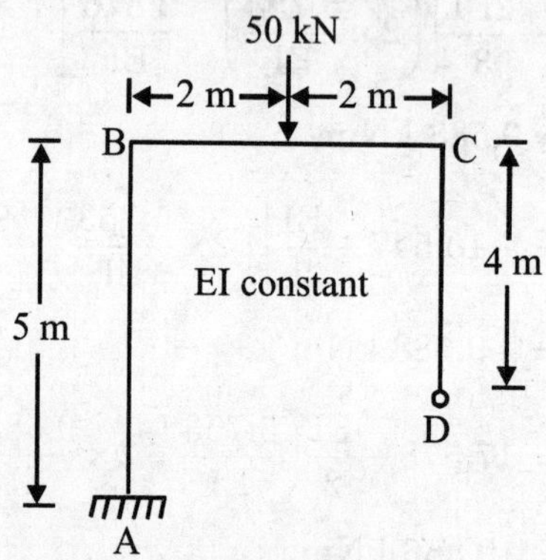

Fig. 2.19

Solution **(a) Fixed end moments :**

Span AB, CD : $M_{FAB} = M_{FBA} = M_{FDC} = M_{FCD} = 0$

Span BC : $M_{FBC} = -\dfrac{Wl}{8} = -\dfrac{50 \times 4}{8}$ $M_{FBC} = -25$ kNm

$M_{FCB} = +\dfrac{Wl}{8} = +\dfrac{50 \times 4}{8}$ $M_{FCB} = +25$ kNm (*b*) **Slope deflection equations :**

The frame is unsymmetrical. Hence the frame will sway. Let B and C move horizontally by δ.

Span AB :

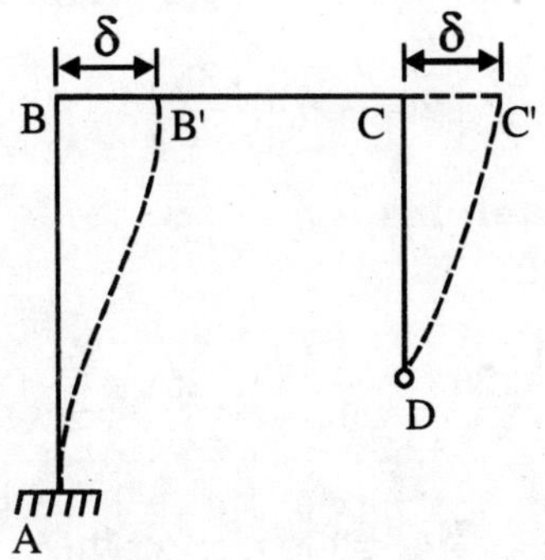

$$M_{AB} = M_{FAB} + 2Ek\left(2\theta_A + \theta_B + \frac{3\delta}{l}\right) = 0 + \frac{2EI}{5}\left(2(0) + \theta_B - \frac{3\delta}{5}\right)$$

$$M_{AB} = \frac{2EI}{5}\left(\theta_B - \frac{3\delta}{5}\right) \qquad ...(1)$$

$$M_{BA} = M_{FBA} + 2Ek\left(2\theta_B + \theta_A + \frac{3\delta}{l}\right) = 0 + \frac{2EI}{5}\left(2\theta_B + 0 - \frac{3\delta}{5}\right)$$

$$M_{BA} = \frac{2EI}{5}\left(2\theta_B - \frac{3\delta}{5}\right) \qquad ...(2)$$

Span BC : $$M_{BC} = M_{FBC} + 2Ek\left(2\theta_B + \theta_C + \frac{3\delta}{l}\right) = -25 + \frac{2EI}{4}(2\theta_B + \theta_C + 0)$$

$$M_{BC} = -25 + \frac{EI}{2}(2\theta_B + \theta_C) \qquad ...(3)$$

$$M_{CB} = M_{FCB} + \frac{2EI}{l}\left(2\theta_C + \theta_B + \frac{3\delta}{l}\right) = 25 + \frac{2EI}{4}(2\theta_C + \theta_B + 0)$$

$$M_{CB} = 25 + \frac{EI}{2}(2\theta_C + \theta_B) \qquad ...(4)$$

Span CD : $$M_{CD} = M_{FCD} + 2Ek\left(2\theta_C + \theta_D + \frac{3\delta}{l}\right) = 0 + \frac{2EI}{4}\left(2\theta_C + \theta_D - \frac{3\delta}{4}\right)$$

$$M_{CD} = \frac{EI}{2}\left(2\theta_C + \theta_D - \frac{3\delta}{4}\right) \qquad ...(5)$$

$$M_{DC} = M_{FDC} + 2Ek\left(2\theta_D + \theta_C + \frac{3\delta}{l}\right)$$

$$= 0 + \frac{2EI}{4}\left(2\theta_D + \theta_C - \frac{3\delta}{4}\right)$$

$$M_{DC} = \frac{EI}{2}\left(2\theta_D + \theta_C - \frac{3\delta}{4}\right) \qquad ...(6)$$

(*c*) **Joint equilibrium equations :**

At Joint B, $M_{BA} + M_{BC} = 0$

From (2) and (3),
$$\frac{2EI}{5}\left(2\theta_B - \frac{3\delta}{5}\right) + (-25) + \frac{EI}{2}(2\theta_B + \theta_C) = 0$$

$$EI\left(\frac{4}{5}\theta_B - \frac{6}{25}\delta + \theta_B + \frac{\theta_C}{2}\right) - 25 = 0$$

$$EI\left(\frac{9}{5}\theta_B - \frac{6}{25}\delta + \frac{\theta_C}{2}\right) = 25$$

$$EI\,(1.8\theta_B + 0.5\theta_C - 0.24\delta) = 25 \qquad ...(7)$$

At joint C,
$$M_{CB} + M_{CD} = 0$$

From (4) and (5),
$$25 + \frac{EI}{2}(2\theta_C + \theta_B) + \frac{EI}{2}\left(2\theta_C + \theta_D - \frac{3\delta}{4}\right) = 0$$

$$\frac{EI}{2}\left(2\theta_C + \theta_B + 2\theta_C + \theta_D - \frac{3\delta}{4}\right) = -25$$

$$EI\left(4\theta_C + \theta_B + \theta_D - \frac{3\delta}{4}\right) = -50 \qquad ...(8)$$

At joint D, $M_{DC} = 0$,
$$\frac{EI}{2}\left(2\theta_D + \theta_C - \frac{3\delta}{4}\right) = 0$$

$$2\theta_D + \theta_C - \frac{3\delta}{4} = 0$$

$$\theta_D = \frac{3\delta}{8} - \frac{\theta_C}{2} \qquad ...(9)$$

Substituting (9) in (8)
$$4\theta_C + \theta_B + \frac{3\delta}{8} - \frac{\theta_C}{2} - \frac{3\delta}{4} = -\frac{50}{EI}$$

$$\theta_B + 3.5\,\theta_C - 0.375\delta = -\frac{50}{EI} \qquad ...(10)$$

(*d*) **Shear equation :**
$$\frac{M_{AB} + M_{BA}}{l_{AB}} + \frac{M_{CD}}{l_{CD}} = 0 \qquad (\because M_{DC} = 0)$$

$$\frac{\frac{2EI}{5}\left(\theta_B - \frac{3\delta}{5}\right) + \frac{2EI}{5}\left(2\theta_B - \frac{3\delta}{5}\right)}{5} + \frac{\frac{EI}{2}\left(2\theta_C + \theta_D - \frac{3\delta}{4}\right)}{4} = 0$$

$$\frac{2EI}{25}\left(3\,\theta_B - \frac{6\delta}{5}\right) + \frac{EI}{8}\left(2\theta_C + \frac{3\delta}{8} - \frac{\theta_C}{2}\right) = 0$$

$$EI\left(\frac{6}{25}\theta_B - \frac{12\delta}{125} + \frac{\theta_C}{4} + \frac{3\delta}{64} - \frac{\theta_C}{16}\right) = 0$$

$$0.24\theta_B + 0.1875\theta_C - 0.143\delta = 0 \qquad ...(11)$$

Solving equations (7), (10), (11)

$$1.8\theta_B + 0.5\theta_C - 0.24\delta = \frac{25}{EI} \qquad ...(7)$$

$$\theta_B + 3.5\,\theta_C - 0.375\delta = -\frac{50}{EI} \qquad ...(10)$$

$(10) \times -1.8 \Rightarrow$ $$-1.8\theta_B - 6.3\theta_C + 0.675\delta = \frac{90}{EI} \qquad ...(12)$$

$(7) \Rightarrow$ $$1.8\theta_B + 1.5\theta_C - 0.24\delta = \frac{25}{EI}$$

Adding the above two,

$$-5.8\theta_C + 0.435\delta \doteq \frac{115}{EI} \qquad ...(13)$$

$(10) \times -0.24 \Rightarrow$ $$-0.24\theta_B - 0.84\theta_C + 0.09\delta = \frac{12}{EI} \qquad ...(14)$$

$(11) + (14) \Rightarrow$ $$-0.6525\theta_C - 0.053\delta = \frac{12}{EI} \qquad ...(15)$$

$(13) \times 0.053 + (15) \times 0.435 \Rightarrow -0.5912\theta_C = 11.315/EI$

$$\therefore \qquad \theta_C = \frac{-19.14}{EI}$$

Substituting θ_C in (13), we get

$$-5.8\left(\frac{-19.14}{EI}\right) + 0.435\delta = \frac{115}{EI}$$

$$\therefore \qquad \delta = \frac{9.21}{EI}$$

Substituting θ_C and δ values in (10)

$$\theta_B + 3.5\theta_C - 0.375\delta = -\frac{50}{EI}$$

$$\theta_B + 3.5\left(-\frac{19.14}{EI}\right) - 0.375\left(\frac{9.21}{EI}\right) = -\frac{50}{EI}$$

$$\therefore \quad \theta_B = \frac{20.43}{EI}$$

Substituting in (9), we get

$$\theta_D = \frac{3\delta}{8} - \frac{\theta_C}{2} = \frac{3 \times 9.21}{8EI} + \frac{19.14}{2EI}$$

$$\therefore \quad \theta_D = \frac{13.02}{EI}$$

$$\theta_B = \frac{20.43}{EI}, \qquad \theta_C = \frac{-19.14}{EI}$$

$$\theta_D = \frac{13.02}{EI}, \qquad \delta = \frac{9.21}{EI}$$

(*e*) **Final moments :**

$$M_{AB} = \frac{2EI}{5}\left(\theta_B - \frac{3\delta}{5}\right) \qquad \text{...(1)}$$

$$M_{AB} = \frac{2EI}{5}\left(\frac{20.43}{EI} - \frac{3}{5}\left(\frac{9.21}{EI}\right)\right) = 5.962 \text{ kNm}$$

$$M_{BA} = \frac{2EI}{5}\left(2\theta_B - \frac{3\delta}{5}\right) \qquad \text{...(2)}$$

$$= \frac{2EI}{5}\left(2\left(\frac{20.43}{EI}\right) - \frac{3}{5}\left(\frac{9.21}{EI}\right)\right) = 14.134 \text{ kNm}$$

$$M_{BC} = -25 + \frac{EI}{2}(2\theta_B + \theta_C) \qquad \text{...(3)}$$

$$= -25 + \frac{EI}{2}\left(2\left(\frac{20.43}{EI}\right) + \left(\frac{-19.14}{EI}\right)\right) = -14.135 \text{ kNm}$$

$$M_{CB} = 25 + \frac{EI}{2}(2\theta_C + \theta_B) \qquad \text{...(4)}$$

$$= +25 + \frac{EI}{2}\left(2\left(\frac{-19.14}{EI}\right) + \left(\frac{20.43}{EI}\right)\right) = +16.080 \text{ kNm}$$

$$M_{CD} = \frac{EI}{2}\left(2\theta_C + \theta_D - \frac{3\delta}{4}\right) \qquad \text{...(5)}$$

$$= \frac{EI}{2}\left(2\left(\frac{-19.14}{EI}\right) + \left(\frac{13.02}{EI}\right) - \frac{3}{4}\left(\frac{9.21}{EI}\right)\right) = -16.080 \text{ kNm}$$

$$M_{DC} = \frac{EI}{2}\left(2\theta_D + \theta_C - \frac{3\delta}{4}\right) \qquad \text{...(6)}$$

$$M_{DC} = \frac{EI}{2}\left(2\left(\frac{13.02}{EI}\right) + \left(\frac{-19.14}{EI}\right) - \frac{3}{4}\left(\frac{9.21}{EI}\right)\right) = 0$$

(*f*) **To draw S.F.D. :**

Span AB :

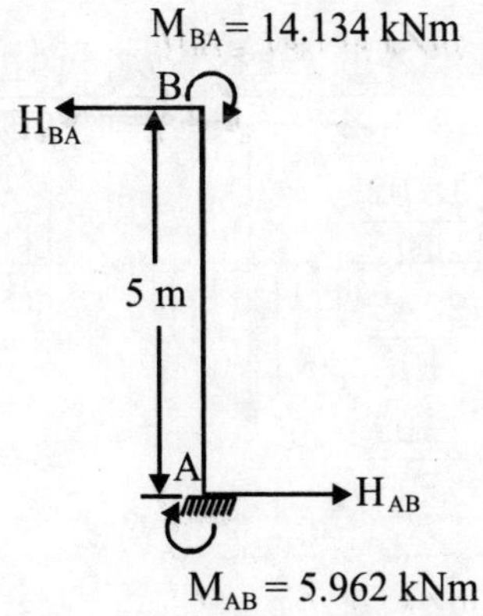

Taking moments about A, $M_{AB} + M_{BA} - H_{BA}(5) = 0$

$$5.962 + 14.134 - H_{BA}(5) = 0$$

$$H_{BA} = 4.0192 \text{ kN} = H_{AB}$$

Span BC :

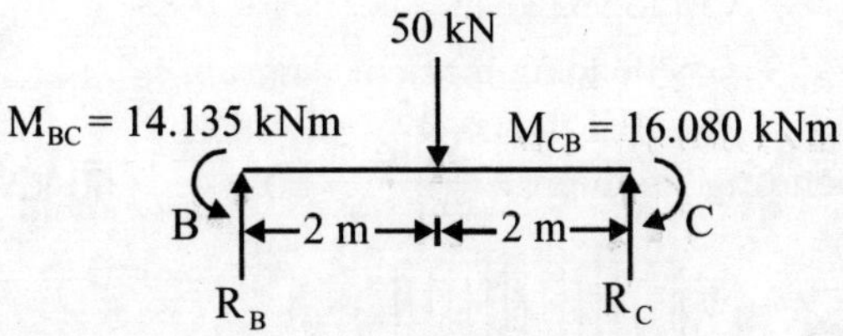

Taking moments about B,

$$M_{CB} + 50(2) - M_{BC} - R_C(4) = 0$$

$$16.080 + 100 - 14.135 - R_C(4) = 0$$

$$R_C = 25.4865 \text{ kN}$$

$$R_B = \text{Total load} - R_C = 50 - 25.4865$$

$$R_B = 24.5135 \text{ kN}$$

Span CD :

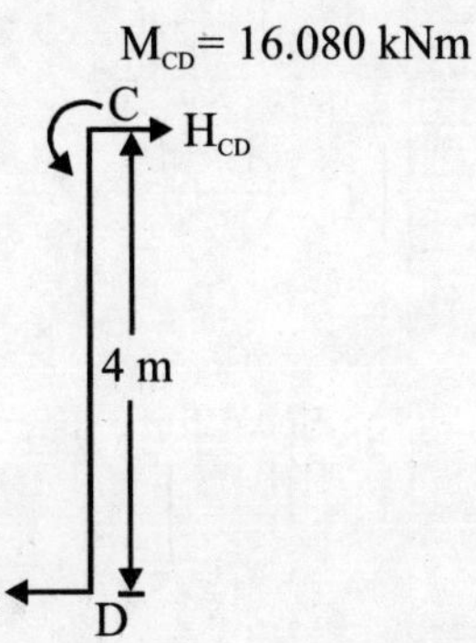

Taking moment about C, $-M_{CD} + H_{DC}(4) = 0$

$$H_{DC} = \frac{(16.080)}{4}$$

$$H_{DC} = 4.02 \text{ kN} = H_{CD}$$

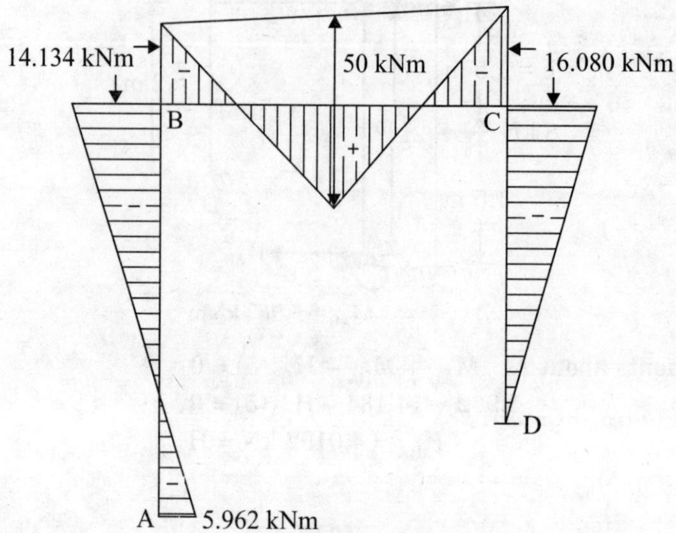

Bending moment diagram

Simply supported bending moment $= \frac{Wab}{l} = 50 \times \frac{4}{4} = 50$ kNm

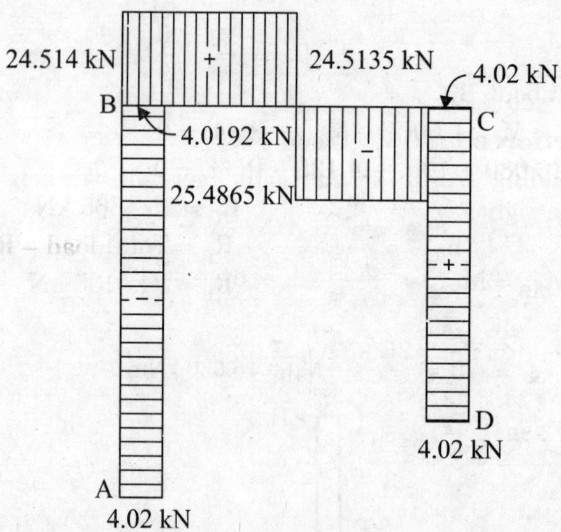

Shear force diagram

Example 2.15 *Analyse the frame shown in Fig. 2.20 by the slope deflection method and draw the bending moment diagram.*

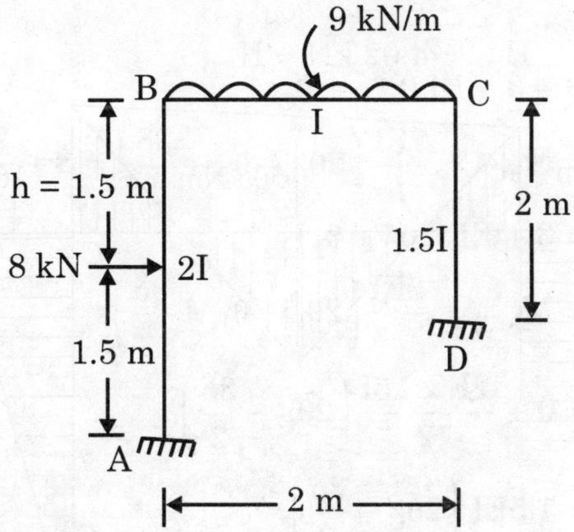

Fig. 2.20

Solution (*a*) **Fixed end moments :**

$$M_{FAB} = -\frac{Wl}{8} = -\frac{8 \times 3}{8} = -3 \text{ kNm}$$

$$M_{FBA} = +\frac{Wl}{8} = +\frac{8 \times 3}{8} = +3 \text{ kNm}$$

$$M_{FBC} = -\frac{wl^2}{12} = -\frac{9 \times 2^2}{12} = -3 \text{ kNm}$$

$$M_{FCB} = +\frac{wl^2}{12} = +\frac{9 \times 2^2}{12} = +3 \text{ kNm}$$

$$M_{FCD} = M_{FDC} = 0 \qquad (\because \text{ No lateral loading on CD})$$

(*b*) **Slope deflection equations :**

The frame and loading are unsymmetrical. Therefore the tops of column AB and DC will sway by δ, say to the right.

$$M_{AB} = M_{FAB} + \frac{2EI}{l}\left(2\theta_A + \theta_B + \frac{3\delta}{l}\right)$$

$$= -3 + \frac{2E \times 2I}{3}\left(\theta_B - \frac{3\delta}{3}\right) \qquad (\because \theta_A = 0)$$

$$= -3 + \frac{4EI}{3}(\theta_B - \delta)$$

$$M_{BA} = M_{FBA} + \frac{2EI}{l}\left(2\theta_B + \theta_A + \frac{3\delta}{l}\right) = 3 + \frac{4EI}{3}\left(2\theta_B - \frac{3\delta}{3}\right)$$

$$= 3 + \frac{4EI}{3}(2\theta_B - \delta)$$

$$M_{BC} = M_{FBC} + \frac{2EI}{l}\left(2\theta_B + \theta_C + \frac{3\delta}{l}\right)$$

$$= -3 + \frac{2EI}{2}(2\theta_B + \theta_C) \qquad (\because \quad \delta = 0 \text{ for beams})$$

$$= -3 + EI(2\theta_B + \theta_C)$$

$$M_{CB} = M_{FCB} + \frac{2EI}{l}\left(2\theta_C + \theta_B + \frac{3\delta}{l}\right) = 3 + \frac{2EI}{2}(2\theta_C + \theta_B)$$

$$= 3 + EI(2\theta_C + \theta_B)$$

$$M_{CD} = M_{FCD} + \frac{2EI}{l}\left(2\theta_C + \theta_D + \frac{3\delta}{l}\right)$$

$$= 0 + \frac{2E \times 1.5I}{2}\left(2\theta_C - \frac{3\delta}{2}\right) \qquad (\because \quad \theta_D = 0)$$

$$= 1.5EI\left(2\theta_C - \frac{3\delta}{2}\right)$$

$$M_{DC} = M_{FDC} + \frac{2EI}{l}\left(2\theta_D + \theta_C + \frac{3\delta}{l}\right) = 0 + \frac{2E \times 1.5I}{2}\left(\theta_C - \frac{3\delta}{2}\right)$$

$$= 1.5EI\left(\theta_C - \frac{3\delta}{2}\right)$$

(*c*) **Equilibrium and shear equations :**

There are 3 unknowns *i.e.*, θ_B, θ_C and δ.

$$M_{BA} + M_{BC} = 0 \qquad ...(1)$$

$$3 + \frac{4EI}{3}(2\theta_B - \delta) - 3 + EI(2\theta_B + \theta_C) = 0$$

$$\frac{14\theta_B EI}{3} - \frac{4EI\delta}{3} + \theta_C EI = 0$$

$$14\theta_B - 4\delta + 3\theta_C = 0 \qquad ...(4)$$

$$M_{CB} + M_{CD} = 0 \qquad ...(2)$$

$$3 + EI(2\theta_C + \theta_B) + 1.5EI\left(2\theta_C - \frac{3\delta}{2}\right) = 0$$

$$EI\theta_B + 5EI\theta_C - 2.25\,\delta\,EI = -3$$

$$\theta_B + 5\theta_C - 2.25\delta = -\frac{3}{EI} \qquad ...(5)$$

$$\frac{M_{AB} + M_{BA} - Ph}{l} + \frac{M_{CD} + M_{DC}}{l} + P = 0 \qquad ...(3)$$

$$\frac{-3 + \frac{4EI}{3}(\theta_B - \delta) + 3 + \frac{4EI}{3}(2\theta_B - \delta) - 12}{3} + \frac{1.5EI\left(2\theta_C - \frac{3\delta}{2}\right) + 1.5EI\left(\theta_C - \frac{3\delta}{2}\right)}{2} + 8 = 0$$

$$\frac{4EI\theta_B - 1.33EI\delta - 1.33EI\delta - 12}{3} + \frac{4.5EI\theta_C - 4.5EI\delta}{2} + 8 = 0$$

$$\frac{8EI\theta_B - 5.32EI\delta - 24 + 13.5EI\theta_C - 13.5EI\delta + 48}{6} = 0$$

$$EI[8\theta_B + 13.5\theta_C - 18.82\delta] + 24 = 0$$

$$8\theta_B + 13.50\theta_C - 18.82\delta = -\frac{24}{EI} \quad ...(6)$$

(5) × 8 ⇒ $8\theta_B + 40\theta_C - 18\delta = -24/EI$ (–)

(6) × 1 ⇒ $8\theta_B + 13.5\theta_C - 18.82\delta = -24/EI$

$\quad - \quad - \quad + \quad +$

$$26.5\theta_C + 0.82\delta = 0 \quad ...(7)$$

(4) × 1 ⇒ $14\theta_B + 3\theta_C - 48\delta = 0$ (–)

(5) × 14 ⇒ $14\theta_B + 70\theta_C - 31.5\delta = -42/EI$

$\quad (-) \quad (-) \quad (+) \quad (+)$

$$-67\theta_C + 27.5\delta = 42/EI \quad ...(8)$$

(7) × 2.53 ⇒ $67\theta_C - 2.07\delta = 0$

(8) × 1 ⇒ $-67\theta_C + 27.5\delta = 42/EI$

$\quad + \quad -$

$$29.57\delta = 42/EI$$

$$\delta = \frac{1.42}{EI}$$

(7) ⇒ $26.5\theta_C = -0.82\delta$

$$\theta_C = -\frac{0.82\delta}{26.5} \times \frac{1.42}{EI}$$

$$\theta_C = -\frac{0.044}{EI}$$

(6) ⇒ $8\theta_B = -\frac{24}{EI} - 13.5 \times \left(-\frac{0.044}{EI}\right) + 18.82\left(\frac{1.42}{EI}\right)$

$$\theta_B = \frac{0.4148}{EI}$$

(*d*) **Final moments :**

$$M_{AB} = -3 + \frac{4EI}{3}(\theta_B - \delta)$$

$$= -3 + \frac{4EI}{3}\left(\frac{0.4148}{EI} - \frac{1.42}{EI}\right) = -4.34 \text{ kNm}$$

$$M_{BA} = 3 + \frac{4EI}{3}(2\theta_B - \delta)$$

$$= 3 + \frac{4EI}{3}\left(2 \times \frac{0.4148}{EI} - \frac{1.42}{EI}\right) = 2.21 \text{ kNm}$$

$$M_{BC} = -3 + EI(2\theta_B + \theta_C)$$

$$= -3 + EI\left(2 \times \frac{0.4148}{EI} - \frac{0.044}{EI}\right) = -2.21 \text{ kNm}$$

$$M_{CB} = 3 + EI(2\theta_C + \theta_B)$$

$$= 3 + EI\left(2 \times \frac{-0.044}{EI} + \frac{0.4148}{EI}\right) = 3.33 \text{ kNm}$$

$$M_{CD} = 1.5EI\left(2\theta_C - \frac{3\delta}{2}\right)$$

$$= 1.5EI\left(2 \times \frac{-0.044}{EI} - \frac{3}{2} \times \frac{1.42}{EI}\right) = -3.33 \text{ kNm}$$

$$M_{DC} = 1.5EI\left(\theta_C - \frac{3\delta}{2}\right)$$

$$= 1.5EI\left(\frac{-0.044}{EI} - \frac{3}{2} \times \frac{1.42}{EI}\right) = -3.26 \text{ kNm}$$

(*e*) **Simply supported moments :**

In AB = $\frac{Wl}{4} = \frac{8 \times 3}{4}$ = 6 kNm

In BC = $\frac{wl^2}{8} = \frac{9 \times 2^2}{8}$ = 4.5 kNm

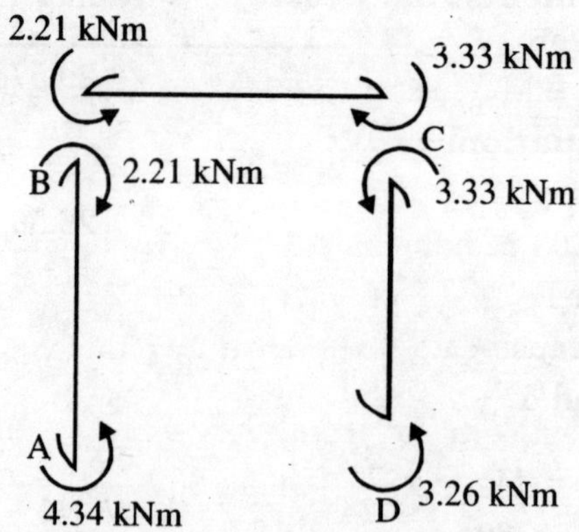

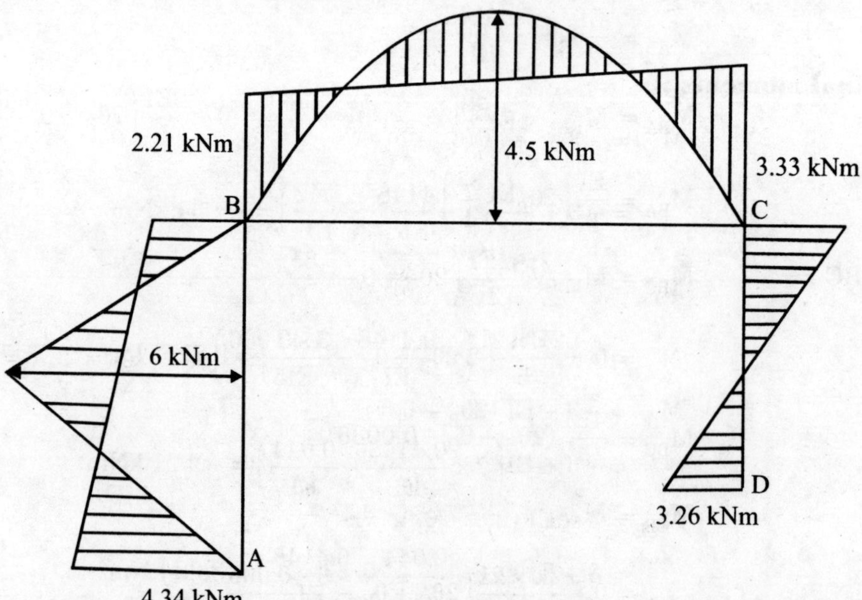

Bending moment diagram

Example 2.16 *The frame shown in Fig. 2.21 undergoes a rotational yield of 0.003 radian clockwise and a vertical yield of 6 mm at D. Taking EI = 3.6 × 10¹² Nmm², analyse the frame by slope deflection method and sketch the B.M.D.*

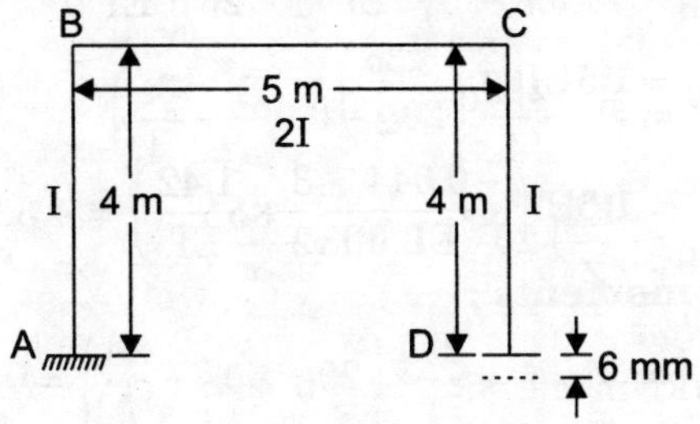

Fig. 2.21

Solution (*a*) **Fixed end moments :** Since there is no loading, the fixed end moments for all the members are zero.

$$\therefore \quad M_{FAB} = M_{FBA} = M_{FBC} = M_{FCB} = M_{FCD} = M_{FDC} = 0$$

(*b*) **Slope deflection equations :**

Known displacements :

Clockwise rotation = 0.003 radians at D

Vertical yield = 6 mm at D

Vertical yield at D will cause C to come down by 6 mm compared to B.

Unknowns are θ_B, θ_C and δ

Span AB :
$$M_{AB} = M_{FAB} + \frac{2EI}{l}\left(2\theta_A + \theta_B + \frac{3\delta}{l}\right) = 0 + \frac{2EI}{4}\left(0 + \theta_B - \frac{3\delta}{4}\right)$$

[$\theta_A = 0 \quad \because$ A is fixed]

$$M_{AB} = \frac{EI}{2}\left(\theta_B - \frac{3\delta}{4}\right) \qquad ...(1)$$

$$M_{BA} = M_{FBA} + \frac{2EI}{l}\left(2\theta_B + \theta_A + \frac{3\delta}{l}\right) = 0 + \frac{2EI}{4}\left(2\theta_B + 0 - \frac{3\delta}{4}\right)$$

$$M_{BA} = \frac{EI}{2}\left(2\theta_B - \frac{3\delta}{4}\right) \qquad ...(2)$$

Span BC :
$$M_{BC} = M_{FBC} + \frac{2FI}{l}\left(2\theta_B + \theta_C + \frac{3\delta}{l}\right)$$

$$= 0 + \frac{2E \times 2I}{5}\left(2\theta_B + \theta_C - \frac{3 \times 0.006}{5}\right) \quad (\because \; I_{BC} = 2I, \delta = 0.006 \text{ m})$$

$$M_{BC} = \frac{4EI}{5}(2\theta_B + \theta_C - 0.0036) \qquad ...(3)$$

$$M_{CB} = M_{FCB} + \frac{2EI}{l}\left(2\theta_C + \theta_B + \frac{3\delta}{l}\right)$$

$$= 0 + \frac{2E \times 2I}{5}\left(2\theta_C + \theta_B - \frac{3 \times 0.006}{5}\right)$$

$$M_{CB} = \frac{4EI}{5}(2\theta_C + \theta_B - 0.0036) \quad \text{...(4)}$$

Span CD :

$$M_{CD} = M_{FCD} + \frac{2EI}{l}\left(2\theta_C + \theta_D + \frac{3\delta}{l}\right)$$

$$= 0 + \frac{2EI}{4}\left(2\theta_C + 0.003 - \frac{3\delta}{4}\right) \quad (\because \ \theta_D = 0.003 \text{ rad})$$

$$M_{CD} = \frac{EI}{2}\left(2\theta_C + 0.003 - \frac{3\delta}{4}\right) \quad \text{...(5)}$$

$$M_{DC} = M_{FDC} + \frac{2EI}{l}\left(2\theta_D + \theta_C + \frac{3\delta}{l}\right) = 0 + \frac{2EI}{4}\left(2 \times 0.003 + \theta_C - \frac{3\delta}{4}\right)$$

$$= \frac{EI}{2}\left(\theta_C + 0.006 - \frac{3\delta}{4}\right) \quad \text{...(6)}$$

(*c*) **Joint equilibrium equations :**

At joint B, $M_{BA} + M_{BC} = 0$

From (2) and (3),

$$\frac{EI}{2}\left(2\theta_B - \frac{3\delta}{4}\right) + \frac{4EI}{5}(2\theta_B + \theta_C - 0.0036) = 0$$

$$EI\left(\theta_B - \frac{3\delta}{8} + \frac{8}{5}\theta_B + \frac{4}{5}\theta_C - 0.00288\right) = 0$$

$$\frac{13}{5}\theta_B + \frac{4}{5}\theta_C - \frac{3\delta}{8} = 0.00288 \quad \text{...(7)}$$

At joint C, $M_{CB} + M_{CD} = 0$

From (4) and (5),

$$\frac{4EI}{5}(\theta_B + 2\theta_C - 0.0036) + \frac{EI}{2}\left(2\theta_C + 0.003 - \frac{3\delta}{4}\right) = 0$$

$$EI\left(\frac{4}{5}\theta_B + \frac{8}{5}\theta_C - 0.00288 + \theta_C + 0.0015 - \frac{3\delta}{8}\right) = 0$$

$$\frac{4}{5}\theta_B + \frac{13}{5}\theta_C - \frac{3\delta}{8} = 0.00138 \quad \text{...(8)}$$

(*d*) **Shear equation :**

$$\frac{M_{AB} + M_{BA}}{l_{AB}} + \frac{M_{CD} + M_{DC}}{l_{CD}} = 0$$

$$\frac{M_{AB} + M_{BA}}{4} + \frac{M_{CD} + M_{DC}}{4} = 0$$

$$\Rightarrow \quad M_{AB} + M_{BA} + M_{CD} + M_{DC} = 0$$

$$\frac{EI}{2}\left(\theta_B - \frac{3\delta}{4}\right) + \frac{EI}{2}\left(2\theta_B - \frac{3\delta}{4}\right) + \frac{EI}{2}\left(2\theta_C + 0.003 - \frac{3\delta}{4}\right) + \frac{EI}{2}\left(\theta_C + 0.006 - \frac{3\delta}{4}\right) = 0$$

$$\frac{EI}{2}\left(\theta_B - \frac{3\delta}{4} + 2\theta_B - \frac{3\delta}{4} + 2\theta_C + 0.003 - \frac{3\delta}{4} + \theta_C + 0.006 - \frac{3\delta}{4}\right) = 0$$

$$3\theta_B + 3\theta_C - 3\delta = -0.009$$

$$\Rightarrow \qquad \theta_B + \theta_C - \delta = -0.003 \qquad \ldots(9)$$

Solving equations (7), (8) and (9), we get

$$\theta_B = 1.645 \times 10^{-3} \text{ rad}$$
$$\theta_C = 8.12 \times 10^{-4} \text{ rad}$$
$$\delta = 5.457 \times 10^{-3} \text{ m}$$

(*e*) **Final moments :**

$$EI = 3.6 \times 10^{12} \text{ Nmm}^2 = \frac{3.6 \times 10^{12}}{10^3 \times 10^6} \text{ kNm}^2 = 3600 \text{ kNm}^2$$

$$(1) \Rightarrow \qquad M_{AB} = \frac{3600}{2}\left(1.645 \times 10^{-3} - \frac{3 \times 5.457 \times 10^{-3}}{4}\right) = -4.406 \text{ kNm}$$

$$(2) \Rightarrow \qquad M_{BA} = \frac{3600}{2}\left[(2 \times 1.645 \times 10^{-3}) - \left(\frac{3 \times 5.457 \times 10^{-3}}{4}\right)\right] = -1.44 \text{ kNm}$$

$$(3) \Rightarrow \qquad M_{BC} = \frac{4 \times 3600}{5}[(2 \times 1.645 \times 10^{-3}) + 8.12 \times 10^{-4} - 0.0036]$$
$$= 1.44 \text{ kNm}$$

$$(4) \Rightarrow \qquad M_{CB} = \frac{4 \times 3600}{5}[(2 \times 8.12 \times 10^{-4}) + 1.645 \times 10^{-3} - 0.0036]$$
$$= -0.95 \text{ kNm}$$

$$(5) \Rightarrow \qquad M_{CD} = \frac{3600}{2}\left[(2 \times 8.12 \times 10^{-4}) + 0.003 - \left(\frac{3 \times 5.457 \times 10^{-3}}{4}\right)\right]$$
$$= 0.95 \text{ kNm}$$

$$(6) \Rightarrow \qquad M_{DC} = \frac{3600}{2}\left[8.12 \times 10^{-4} + 0.006 - \left(\frac{3 \times 5.457 \times 10^{-3}}{4}\right)\right]$$
$$= 4.895 \text{ kNm}$$

(*f*) **Simply supported bending moments :** Since there is no loading on spans AB, BC and CD, bending moments are zero.

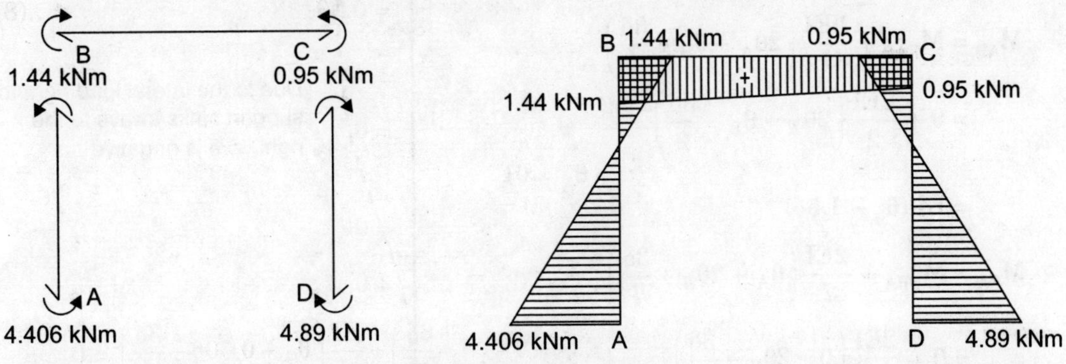

Bending moment diagram

Example 2.17 *Draw the moment diagram for the frame shown in Fig. 2.22. EI is constant for all the members. Use the slope deflection method for analysis.*

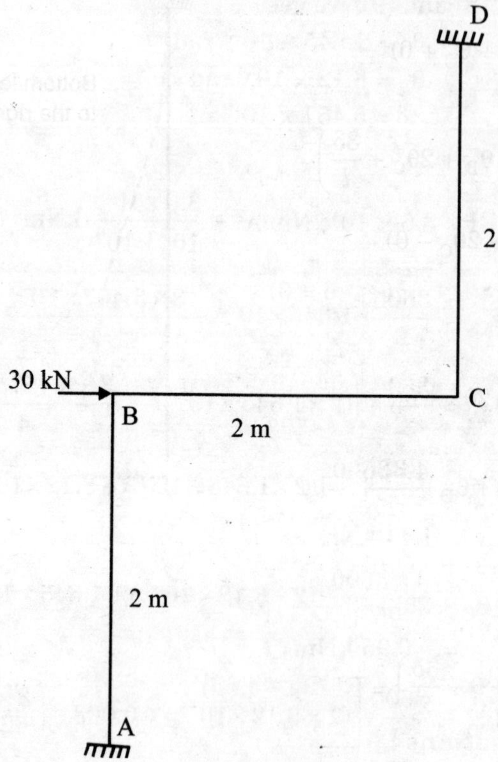

Fig. 2.22

Solution Unknowns θ_B, θ_C, δ

(*a*) **Fixed end moments:**

Since, there is no span loading in any of the member, fixed moment for all the members is zero.

(*b*) **Slope deflection equations :**

$$M_{AB} = M_{FAB} + \frac{2EI}{l}\left(2\theta_A + \theta_B + \frac{3\delta}{l}\right)$$

$$= 0 + \frac{2EI}{2}\left(2\theta_A + \theta_B - \frac{3\delta}{2}\right) \quad (\because \quad \theta_A = 0)$$

$$= EI\,(\theta_B - 1.5\delta)$$

$$M_{BA} = M_{FBA} + \frac{2EI}{l}\left(\theta_A + 2\theta_B + \frac{3\delta}{l}\right)$$

$$= 0 + \frac{2EI}{2}\left(0 + 2\theta_B - \frac{3\delta}{2}\right)$$

$$= EI\,(2\theta_B - 1.5\delta)$$

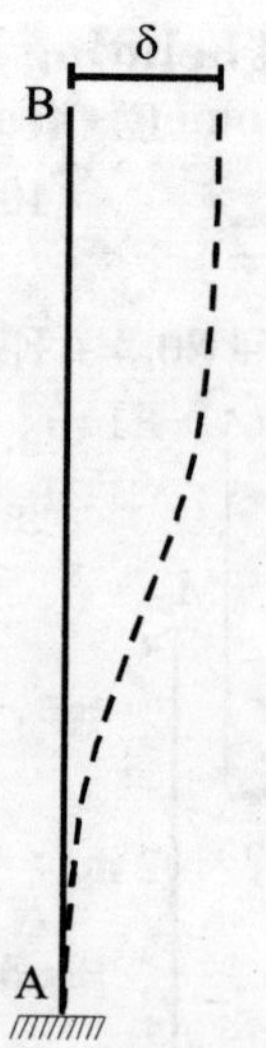

Due to the lateral load right/top support sinks/sways to the right, ∴ δ is negative

$$M_{BC} = M_{FBC} + \frac{2EI}{l}\left(2\theta_B + \theta_C + \frac{3\delta}{l}\right)$$

$$= 0 + \frac{2EI}{2}(2\theta_B + \theta_C + 0)$$

$$= EI(2\theta_B + \theta_C)$$

$$M_{CB} = M_{FCB} + \frac{2EI}{l}\left(\theta_B + 2\theta_C + \frac{3\delta}{l}\right)$$

$$= 0 + \frac{2EI}{2}(\theta_B + 2\theta_C - 0)$$

$$(\because \quad \delta = 0)$$

$$= EI(\theta_B + 2\theta_C)$$

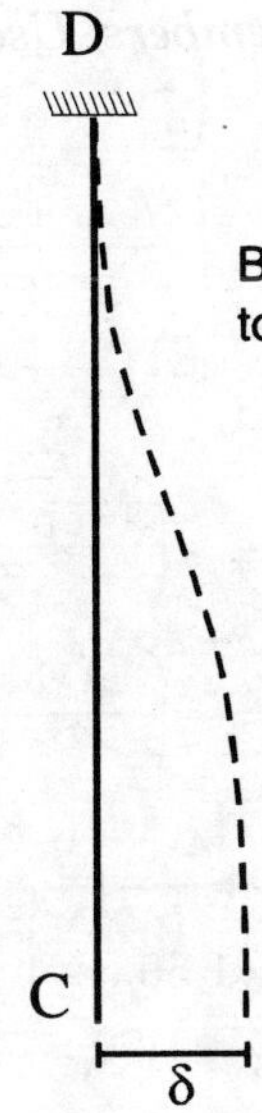

Bottom/left support sinks/moves to the right, ∴ δ is positive

$$M_{CD} = M_{FCD} + \frac{2EI}{l}\left(2\theta_C + \theta_D + \left(\frac{3\delta}{l}\right)\right)$$

$$= 0 + \frac{2EI}{2}\left(2\theta_C + \theta_D + \frac{3\delta}{2}\right)$$

$$= EI(2\theta_C + 1.5\delta)$$

$$M_{DC} = M_{FDC} + \frac{2EI}{l}\left(\theta_C + 2\theta_D + \left(\frac{+3\delta}{l}\right)\right)$$

$$= 0 + \frac{2EI}{2}\left(\theta_C + 0 + \frac{3\delta}{2}\right) = EI(\theta_C + 1.5\delta)$$

(*c*) **Equilibrium equations :**

$$M_{BA} + M_{BC} = 0$$

$$M_{CB} + M_{CD} = 0$$

$$\Sigma H = 0 \Rightarrow H_A + H_D + 30 = 0$$

Substituting in $M_{BA} + M_{BC} = 0$

$$\Rightarrow \quad EI(2\theta_B - 1.5\delta) + EI(2\theta_B + \theta_C) = 0$$

$$EI(4\theta_B + \theta_C - 1.5\delta) = 0$$

$$4\theta_B + \theta_C - 1.5\delta = 0 \quad \text{...(1)}$$

Substituting in $M_{CB} + M_{CD} = 0$

$$\Rightarrow \quad EI(\theta_B + 2\theta_C) + EI(2\theta_C + 1.5\delta) = 0$$

$$EI(\theta_B + 4\theta_C + 1.5\delta) = 0$$

$$\theta_B + 4\theta_C + 1.5\delta = 0 \quad \text{...(2)}$$

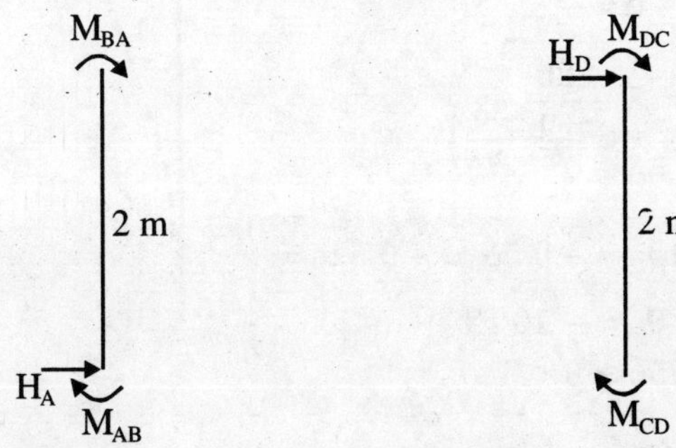

$$2H_A = M_{AB} + M_{BA} \qquad 2H_D + M_{CD} + M_{DC} = 0$$

$$H_A = \left(\frac{M_{AB} + M_{BA}}{2}\right) \qquad H_D = -\left(\frac{M_{CD} + M_{DC}}{2}\right)$$

$$H_A = \left(\frac{M_{AB} + M_{BA}}{2}\right) = \frac{EI\,(\theta_B - 1.5\delta) + EI\,(2\theta_B - 1.5\delta)}{2}$$

$$= \frac{EI\,(3\theta_B - 3\delta)}{2} = EI\,(1.5\theta_B - 1.5\delta)$$

$$H_D = -\left(\frac{M_{CD} + M_{DC}}{2}\right) = -\left(\frac{EI\,(2\theta_C + 1.5\delta) + EI\,(\theta_C + 1.5\delta)}{2}\right)$$

$$= -\left(\frac{EI\,(3\theta_C + 3\delta)}{2}\right) = -\,EI\,(1.5\theta_C + 1.5\delta)$$

$$H_A + H_D + 30 = 0$$

Substituting in $H_A + H_D = -\,30$

$$EI\,(1.5\theta_B - 1.5\delta) - EI\,(1.5\theta_C + 1.5\delta) = -\,30$$

$$EI\,(1.5\theta_B - 1.5\theta_C - 3\delta) = -\,30$$

$$1.5\theta_B - 1.5\theta_C - 3\delta = \frac{(-\,30)}{EI}$$

Dividing 1.5,
$$\theta_B - \theta_C - 2\delta = \frac{(-\,20)}{EI} \qquad \text{...(3)}$$

$$\begin{bmatrix} 4 & 1 & -1.5 \\ 1 & 4 & 1.5 \\ 1 & -1 & -2 \end{bmatrix} \begin{Bmatrix} \theta_B \\ \theta_C \\ \delta \end{Bmatrix} = \begin{Bmatrix} 0 \\ 0 \\ \dfrac{-\,20}{EI} \end{Bmatrix}$$

Using Gaussian elimination method,

$$\left[\begin{array}{ccc:c} 1 & 0.25 & -0.375 & 0 \\ 0 & 3.75 & 1.875 & 0 \\ 0 & -1.25 & -1.625 & \dfrac{-\,20}{EI} \end{array}\right] \begin{Bmatrix} \theta_B \\ \theta_C \\ \delta \end{Bmatrix}$$

$$\left[\begin{array}{ccc:c} 1 & 0.25 & -0.375 & 0 \\ 0 & 1 & 0.5 & 0 \\ 0 & 0 & -1 & \dfrac{-\,20}{EI} \end{array}\right] \begin{Bmatrix} \theta_B \\ \theta_C \\ \delta \end{Bmatrix}$$

$$-\,\delta = -\,\frac{20}{EI}$$

$$\delta = \frac{20}{EI}$$

$$\theta_C + 0.5\delta = 0$$

$$\theta_C = -\,0.5\delta = -\,0.5 \times \frac{(20)}{EI}$$

$$\theta_C = -\,10/EI$$

$$\theta_B + 0.25\theta_C - 0.375\delta = 0$$

$$\theta_B = -0.25\theta_C + 0.375\delta = -0.25\frac{(-10)}{EI} + 0.375\frac{(+20)}{EI}$$

$$\theta_B = \frac{10}{EI}$$

$$M_{AB} = EI(\theta_B - 1.5\delta)$$

$$= EI\left(+\frac{10}{EI} - 1.5\left(\frac{20}{EI}\right)\right) = -20 \text{ kNm}$$

$$M_{BA} = EI(2\theta_B - 1.5\delta)$$

$$= EI\left(2\left(+\frac{10}{EI}\right) - 1.5\left(\frac{20}{EI}\right)\right) = -10 \text{ kNm}$$

$$M_{BC} = EI(2\theta_B - 5\theta_C)$$

$$= EI\left(2\left(+\frac{10}{EI}\right) + \left(-\frac{10}{EI}\right)\right) = +10 \text{ kNm}$$

$$M_{CB} = EI(\theta_B - 2\theta_C)$$

$$= EI\left(\left(+\frac{10}{EI}\right) + 2\left(-\frac{10}{EI}\right)\right) = -10 \text{ kNm}$$

$$M_{CD} = EI(2\theta_C - 1.5\delta)$$

$$= EI\left(2\left(-\frac{10}{EI}\right) + 1.5\left(\frac{20}{EI}\right)\right) = 10 \text{ kNm}$$

$$M_{DC} = EI(\theta_C + 1.5\delta)$$

$$= EI\left(\left(-\frac{10}{EI}\right) + 1.5\left(\frac{20}{EI}\right)\right) = 20 \text{ kNm}$$

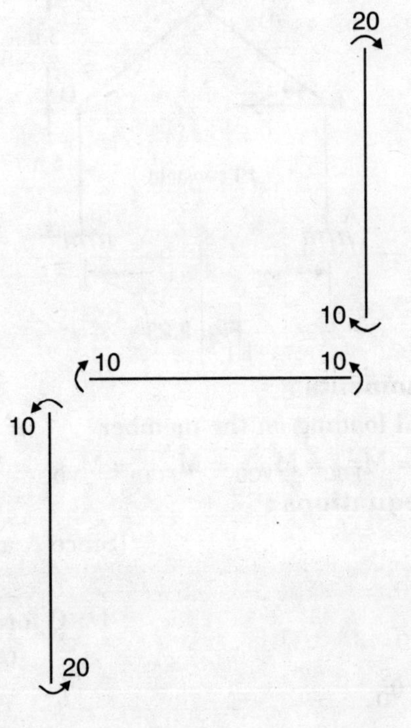

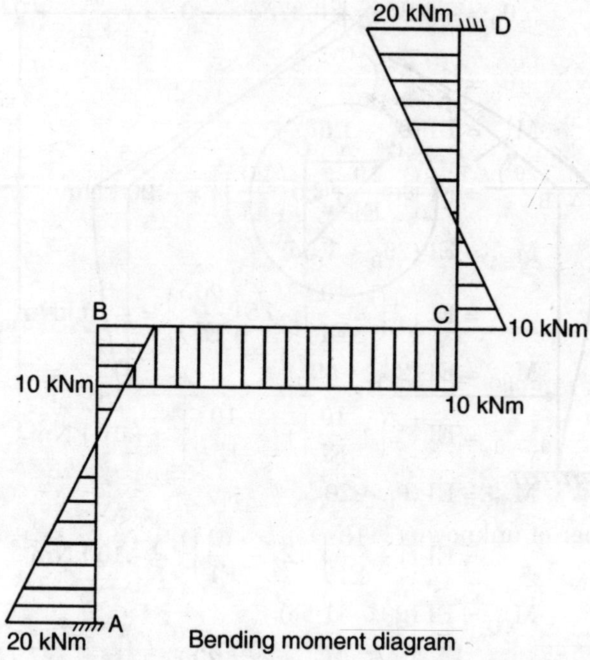

Bending moment diagram

Example 2.18 *Analyse the frame shown in Fig. 2.23 by slope deflection method and sketch the bending moment diagram.*

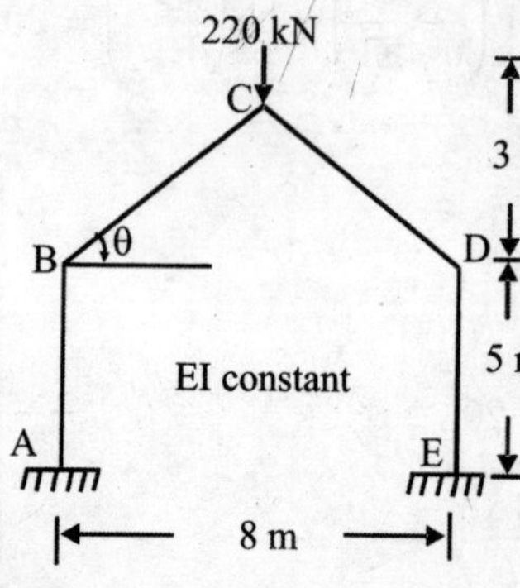

Fig. 2.23

Solution (*a*) **Fixed end moments :**

Since there is no lateral loading on the member

$$M_{FAB} = M_{FBA} = M_{FBC} = M_{FCB} = M_{FCD} = M_{FDC} = M_{FDE} = M_{FED} = 0$$

(*b*) **Slope deflection equations :**

$\theta_A = 0$ [Since A and E are fixed ends $\theta_A = \theta_E = 0$]

$\theta_E = 0$

$\theta_C = 0$ [∵ C forms the line of symmetry as the frame and load are symmetrical]

$\theta_B = -\theta_D$

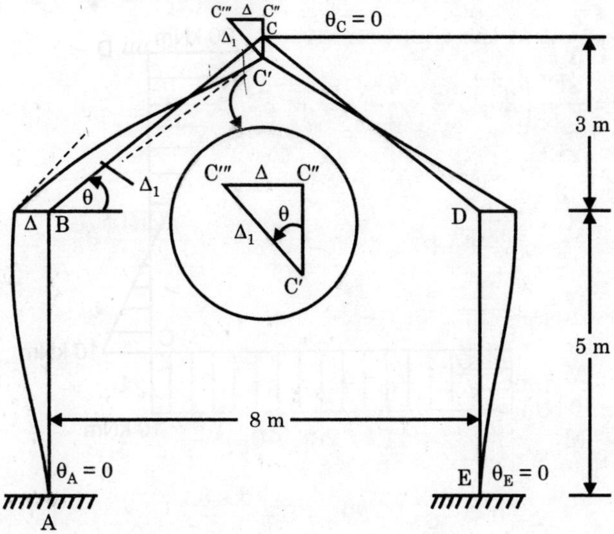

Hence the number of unknowns gets reduced to two *i.e.*, θ_B $(= -\theta_D)$ and Δ

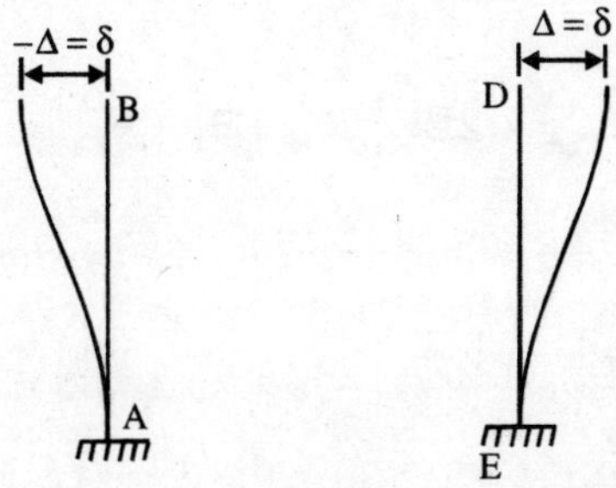

$$M_{AB} = M_{FAB} + 2Ek\left(2\theta_A + \theta_B + \frac{3\delta}{l}\right) = 0 + \frac{2EI}{5}\left(0 + \theta_B + \frac{3\Delta}{l}\right)$$

$$= \frac{2EI}{5}\left(\theta_B + \frac{3\Delta}{l}\right) \quad ...(1)$$

$$M_{BA} = M_{FBA} + 2Ek\left(2\theta_B + \theta_A + \frac{3\delta}{l}\right) = 0 + \frac{2EI}{5}\left(2\theta_B + 0 + \frac{3\Delta}{l}\right)$$

$$= \frac{2EI}{5}\left(2\theta_B + \frac{3\Delta}{l}\right) \quad ...(2)$$

$$M_{BC} = M_{FBC} + 2Ek\left(2\theta_B + \theta_C - \frac{3\delta}{l}\right)$$

$$= 0 + \frac{2EI}{5}\left(2\theta_B + 0 - \frac{3\Delta}{l}\right) \quad ...(3) \quad \left[\therefore\ k = \frac{I}{l};\ l = \sqrt{3^2 + 4^2} = 5;\ k = \frac{I}{5}\right]$$

Settlement (or) Displacement of C = $\delta = \Delta_1$

From the triangle C′ C″ C‴

$$\sin\theta = \frac{\Delta}{\Delta_1} = \frac{3}{5} \quad \left[\text{From the } \Delta\text{OCB } \sin\theta = \frac{3}{5}\right]$$

$$\Delta_1 = \frac{5}{3}\Delta$$

Substitute $\Delta_1 = \frac{5}{3}\Delta$ in the equation (3)

$$M_{BC} = \frac{2EI}{3}\left(2\theta_B - 3\times\frac{5}{3}\Delta\times\frac{1}{5}\right)$$

$$M_{BC} = \frac{2EI}{3}(2\theta_B - \Delta) \quad ...(4)$$

$$M_{CB} = M_{FCB} + 2Ek\left(\theta_B + 2\theta_C + \frac{3\delta}{l}\right) = 0 + \frac{2EI}{3}\left(\theta_B + 2\times 0 - 3\times\frac{5}{3}\times\frac{\Delta}{5}\right)$$

$$M_{CB} = \frac{2EI}{3}(\theta_B - \Delta) \quad ...(5)$$

$$M_{CD} = \frac{2EI}{3}(-\theta_B - \Delta) \quad ...(6)$$

$$M_{DC} = M_{FDC} + 2Ek\left(2\theta_D + \theta_C + \frac{3\delta}{l}\right)$$

$$= 0 + \frac{2EI}{5}\left(-2\theta_B + 0 + \frac{3\times 5\Delta}{3\times 5}\right) \quad [\because \theta_D = -\theta_B;\ \therefore \delta = \Delta]$$

$$M_{DC} = \frac{2EI}{5}(-2\theta_B + \Delta) \quad ...(7)$$

$$M_{DE} = M_{FDE} + 2Ek\left(2\theta_D + \theta_E + \frac{3\delta}{l}\right) = 0 + \frac{2EI}{5}\left(-2\theta_B + 0 - \frac{3\Delta}{l}\right)$$

$$M_{DE} = \frac{2EI}{5}\left(-2\theta_B - \frac{3\Delta}{l}\right) \quad ...(8)$$

$$M_{ED} = M_{FED} + 2Ek\left(2\theta_E + \theta_D + \frac{3\delta}{l}\right) = 0 + \frac{2EI}{5}\left(2\times 0 - \theta_B - \frac{3\Delta}{l}\right)$$

$$M_{DE} = \frac{2EI}{5}\left(-\theta_B - \frac{3\Delta}{l}\right) \quad ...(9)$$

(*c*) **Joint equilibrium equations :**

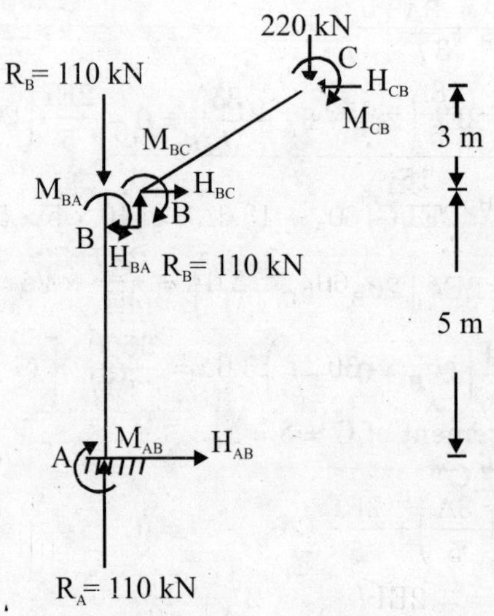

$$\Sigma V = 0$$

$$R_A + R_E = 220 \text{ kN}$$

Since the load is symmetrical $R_A = R_E = (220/2) = 110$ kN

Consider all the moments to be in the clockwise (+ve) direction.

Considering the free body diagram of AB and taking moments about B, $\Sigma M = 0$

$$M_{AB} + M_{BA} - H_{AB} \times 5 = 0$$

$$H_{AB} = \frac{M_{AB} + M_{BA}}{5} \qquad ...(10)$$

Considering the free body diagram of BC and taking moments about C,

$$M_{CB} + M_{BC} + 110 \times 4 - H_{BC} \times 3 = 0$$

$$H_{BC} = \frac{M_{CB} + M_{BC}}{3} + \frac{440}{3} \qquad ...(11)$$

Substituting equations (4) and (5) in eqn. (11)

$$H_{BC} = \frac{\frac{2EI}{5}(2\theta_B - \Delta) + \frac{2EI}{3}(\theta_B - \Delta)}{3} + \frac{440}{3}$$

$$H_{BC} = \frac{\frac{2EI}{5}(3\theta_B - 2\Delta) + 440}{3}$$

Since $\qquad H_{BC} = H_{BA} = H_{AB}$

From eqn. (10), $\qquad H_{BC} = \frac{M_{AB} + M_{BA}}{5} = \frac{\frac{2EI}{5}(3\theta_B - 2\Delta)}{3} + \frac{440}{3}$

Substitute equations (1) and (2) in the above equation,

$$\frac{\frac{2EI}{5}\left(\theta_B + \frac{3\Delta}{5}\right) + \frac{2EI}{5}\left(2\theta_B + \frac{3\Delta}{5}\right)}{5} = \frac{\frac{2EI}{5}(3\theta_B - 2\Delta)}{3} + \frac{440}{3}$$

$$\frac{\frac{2EI}{5}\left(3\theta_B + \frac{6\Delta}{5}\right)}{3} = \frac{\frac{2EI}{5}(3\theta_B - 2\Delta)}{3} + \frac{440}{3}$$

$$\frac{\frac{2EI}{5}\left(9\theta_B + \frac{18\Delta}{5} - 15\theta_B + 10\Delta\right)}{15} = \frac{440}{3}$$

$$2EI\,(-6\theta_B + 13.6\Delta) = 440 \times 5 \times 5$$

$$-6\theta_B + 13.6\Delta = \frac{220}{EI} \times 25 = \frac{5500}{EI}$$

$$-6\theta_B + 13.6\Delta = \frac{5500}{EI} \qquad ...(12)$$

At joint B, $\qquad M_{BC} + M_{BA} = 0$

From eqns. (2) and (4),

$$\frac{2EI}{5}\left(2\theta_B + \frac{3\Delta}{5}\right) + \frac{2EI}{5}(2\theta_B - \Delta) = 0$$

$$\frac{2EI}{5}\left(4\theta_B - \frac{2}{5}\Delta\right) = 0$$

$$4\theta_B - 0.4\Delta = 0 \qquad ...(13)$$

By solving (12) and (13), the values of θ_B and Δ can be obtained

$$(13) \times 1.5 \quad \Rightarrow \quad 6\theta_B - 0.6\Delta = 0$$

$$(12) \times 1 \quad \Rightarrow \quad -6\theta_B + 13.6\Delta = \frac{5500}{EI}$$

$$13.0\Delta = \frac{5500}{EI}$$

$$\Delta = \frac{423.08}{EI}$$

Substitute in (13) $4\theta_B = \dfrac{423.08}{EI} \times 0.4$

$$\theta_B = \frac{42.308}{EI}$$

(*d*) **Final moments :**

$$M_{AB} = \frac{2EI}{5}\left(\theta_B + \frac{3\Delta}{5}\right) = \frac{2EI}{5}\left(\frac{42.308}{EI} + \frac{3}{5}\left(\frac{423.076}{EI}\right)\right)$$

$$M_{AB} = 118.461 \text{ kNm}$$

$$M_{BA} = \frac{2EI}{5}\left(2\theta_B + \frac{3\Delta}{5}\right)$$

$$M_{BA} = \frac{2EI}{5}\left(2\left(\frac{42.308}{EI}\right) + \frac{3}{5}\left(\frac{423.076}{EI}\right)\right) \qquad M_{BA} = 135.385 \text{ kNm}$$

$$M_{BC} = \frac{2EI}{5}(2\theta_B - \Delta) = \frac{2EI}{5}\left(2\left(\frac{42.308}{EI}\right) - \left(\frac{423.076}{EI}\right)\right)$$

$$M_{BC} = -135.385 \text{ kNm}$$

$$M_{CB} = \frac{2EI}{5}(\theta_B - \Delta) = \frac{2EI}{5}\left(\frac{42.308}{EI} - \frac{423.076}{EI}\right)$$

$$M_{CB} = -152.307 \text{ kNm}$$

$$M_{CD} = \frac{2EI}{5}(-\theta_B + \Delta) = \frac{2EI}{5}\left(-\frac{42.308}{EI} + \frac{423.076}{EI}\right)$$

$$M_{CD} = +152.307 \text{ kNm}$$

$$M_{DC} = \frac{2EI}{5}(-2\theta_B - \Delta) = \frac{2EI}{5}\left(-2\left(\frac{42.308}{EI}\right) + \frac{423.076}{EI}\right)$$

$$M_{DC} = +135.385 \text{ kNm}$$

$$M_{DE} = \frac{2EI}{5}\left(-2\theta_B - \frac{3\Delta}{l}\right) = \frac{2EI}{5}\left(-2\left(\frac{42.308}{EI}\right) - \frac{3}{5}\left(\frac{423.076}{EI}\right)\right)$$

$$M_{DE} = -135.385 \text{ kNm}$$

$$M_{ED} = \frac{2EI}{5}\left(-\theta_B - \frac{3\Delta}{l}\right) = \frac{2EI}{5}\left(-\frac{42.308}{EI} - \frac{3}{5}\left(\frac{423.076}{EI}\right)\right)$$

$$M_{ED} = -118.461 \text{ kNm}$$

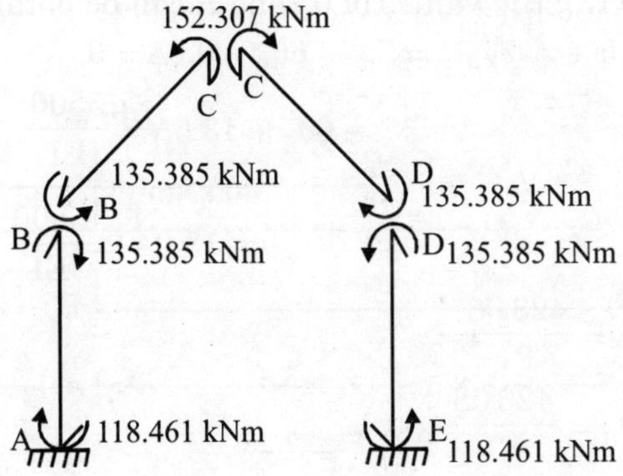

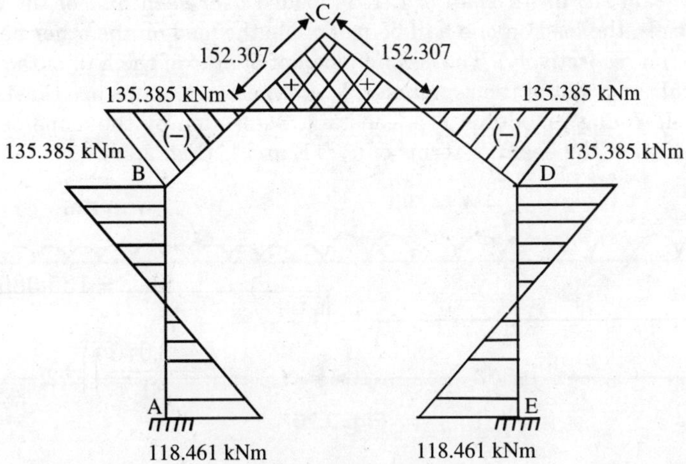

Bending moment diagram

PROBLEMS

1. A continuous beam is built in at A and is supported over rollers at B and C, as shown in Fig. 2.24 AB = BC = 12 m. The beam carries a uniformly distributed load of 30 kN/m over AB and a point load of 240 kN at a distance of 4 m from B on span BC. B has a settlement of 30 mm. E = 2 × 10^5 N/mm^2 ; I = 2 × 10^9 mm^4. Analyse the beam by slope deflection method.

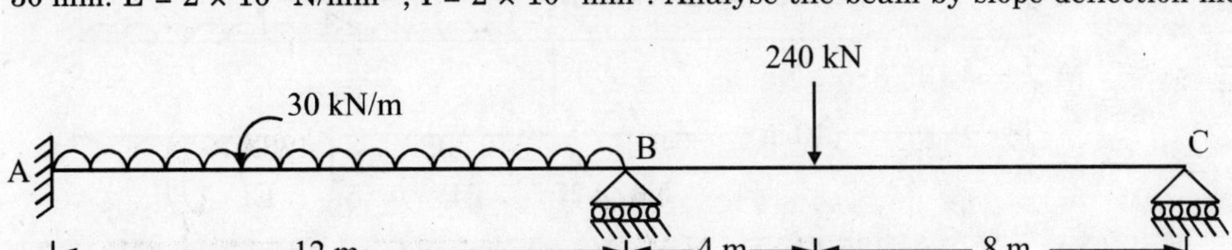

Fig. 2.24

[**Ans.** M_{AB} = – 739 kNm ; M_{BA} = + 102 kNm ; M_{BC} = – 102 kNm ; M_{CB} = 0]

2. Analyse the nonprismatic fixed beam shown in Fig. 2.25 by the slope deflection method and sketch the bending moment diagram.

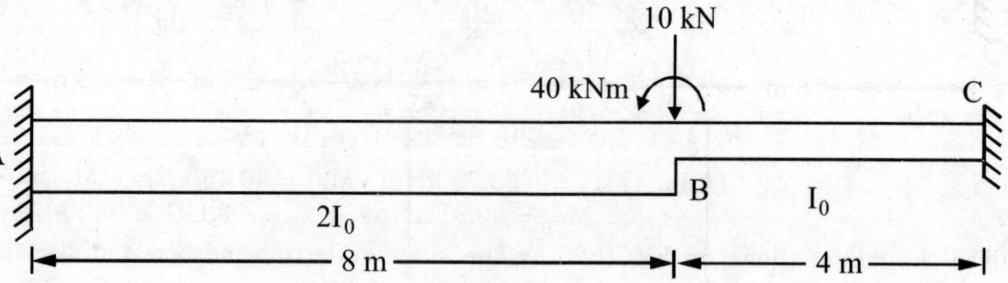

Fig. 2.25

[**Ans.** $M_{AB} = -24.86$ kNm ; $M_{BA} = -37.84$ kNm ; $M_{BC} = -2.16$ kNm ; $M_{CB} = 10.81$ kNm]

3. An encastre beam AB has a span of l. It is loaded over each half of the span with uniformly distributed loads, the load on one half being double the load on the other half. (The loads are $2w$ and w per m run respectively). The second moment of area of the half of the beam supporting $2w$ is equal to $1.5I_0$ and that of the remainder, I_0. Both cross-sections are constant along the respective lengths. If Young's modulus E is constant, determine by the slope deflection method, the central deflection of the beam in terms of w, l, E and I_0 (Fig. 2.26).

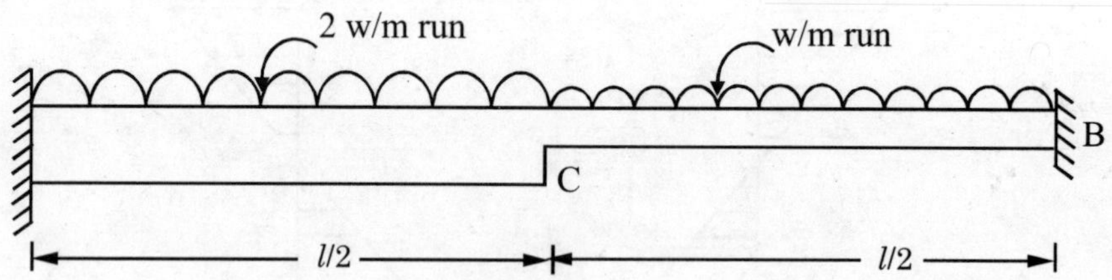

Fig. 2.26

$$\left[\textbf{Ans.}\quad \delta = 0.0086\,\frac{wl^4}{EI}\right]$$

4. ABC is a continuous beam with constant EI throughout its length (Fig. 2.27). The end supports A and C are fixed and the beam is continuous over middle support B. Span BC is uniformly loaded with 10 kN per metre length while a concentrated vertical down ward load of 100 kN acts at the midspan of AB. Calculate the moments by slope deflection method.

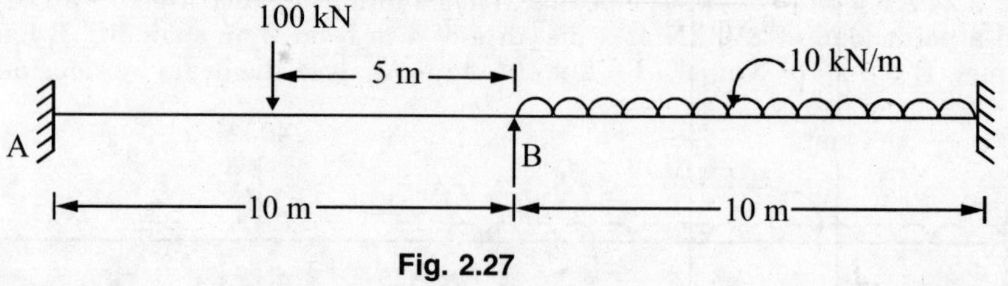

Fig. 2.27

[**Ans.** $M_{AB} = -135.4$ kNm ; $M_{BA} = 104.2$ kNm ; $M_{BC} = -104.2$ kNm ; $M_{CB} = 72.9$ kNm]

5. A continuous beam ABCD consists of three spans and is loaded as shown in Fig. 2.28. Ends A and D are fixed. Using the slope deflection method, determine the bending moments at the supports and sketch the bending moments at the supports. Sketch the bending moment diagram.

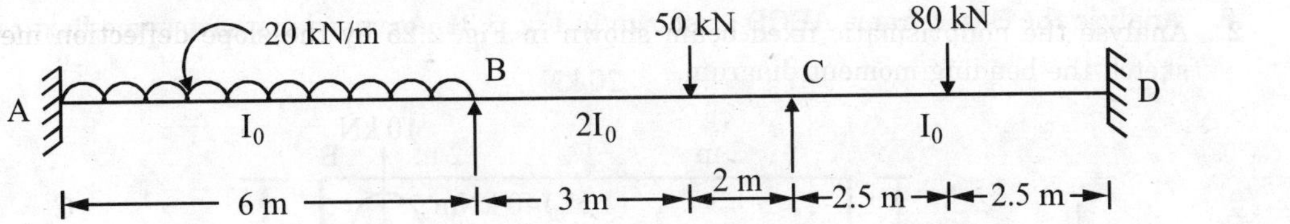

Fig. 2.28

[**Ans.** $M_{AB} = -66.78$ kNm ; $M_{BA} = 46.45$ kNm ; $M_{BC} = -46.45$ kNm ;
$M_{CB} = 39.91$ kNm ; $M_{CD} = -33.91$ kNm ; $M_{DC} = 55.04$ kNm]

6. Analyse the frame shown in Fig. 2.29 by the slope deflection method and sketch the bending moment diagram.

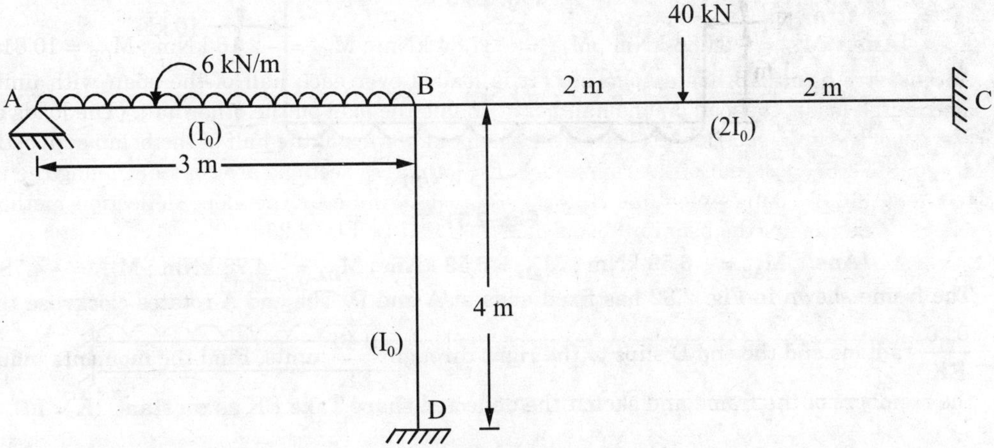

Fig. 2.29

[**Ans.** $M_{AB} = 0$; $M_{BA} = 10.0625$ kNm ; $M_{BC} = -13.375$ kNm ; $M_{CB} = 23.3125$ kNm ;
$M_{BD} = 3.3125$ kNm ; $M_{DB} = 1.6563$ kNm]

7. Find the moments at A, B and C in the frame shown in Fig. 2.30. Find also the reactions at the supports.

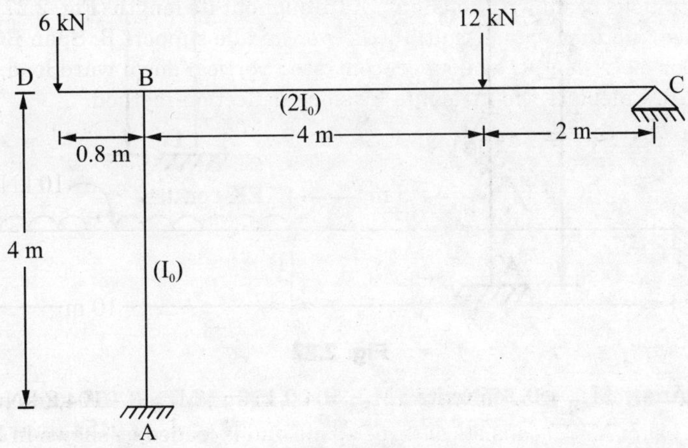

Fig. 2.30

[**Ans.** $M_{AB} = 1.47$ kNm ; $M_{BA} = 2.94$ kNm ; $M_{BC} = -7.74$ kNm ; $M_{BD} = 4.80$ kNm ;
$H_A = 1.1025$ kN($\rightarrow$) ; $H_C = 1.1025$ kN($\leftarrow$) ; $V_A = 5.29$ kN ; $V_C = 6.71$ kN]

8. Analyse the closed frame ABCD as shown in Fig. 2.31. All members have same flexural rigidity.

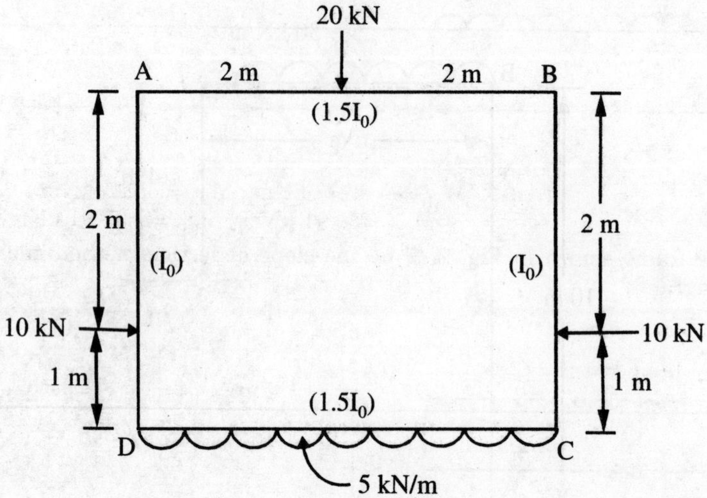

Fig. 2.31

[**Ans.** $M_{AB} = -6.59$ kNm ; $M_{AD} = 6.59$ kNm ; $M_{DA} = -4.78$ kNm ; $M_{DC} = -4.78$ kNm]

9. The frame shown in Fig. 2.32 has fixed ends at A and D. The end A rotates clockwise through $\frac{0.20}{EK}$ radians and the end D slips to the right through $\frac{0.40}{EK}$ units. Find the moments induced in the members of the frame and sketch the deflected shape.Take EK as constant. ($K = I/l$).

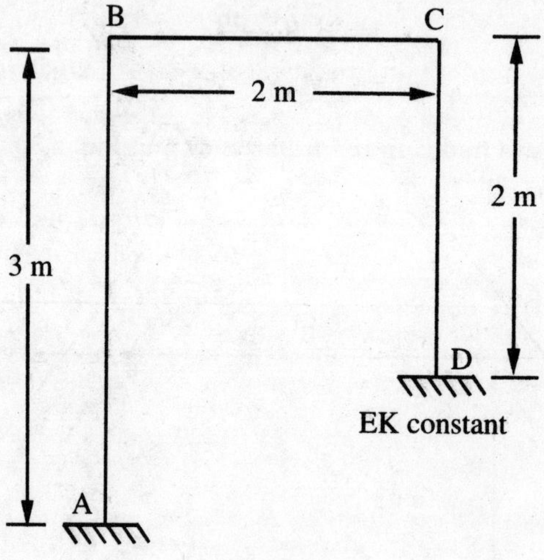

Fig. 2.32

[**Ans.** $M_{AB} = 0.145$ units ; $M_{BA} = -0.156$; $M_{BC} = 0.156$ units ; $M_{CB} = 0.0154$ units ; $M_{CD} = -0.0154$ units ; $M_{DC} = 0.025$ units.]

10. Analyse the portal frame shown in Fig. 2.33 E is constant, and $I_1 : I_2 : I_3 = 3 : 2 : 1$

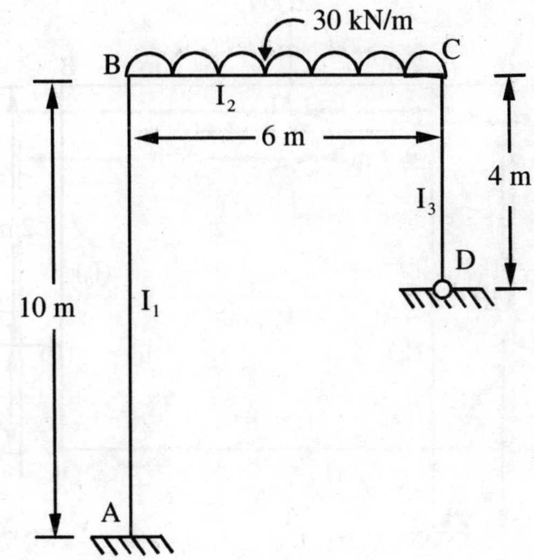

Fig. 2.33

[**Ans.** M_{AB} = 35.528 kNm ; M_{BA} = 65.152 kNm ; M_{BC} = – 65.416 kNm ; M_{CB} = – 40.42 kNm ; M_{CD} = –40.305 kNm ; M_{DC} = 0]

11. Analyse the portal frame shown in Fig. 2.34 by the slope deflection method and sketch the bending moment diagram.

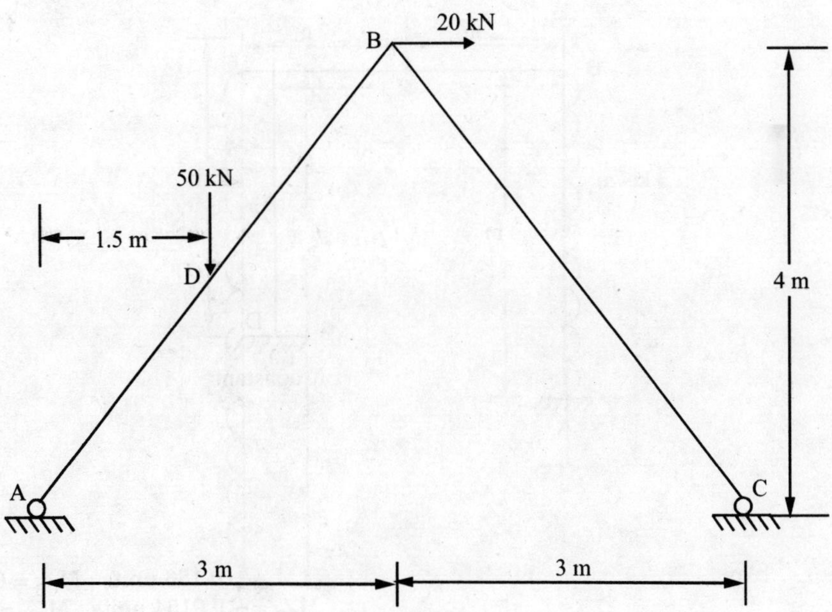

Fig. 2.34

[**Ans.** M_{BA} = 14.05 kNm]

12. Find the moments at A, B, C, D and the reactions at the supports for the portal frame shown in Fig. 2.35.

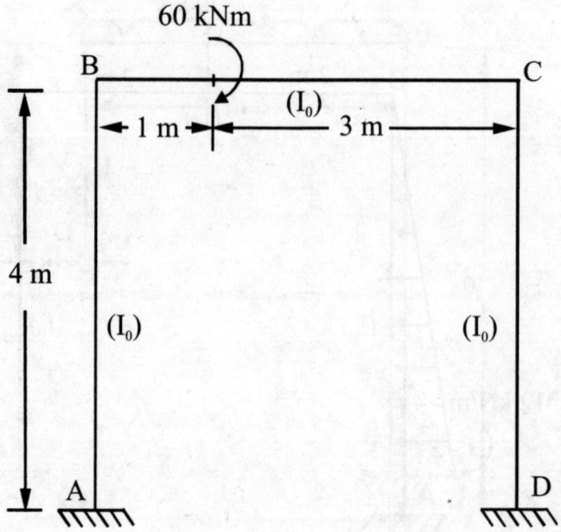

Fig. 2.35

[**Ans.** M_{AB} = 5.29 kNm ; M_{BA} = 9.71 kNm ; M_{BC} = – 9.71 kNm ; M_{CB} = 10.29 kNm ; M_{CD} = – 10.29 kNm ; M_{DC} = – 4.71 kNm ; H_A = 3.75 kN ; H_D = – 3.75 kN ; V_A = – 15.145 kN ; V_D = 15.145 kN]

13. Analyse the portal frame shown in Fig. 2.36.

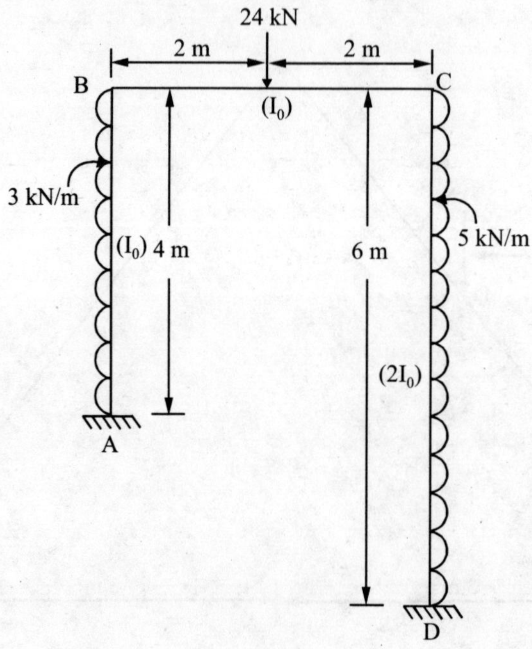

Fig. 2.36

[**Ans.** M_{AB} = 19.42 kNm ; M_{BC} = 22.01 kNm ; M_{BC} = – 22.01 kNm ; M_{CB} = 8.14 kNm ; M_{CD} = – 8.14 kNm]

14. Analyse the portal frame shown in Fig. 2.37 by the slope deflection method and sketch the bending moment diagram.

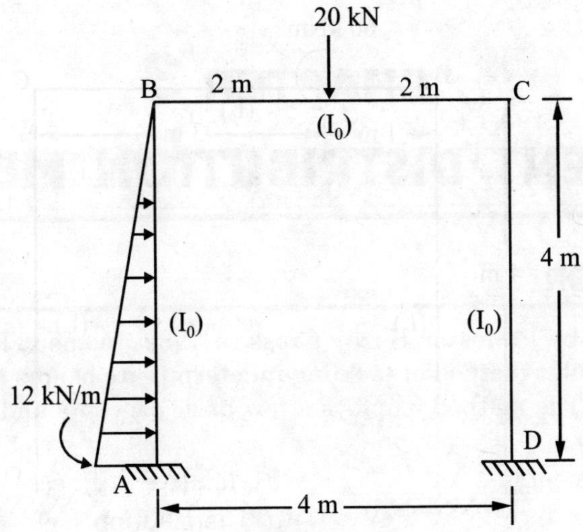

Fig. 2.37

[**Ans.** M_{AB} = – 12.42 kNm ; M_{BA} = 1.86 kNm ; M_{BC} = – 1.86 kNm ; M_{CB} = 8.90 kNm ; M_{CD} = – 8.90 kNm ; M_{DC} = – 0.52 kNm]

CHAPTER 3

MOMENT DISTRIBUTION METHOD

3.1 INTRODUCTION

First introduced by Professor Hardy Cross in 1932, Moment Distribution method has remained the most popular method of tackling indeterminate beams and rigid frames. It uses an iterative technique. The method employs a few basic concepts and a few specialized terms such as

(*a*) Fixed end moments (*b*) Relative stiffness
(*c*) Carry over (*d*) Distribution factor.

3.2 FUNDAMENTAL PROPOSITIONS

(*a*) **Fixed End Moments.** All members of a given frame are initially assumed fixed at both ends. The loads acting on these fixed beams produce fixed end moments at the ends. These are the moments exerted by the supports on the beam-ends. These (non-existent) moments keep the rotations at the ends of each member zero.

Table A1 gives the fixed end moments in prismatic fixed beams due to various load types. These moments are treated as positive if clockwise and negative if anti-clockwise.

(*b*) **Relative Stiffness**

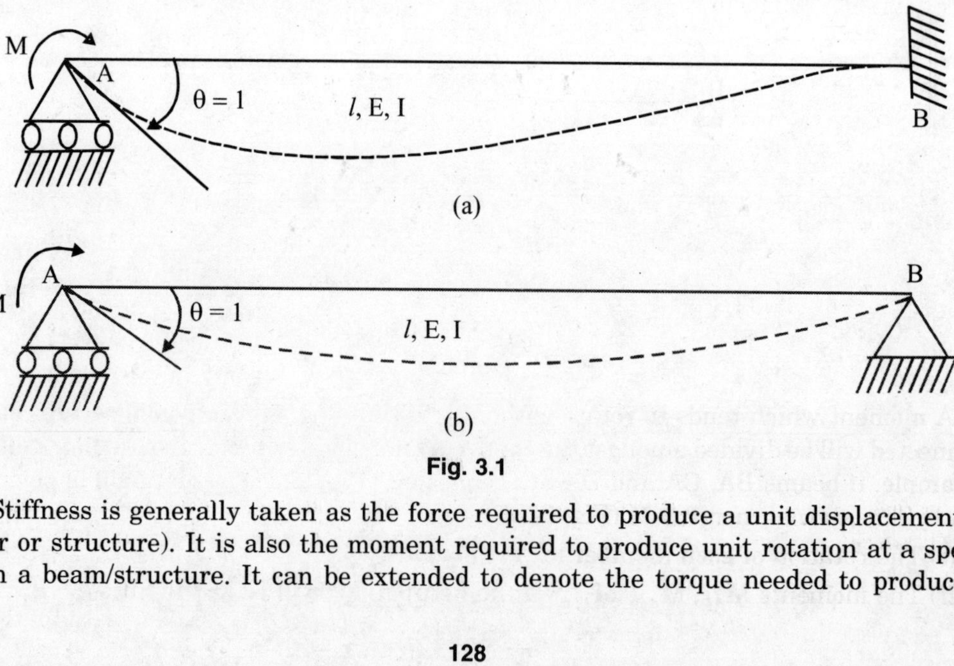

Fig. 3.1

Stiffness is generally taken as the force required to produce a unit displacement (in a member or structure). It is also the moment required to produce unit rotation at a specified point in a beam/structure. It can be extended to denote the torque needed to produce unit twist.

In moment distribution method we are interested to know the moment required to produce a unit clockwise rotation at a support point in a beam [say support A in Figs. 3.1 (*a*) and (*b*)]

Stiffness is usually denoted as k and can be proved to be given by

$$\mathbf{k} = 4\ \mathrm{EI}/l \qquad \text{[Far end fixed ; Fig. 3.1 (a)]}$$
$$\mathbf{k} = 3\ \mathrm{EI}/l \qquad \text{[Far end hinged ; Fig. 3.1 (b)]}$$

It is easy to realize that **k** will be large if E or/and I are large. **k** is less when l is more.

(*c*) **Carry Over.** In moment distribution method, we have to analyse the effects of applying imaginary moments at a specified point.

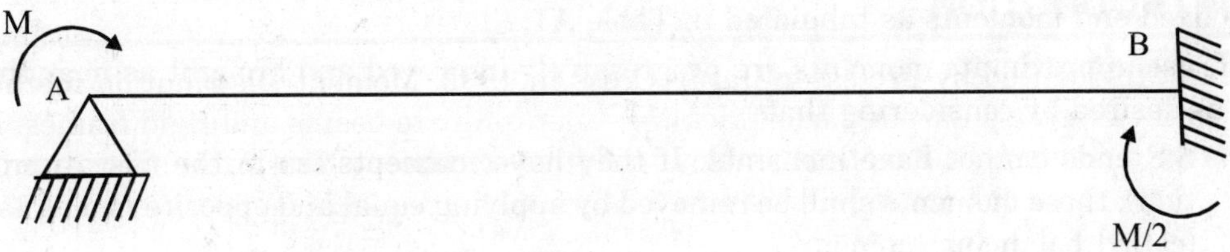

Fig. 3.2

The beam in Fig. 3.2 when it receives a moment M at A, will develop at B a moment of M/2. This M/2 is called the carry over moment. The factor $\frac{1}{2}$ is then the "Carry over factor" (C.O.F.). If the far end B were hinged, the C.O.F. will be zero.

(*d*) **Distribution Factor**

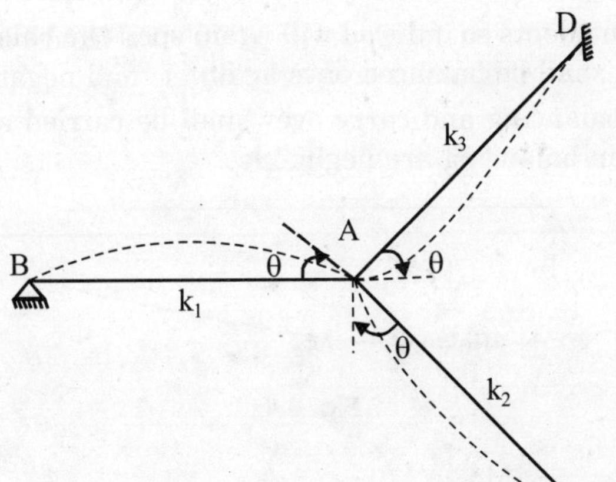

Fig. 3.3

A moment which tends to rotate without translation a joint to which several members are connected will be divided amongst the connected members in proportion to their stiffnesses. For example, if beams BA, CA and DA are connected at A, and if a rotation θ is given to the joint A by an external moment M (Fig. 3.3).

(*i*) The rotation of each member at A is obviously θ

(*ii*) The moments M_{AB}, M_{AC}, M_{AD} (summing up to M) will be in the ratio $k_1 : k_2 : k_3$.

Hence, $M_{AB} = (k_1/\Sigma k) \cdot M$

$M_{AC} = (k_2/\Sigma k) \cdot M$

$M_{AD} = (k_3/\Sigma k) \cdot M$

The factors $k_i/\Sigma k$ are called distribution factors.

3.3 MOMENT DISTRIBUTION PROCEDURE

Moment distribution method is an iterative procedure. To start with, all joints in a beam or frame are considered fixed, and the given loading applied on this modified structure. Under this (approximate) condition all the joint moments can be considered as known. They are the fixed end moments as tabulated in Table A1.

These approximate moments are progressively improved and brought as near the true values as desired by considering that :

(*i*) S.S ends cannot have moments. If they have moments (as in the first approximation), these moments shall be removed by applying equal and opposite moments there (called balancing moments).

(*ii*) Interior joints cannot have any net moment. If the sum of the moments (ΣM) at an interior joint is not zero, it is an error. To offset this error, a balancing moment of – ΣM shall be applied at the joint.

(*iii*) The balancing moment will be shared by the members connected to the joint in the ratio of their stiffnesses.

(*iv*) Now each of the balancing moments introduced in members (in steps (*i*), (*ii*), and (*iii*) above) will induce moments at the opposite end of the members as carry over moments.

(*v*) The carry over moments so induced will again spoil the balance achieved in steps (*i*), (*ii*) and (*iii*) and shall be balanced once again.

(*vi*) The process of balancing and carry over shall be carried out till the values of moments involved in balancing are negligible.

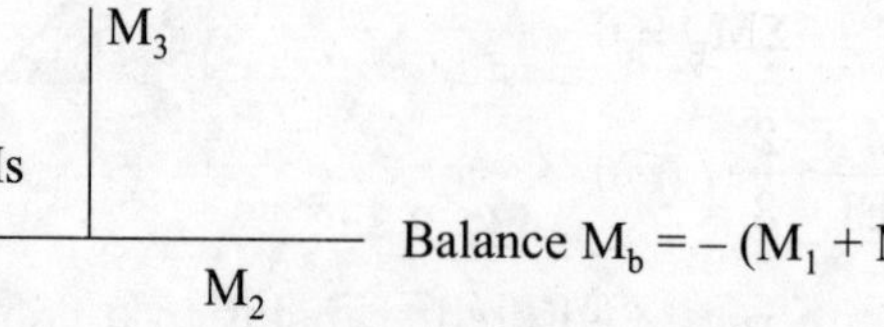

Fig. 3.4

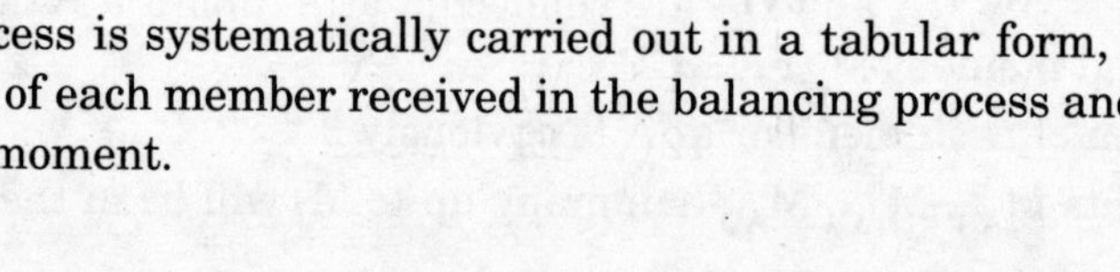

Fig. 3.5

The whole process is systematically carried out in a tabular form, and the sum of the moments at each end of each member received in the balancing process and carry over process is taken as the final moment.

Stiffness and carry over factors: Consider a prismatic beam AB, simply supported at one end and fixed at the other as shown in Fig. 3.6 (*a*).

Let us apply a clockwise moment (M_A) at the left end of the member.

The moment required to cause a rotation θ_A at the simply supported (near) end and the moment M_B carried over to the fixed far end B are required.

Let us use the consistent deformation method. The required geometry in this case is

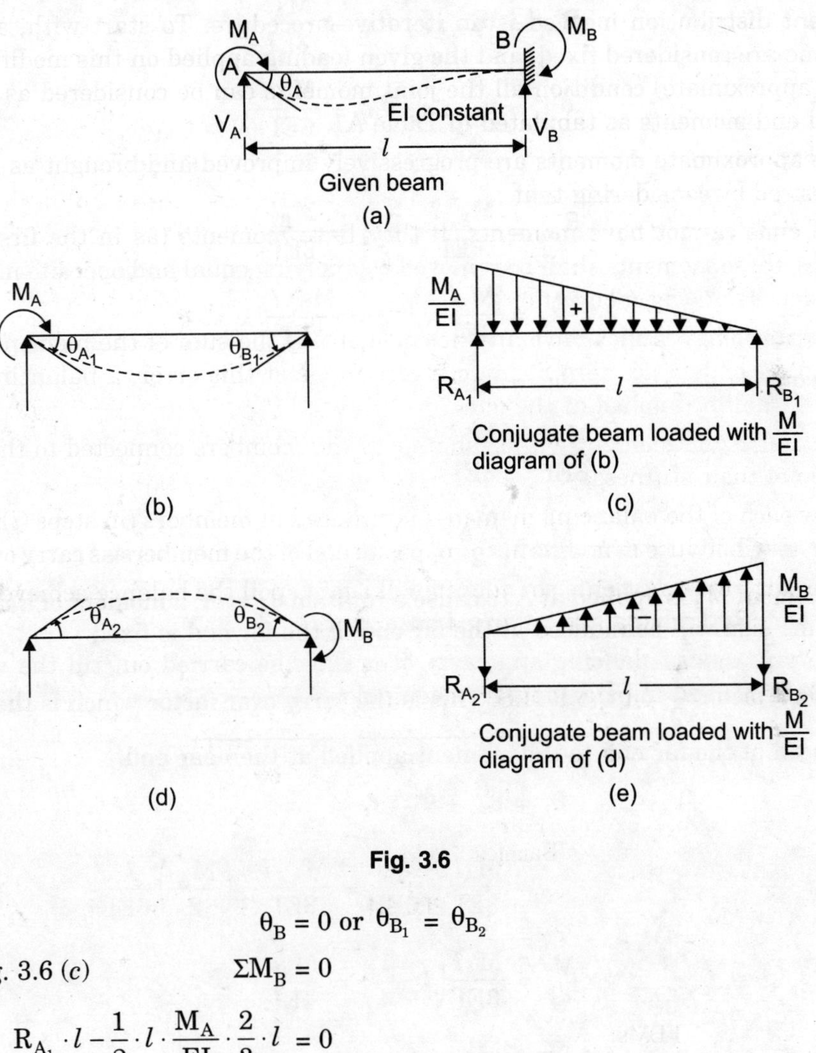

Fig. 3.6

$$\theta_B = 0 \text{ or } \theta_{B_1} = \theta_{B_2} \qquad \ldots(1)$$

In Fig. 3.6 (*c*)

$$\Sigma M_B = 0$$

$$R_{A_1} \cdot l - \frac{1}{2} \cdot l \cdot \frac{M_A}{EI} \cdot \frac{2}{3} \cdot l = 0$$

$$R_{A_1} = \frac{M_A \cdot l}{3EI} \qquad \ldots(2)$$

$$\Sigma V = 0$$

$$R_{A_1} + R_{B_1} = \frac{1}{2} \cdot l \cdot \frac{M_A}{EI}$$

$$\frac{M_A \cdot l}{3EI} + R_{B_1} = \frac{M_A l}{2EI}$$

$$\therefore \quad R_{B_1} = \frac{M_A l}{6EI} \qquad \text{...(3)}$$

We know that, the S.F. at A of the conjugate beam is the slope at A of the original beam

i.e., $$R_{A_1} = \theta_{A_1} \; i.e., \; \theta_{A_1} = \frac{M_A l}{3EI} \qquad \text{...(4) [from (2)]}$$

and $$R_{B_1} = \theta_{B_1} \; i.e., \; \theta_{B_1} = \frac{M_A l}{6EI} \qquad \text{...(5) [from (3)]}$$

From Fig. 3.6 (*e*)

$$R_{A_2} = \frac{M_B \cdot l}{6EI}; \; R_{B_2} = \frac{M_B l}{3EI} \qquad \text{(as above)}$$

$$\therefore \quad \theta_{A_2} = \frac{M_B l}{6EI}; \; \theta_{B_2} = \frac{M_B \cdot l}{3EI}$$

From eqn. (1) $$\theta_{B_1} = \theta_{B_2}$$

$$\frac{M_A l}{6EI} = \frac{M_B l}{3EI}$$

$$\therefore \quad M_B = \frac{M_A}{2} \qquad \text{...(6)}$$

i.e., when a moment M_A is applied at A to cause a rotation θ_A at A, a moment of half of its value and of the same sign will be induced at the far end, if the far end is fixed.

There is a factor $\frac{1}{2}$ on the R.H.S. This is the carry over factor which is the ratio of the moment induced at the far end to the moment applied at the near end.

Also, $$\theta_A = \theta_{A_1} - \theta_{A_2}$$

$$= \frac{M_A l}{3EI} - \frac{M_B l}{6EI} = \frac{M_A \cdot l}{3EI} - \left(\frac{M_A}{2}\right)\frac{l}{6EI}$$

$$= \frac{M_A l}{3EI}\left(1 - \frac{1}{4}\right) = \frac{M_A l}{4EI}$$

$$\therefore \quad \theta_A = \frac{l}{4EI} \cdot M_A \text{ or } M_A = \frac{4EI}{l} \cdot \theta_A \qquad \text{...(7)}$$

$$\frac{M_A}{\theta_A} = k = \frac{4EI}{l}$$

$$k = \frac{4EI}{l} = \text{Stiffness factor}$$

Stiffness factor (k) is defined as the moment required to be applied at the end under consideration of a member to cause unit rotation at it.

Here the moment k is required to rotate the near end of the prismatic beam through unit angle, without translation, the far end being fixed, is given by $k = \frac{4EI}{l}$.

If the beam is simply supported at B as shown in Fig. 3.6 (b) and (c), θ_A caused by M_A,

$$\theta_A = \theta_{A_1} = \frac{M_A l}{3EI} \qquad ...(8)$$

Hence the moment required to be applied at A, to cause a rotation of 1 radian at A of a simply supported beam is $\frac{3EI}{l}$ and the carry over factor is zero in such beams.

Example 3.1 *Determine the fixed end moment at each end of each member in the continuous beam in Fig. 3.7 for the loading shown.*

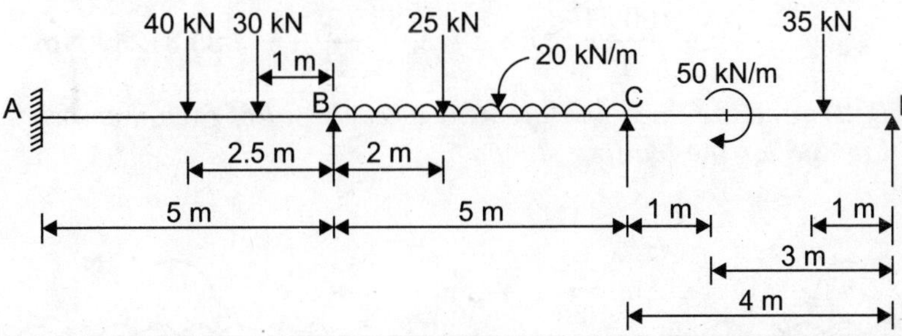

Fig. 3.7

Solution

$$M_{FAB} = -\left[\frac{W_1 l}{8} + \frac{W_2 ab^2}{l^2}\right]$$

$$= -\left[\frac{40 \times 5}{8} + \frac{30 \times 4 \times 1^2}{5^2}\right]$$

$$= -[25 + 4.8] = -29.8 \text{ kNm}$$

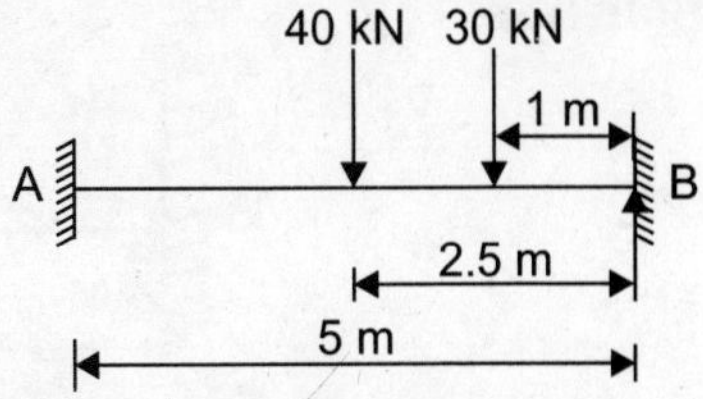

$$M_{FBA} = +\left[\frac{W_1 l}{8} + \frac{W_2 a^2 b}{l^2}\right]$$

$$= \frac{40 \times 5}{8} + \frac{30 \times 4^2 \times 1}{5^2}$$

$$= 25 + 19.2 = 44.2 \text{ kNm}$$

$$M_{FBC} = -\left[\frac{wl^2}{12} + \frac{Wab^2}{l^2}\right]$$

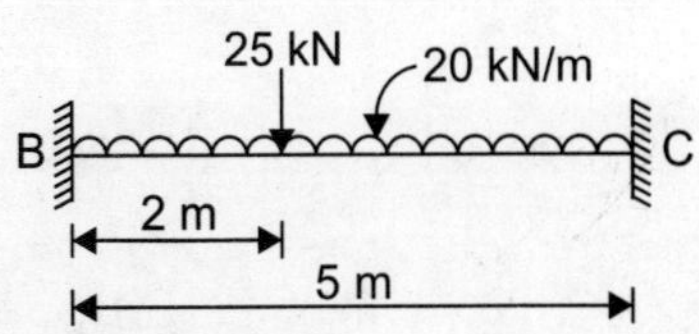

$$= -\left[\frac{20 \times 5^2}{12} + \frac{25 \times 2 \times 3^2}{5^2}\right]$$

$$= -[41.67 + 18] = -59.67 \text{ kNm}$$

$$M_{FCB} = +\left[\frac{wl^2}{12} + \frac{Wa^2b}{l^2}\right]$$

$$= +\left[\frac{20 \times 5^2}{12} + \frac{25 \times 2^2 \times 3}{5^2}\right]$$

$$= 41.67 + 12 = 53.67 \text{ kNm}$$

$$M_{FCD} = \frac{M \cdot b}{l^2}(3a - l) - \frac{Wab^2}{l^2}$$

$$= \frac{50 \times 3}{4^2}(3 \times 1 - 4) - \frac{35 \times 3 \times 1^2}{4^2}$$

$$= -\ 15.9375 \text{ kNm}$$

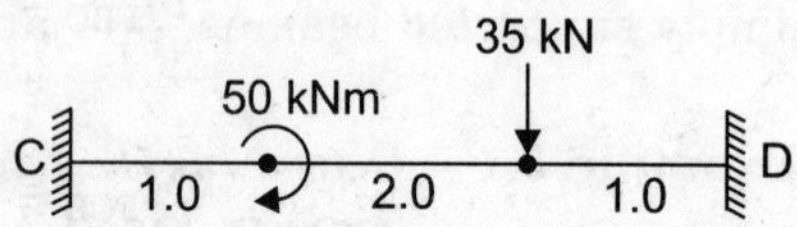

$$M_{FDC} = \frac{M \,.\, a}{l^2}(3b - l) + \frac{Wa^2b}{l^2}$$

$$= \frac{50 \times 1}{4^2}(3 \times 3 - 4) + \frac{35 \times 3^2 \times 1}{4^2} = 35.3125 \text{ kNm}$$

Example 3.2 *Determine the fixed end moment at each end of each member in the frame in Fig. 3.8 for the loading shown.*

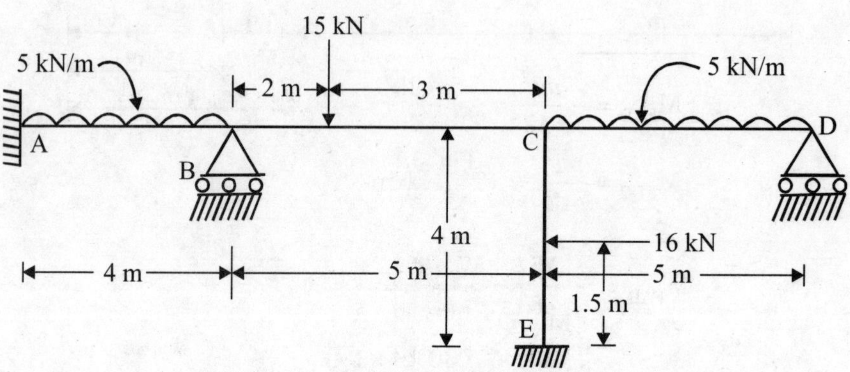

Fig. 3.8

Solution

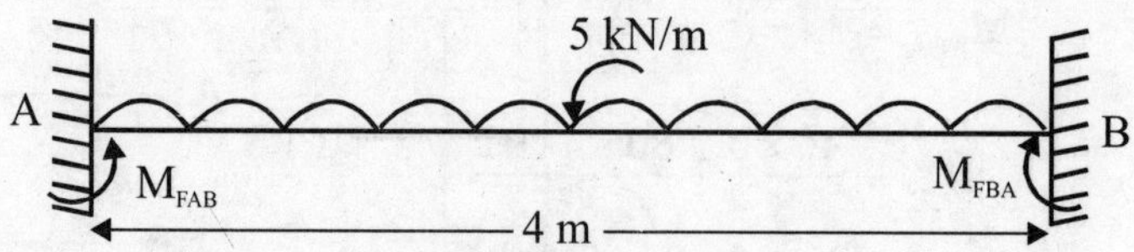

Fig. 3.9

$$M_{FAB} = -\frac{wl^2}{12} = -\frac{5 \times 4^2}{12} = -\ 6.67 \text{ kNm}$$

$$M_{FBA} = \frac{wl^2}{12} = \frac{5 \times 4^2}{12} = 6.67 \text{ kNm}$$

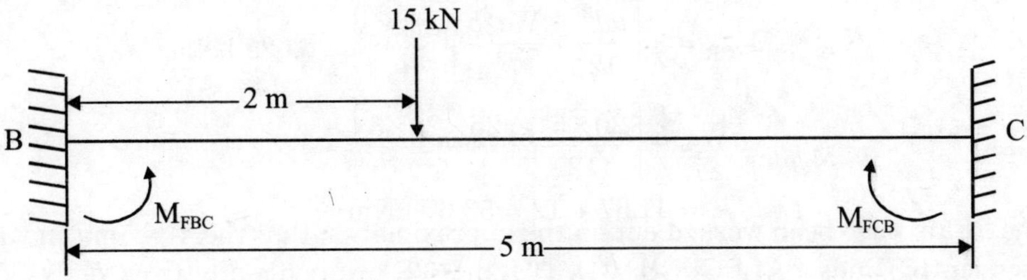

Fig. 3.10

$$M_{FBC} = -\frac{Wab^2}{l^2} = -\frac{15 \times 2 \times 3^2}{5^2} = -10.8 \text{ kNm}$$

$$M_{FCB} = +\frac{Wa^2b}{l^2} = \frac{15 \times 2^2 \times 3}{5^2} = 7.2 \text{ kNm}$$

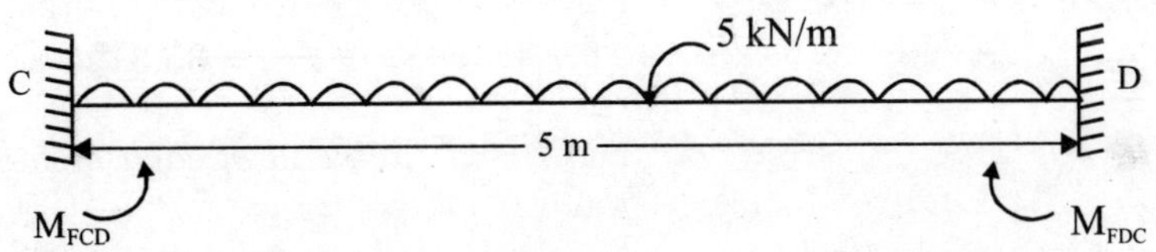

Fig. 3.11

$$M_{FCD} = -\frac{wl^2}{12} = -\frac{5 \times 25}{12} = -10.42 \text{ kNm}$$

$$M_{FDC} = \frac{wl^2}{12} = 10.42 \text{ kNm}$$

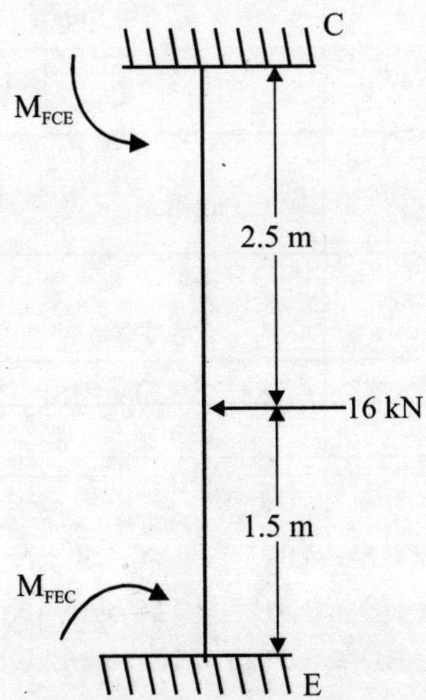

Fig. 3.12

$$M_{FCE} = -\frac{Wab^2}{l^2} = -\frac{16 \times 2.5 \times 1.5^2}{4^2} = -5.625 \text{ kNm}$$

$$M_{FEC} = \frac{Wa^2b}{l^2} = \frac{16 \times 2.5^2 \times 1.5}{4^2} = 9.375 \text{ kNm}$$

[The FEMs have been worked out on the approximation that the 4 'beams' in the frame act independently. Ends BA, BC, CB, CD, DC and CE are not really fixed ends. They are assumed to be, as a first approximation in the moment distribution process.]

Example 3.3 *Find the moments shared by the members BA, BC and BD if the joint B is subjected to a clockwise moment of 144 kNm as shown in Fig. 3.13.*

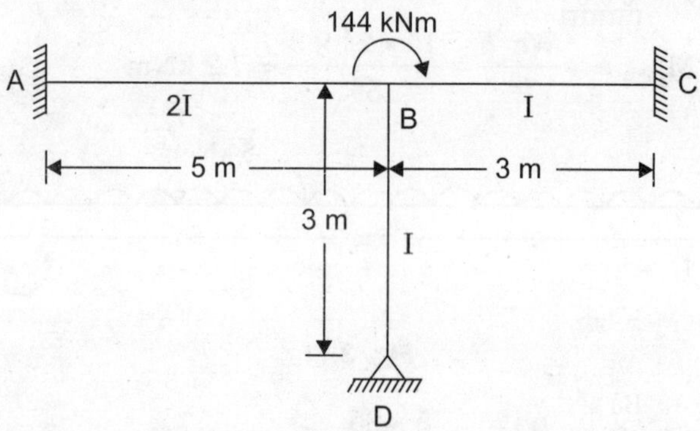

Fig. 3.13

Solution Distribution factors :

Joint	*Member*	*Relative Stiffness (R.S.)*	*Sum of Stiffness (T.S.)*	*Distribution Factor* = $\frac{R.S.}{T.S.}$
	BA	$\frac{I}{l} = \frac{2I}{5}$		$\frac{2I}{5} \Big/ \frac{16I}{15} = \frac{2I}{5} \times \frac{15}{16I} = 0.375$
B	BC	$\frac{I}{l} = \frac{I}{3}$	$\frac{2I}{5} + \frac{I}{3} + \frac{I}{3}$	$\frac{I}{3} \times \frac{15}{16I} = 0.3125$
	BD	$\frac{I}{l} = \frac{I}{3}$	$\frac{6I + 5I + 5I}{15} = \frac{16I}{15}$	$\frac{I}{3} \times \frac{15}{16I} = 0.3125$

Hence,

Moment shared by member BA = 0.375 × 144 = 54 kNm

Moment shared by member BC = 0.3125 × 144 = 45 kNm

and moment shared by member BD = 0.3125 × 144 = 45 kNm

Example 3.4 *Three flexure members (Fig. 3.14) are connected at joint J. Their section properties are given under. Find the moments induced in each end of each beam, if an external moment of 24 kNm acts on the joint.*

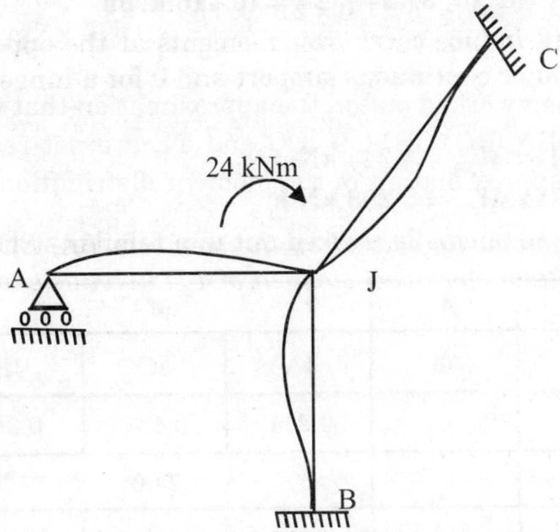

Fig. 3.14

Member	Length (m)	I (mm^4)
AJ	4	3×10^8
BJ	4	4×10^8
CJ	5	6×10^8

Solution The applied moment will be shared by JA, JB and JC in proportion to their stiffnesses. [The reason is, the rotation θ of the joint J is the same as the rotation of JA , JB or JC]. The stiffness of the joint J is actually the sum of the stiffnesses of JA, JB and JC.

$$k_J = k_{JA} + k_{JB} + k_{JC}$$

$$k_{JA} = \frac{3EI}{l} = \frac{3 \times E \times 3 \times 10^8}{4 \times 10^3} = \frac{9}{4 \times 10^3} \times 10^8 \times E = 2.25 \times 10^5\ E$$

$$k_{JB} = \frac{4EI}{l} = \frac{4 \times E \times 4 \times 10^8}{4 \times 10^3} = 4 \times 10^5\ E$$

$$k_{JC} = \frac{4EI}{l} = \frac{4 \times E \times 6 \times 10^8}{5 \times 10^3} = \frac{24}{5} \times 10^5\ E = 4.8 \times 10^5\ E$$

$$k_J = 11.05 \times 10^5\ E$$

The distribution factors are therefore

$$DF_{JA} = \frac{2.25}{11.05} = 0.204$$

$$DF_{JB} = \frac{4}{11.05} = 0.362$$

$$DF_{JC} = \frac{4.8}{11.05} = 0.434$$

The moments shared by the members are

$$M_{JA} = 0.204 \times 24 = 4.896 \text{ kNm}$$
$$M_{JB} = 0.362 \times 24 = 8.688 \text{ kNm}$$
$$M_{JC} = 0.434 \times 24 = 10.416 \text{ kNm}$$

These moments will induce carry over moments at the opposite ends. The carry over factor is 1/2 for a fixed end or continuous support and 0 for a hinged end.

Hence, $M_{AJ} = 0$

$$M_{BJ} = (1/2) \times M_{JB} = 4.344 \text{ kNm}$$
$$M_{CJ} = (1/2) \times M_{JC} = 5.208 \text{ kNm}$$

The above results are normally worked out in a tabular structure as under :

Joint	***A***	***J***			***B***	***C***
Member	AJ	JA	JC	JB	BJ	CJ
DF(Distribution Factor)	–	0.204	0.434	0.362	–	–
Applied moment	–		24.0		–	–
Distribution		4.896	10.416	8.688		
Carry over	0				4.344	5.208

Example 3.5 *A propped cantilever carries a load of 20 kN as shown in Fig. 3.15. Find the moment at the fixed end. Sketch the bending moment diagram.*

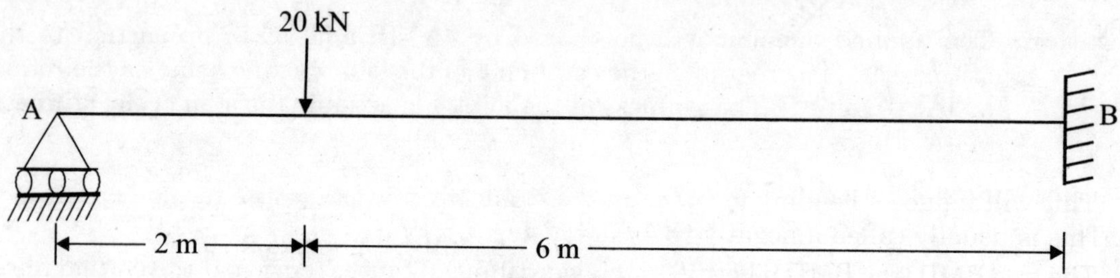

Fig. 3.15

Solution **Fixed end moments :**

$$M_{FAB} = -\frac{Wab^2}{l^2} = -\frac{20 \times 2 \times 6^2}{8^2} = -22.5 \text{ kNm}$$

$$M_{FBA} = \frac{Wa^2b}{l^2} = \frac{20 \times 2^2 \times 6}{8^2} = 7.5 \text{ kNm}$$

Member AB

Joint	A	B
FEM	– 22.5	7.5
Balance	+ 22.5	–
Carry over	–	11.25
Final moments	**0**	**18.75**

There is only one member. There is no question of distribution or distribution factor. The first step assumes a moment of – 22.5 at the hinge A. This is obviously wrong. To restore zero moment at the hinge, we must balance it. It means we must apply a moment of + 22.5 at A. This applied moment induces half that value at the opposite end B, *i.e.*, a value of 11.25. This is called carry over. Since balance is restored at A, the process is over and we get the final moments by adding the moments under A and B separately.

The bending moment diagram is made up of 2 parts, the BMD due to support moments and the free bending moment diagram. These two are superposed to get the final BMD.

The value 18.75 in the last line of the table is the moment at the fixed end B. It is the moment applied by the fixed support B on the beam and is positive meaning clockwise.

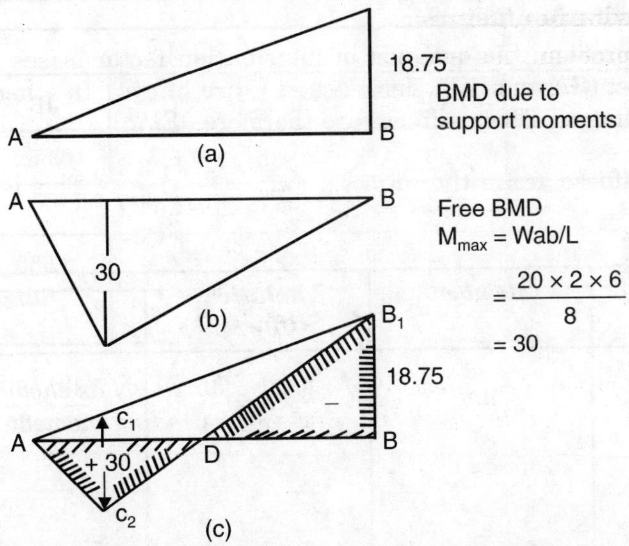

Fig. 3.16

This will induce a bending moment on the beam which would cause tension on the top fibre. This is usually called a negative bending moment.

The free BMD (the BMD if both A and B were hinged) causes tension at the bottom fibre of AB. Hence positive bending moment. These two BMDs are best combined as in Fig. 3.16 (*c*). The positive BMD is 'suspended' from the negative BMD to get the net BMD.

Example 3.6 *Analyse the continuous beam shown in Fig. 3.17 by the method of moment distribution.*

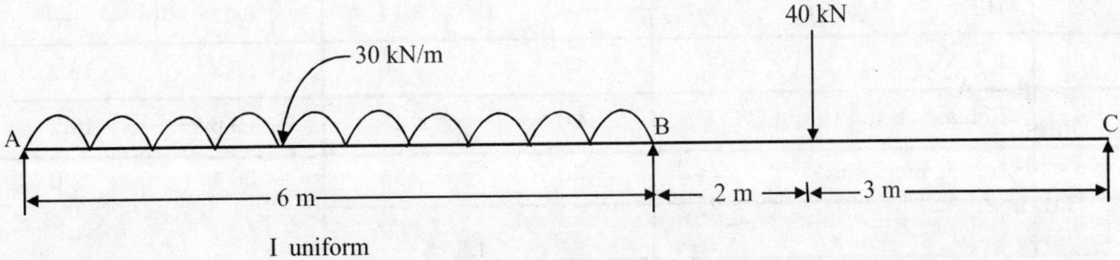

Fig. 3.17

Solution **(*a*) Fixed end moments :**

$$M_{FAB} = -\frac{wl^2}{12} = -\frac{30 \times 6^2}{12} = -90 \text{ kNm}$$

$$M_{FBA} = +\frac{wl^2}{12} = \frac{30 \times 6^2}{12} = +90 \text{ kNm}$$

$$M_{FBC} = -\frac{Wab^2}{l^2} = \frac{40 \times 2 \times 3^2}{5^2} = -28.8 \text{ kNm}$$

$$M_{FCB} = +\frac{Wa^2b}{l^2} = +\frac{40 \times 2^2 \times 3}{5^2} = +19.2 \text{ kNm}$$

(*b*) Distribution factors :

In this problem, the question of distribution factor arises only at support B where 2 members meet. (BA and BC). Here, A and C are hinged. In other words, BA and BC have their far ends hinged. Their stiffness are therefore 3EI/*l*.

Their stiffness are in the ratio $k_{BA} : k_{BC} = \frac{3}{4}\frac{I}{l} : \frac{3}{4}\frac{I}{l}$. This is made use of in the table below.

Joint	***Member***	***Relative Stiffness, k***	***Sum***	***Distribution Factor***
B	BA	$\frac{3}{4} \times \frac{I}{6} = \frac{3I}{24}$	$\frac{3I}{24} + \frac{3I}{20}$	$\left(\frac{3I}{24}\right) / \left(\frac{33I}{120}\right)$
			$= \frac{15I + 18I}{120}$	$= \frac{3I}{24} \times \frac{120}{33I} = \frac{5}{11}$
	BC	$\frac{3}{4} \times \frac{I}{5} = \frac{3I}{20}$	$= \frac{33I}{120}$	$\left(\frac{3I}{20}\right) / \left(\frac{33I}{120}\right) = \frac{6}{11}$

Any external moment applied to B would be shared by BA and BC in the ratio of their stiffnesses, namely 3EI/L.

(*c*) Moment distribution :

Line	***Joint***	***A***	***B***		***C***
1	Member	AB	BA	BC	CB
2	D.F.	—	5/11	6/11	—
3	F.E.M.	– 90	+ 90	– 28.8	19.2
4	Release A and C and C/O	+ 90 ⟶	+ 45	– 9.6 ⟵	– 19.2
5	Initial moments	0	135	– 38.4	0
6	Balance B	—	– 43.9	– 52.7	—
7	Final moments	0	91.1	– 91.1	0

M_{BA} is + 91.1 kNm. This is the moment applied by the support system on BA. M_{BC} is – 91.1 kNm. The BMD is drawn by combining the support moment diagram with the span BMD (for simply support condition) of BA and BC.

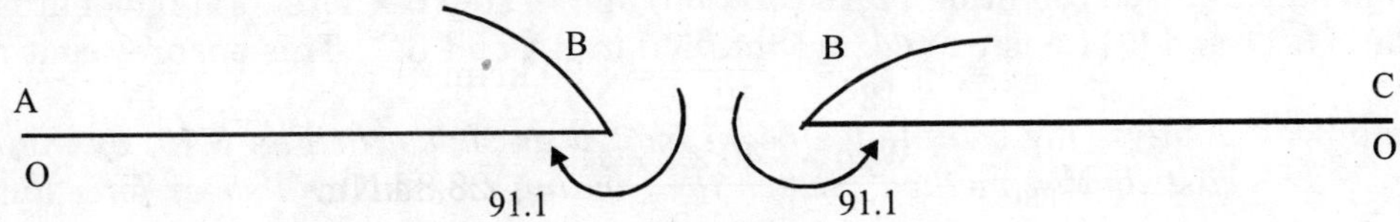

Fig. 3.18 (*a*)

Simply supported moments :

In AB : $\frac{wl^2}{8} = \frac{30 \times 6^2}{8} = 135$ kNm

In BC : $\frac{Wab}{l} = \frac{40 \times 2 \times 3}{5} = 48$ kNm

Some lines in the moment distribution table above need explaining :

Line 2. The distribution factors (D.F.), 5/11 and 6/11 are the factors by which any balancing moment at B would be multiplied to give the share of moments to BA and BC.

Line 3. None of the supports A, B or C is fixed. But all of them are assumed to be fixed in this step and the resulting moments are written.

Line 4. The hinged conditions at A and C are realised by removing moments – 90 at A and 19.2 at C (by providing their opposites here). These opposite moments 90 and – 19.2 induce at the far end (B) carry over moments of half the value (45 and – 9.6).

Line 5. Gives the situation when the restoration of hinged condition at A and C is complete.

Line 6. Support B in now unbalanced. The moment on BA is 135 and the moment on BC is – 38.4. The sum should be zero and it is not. The error is 135 – 38.4 = 96.6. This moment is balanced in this step by applying – 96.6 at B. This obviously would be shared by BA and BC in the ratio 5/11 and 6/11 (giving – 43.9 to BA and – 52.7 to BC).

Line 7. Gives the final support moments.

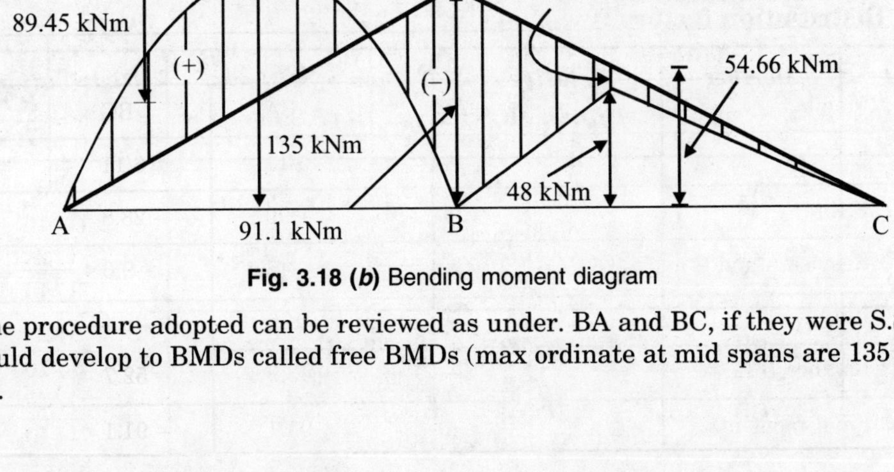

Fig. 3.18 (*b*) Bending moment diagram

The procedure adopted can be reviewed as under. BA and BC, if they were S.S. at both ends, would develop to BMDs called free BMDs (max ordinate at mid spans are 135 kNm and 48 kNm).

Since the beam is continuous at B, BC will exert a moment of 91.1 on BA. BA would exert – 91.1 on BC. The net moment is of course zero (equilibrium).

Since the beam is continuous at B, the rotation of BA and BC would be same. This is ensured in step 3 (Fixed condition ; zero rotation) and in step 6 where total moment is shared in the ratio 5/11 and 6/11 ensuring equal rotation to BA and BC . This ensures continuity.

Example 3.7 *Analyse the continuous beam loaded as shown in Fig. 3.19, by the moment distribution method. Sketch the bending moment and shear force diagrams.*

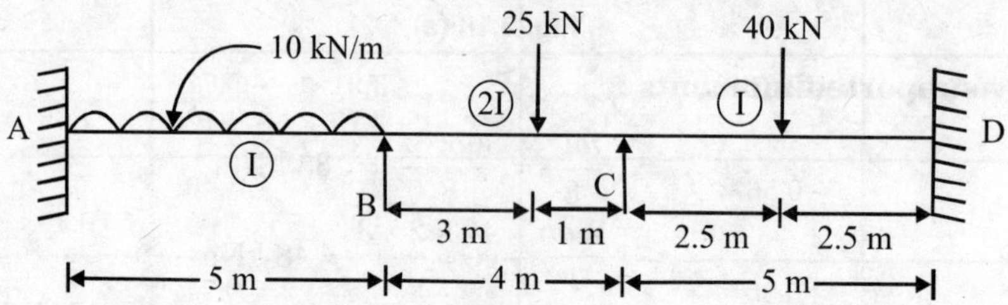

Fig. 3.19

Solution (*a*) **Fixed end moments :**

Span AB : $$M_{FAB} = -\frac{wl^2}{12} = -\frac{10\times 5^2}{12} = -20.83 \text{ kNm}$$

$$M_{FBA} = +\frac{wl^2}{12} = +\frac{10\times 5^2}{12} = +20.83 \text{ kNm}$$

Span BC : $$M_{FBC} = -\frac{Wab^2}{l^2} = -\frac{25\times 3\times (1)^2}{(4)^2} = -4.69 \text{ kNm}$$

$$M_{FCB} = +\frac{Wa^2b}{l^2} = +\frac{25\times (3)^2\times 1}{(4)^2} = +14.06 \text{ kNm}$$

Span CD : $$M_{FCD} = -\frac{Wl}{8} = -\frac{40\times 5}{8} = -25 \text{ kNm}$$

$$M_{FDC} = +\frac{Wl}{8} = +\frac{40\times 5}{8} = +25 \text{ kNm}$$

(*b*) **Distribution factors :**

Joint	*Member*	*Relative stiffness (R.S.)*	*Sum of stiffness (T.S.)*	*Distribution factor* $= \frac{R.S.}{T.S.}$
B	BA	I/5	I/5 + I/2 = 7I/10	$\frac{I/5}{7I/10} = 2/7$
	BC	2I/4 = I/2		$\frac{I/2}{7I/10} = 5/7$
C	CB	2I/4 = I/2	I/2 + I/5 = 7I/10	$\frac{I/2}{7I/10} = 5/7$
	CD	I/5		$\frac{I/5}{7I/10} = 2/7$

(*c*) **Moment distribution :**

Joint	*A*	*B*		*C*		*D*
Members	AB	BA	BC	CB	CD	DC
Distribution factor	—	2/7	5/7	5/7	2/7	—
Fixed end moments	– 20.83	+ 20.83	– 4.69	14.06	– 25	+ 25
Balancing		– 4.61	– 11.53	7.81	3.13	
Carry over	– 2.31		3.91	– 5.77		1.57
Balancing		– 1.12	– 2.79	4.12	1.65	
Carry over	– 0.56		2.06	– 1.40		0.83
Balancing		– 0.59	– 1.47	1.00	0.40	
Carry over	– 0.30		0.50	– 0.74		0.20
Balancing		– 0.14	– 0.36	+ 0.53	0.21	
Carry over Balancing	– 0.07		+ 0.27	– 0.18		0.11
and Carry over	– 0.04	– 0.08	– 0.19	0.13	0.05	0.03
Final moments	– 24.11	14.29	– 14.29	19.56	– 19.56	27.74

(*d*) **Model calculations :**

Joint B. (Balancing). $M_{FBA} + M_{FBC}$ = 20.83 – 4.69 = 16.14 (unbalanced moment). For equilibrium of joint B, this algebraic sum must be zero. Hence for equilibrium a moment of – 16.14 kNm is applied at joint B. This is the balancing moment. This has to be applied to members BA and BC in proportion to their distribution factors.

Moment to be applied at B of member BA = (2/7) × (–16.14) = – 4.61 kNm.

Moment to be applied at B of member BC = (5/7) × (–16.14) = – 11.53 kNm.

Similarly, calculations are done for joint C. This process will keep the joints B and C under equilibrium. (Since they are balanced)

Joint B : (Carry over). The moment applied at B in BA will induce a carry over moment of half of its value and of the same sign at end A.

i.e.,

$$-\frac{4.61}{2} = -2.31 \text{ kNm to A}$$

Similarly from B of BC, there will be a carry over moment of (– 11.53/2) = – 5.77 kNm to joint C in CB. Similar calculations are done at joint C. These moments, carried over from C to B and B to C will again unbalance the joints B and C., *i.e.,* the moment of 3.91 kNm carried over from C to B is balanced by applying – 3.91 kNm and distributing the same to members BA and BC as given under 'balancing.'

This process of balancing and carry over are to be repeated till the balancing moments become very small. In the last cycle, the moments are carried over only to the ends. The final moments are obtained by algebraically adding each column.

(*e*) **To draw S.F.D. :**

Span AB :

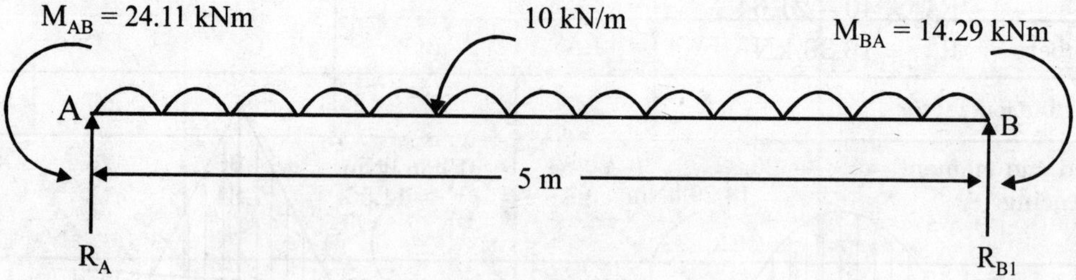

Fig. 3.20

Taking moments about A,

$$M_{BA} + \frac{wl^2}{2} - M_{AB} - R_{B1}(l) = 0$$

$$14.29 + \frac{10 \times 25}{2} - 24.11 - R_{B1}(5) = 0$$

$$R_{B1} = 23.04 \text{ kN}$$

$$R_A = \text{Total load} - R_{B1} = 10 \times 5 - 23.04 = 26.96 \text{ kN}$$

Span BC :

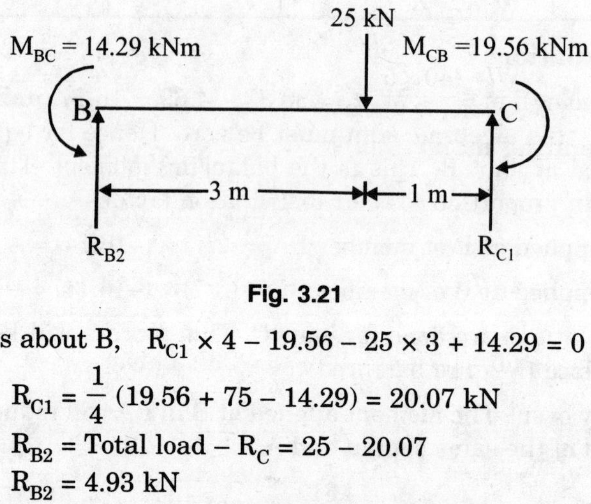

Fig. 3.21

Taking moments about B, $R_{C1} \times 4 - 19.56 - 25 \times 3 + 14.29 = 0$

$$R_{C1} = \frac{1}{4}(19.56 + 75 - 14.29) = 20.07 \text{ kN}$$

$$R_{B2} = \text{Total load} - R_C = 25 - 20.07$$

$$R_{B2} = 4.93 \text{ kN}$$

Span CD :

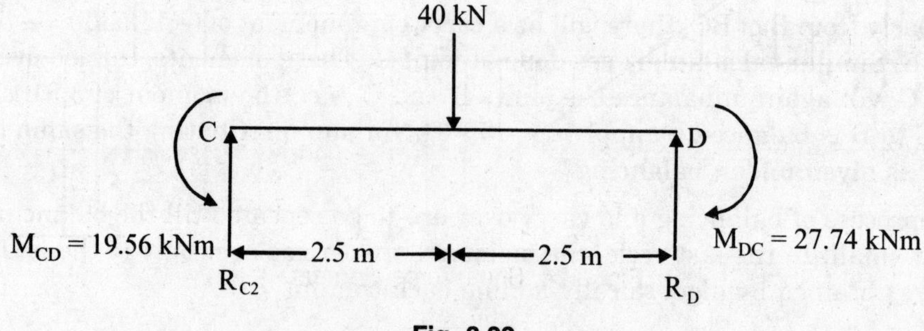

Fig. 3.22

Taking moments about C, $27.74 + 40 (2.5) - 19.56 - R_D (5) = 0$

$R_D = 21.64$ kN

$R_{C2} = 40 - 21.64$

$R_{C2} = 18.36$ kN

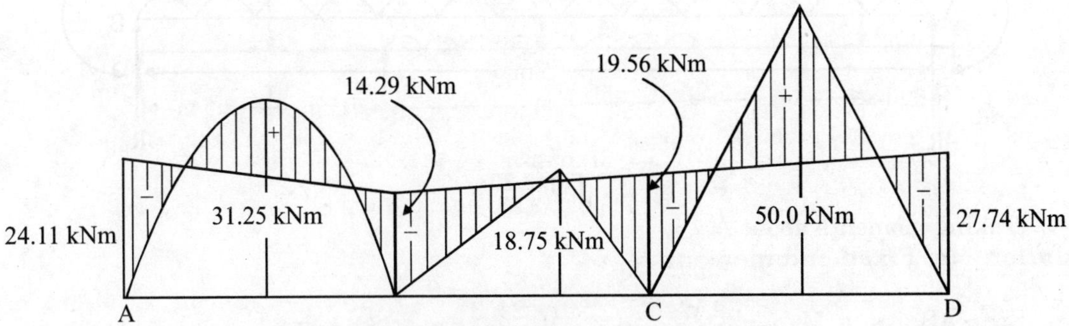

Fig. 3.23. Bending moment diagram.

Simply supported beam moment :

Span AB : $M = \frac{wl^2}{8} = \frac{10 \times (5)^2}{8} = 31.25$ kNm

Span BC : $M = \frac{Wab}{l} = \frac{25 \times 3 \times 1}{4} = 18.75$ kNm

Span CD : $M = \frac{Wl}{4} = \frac{40 \times 5}{4} = 50$ kNm

Net positive bending moments :

Span AB : $M_{+ve} = 31.25 - \left(\frac{24.11 + 14.29}{2}\right) = 12.05$ kNm

Span BC : $M_{+ve} = 18.75 - \left\{14.29 + \frac{(19.56 - 14.29)}{4} \times 3\right\} = 0.51$ kNm

Span CD : $M_{+ve} = 50 - \left(\frac{19.56 + 27.74}{2}\right) = 26.35$ kNm

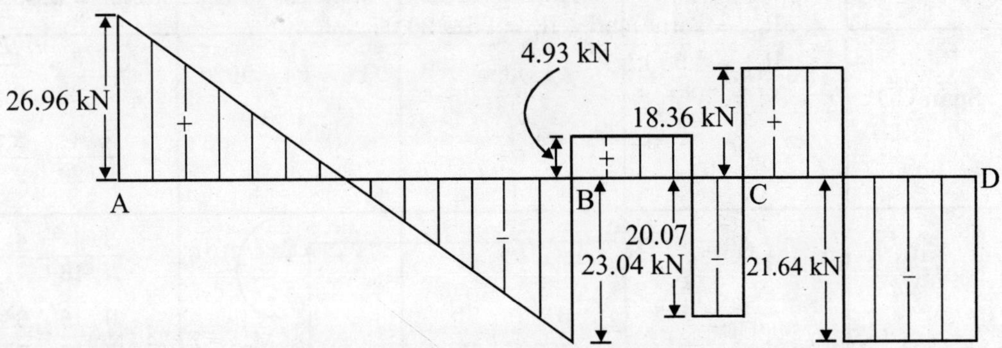

Fig. 3.24. Shear force diagram.

Example 3.8 *Analyse the continuous beam loaded as shown in Fig. 3.25 by the method of moment distribution. Sketch the bending moment and shear force diagrams.*

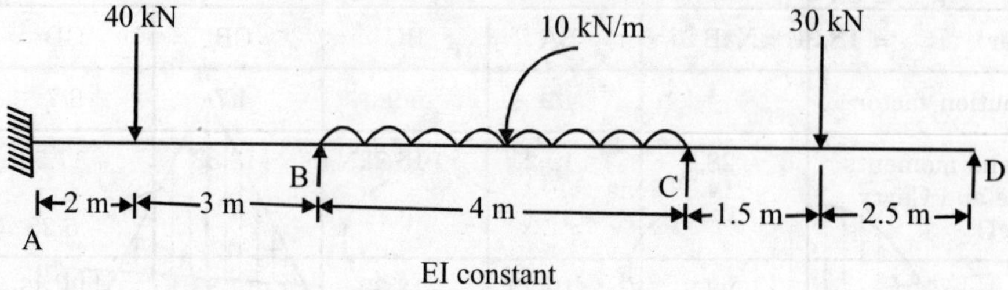

Fig. 3.25

Solution (*a*) **Fixed end moments :**

Span AB :

$$M_{FAB} = -\frac{Wab^2}{l^2} = -\frac{40 \times 2 \times 3^2}{5^2} = -28.8 \text{ kNm}$$

$$M_{FBA} = \frac{Wa^2b}{l^2} = \frac{40 \times 2^2 \times 3}{5^2} = 19.2 \text{ kNm}$$

Span BC :

$$M_{FBC} = -\frac{wl^2}{12} = -\frac{10 \times (4)^2}{12} = -13.33 \text{ kNm}$$

$$M_{FCB} = +\frac{wl^2}{12} = +\frac{10 \times (4)^2}{12} = +13.33 \text{ kNm}$$

Span CD :

$$M_{FCD} = -\frac{Wab^2}{l^2} = -\frac{30 \times 1.5 \times (2.5)^2}{(4)^2} = -17.58 \text{ kNm}$$

$$M_{FDC} = +\frac{Wa^2b}{l^2} = +\frac{30 \times 1.5^2 \times (2.5)}{(4)^2} = +10.55 \text{ kNm}$$

(*b*) **Distribution factor :**

Joints	***Members***	***Relative stiffness (R.S.)***	***Total stiffness (T.S.)***	***Distribution factor = R.S./T.S.***
B	BA	I/5	I/5 + I/4 = 9I/20	$\frac{I/5}{9I/20} = \frac{4}{9}$
	BC	I/4		$\frac{I/4}{9I/20} = \frac{5}{9}$
C	CB	I/4	I/4 +3I/ 16 = 7I/16	$\frac{I/4}{7I/16} = \frac{4}{7}$
	CD	$\frac{3}{4} \times \frac{I}{4} = \frac{3I}{16}$		$\frac{3I/16}{7I/16} = \frac{3}{7}$

(*c*) **Moment distribution :**

Joint	*A*	*B*		*C*		*D*
Member	AB	BA	BC	CB	CD	DC
Distribution factor	–	4/9	5/9	4/7	3/7	–
Fixed end moments	– 28.8	+19.2	– 13.33	+13.33	– 17.58	+ 10.55
Release and Carry over to C					– 5.28	– 10.55
Initial moments	– 28.8	+19.2	– 13.33	+13.33	– 22.86	0
Balancing		– 2.61	– 3.26	5.45	4.08	
Carry over	– 1.31		2.73	– 1.63		
Balancing		– 1.21	– 1.52	0.93	0.70	
Carry over	– 0.61		0.47	– 0.76		
Balancing		– 0.21	– 0.26	0.43	0.33	
Carry over	– 0.11		0.22	– 0.13		
Balancing		– 0.10	– 0.12	0.07	0.06	
Carry over	– 0.05		0.04	– 0.06		
Balancing		– 0.02	– 0.02	0.03	0.03	
Final moments	– 30.88	15.05	– 15.05	17.66	– 17.66	0

(*d*) **To draw S.F.D. :**

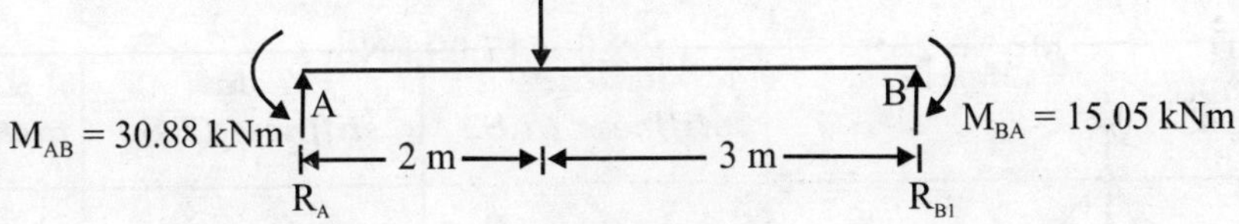

Fig. 3.26

Span AB :

Taking moments about A, 150.5 + 40(2) – 30.88 – R_{B1} (5) = 0

$$R_{B1} = 12.83 \text{ kN},\quad R_A = 40 - R_{B1},\quad R_A = 27.17 \text{ kN}$$

Span BC :

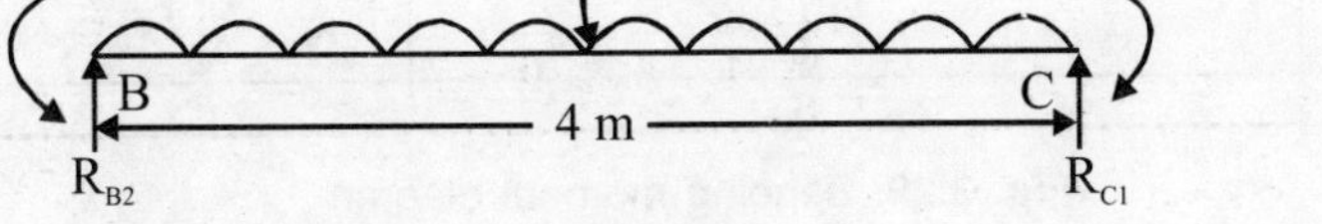

Fig. 3.27

Taking moments about B,

$$17.66 + \frac{10 \times (4)^2}{2} - 15.05 - R_{C1}(4) = 0$$

$$R_{C1} = 20.65 \text{ kN}, \quad R_{B2} = 10 \times 4 - 20.65, \quad R_{B2} = 19.35 \text{ kN}$$

Span CD :

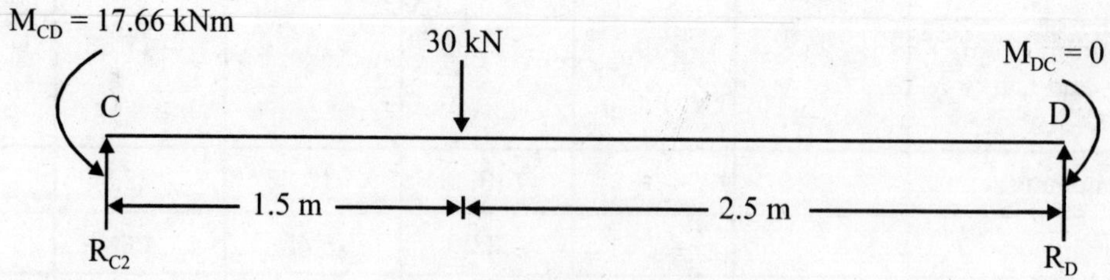

Fig. 3.28

Taking moments about C, $0 + 30(1.5) - 17.66 - R_D(4) = 0$

$$R_D = 6.84 \text{ kN}, \quad R_{C2} = 30 - 6.84, \quad R_{C2} = 23.16 \text{ kN}$$

Simply supported bending moments :

Span AB : $$M = \frac{Wab}{l} = \frac{40 \times 2 \times 3}{5} = 48 \text{ kNm}$$

Span BC : $$M = \frac{wl^2}{8} = \frac{10 \times (4)^2}{8} = 20 \text{ kNm}$$

Span CD : $$M = \frac{Wab}{l} = \frac{30 \times 1.5 \times 2.5}{4} = 28.13 \text{ kNm}$$

Net positive bending moments :

Span AB : $$M_{+ve} = 48 - \left\{15.05 + \frac{(30.88 - 15.05)}{5} \times 3\right\} = 23.46 \text{ kNm}$$

Span BC : $$M_{+ve} = 20 - \left\{15.05 + \frac{(17.66 - 15.05)}{2}\right\} = 3.63 \text{ kNm}$$

Span CD : $$M_{+ve} = 28.13 - \frac{17.66}{4} \times 2.5 = 17.09 \text{ kNm}$$

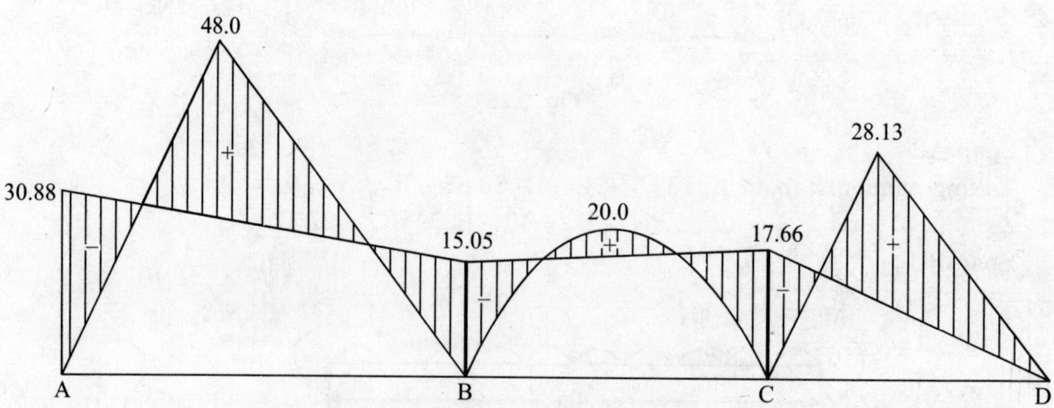

Fig. 3.29. Bending moment diagram.

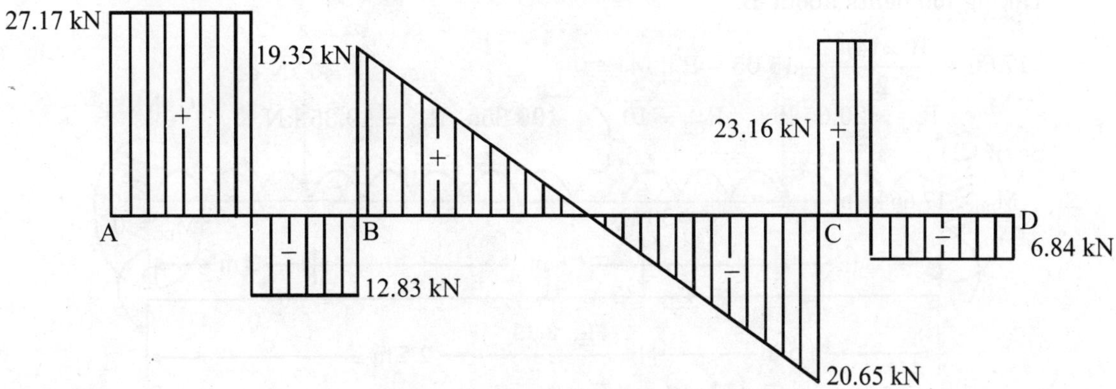

Fig. 3.30. Shear force diagram.

Example 3.9 *Analyse the continuous beam loaded as shown in Fig. 3.31 by the moment distribution method. Sketch the bending moment diagram and shear force diagram.*

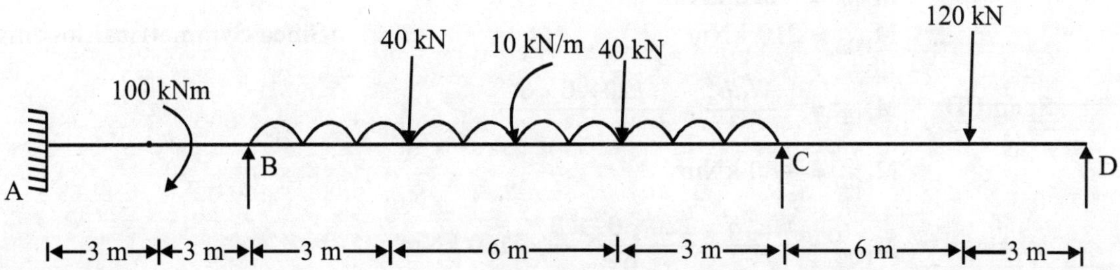

Fig. 3.31

Solution (*a*) **Fixed end moments :**

Span AB :

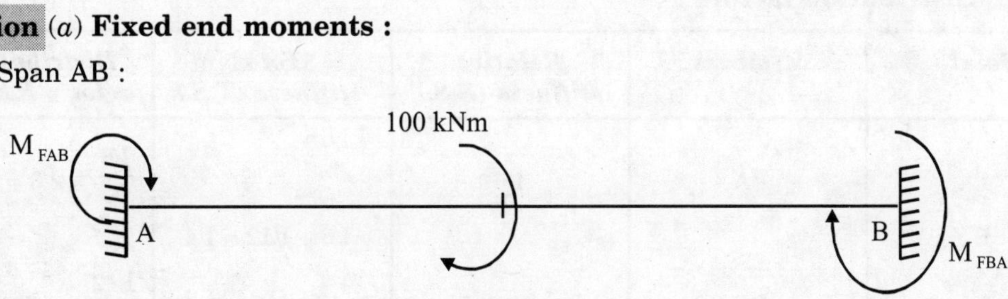

Fig. 3.32

$$M_{FAB} = +\frac{M_0}{4} \text{ where } M_0 = +100 \text{ kNm}$$

$$M_{FAB} = +\frac{100}{4}$$

$$M_{FAB} = +25 \text{ kNm}$$

$$M_{FBA} = +\frac{100}{4} = +25 \text{ kNm}$$

Span BC :

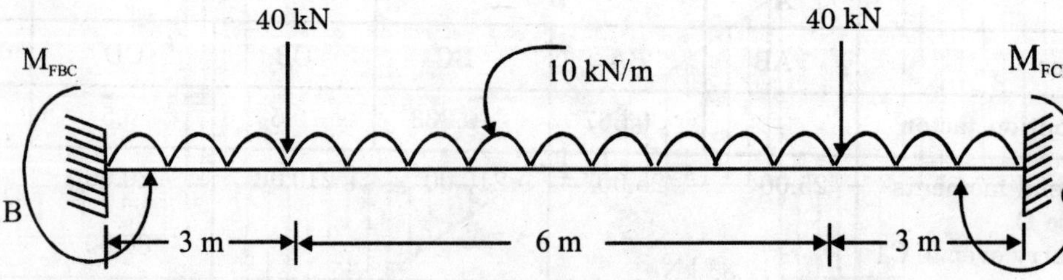

Fig. 3.33

$$M_{FBC} = -\frac{wl^2}{12} - \frac{W_1 a_1 b_1^{\,2}}{l^2} - \frac{W_2 a_2 b_2^{\,2}}{l^2}$$

$$= -\frac{10 \times 12^2}{12} - \frac{40 \times 3 \times 9^2}{12^2} - \frac{40 \times 9 \times 3^2}{12^2} = -120 - 67.5 - 22.5$$

$M_{FBC} = -210$ kNm

$M_{FCB} = 210$ kNm (Since Symmetrical loading)

Span CD : $M_{FCD} = -\frac{Wab^2}{l^2} = -\frac{120 \times 6 \times 3^2}{9^2}$

$M_{FCD} = -80$ kNm

$$M_{FDC} = \frac{Wa^2b}{l^2} = \frac{120 \times 6^2 \times 3}{9^2} = 160 \text{ kNm}$$

(*b*) **Distribution factors :**

Joint	***Members***	***Relative stiffness (R.S.)***	***Total stiffness (T.S.)***	***Distribution factor = R.S./T.S.***
B	BA	I/6	I/6 + I/12 = I/4	$\frac{I/6}{I/4}$ = 2/3 = 0.667
	BC	I/12		$\frac{I/12}{I/4}$ = I/3 = 0.333
C	CB	I/12	I/12 + I/12 = I/6	$\frac{I/12}{I/6}$ = 1/2 = 0.5
	CD	3/4 × I/9 = I/12		$\frac{I/12}{I/6}$ = 1/2 = 0.5

(*c*) **Moment distribution :**

Joint	A	B		C		D
Member	AB	BA	BC	CB	CD	DC
Distribution factor	–	0.667	0.333	0.5	0.5	–
Fixed end moments Release D and Carry over to C	25.00	25.00	– 210.00	+ 210.00	– 80.00 – 80.00	160.00 – 160.00
Initial moments Balance	25.00	25.00 123.395	– 210.00 61.605	210.00 – 25.00	– 160.00 – 25.00	0
Carry over Balance	61.698	 8.338	– 12.50 4.163	30.803 – 15.402	 – 15.402	 –
Carry over Balance	4.169	 5.137	– 7.701 2.564	2.082 – 1.041	 – 1.041	 –
Carry over Balance	2.569	 0.348	– 0.521 0.173	1.282 – 0.6410	 – 0.641	 –
Carry over Balance	0.174	 0.214	– 0.321 0.107	0.086 – 0.044	 – 0.044	
Carry over Balance and C.O. to ends	0.107 0.008	 0.015	– 0.022 0.007	0.054 – 0.027	 – 0.027	 –
Final moments	93.725	162.447	– 162.446	202.153	– 202.155	0

(*d*) **To draw S.F.D. :**

Span AB :

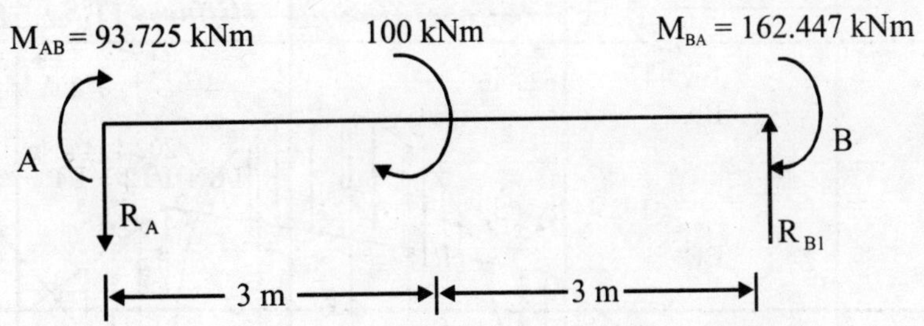

Taking moments about A

$$R_{B1} \times 6 - 100 - 162.447 - 93.725 = 0$$

$$R_{B1} = 59.362 \text{ kN } (\uparrow)$$

$$R_A = -R_{B1} = -59.362 \text{ kN}$$

i.e., $$R_A = 59.362 \text{ kN } (\downarrow)$$

Span BC :

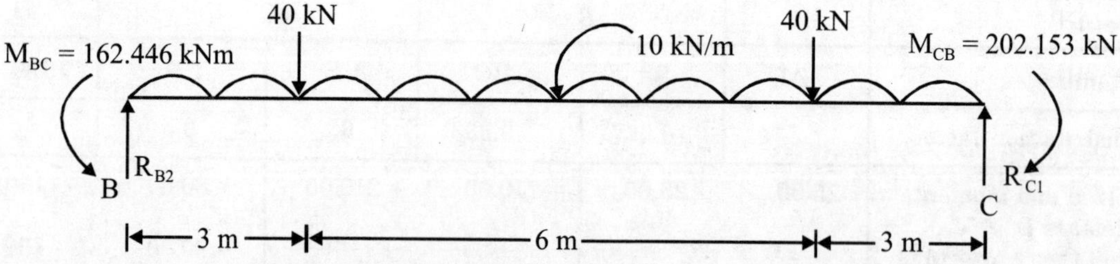

Taking moments about B,

$$R_{C1} \times 12 - 202.153 - 40 \times 9 - 40 \times 3 - 10 \times 12 \times 6 + 162.446 = 0$$

$$R_{C1} = 103.309 \text{ kN}$$

$$R_{B2} = 40 + 40 + 12 \times 10 - 103.309 = 96.691 \text{ kN}$$

Span CD :

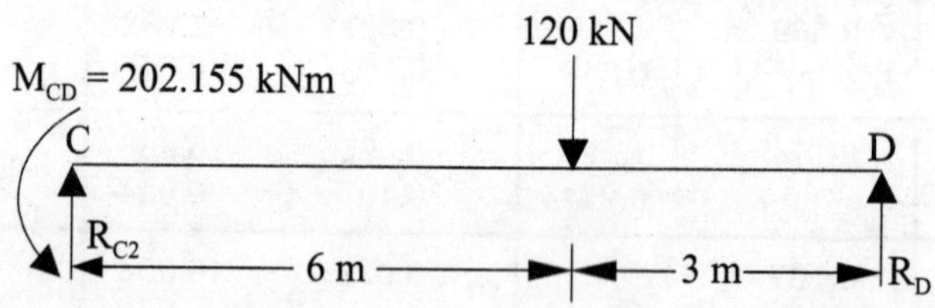

Taking moments about C, $R_D \times 9 - 120 \times 6 + 202.155 = 0$

$$R_D = 57.538 \text{ kN}$$

$$R_{C2} = 120 - 57.538 = 62.462 \text{ kN}$$

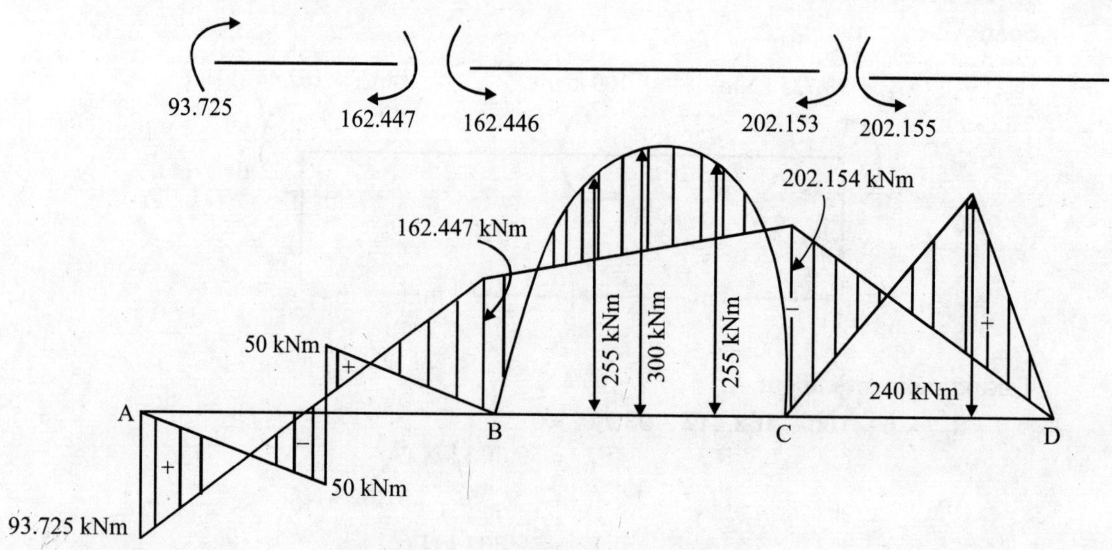

Fig. 3.34. Bending moment diagram.

Simply supported bending moments :

Span AB = 100 kNm

Span CD, $\frac{Wab}{l} = \frac{120 \times 6 \times 3}{9} = 240$ kNm

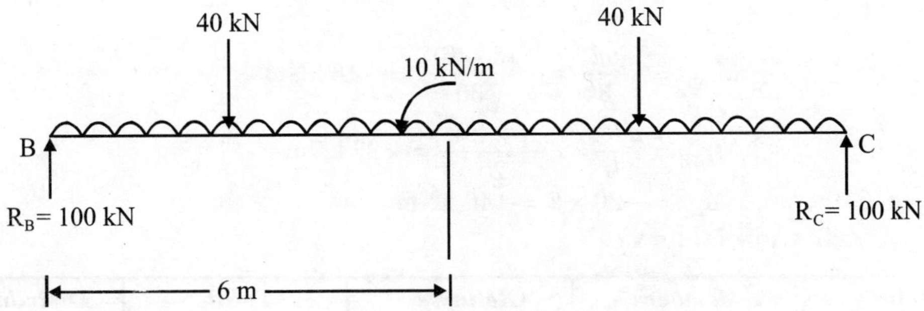

Fig. 3.35

Bending moment at midspan = + 100(6) – 40(3) – 10 × 6 × 3 = 300 kNm

Bending moment at quarter span = 100 (3) – 10 × 3 × 1.5 = 255 kNm.

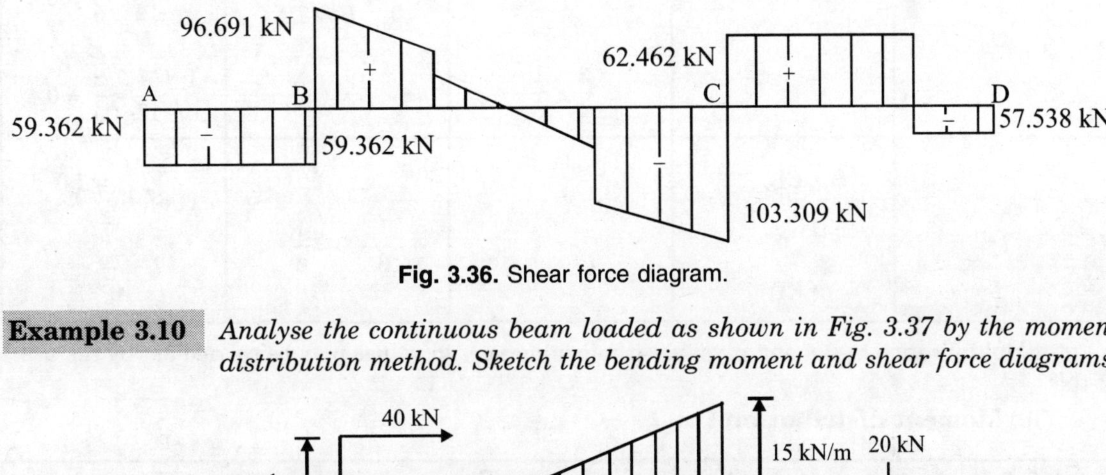

Fig. 3.36. Shear force diagram.

Example 3.10 *Analyse the continuous beam loaded as shown in Fig. 3.37 by the moment distribution method. Sketch the bending moment and shear force diagrams.*

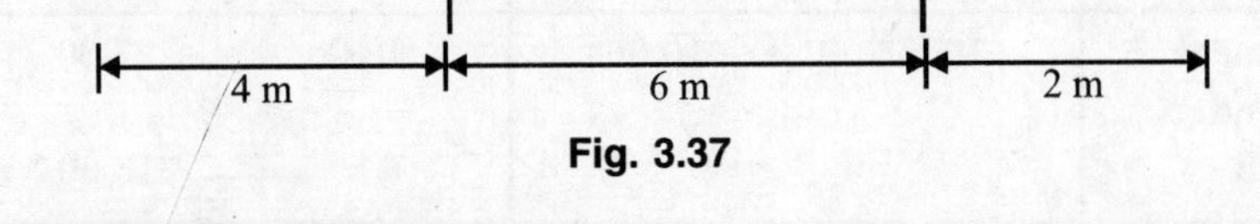

Fig. 3.37

Solution (*a*) **Fixed end moments :**

Span AB : $M_{FAB} = M_{FBA} = \frac{M_0}{4}$ (For centrally applied moment)

$M_0 = 40 \times 1 = 40$ kNm

$$M_{FAB} = \frac{40}{4} = 10 \text{ kNm}$$

$$M_{FBA} = \frac{40}{4} = 10 \text{ kNm}$$

Span BC :

$$M_{FBC} = -\frac{wl^2}{30} = -\frac{15 \times (6)^2}{30} = -18 \text{ kNm}$$

$$M_{FCB} = \frac{wl^2}{20} = +\frac{15 \times (6)^2}{20} = +27 \text{ kNm}$$

Span CD : $M_{CD} = -20 \times 2 = -40$ kNm

(*b*) **Distribution factors :**

Joint	***Members***	***Relative stiffness (R.S.)***	***Total stiffness (T.S.)***	***Distribution factor =*** $\frac{R.S.}{T.S.}$
B	BA	$\frac{I}{4}$	$\frac{I}{4} + \frac{I}{6} = \frac{5I}{12}$	$\frac{I}{4} \Big/ \frac{5I}{12} = \frac{3}{5} = 0.6$
	BC	$^*\frac{3}{4} \times \frac{I}{6}$		$\frac{I}{6} \Big/ \frac{5I}{12} = \frac{2}{5} = 0.4$
C	CB	$\frac{I}{6}$	$\frac{I}{6} + 0 = \frac{I}{6}$	$\frac{I}{6} \Big/ \frac{I}{6} = 1.0$
	CD	–		0

*Multiplication by 3/4 and consequent minor changes in values may be carried out by the users of this.

(*c*) **Moment distribution :**

Joint	***A***	***B***		***C***	
Members	AB	BA	BC	CB	CD
Distribution factors	–	0.6	0.4	1.0	–
Fixed end moment	10.00	10.00	– 18.00	27.00	– 40.00
Balancing C and Carry over to B			6.5 ⟵	13.00	
Initial moments	10.0	10.00	– 11.5	40.00	– 40.00
Balance B and Carry over to A	0.45 ⟵	0.90	0.60		
Final moments	10.45	10.90	– 10.90	40.00	– 40

(*d*) **To draw S.F.D. :**

Span AB :

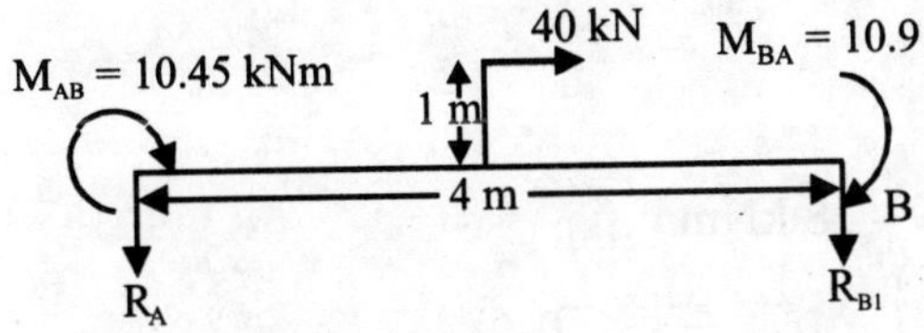

Span BC :

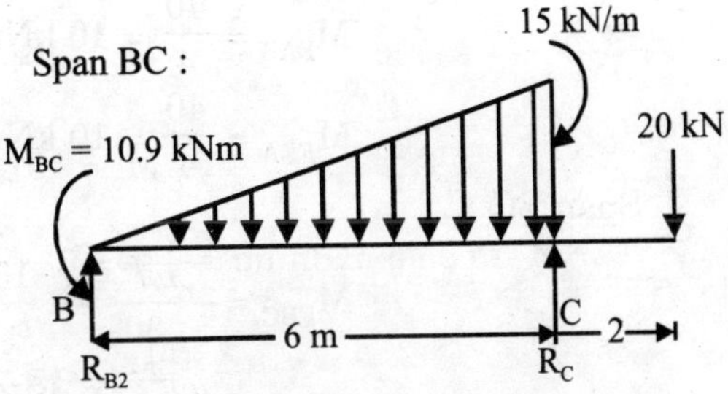

Taking moments about A,

$$R_B \times 4 + 10.9 + 40 + 10.45 = 0$$

$$R_{B1} = -15.34 \text{ kN}$$
$$R_A = \text{Total load} - R_B = 0 - R_{B1}$$
$$R_A = -15.34 \text{ kN}(\downarrow)$$

Taking moments about B,

$$R_C \times 6 - 20 \times 8 - \frac{1}{2} \times 6 \times 15 \times \frac{2}{3} \times 6 + 10.9 = 0$$

$$R_{C1} = 54.85 \text{ kN}$$
$$R_{B2} = 10.15 \text{ kN}$$

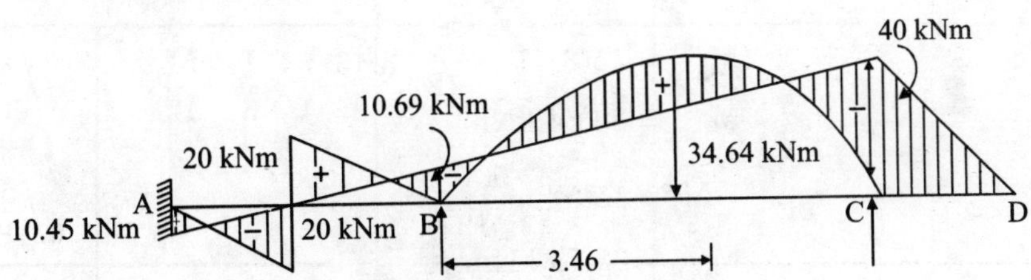

Fig. 3.38. Bending moment diagram.

Simply supported bending moments :

Span BC :

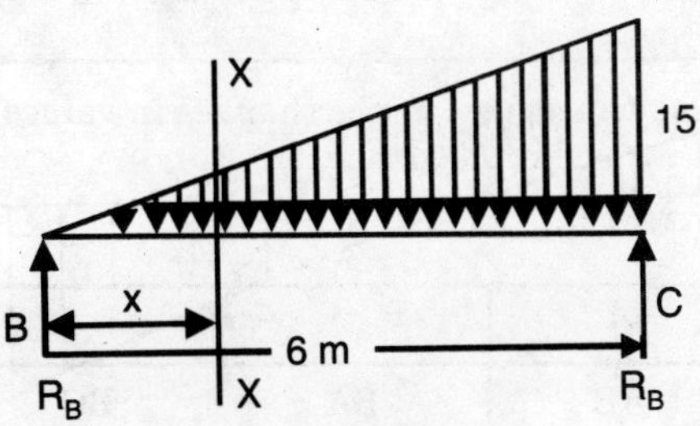

Fig. 3.39

at XX, $\dfrac{\text{Load ordinate}}{x} = \dfrac{15}{6}$

Load ordinate = $2.5x$

Taking moments about B, $R_C \times 6 - \dfrac{1}{2} \times 6 \times 15 \times \dfrac{2}{3} \times 6 = 0$

$$R_C = 30 \text{ kN}$$

$$R_B = 15 \times \frac{6}{2} - 30$$

$$R_B = 15 \text{ kN}$$

$$M_x = 15x - \frac{1}{2} \times x \times (2.5x) \times \frac{x}{3} = 15x - \frac{5}{4} \times \frac{x^3}{3}$$

$$M_x = 15x - \frac{5x^3}{12}$$

For M_x to be maximum $\frac{dM_x}{dx} = 0$

$$\frac{dM_x}{dx} = 15 - \frac{5}{4}\left(\frac{3x^2}{3}\right) = 15 - \frac{5}{4}x^2 = 0$$

$$x^2 = 15 \times \frac{4}{5}$$

$$x^2 = 12$$

$$x = 3.46 \text{ m from B}$$

$$M_x = 15x - \frac{5x^3}{12}$$

$$M_{max} = 15(3.46) - \frac{5}{12}(3.46)^3$$

$$M_{max} = 34.64 \text{ kNm}$$

$$M_x = 15(3) - \frac{5(3)^3}{12}$$

$$M_{mid} = 33.75 \text{ kNm.}$$

or
$$W = \frac{1}{2} \times 6 \times 15 = 45$$

From formula $M_{max} = 0.128\ Wl = 0.128(45) \times 6$

$= 34.56$ kNm. Occurs at $0.577(l) = 3.462$ m from B

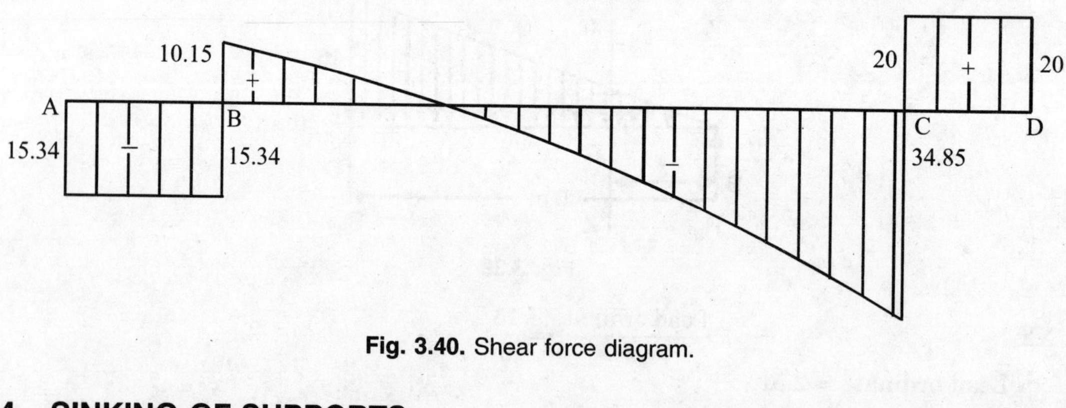

Fig. 3.40. Shear force diagram.

3.4 SINKING OF SUPPORTS

In practical structures such as in Rigid frame bridges, Continuous beam bridges, building frames and frames for industrial structures, 'loading' can be changed even by sinking of supports or rotation of 'fixed' ends. Supports sink mostly due to soil settlement. Rotation of 'fixed' ends can happen either because of soil settlement or upheaval of horizontal or inclined fixed ends. Fixed end moments induced in beam ends because of settlement or rotation of supports are tabulated in appendix A1 (items 10, 11, 12 and 13). These moments will get superposed on moments due to conventional loads. Examples 3.11 and 3.12 deal with cases of continuous beams due to sinking of supports.

Example 3.11 *A continuous beam ABCD, 20 m long is simply supported at its ends and is propped at the same level at B and C as shown. It is loaded as shown in Fig. 3.41.*

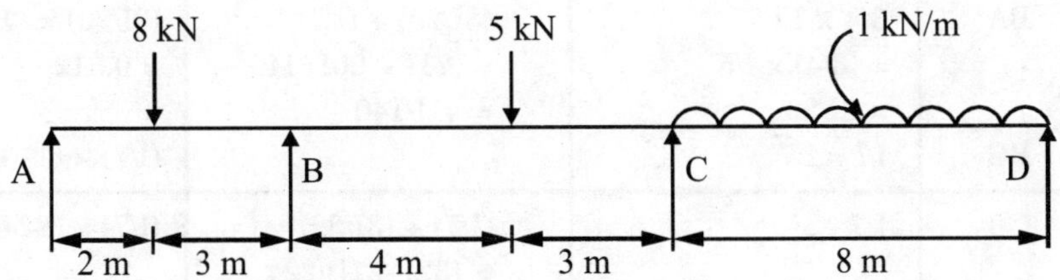

Fig. 3.41

If support B sinks by 10 mm, analyse the beam by moment distribution method and sketch the bending moment diagram. Take E = 2.1 × 10⁵ N/mm² and I = 85 × 10⁵ mm⁴.

Solution **Fixed end moments :**

$$M_{FAB} = -\frac{Wab^2}{l^2} - \frac{6EI\delta}{l^2}$$

$$= -\frac{8 \times 2 \times 3^2}{5^2} - \frac{6 \times 2.1 \times 10^8 \times 85 \times 10^{-7}}{5^2} \times \frac{10}{1000}$$

$$= -5.76 - 4.284 = -10.044 \text{ kNm}$$

$$\left[E = \frac{2.1 \times 10^5}{1000} \times 1000^2 = 2.1 \times 10^8 \text{ kN/m}^2 \right.$$

$$I = 85 \times 10^5 \text{ mm}^4 = \frac{85 \times 10^5}{(1000)^4} = 85 \times 10^{-7} \text{ m}^4$$

$$\left. \delta = \frac{10}{1000} \text{ m} \right]$$

$$M_{FBA} = +\frac{Wa^2b}{l^2} - \frac{6EI\delta}{l^2} = \frac{8 \times 2^2 \times 3}{5^2} - 4.284 = -0.444 \text{ kNm}$$

$$M_{FBC} = -\frac{Wab^2}{l^2} + \frac{6EI\delta}{l^2} = -\frac{5 \times 4 \times 3^2}{7^2} + \frac{6 \times 2.1 \times 10^8 \times 85 \times 10^{-7}}{7^2 \times 1000} \times 10$$

[$\because$ In BC, left support sinks, $\delta = -10$ mm]

$$= -3.673 + 2.186 = -1.487 \text{ kNm}$$

$$M_{FCB} = +\frac{Wa^2b}{l^2} + \frac{6EI\delta}{l^2} = +\frac{5 \times 4^2 \times 3}{7^2} + \frac{6 \times 2.1 \times 10^8 \times 85 \times 10^{-7} \times 10}{7^2 \times 1000}$$

$$= 4.898 + 2.186 = +7.084 \text{ kNm}$$

$$M_{FCD} = -\frac{wl^2}{12} = -\frac{1 \times 8^2}{12} = -5.33 \text{ kNm}$$

$$M_{FDC} = +\frac{wl^2}{12} = +\frac{1\times 8^2}{12} = +5.33 \text{ kNm}$$

Distribution factor :

Joint	*Member*	*Relative stiffness*	*Sum*	*Distribution factor*
B	BA	3/4 × I/*l* = (3/4) × I/5 = 3I/20	(3I/20) + (I/7) = (21I + 20I)/140 = 41I/140	(3I/20) × (140/41I) = 0.512
	BC	I/*l* = I/7		(I/7) × (140/41I) = 0.488
C	CB	I/7	(I/7) + (3I/32) = (32I+21I)/224 = 53I/224	(I/7) × (224/53I) = 0.604
	CD	(3/4) × (I/8) = (3I/32)		(3I/32) × (224/53I) = 0.396

Moment distribution :

Joint	*A*	*B*		*C*		*D*
Member	AB	BA	BC	CB	CD	DC
Distribution factor	–	0.512	0.488	0.604	0.396	–
FEM	– 10.044	– 0.444	– 1.487	7.084	– 5.33	+ 5.33
Release A, D and Carry over	+ 10.044	5.022			– 2.665	– 5.33
Initial moments	0	+ 4.578	– 1.487	+ 7.084	– 7.995	0
Balance	–	– 1.583	– 1.508	+ 0.550	+ 0.361	–
Carry over	–		0.275	– 0.754	–	–
Balance	–	– 0.141	– 0.134	0.455	0.299	–
Carry over	–	–	0.228	– 0.067	–	–
Balance	–	– 0.117	– 0.111	0.040	0.027	–
Final moments	0	+ 2.737	– 2.737	+ 7.308	– 7.308	0

Simply supported bending moments :

$$M_{AB} = \frac{Wab}{l} = \frac{8\times 2\times 3}{5} = 9.6 \text{ kNm}$$

$$M_{BC} = \frac{Wab}{l} = \frac{5\times 4\times 3}{7} = 8.6 \text{ kNm}$$

$$M_{CD} = \frac{wl^2}{8} = \frac{1\times 8^2}{8} = 8 \text{ kNm}$$

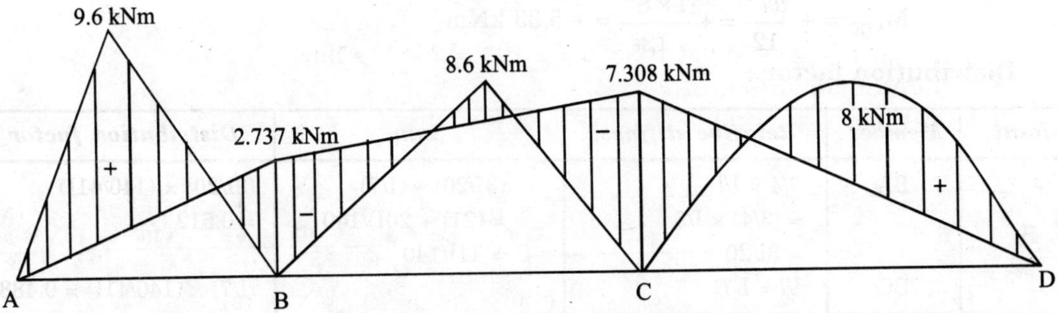

Fig. 3.42. Bending moment diagram.

S.F. Calculations :

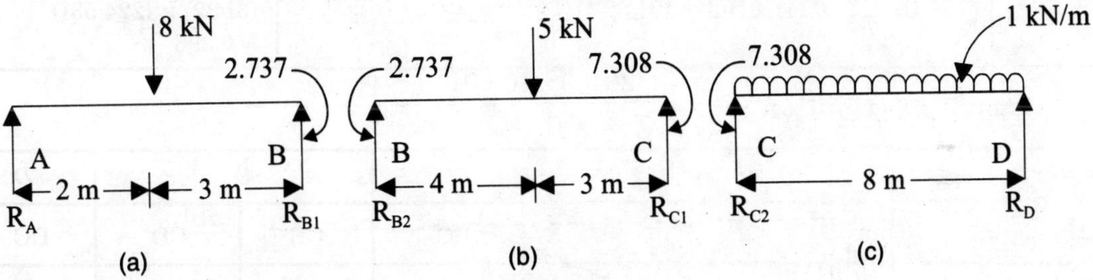

Fig. 3.43

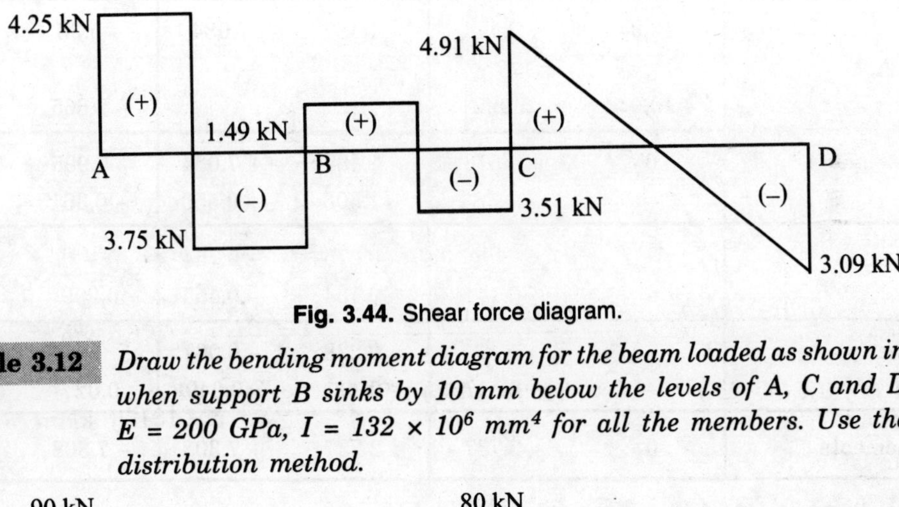

Fig. 3.44. Shear force diagram.

Example 3.12 *Draw the bending moment diagram for the beam loaded as shown in Fig. 3.45 when support B sinks by 10 mm below the levels of A, C and D. Assume E = 200 GPa, I = 132 × 10⁶ mm⁴ for all the members. Use the moment distribution method.*

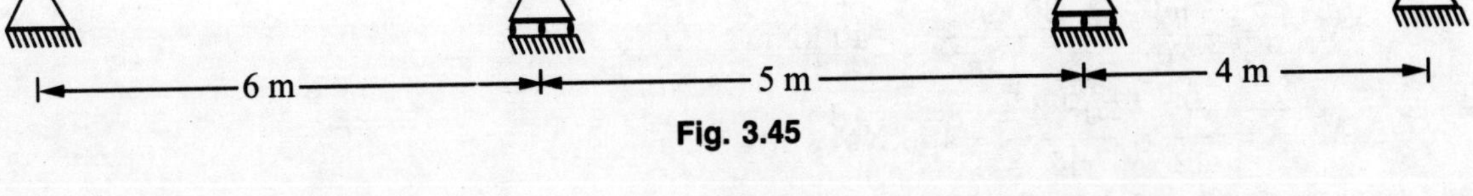

Fig. 3.45

Solution

$$\delta_B = 10 \text{ mm}$$

$$EI = 200 \times 10^6 \times 132 \times 10^{-6} = 26400 \text{ kNm}^2$$

Fixed end moments :

Span AB :

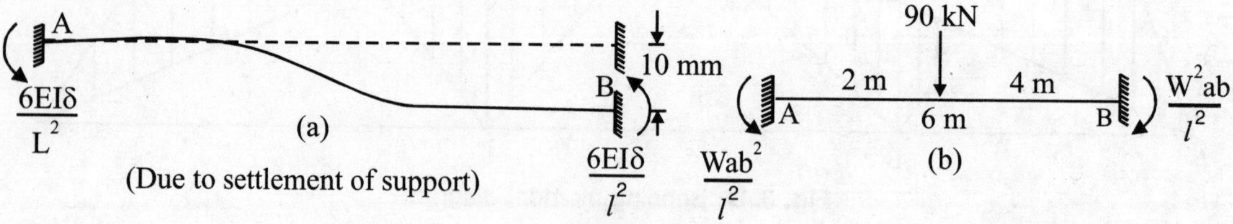

Fig. 3.46

$$M_{FAB} = -\frac{6EI\delta}{l^2} - \frac{Wab^2}{l^2} = -\frac{6 \times 26400 \times 10 \times 10^{-3}}{6^2} - \frac{90 \times 2 \times 4^2}{6^2} = -124 \text{ kNm}$$

$$M_{FBA} = -\frac{6EI\delta}{l^2} + \frac{Wa^2b}{l^2} = -\frac{6 \times 26400 \times 10 \times 10^{-3}}{6^2} + \frac{90 \times 2^2 \times 4}{6^2} = -4 \text{ kNm}$$

Span BC :

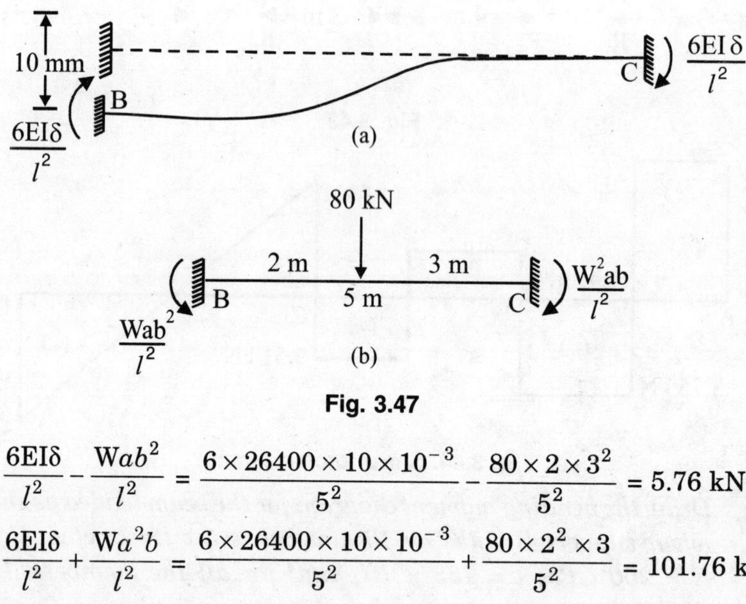

Fig. 3.47

$$M_{FBC} = \frac{6EI\delta}{l^2} - \frac{Wab^2}{l^2} = \frac{6 \times 26400 \times 10 \times 10^{-3}}{5^2} - \frac{80 \times 2 \times 3^2}{5^2} = 5.76 \text{ kNm}$$

$$M_{FCB} = \frac{6EI\delta}{l^2} + \frac{Wa^2b}{l^2} = \frac{6 \times 26400 \times 10 \times 10^{-3}}{5^2} + \frac{80 \times 2^2 \times 3}{5^2} = 101.76 \text{ kNm}$$

Span CD :

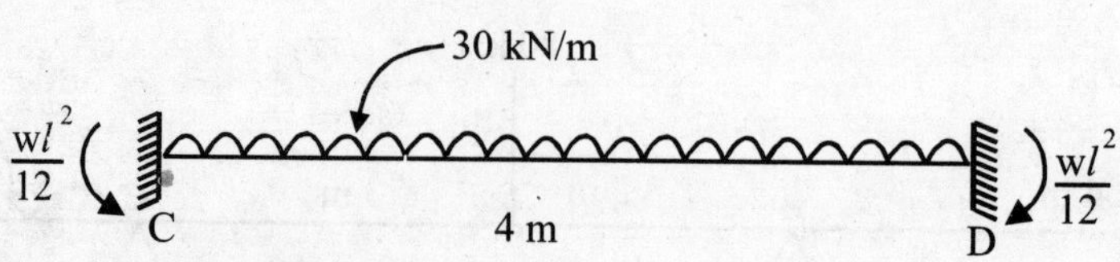

$$M_{FCD} = -\frac{wl^2}{12} = -\frac{30 \times 4^2}{12} = -40 \text{ kNm}$$

$$M_{FDC} = \frac{wl^2}{12} = \frac{30 \times 4^2}{12} = 40 \text{ kNm}$$

Distribution factors :

Joint	*Member*	*Relative stiffness*	*Sum of stiffness*	*Distribution factor*
B	BA	$\frac{3}{4} \times \frac{I}{l} = \frac{3}{4} \times \frac{I}{6} = \frac{I}{8}$	$\frac{I}{8} + \frac{I}{5} = \frac{5I + 8I}{40} = \frac{13I}{40}$	$\frac{I/8}{13I/40} = \frac{I}{8} \times \frac{40}{13I} = 0.385$
	BC	$\frac{I}{l} = \frac{I}{5} = \frac{I}{5}$		$\frac{I/5}{13I/40} = \frac{I}{5} \times \frac{40}{13I} = 0.615$
C	CB	$\frac{I}{l} = \frac{I}{5} = \frac{I}{5}$	$\frac{I}{5} + \frac{3I}{16} = \frac{16I + 15I}{80} = \frac{31I}{80}$	$\frac{I/5}{13I/80} = \frac{I}{5} \times \frac{80}{31I} = 0.516$
	CD	$\frac{3}{4} \times \frac{I}{l} = \frac{3}{4} \times \frac{I}{4} = \frac{3I}{16}$		$\frac{3I/16}{31I/80} = \frac{3I}{5} \times \frac{80}{31I} = 0.484$

Moment distribution :

Joint	*A*	*B*		*C*		*D*
Member	AB	BA	BC	CB	CD	DC
Distribution factor		0.385	0.615	0.516	0.484	
FEM	– 124	– 4	5.76	101.76	– 40	40
Balance 'A' 'D'	124					– 40
Carry over		62			– 20	
Initial moments	0	58	5.76	101.761	– 60	0
Balance 'B' 'C'		– 24.548	– 39.212	– 21.548	– 20.212	
Carry over		– 10.774	– 19.606			
Balance 'B' 'C'		4.148	6.626	10.117	9.489	
Carry over			5.059	3.313		
Balance 'B' 'C'		– 1.948	– 3.111	– 1.709	– 1.604	
Carry over			– 0.855	– 1.556		
Balance 'B' 'C'		0.329	0.526	0.803	0.753	
Carry over			0.402	0.263		
Balance 'B' 'C'		– 0.155	– 0.247	– 0.136	– 0.127	
Carry over			– 0.068	– 0.124		
Balance 'B' 'C'		0.026	0.042	0.064	0.06	
Carry over			0.032	0.021		
Balance 'B' 'C'		– 0.012	– 0.019	– 0.011	– 0.01	
Carry over			– 0.006	– 0.009		
Balance 'B' 'C'		0.002	0.004	0.005	0.004	
Carry over			0.003	0.002		
Balance 'B' 'C'		– 0.001	– 0.002	– 0.001	– 0.001	
Net moment	0	35.841	– 35.84	71.648	– 71.648	0

$M_{AB} = 0$
$M_{BA} = 35.841$ kNm
$M_{BC} = -35.841$ kNm
$M_{CB} = 71.648$ kNm
$M_{CD} = -71.648$ kNm
$M_{DC} = 0$

Simply supported bending moments :

$$M_{AB} = \frac{Wab}{l} = \frac{90 \times 2 \times 4}{6} = 120 \text{ kNm}$$

$$M_{BC} = \frac{Wab}{l} = \frac{80 \times 2 \times 3}{5} = 96 \text{ kNm}$$

$$M_{CD} = \frac{wl^2}{8} = \frac{30 \times 4^2}{8} = 60 \text{ kNm}$$

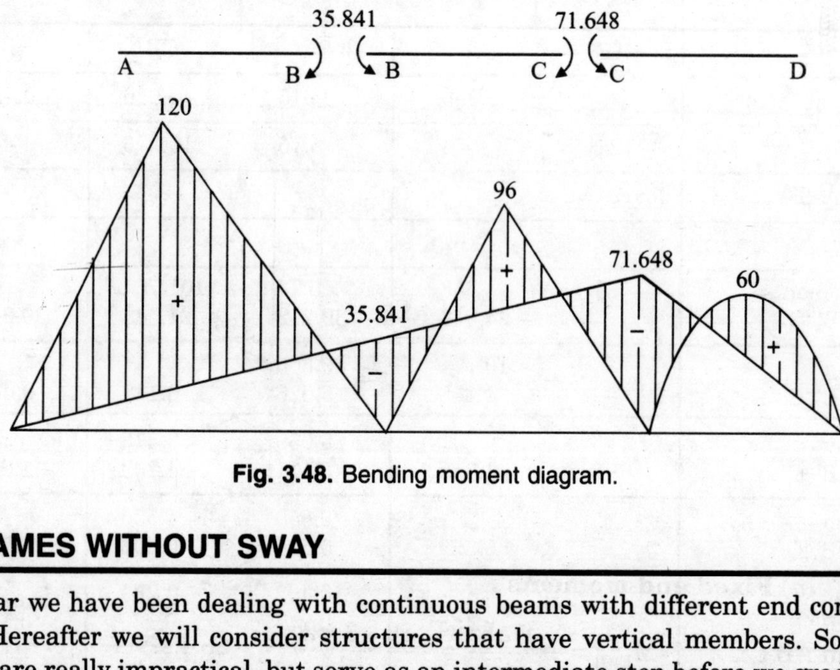

Fig. 3.48. Bending moment diagram.

3.5 FRAMES WITHOUT SWAY

So far we have been dealing with continuous beams with different end conditions and loadings. Hereafter we will consider structures that have vertical members. Some of these structures are really impractical, but serve as an intermediate step before we graduate to real frames which are capable of "swaying". 'Swaying' is the lateral movement of column members in a structural frame. Frames in examples 3.13 to 3.19 do not sway because the frame is prevented from horizontal movement at the top level of column members.

Frames in examples 3.22 to 3.24 are symmetrical about their vertical centre lines and are symmetrically loaded. Hence there is no reason for them to sway to the left or to the right. All these problems are solvable in much the same way as the beams we solved hitherto. The additional feature is that in frames we can have joints with more than two members. Accordingly the relative stiffnesses and distribution factors have to be apportioned.

3.6 BENDING MOMENT DIAGRAMS OF STRUCTURAL FRAMES

Usually, while drawing BMD for continuous beams we draw the free bending moments (positive BMs) **below** the beam centre line and support moments (negative BMs) **above** the beam. This procedure has the advantage of showing the BMD on the tension side of beam.

In vertical or inclined members in frame it is difficult to define positive or negative bending moments ; nor hogging and sagging moments. This difficulty can be overcome by adopting the bending moment diagram on the **tension** side of the member, be it beam or column. This procedure is particularly meaningful, when we analyse reinforced concrete frames. Herein the BMD readily indicates which side of the member needs tension reinforcement.

Example 3.13 *Analyse the structure loaded as shown in Fig. 3.49 by the moment distribution method and sketch the bending moment and shear force diagrams.*

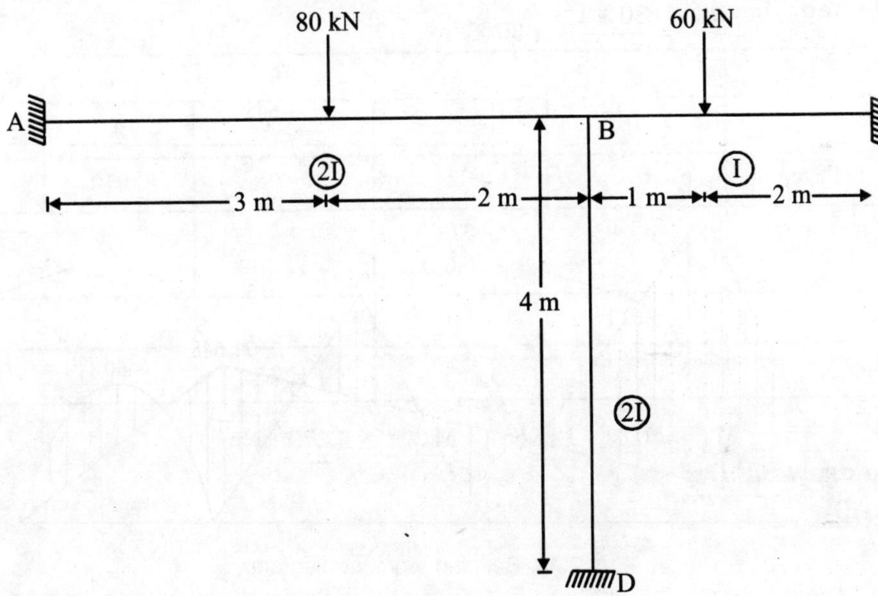

Fig. 3.49

Solution (*a*) **Fixed end moments :**

Span AB : $M_{FAB} = -\dfrac{Wab^2}{l^2} = -\dfrac{80 \times 3 \times 2^2}{5^2} = -38.4$ kNm

$M_{FBA} = \dfrac{Wa^2b}{l^2} = +\dfrac{80 \times 3^2 \times 2}{5^2} = +57.6$ kNm

Span BC : $M_{FBC} = -\dfrac{Wab^2}{l^2} = -\dfrac{60 \times 1 \times 2^2}{3^2} = -26.67$ kNm

$M_{FCB} = +\dfrac{Wa^2b}{l^2} = +\dfrac{60 \times 1^2 \times 2}{3^2} = +13.33$ kNm

Span BD : $M_{FBD} = 0 = M_{FDB}$

(*b*) **Distribution factors :**

Joint	*Members*	*Relative stiffness (R.S.)*	*Total stiffness (T.S.)*	*Distribution factor = (R.S.)/(T.S.)*
B	BA	$\frac{2I}{5}$	$\frac{2I}{5}+\frac{I}{2}+\frac{I}{3}=\frac{37I}{30}$	$\frac{2I/5}{37I/30}=0.324$
	BD	$\frac{2I}{4}=\frac{I}{2}$		$\frac{I/2}{37I/30}=0.406$
	BC	$\frac{I}{3}$		$\frac{I/3}{37I/30}=0.270$

(*c*) **Moment distribution :**

Joint	*A*	*B*			*C*
Members	AB	BA	BD	BC	CB
Distribution factor	–	0.324	0.406	0.270	–
Fixed end moments Balancing	– 38.4	+ 57.6 – 10.021	0 – 12.558	– 26.67 – 8.351	+ 13.33
Carry over	– 5.011				– 4.176
Final moments	– 43.411	47.579	– 12.558	– 35.021	+ 9.154

$$M_{DB} = 1/2\,(-12.558), \quad M_{DB} = -6.279 \text{ kNm}$$

(*d*) **To draw S.F.D. :**

Span AB :

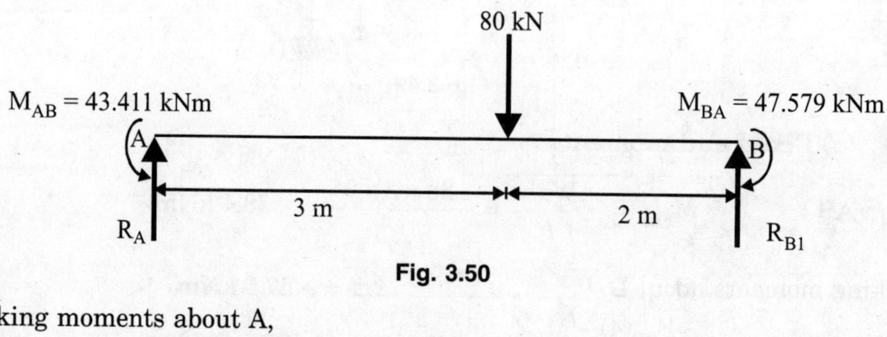

Fig. 3.50

Taking moments about A,

$$R_B(5) + M_{AB} - W(a) - M_{BA} = 0$$
$$R_B(5) + 43.411 - 80 \times 3 - 47.579 = 0$$
$$R_{B1} = 48.834 \text{ kN}$$
$$R_A = \text{Total load} - R_B = 80 - 48.834$$
$$R_A = 31.166 \text{ kN}$$

Span BC :

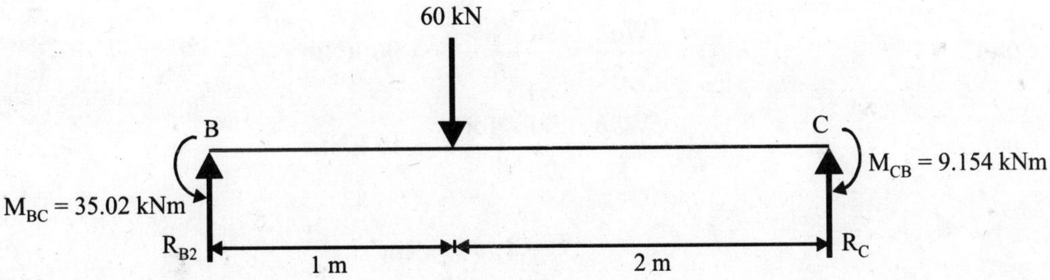

Fig. 3.51

Taking moments about B,

$$R_C(3) + M_{BC} - M_{CB} - W(a) = 0$$

$$R_C(3) + 35.021 - 9.154 - 60(1) = 0$$

$$R_C = 11.378 \text{ kN}, \quad R_{B2} = 60 - 11.378, \quad R_{B2} = 48.622 \text{ kN}$$

Span BD :

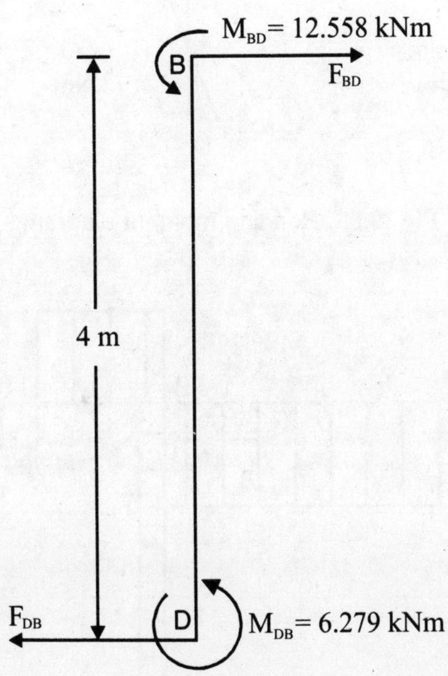

Fig. 3.52

Taking moments about D,

$$F_{BD}(4) - M_{BD} - M_{DB} = 0$$

$$F_{DB}(4) - 12.558 - 6.279 = 0$$

$$F_{BD} = 4.709 \text{ kN} \quad (\rightarrow)$$

$$F_{DB} = 4.709 \text{ kN} \quad (\leftarrow)$$

Simply supported bending moments :

Span $AB = \frac{Wab}{l} = \frac{80 \times 3 \times 2}{5} = 96$ kNm

Span $BC = \frac{Wab}{l} = \frac{60 \times 1 \times 2}{3} = 40$ kNm

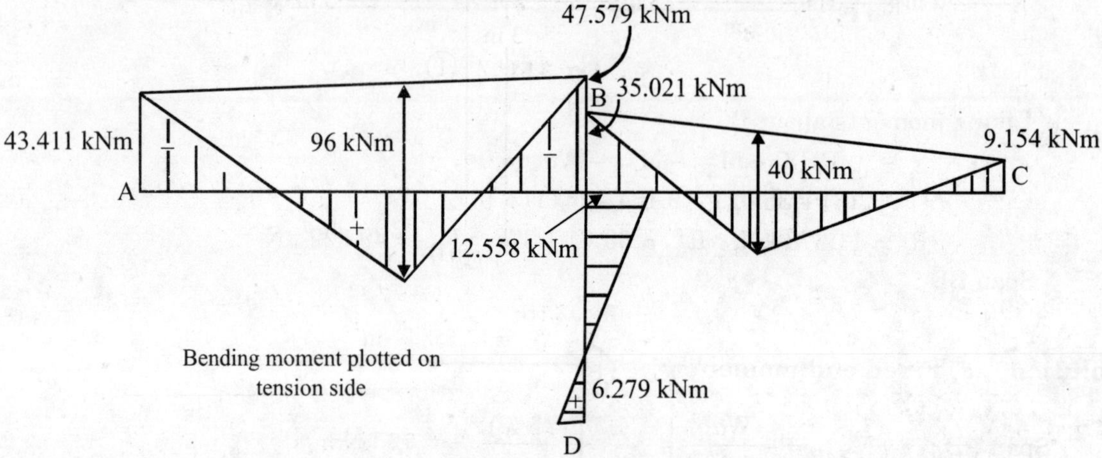

Fig. 3.53. Bending moment diagram.

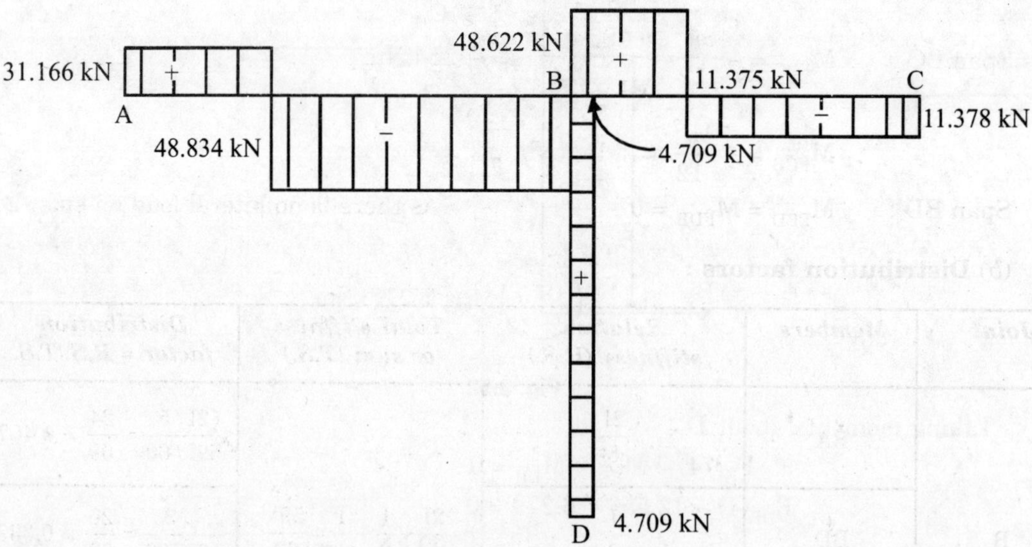

Fig. 3.54. Shear force diagram.

Example 3.14 *A continuous beam ABC is supported on an elastic column BD and is loaded as shown in Fig. 3.55. Treating joint B as rigid, analyse the frame and the B.M.D. and sketch the deflected shape of the structure.*

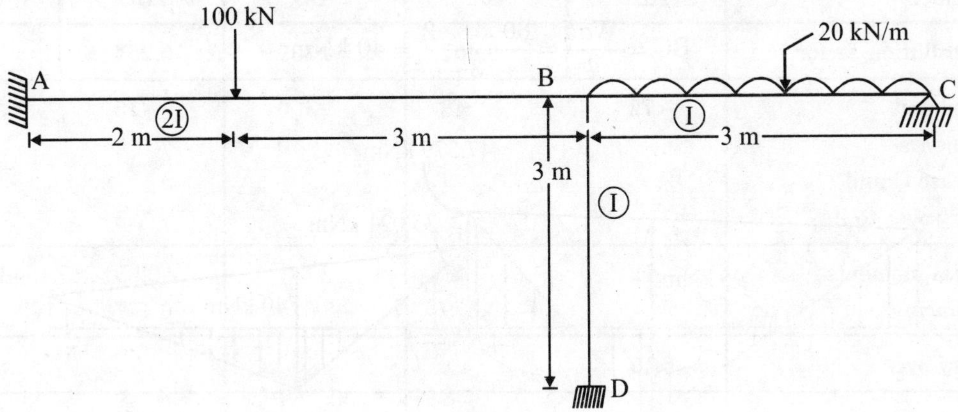

Fig. 3.55

Solution (*a*) **Fixed end moments :**

Span AB : $M_{FAB} = -\dfrac{Wab^2}{l^2} = -\dfrac{100 \times 2 \times (3)^2}{5^2} = -72$ kNm

$M_{FBA} = +\dfrac{Wa^2b}{l^2} = +\dfrac{100 \times 2^2 \times 3}{5^2} = +48$ kNm

Span BC : $M_{FBC} = -\dfrac{wl^2}{12} = -\dfrac{20 \times (3)^2}{12} = -15$ kNm

$M_{FCB} = +\dfrac{wl^2}{12} = +15$ kNm

Span BD : $M_{FBD} = M_{FDB} = 0$ (As there is no lateral load on span BD)

(*b*) **Distribution factors :**

Joint	*Members*	*Relative stiffness (R.S.)*	*Total stiffness or sum (T.S.)*	*Distribution factor = R.S./T.S.*
B	BA	$\dfrac{2I}{5}$	$\dfrac{2I}{5} + \dfrac{I}{3} + \dfrac{I}{4} = \dfrac{59I}{60}$	$\dfrac{2I/5}{59I/60} = \dfrac{24}{59} = 0.407$
	BD	$\dfrac{I}{3}$		$\dfrac{I/3}{59I/60} = \dfrac{20}{59} = 0.339$
	BC	$\dfrac{3}{4} \times \dfrac{I}{3} = \dfrac{I}{4}$		$\dfrac{I/4}{59I/60} = \dfrac{15}{59} = 0.254$

(*c*) **Moment distribution :**

Joint	*A*	*B*			*C*
Member	AB	BA	BD	BC	CB
Distribution factor	–	0.407	0.339	0.254	–
Fixed end	– 72	48	0	– 15	15
Moments			0		
Release C and					
Carry over to B				– 7.5	– 15
Initial moments	– 72	48		– 22.5	0
Balancing		– 10.379	– 8.645	– 6.477	
Carry over	– 5.19				
Final moments	– 77.19	37.621	– 8.645	– 28.977	0

We know that, A moment which rotates the near end of a prismatic beam without translation, the far end being fixed induces at the far end, a moment of one half its magnitude and in the same direction (*i.e.,* of the same sign)

$$M_{DB} = \frac{1}{2} M_{BD} = \frac{1}{2} (-8.645) = -4.323 \text{ kNm}$$

(*d*) **To draw S.F.D. :**

Span AB :

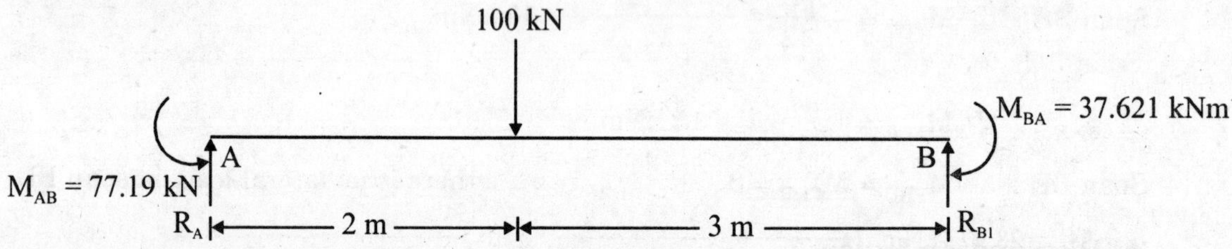

Fig. 3.56

Taking moments about B,

$$R_A(5) + M_{BA} - 77.19 - 100(3) = 0$$

$$R_A(5) + 37.621 - 77.19 - 300 = 0$$

$$R_A = 67.914 \text{ kN}$$

$$R_{B1} = \text{Total load} - 67.914 = 100 - 67.914$$

$$R_{B1} = 32.086 \text{ kN}$$

Span BD :

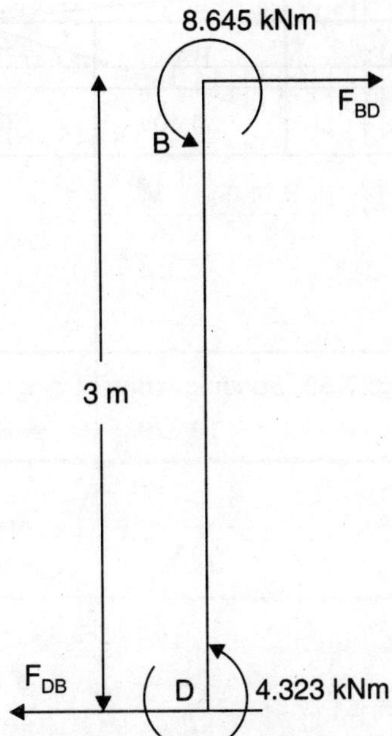

Fig. 3.57

Taking moments about D, $8.645 + 4.323 - F_{BD}(3) = 0$

$$F_{BD} = 4.323 \text{ kN } (\longrightarrow)$$

$$F_{DB} = 4.323 \text{ kN } (\longleftarrow)$$

Span BC :

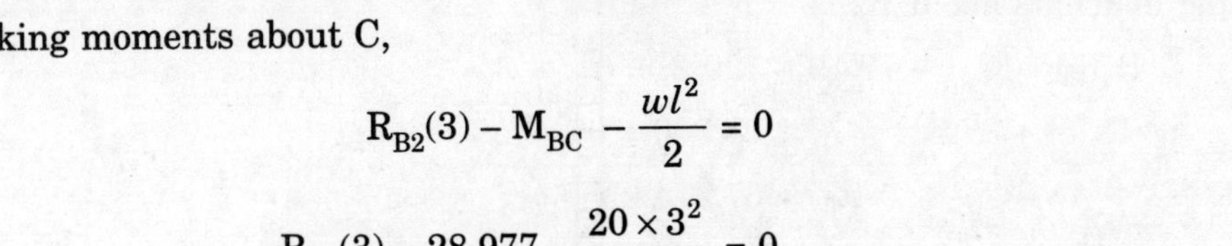

Fig. 3.58

Taking moments about C,

$$R_{B2}(3) - M_{BC} - \frac{wl^2}{2} = 0$$

$$R_{B2}(3) - 28.977 - \frac{20 \times 3^2}{2} = 0$$

$$R_{B2} = 39.659 \text{ kN}$$

$$R_C = \text{Total load} - R_{B2} = 20 \times 3 - 39.659 = 20.341 \text{ kN}$$

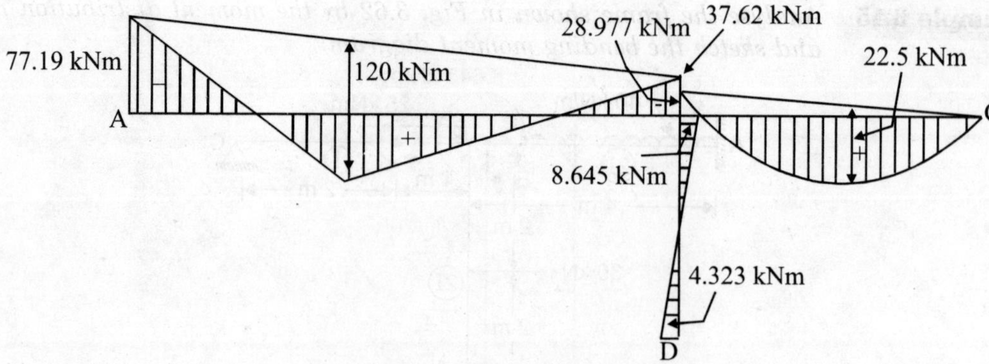

Fig. 3.59. Bending moment diagram.

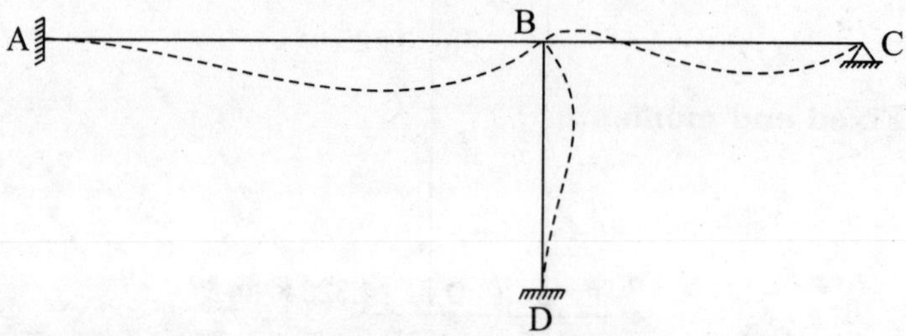

Fig. 3.60. Deflected shape.

Simply supported bending moments :

$$\text{Span AB} = \frac{Wab}{l} = \frac{100 \times 2 \times 3}{5} = 120 \text{ kNm}$$

$$\text{Span BD} = 0$$

$$\text{Span BC} = \frac{wl^2}{8} = \frac{20 \times 3^2}{8} = 22.5 \text{ kNm}$$

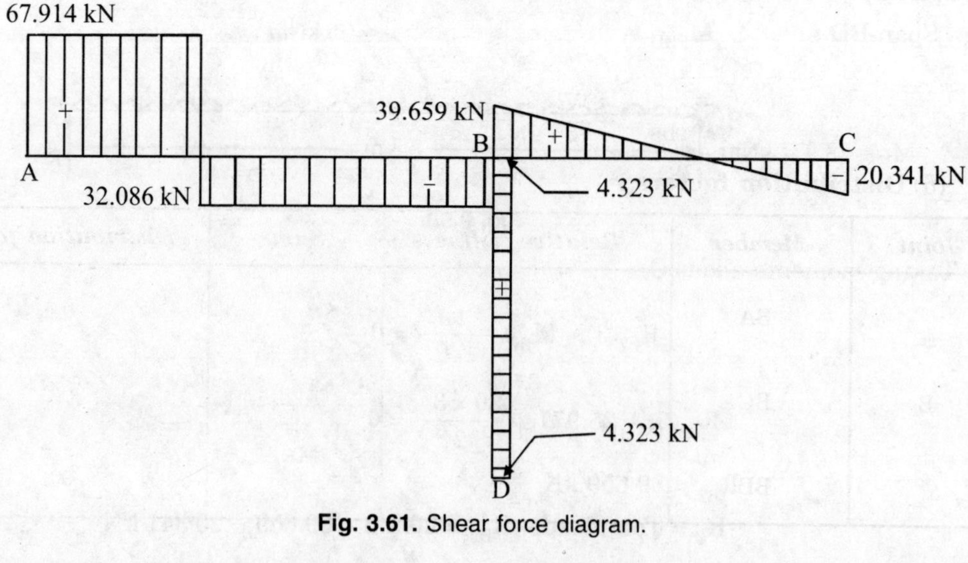

Fig. 3.61. Shear force diagram.

Example 3.15 *Analyse the frame shown in Fig. 3.62 by the moment distribution method and sketch the bending moment diagram.*

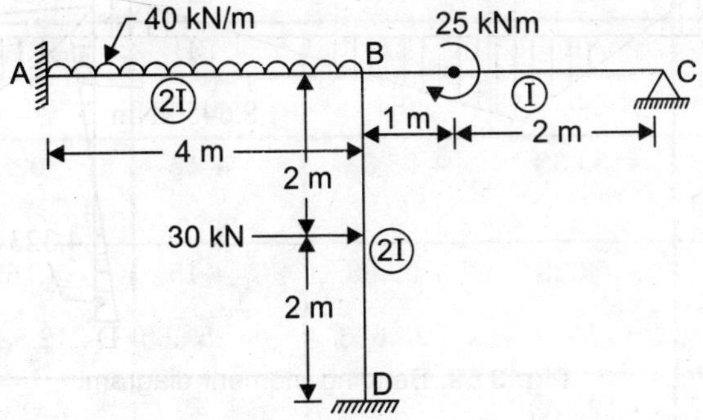

Fig. 3.62

Solution (*a*) **Fixed end moments :**

Span AB :

$$M_{FAB} = -\frac{wl^2}{12} = -\frac{40\times4^2}{12} = -53.33 \text{ kNm}$$

$$M_{FBA} = +\frac{wl^2}{12} = \frac{40\times4^2}{12} = +53.33 \text{ kNm}$$

Span BC :

$$M_{FBC} = \frac{Mb}{l^2}(3a-l) = \frac{25\times2}{3^2}(3\times1-3) = 0$$

$$M_{FCB} = \frac{Ma}{l^2}(3b-l) = \frac{25\times1}{3^2}(3\times2-3) = 8.33 \text{ kNm}$$

Span BD :

$$M_{FBD} = +\frac{Wl}{8} = +\frac{30\times4}{8} = +15 \text{ kNm}$$

$$M_{FDB} = -\frac{Wl}{8} = -\frac{30\times4}{8} = -15 \text{ kNm}$$

(*b*) **Distribution factors :**

Joint	*Member*	*Relative stiffness*	*Sum*	*Distribution factor*
B	BA	$\frac{2I}{4}$	1.25I	0.4
	BC	$\frac{3}{4}\times\frac{I}{3}$		0.2
	BD	$\frac{2I}{4}$		0.4

(*c*) **Moment distribution :**

Joint	***A***		***B***		***C***	***D***
Member	AB	BA	BD	BC	CB	DB
Distribution factors	–	0.4	0.4	0.2	–	–
FEM	– 53.33	+ 53.33	+ 15	0	8.33	– 15
Release C and C/O to B				– 4.165 ⟵	– 8.33	
Initial moments	– 53.33	+ 53.33	+ 15	– 4.165	0	– 15
Balancing		– 25.666	– 25.666	– 12.833		
Carry over	– 12.833	–	–	–		– 12.833
Balancing	–	–	–	–		–
Final moments	– 66.163	27.664	– 10.666	– 16.998		– 27.833

Simply supported moments :

Span AB : $M_{max} = +\frac{wl^2}{8} = +\frac{40 \times 4^2}{8} = 80$ kNm

Span BD : $M_{max} = \frac{Wl}{4} = \frac{30 \times 4}{4} = 30$ kNm

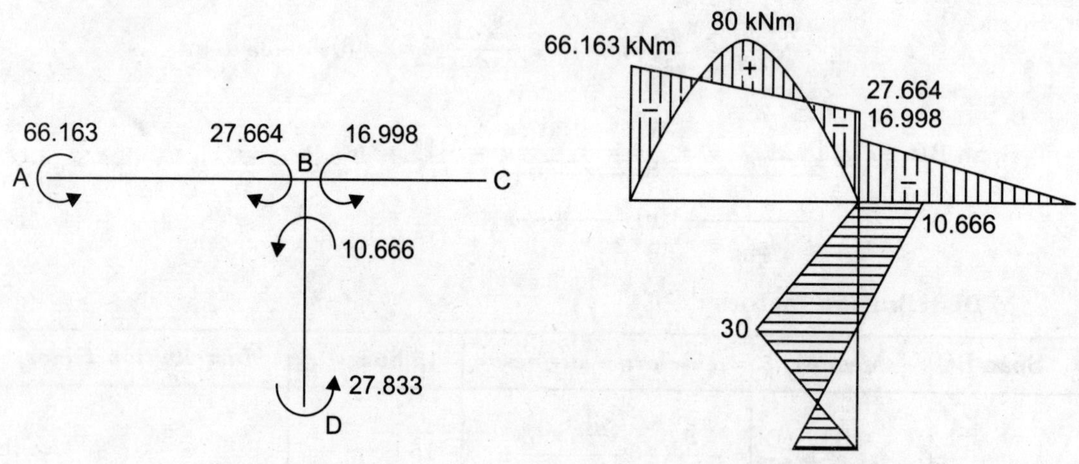

Free body diagram Bending moment diagram

Fig. 3.63

Example 3.16 *Analyse the structure loaded as shown in Fig. 3.64 by the moment distribution method and sketch bending moment and shear force diagrams.*

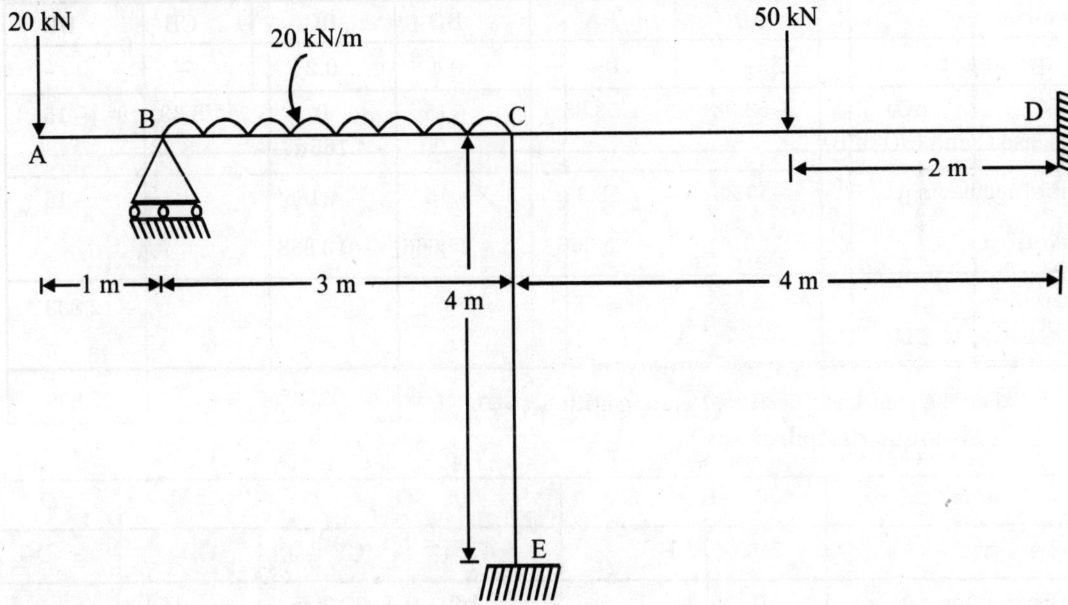

Fig. 3.64

Solution (*a*) **Fixed end moments :**

Span AB : $M_{BA} = 20 \times 1 = 20$ kNm

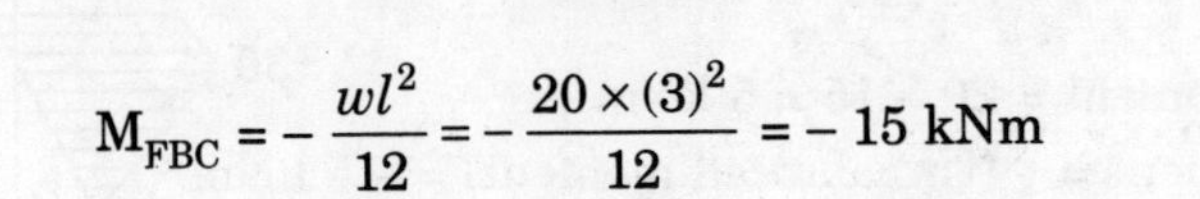

Fig. 3.65

Span BC : $M_{FBC} = -\dfrac{wl^2}{12} = -\dfrac{20 \times (3)^2}{12} = -15$ kNm

$M_{FCB} = +\dfrac{wl^2}{12} = +\dfrac{20 \times (3)^2}{12} = +15$ kNm

Span CE : $M_{FCE} = 0$

$M_{FEC} = 0$

Span CD : $M_{FCD} = -\dfrac{Wl}{8} = -\dfrac{50 \times 4}{8}$

$M_{FCD} = -25$ kNm

$M_{FDC} = +25$ kNm

(*b*) **Distribution factors :**

Joint	Member	Relative stiffness	Sum	Distribution factor
B	BA	0*	$\frac{I}{3}$	0
	BC	$\frac{I}{3}$		$\frac{I/3}{I/3} = 1$
C	CB	$\frac{3}{4} \times \frac{I}{3} = \frac{I}{4}$	$\frac{I}{4} + \frac{I}{4} + \frac{I}{4} = \frac{3I}{4}$	$\frac{I/4}{3I/4} = \frac{1}{3}$
	CE	$\frac{I}{4}$		$\frac{I/4}{3I/4} = \frac{1}{3}$
	CD	$\frac{I}{4}$		$\frac{I/4}{3I/4} = \frac{1}{3}$

*For the cantilever portion, Relative stiffness = 0.

(*c*) **Moment distribution :**

Joint	*B*		*C*			*D*
Member	BA	BC	CB	CE	CD	DC
Distribution factor	0	1	1/3	1/3	1/3	–
Fixed end moments	20	– 15	+15	0	– 25	+ 25
Balancing (1)		– 5	3.333	3.333	3.333	–
Carry over (1)		–	– 2.5	–	–	1.665
Balancing (2)		–	+ 0.833	+ 0.833	0.833	–
Carry over (2)		–				0.4165
Final moments	20	– 20	16.667	4.167	– 20.834	27.082

Calculations

Balancing (1) :

Joint B :

Unbalanced moment = 20 – 15 = 5 kNm

Balancing moment = – (Unbalanced moment) = – 5 kNm

Moment contributed by BA = 0

Moment contributed by BC = – 5 × DF = – 5 × 1 = – 5 kNm

Joint C :

Unbalanced moment = +15 – 25 = – 10 kNm

Balancing moment = +10 kNm

Moment contribution by CB = Balancing moment × Distribution factor of CB

= +10 × (1/3) = + 3.333 kNm

Moment contribution of CE = +10 × (1/3) = + 3.333 kNm

Moment contribution of CD = +10 × (1/3) = + 3.333 kNm

Carry over (1) :

From B to C. Half of the balancing moment in M_{BC} will be carried over to M_{CB}

$$\text{Moment carried over} = \frac{\text{Balancing moment at joint B}}{2}$$

$$= -\frac{5}{2} = -\ 2.5 \text{ kNm}$$

From C to D. $\text{Moment carried over} = \dfrac{(\text{Balancing moment contributed by CD})}{2}$

$$= 3.333/2 = 1.665 \text{ kNm}$$

Balancing (2) :

Unbalanced moment = – 2.5 kNm,

Joint C Balancing moment = + 2.5 kNm

Moment contribution by CB = + 2.5 × (1/3) = 0.833 kNm

Moment contribution by CE = + 2.5 × (1/3) = + 0.833 kNm

Moment contribution by CD = + 2.5 × (1/3) = + 0.833 kNm

Carry over (2) :

From C to D. $\text{Moment carried over} = \dfrac{(\text{Balancing moment contributed by CD})}{2}$

$$= 0.833/2 = 0.4166 \text{ kNm}$$

Final moments :

$M_{BA} = 20$ kNm

$M_{BC} = -15 - 5 = -20$ kNm

$M_{CB} = +15 + 3.333 + (-2.5) + 0.833 = +16.667$ kNm

$M_{CE} = +4.167$ kNm

$M_{EC} = +4.167/2 = +2.0835$ kNm

$M_{CD} = -25 + 3.333 + 0.833 = -20.834$ kNm

$M_{DC} = +25 + 1.665 + 0.4165 = +27.082$ kNm

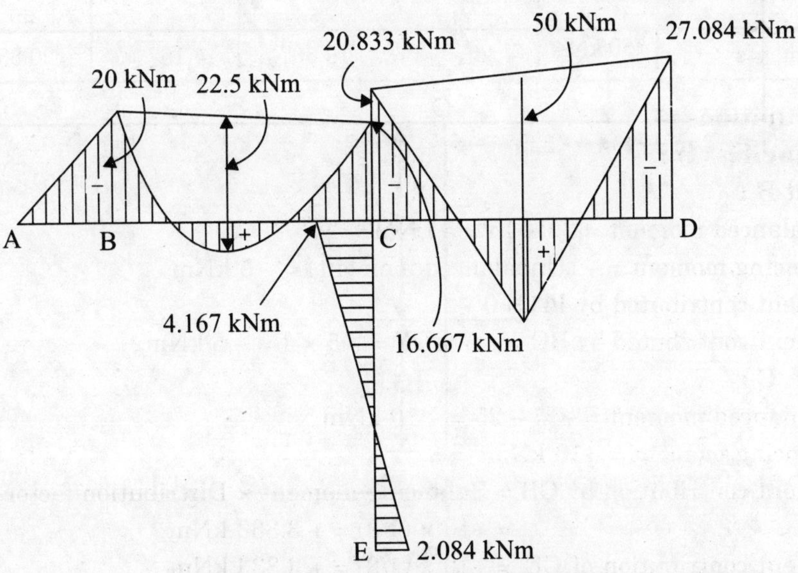

Fig. 3.66. Bending moment diagram.

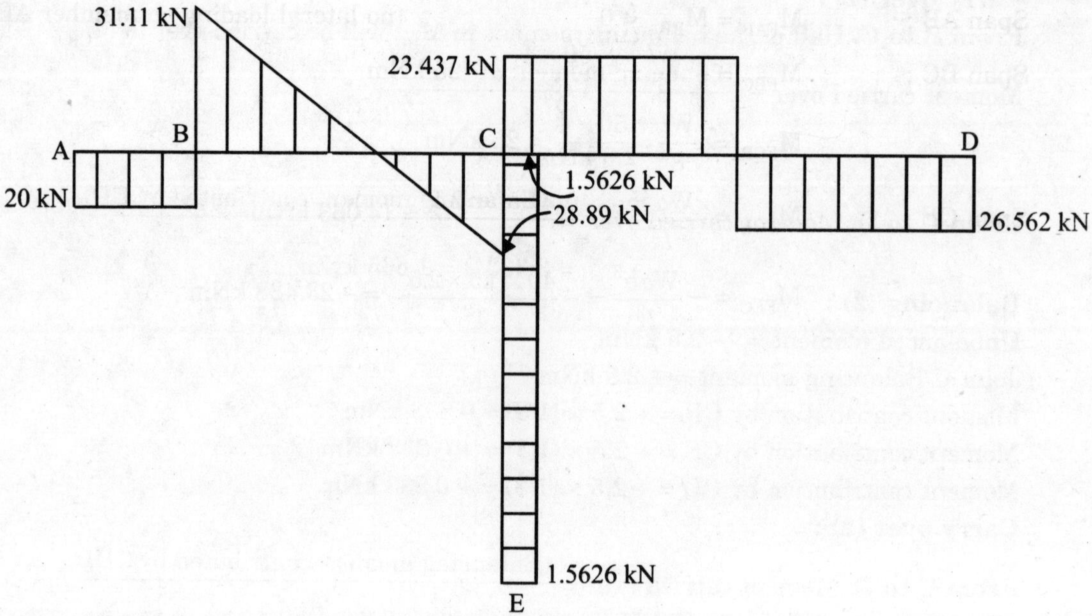

Fig. 3.67. Shear force diagram.

Example 3.17 *Analyse the frame shown in Fig. 3.68 by the moment distribution method and draw the bending moment diagram.*

$$I_{AB} = I\ ;\ I_{BC} = I_{CD} = 2I\ ;\ I_{CE} = 1.5I.$$

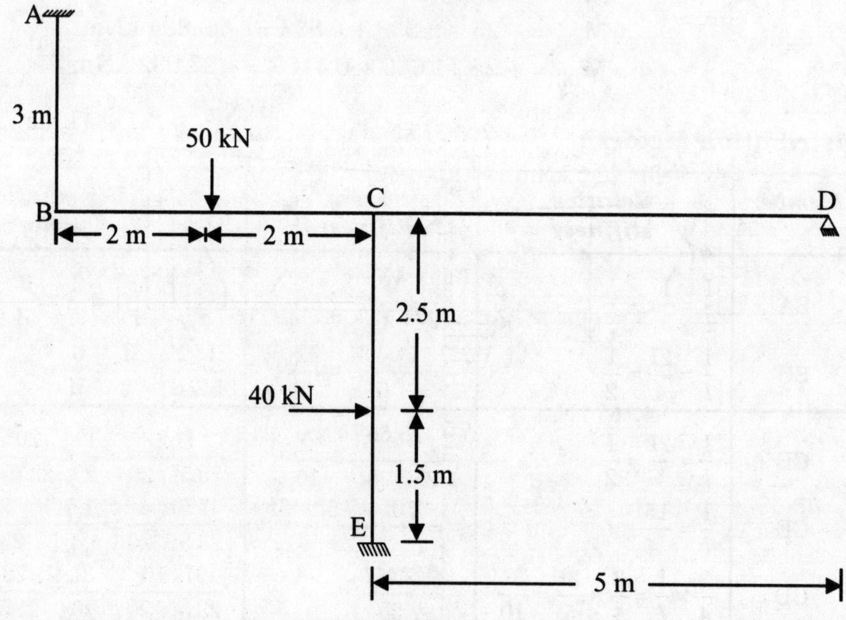

Fig. 3.68

Solution (*a*) **Fixed end moments :**

Span AB : $M_{FAB} = M_{FBA} = 0$ (no lateral loading on member AB)

Span BC : $M_{FBC} = -\frac{Wl}{8} = -\frac{50 \times 4}{8} = -25$ kNm

$M_{FCB} = \frac{Wl}{8} = \frac{50 \times 4}{8} = 25$ kNm

Span CE : $M_{FCE} = \frac{Wa^2b}{l^2} = \frac{40 \times 1.5^2 \times 2.5}{4^2} = 14.063$ kNm

$M_{FEC} = -\frac{Wab^2}{l^2} = -\frac{40 \times 1.5 \times 2.5^2}{4^2} = -23.438$ kNm

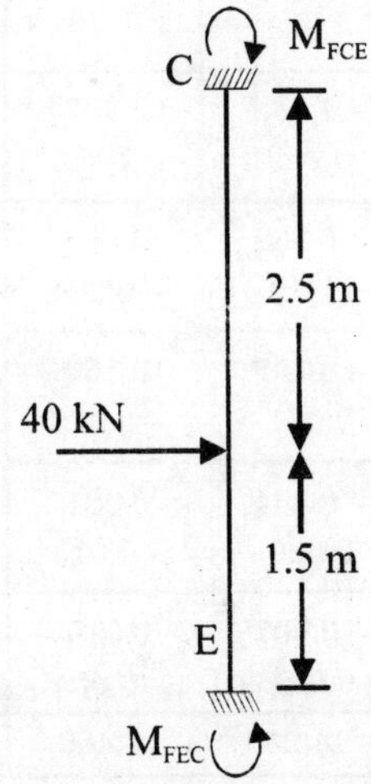

Fig. 3.69

Span CD : $M_{FCD} = M_{FDC} = 0$ (no lateral loading on member CD)

(*b*) **Distribution factors :**

Joint	*Member*	*Relative stiffness*	*Sum of stiffness*	*Distribution factor*
B	BA	$\frac{I}{l} = \frac{I}{3}$	$\frac{I}{3} + \frac{I}{2}$	$\left(\frac{I}{3}\right)/\left(\frac{5I}{6}\right) = \frac{I}{3} \times \frac{6}{5I} = 0.4$
	BC	$\frac{I}{l} = \frac{2I}{4} = \frac{I}{2}$	$= \frac{2I + 3I}{6} = \frac{5I}{6}$	$\frac{I/2}{5I/6} = \frac{I}{2} \times \frac{6}{5I} = 0.6$
C	CB	$\frac{I}{l} = \frac{2I}{4} = \frac{I}{2}$	$\frac{I}{2} + \frac{1.5I}{4} + \frac{3I}{10}$	$\frac{I/2}{23.5I/20} = \frac{I}{2} \times \frac{20}{23.5I} = 0.426$
	CE	$\frac{I}{l} = \frac{1.5I}{4}$	$= \frac{10I + 7.5I + 6I}{20}$	$\frac{1.5I/4}{23.5I/20} = \frac{1.5I}{4} \times \frac{20}{23.5I} = 0.319$
	CD	$\frac{3}{4} \times \frac{I}{l} = \frac{3}{4} \times \frac{2I}{5} = \frac{3I}{10}$	$= \frac{23.5I}{20}$	$\frac{3I/10}{23.5I/20} = \frac{3I}{10} \times \frac{20}{23.5I} = 0.255$

(*c*) **Moments distribution :**

Joint	*A*	*B*		*C*			*D*	*E*
Member	AB	BA	BC	CB	CE	CD	DC	EC
Distribution factors	–	0.4	0.6	0.426	0.319	0.255	–	1
Fixed end moments	0	0.00	– 25.00	25.00	14.063	0.00	0.00*	– 23.438
Balance		10.00	15.00	– 16.641	– 12.461	– 9.961		
Carry over	5.00		– 8.321	7.50			No carry to over this since D is simply supported	– 6.231
Balance		3.328	4.993	– 3.195	– 2.393	– 1.913		
Carry over	1.664		– 1.598	2.497			–	– 1.197
Balance		0.639	0.959	– 1.064	– 0.797	– 0.637		
Carry over	0.320		– 0.532	0.480			–	– 0.399
Balance		0.213	0.319	– 0.204	– 0.153	– 0.122		
Carry over	0.107		– 0.102	0.160			–	– 0.077
Balance		0.041	0.061	– 0.068	– 0.051	– 0.041		
Carry over	0.021		– 0.034	0.031			–	– 0.026
Balance		0.014	0.020	– 0.013	– 0.010	– 0.008		
Carry over	0.007		– 0.007	0.010			–	– 0.005
Balance and Co.	0.002 ←	0.003	0.004	– 0.004	– 0.003	– 0.003		– 0.002
Final moments	7.121	14.238	– 14.238	14.489	– 1.804	– 12.685	–	– 31.375

*Since there is no loading on CD, the fixed end moment at D = 0. However, if there is loading on CD, there will be M_{FDC}. Since D is simply supported, to restore the realistic support condition, D is to be first released and carried over to C of CD. Then initial moments are to be obtained and then the routine moment distribution is to be done without carrying over to D.

M_{AB} = 7.121 kNm $\quad$ M_{BA} = 14.238 kNm $\quad$ M_{BC} = – 14.238 kNm

M_{CB} = 14.489 kNm $\quad$ M_{CD} = – 12.685 kNm $\quad$ M_{CE} = – 1.804 kNm

M_{EC} = – 31.377 kNm $\quad$ M_{DC} = 0

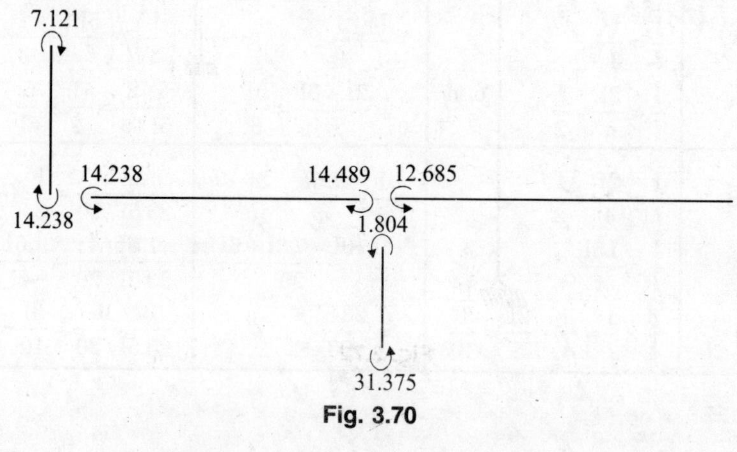

Fig. 3.70

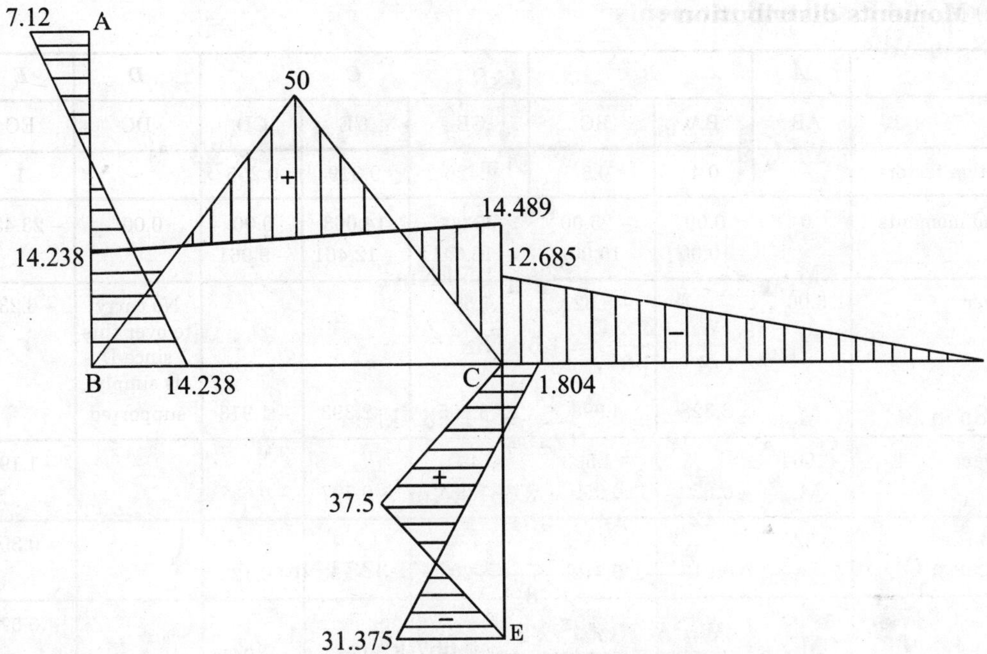

Fig. 3.71. Bending moment diagram.

Example 3.18 *Analyse the frame shown in Fig. 3.72 and draw the bending moment diagram.*

$$I_{AB} = I_{BE} = 2I \; ; I_{BC} = I_{CD} = I_{CF} = I.$$

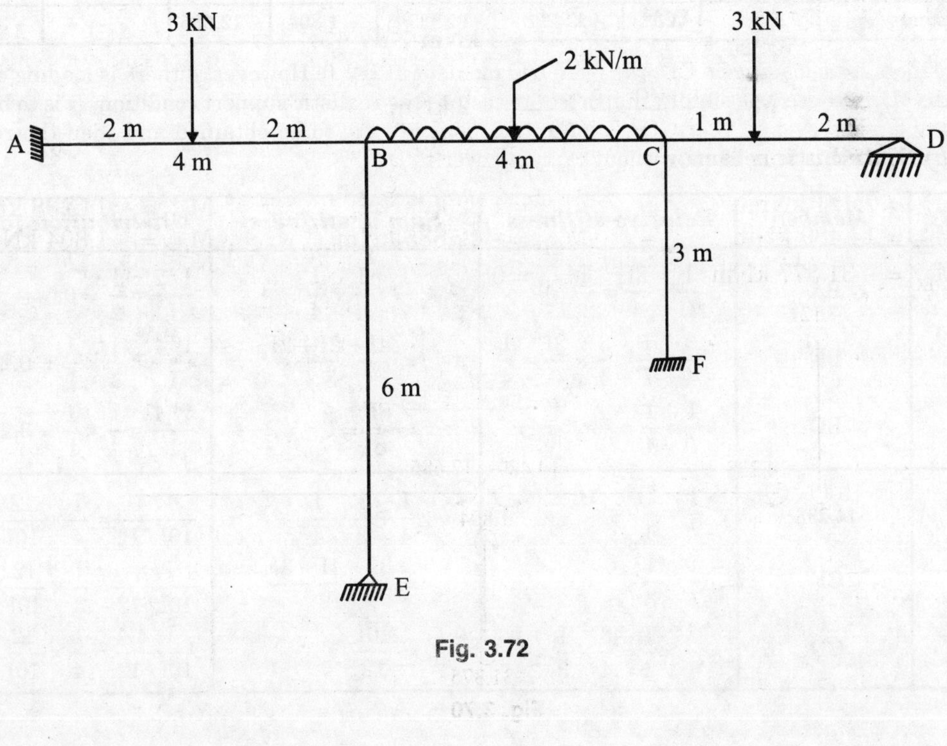

Fig. 3.72

Solution (*a*) **Fixed end moments :**

Span AB :

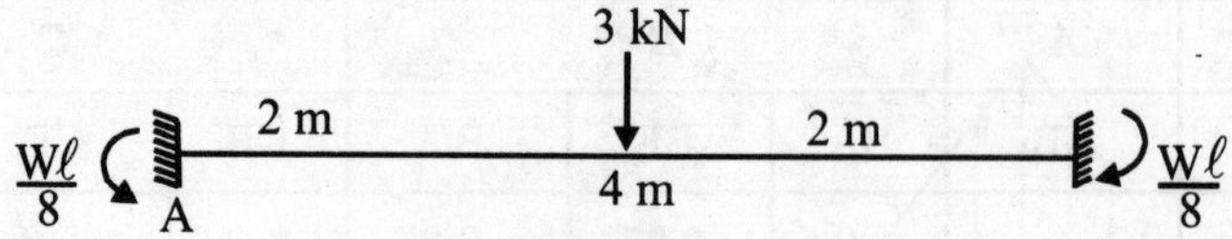

Fig. 3.73

$$M_{FAB} = -\frac{Wl}{8} = -\frac{3 \times 4}{8} = -1.5 \text{ kNm}$$

$$M_{FBA} = \frac{Wl}{8} = \frac{3 \times 4}{8} = 1.5 \text{ kNm}$$

Span BC : $$M_{FBC} = -\frac{wl^2}{12} = -\frac{2 \times 4^2}{12} = -2.667 \text{ kNm}$$

$$M_{FCB} = \frac{wl^2}{12} = \frac{2 \times 4^2}{12} = 2.667 \text{ kNm}$$

Span CD : $$M_{FCD} = -\frac{Wab^2}{l^2} = -\frac{3 \times 1 \times 2^2}{3^2} = -1.333 \text{ kNm}$$

$$M_{FDC} = \frac{Wa^2b}{l^2} = \frac{3 \times 1^2 \times 2}{3^2} = 0.667 \text{ kNm}$$

$$M_{FBE} = M_{FEB} = M_{FCF} = M_{FFC} = 0$$

Simply supported BM :

$$M_{AB} = \frac{Wl}{4} = \frac{3 \times 4}{4} = 3 \text{ kNm}$$

$$M_{BC} = \frac{Wl^2}{8} = \frac{2 \times 4^2}{8} = 4 \text{ kNm}$$

$$M_{CD} = \frac{Wab}{l} = \frac{3 \times 1 \times 2}{3} = 2 \text{ kNm}$$

(*b*) **Distribution factor :**

Joint	*Member*	*Relative stiffness*	*Sum of stiffness*	*Distribution factor*
B	BA	$\frac{I}{L} = \frac{2I}{4} = \frac{I}{2}$	$\frac{I}{2} + \frac{I}{4} + \frac{I}{4}$	$\frac{I/2}{I} = \frac{I}{2} = 0.5$
	BE	$\frac{3}{4} \times \frac{I}{l} = \frac{3}{4} \times \frac{2I}{6} = \frac{I}{4}$	$= \frac{4I + 2I + 2I}{8}$	$\frac{I/4}{I} = \frac{I}{4} \times \frac{1}{I} = 0.25$
	BC	$\frac{I}{l} = \frac{I}{4}$	$= \frac{8I}{8} = I$	$\frac{I/4}{I} = \frac{I}{4} \times \frac{1}{I} = 0.25$
C	CB	$\frac{I}{l} = \frac{I}{4}$	$\frac{I}{4} + \frac{I}{3} + \frac{I}{4}$	$\frac{I/4}{10I/12} = \frac{I}{4} \times \frac{12}{10I} = 0.3$
	CF	$\frac{I}{l} = \frac{I}{3}$	$= \frac{3I + 4I + 3I}{12}$	$\frac{I/3}{10I/12} = \frac{I}{3} \times \frac{12}{10I} = 0.4$
	CD	$\frac{3}{4} \times \frac{I}{l} = \frac{3}{4} \times \frac{I}{3} = \frac{I}{4}$	$= \frac{10I}{12}$	$\frac{I/4}{10I/12} = \frac{I}{4} \times \frac{12}{10I} = 0.3$

(*c*) **Moment distribution :**

Joint	A	B			C			D
Member	AB	BA	BE	BC	CB	CF	CD	DC
Distribution factors	—	0.5	0.25	0.25	0.3	0.4	0.3	—
FEM	– 1.5	1.5	—	– 2.667	2.667	—	– 1.333	0.667
Balance 'D'	—	—	—	—	—	—	—	– 0.667
Carry over	—	—	—	—	—	—	– 0.333	—
Initial moments	– 1.5	1.5	—	– 2.667	2.667	—	– 1.666	0
Balance 'B' 'C'	—	0.584	0.292	0.292	– 0.3	– 0.4	– 0.3	—
Carry over	0.292	—	—	– 0.15	0.146	—	—	—
Balance 'B' 'C'		0.075	0.038	0.038	– 0.044	– 0.058	– 0.044	—
Carry over	0.038	—	—	– 0.022	0.019	—	—	—
Balance 'B' 'C'	—	0.011	0.006	0.006	– 0.006	– 0.008	– 0.006	—
Carry over	0.006	—	—	– 0.003	0.003	—	—	—
Balance 'B' 'C'	—	0.002	0.0008	0.0008	– 0.0009	– 0.0012	– 0.0009	—
Final moments	– 1.164	2.172	0.337	– 2.505	2.484	– 0.467	– 2.017	0

$$M_{FC} = \frac{1}{2} M_{CF} = \frac{1}{2} \times (-0.467) = -0.233 \text{ kNm}$$

$M_{EB} = 0$ ($\because$ E is hinged)

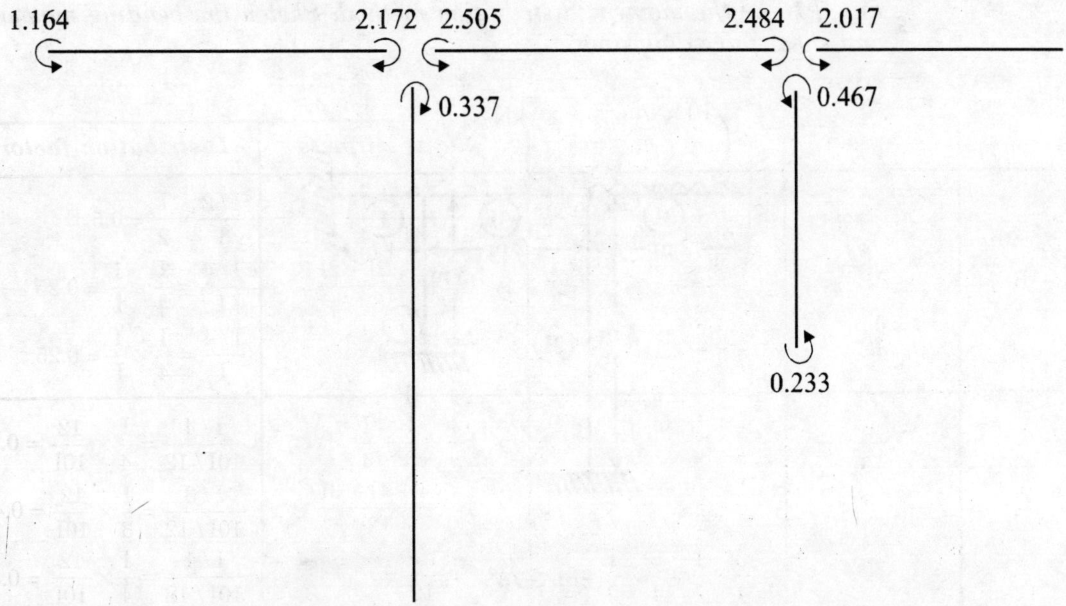

Fig. 3.74

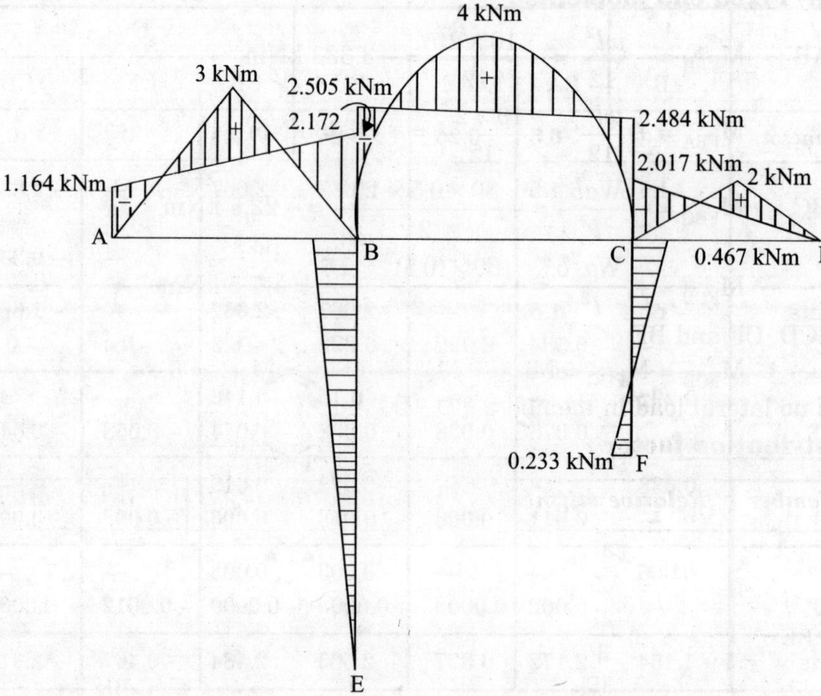

Fig. 3.75. Bending moment diagram.

Example 3.19 *Analyse the continuous beam supported by columns and loaded as shown in Fig. 3.76 by the moment distribution method. Sketch the bending moment and shear force diagrams.*

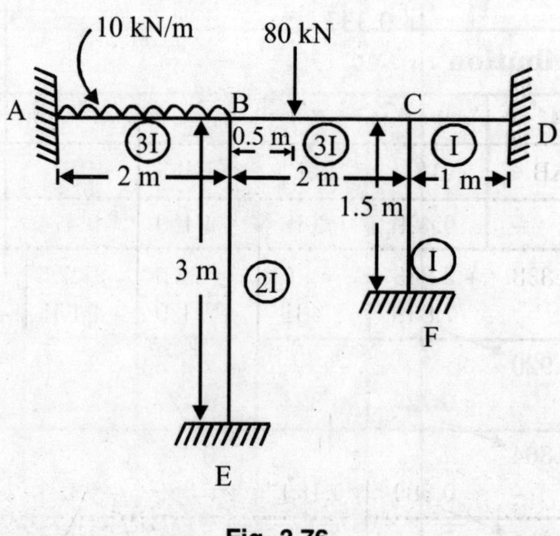

Fig. 3.76

Solution (*a*) **Fixed end moments :**

Span AB : $M_{FAB} = -\dfrac{wl^2}{12} = -\dfrac{10 \times 2^2}{12} = -3.333$ kNm

$M_{FBA} = +\dfrac{wl^2}{12} = +\dfrac{10 \times 2^2}{12} = +3.333$ kNm

Span BC : $M_{FBC} = -\dfrac{Wab^2}{l^2} = -\dfrac{80 \times 0.5 \times 1.5^2}{2^2} = -22.5$ kNm

$M_{FCB} = +\dfrac{Wa^2b}{l^2} = +\dfrac{80 \times (0.5)^2 \times 1.5}{2^2} = +7.5$ kNm

Spans CD, CF and BE :

$M_{FCD} = M_{FDC} = M_{FBE} = M_{FEB} = M_{FCF} = M_{FFC} = 0,$

since there is no lateral load in members CD, CF and BE.

(*b*) **Distribution factor :**

Joint	*Member*	*Relative stiffness*	*Total stiffness*	*Distribution factor*
B	BA	$\frac{3I}{2}$	$\frac{3I}{2}+\frac{2I}{3}+\frac{3I}{2}=\frac{11I}{3}$	$\frac{3I/2}{11I/3}=\frac{9}{22}=0.409$
	BE	$\frac{2I}{3}$		$\frac{2I/3}{11I/3}=\frac{2}{11}=0.182$
	BC	$\frac{3I}{2}$		$\frac{3I/2}{11I/3}=\frac{9}{22}=0.409$
C	CB	$\frac{3I}{2}$	$\frac{3I}{2}+\frac{I}{1.5}+\frac{I}{1}$	$\frac{3I/2}{19I/6}=\frac{9}{19}=0.474$
	CF	$\frac{I}{1.5}$	$=\frac{9I+4I+6I}{6}$	$\frac{I/1.5}{19I/6}=\frac{4}{19}=0.210$
	CD	$\frac{I}{1}$	$=\frac{19I}{6}$	$\frac{I/1}{19I/6}=\frac{6}{19}=0.316$

(*c*) **Moments distribution :**

Joints	*A*	*B*			*C*			*D*
Member	AB	BA	BE	BC	CB	CF	CD	DC
Distribution factors	–	0.409	0.182	0.409	0.474	0.210	0.316	–
Fixed end moments	– 3.333	+ 3.333	0	– 22.5	+ 7.5	–	–	–
Balance		7.839	3.488	7.839	– 3.555	– 1.575	– 2.370	–
Carry over	3.920			– 1.778	3.920			– 1.185
Balance		0.727	0.324	0.727	– 1.858	– 0.823	– 1.239	
Carry over	0.364			– 0.929	0.364			– 0.620
Balance	–	0.380	0.169	+ 0.380	– 0.173	– 0.076	– 0.115	
Carry over	0.190			– 0.087	0.190			– 0.058
Balance and C.O.	0.018 ←	0.036	0.016	0.036	– 0.090	– 0.040	– 0.060 →	– 0.030
Final moments	1.159	12.315	3.997	– 16.312	6.298	– 2.514	– 3.784	– 1.893

(*d*) **To draw S.F.D. :**

Span AB :

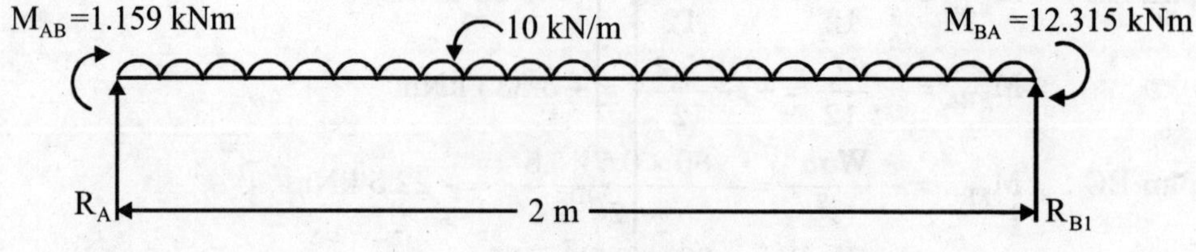

Fig. 3.77

Taking moment about A, $R_{B1}(2) - \frac{10(2)^2}{2} - 1.159 - 12.315 = 0$

$$R_{B1} = 16.737 \text{ kN}$$

$$R_A = \text{Total load} - R_B = 10 \times 2 - 16.737$$

$$R_A = 3.263 \text{ kN}$$

Span BC :

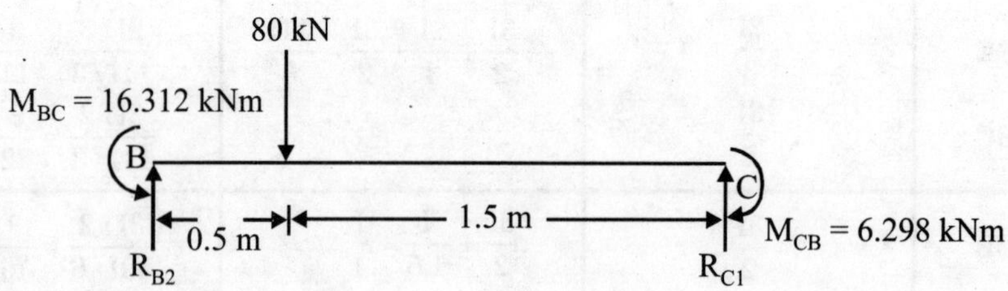

Fig. 3.78

Taking moment about B, $R_{C1} \times 2 - 6.298 - 80 \times 0.5 + 16.312 = 0$

$$R_{C1} = 14.993 \text{ kN}$$

$$R_{B2} = 80 - 14.993$$

$$R_{B2} = 65.007 \text{ kN}$$

Span CD :

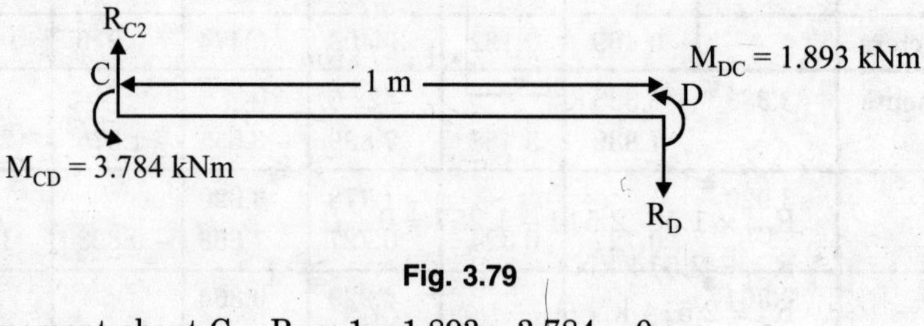

Fig. 3.79

Taking moment about C, $R_D \times 1 - 1.893 - 3.784 = 0$

$$R_D = 5.667 \text{ kN } (\downarrow)$$

$$R_{C2} = 5.667 \ (\uparrow)$$

Span BE :

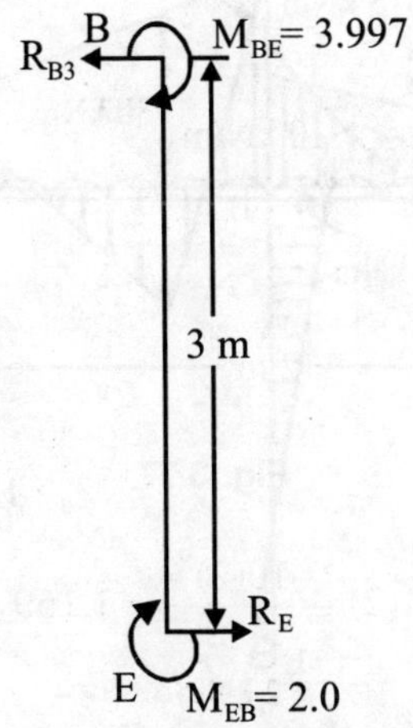

Fig. 3.80

Taking moments about E, $R_{B3} \times 3 - 3.997 - 2 = 0$

$R_{B3} = 1.999$ kN (⟵)

$R_E = 1.999$ kN (⟶)

Span CF :

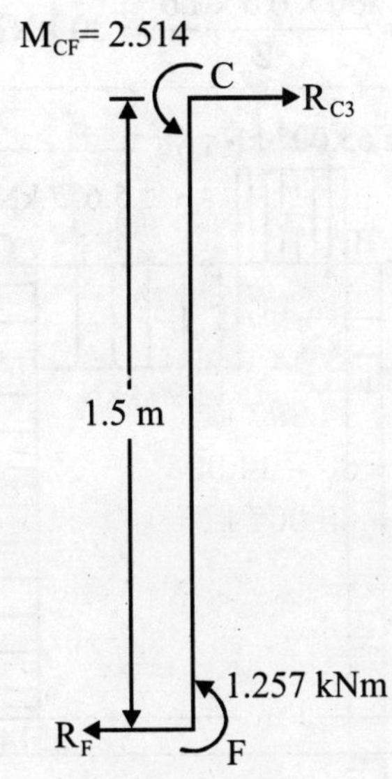

Fig. 3.81

$R_{C3} \times 1.5 - 2.514 - 1.257 = 0$

$R_{C3} = 2.514$ kN (⟶)

$R_F = 2.514$ kN (⟵)

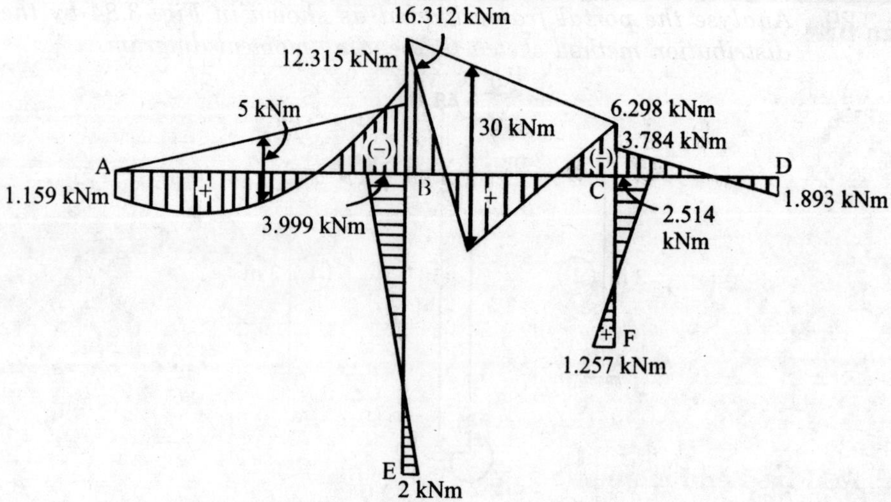

Fig. 3.82. Bending moment diagram.

Simply supported bending moments :

In span $AB = \frac{wl^2}{8} = \frac{10 \times 2^2}{8} = 5$ kNm

In span $BC = \frac{Wab}{l} = \frac{80 \times 0.5 \times 1.5}{2} = 30$ kNm

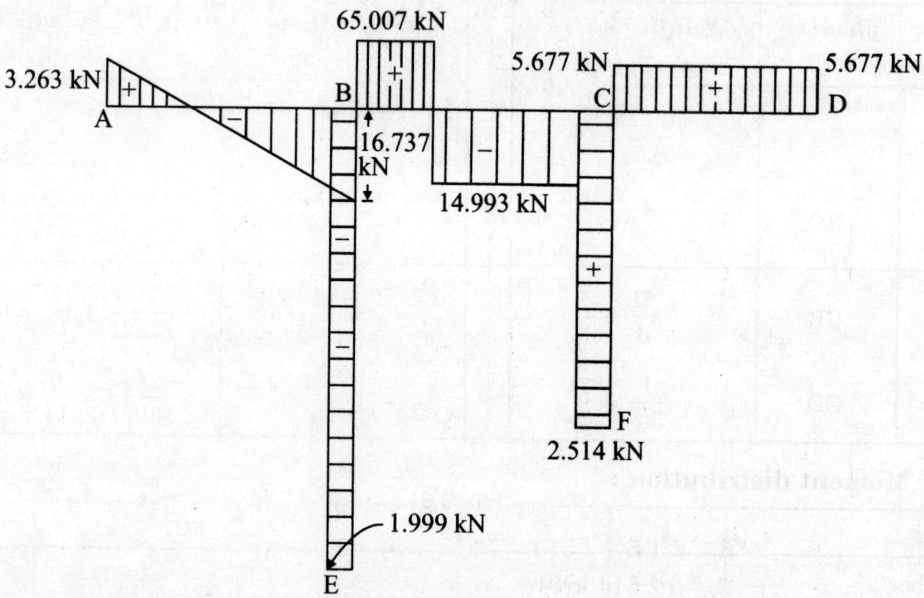

Fig. 3.83. Shear force diagram.

Example 3.20 *Analyse the portal frame loaded as shown in Fig. 3.84 by the moment distribution method sketch the bending moment diagram.*

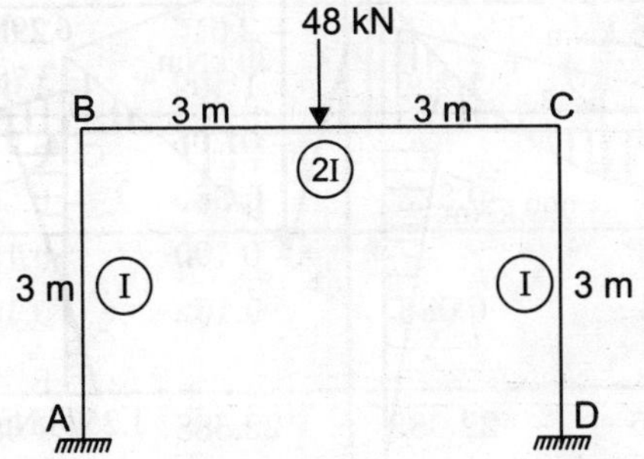

Fig. 3.84

Solution (*a*) **Fixed end moments :**

Span AB : $M_{FAB} = M_{FBA} = 0$

Span BC : $M_{FBC} = -\dfrac{Wl}{8} = -\dfrac{48 \times 6}{8} = -36$ kNm

$M_{FCB} = \dfrac{Wl}{8} = \dfrac{48 \times 6}{8} = 36$ kNm

Span CD : $M_{FCD} = M_{FDC} = 0$

(*b*) **Distribution factors :**

Joint	*Member*	*Relative stiffness*	*Total stiffness*	*Distribution factor*
B	BA	$\dfrac{I}{3.5} = \dfrac{2I}{7}$	$\dfrac{2I}{7} + \dfrac{I}{3} = \dfrac{6I + 7I}{21}$	$\dfrac{2I/7}{13I/21} = \dfrac{6}{13} = 0.461$
	BC	$\dfrac{2I}{6} = \dfrac{I}{3}$	$= \dfrac{13I}{21}$	$\dfrac{I/3}{13I/21} = \dfrac{7}{13} = 0.539$
C	CB	$\dfrac{2I}{6} = \dfrac{I}{3}$	$\dfrac{2I}{7} + \dfrac{I}{3} = \dfrac{6I + 7I}{21}$	$\dfrac{I/3}{13I/21} = \dfrac{7}{13} = 0.539$
	CD	$\dfrac{I}{3.5} = \dfrac{2I}{7}$	$= \dfrac{13I}{21}$	$\dfrac{2I/7}{13I/21} = \dfrac{6}{13} = 0.461$

(*c*) **Moment distribution :**

Joint	*A*	*B*		*C*		*D*
Member	AB	BA	BC	CB	CD	DC
Distribution factor	–	0.461	0.539	0.539	0.461	
Fixed end moments	–	–	– 36	36	–	–
Balancing		16.596	19.404	– 19.404	– 16.596	

Carry over	8.298		– 9.702	9.702		– 8.298
Balancing		4.473	5.229	– 5.229	– 4.473	
Carry over	2.237		– 2.615	2.615		– 2.237
Balancing		1.206	1.409	– 1.409	– 1.206	
Carry over	0.603		– 0.705	0.705		– 0.603
Balancing		0.325	0.380	– 0.380	– 0.325	
Carry over	0.163		– 0.190	0.190		– 0.163
Balancing and C.O. to the ends	0.044	0.088	0.102	– 0.102	– 0.088	– 0.044
Final moments	11.345	22.688	– 22.688	22.688	– 22.688	– 11.345

Simply supported bending moments :

Span AB : $\mu_{max} = 0$

Span BC : $\mu_{max} = \dfrac{Wl}{4} = \dfrac{48 \times 6}{4} = 72 \text{ kNm}$

Span CD : $\mu_{max} = 0$

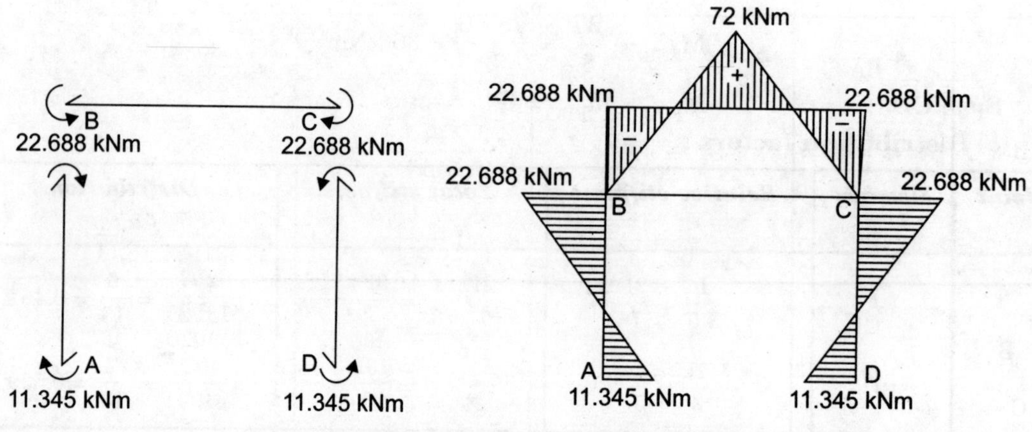

Fig. 3.85. Bending moment diagram.

Example 3.21 *Analyse the portal frame loaded as shown in Fig. 3.86 by moment distribution method and sketch the BMD.*

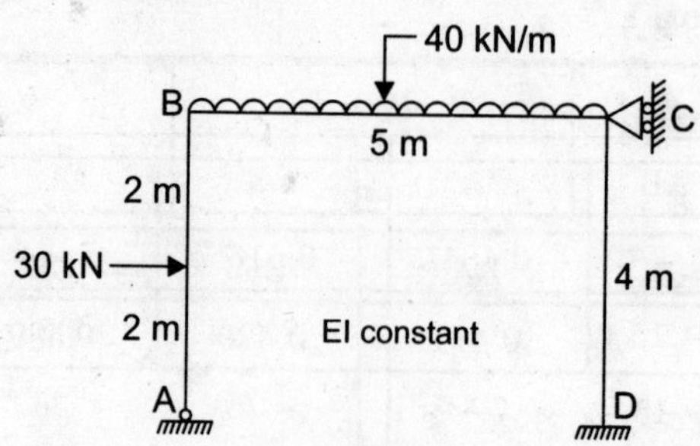

Fig. 3.86

Solution Here, the frame and loading are unsymmetrical. However because of the support at C, the frame cannot sway. Hence analysed as a non-sway frame.

(*a*) **Fixed end moments :**

Span AB : $M_{FAB} = -\frac{Wl}{8} = -\frac{30 \times 4}{8} = -15$ kNm

$M_{FBA} = \frac{Wl}{8} = \frac{30 \times 4}{8} = 15$ kNm

Span BC : $M_{FBC} = -\frac{wl^2}{12} = -\frac{40 \times 5^2}{12} = -83.33$ kNm

$M_{FCB} = \frac{wl^2}{12} = \frac{40 \times 5^2}{12} = 83.33$ kNm

Span CD : $M_{FCD} = M_{FDC} = 0$

(*b*) **Distribution factor :**

Joint	*Member*	*Relative stiffness*	*Total stiffness*	*Distribution factor*
B	BA	$\frac{3}{4}\frac{I}{L} = \frac{3I}{16}$	$\frac{3I}{16} + \frac{I}{5} = \frac{31I}{80}$	$\frac{3I/16}{31I/80} = 0.484$
	BC	$\frac{I}{L} = \frac{I}{5}$		$\frac{I/5}{31I/80} = 0.516$
C	CB	$\frac{I}{L} = \frac{I}{5}$	$\frac{I}{5} + \frac{I}{4} = \frac{9I}{20}$	$\frac{I/5}{9I/20} = 0.444$
	CD	$\frac{I}{L} = \frac{I}{4}$		$\frac{I/4}{9I/20} = 0.556$

(*c*) **Moment distribution :**

Joint	*A*	*B*		*C*		*D*
Member	AB	BA	BC	CB	CD	DC
D.F.	–	0.484	0.516	0.444	0.556	
Fixed end moments	– 15	15	– 83.33	83.33	0	0
Release A and carry over	15 →	7.5				

Initial moments	–	22.5	– 83.33	83.33	–	–
Balancing		29.442	31.388	– 36.999	– 46.331	
Carry over			– 18.50	15.694		– 23.166
Balancing		8.954	9.546	– 6.968	– 8.726	
Carry over			– 3.484	4.773		– 4.363
Balancing		1.686	1.798	– 2.119	– 2.654	
Carry over			– 1.056	0.899		– 1.327
Balancing		0.511	0.545	– 0.399	– 0.500	
Carry over			– 0.200	+ 0.273		–0.250
Balancing and C.O.	–	0.097	0.103	– 0.121	– 0.152	– 0.076
Final moments	0	63.19	– 63.19	58.363	– 58.363	– 29.182

Simply supported bending moments :

Span AB : $\mu_{max} = \dfrac{Wl}{4} = \dfrac{30 \times 4}{4} = 30$ kNm

Span BC : $\mu_{max} = \dfrac{wl^2}{8} = \dfrac{40 \times 5^2}{8} = 125$ kNm

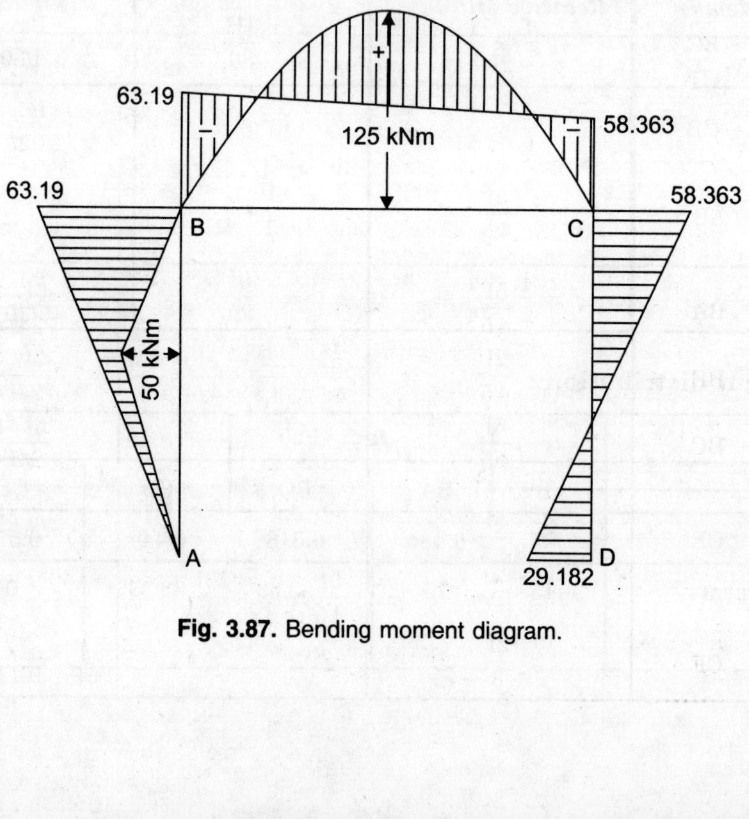

Fig. 3.87. Bending moment diagram.

Example 3.22 *Analyse the frame loaded as shown in Fig. 3.88 by the moment distribution method. Sketch bending moment and shear force diagrams.*

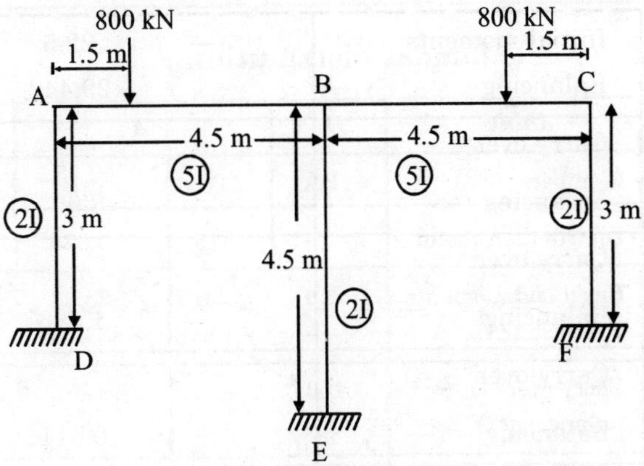

Fig. 3.88

Solution (*a*) **Fixed end moments :**

Span AB :

$$M_{FAB} = -\frac{Wab^2}{l^2} = -\frac{800 \times 1.5 \times 3^2}{4.5^2}$$

$$= -533.33 \text{ kNm}$$

$$M_{FBA} = +\frac{Wa^2b}{l^2} = +\frac{800 \times 1.5^2 \times 3}{4.5^2}$$

$$= +266.67 \text{ kNm}$$

Span BC :

$$M_{FBC} = -\frac{Wab^2}{l^2} = -\frac{800 \times 3 \times 1.5^2}{4.5^2} = -266.67 \text{ kNm}$$

$$M_{FBC} = -266.67 \text{ kNm}$$

$$M_{FCB} = +\frac{Wa^2b}{l^2} = +\frac{800 \times 3^2 \times 1.5}{4.5^2} = +533.33 \text{ kNm}$$

Span AD, BE and CF :

$$M_{FAD} = M_{FDA} = M_{FBE} = M_{FEB} = M_{FCF} = M_{FFC} = 0$$

(*b*) **Distribution factors :**

Joint	*Member*	*Relative stiffness*	*Total stiffness*	*Distribution factor*
A	AD	$\frac{2I}{3}$	$\frac{2I}{3} + \frac{5I}{4.5} = \frac{16I}{9}$	$\frac{2I/3}{16I/9} = 0.375$
	AB	$\frac{5I}{4.5}$		$\frac{5I/4.5}{16I/9} = 0.625$
B	BA	$\frac{5I}{4.5}$	$\frac{5I}{4.5} + \frac{2I}{4.5} + \frac{5I}{4.5} = \frac{24I}{9}$	$\frac{5I/4.5}{24I/9} = 0.417$
	BE	$\frac{2I}{4.5}$		$\frac{2I/4.5}{24I/9} = 0.167$
	BC	$\frac{5I}{4.5}$		$\frac{5I/4.5}{24I/9} = 0.417$
C	CB	$\frac{5I}{4.5}$	$\frac{5I}{4.5} + \frac{2I}{3} = \frac{16I}{9}$	$\frac{5I/4.5}{16I/9} = 0.625$
	CF	$\frac{2I}{3}$		$\frac{2I/3}{16I/9} = 0.375$

(*c*) **Moment distribution :**

Joint	D	A		B			C		F
Members	DA	AD	AB	BA	BE	BC	CB	CF	FC
Distribution factor	–	0.375	0.625	0.417	0.167	0.417	0.625	0.375	–
Fixed end moments	0	0	– 533.33	266.67	0	– 266.67	533.33	0	0
Balancing (1)	–	+ 200	+ 333.33	–	–	–	– 333.33	– 200	–
Carry over (1)	+ 100	–	–	166.67		– 166.67	–		– 100
Balancing (2)	–	–	–	–	–	–	–	–	–
Final moments	+ 100	+ 200	– 200	433.34	0	– 433.34	200	– 200	– 100

We know from basic properties that, $M_{EB} = \frac{1}{2} M_{BE} = \frac{1}{2} (0) = 0$

Simply supported bending moments :

Span AB : $\frac{Wab}{l} = \frac{800 \times 1.5 \times 3}{4.5} = 800 \text{ kNm}$

Span BC : $\frac{Wab}{l} = \frac{800 \times 3 \times 1.5}{4.5} = 800 \text{ kNm}$

(*d*) **To draw S.F.D. :**

Span AB :

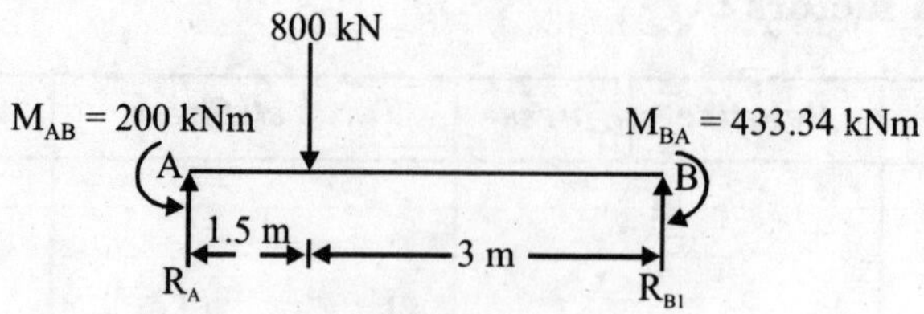

Fig. 3.89

Taking moment about A,

$$433.34 + 800\,(1.5) - 200 - R_{B1}\,(4.5) = 0$$

(↻) (↻) (↺) (↺)

$R_{B1} = 318.52$ kN

$R_A = 800 - 318.52 = 481.48$ kN

Span BC :

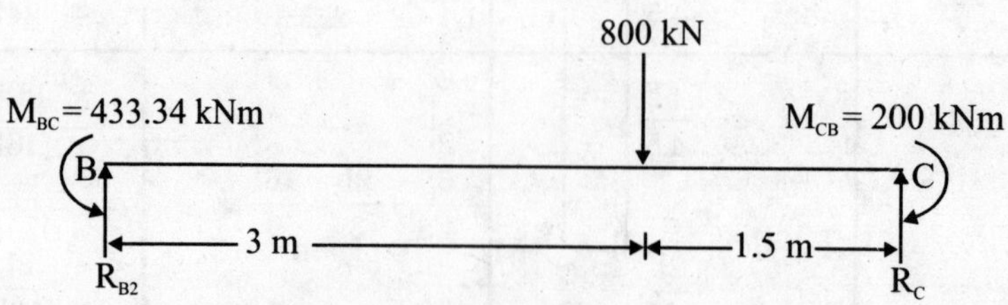

Fig. 3.90

Taking moment about B,

$$200 + 800 \times 3 - 433.34 - R_C \times 4.5 = 0$$

(↻) (↻) (↺) (↺)

$$R_C = 481.48 \text{ kN}$$
$$R_{B2} = 318.52 \text{ kN}$$

Member AD :

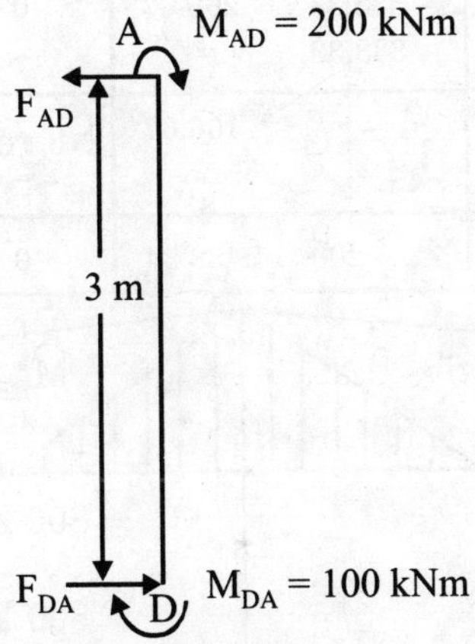

Fig. 3.91

Taking moments about D, $200 + 100 - F_{AD}(3) = 0$

(↻) (↻) (↺)

$$F_{AD} = 100 \text{ kNm } (\longleftarrow)$$
$$F_{DA} = 100 \text{ kNm } (\longrightarrow)$$

Member CF :

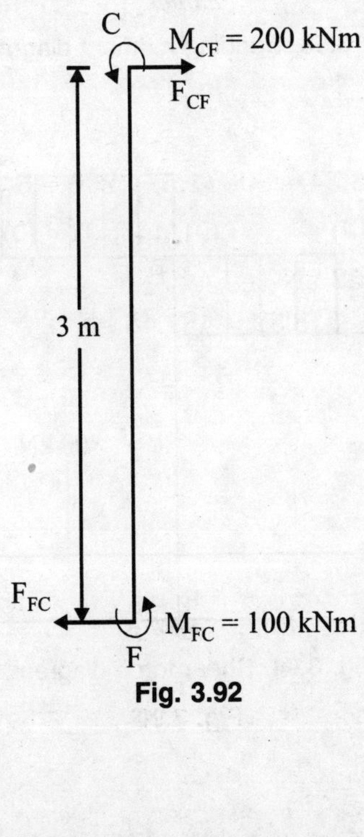

Fig. 3.92

Taking moments about F, $-200 - 100 + F_{CF} \times 3 = 0$

(↻) (↻) (↺)

$F_{CF} = 100$ kNm(→)

$F_{FC} = 100$ kNm(←)

Member BE :

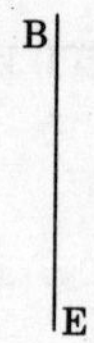

No reactions.

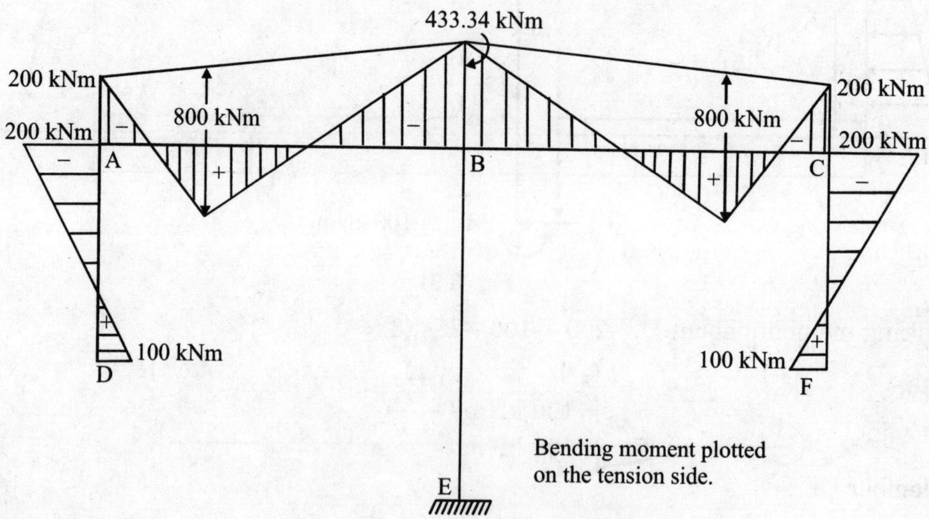

Fig. 3.93. Bending moment diagram.

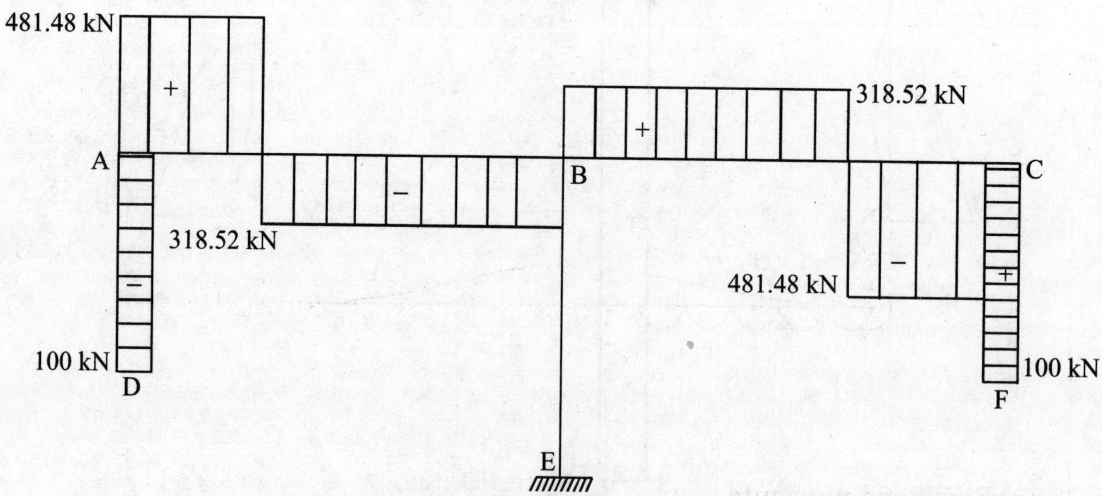

Fig. 3.94. Shear force diagram.

Example 3.23 *Analyse the box culvert loaded as shown in Fig. 3.95 by the moment distribution method. Sketch the bending moment and shear force diagrams.*

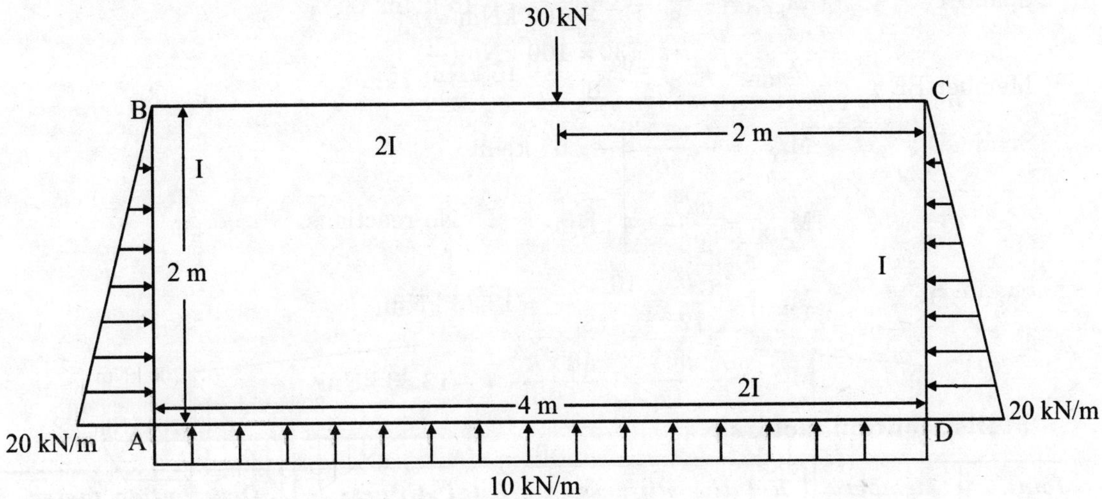

Fig. 3.95

Solution

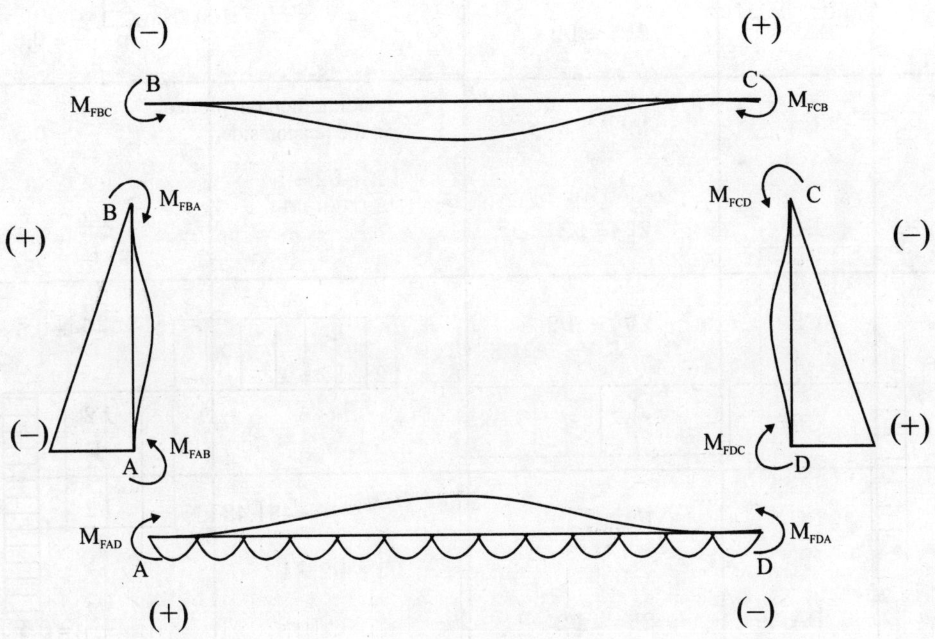

Fig. 3.96

(*a*) **Fixed end moments :**

Span AB : $M_{FAB} = -\dfrac{wl^2}{20} = -\dfrac{20 \times 2^2}{20} = -4$ kNm

$$M_{FBA} = +\frac{wl^2}{30} = \frac{20 \times 2^2}{30} = 2.67 \text{ kNm}$$

Span BC : $$M_{FBC} = -\frac{wl}{8} = -\frac{30 \times 4}{8} = -15 \text{ kNm}$$

$$M_{FCB} = +\frac{wl}{8} = \frac{30 \times 4}{8} = 15 \text{ kNm}$$

Span CD : $$M_{FCD} = -\frac{wl^2}{30} = -2.67 \text{ kNm}$$

$$M_{FDC} = \frac{wl^2}{20} = 4 \text{ kNm}$$

Span AD : $$M_{FAD} = \frac{wl^2}{12} = \frac{10 \times 4^2}{12} = 13.33 \text{ kNm}$$

$$M_{FDA} = -\frac{wl^2}{12} = -\frac{10 \times 4^2}{12} = -13.33 \text{ kNm}$$

(*b*) **Distribution factors :**

Joint	***Members***	***Relative stiffness***	***Total stiffness***	***Distribution factor***
A	AB	I/2	I/2 + I/2 = I	$\frac{I/2}{I} = 0.5$
	AD	2I/4 = I/2		$\frac{I/2}{I} = 0.5$
B	BA	I/2	I/2 + I/2 = I	$\frac{I/2}{I} = 0.5$
	BC	2I/4= I/2		$\frac{I/2}{I} = 0.5$
C	CB	2I/4 = I/2	I/2 + I/2 = I	$\frac{I/2}{I} = 0.5$
	CD	I/2		$\frac{I/2}{I} = 0.5$
D	DC	I/2	I/2 + I/2 = I	$\frac{I/2}{I} = 0.5$
	DA	2I/4 = I/2		$\frac{I/2}{I} = 0.5$

(*c*) **Moment distribution :**

Joint	*A*		*B*		*C*		*D*	
Members	AD	AB	BA	BC	CB	CD	DC	DA
Distribution factor	0.5	0.5	0.5	0.5	0.5	0.5	0.5	0.5
Fixed end moments	13.33	– 4	+ 2.67	– 15	+ 15	– 2.67	4	– 13.33
Balancing (1)	– 4.67	– 4.67	6.17	6.17	– 6.17	– 6.17	4.67	4.67
Carry over to 'DA' (1)	2.34	3.08	– 2.34	– 3.08	3.08	2.34	– 3.08	– 2.34
Balancing (2)	– 2.71	– 2.71	+ 2.71	+ 2.71	– 2.71	– 2.71	+ 2.71	+ 2.71
Carry over (2)	1.36	1.36	– 1.36	– 1.36	1.36	1.36	– 1.36	– 1.36
Balancing (3)	– 1.36	– 1.36	+ 1.36	+ 1.36	– 1.36	– 1.36	+ 1.36	+ 1.36
Carry over (3)	0.68	0.68	– 0.68	– 0.68	0.68	0.68	– 0.68	– 0.68
Balancing (4)	– 0.68	– 0.68	+ 0.68	+ 0.68	– 0.68	– 0.68	+ 0.68	+ 0.68
Final moments	8.29	– 8.30	9.21	– 9.20	9.20	– 9.21	8.3	– 8.29

Simply supported bending moments :

Span AB : $M_{max} = 0.128\ Wl = 0.128 \times 20 \times 2$

$M_{max} = 5.12$ kNm occurs at $0.577l$ from B

Occurs at 1.154 m from B

Span BC : $M_{max} = \frac{Wl}{4} = \frac{30 \times 4}{4} = 30$ kNm

Span CD : $M_{max} = 0.128\ Wl = 5.12$ kNm occurs at 1.15 m from C

Span AD : $M_{max} = \frac{Wl^2}{8} = \frac{10 \times 4^2}{8} = 20$ kNm

(*d*) **To draw S.F.D. :**

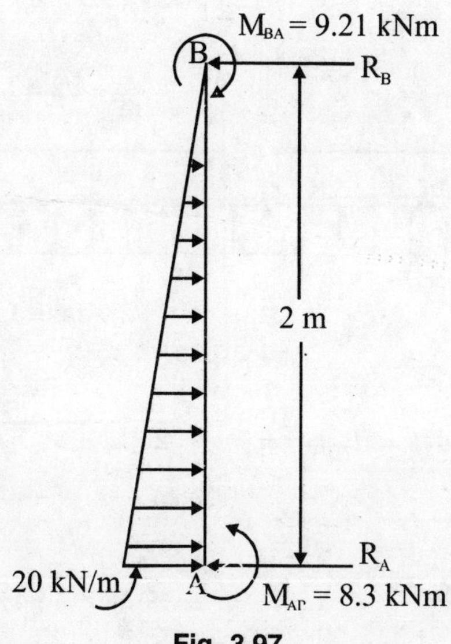

Fig. 3.97

Taking moments about A,

$$M_{BA} + \left(\frac{1}{2} \times 2 \times 20\right) \times \frac{1}{3} \times 2 - R_B(2) - M_{AB} = 0$$

$$9.21 + 13.33 - 2R_B - 8.3 = 0$$

$$R_B = 7.12 \text{ kN}$$

$$R_A = \text{Total load} - R_B = \left(\frac{1}{2} \times 2 \times 20\right) - 7.12$$

$$R_A = 12.88 \text{ kN}$$

Span BC :

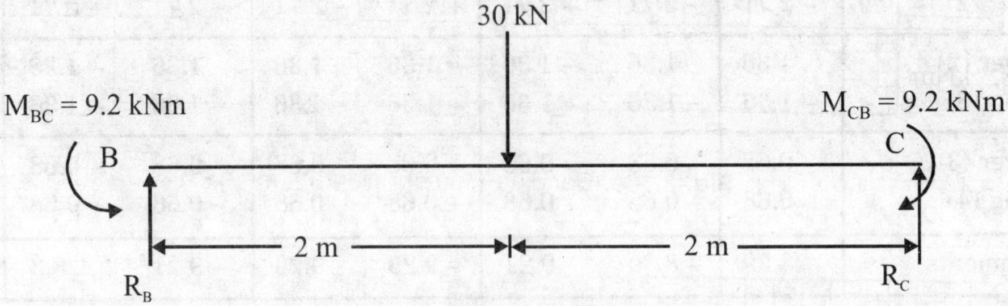

Fig. 3.98

Taking moments about B, $M_{BC} = - M_{CB}$

∴ This will not cause any change in reaction

∴ For centrally applied load, $R_B = R_C = 30/2 = 15$ kN

Span CD :

Loads and moments similar to span AB. Hence ;

$$R_C = 7.12 \text{ kN } (\rightarrow)$$

$$R_D = 12.88 \text{ kN } (\rightarrow)$$

Span AD :

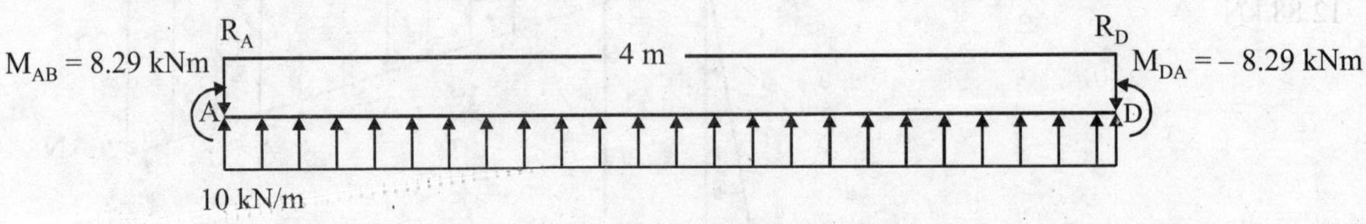

Fig. 3.99

$$R_A = R_D = \frac{10 \times 4}{2} = 20 \text{ kN}$$

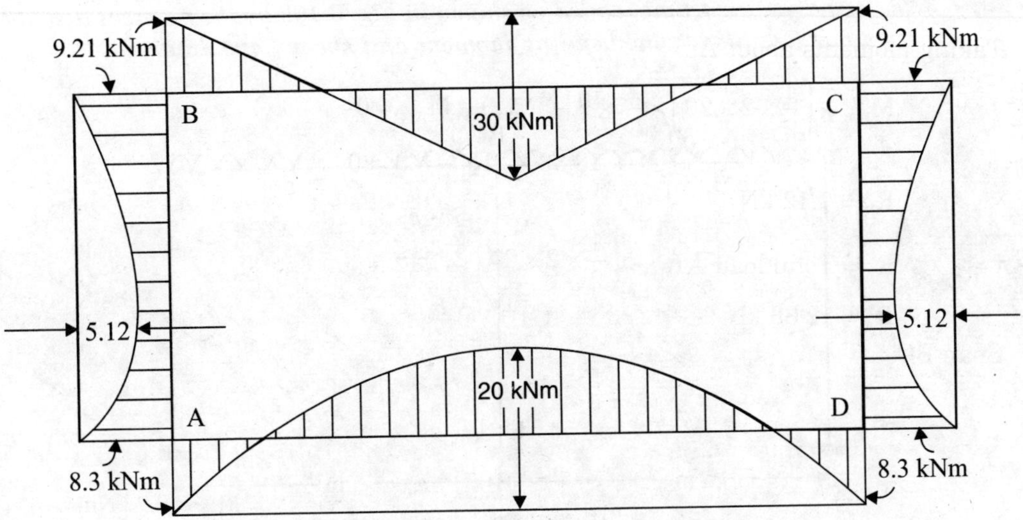

Fig. 3.100. Bending moment diagram.

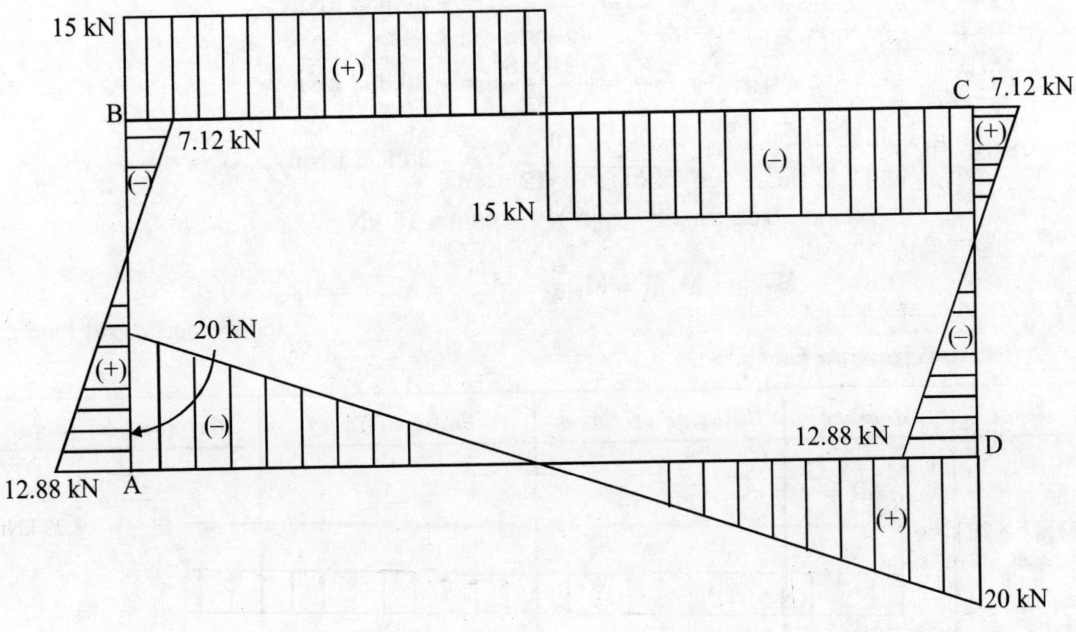

Fig. 3.101. Shear force diagram.

Example 3.24 *Analyse the frame loaded as shown in Fig. 3.102 by the moment distribution method. Sketch the bending moment and shear force diagrams.*

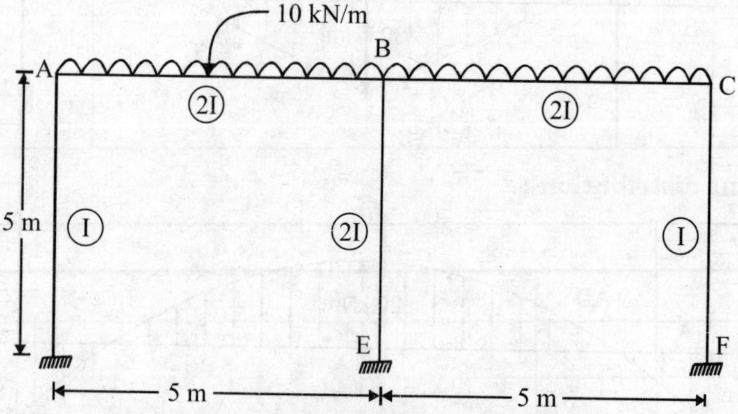

Fig. 3.102

Solution (*a*) **Fixed end moments :**

Span AB : $M_{FAB} = -\dfrac{wl^2}{12} = -\dfrac{10 \times 5^2}{12} = -20.833$ kNm

$M_{FBA} = +\dfrac{wl^2}{12} = +\dfrac{10 \times 5^2}{12} = +20.833$ kNm

Span BC : $M_{FBC} = -\dfrac{wl^2}{12} = -\dfrac{10 \times 5^2}{12} = -20.833$ kNm

$M_{FCB} = +20.833$ kNm

Spans AD, BE, CF :

$$M_{FAD} = M_{FDA} = M_{FBE} = M_{FEB} = M_{FCF} = M_{FFC} = 0$$

(Since no lateral loading)

(*b*) **Distribution factors :**

Joint	*Member*	*Relative stiffness*	*Total stiffness*	*Distribution factor*
A	AD	$\frac{I}{5}$	$\frac{3I}{5}$	$\frac{I/5}{3I/5} = \frac{1}{3}$
	AB	$\frac{2I}{5}$		$\frac{2I/5}{3I/5} = \frac{2}{3}$
B	BA	$\frac{2I}{5}$	$\frac{6I}{5}$	$\frac{2I/5}{6I/5} = \frac{1}{3}$
	BE	$\frac{2I}{5}$		$\frac{2I/5}{6I/5} = \frac{1}{3}$
	BC	$\frac{2I}{5}$		$\frac{2I/5}{6I/5} = \frac{1}{3}$

C	CB	$\frac{2I}{5}$	$\frac{3I}{5}$	$\frac{2I/5}{3I/5} = \frac{2}{3}$
	CF	$\frac{I}{5}$		$\frac{I/5}{3I/5} = \frac{1}{3}$

(*c*) **Moment distribution :**

Joint	***A***		***B***			***C***	
Members	AD	AB	BA	BE	BC	CB	CF
Distribution factors	1/3	2/3	1/3	1/3	1/3	2/3	1/3
Fixed end moments	0	– 20.833	+ 20.833	0	– 20.833	+ 20.833	–
Balancing	6.944	13.889	–	–	–	– 13.889	– 6.944
Carry over	–	–	6.445	–	– 6.445	–	–
Final moments	6.944	– 6.944	27.278	0	– 27.278	+ 6.944	– 6.944

$$M_{FDA} = \frac{6.944}{2} = 3.472 \text{ kNm}$$

$$M_{FFC} = -\frac{6.944}{2} = -3.472 \text{ kNm}$$

(*d*) **To draw S.F.D. :**

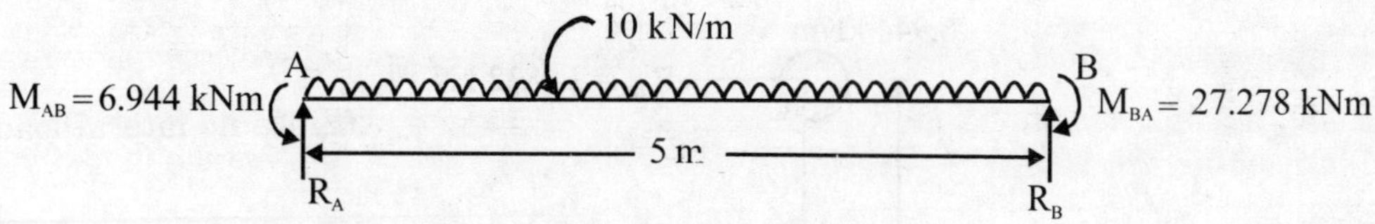

Fig. 3.103

Span AB : Taking moments about A,

$$-R_B \times 5 - 6.944 + 27.278 + 10 \times 5 \times \frac{5}{2} = 0$$

(↺) (↺) (↻) (↻)

$R_B = 29.067$ kN

$R_A = 10 \times 5 - 29.067$

$R_A = 20.933$ kN

Span BC : As the frame and loading are symmetrical,

$R_B = 29.067$ kN

$R_C = 20.933$ kN

Span AD :

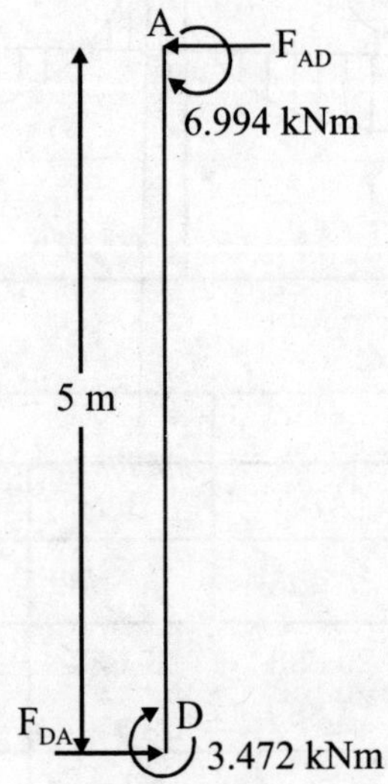

Fig. 3.104

Taking moments about D, $\quad - F_{AD} \times 5 + M_{DA} + M_{AC} = 0$

(↻) (↺) (↺)

$$3.472 + 6.944 - F_{AD}\,(5) = 0$$
$$F_{AD} = 2.0832 \text{ kN } (\rightarrow)$$
$$F_{DA} = 2.0832 \text{ kN } (\leftarrow)$$

Span CF :

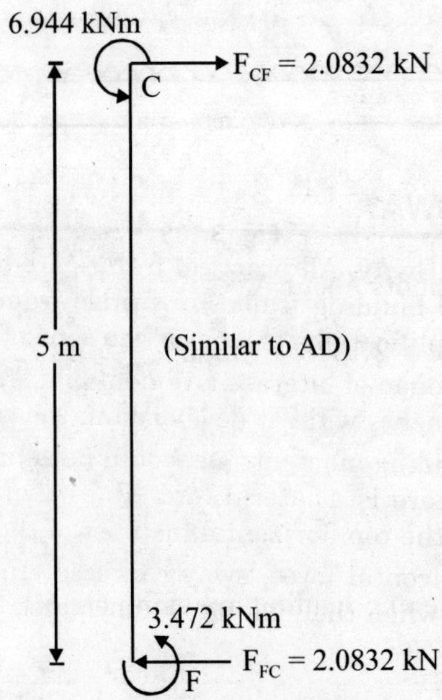

Fig. 3.105

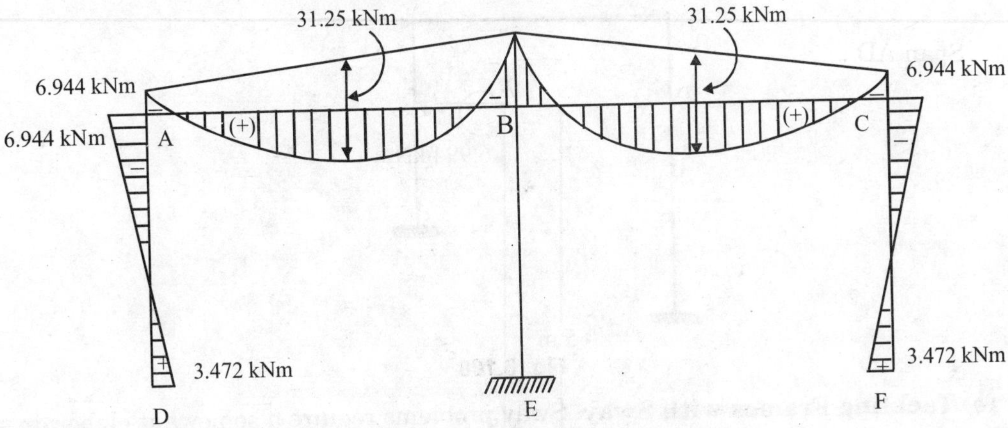

Fig. 3.106. Bending moment diagram.

$$\text{Simply supported bending moment} = \frac{10 \times (5)^2}{8} = 31.25 \text{ kNm}$$

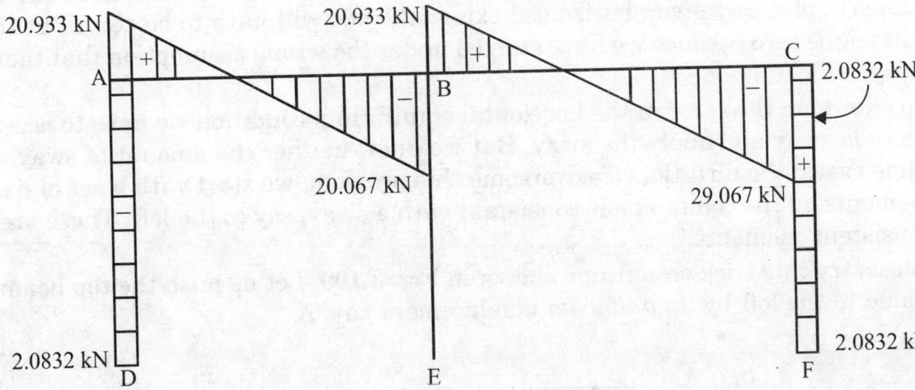

Fig. 3.107. Shear force diagram.

3.7 FRAMES WITH SWAY

(*a*) **The Concept of Sway.** All practical frames are liable to sway, be they multistoried building frames, industrial building frames or bridge frames. Non-sway cases that we dealt with hitherto are only simplifications or approximations of real problems.

Sway is the phenomenon of lateral movement of a structural frame (Fig. 3.108). Due to sway, a set of bending moments will get developed in the columns and beams in a frame.

To sustain these bending moments we would need a lateral force at the top level of the columns. Conversely, if there is a lateral force F_H, it will give rise to a certain horizontal displacement (or sway) of the top horizontal member of the frame.

Even without a horizontal force, sway can occur in a frame. This is when the frame itself is unsymmetrical or when the loading is unsymmetrical.

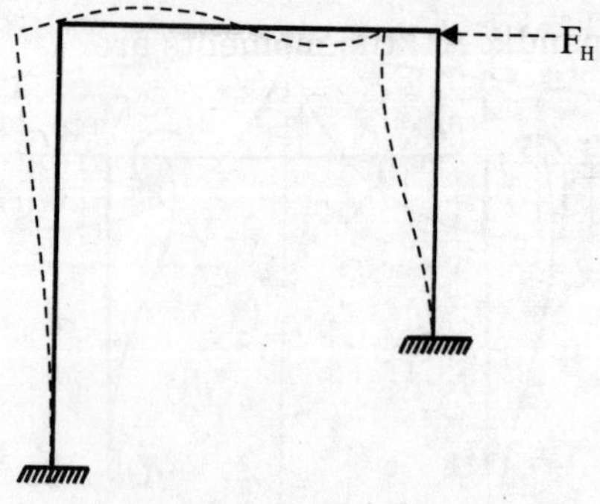

Fig. 3.108

(*b*) **Tackling Frames with Sway.** Sway problems require a somewhat elaborate solution process.

(*i*) As a first step, the given frame is analysed assuming that there is no sway. At the end of this step we compute the moments and reactions at the supports. When we consider the horizontal reactions at the base of column members we would come across the first sign of trouble. The sum of the horizontal reactions at the column supports plus any given horizontal external loads will have to be zero. But the sum will not be zero because we have started under the wrong assumption that there is no sway.

(*ii*) To eliminate the error in the horizontal equilibrium condition we have to accept that there is sway and apply the sway. But we know neither the amount of sway, nor (in some cases) the direction. To overcome this problem, we start with a set of fixed end moments on the column ends consistent with a sway, say to the left. These are called consistent moments.

(*iii*) Let us try this trick on a frame shown in Fig. 3.109. Let us push the top beam of the frame to the left by an *unknown* displacement say, Δ.

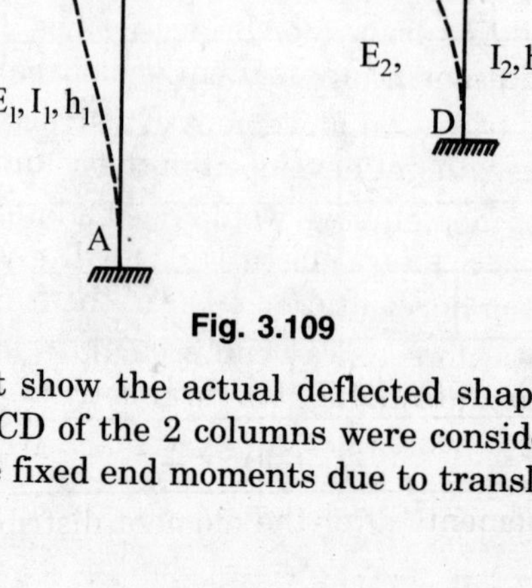

Fig. 3.109

The dotted lines do not show the actual deflected shape, but they represent the deflected shape if ends BA and CD of the 2 columns were considered fixed. The corresponding moments at A, B, C and D are fixed end moments due to translation of fixed ends.

Referring to case 12 of appendix A1, the moments are

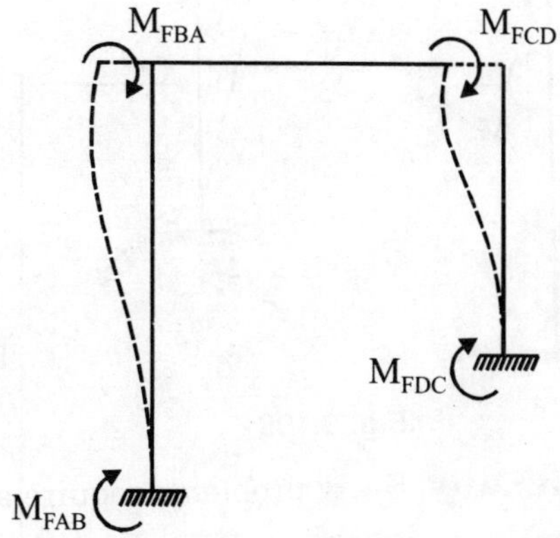

Fig. 3.110

$$M_{FAB} = M_{FBA} = +\frac{6E_1I_1\Delta}{h_1^2}$$

$$M_{FCD} = M_{FDC} = +\frac{6E_2I_2\Delta}{h_2^2}$$

Since we do not know Δ, the actual values of these moments are unimportant. It is the ratio of these moments that is important.

Hence we need $M_{FAB} / M_{FDC} = (E_1I_1/h_1^2) / (E_2I_2/h_2^2)$

Since E_1 and E_2 cannot usually be different $M_{FAB} / M_{FDC} = (I_1/h_1^2) / (I_2/h_2^2)$.

(*c*) **Sway Moment Distribution :**

For the frame in Fig. 3.109, the stiffness ratios are as under:

Joint	***Member***	k_i	Σk_i	$DF = \frac{k_i}{\Sigma k_i}$
B	BA BC	$k_1 = 4E_1I_1/h_1$ $k_2 = 4E_3I_3/l_3$	$(k_1 + k_2)$	$D_1 = k_1/(k_1 + k_2)$ $D_2 = k_2/(k_1 + k_2)$
C	CB CD	$k_2 = 4E_3I_3/l_3$ $k_3 = 4EI/h_2$	$(k_2 + k_3)$	$D_3 = k_2/(k_2 + k_3)$ $D_4 = k_3/(k_2 + k_3)$

The moment distribution process will be as under :

Joint	***A***	***B***	***B***	***C***	***C***	***D***
Member	AB		BC		CD	
DF		D_1	D_2	D_3	D_4	
FEM	M_{FAB}	M_{FAB}			M_{FDC}	M_{FDC}
Final moments	M_{AB}	M_{BA}	$-M_{BA}$	$-M_{CD}$	M_{CD}	M_{DC}

The last line indicates the moments after the moment distribution process.

(*d*) **Determining the Force Imbalance :**

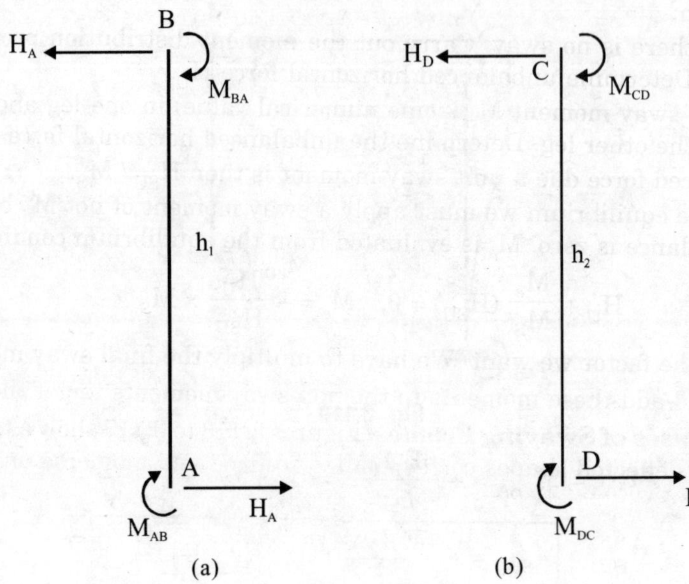

Fig. 3.111

Figure 3.111 (*a*) and (*b*) are the free bodies of the two columns of the portal frame. These will be in equilibrium with the possible horizontal forces at A, B, C and D. We can guess that the horizontal forces H_A and H_B will be equal and opposite. Likewise H_D and H_C.

(*e*) **From Moment Equilibrium Conditions:**

$$H_A \times h_1 = M_{AB} + M_{BA} \ ; \ H_A = (M_{AB} + M_{BA})/h_1$$

$$H_D = (M_{DC} + M_{CD}) \ / \ h_2$$

It is very likely that H_A and H_D do not balance each other. In that case $H_A + H_D$ is the horizontal imbalance.

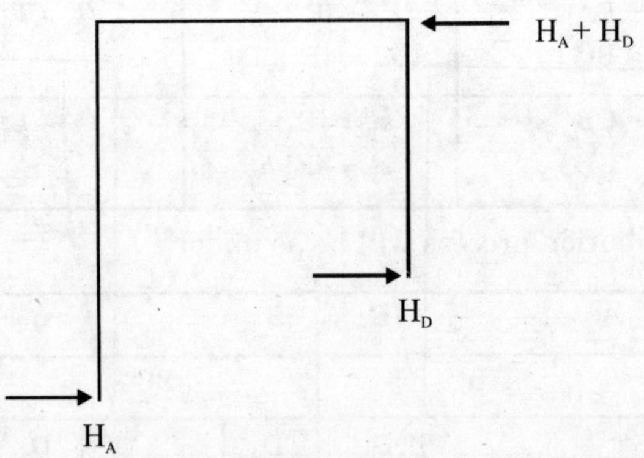

Fig. 3.112

Finding ($H_A + H_D$) the imbalance corresponding to the arbitrary set of consistent fixed end moments we started with in section 3.7 (*b*), (*c*) (M_{FAB} and M_{FDC}) finishes our quest to determine the imbalance which is the end result of sway moment distribution.

Thus we can evolve here a step by step method of tackling likely-to-sway frames loaded in any give fashion.

(*i*) Assume there is no sway. Carry out the moment distribution process for the given loading. Determine unbalanced horizontal force H_U.

(*ii*) Choose a sway moment M_S (some numerical value) in one leg and a consistent moment in the other leg. Determine the unbalanced horizontal force H_{SU}.

(*iii*) Unbalanced force due a unit sway moment is then H_{SU} / M_S.

(*iv*) To ensure equilibrium we must apply a sway moment of not M_S but M_e, so that the net imbalance is zero. M_e is evaluated from the equilibrium condition.

$$H_U + \frac{M_e}{M_S}(H_{SU}) = 0, \quad M_e = -\frac{H_U}{H_{SU}} \times M_S$$

H_U/H_{SU} is the factor we want. We have to multiply the final sway moments by H_U/H_{SU}, and add these moments to the non-sway moments to get the final moments.

(*f*) **Some Cases of Swaying Frames:** Figures 3.113 to 3.117 show a few typical swaying patterns, and the deflected shapes corresponding to fixed end moments on frame columns.

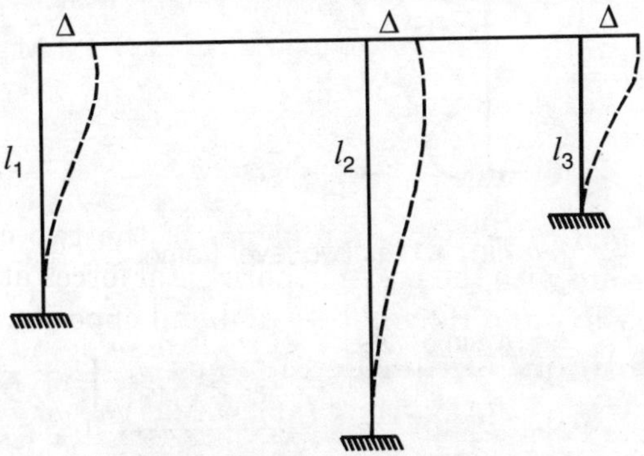

Fig. 3.113. Two-bay portal.

The FEMs are $-\dfrac{6EI\Delta}{l_1^2}, -\dfrac{6EI\Delta}{l_2^2}$ and $-\dfrac{6EI\Delta}{l_3^2}$

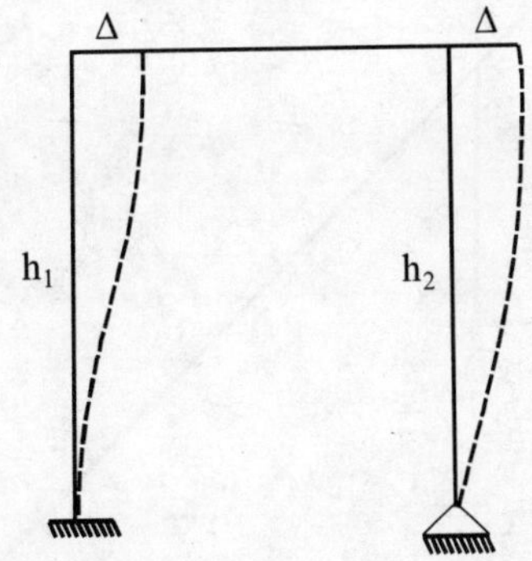

Fig. 3.114. Fixed-hinged portal.

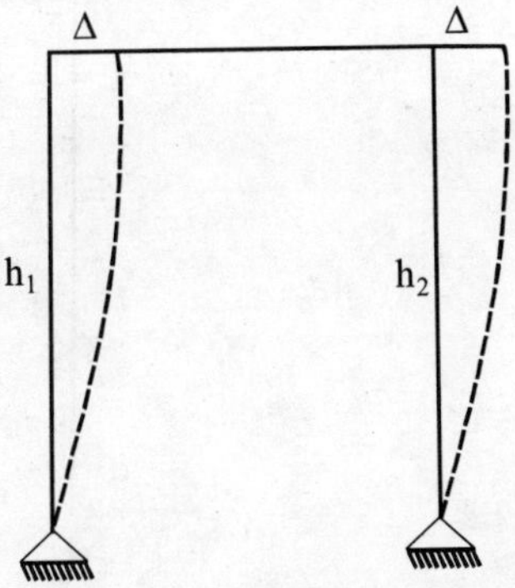

Fig. 3.115. Hinged-hinged portal.

For Fig. 3.114, the FEMs are $\dfrac{6EI\Delta}{h_1^2}; \dfrac{3EI\Delta}{h_2^2}$.

For Fig. 3.115, the FEMs are $\dfrac{3EI\Delta}{h_1^2}; \dfrac{3EI\Delta}{h_2^2}$.

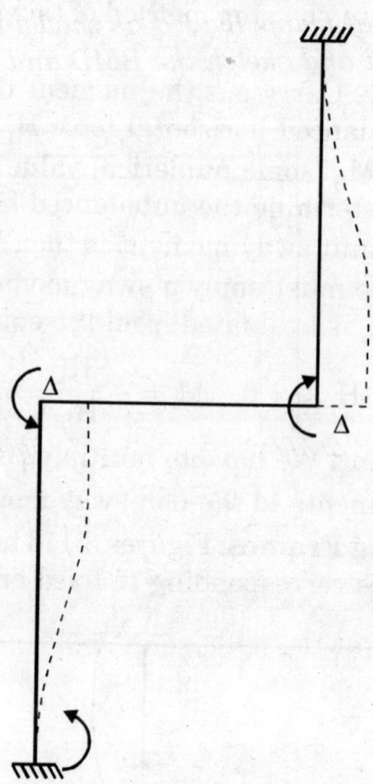

Fig. 3.116. Two-level frame.

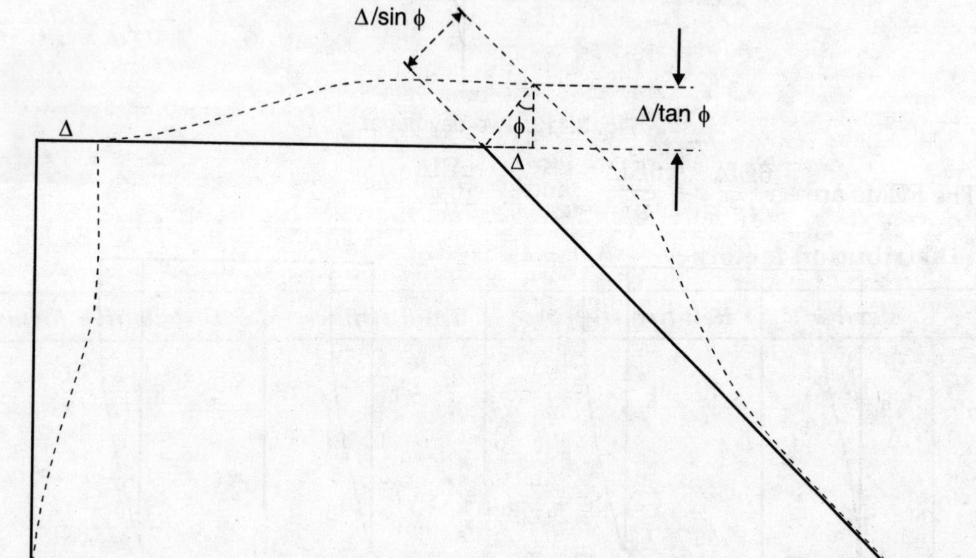

Fig. 3.117. Portal with a sloping leg.

Example 3.25 *Analyse the portal frame loaded as shown in Fig. 3.118 by the moment distribution method and sketch the BMD and SFD.*

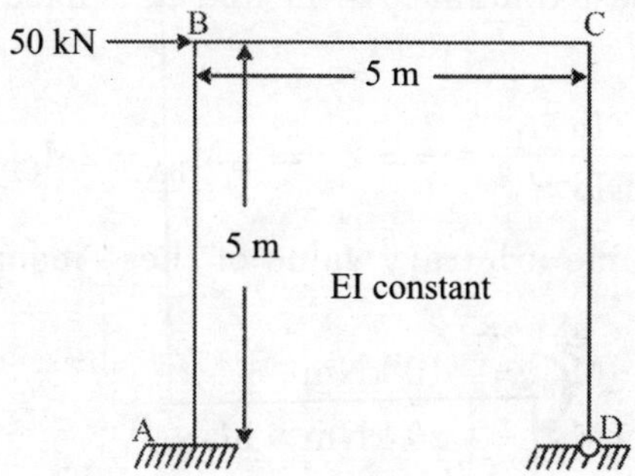

Fig. 3.118

Solution (*a*) **Fixed end moments :**

Span AB, BC, CD : $M_{FAB} = M_{FBA} = M_{FBC} = M_{FCB} = M_{FCD} = M_{FDC} = 0$.

Since the load is acting only at the joint, there will be no fixed end moments. However, due to side sway, moments will be induced at joints A, B and C.

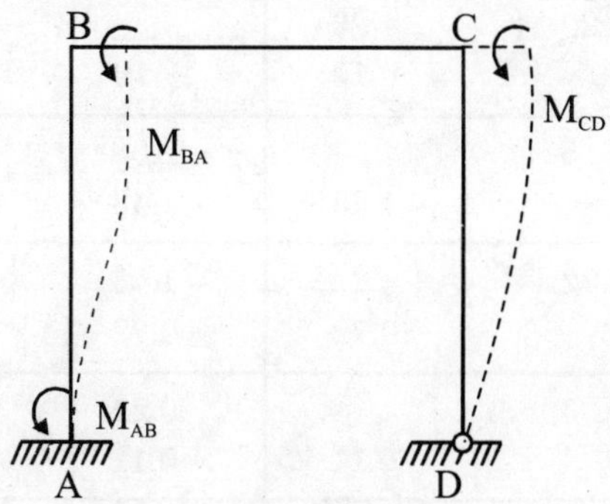

Fig. 3.119

(*b*) **Distribution factors :**

Joint	*Member*	*Relative stiffness*	*Total stiffness*	*Distribution factor*
B	BA	$\frac{I}{5}$	$\frac{I}{5}+\frac{I}{5}=\frac{2I}{5}$	$\frac{I/5}{2I/5}=\frac{1}{2}$
	BC	$\frac{I}{5}$		$\frac{I/5}{2I/5}=\frac{1}{2}$
C	CB	$\frac{I}{5}$	$\frac{I}{5}+\frac{3I}{20}=\frac{7I}{20}$	$\frac{I/5}{7I/20}=\frac{4}{7}$
	CD	$\frac{3}{4}\times\frac{I}{5}=\frac{3I}{20}$		$\frac{3I/20}{7I/20}=\frac{3}{7}$

(*c*) **Side sway:** Under the action of the 50 kN load, there will be side sway to the right and the columns AB and CD will rotate in a clockwise direction. Thus negative moments will be induced at A, B and C in these columns. As the end 'A' is fixed and 'D' is hinged, the ratio of moments will be,

$$\frac{M_{BA}}{M_{CD}} = \frac{6EI\delta / l_1^2}{3EI\delta / l_2^2} = \frac{2}{1} = 2 \Rightarrow M_{BA} = 2M_{CD} \qquad (I_1 = I_2 = I \,;\, l_1 = l_2 = l)$$

Let us, first of all assume arbitrary value of these moments and find out the corresponding sway force.

Let $M_{CD} = -10$ kNm

$M_{BA} = -20$ kNm $= M_{AB}$

(*d*) **Moment distribution :**

Joint	***A***	***B***		***C***		***D***
Member	AB	BA	BC	CB	CD	DC
Distribution factors	–	1/2	1/2	4/7	3/7	–
Fixed end moments	– 20	– 20	–	–	– 10	–
Balancing		10	10	5.71	4.29	
Carry over	5		2.86	5		
Balancing		– 1.43	– 1.43	– 2.86	– 2.14	
Carry over	– 0.72		– 1.43	– 0.72		
Balancing		0.72	0.72	0.41	0.31	
Carry over	0.36		0.21	0.36		
Balancing		– 0.11	– 0.11	– 0.21	– 0.15	
Carry over	– 0.06		– 0.11	– 0.06		
Balancing	0.06	0.05	0.06	0.03	0.03	
Final moments	– 15.42	– 10.76	10.77	7.66	– 7.66	0

(*e*) **Sway forces :**

Balancing moment for AB = 10.76 + 15.42 = 26.18 kNm

Horizontal reaction at B = (26.18) / 5 = 5.24 kN (→) = H_{BA}

Horizontal reaction at A = 5.24 kN (←) = H_{AB}

Balancing moment for CD = 7.66 kNm

Horizontal reaction at C = (7.66) / 5 = 1.53 kN (→)

Horizontal reaction at D = 1.53 kN (←)

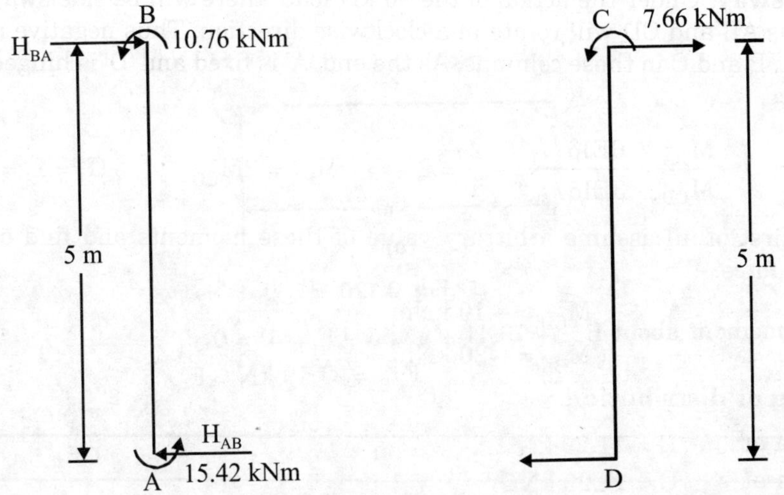

The sway force causing the assumed moments = 5.24 + 1.53 = 6.77 kN (←)

But actual sway force is 50 kN ; hence the moments be increased proportionately in the ratio of (50) / (6.77) = 7.38.

Joints	*A*	*B*		*C*		*D*
H_{sway} = 6.77 kN	– 15.42	– 10.76	10.76	7.66	– 7.66	0
H_{sway} = 50 kN	– 113.8	– 79.41	79.41	56.53	– 56.53	0

Horizontal reaction at A = 5.24 × 7.38 = 38.67 kN (←)

Horizontal reaction at D = 1.53 × 7.38 = 11.29 kN (←)

(*f*) **To draw S.F.D.:**

Span AB :

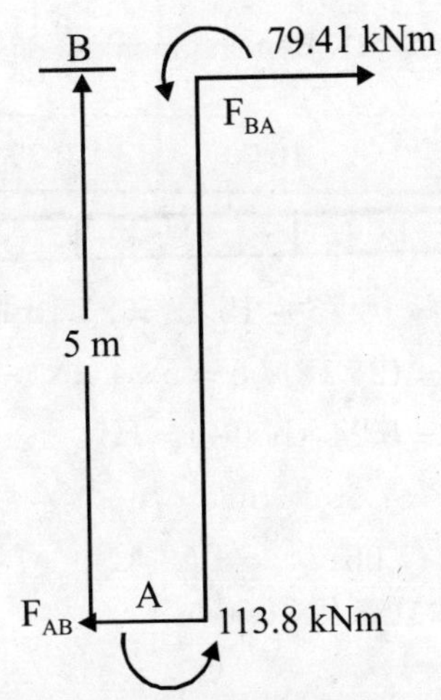

(a)

Taking moments about A, $\quad -79.41 - 113.8 + F_{BA}(5) = 0$

$$F_{BA} = 38.642 \text{ kN} = F_{AB}$$

Span BC :

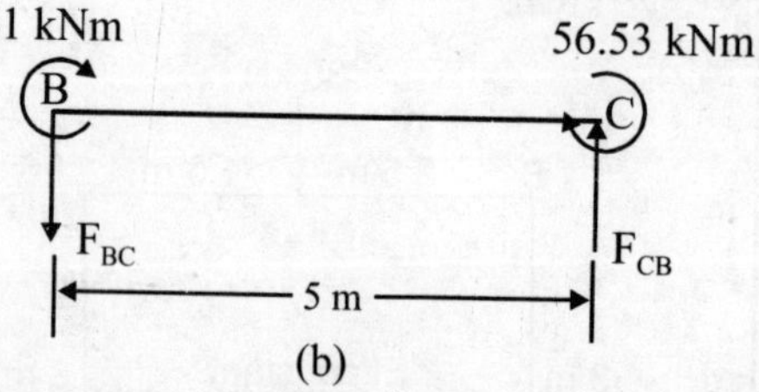

(b)

Fig. 3.120

Taking moment about B, $+ 79.41 + 56.53 - F_{CB}(5) = 0$

$$F_{CB} = 27.19 \text{ kN} = F_{BC}$$

Span CD :

Reaction at D = 11.29 kN ↓

Reaction at C = 11.29 kN ↑

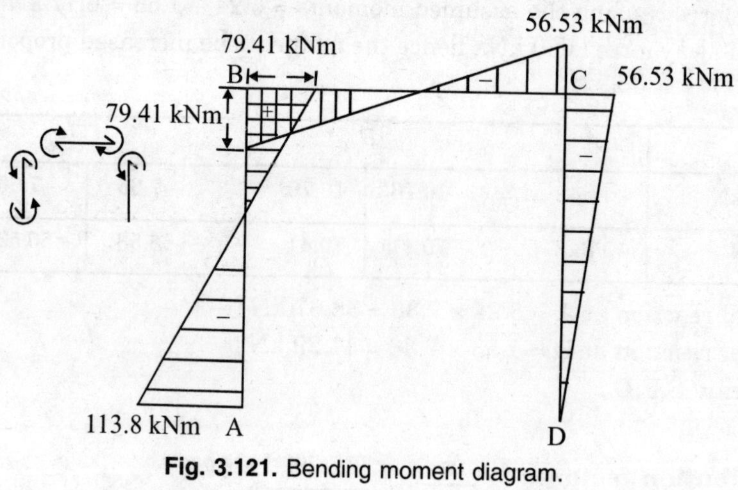

Fig. 3.121. Bending moment diagram.

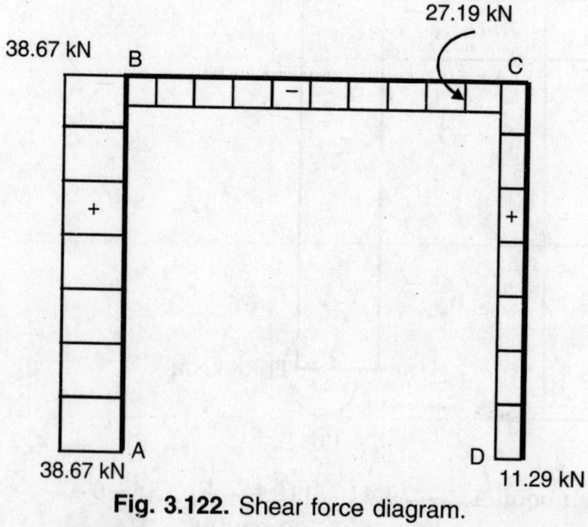

Fig. 3.122. Shear force diagram.

Example 3.26 *Analyse the portal frame loaded as shown in Fig. 3.123 by moment distribution method and sketch the bending moment diagram.*

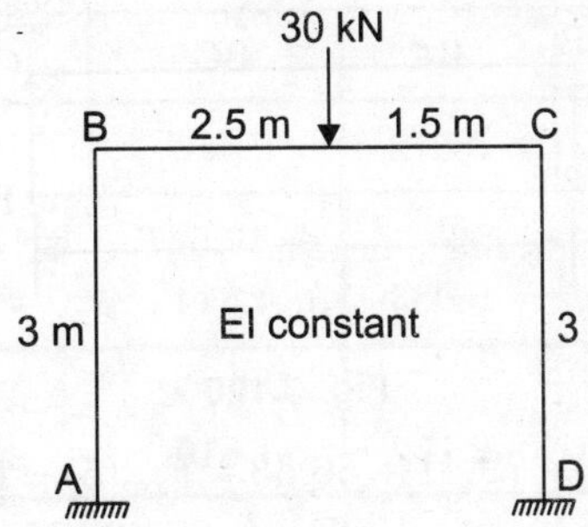

Fig. 3.123

Solution The frame is subjected to unsymmetrical loading. The frame will be subjected to side sway.

(A) **Non-sway Analysis :**

(*a*) **Fixed end moments :**

Span AB : $M_{FAB} = M_{FBA} = 0$

Span BC : $M_{FBC} = -\frac{Wab^2}{l^2} = -\frac{30 \times 2.5 \times 1.5^2}{4^2}$

$= -10.547$ kNm

$M_{FCB} = \frac{Wa^2b}{l^2} = \frac{30 \times 2.5^2 \times 1.5}{4^2}$

$= 17.578$ kNm

Span CD : $M_{FCD} = M_{FDC} = 0$

(*b*) **Distribution factors :**

Joint	*Member*	*Relative stiffness*	*Total stiffness*	*Distribution factor*
B	BA	$\frac{I}{L} = \frac{I}{3}$	$\frac{I}{3} + \frac{I}{4}$	$\frac{I/3}{7I/12} = 0.572$
	BC	$\frac{I}{L} = \frac{I}{4}$	$= \frac{7I}{12}$	$\frac{I/4}{7I/12} = 0.428$
C	CB	$\frac{I}{L} = \frac{I}{4}$	$\frac{I}{3} + \frac{I}{4}$	$\frac{I/4}{7I/12} = 0.428$
	CD	$\frac{I}{L} = \frac{I}{3}$	$= \frac{7I}{12}$	$\frac{I/3}{7I/12} = 0.572$

(*c*) **Moment distribution :**

Joint	*A*	*B*		*C*		*D*
Member	AB	BA	BC	CB	CD	DC
Distribution factor	–	0.572	0.428	0.428	0.572	
Fixed end moments	–	–	– 10.547	17.578	–	–
Balancing		6.033	4.514	– 7.523	– 10.055	
Carry over	3.017		– 3.762	2.257		– 5.028
Balancing		2.152	1.610	– 0.966	– 1.291	
Carry over	1.076		– 0.483	0.805		– 0.646
Balancing		0.276	0.207	– 0.345	– 0.460	
Carry over	0.138		– 0.173	0.103		– 0.23
Balancing and C.O.	0.050	0.099	0.074	– 0.044	– 0.059	– 0.03
Final moments	4.281	8.56	– 8.56	11.865	– 11.865	– 5.934

(*d*) **Shear force :**

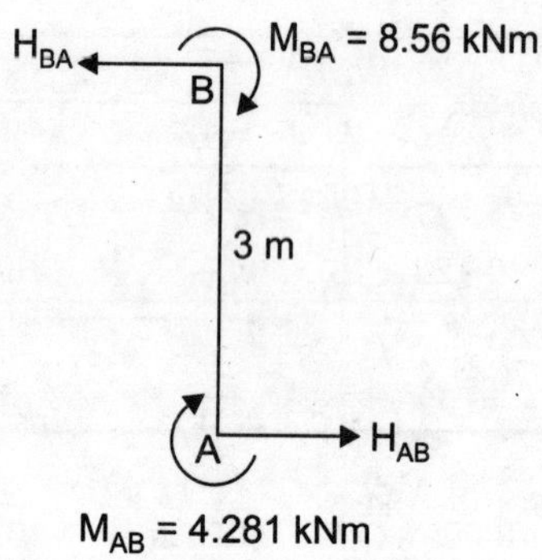

Taking moment about B,

$$-H_{AB} \times 3 + 4.281 + 8.56 = 0$$

$$H_{AB} = 4.28 \text{ kN } (\rightarrow)$$

$$H_{BA} = 4.28 \text{ kN } (\leftarrow)$$

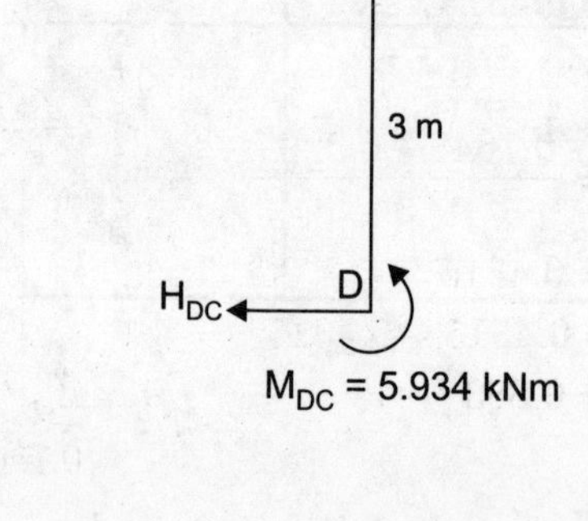

Taking moment about C,

$$H_{DC} \times 3 - 5.934 - 11.865 = 0$$

$$H_{DC} = 5.933 \text{ kN } (\leftarrow)$$

$$H_{CD} = 5.933 \text{ kN } (\rightarrow)$$

Value of force 'P' preventing sway = $H_{AB} \sim H_{DC}$

$$= 4.280 \sim 5.933 = 1.653 \text{ kN } (\rightarrow)$$

(B) **Side Sway Analysis :** Now, let a sway force of 1.653 kN (←) be applied at point C. This will induce clockwise moments at A and D.

$$\frac{M_{BA}}{M_{CD}} = \frac{I_1/l_1^2}{I_2/l_2^2} = \frac{I/3^2}{I/3^2} = \frac{1}{1}$$

Assume $M_{BA} = M_{AB} = +1$ kNm

$$M_{CD} = M_{DC} = +1 \text{ kNm}$$

Sway moment distribution:

Joint	***A***	***B***		***C***		***D***
Member	AB	BA	BC	CB	CD	DC
D.F.		0.572	0.428	0.428	0.572	
FEM	1.0	1.0			1.0	1.0
Balancing		– 0.572	– 0.428	– 0.428	– 0.572	
Carry over	– 0.286		– 0.214	– 0.214		– 0.286
Balancing		0.122	0.092	0.092	0.122	
Carry over	0.061		0.046	0.046		0.061
Balancing		– 0.026	– 0.02	– 0.02	– 0.026	
Carry over	– 0.013		– 0.01	– 0.01		– 0.013
Balancing and C.O.	0.0028	0.0057	0.0043	0.0043	0.0057	0.0028
Final moments	0.7648	0.5297	– 0.5297	– 0.5297	0.5297	0.7648

Sway force corresponds to the assumed moments:

For span AB:

Horizontal reaction, $H_A = \dfrac{0.7648 + 0.5297}{3.0}$

$$= 0.4315 \; (\rightarrow)$$

For span CD:

Horizontal reaction, $H_D = \dfrac{0.7648 + 0.5297}{3}$

$$= 0.4315 \; (\rightarrow)$$

Total horizontal reaction $= 0.4315 + 0.4315$

$$= 0.863 \; (\rightarrow)$$

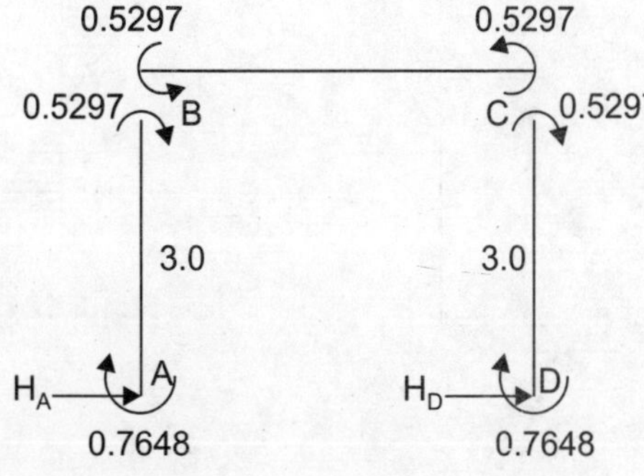

But the actual sway force = 1.653 kN

$$\therefore \quad \text{Correction factor} = \frac{1.653}{0.863} = 1.9154$$

To obtain correct sway moments, multiply the sway moments by 1.9154.

	Moments	M_{AB}	M_{BA}	M_{BC}	M_{CB}	M_{CD}	M_{DC}
1.	Non-sway moments	4.281	8.56	– 8.56	11.865	– 11.865	– 5.934
2.	Sway moment after correction	0.7648 × 1.9154 = 1.4649	0.5297 × 1.9154 = 1.0146	– 0.5297 × 1.9154 = – 1.0146	– 0.5297 × 1.9154 = – 1.0146	0.5297 × 1.9154 = 1.0146	0.7648 × 1.9154 = 1.4649
3.	Final moments (1) + (2)	5.7459	9.5746	– 9.5746	10.8504	– 10.8504	– 4.4691

$$H_A = \frac{5.7459 + 9.5746}{3}$$

$$= 5.1068\ (\rightarrow)$$

$$H_D = \frac{4.4691 + 10.8504}{3}$$

$$= 5.1068\ (\leftarrow)$$

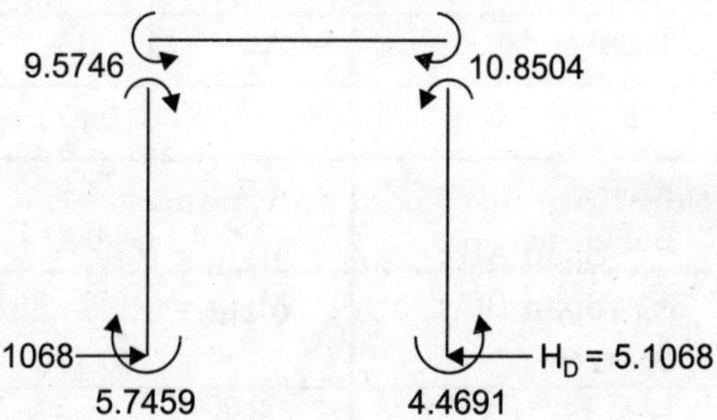

Free bending moment:

Span AB and CD = 0

$$\text{Span BC} = \frac{Wab}{l} = \frac{30(2.5)(1.5)}{4}$$

$$= 28.125 \text{ kNm}$$

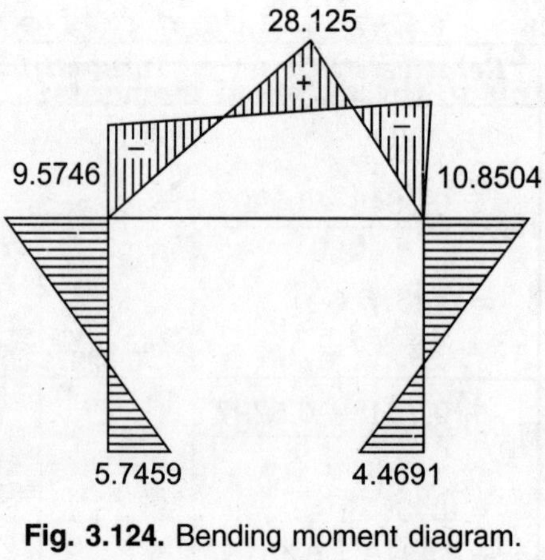

Fig. 3.124. Bending moment diagram.

Example 3.27 *Analyse the portal frame loaded as shown in Fig. 3.125 by the moment distribution method and sketch the BMD and SFD.*

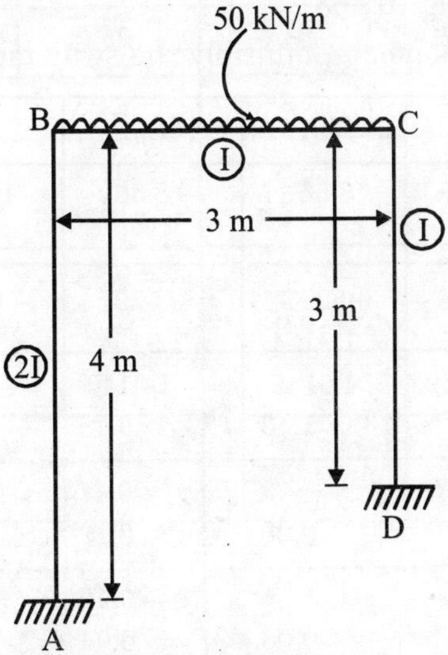

Fig. 3.125

Solution (*a*) **Fixed end moments :**

Span AB : $M_{FAB} = M_{FBA} = 0$

Span BC : $M_{FBC} = -M_{FCB}$

$$M_{FBC} = -\frac{wl^2}{12} = -\frac{50 \times 3^2}{12} = -37.5 \text{ kNm}$$

$$M_{FCB} = +\frac{wl^2}{12} = +\frac{50 \times 3^2}{12} = +37.5 \text{ kNm}$$

Span CD : $M_{FCD} = M_{FDC} = 0$

(*b*) **Distribution factors :**

Joint	*Members*	*Relative stiffness*	*Total stiffness*	*Distribution factor*
B	BA	$\frac{2I}{4} = \frac{I}{2}$	$\frac{I}{2} + \frac{I}{3} = \frac{5I}{6}$	$\frac{I/2}{5I/6} = \frac{3}{5}$
	BC	$\frac{I}{3}$		$\frac{I/3}{5I/6} = \frac{2}{5}$
C	CB	$\frac{I}{3}$	$\frac{I}{3} + \frac{I}{3} = \frac{2I}{3}$	$\frac{I/3}{2I/3} = \frac{1}{2}$
	CD	$\frac{I}{3}$		$\frac{I/3}{2I/3} = \frac{1}{2}$

(*c*) **Moment distribution (Non-sway):**

Joints	***A***	***B***		***C***		***D***
Members	AB	BA	BC	CB	CD	DC
Distribution factors	–	3/5	2/5	1/2	1/2	–
Fixed end moments			– 37.5	+ 37.5		
Balancing		22.5	15.0	– 18.75	– 18.75	
Carry over	11.25		– 9.38	7.5		– 9.38
Balancing		5.63	3.75	– 3.75	– 3.75	
Carry over	2.82		– 1.88	1.88		– 1.88
Balancing		1.13	0.75	– 0.94	– 0.94	
Carry over	0.57		– 0.47	0.38		– 0.47
Balancing		0.28	+ 0.19	– 0.19	– 0.19	
Carry over	0.14		– 0.10	0.10		– 0.10
Balancing		0.06	0.04	– 0.05	– 0.05	
Final moments	14.78	29.60	– 29.60	23.68	– 23.68	– 11.83

Span AB :

Moment to be balanced = 14.78 + 29.60 = 44.38 kNm

Horizontal reaction at A $= \frac{44.38}{l} = \frac{44.38}{4} = 11.10$ kN (→)

Span CD :

Moment to be balanced = – 23.68 – 11.83 = – 35.51 kNm

Horizontal reaction at D $= \frac{35.51}{3} = 11.84$ kN (←)

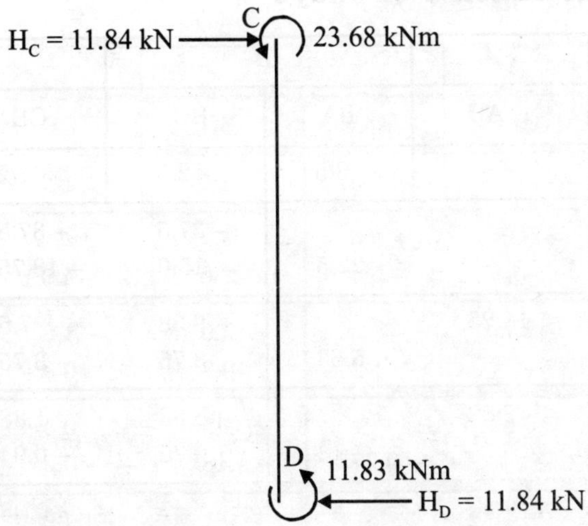

Value of force 'P' preventing sway = 11.1 ~ 11.84 = 0.74 kN (→)

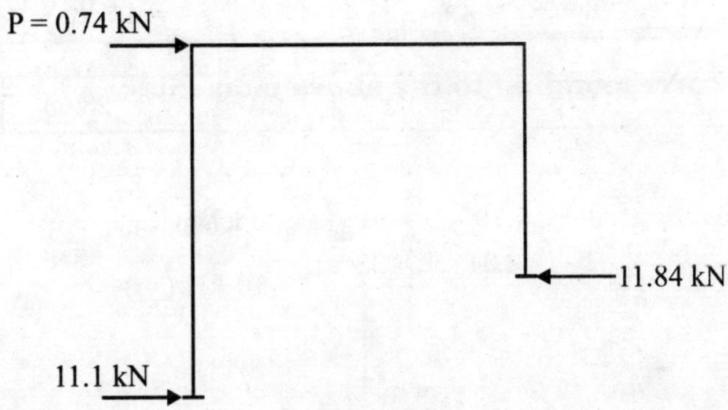

Fig. 3.126

(*d*) **Side sway:** Now, let a sway force of 0.74 kN (⟵) be applied at C. This will induce clockwise moments at A and D.

$$\frac{M_{BA}}{M_{CD}} = \frac{I_1 / l_1^2}{I_2 / l_2^2} = \frac{2I / 4^2}{I / 3^2} = \frac{2}{16} \times 9$$

$$\frac{M_{BA}}{M_{CD}} = \frac{9}{8}$$

Assume $M_{BA} = +9$ kNm ; $M_{AB} = M_{BA} = 9$ kNm

$M_{CD} = 8 \times (M_{BA} / 9) = +8$ kNm ;

$M_{DC} = M_{CD} = 8$ kNm

These are the arbitrary f.e.m.s for sway analysis.

(*e*) **Sway moment distribution :**

Joints	*A*	*B*		*C*		*D*
Members	AB	BA	BC	CB	CD	DC
Distribution factors	–	3/5	2/5	1/2	1/2	–
Fixed end moments	+ 9.0	+ 9.0	–	–	+ 8.0	+ 8.0
Balancing		– 5.4	– 3.6	– 4.0	– 4.0	
Carry over	– 2.7		– 2.0	– 1.8		– 2.0
Balancing		1.2	0.8	0.9	0.9	
Carry over	0.6		0.45	0.4		0.45
Balancing		– 0.27	– 0.18	– 0.2	– 0.2	
Carry over	– 0.14		– 0.1	– 0.09		– 0.1
Balancing		0.06	0.04	0.05	0.05	
Final moments	6.76	4.59	– 4.59	– 4.74	4.75	6.35

Sway forces corresponding to the above moments :

Span AB :

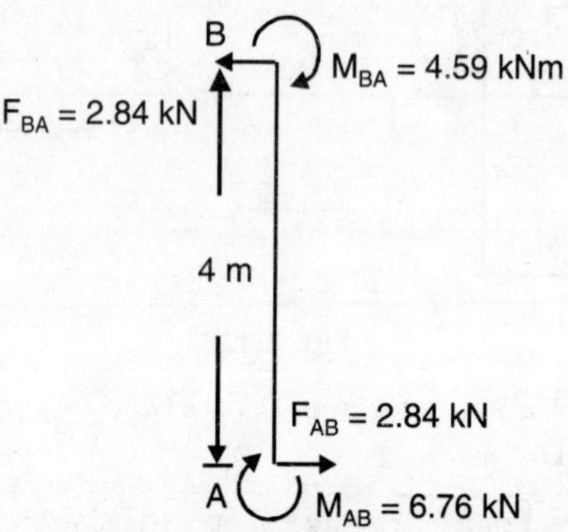

Fig. 3.127

Taking moments about A, $-F_{BA}(4) + M_{BA} + M_{AB} = 0$

$$F_{BA} = (4.59 + 6.76)/4$$

$$F_{BA} = 2.84 \text{ kN} = F_{AB} (\longrightarrow)$$

Span CD :

Taking moments about D, $-F_{CD}(3) + M_{CD} + M_{DC} = 0$.

$$F_{CD} = \frac{(4.75 + 6.35)}{3} = 3.7 \text{ kN} = F_{DC} (\longrightarrow)$$

Total horizontal reaction at support = $F_{AB} + F_{DC}$ = 2.84 (→) + 3.7(→) = 6.54 kN (→)

But the actual sway force is 0.74 kN. Hence, the correction factor = $\frac{0.74}{6.54}$ = 0.113.

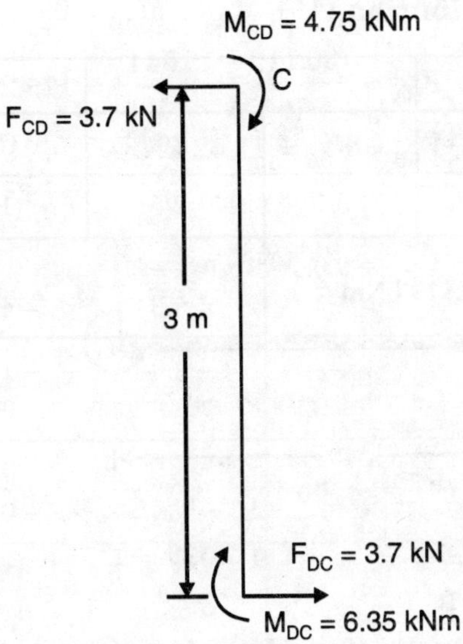

The moments have to be multiplied by 0.113. The following table shows the non-sway moments, sway moments after correction and the final moments.

Moments	M_{AB}	M_{BA}	M_{BC}	M_{CB}	M_{CD}	M_{DC}
1. Non-sway moments	14.78	29.60	– 29.60	23.68	– 23.68	– 11.83
2. Sway moments after correction	6.76 × 0.113 = 0.764	4.59 × 0.113 = 0.519	– 4.59 × 0.113 = – 0.519	– 4.75 × 0.113 = – 0.537	4.75 × 0.113 = 0.537	6.35 × 0.113 = 0.718
3. Final moments (1) + (2)	15.544	30.119	– 30.119	23.143	– 23.143	– 11.112

(*f*) **To draw S.F.D.:**

Span AB :

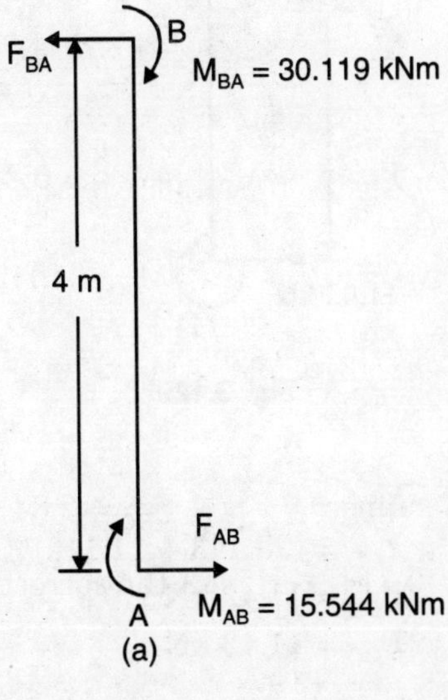

(a)

Taking moments about A, $-F_{BA}(4) + M_{BA} + M_{AB} = 0$.

$$F_{BA} = \frac{(30.119 + 15.544)}{4} = 11.42 \text{ kN}$$

$$F_{AB} = F_{BA} = 11.42 \text{ kN}$$

Span BC :

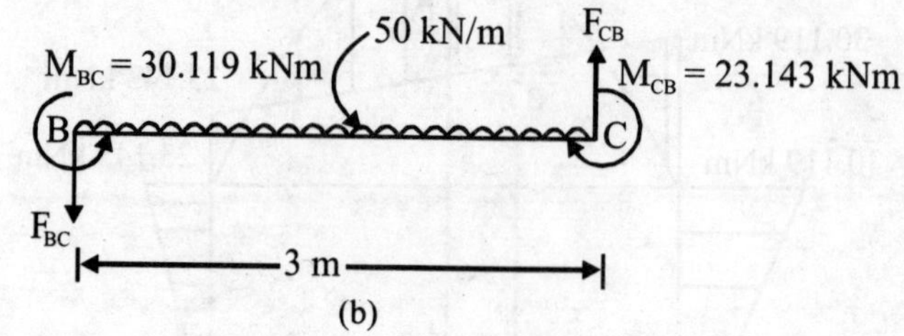

Fig. 3.128

Taking moments about B,

$$50 \times (3^2/2) - F_{CB}(3) + M_{CB} - M_{BC} = 0$$

$$50 \times (9/2) - F_{CB}(3) + 23.143 - 30.119 = 0$$

$$F_{CB}(3) = +\ 218.024$$

$$F_{CB} = 72.67 \text{ kN}$$

$$F_{BC} = 77.33 \text{ kN}$$

Span CD :

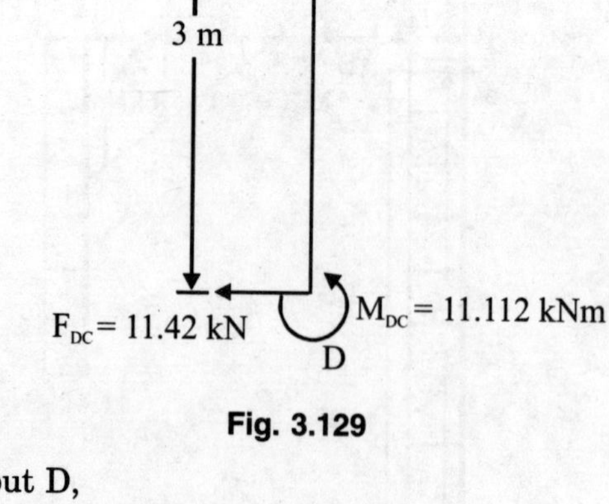

Fig. 3.129

Taking moments about D,

$$F_{CD}(3) - M_{CD} - M_{DC} = 0$$

$$F_{CD} = (23.143 + 11.112)/3$$

$$F_{CD} = 11.42 \text{ kN}$$

$$F_{DC} = 11.42 \text{ kN}$$

Simply supported bending moment in the span BC = $\frac{wl^2}{8} = \frac{50 \times 3^2}{8}$ = 56.25 kN

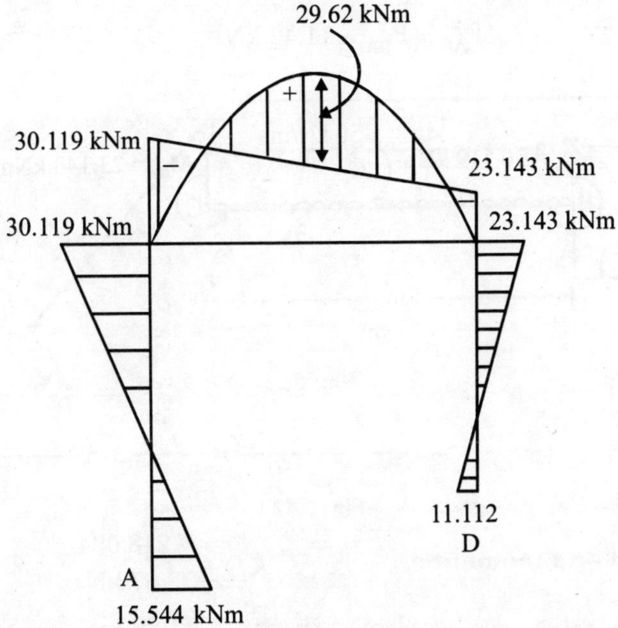

Fig. 3.130. Bending moment diagram.

Positive moment ordinate at middle of the span BC

$$= 56.25 - \left[\frac{23.143 + 30.119}{2}\right] = 29.62 \text{ kNm}$$

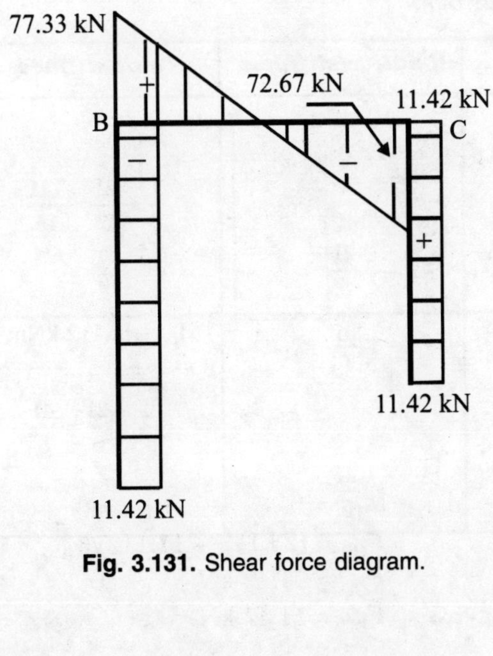

Fig. 3.131. Shear force diagram.

Example 3.28 *Analyse the portal frame shown in Fig. 3.132 by the moment distribution method and sketch the BMD and SFD.*

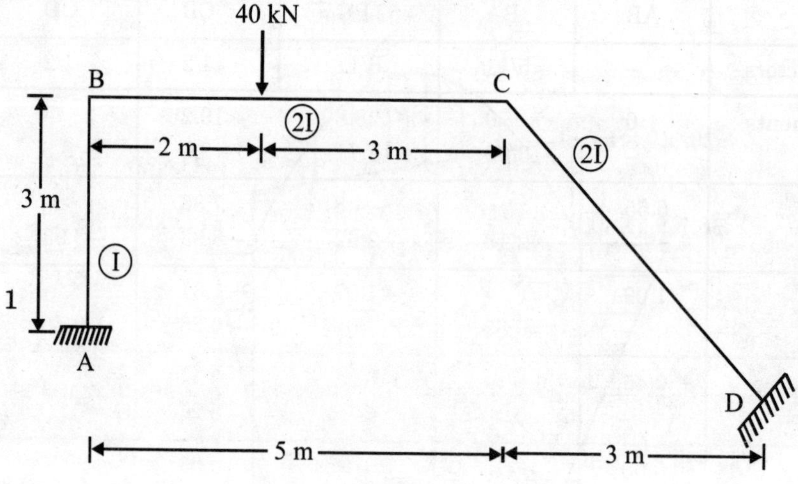

Fig. 3.132

Solution (*a*) **Fixed end moments:**

Span AB, CD :

$$M_{FAB} = M_{FBA} = M_{FCD} = M_{FDC} = 0$$

Span BC :

$$M_{FBC} = -\frac{Wab^2}{l^2} = \frac{40 \times 2 \times 3^2}{5^2} = -28.8 \text{ kNm}$$

$$M_{FCB} = +\frac{Wa^2b}{l^2} = \frac{40 \times 2^2 \times 3}{5^2} = +19.2 \text{ kNm}$$

(*b*) **Distribution factors:**

Joints	*Members*	*Relative stiffness*	*Total stiffness*	*Distribution factor*
B	BA	$\frac{I}{3}$	$\frac{I}{3}+\frac{2I}{5}=\frac{11I}{15}$	$\frac{I/3}{11I/15}=\frac{5}{11}$
	BC	$\frac{2I}{5}$		$\frac{2I/5}{11I/15}=\frac{6}{11}$
C	CB	$\frac{2I}{5}$	$\frac{2I}{5}+\frac{2I}{5}=\frac{4I}{5}$	$\frac{2I/5}{4I/5}=\frac{1}{2}$
	CD	$\frac{2I}{5}$		$\frac{2I/5}{4I/5}=\frac{1}{2}$

(*c*) **(Non-sway) moment distribution:**

Joints	***A***	***B***		***C***		***D***
Members	AB	BA	BC	CB	CD	DC
Distribution factors	–	5/11	6/11	1/2	1/2	–
Fixed end moments	0	0	– 28.8	+ 19.2	–	–
Balance	–	13.09	15.71	– 9.6	– 9.6	–
Carry over	6.55	–	– 4.8	7.86	–	– 4.8
Balancing	–	2.18	2.62	– 3.93	– 3.93	–
Carry over	1.09	–	– 1.97	1.31	–	– 1.97
Balancing	–	0.89	1.08	– 0.66	– 0.66	–
Carry over	0.45	–	– 0.33	0.54	–	– 0.33
Balancing	–	0.15	0.18	– 0.27	– 0.27	–
Carry over	0.08	–	– 0.14	0.09	–	– 0.14
Balancing	–	0.06	0.08	– 0.05	– 0.05	–
Final moments	8.17	16.37	– 16.37	14.49	– 14.49	– 7.24

The above moments act but assuming a restraining force of P_1 at B.

The restraining force P_1 may be found by taking moments about O, the point of intersection of AB and CD.

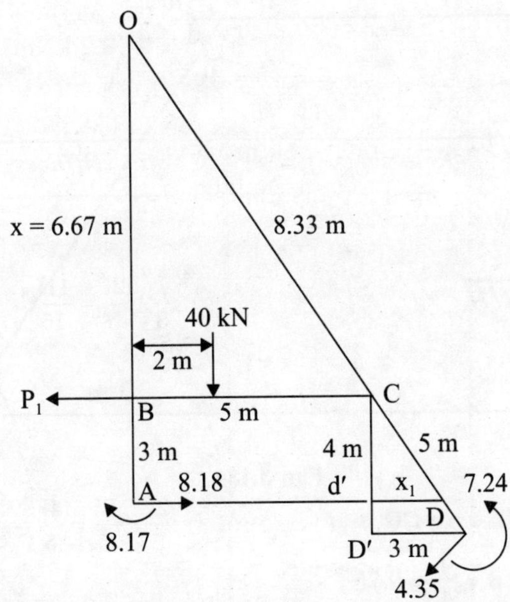

Fig. 3.133

In ΔC D′D and ΔCd′d, $\frac{4}{3}=\frac{3}{x_1}$ $x_1 = \frac{9}{4} = 2.25$ m ; Ad = 7.25 m

In ΔOAd, ΔOBC

$$\frac{7.25}{x+3}=\frac{5}{x}$$

$$7.25x = 5x + 15$$

$$2.25x = 15$$

$$x = 6.67 \text{ m}$$

$$H_{AB} = \frac{M_{AB} + M_{BA}}{l_{AB}} = \frac{16.37 + 8.17}{3}$$

$$H_{AB} = 8.18 \text{ kN}$$

$$H_{DC} = \frac{M_{DC} + M_{CD}}{l_{CD}} = \frac{7.24 + 14.49}{5} = 4.35 \text{ kN}$$

Taking moments about O,

$$P_1(6.67) + 40(2) + 8.17 + 4.35(13.33) - 8.18(9.67) - 7.24 = 0$$

$$P_1 = -8.97 \text{ kN}$$

(*d*) **Sway solution:** Remove the load from the frame, fix all the joints and allow the frame to sway to the right through a horizontal distance d. The new positions of B and C are B′ and C′. Notice that C move to the right and upwards so that $CC_1 = d$ and CC′ is perpendicular to DC.

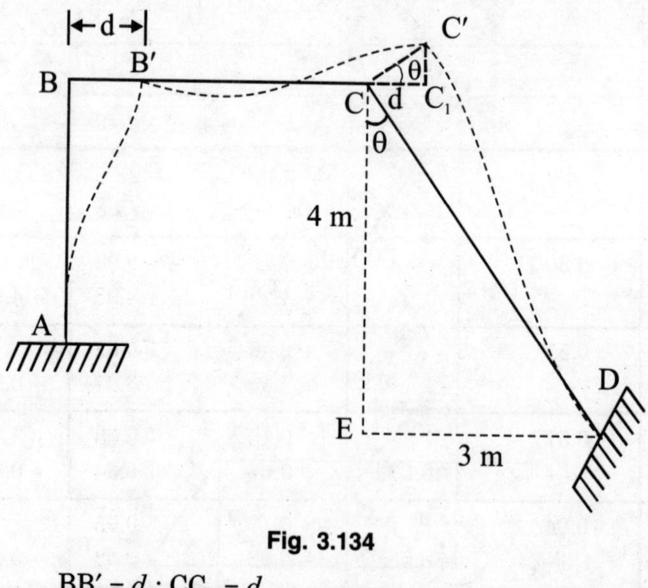

Fig. 3.134

$$BB' = d \; ; \; CC_1 = d$$

In ΔCDE, $\tan\theta = \frac{3}{4} = 0.75$ ...(1)

In ΔCC′C_1, $\tan\theta = \frac{C'C_1}{CC_1} = \frac{C'C_1}{d} = 0.75$ [From (1)]

$$C'C_1 = 0.75d \quad ...(2)$$

$$CC' = \sqrt{d^2 + (0.75d)^2} = 1.25d \quad ...(3)$$

$$CC_1 = d \quad ...(4)$$

(*e*) **Fixed end moments :**

Span AB :

This member undergoes a deflection of '*d*' units. Thereby,

$$M_{FBA} = M_{FAB} = \frac{-6EI\delta}{l^2} = \frac{-6EId}{3^2} = \frac{-2}{3} EId$$

Span BC :

This member undergoes a deflection of '0.75*d*' units.

$$M_{FBC} = M_{FCB} = \frac{6EI\delta}{l^2} = \frac{6EI(0.75d)}{3^2} = 0.18 EId$$

Span CD :

This member undergoes a deflection of '1.25*d*' units

$$M_{FCD} = M_{FDC} = \frac{-6EI\delta}{l^2} = \frac{-6EI(1.25d)}{(5)^2} = -0.3 EId$$

Assuming, $M_{FAB} = M_{FBA} = 50$ kNm $= -0.67$ EI*d* (EI*d* = – 75 units)

$M_{BC} = M_{CB} = 0.18EId = 0.18(-75) = -13.5$ kNm

$M_{CD} = M_{DC} = -0.3 EId = -0.3(-75) = 22.5$ kNm

(*f*) **Sway moment distribution :**

Joints	***A***	***B***		***C***		***D***
Members	AB	BA	BC	CB	CD	DC
Distribution factors	–	5/11	6/11	1/2	1/2	–
Fixed end moments	50	50	– 13.5	– 13.5	22.5	22.5
Balancing	–	– 16.59	– 19.91	– 4.5	– 4.5	–
Carry over	– 8.30	–	– 2.25	– 9.96	–	– 2.25
Balancing	–	1.02	+ 1.23	+ 4.98	+ 4.98	–
Carry over	0.51	–	+ 2.49	+ 0.62	–	+ 2.49
Balancing	–	– 1.13	– 1.36	– 0.31	– 0.31	–
Carry over	– 0.57	–	– 0.16	– 0.68	–	– 0.16
Balancing	–	0.07	0.09	+ 0.34	+ 0.34	–
Carry over	0.04	–	+ 0.17	0.05	–	+ 0.17
Balancing	–	– 0.08	– 0.09	– 0.03	– 0.02	–
Carry over	– 0.04	–	– 0.02	– 0.05	–	– 0.01
Balancing	–	0.01	0.01	+ 0.02	+ 0.03	–
Final moments	41.64	33.3	– 33.3	– 23.02	+ 23.02	22.74

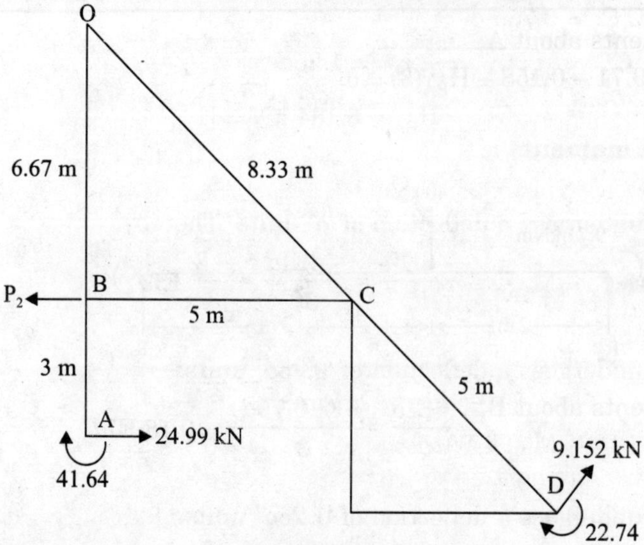

Fig. 3.135

$$F_{AB} = \frac{41.64 + 33.33}{3} = 24.99 \text{ kN}, \quad F_{DC} = \frac{23.02 + 22.74}{5} = 9.152 \text{ kN}$$

Taking moments about O,

$$P_2(6.67) + 41.64 + 22.74 - 24.99(9.670) - 9.152(13.33) = 0$$

$$P_2 = + 44.87 \text{ kN}$$

(*g*) **Correction factor and the final moments :**

$$\text{C.F.} = (-8.97)/\ 44.87 = -0.199 = -0.2$$

$$M_{AB} = (\text{Non-sway moment})_{AB} + \text{C.F.} \times (\text{Sway moment})_{AB}$$

$$M_{AB} = 8.17 + (-0.2) \times 41.64 = -0.158 \text{ kNm}$$

$$M_{BA} = 16.37 + (-0.2) \times 33.30 = 9.71 \text{ kNm}$$

$$M_{BC} = -16.37 + (-0.2) \times (-33.30) = -9.71 \text{ kNm}$$

$$M_{CB} = 14.49 + (-0.2) \times (23.02) = 19.09 \text{ kNm}$$

$$M_{CD} = -14.49 + (-0.2) \times (23.02) = -19.09 \text{ kNm}$$

$$M_{DC} = -7.24 + (-0.2) \times (22.740) = -11.79 \text{ kNm}$$

$$\text{Simply supported bending moment} = \frac{Wab}{l} = \frac{40 \times 2 \times 3}{5} = 48 \text{ kNm}$$

(*h*) **To draw S.F.D. :**

Span AB :

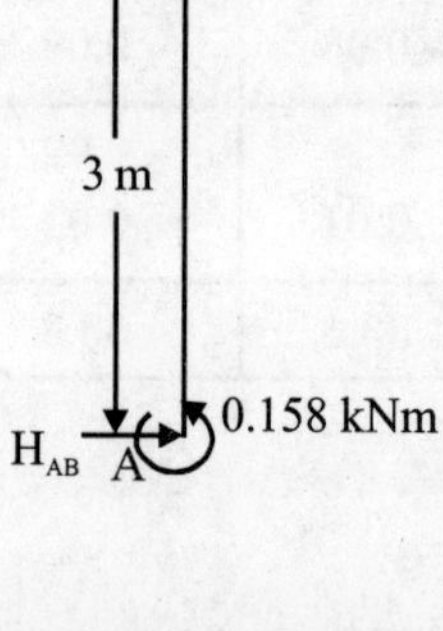

Taking moments about A,

$$9.71 - 0.158 - H_{BA}(3) = 0$$

$$H_{BA} = 3.184 \text{ kN} = H_{AB}$$

Span BC :

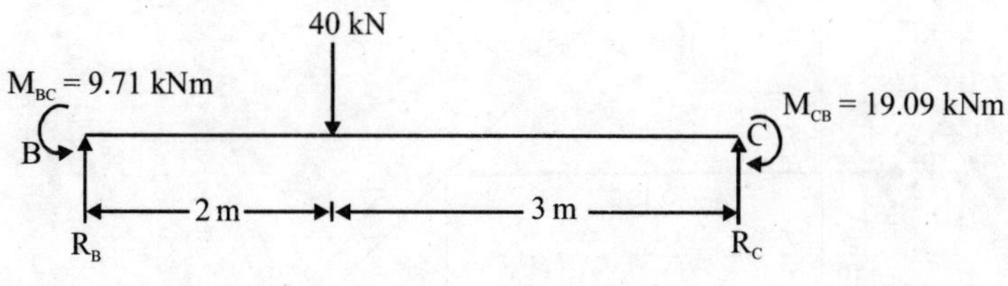

Taking moments about B,

$$M_{CB} + W(a) - M_{BC} - R_C(l) = 0$$

$$19.09 + 40(2) - 9.71 - R_C(5) = 0$$

$$R_C = 17.88 \text{ kN}$$

$$R_B = \text{Total load} - R_C = 40 - 17.88$$

$$R_B = 22.12 \text{ kN}$$

Span CD :

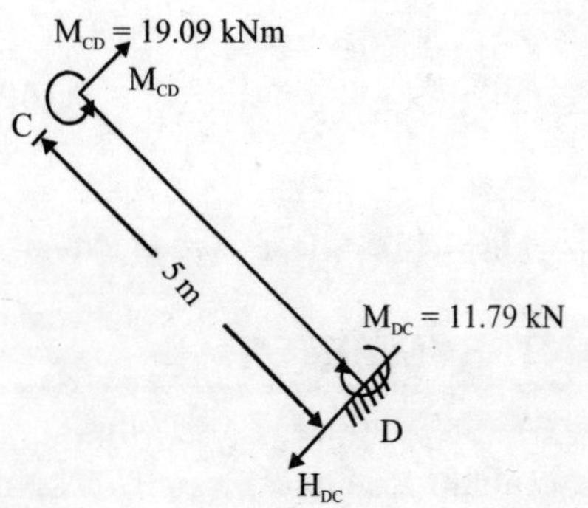

Taking moments about C,

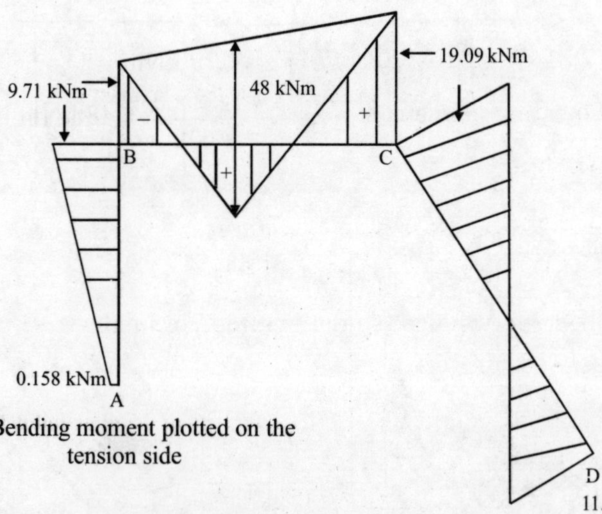

Fig. 3.136. Bending moment diagram.

$$H_{DC}(5) - M_{DC} - M_{CD} = 0$$
$$H_{DC}(5) - 11.79 - 19.09 = 0$$
$$H_{DC} = 6.18 \text{ kN}$$
$$H_{CD} = 6.18 \text{ kN}$$

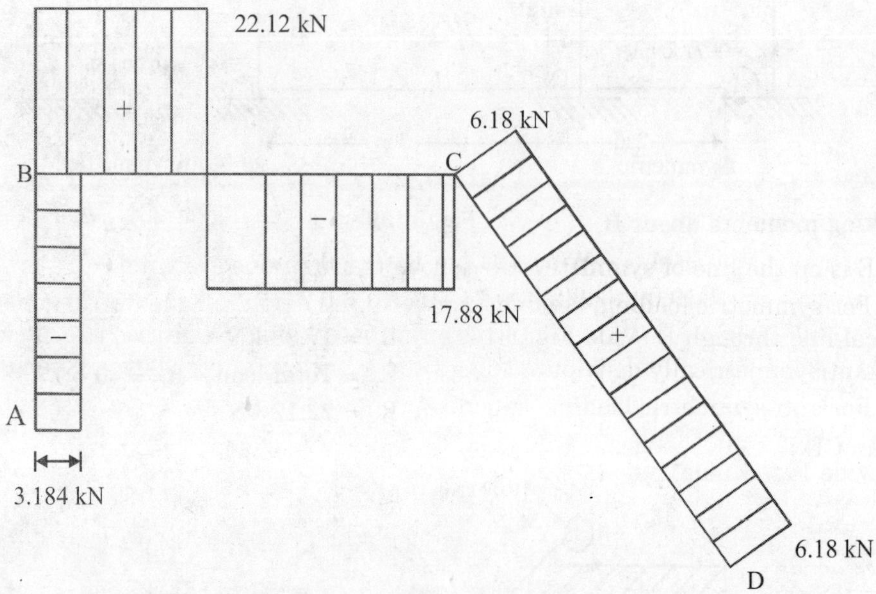

Fig. 3.137. Shear force diagram.

3.8 SYMMETRY AND ANTI-SYMMETRY

In many cases structures have symmetry about one or more axes. Figure 3.138 shows structures that are symmetrical about a vertical axis.

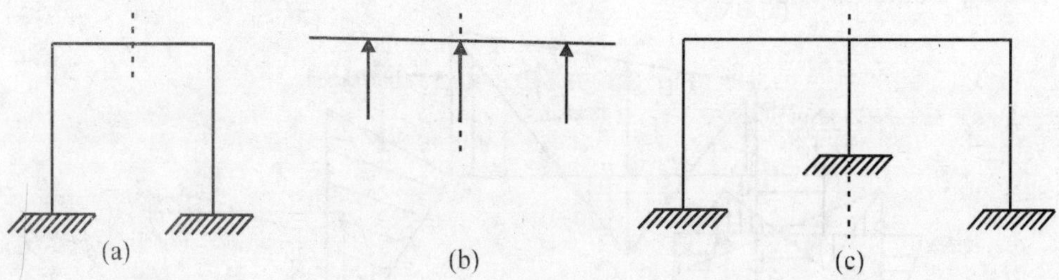

Fig. 3.138

Such symmetrical structures can have symmetric loads, anti-symmetric loads or unsymmetric loads (Fig. 3.139).

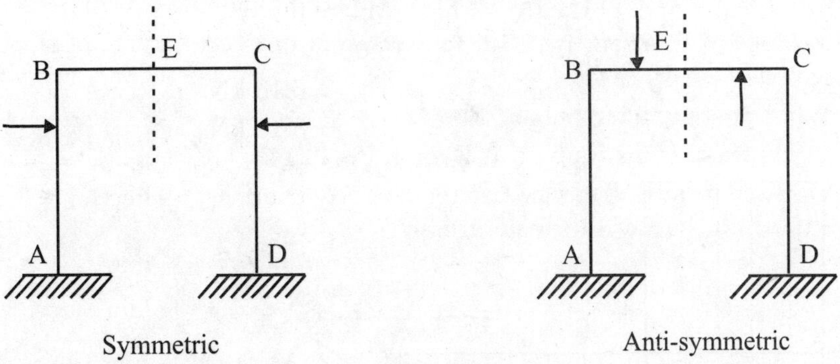

Fig. 3.139

If E is on the line of symmetry, the following conditions are valid.

(*i*) For symmetric loading, bending moments and displacements are symmetrical about the vertical line through E (Vide Fig. 3.140). Anti-symmetric quantities like slope and shear force are antisymmetrically distributed about the axis through E. They will be zero at E.

(*ii*) For anti-symmetric loading, symmetric quantities will be zero at E. *i.e.*, bending moment at E will be zero. Displacements at E will also be zero. In other words, E will function as a hinge (Vide Fig. 3.141)

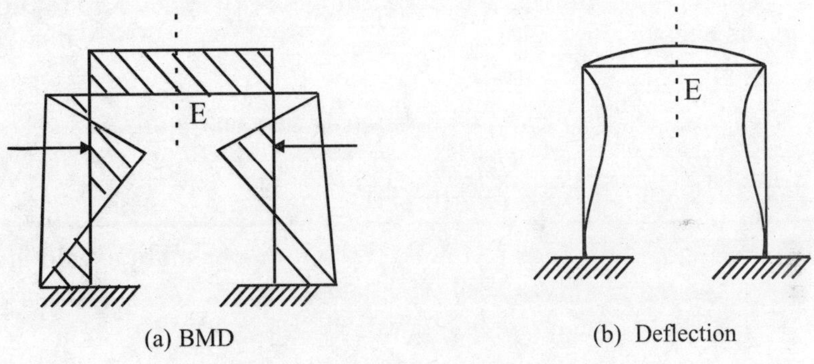

Fig. 3.140. Symmetrical loading.

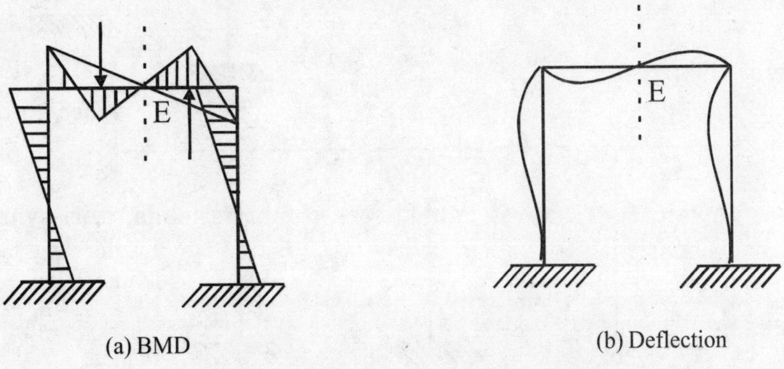

Fig. 3.141. Anti-symmetrical loading.

Observing Figs. 3.140 and 3.141 we can draw the following conclusions.

(*i*) For symmetrical cases, the point of symmetry has zero change of slope.

This gives rise to two cases.

(*a*) When there is a support at E

(*b*) When there is no support at E as in Fig. 3.140.

If there is support at E, E can be treated as a fixed support (deflection = 0, slope = 0) and we can ignore the right half of the structure.

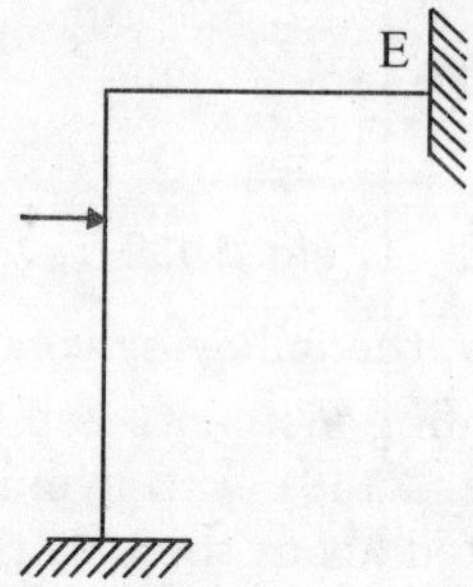

If there is no support at E, We can make a simplification in the moment distribution process. This is described a bit later.

(*ii*) For anti-symmetrical cases the point of anti-symmetry has zero deflection and zero bending moment. This is like a hinge. Hence we can ignore the structure to the right of E and treat E as a hinge as below.

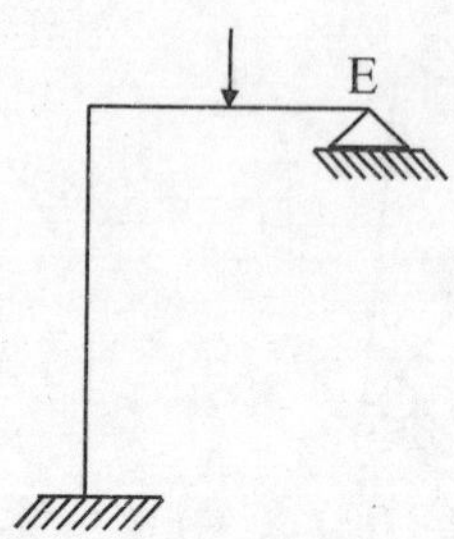

With the above simplification we can tackle examples 3.29 to 3.34.

Example 3.29 *Determine the bending moments at strategic points in the frame shown in Fig. 3.142.*

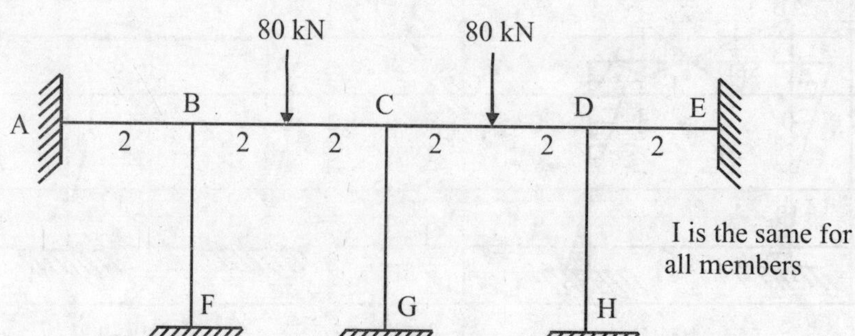

Fig. 3.142

Solution We can take advantage of the symmetry in the structure and the loading and solve the reduced structure as shown in Fig. 3.143.

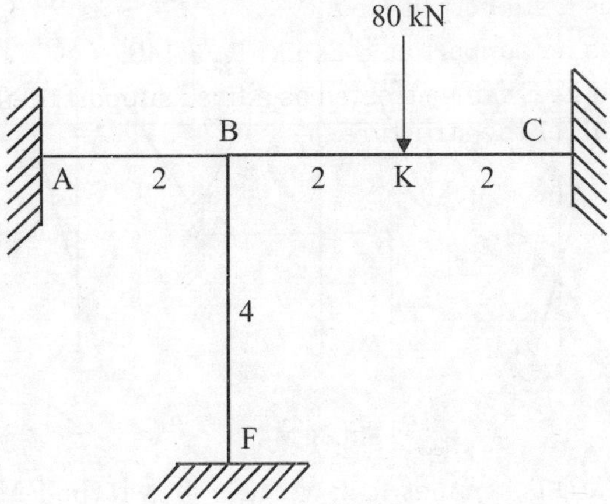

Fig. 3.143

(*a*) **Fixed end moments :**

$$M_{FAB} = M_{FBA} = M_{FBF} = M_{FFB} = 0$$

$$M_{FBC} = -\frac{Wl}{8} = -\frac{80 \times 4}{8} = -40 \text{ kNm}$$

$$M_{FCB} = +\frac{Wl}{8} = +\frac{80 \times 4}{8} = +40 \text{ kNm}$$

(*b*) **Free BM :** $M_K = \frac{Wl}{4} = 80$ kNm

(*c*) **Distribution factors :**

$$k_{BA} : k_{BC} : k_{BF} = 1/2 : 1/4 : 1/4 = 2 : 1 : 1$$

$$DF_{BA} = 0.5 \qquad DF_{BC} = DF_{BF} = 0.25$$

(*d*) **Moment distribution :**

Joint	***A***		***B***		***C***	
Member	AB	BA	BF	BC	CB	FB
DF		0.5	0.25	0.25		
FEM	0	0	0	– 40	40	0
Balancing		20	10	10		
Carry over	10				5	5
Final	10	20	10	– 30	45	5

(*e*) **Bending moment diagram :**

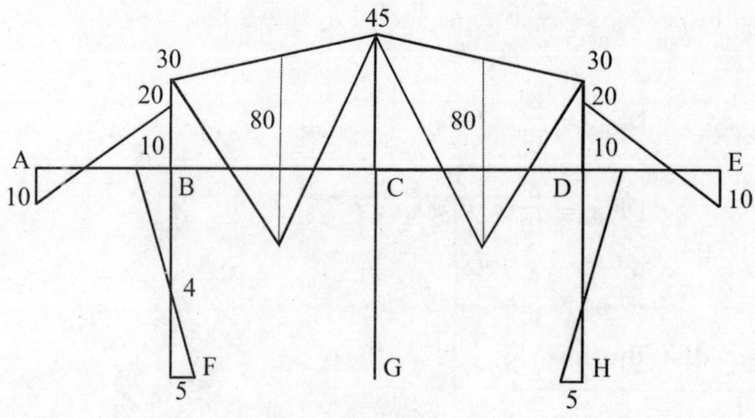

Fig. 3.144

Knowing that the BMD is symmetrical, we have drawn the BMD for the left half of the frame and drawn its mirror reflection for the right half.

Our next example will consider the same frame loaded anti-symmetrically.

Example 3.30 *Analyse the frame in Fig. 3.142 when the 80 kN load on CD acts upwards instead of downwards.*

Solution This is an anti-symmetric loading case. There will be no bending moment at C (and of course no deflection either). Hence the frame reduces to the following Fig. 3.145.

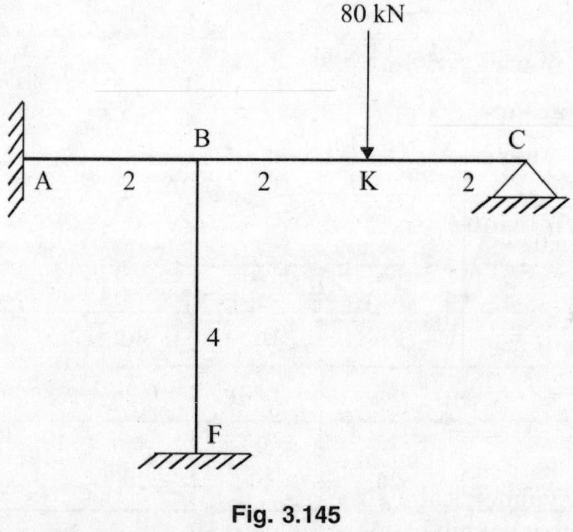

Fig. 3.145

(*a*) **Fixed end moments :**

$$M_{FBC} = -40 \text{ kNm}, M_{FCB} = 40 \text{ kNm}$$

(*b*) **Free BM :** $M_k = 80$ kNm

(*c*) **Distribution factors :**

$$k_{BA} : k_{BC} : k_{BF} = \frac{1}{2} : \frac{3}{4} \times \frac{1}{4} : \frac{1}{4} = 8 : 3 : 4$$

$$DF_{BA} = \frac{8}{15} = 0.533$$

$$DF_{BC} = \frac{3}{15} = 0.200$$

$$DF_{BF} = \frac{4}{15} = 0.267$$

(*d*) **Moment distribution :**

Joint	***A***		***B***		***C***	
Member	AB	BA	BF	BC	CB	FB
DF		0.533	0.267	0.200		
FEM Bal C	0	0	0	– 40	40 – 40	0
Carry over to B Bal B		31.98	16.02	– 20 +12		
Carry over	15.99					8.01
Final	15.99	31.98	16.02	– 48.0	0	8.01

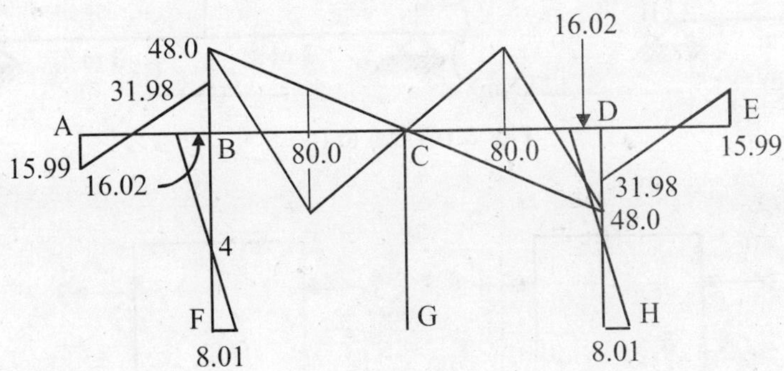

Fig. 3.146

The BMD is anti-symmetrical. At the point of symmetry C, the BM is zero. The column CG at the line of symmetry has no bending moment (It will have no thrust either).

Once we have a solution to symmetrical and anti-symmetrical loading cases, we are in a position to tackle any loading case.

Example 3.31 *Express the six general loading cases in Fig. 3.147 (a) to (f) as combinations of symmetrical and anti-symmetrical loading cases.*

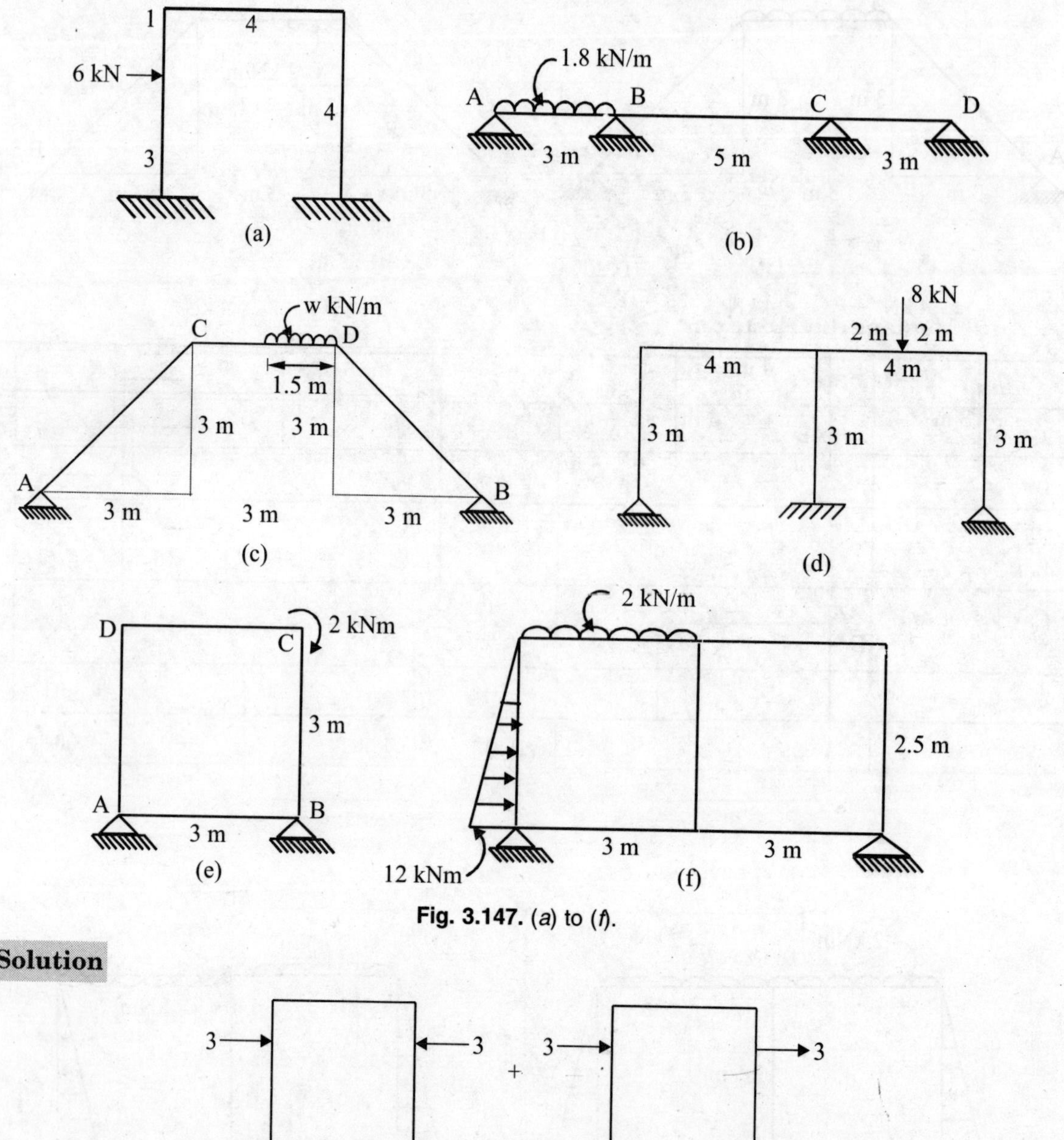

Fig. 3.147. (*a*) to (*f*).

Solution

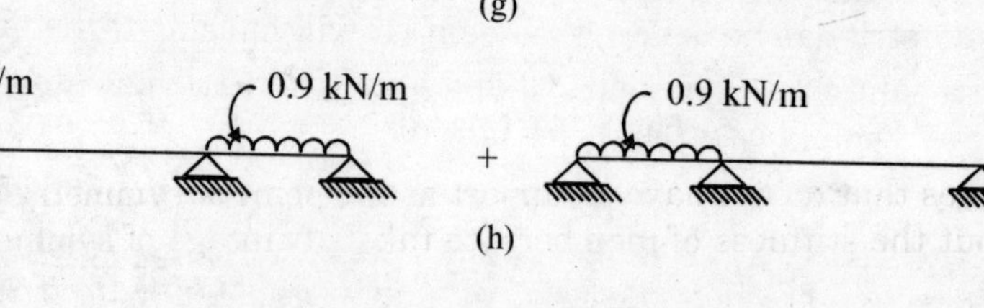

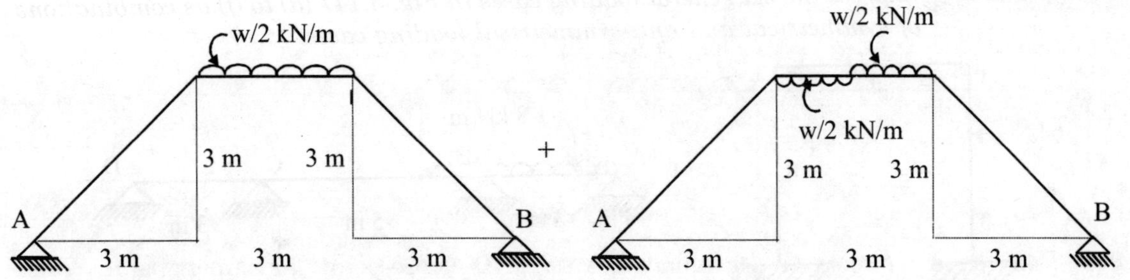

(i)

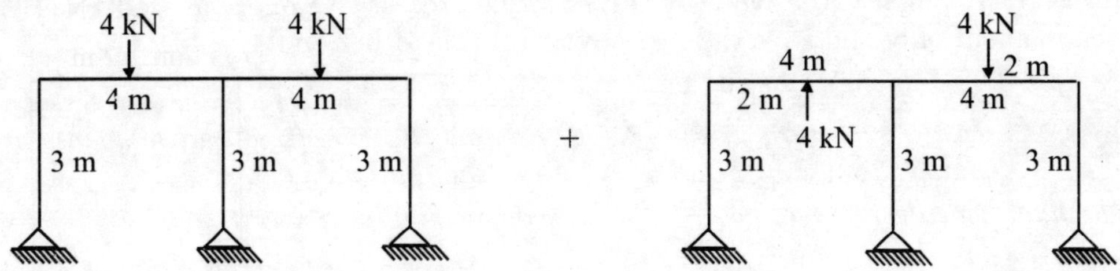

(j)

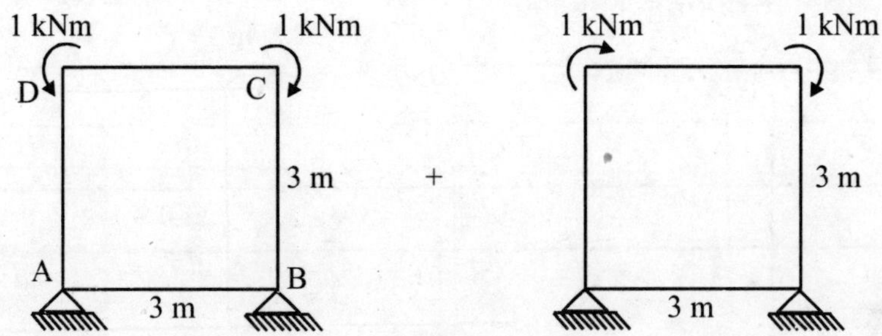

(k)

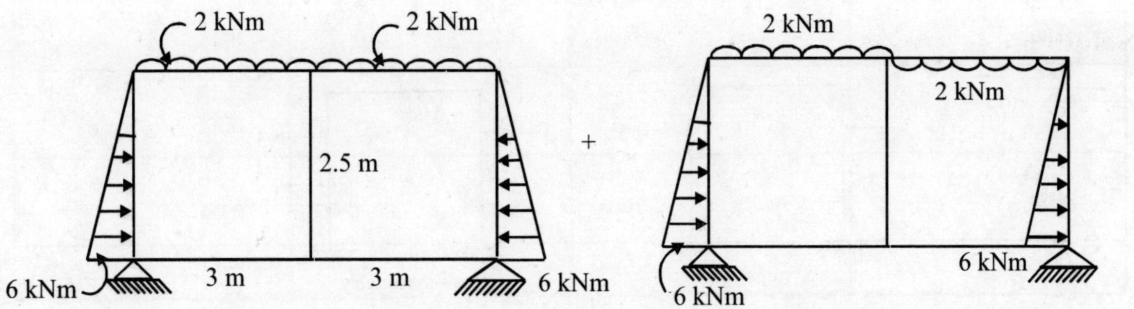

(*l*)

Fig. 3.147. (*g*) to (*l*)

In cases of frames that do not have a support at the point of symmetry, some modification can be made about the stiffness of members, to take advantage of symmetry.

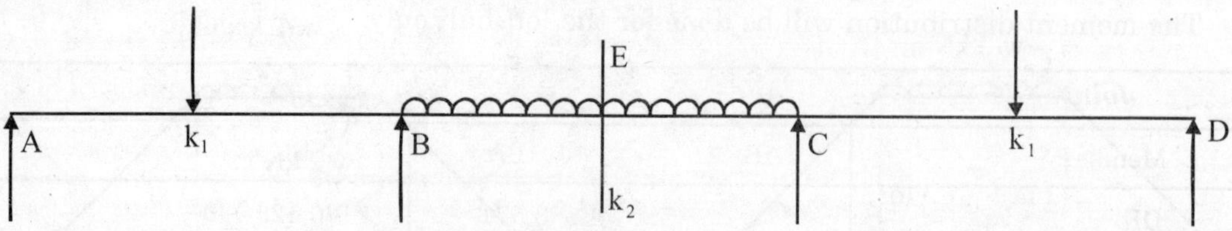

Fig. 3.148

Consider a symmetrical continuous beam ABCD whose point of symmetry is E. When we carry out moment distribution for the structure, every time we balance B and C, BC and CB get equal (and opposite) contributions. In the next step half the moment would go to CB and BC respectively. Their values will get halved in the process. We might as well give BC and CB half the moment only and save the carry over from the other end.

Example 3.32 demonstrates the process.

Example 3.32 *In the structure in Fig. 3.148, assume that the fixed end moments are as under. Take $k_1 = 2$ and $k_2 = 3$. Find the final moments.*

(i) without considering symmetry *(ii) considering symmetry.*

$M_{FAB} = -24$ $M_{FBA} = 24$ $M_{FBC} = -32$ $M_{FCB} = 32$

$M_{FCD} = -24$ $M_{FDC} = 24.$

Solution *(i)* **Without taking symmetry:**

Joint	***A***	***B***		***C***		***D***
Member	AB	BA	BC	CB	CD	DC
DF		0.4	0.6	0.6	0.4	
FEM	– 24	24	– 32	32	– 24	24
Bal and C.O.	24	12			– 12	– 24
Bal B, C		– 1.6	– 2.4	2.4	1.6	
C.O.			1.2	– 1.2		
Bal		– 0.48	– 0.72	0 .72	+ 0.48	
C.O.			0.36	– 0.36		
Bal		– 0.14	– 0.22	– 0.22	– 0.14	
C.O.			0.11	– 0.11		
Bal		– 0.04	– 0.07	0.07	0.04	
Final	0	33.74	– 33.74	33.74	– 33.74	0

(ii) **Taking symmetry into account:**

Here we will modify the stiffness of BC into $K_m = k_2 \div 2 = 1.5$.

Hence the distribution factors at B are

$$DF_{BA} = \frac{2}{2+1.5} = 0.571\ ; \quad DF_{BC} = \frac{1.5}{2+1.5} = 0.429$$

The moment distribution will be done for the left half only.

Joint	*A*	*B*	
Member	AB	BA	BC
DF		0.571	0.429
FEM	– 24	24	– 32
Bal and C.O.	24	12	
Bal		– 2.28	– 1.72
Final	0	33.72	– 33.72

It is needless to say that the idea of taking advantage of symmetry saves a lot of avoidable arithmetics.

When the loading on the structure is anti-symmetrical, as shown in Fig. 3.149, we can tackle the moment distribution with appropriate modifications. This is carried out in example 3.33.

Example 3.33 *Find the final moments in the continuous beam in Fig. 3.149. The fixed end moments are (in kNm)*

$M_{FAB} = -24$ $M_{FBA} = 24$

$M_{FBC} = -16$ $M_{FCB} = -16$

$M_{FCD} = 24$ $M_{FDC} = -24$

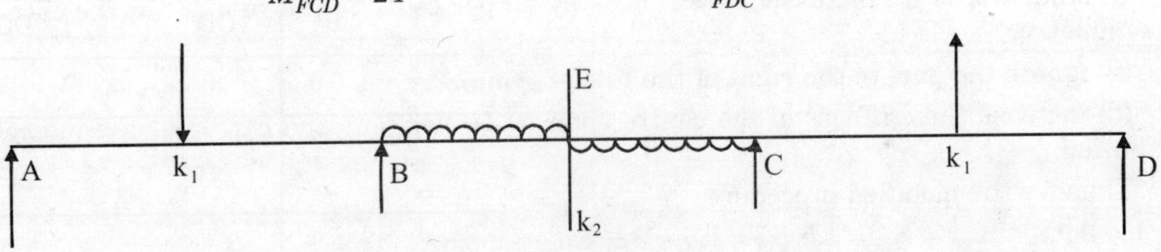

Fig. 3.149

Solution (*i*) **Not considering anti-symmetry:**

Joint	*A*	*B*		*C*		*D*
Member	AB	BA	BC	CB	CD	DC
DF		0.4	0.6	0.6	0.4	
FEM	– 24	24	– 16	– 16	24	– 24
Bal and C.O.	24	12			12	24
Bal		– 8	– 12	– 12	– 8	
C.O.			– 6	– 6		
Bal		2.4	3.6	3.6	2.4	
C.O.			1.8	1.8		
Bal		– 0.72	– 1.08	– 1.08	– 0.72	

C.O.			– 0.54	– 0.54		
Bal		0.22	0.32	0.32	0.22	
C.O.			0.16	0.16		
Bal		– 0.06	– 0.10	– 0.10	0.06	
Final	0	29.84	– 29.84	– 29.84	29.84	0

Note. This is an anti-symmetric loading case. But the moment distribution table is perfectly symmetrical about E. The reason is this. In a beam AB if the FEMs at A and B are equal and opposite in sign, it is a symmetric case. If the moments are equal and of the same sign it is an anti-symmetric case.

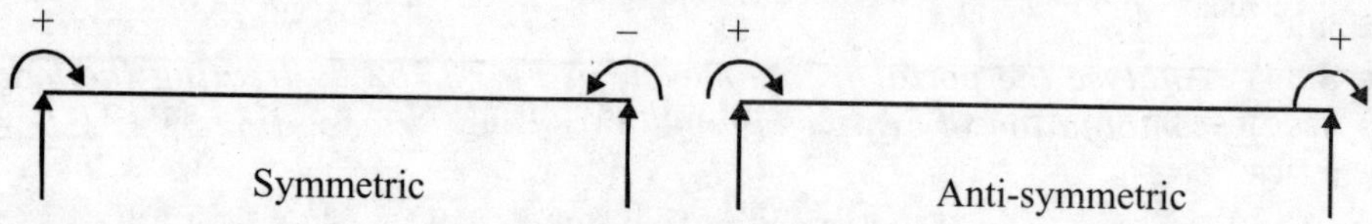

Fig. 3.150

(*ii*) **Taking advantage of anti-symmetry:**

When we study the moment distribution table above, we find that every time BC gets a contribution after balancing, it gets half that contribution immediately by carry over from CB (and vice versa). We might as well make 1.5 times the contribution in the first instance itself and forget about the right half.

In other words we increase the stiffness by 50 per cent. So the rule is : in the case of anti-symmetry.

(*a*) ignore the part to the right of the line of symmetry.

(*b*) increase the stiffness of the centre span by 50 per cent and carry out the moment distribution.

Thus for the modified procedure, $k_{BA} = k_1 = 2$

$$K_{BC} = K_m = 1.5\ K_2 = 1.5 \times 3 = 4.5$$

$$D_{FBA} = \frac{2}{6.5} = 0.308$$

$$D_{FBC} = \frac{4.5}{6.5} = 0.692$$

Joint	***A***	***B***		***E***
Member	AB	BA	BC	
DF		0.308	0.692	
FEM	– 24	24	– 16	
Bal and C.O.	24	12		
Bal		– 6.16	– 13.84	
Final	0	29.84	– 29.84	

Thus by the modified procedure we get at the final moments without too many hassles.

Note. The idea of augmenting the stiffness by 50 per cent can be arrived at by a different reasoning process also. The point E in Fig. 3.151 is a virtual hinge (Zero moment, Zero displacement). Hence we have the modified structure as under:

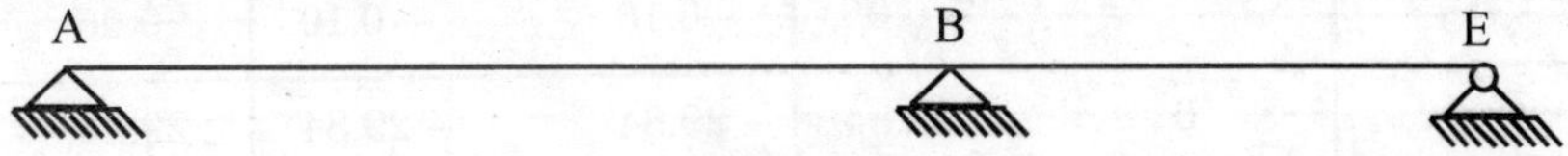

Fig. 3.151

$k_{BC} = k_2$. Hence $k_{BE} = 2k_2$ if E were fixed.

$$k_{BE} = \frac{3}{4} \times 2k_2 = \frac{3}{2} \times k_2 \text{ since E is hinged.}$$

Example 3.34 *Analyse the portal frame shown in Fig. 3.152 by treating the loading as a combination of symmetric and anti-symmetric loadings.*

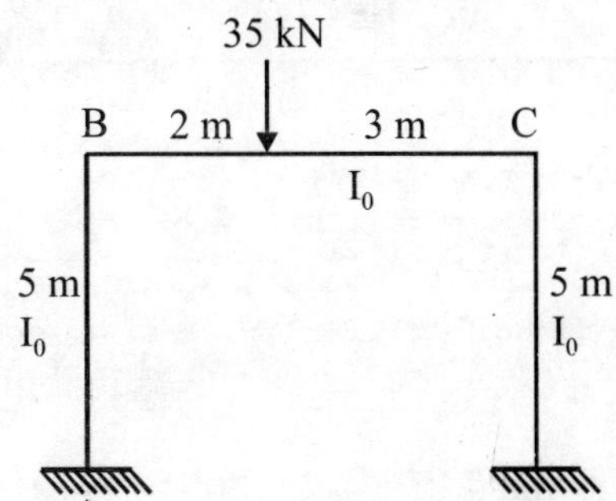

Fig. 3.152

Solution The given problem is treated as a combination of cases 1 and 2 below :

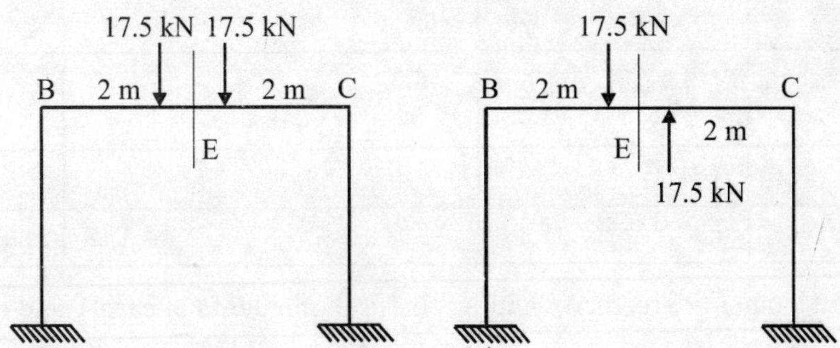

Case 1: Symmetric loading

$$M_{FBC} = -\left[\frac{Wab^2}{l^2} + \frac{Wa^2b}{l^2}\right] = -\frac{Wab}{l^2}(a+b) = -\frac{Wab}{l}$$

$$= \frac{17.5 \times 2 \times 3}{5} = -21 \text{ kNm}$$

$$k_{BA} = \frac{1}{5} \qquad k_{BC} = \frac{1}{2} \times \frac{1}{5} = \frac{1}{10}$$

$$DF_{BA} = \frac{2}{3} \qquad DF_{BC} = \frac{1}{3}$$

Joint	*A*	*B*		*C*		*D*
Member	AB	BA	BC			
DF		0.67	0.33			
FEM Bal	0	0 14	– 21 7			
C.O.	7					
Final	7	14	– 14	14	– 14	– 7

Case 2: Anti-symmetric loadings

$$M_{FBC} = -\frac{Wab^2}{l^2} + \frac{Wa^2b}{l^2} = \frac{17.5}{25}[-2 \times 3 \times 3 + 2 \times 2 \times 3] = -4.2 \text{ kNm}$$

$$k_{BA} = \frac{1}{5}\ ;\ k_{BC} \text{ (modified)} = 1.5 \times \frac{1}{5} = \frac{3}{10}$$

$k_{BA} : k_{BC} = 2 : 3.$

$DF_{BA} = 0.4 \qquad DF_{BC} = 0.6.$

Joint	*A*	*B*		*E*	*C*		*D*
Member	AB	BA	BC				
DF		0.4	0.6				
FEM Bal		 1.68	– 4.2 2.52				
C.O.	0.84						
Final	0.84	1.68	– 1.68		– 1.68	1.68	0.84

Non-sway moments are got by adding the final moments of case 1 and case 2.

$M_{AB} = 7.84 \qquad M_{BA} = 15.68 \qquad M_{BC} = -15.68$

$M_{DC} = -6.16 \quad M_{CD} = -12.32 \quad M_{CB} = 12.32$

Non-sway shears at A and D are obtained from the above moments.

$$H_A = \frac{M_{AB} + M_{BA}}{5} = \frac{7.84 + 15.68}{5} = 4.704 \ (\longrightarrow)$$

$$H_D = \frac{M_{DC} + M_{CD}}{5} = -\frac{6.16 - 12.32}{5} = -3.696 \ (\longleftarrow)$$

Unbalanced horizontal force $H_U = 1.008 \ (\longrightarrow)$

To off-set this imbalance we have to introduce a set of consistent moments at A, B, C and D on the columns AB and CD that would develop a net horizontal force H′ of 1.008 ←

We will start with trial moments of 10 at A, B, C and D corresponding to an arbitrary horizontal force H_a at C.

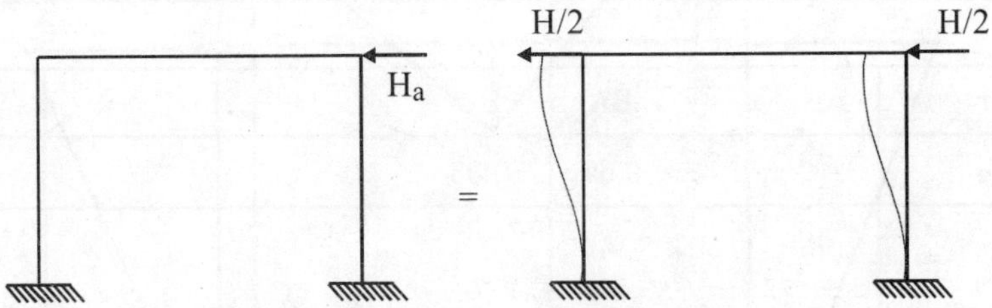

This is an anti-symmetrical case and is analysed below :

Joint	***A***	***B***		***E***	***C***		***D***
Member	AB	BA	BC				
DF		0.4	0.6				
FEM	10	10	0				
Bal		– 4	– 6				
C.O.	– 2						
Final	8	6	– 6		– 6	6	8

$$H'_A = \frac{8+6}{5} = 2.8$$

$$H'_D = \frac{8+6}{5} = 2.8$$

$$H_a = 2.8 + 2.8 = 5.6 \rightarrow$$

we need $H' = 1.008 \leftarrow$

Hence the corresponding column moments are modified by multiplying the final moments of sway moment distribution table by $-\frac{1.008}{5.6} = -0.18$

Sway moments are

$M_{AB} = -1.44$ $M_{BA} = -1.08$ $M_{BC} = +1.08$

$M_{DC} = -1.44$ $M_{CD} = -1.08$ $M_{CD} = +1.08$

The final moments are

$M_{AB} = 7.84 - 1.44 = 6.40$ $M_{BA} = 15.68 - 1.08 = 14.60$

$M_{BC} = -15.68 + 1.08 = -14.60$ $M_{DC} = -6.16 - 1.44 = -7.60$

$M_{CD} = -12.32 - 1.08 = -13.40$ $M_{CB} = 12.32 + 1.08 = 13.40$

The final moments are plotted below :

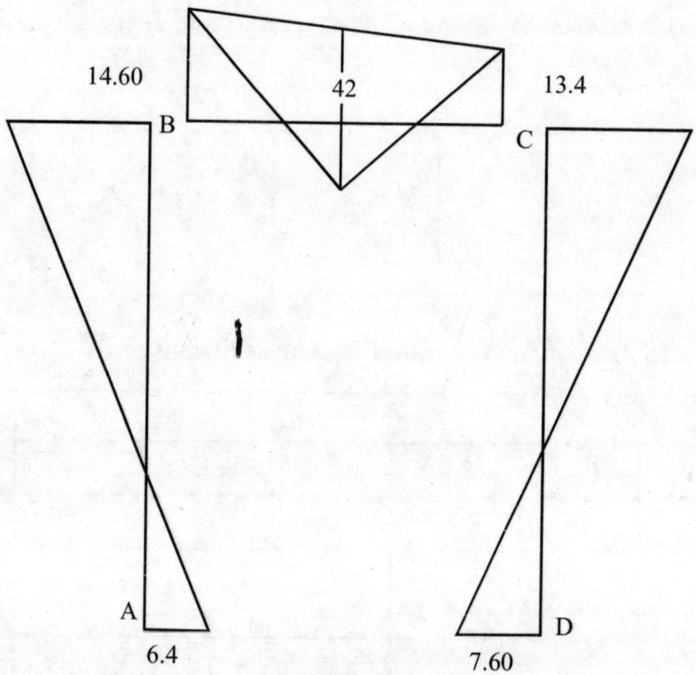

When we review example 3.34, a few things stand out.

1. The symmetric moment distribution results in considerable simplification. In addition, the column moments result in no unbalanced horizontal force.

$$H = \frac{7+14}{5} + \frac{-7-14}{5} = 0.$$

2. The imbalance in H is only due to anti-symmetric loading part. The H_U of 1.008 is actually the result of anti-symmetric moment distribution.

$$H_U = \frac{0.84 + 1.68}{5} + \frac{1.68 + 0.84}{5} = 1.008.$$

3. Probably we can altogether avoid the sway moment distribution process if we can modify our anti-symmetric moment distribution procedure in such a way that would keep the net horizontal shear at A and D zero. This modification is dealt with in the next section (3.9), Cantilever Moment Distribution.

Example 3.35 *Solve the problem in Fig. 3.152 by Naylor's method.*

Solution The symmetric loading part of the problem remains unchanged (same as case 1 of example 3.34)

$$M_{AB} = 7\ ; \qquad M_{BA} = 14\ ; \qquad M_{BC} = -14.$$

For the anti-symmetric part (Case 2 of example 3.28) we have to work out the FEMs and distribution factors

FEM : $M_{FBF} = 0$

$$M_{FBE} = -\frac{Wab^2}{l^2} + \frac{Wa^2b}{l^2} = \frac{17.5}{25}(-2 \times 3 \times 3 + 2 \times 2 \times 3) = -4.2$$

Distribution factors (DF)

$$k_{BF} = \frac{EI}{l} \qquad k_{BE} = \frac{6 \times EI}{l}$$

$$k_{BF} : k_{BE} = 1 : 6$$

$$DF_{BF} = \frac{1}{7} \qquad DF_{BE} = \frac{6}{7}$$

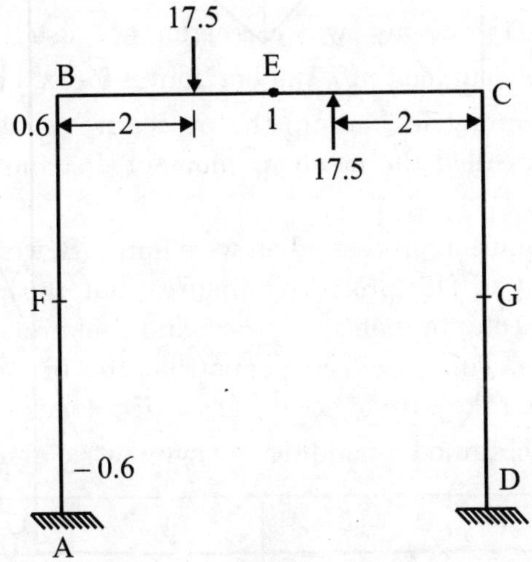

Fig. 3.153

Naylor's moment distribution :

Joint	*A*	*B*		*E*	*C*		*D*
Member	AB	BA	BE				
DF		$\frac{1}{7}$	$\frac{6}{7}$				
FEM Bal		 0.6	– 4.2 3.6				
C.O.	– 0.6						
Final	– 0.6	0.6	– 0.6		– 0.6	0.6	– 0.6

The final moments are

$$M_{AB} = 7 - 0.6 = 6.4 ;$$

$$M_{BA} = 14 + 0.6 = 14.60$$

$$M_{BC} = -14 - 0.6 = -14.6 ;$$

$$M_{CB} = 14 - 0.6 = 13.4$$

$$M_{CD} = -14 + 0.6 = -13.4 ;$$

$$M_{DC} = -7 - 0.6 = -7.6$$

These moments do not leave any unbalanced horizontal force, since

$$\frac{6.4 + 14.6}{5} + \frac{-7.6 - 13.4}{5} = 0$$

And we did not carry out any sway correction analysis. This sounds like too good to be true. The fact is Naylor's method is remarkable for its efficiency and simplicity. It has not received the recognition due to it.

We do not have to carry out any sway correction because

I. In step 1, (Symmetrical loading), the horizontal forces balance. (There is no sway)

II. In step 2, (anti-symmetric loading) the procedure leaves no shear on the column. This is sometimes called the no-shear moment distribution. The Carry over of – 1 takes care of this.

In this moment distribution process when we balance B, we need + 0.6 on BA. The carry over on AB would then be – 0.6. This procedure ensures that, during every balancing plus C.O stage, the net increment in column moment is zero and this prevents the introduction of any unbalanced horizontal force. This in effect means that the frame is allowed to sway to the required amount in order to take care of eccentric vertical loads or horizontal forces.

The final solution is obtained by adding the symmetric and anti-symmetric solutions :

Member	AB	BA	BC	C B	CD	DC
Symmetric	7	14	– 14	14	– 14	– 7
Anti-symmetric	– 0.6	0.6	– 0.6	– 0.6	0.6	– 0.6
Final	6.4	14.6	– 14.6	13.4	– 13.4	– 7.6

The bending moment diagram is as under:

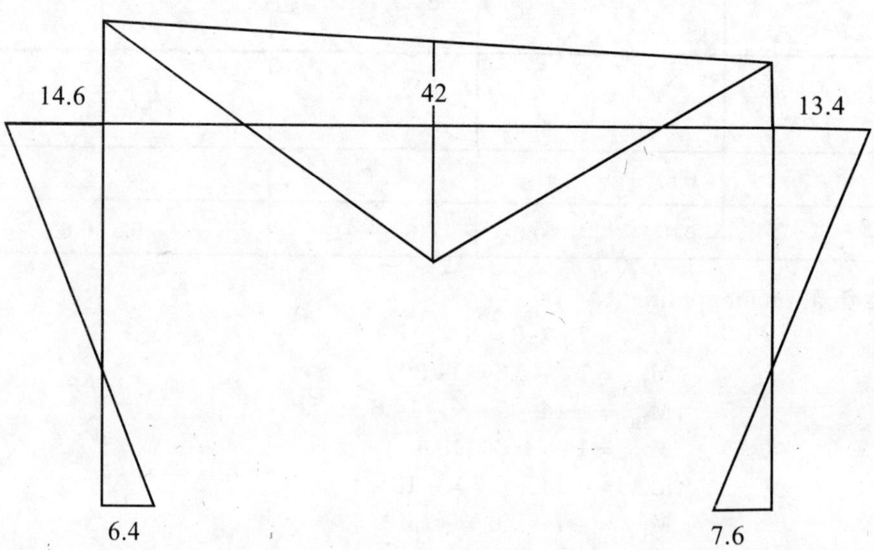

Example 3.36 *Analyse the portal frame loaded as shown is Fig. 3.154, using Naylor's method*

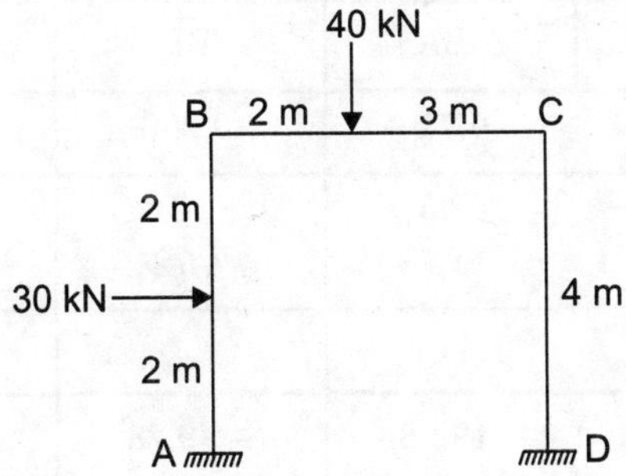

Fig. 3.154

Solution

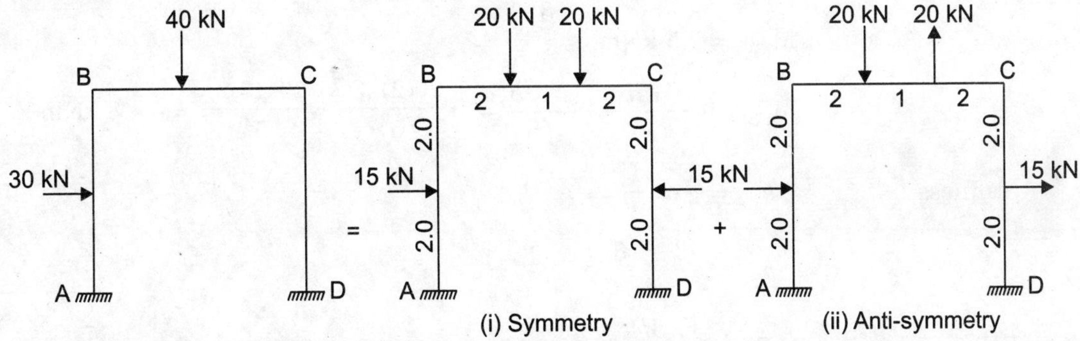

Fig. 3.155

Case (*i*) : Symmetry

FEM : $M_{FAB} = -\frac{Wl}{8} = -\frac{15 \times 4}{8} = -7.5 \text{ kNm}$

$M_{FBA} = 7.5 \text{ kNm}$

$$M_{FBC} = -\frac{W \cdot ab^2}{l^2} - \frac{W \cdot a^2 b}{l^2} = -\frac{20(2) \cdot 3^2}{5^2} - \frac{20(3) \cdot 2^2}{5^2} = -24 \text{ kNm}$$

Stiffness : $k_{BA} = \frac{I}{l} = \frac{I}{4}$

$$k_{BC} = \frac{1}{2} \cdot \frac{I}{l} = \frac{1}{2} \cdot \frac{I}{5} = \frac{I}{10}$$

DF : $BA = \frac{I/4}{I/4 + I/10} = 0.714$

$$BC = \frac{I/10}{I/4 + I/10} = 0.286$$

Joint	*A*	*B*		*E*	*C*	*D*
Member	AB	BA	BE	CE	CD	DC
DF		0.714	0.286			
FEM Balancing	– 7.5	7.5 11.78	– 24 4.72			
C.O.	5.89					
Total	– 1.61	19.28	– 19.28	19.28	– 19.28	1.61

Case (*ii*) : Anti-symmetry

FEM :

$$M_{FAB} = -\frac{Wl}{8} = -\frac{15 \times 4}{8} = -7.5 \text{ kNm}$$

$$M_{FBA} = 7.5 \text{ kNm}$$

$$M_{FBC} = -\frac{Wab^2}{l^2} + \frac{Wa^2b}{l^2} = -\frac{20(2) \cdot 3^2}{5^2} + \frac{20 \cdot (3)2^2}{5^2} = -4.8 \text{ kNm}$$

Stiffness :

$$k_{BA} = \frac{I}{l} = \frac{I}{4}$$

$$k_{BC} = \frac{6I}{l} = \frac{6 \cdot I}{5}$$

DF :

$$BA = \frac{I/4}{I/4 + 6I/5} = 0.172$$

$$BC = \frac{6I/5}{I/4 + 6I/5} = 0.828$$

Joint	*A*	*B*		*E*	*C*	*D*
Member	AB	BA	BE	CE	CD	DC
DF		0.172	0.828			
FEM Balancing	– 7.5	7.5 – 0.46	– 4.8 – 2.24			
C.O.	0.46					
Total	– 7.04	7.04	– 7.04	– 7.04	7.04	– 7.04

Free BM :

$$\mu_{AB} = \frac{30 \times 4}{4} = 30 \text{ kNm}$$

$$\mu_{BC} = \frac{40 \times 2 \times 3}{5} = 48 \text{ kNm}$$

Final moments :

Member	AB	BA	BC	CB	CD	DC
(*i*) Symmetry	– 1.61	19.28	– 19.28	19.28	– 19.28	1.61
(*ii*) Anti-symmetry	– 7.04	7.04	– 7.04	– 7.04	7.04	– 7.04
Final moments (*i*) + (*ii*)	– 8.65	26.32	– 26.32	12.24	– 12.24	– 5.43

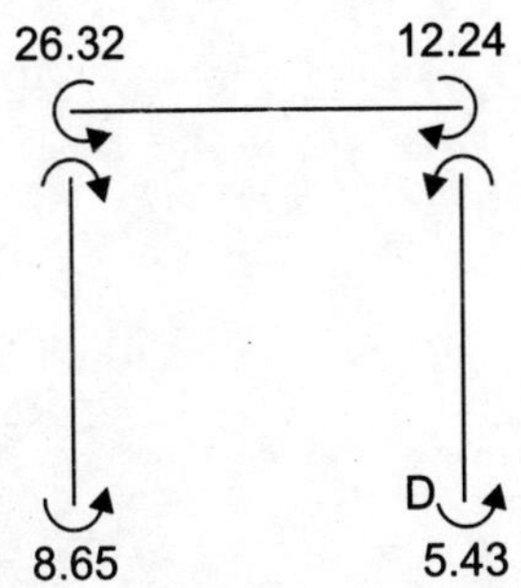

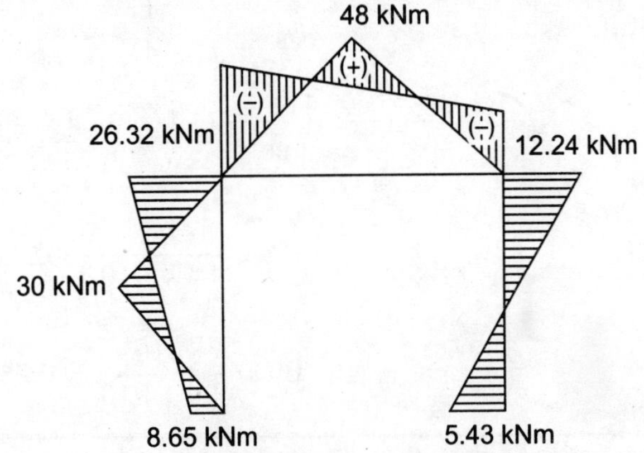

Bending moment diagram

Example 3.37 *Analyse the portal frame loaded as shown in Fig. 3.156, using Naylor's method.*

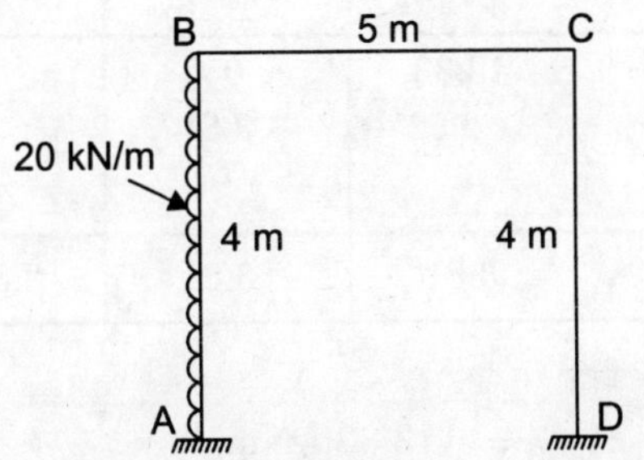

Fig. 3.156

Solution

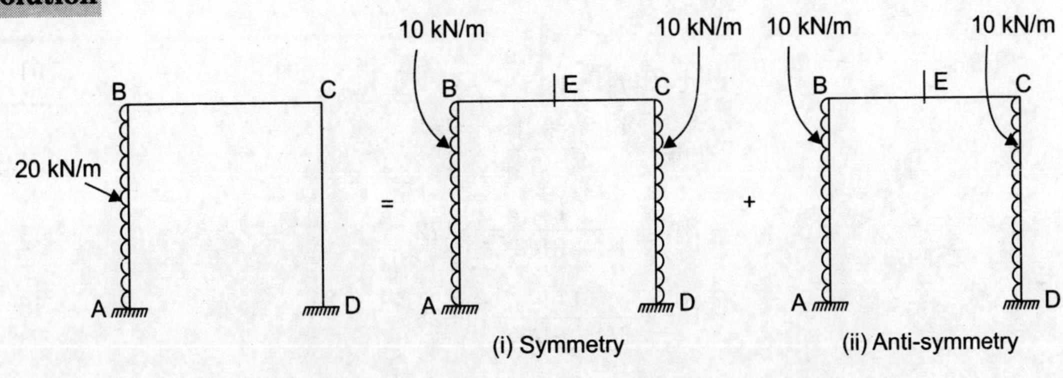

Case (*i*) : Symmetry

FEM :

$$M_{FAB} = -\frac{wl^2}{12} = -\frac{10 \times 4^2}{12} = -13.33 \text{ kNm}$$

$$M_{FBA} = \frac{wl^2}{12} = \frac{10 \times 4^2}{12} = 13.33 \text{ kNm}$$

$$M_{FBC} = 0$$

Stiffness :

$$k_{BA} = \frac{I}{l} = \frac{I}{4}$$

$$k_{BE} = \frac{1}{2} \cdot \frac{I}{l} = \frac{1}{2} \cdot \frac{I}{5} = \frac{I}{10}$$

DF :

$$BA = \frac{I/4}{I/4 + I/10} = 0.714$$

$$BC = \frac{I/10}{I/4 + I/10} = 0.286$$

Joint	***A***	***B***		***E***	***C***	***D***
Member	AB	BA	BE	CE	CD	DC
DF		0.714	0.286			
FEM	– 13.33	13.33	0			
Balancing		– 9.52	– 3.81			
C.O.	– 4.76					
Final	– 18.09	3.81	– 3.81	3.81	– 3.81	18.09

Case (*ii*) : Anti-symmetry

FEM :

$$M_{FAB} = -13.33 \text{ kNm}$$

$$M_{FBA} = 13.33 \text{ kNm}$$

$$M_{FBC} = 0$$

Stiffness :

$$k_{BA} = \frac{I}{l} = \frac{I}{4}$$

$$k_{BE} = \frac{6I}{l} = \frac{6 \cdot I}{5}$$

DF :

$$BA = \frac{I/4}{I/4 + 6I/5} = 0.172$$

$$BC = \frac{6I/5}{I/4 + 6I/5} = 0.828$$

Joint	*A*	*B*		*E*	*C*	*D*
Member	AB	BA	BE	CE	CD	DC
DF		0.172	0.828			
FEM Balancing C.O.	– 13.33 2.29	13.33 – 2.29	0 – 11.04			
Final	– 11.04	11.04	– 11.04	– 11.04	11.04	– 11.04

Final moments :

	Member	*AB*	*BA*	*BC*	*CB*	*CD*	*DC*
(*i*)	Symmetry	– 18.09	3.81	– 3.81	3.81	– 3.81	18.09
(*ii*)	Anti-symmetry	– 11.04	11.04	– 11.04	– 11.04	11.04	– 11.04
	Final moments (*i*) + (*ii*)	– 29.13	14.85	– 14.85	– 7.23	7.23	7.05

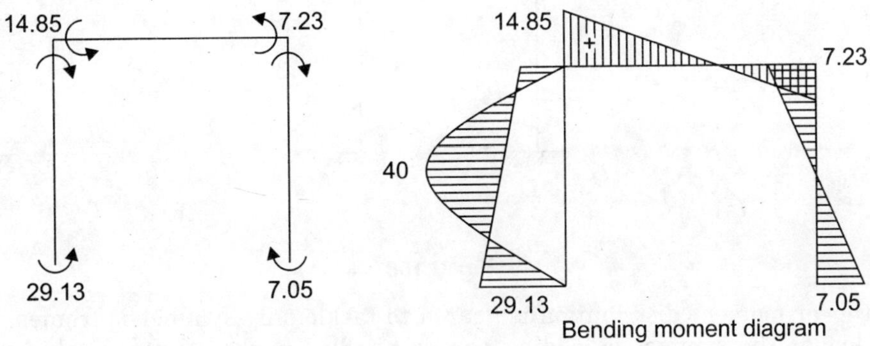

Fig. 3.157

3.9 CANTILEVER MOMENT DISTRIBUTION (NAYLOR'S METHOD)

3.9.1 Introduction

We have already seen that moment distribution method starts with a wrong assumption—that all the joints are fixed. We progressively correct this mistake by realising balanced condition in each joint. While balancing, we are forced to introduce some moment at one end of any given member. This created a **Carry over** to the far end. We also saw that in the process of balancing any joint with several members, we

(*i*) first compute the total imbalance in the joint

(*ii*) apply a balancing moment to the joint equal and opposite to the unbalanced moment (*iii*) apportion this total balancing moment to each member at the joint in proportion to their **stiffnesses**.

The process of moment distribution ensures that all joints originally (wrongly) assumed to be fixed are allowed to rotate sufficiently to let each joint stay in moment equilibrium. However, we do not **compute** the actual rotation of any joint.

Cantilever moment distribution is a method suited to frames, especially multistorey frames. Herein, instead of starting with the wrong assumption that joints are fixed, we start with another wrong assumption that the vertical members (columns) are free to rotate and translate at their top ends. Thus each column acts as a cantilever, fixed at the bottom and free at the top end.

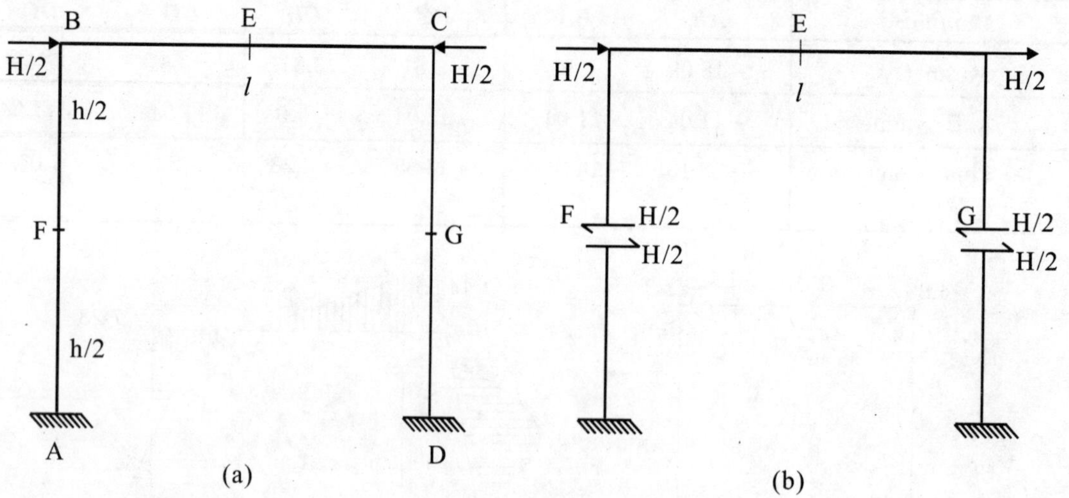

Fig. 3.158

Cantilever moment distribution is geared to tackle only symmetric frames, with anti-symmetric loads. The symmetric loading part is handled in the conventional way. In other words, the method can handle symmetric frames with any loading since it can be split into symmetric and anti-symmetric loading components. In this method sway analysis is completely avoided.

3.9.2 Redefining the Basic Terms

The terms 'fixed end moment', 'stiffness' and 'carry over' need to be redefined in the case of Cantilever moment distribution method.

(*i*) The fixed end moments on columns are computed from the justifiable assumption that columns have zero bending moments at mid heights. Hence the total shear above the mid height section would be shared by the columns at that level. The FEM at BA will be $-\frac{H}{2}\cdot\frac{h}{2}$, where H/2 is the shear on the column. Obviously this is the moment at the fixed end of the Cantilever BF. If there are more storeys, the shear S at F can be different from H/2, and the FEM has to be $-S\cdot\frac{h}{2}$.

(*ii*) Carry over

We were all along dealing with a moment applied at B which was hinged. In Cantilever method, B will be free (Fig. 3.159 (*b*)) and M applied at B would produce – M at A. The Carry over factor is thus – 1 (instead of ½)

(*iii*) **Stiffness.** Again considering Fig. 3.159 (*b*), M at B produces θ_B given by $\frac{Ml}{EI}$. Hence the stiffness of BA is $\frac{EI}{l}$ (compare this with $\frac{4EI}{l}$ when near end is hinged and far end is fixed ; and $\frac{3EI}{l}$ when both near end and far end are hinged).

[Guess what would be the stiffness when B were free and A were hinged]

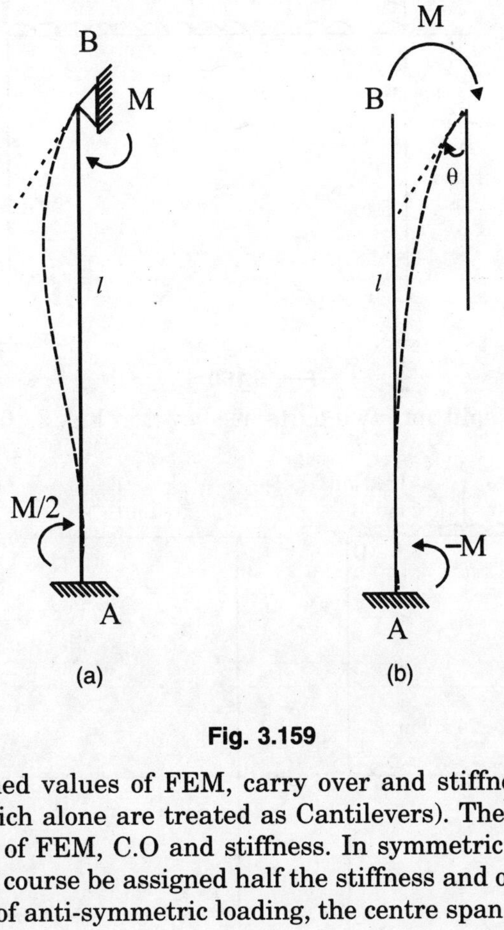

Fig. 3.159

The above redefined values of FEM, carry over and stiffness apply only for column members of a frame (which alone are treated as Cantilevers). The beam members will carry the same old definitions of FEM, C.O and stiffness. In symmetric loading cases, the central horizontal span would of course be assigned half the stiffness and one half of the frame would be analysed. In the case of anti-symmetric loading, the centre span would be assumed to have a hinge at mid span. Its stiffness would be $\frac{6EI}{l}$ (1.5 times $4EI/l$).

Armed with these concepts we will now tackle the next example.

Example 3.38 *Analyse the two storey frame shown in Fig. 3.160 by Naylor's method and tabulate the final joint moments.*

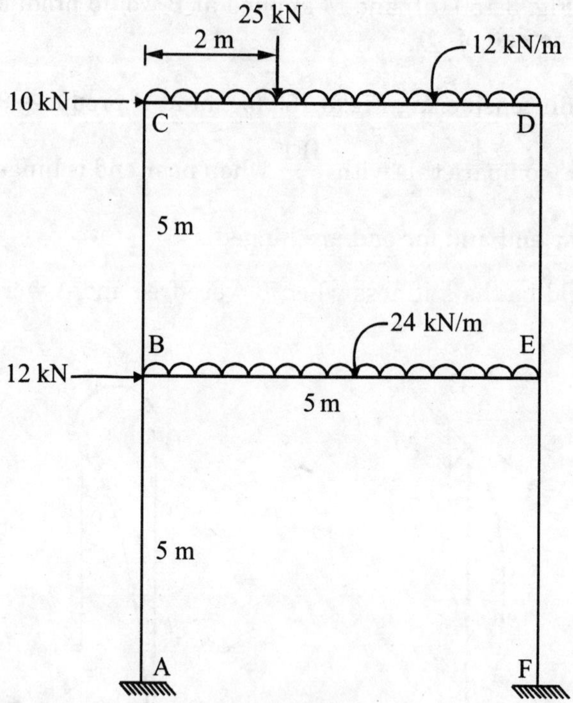

Fig. 3.160

Solution The loading is split into two parts as shown in Fig. 3.161

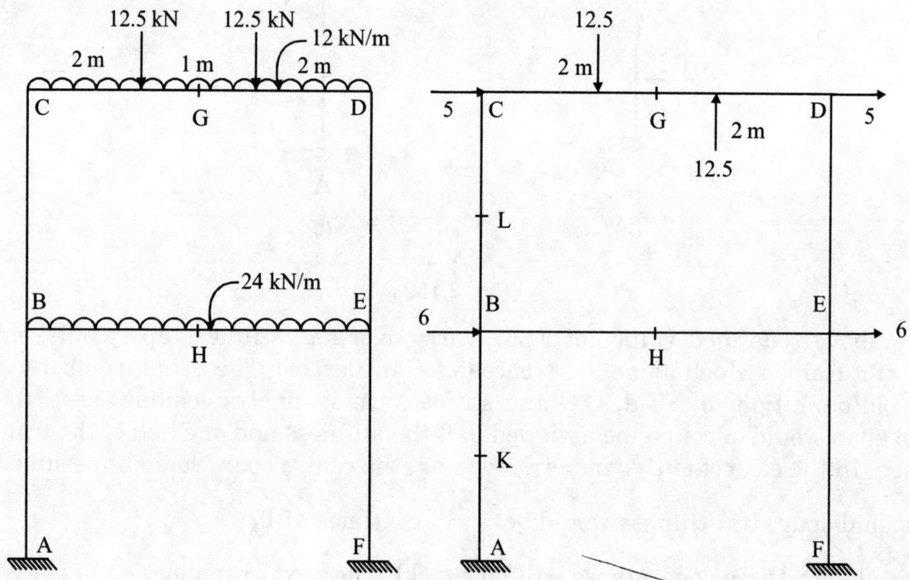

Fig. 3.161

Part I. Symmetric solution

FEM : $M_{FAB} = M_{FBA} = M_{FCB} = M_{FBC} = 0$

$$M_{FBE} = -\frac{wl^2}{12} = -\frac{24 \times 25}{12} = -50 \text{ kNm}$$

$$M_{FCD} = -\frac{wl^2}{12} - \frac{Wab^2}{l^2} - \frac{Wa^2b}{l^2} = -25 - 9 - 6 = -40 \text{ kNm}$$

Stiffnesses : taking I = 1

$$K_{BA} = \frac{1}{5}; \quad K_{BC} = \frac{1}{5}; K_{BE} \text{ (modified)} = \frac{1}{2} \times \frac{1}{5} = \frac{1}{10}$$

$$K_{CD} \text{ (modified)} = \frac{1}{10}$$

$DF_{BA} = 0.4$ $\quad DF_{BC} = 0.4$ $\quad DF_{BE} = 0.2$

$DF_{CB} = \frac{2}{3}$ $\quad DF_{CD} = \frac{1}{3}$

Symmetric moment distribution:

Joint	*A*	*B*			*C*	
Member	AB	BA	BE	BC	CB	CD
DF		0.4	0.2	0.4	0.667	0.333
FEM			– 50			– 40
Bal		20	10	20	26.67	13.33
C.O.	10			13.33	10	
Bal		– 5.33	– 2.67	– 5.33	– 6.67	– 3.33
C.O.	– 2.67		0.67	– 3.33	– 2.67	
Bal		1.33		1.33	1.78	0.89
C.O.	+ 0.67			+ 0.87	+ 0.67	
Bal		– 0.35	– 0.17	– 0.35	– 0.45	– 0.22
Final	8.00	15.65	– 42.17	26.52	29.33	– 29.33

By symmetry,
$M_{DC} = 29.33$
$M_{DE} = -29.33$
$M_{ED} = -26.52$
$M_{EB} = 42.17$
$M_{EF} = -15.65$
$M_{FE} = -8.00$

Part II. Anti-symmetric loading. (Horizontal loads can be apportioned equally to the left and right side of the frame. The symmetric part can be ignored as it produces only a thrust on BE and CD)

FEMs. To get the FEMs on columns let us work out the shears on CB and BA (see figure below)

$$M_{FBA} = 11 \times 2.5 = -27.5 \text{ kNm}, \quad M_{FCB} = 5 \times 2.5 = -12.5$$
$$M_{FBC} = -12.5, \quad M_{FAB} = -27.5$$

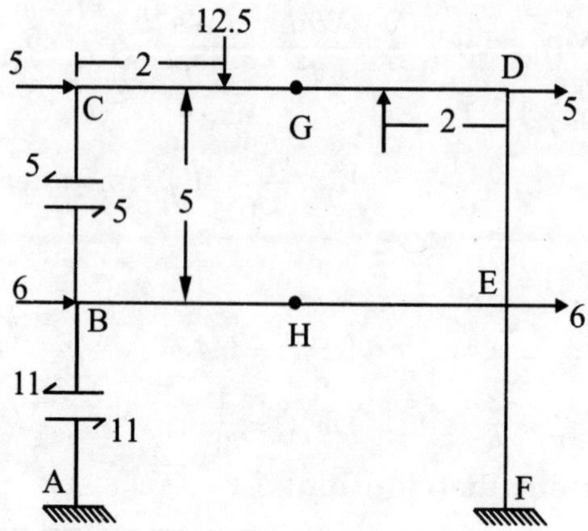

FEMs on beams is considered in the conventional method.

$$M_{FCD} = -\frac{Wab^2}{l^2} + \frac{Wa^2b}{l^2} = -3.0 \text{ kNm}, \quad M_{FBE} = 0$$

Stiffnesses $k_{BA} = \frac{1}{5}$ $k_{BC} = \frac{1}{5}$ $k_{BH} = \frac{6}{5}$

Hence, $DF_{BA} = 0.125$ $DF_{BC} = 0.125$ $DF_{BH} = 0.750$

$$k_{CB} = \frac{1}{5} \qquad k_{CG} = \frac{6}{5}$$

$$DF_{CB} = \frac{1}{7} = 0.143, \quad DF_{CG} = \frac{6}{7} = 0.857$$

Joint	***A***	***B***			***C***	
Member	AB	BA	BE	BC	CB	CD
DF		0.125	0.750	0.125	0.143	0.857
FEM Bal	– 27.5	– 27.5 5	0 30	– 12.5 5	– 12.5 2.21	– 3.00 13.29
C.O. Bal	– 5	 0.28	 1.66	– 2.21 0.27	– 5 0.71	 4.29
C.O. Bal	– 0.28	 0.09	 0.53	– 0.71 0.09	– 0.27 0.04	 0.23
C.O. Bal	– 0.09		 0.04	– 0.04	– 0.09 0.01	 0.08
Final	– 32.87	– 22.13	32.23	– 10.10	– 14.89	14.89

Member	AB	BA	BE	BC	CB	CD	DC	DE	ED	EB	EF	FE
Symm	8.00	15.65	– 42.17	26.52	29.33	– 29.33	29.33	– 29.33	– 26.52	42.17	– 15.65	– 8.00
A / S	– 32.87	– 22.13	32.23	– 10.10	14.89	14.89	14.89	– 14.89	– 10.10	32.23	– 22.13	– 32.84
Final moments	– 24.87	– 6.48	– 9.94	16.42	14.44	– 14.44	44.22	– 44.22	– 36.62	74.40	– 37.78	– 40.84

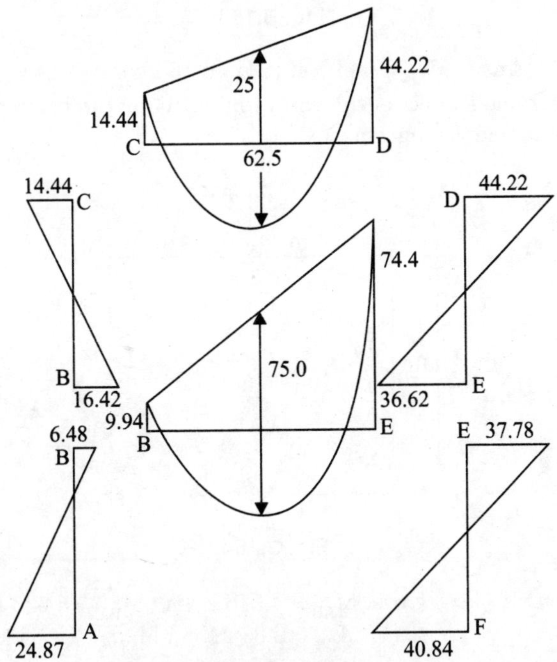

Bending moment diagram

PROBLEMS

1. A continuous beam is shown in Fig. 3.162. The support B sinks by 10 mm relative to A and C. Using moment distribution, calculate the support moments and draw bending moment diagram, EI = 6000 kNm^2.

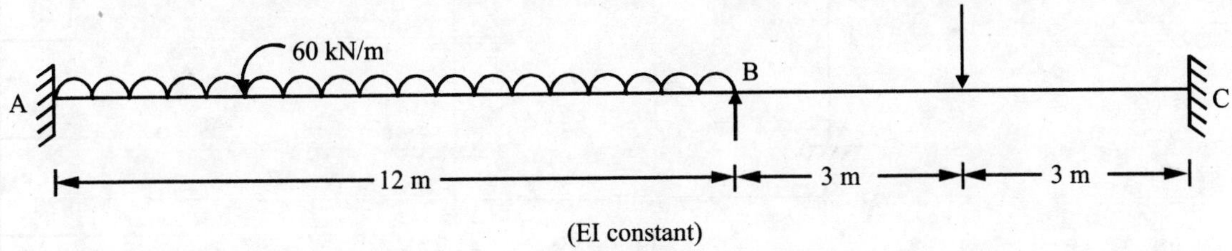

Fig. 3.162

[**Ans.** M_{AB} = – 753.719 kNm ; M_{BA} = 655.062 kNm ; M_{BC} = – 655.063 kNm ; M_{CB} = 487.469 kNm]

2. A continuous beam ABC fixed at the ends is loaded as shown in Fig. 3.163. Find the reactions and support moments using the method of moment distribution. Draw also the BM and SF diagrams.

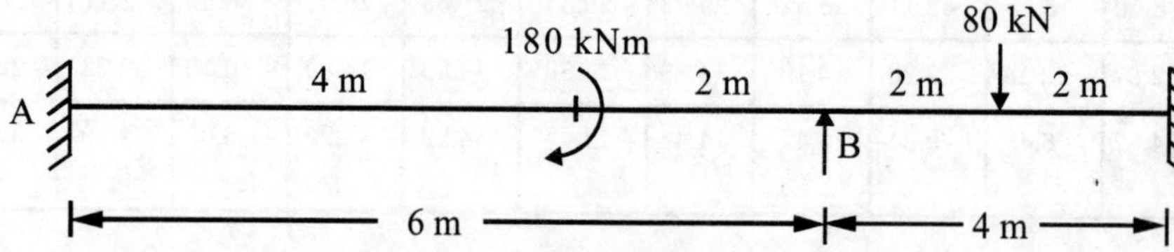

Fig. 3.163

[**Ans.** M_{AB} = 68 kNm ; M_{BC} = 16 kNm ; M_{BC} = – 16 kNm ; M_{CB} = 52 kNm]

3. Analyse the portal frame loaded as shown in Fig. 3.164 by the moment distribution method and sketch the bending moment diagram.

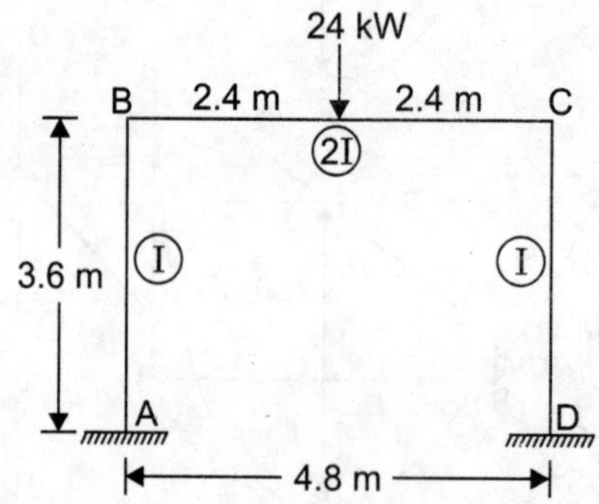

Fig. 3.164

[**Ans.** M_{AB} = 4.315 kNm ; M_{BA} = 8.630 kNm ; M_{BC} = – 8.63 kNm ;
M_{CB} = 8.63 kNm ; M_{CD} = – 8.63 kNm ; M_{DC} = – 4.315 kNm]

4. Analyse the portal frame with fixed base shown in Fig. 3.165 using moment distribution method, given $I_{AB} = 3I_0$; $I_{BC} = 2I_0$; $I_{CD} = 3I_0$.

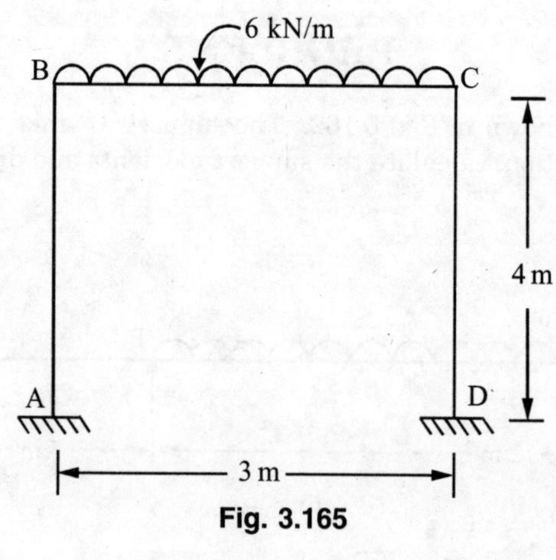

Fig. 3.165

[**Ans.** M_{AB} = 1.57 kNm ; M_{BA} = 3.11 kNm ; M_{BC} = – 3.11 kNm ;
M_{CB} = 3.11 kNm ; M_{CD} = – 3.11 kNm ; M_{DC} = – 1.57 kNm]

5. Analyse the frame shown in Fig. 3.166, using the method of moment distribution and draw the bending moment diagram.

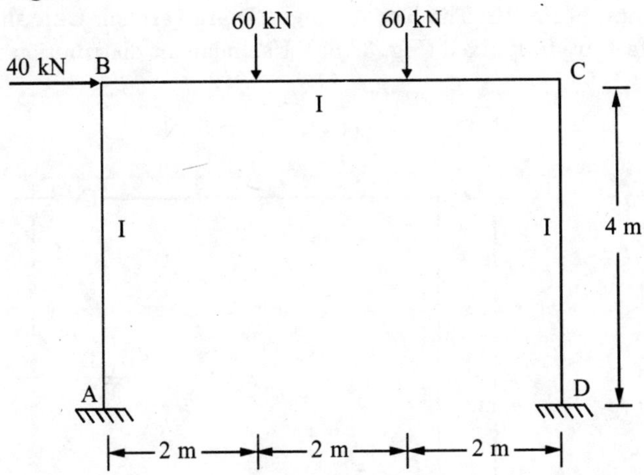

Fig. 3.166

[**Ans.** M_{AB} = 18 kNm ; M_{BA} = – 28 kNm ; M_{BC} = 28 kNm ; M_{CB} = – 92 kNm ; M_{CD} = 92 kNm ; M_{DC} = 78 kNm]

6. The simple rectangular frame shown in Fig. 3.167 is asymmetrically loaded. EI is constant for the frame. Determine the final moments and horizontal reactions, using the moment distribution method.

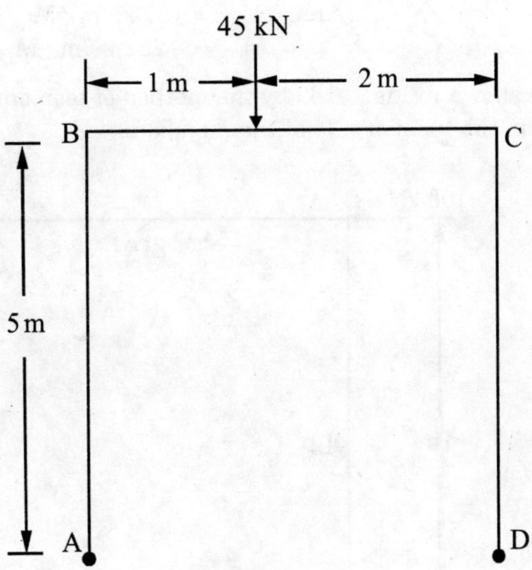

Fig. 3.167

[**Ans.** M_{BA} = 7 kNm ; M_{BC} = – 7 kNm ; M_{CB} = 7 kNm ; M_{CD} = – 7 kNm ; H_A = 1.40 kN(⟶) ; H_D = 1.40 kN(⟵)]

7. A stiff jointed frame ABCD is of constant section throughout and has the following dimensions. AB = 4.5 m, BC = 3 m and CD = 1.8 m. It carries loads of 600 kN each at each of the third points of the horizontal beam BC. The legs AB and CD are vertical. Calculate the moments at the joints if the feet are fully fixed (Fig. 3.168). Use moment distribution method.

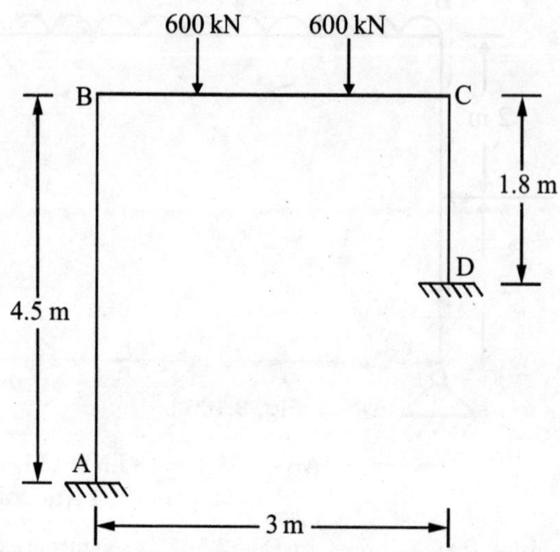

Fig. 3.168

[**Ans.** M_{AB} = 157 kNm ; M_{BA} = 259 kNm ; M_{BC} = – 259 kNm ; M_{CB} = – 222 kNm ; M_{CD} = – 222 kNm ; M_{DC} = 66 kNm]

8. Analyse the frame shown in Fig. 3.169 by the method of moment distribution. Draw the bending moment diagram on the tension side of the members.

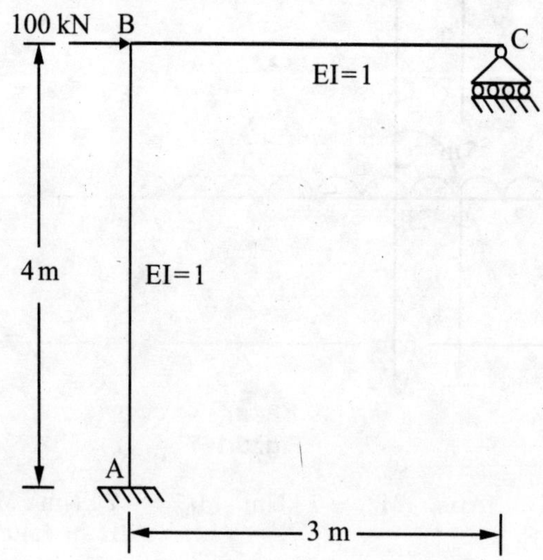

Fig. 3.169

[**Ans.** M_{AB} = – 240 kNm ; M_{BA} = – 160 kNm ; M_{BC} = 160 kNm ; M_{CB} = 0]

9. Find the end moments of the members of the portal frame shown in Fig. 3.170, using the moment distribution method.

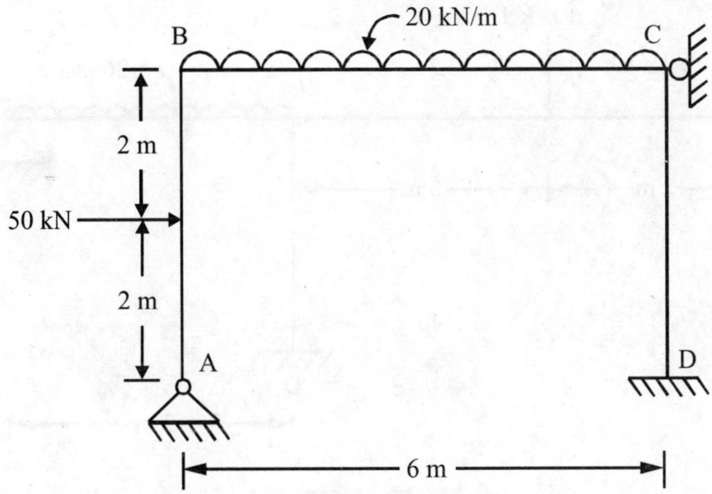

Fig. 3.170

[**Ans.** $M_{BA} = -54.83$ kNm ; $M_{BC} = 54.83$ kNm ; $M_{CB} = -35.72$ kNm ; $M_{CD} = 35.72$ kNm ; $M_{DC} = 17.86$ kNm]

10. Determine the moments at A and D of the frame shown in Fig. 3.171 using moment distribution method. Assume E to be uniform.

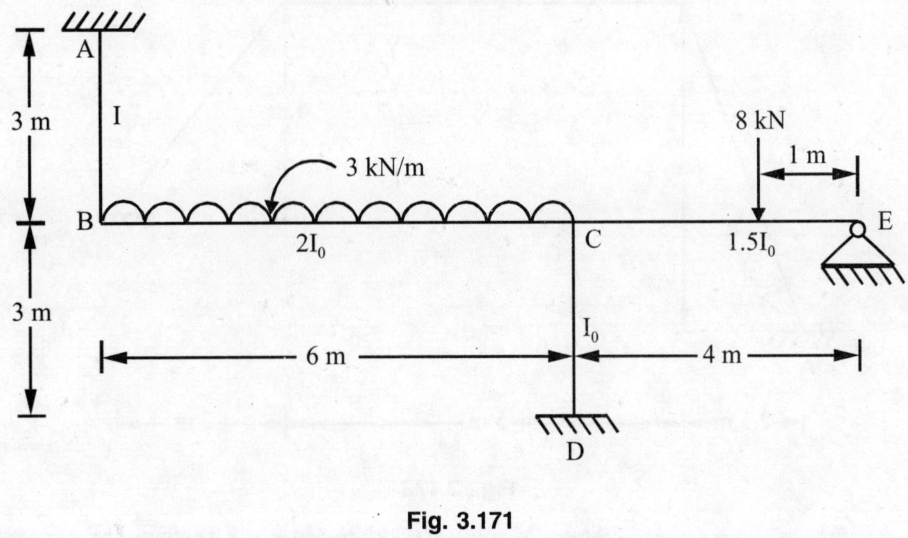

Fig. 3.171

[**Ans.** $M_A = 2.59$ kNm ; $M_D = 1.39$ kNm]

11. Analyse the frame shown in Fig. 3.172 by moment distribution method. Draw the bending moment diagram. The second moment of the area is indicated in the figure.

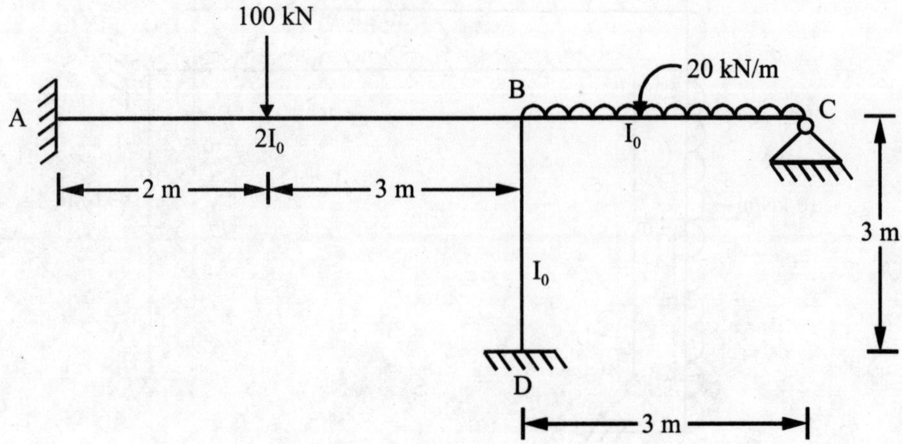

Fig. 3.172

[**Ans.** M_{AB} = – 77.23 kNm ; M_{BA} = – 37.54 kNm ; M_{BC} = – 28.875 kNm ; M_{CB} = 0 ; M_{BD} = – 8.67 kNm ; M_{DB} = – 4.35 kNm]

12. Analyse the portal frame shown in Fig. 3.173, using the method of moment distribution. All members have same flexural rigidity.

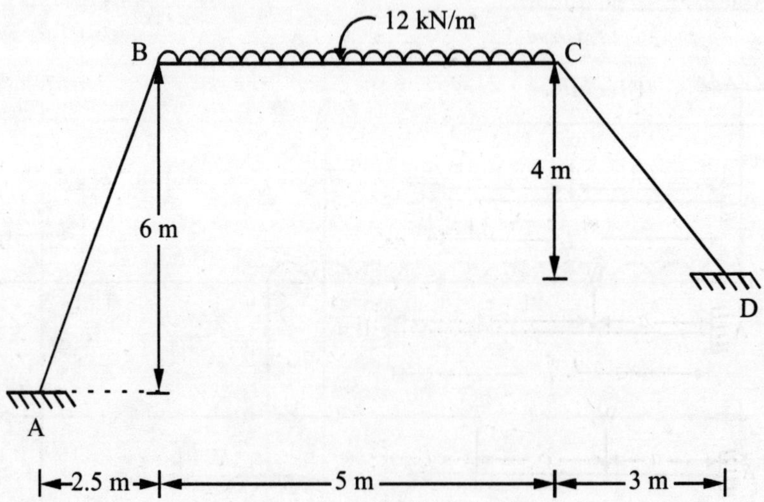

Fig. 3.173

[**Ans.** M_{AB} = – 0.03 kNm ; M_{BA} = 8.79 kNm ; M_{BC} = – 8.79 kNm ; M_{CB} = 1.18 kNm ; M_{CD} = – 1.18 kNm ; M_{DC} = – 0.09 kNm]

13. Analyse the culvert ABCD shown in Fig. 3.174, using the moment distribution method. All members have the same flexural rigidity.

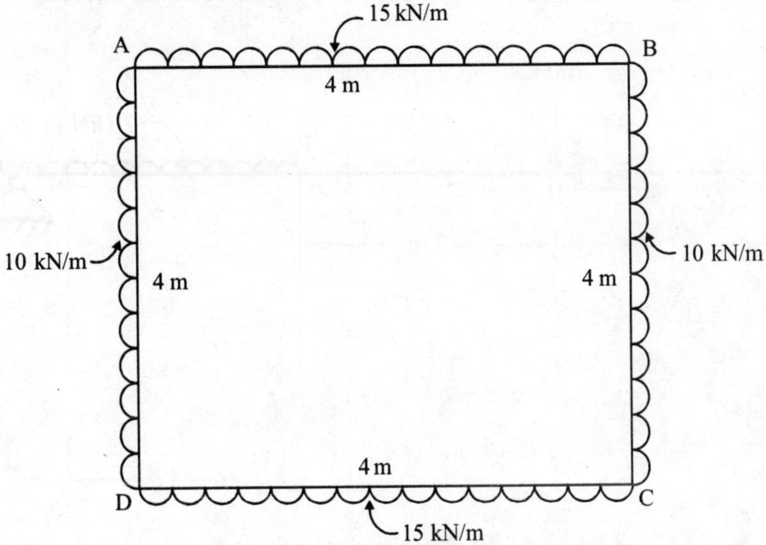

Fig. 3.174

[**Ans.** $M_{AB} = -16.67$ kNm ; $M_{AD} = 16.67$ kNm ; $M_{CD} = -16.67$ kNm ; $M_{CB} = +16.67$ kNm]

Table A1
Standard Cases of Fixed end Moments

S.No.	*Case* (*Clockwise : +ve*)	*Both ends fixed* M_{FAB}	M_{FBA}
1.	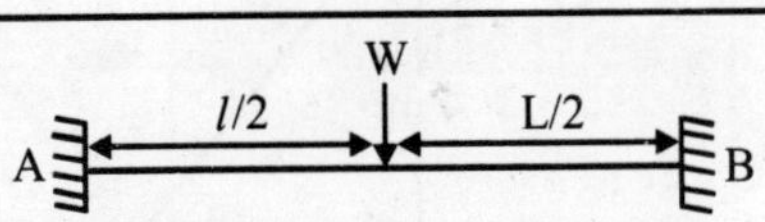	$-\dfrac{Wl}{8}$	$+\dfrac{Wl}{8}$
2.	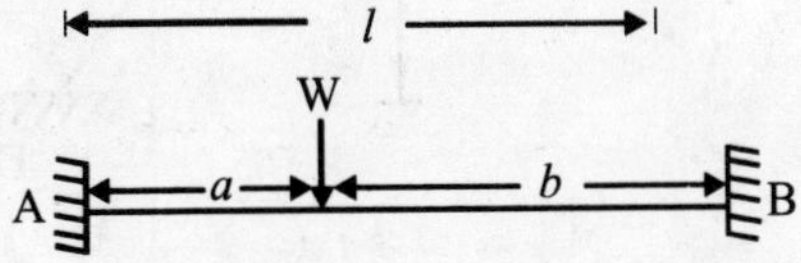	$-\dfrac{Wab^2}{l^2}$	$+\dfrac{Wa^2b}{l^2}$
3.	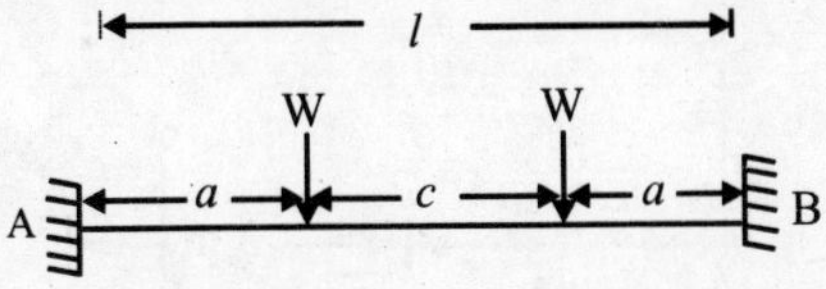	$-\dfrac{Wa(a+c)}{l}$	$+\dfrac{Wa(a+c)}{l}$
4.	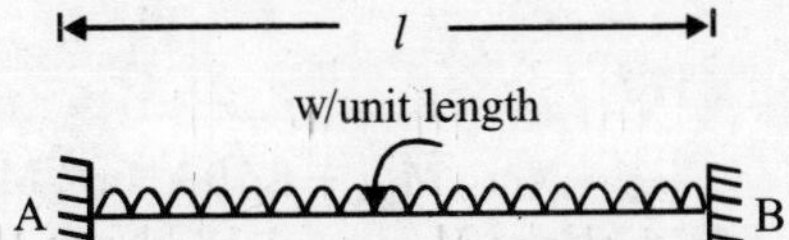	$-\dfrac{wl^2}{12}$	$+\dfrac{wl^2}{12}$

5.	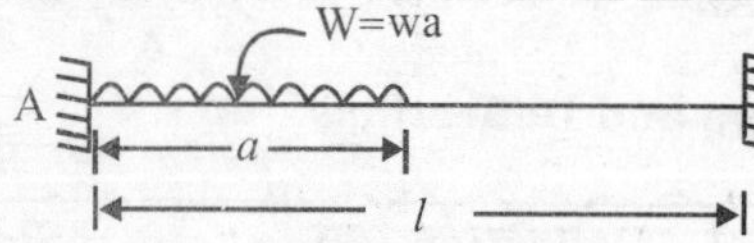	$-\frac{Wa}{12l^2}(6l^2 - 8al + 3a^2)$	$+\frac{Wa^2}{12l^2}(4l - 3a)$
6.	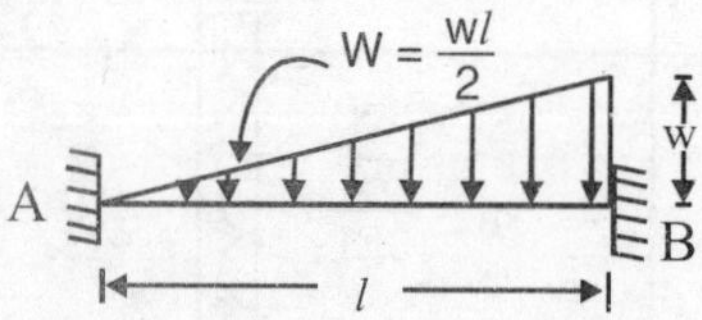	$-\frac{Wl}{15}$ or $\frac{-wl^2}{30}$	$+\frac{Wl}{10}$ or $\frac{+wl^2}{20}$
7.	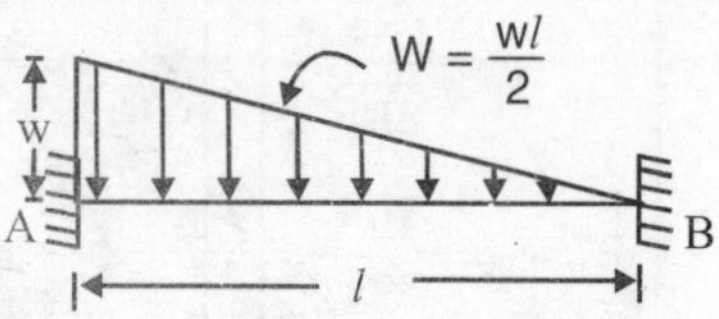	$-\frac{Wl}{10}$ or $\frac{-wl^2}{20}$	$+\frac{Wl}{15}$ or $\frac{wl^2}{30}$
8.	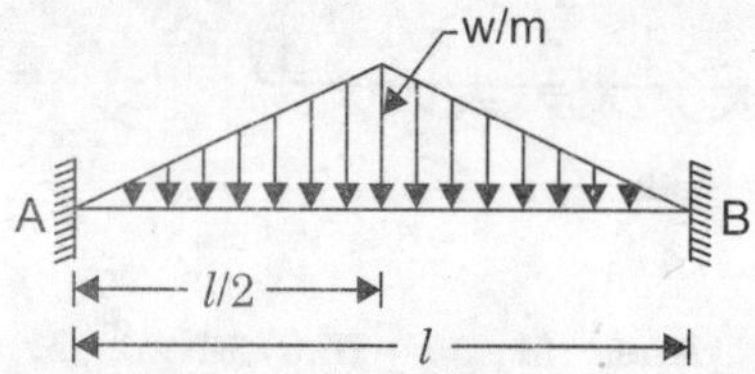	$\frac{-5}{96}wl^2$ or $\frac{-5}{48}Wl$	$\frac{5}{96}wl^2$ or $\frac{5}{48}Wl$
9.	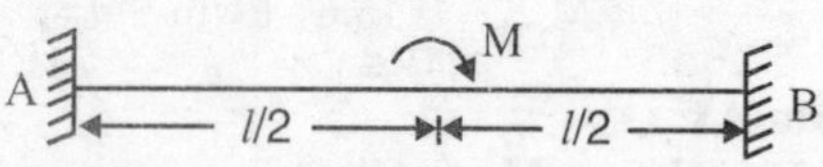	$+\frac{M}{4}$	$+\frac{M}{4}$
10.	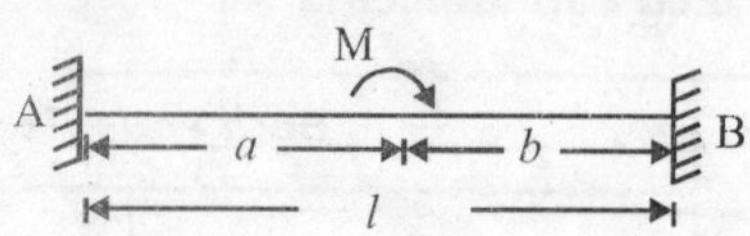	$\frac{Mb}{l^2}(3a - l)$	$\frac{Ma}{l^2}(3b - l)$
11.		$\frac{4EI}{l}\theta_A$	$\frac{2EI}{l}\theta_A$
12.	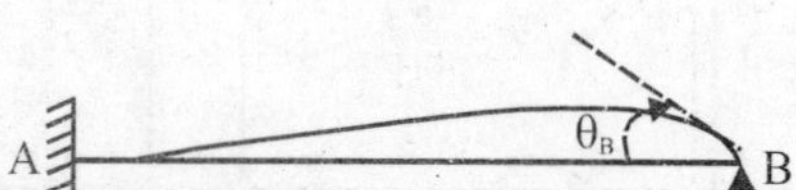	$\frac{2EI}{l}\theta_B$	$\frac{4EI}{l}\theta_B$
13.	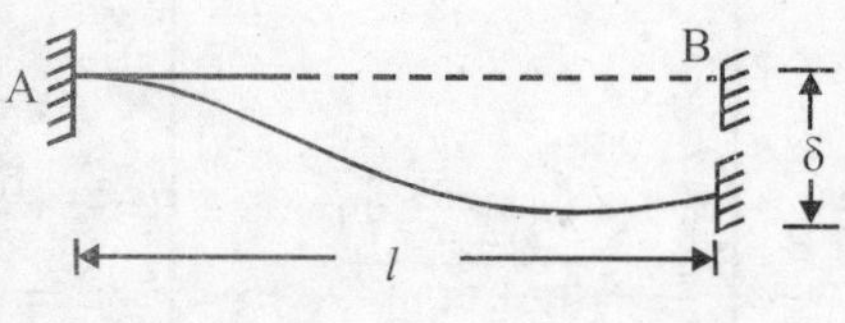	$-\frac{6EI\delta}{l^2}$	$-\frac{6EI\delta}{l^2}$
14.	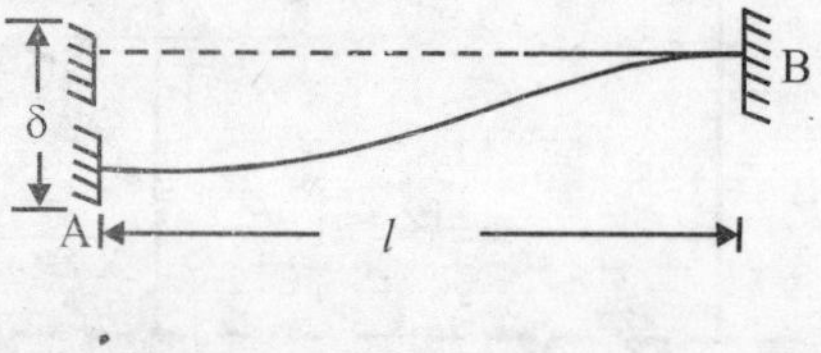	$+\frac{6EI\delta}{l^2}$	$+\frac{6EI\delta}{l^2}$

Table A2

Standard Cases of Slope and Deflections

S.No.	*Cases*	*Deflection (y)*	*Slope (i)*
1.	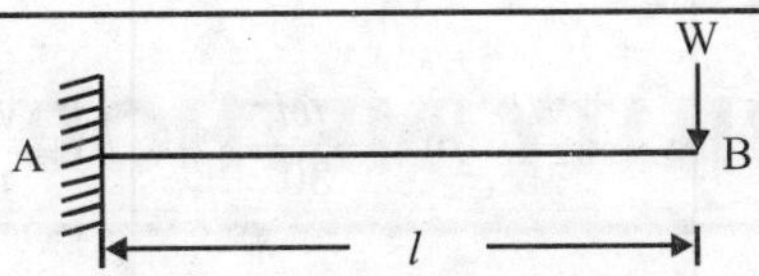	$y_B = \frac{Wl^3}{3EI}$	$i_B = \frac{Wl^2}{2EI}$
2.	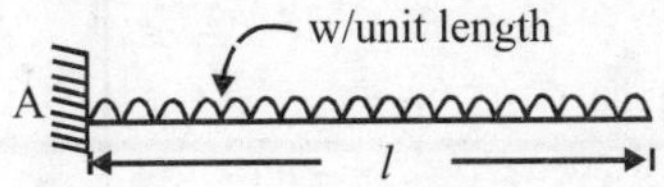	$y_B = \frac{wl^4}{8EI}$	$i_B = \frac{wl^3}{6EI}$
3.	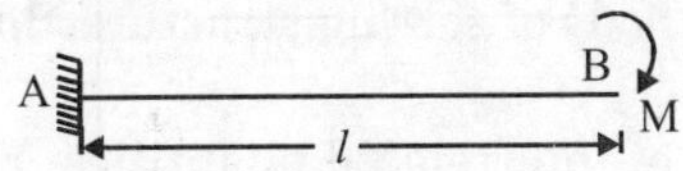	$y_B = \frac{Ml^2}{2EI}$	$i_B = \frac{Ml}{EI}$
4.	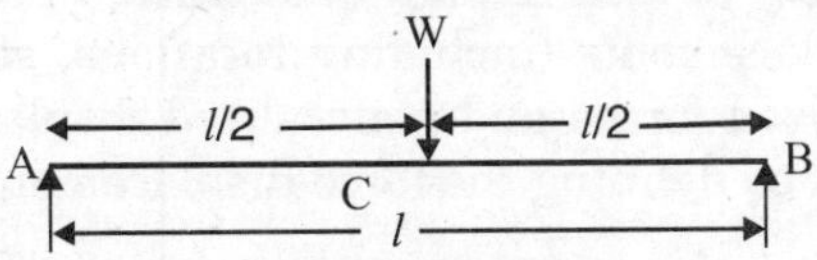	$y_C = \frac{Wl^3}{48EI}$	$i_A = \frac{Wl^2}{16EI}$, $i_B = \frac{-Wl^2}{16EI}$
5.	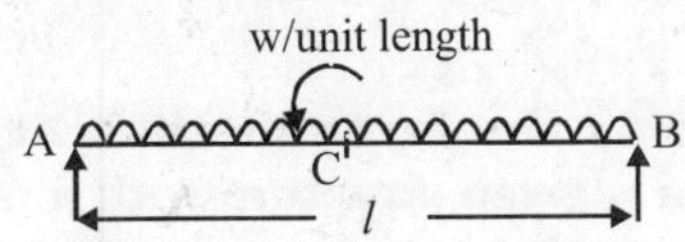	$y_C = \frac{5wl^4}{384EI}$	$i_A = \frac{wl^3}{24EI}$, $i_B = \frac{-wl^3}{24EI}$
6.	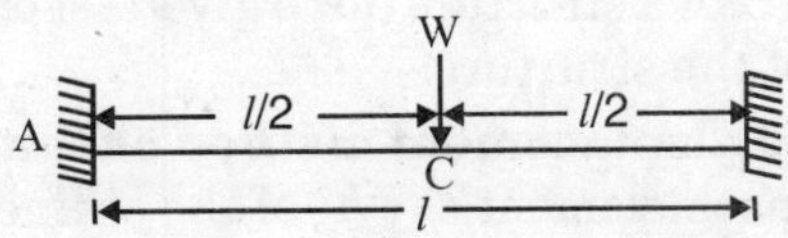	$y_C = \frac{Wl^3}{192EI}$	$i_A = i_B = 0$
7.	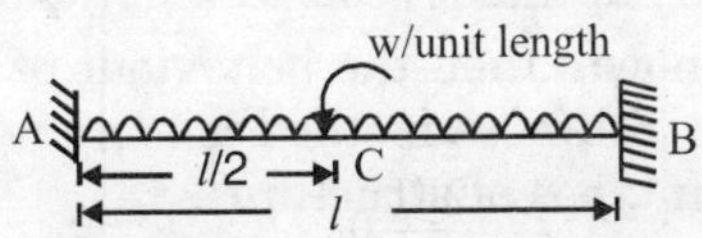	$y_C = \frac{wl^4}{384EI}$	$i_A = i_B = 0$

CHAPTER 4

MATRIX FLEXIBILITY METHOD

4.1 MATRIX METHODS

With the advent of computers, matrix methods of solving structures have become very popular.

In an elastic structure there are two sets of interrelated quantities, Forces (including moments, stresses, reactions, etc.,) and Displacements (including rotations, strains, twists, etc.). They are sometimes termed as generalised forces and generalised displacements. The behaviour of a structure can largely be defined by defining the force-displacement relationship in the form of a matrix.

There are two kinds of matrix methods :

1. The flexibility method also known as the force method or compatibility method. Herein we define a flexibility matrix for a given structure with a pre-defined set of co-ordinates. Each element of a flexibility matrix ($[a]$ matrix or $[\alpha]$ matrix) represents a displacement at a co-ordinate (i) due to unit force at a co-ordinate (j). That element of (α) can be called flexibility coefficient, α_{ij}. If the $[\alpha]$ matrix of a structure (for a given set of co-ordinates) is known, we can say we know the behaviour of the structure.

2. The stiffness method, also called the displacement method or the equilibrium method. For a pre-defined set of co-ordinates, each element say k_{ij} of the stiffness matrix $[k]$, called the stiffness coefficient is the force at co-ordinate at i due to unit displacement at co-ordinate at j with the condition that no other displacement occurs except the one stated at j.

If the stiffness matrix of a structure is known, then the behaviour of the structure is deemed to be known. Among these two methods, Stiffness method is more popular, although Flexibility method proves to be easier for certain types of structures.

Both methods involve the calculation of displacements due to forces or forces consistent with displacements. To do this we have to employ any one of the classical methods, such as Mohr's theorems, Conjugate beam method, Strain energy methods, etc.

4.2 ELEMENTARY CONCEPTS

4.2.1 Co-ordinates

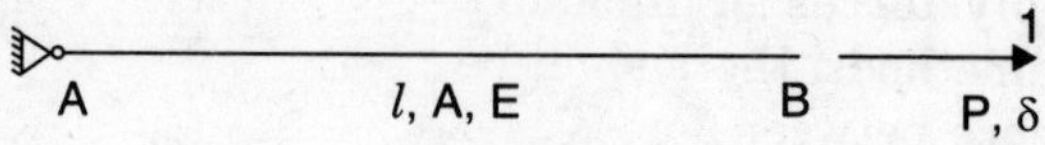

Fig. 4.1

Consider at bar AB held in position at A and free to move at B.

If a force P acts at B as shown

(*i*) Hinge A will develop a force P towards the left of A.

(*ii*) Bar AB will stretch by a value $\delta = \frac{Pl}{AE}$ and so B will move to the right by $\frac{Pl}{AE}$. In short, at B the stiffness of the member (force per unit displacement) $k = \frac{AE}{l}$ and the flexibility (displacement per unit force) $a = \frac{l}{AE}$.

The arrow at B defines a **"co-ordinate"** and it indicates the point and the direction of either the force P or the displacement δ, based on which flexibility a or stiffness k are defined.

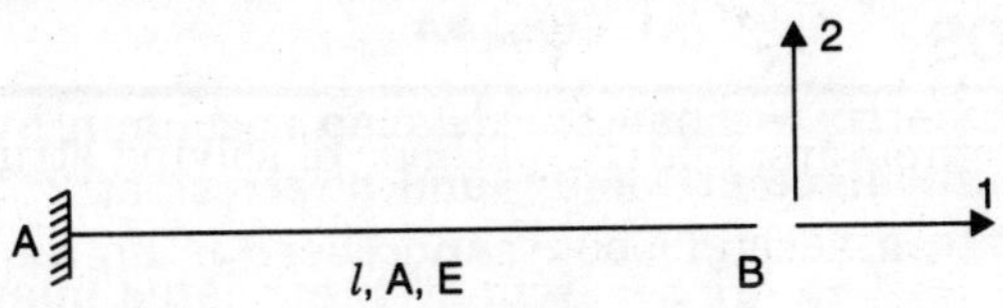

Fig. 4.2

Figure 4.2 is a 2 co-ordinate system definable for the beam AB. The beam is fixed at A. Hence there will be no displacement there. You cannot apply any desired force there either, except what can be developed there. Hence there is no question of defining co-ordinates at A.

4.2.2 Stiffness and Flexibility Matrices

We can apply unit displacements at 1 and / or 2. The corresponding forces sustaining these displacements are called stiffness terms. The displacements at 1 and 2 will form a vector $\begin{Bmatrix}\delta_1\\ \delta_2\end{Bmatrix}$ and the forces, a vector $\begin{Bmatrix}P_1\\ P_2\end{Bmatrix}$. Thus the stiffness of the system would be a matrix $[k] = \begin{bmatrix}k_{11} & k_{12}\\ k_{21} & k_{22}\end{bmatrix}$ and the flexibility matrix $[\alpha] = \begin{bmatrix}\alpha_{11} & \alpha_{12}\\ \alpha_{21} & \alpha_{22}\end{bmatrix}$ and obviously.

$$\{\delta\} = [\alpha]\{P\} \qquad ...(4.1)$$

and

$$\{P\} = [k]\{\delta\} \qquad ...(4.2)$$

Now we can concentrate on the terms α_{11}, α_{12}, α_{21} and α_{22}.

α_{11} is the displacement at 1 due to a unit force at 1.

α_{21} is the displacement at 2 due to a unit force at 1, and generally

α_{ij} is the displacement at i due to a unit force at j only.

Likewise k_{ij} is the force at i due to a unit displacement at j only (and zero displacement at all other co-ordinates).

4.2.3 One to One Correspondence

The arrows in Fig. 4.2 are not forces, nor displacements. They are co-ordinates along which only, we can apply forces or measure displacements. In matrix methods, there shall be a one to one correspondence between forces and displacements.

Figure 4.2 can be made a bit more general by defining one more co-ordinate as in Fig. 4.3.

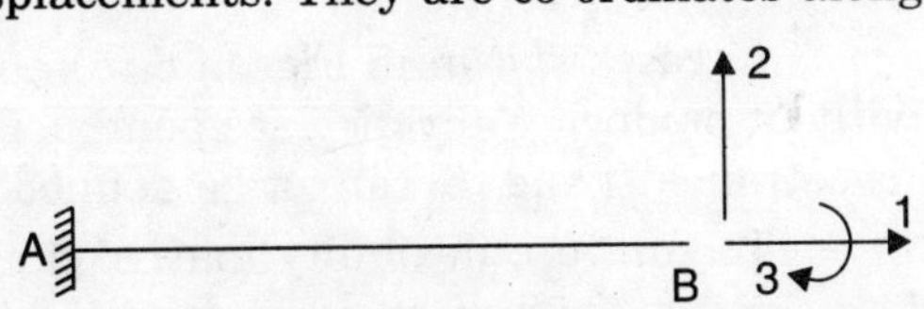

Fig. 4.3

Here 3 is the co-ordinate for the moment at B and the corresponding rotation. The stiffness and the flexibility matrices for this system would be of size 3 × 3.

4.2.4 Existence of [α] Matrix

[α] matrix for a given co-ordinate system need not necessarily exist. We will now look at a 4 co-ordinate system (Fig. 4.4).

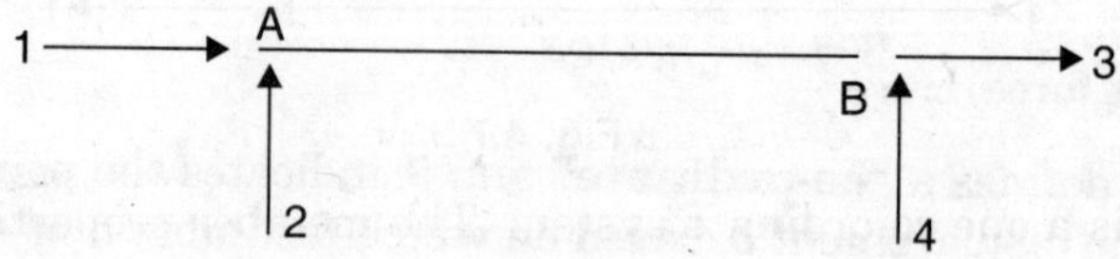

Fig. 4.4

For finding the [α] matrix we have to develop it column by column. To get the first column we shall apply a unit force at 1 only (and no forces at 2, 3 or 4). If we do that, the member AB will accelerate as a rigid body and there is no question of determining its displacements.

Hence the [α] matrix, made up of displacements at 1, 2, 3 and 4 cannot exist.

Generalising the above example, a system in which certain co-ordinates are specified shall be stable without any forces at the co-ordinates. Then only an [α] matrix will be derivable.

4.2.5 Existence of [k] Matrix

For the existence of [k] matrix the requirement of the co-ordinate system is roughly opposite to that of the requirement for [α] matrix to exist.

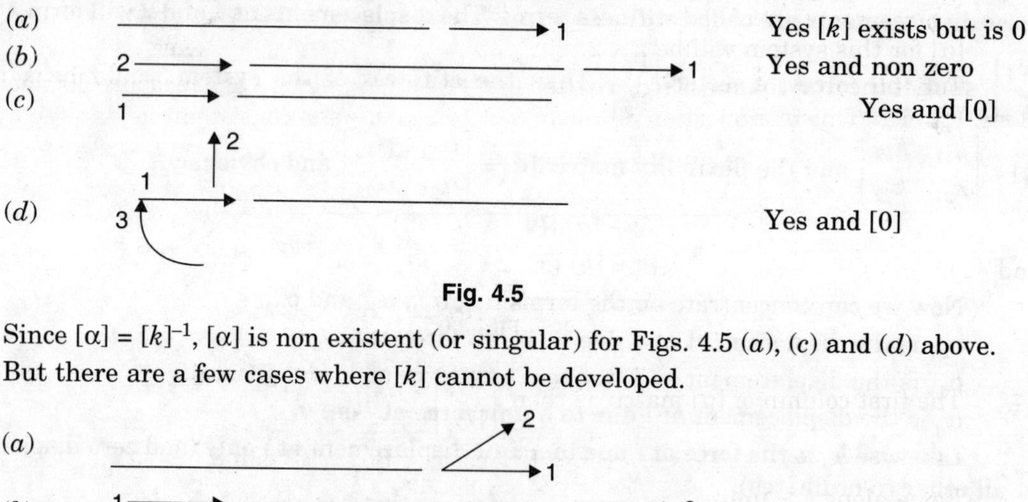

Fig. 4.5

Since [α] = [k]$^{-1}$, [α] is non existent (or singular) for Figs. 4.5 (a), (c) and (d) above. But there are a few cases where [k] cannot be developed.

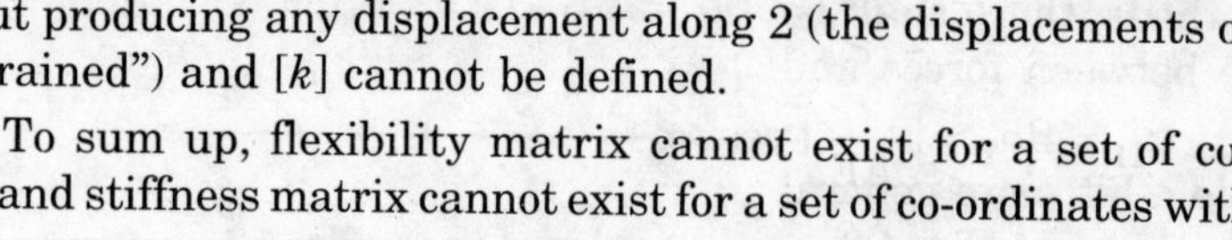

Fig. 4.6

In cases shown in Figs. 4.6 (a) and (b) you cannot apply a unit displacement along 1 without producing any displacement along 2 (the displacements on 1 and 2 are interrelated or "constrained") and [k] cannot be defined.

To sum up, flexibility matrix cannot exist for a set of co-ordinates with constrained forces and stiffness matrix cannot exist for a set of co-ordinates with constrained displacements.

[Can [F] exist for the case in Fig. 4.5 (*b*) ?]*

4.2.6 Flexibility of a Truss Member

A truss member is assumed to be capable of taking only tension and compression and incapable of taking bending and/or shear.

Fig. 4.7

AB in Fig. 4.7 is a one co-ordinate system. The member properties are l, A and E.

To get its flexibility matrix which is of course 1 × 1, we have to apply a unit force at 1 and find the displacement along 1. Obviously

$$[\alpha] = \frac{l}{AE}$$

Here we have obtained [α] for the **local co-ordinate** (*i.e.*, co-ordinate along the member axis).

Let us consider the system in Fig. 4.8.

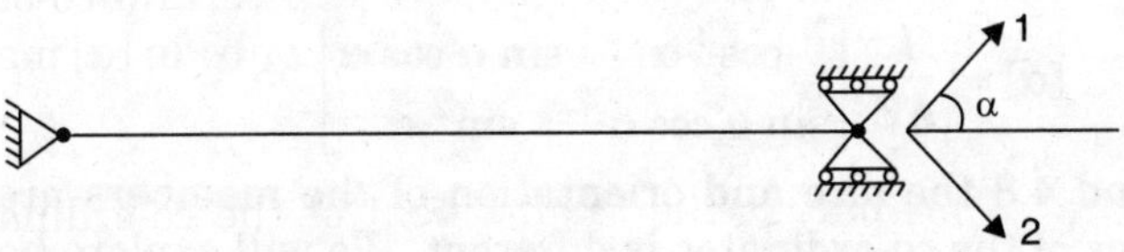

Fig. 4.8

[α] for this system will be 2 × 2.

To get the first column of [α], we shall apply unit force along 1 and measure displacements along 1 and 2. This is a question of kinematics (calculation of consistent displacements).

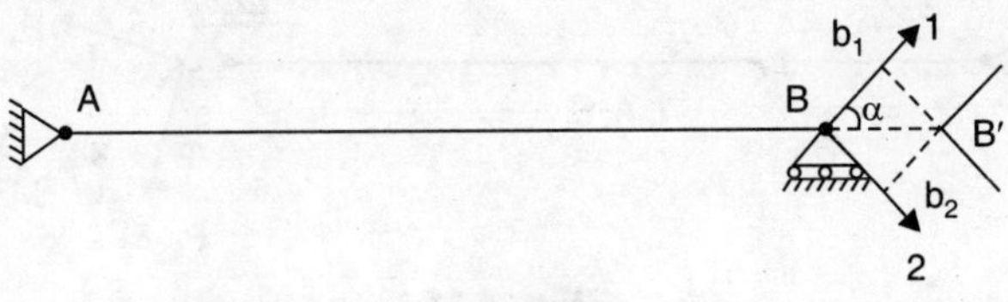

Fig. 4.9

The first column of [α] matrix is then $\begin{Bmatrix} Bb_1 \\ Bb_2 \end{Bmatrix}$

$$Bb_1 = BB' \cos \alpha, \quad Bb_2 = BB' \sin \alpha$$

Force along 1 = unity

Its axial component is 1 . cos α

BB′ is the displacement due to 1 . cos α

$$BB' = \left(\frac{l}{AE}\right) \cos \alpha \; ; \; Bb_1 = BB' \cos \alpha \; ; \; Bb_2 = BB' \sin \alpha$$

Hence, $$\alpha_{11} = Bb_1 = \left(\frac{l}{AE}\right) \cos^2 \alpha$$

*For Fig. 4.5 (*b*) too, [α] cannot exist, since the forces at 1 and 2 depend on each other.

$$\alpha_{12} = Bb_2 = \left(\frac{l}{AE}\right) \cos \alpha \sin \alpha$$

To get the second column of [α] we shall apply a unit force along 2 only.

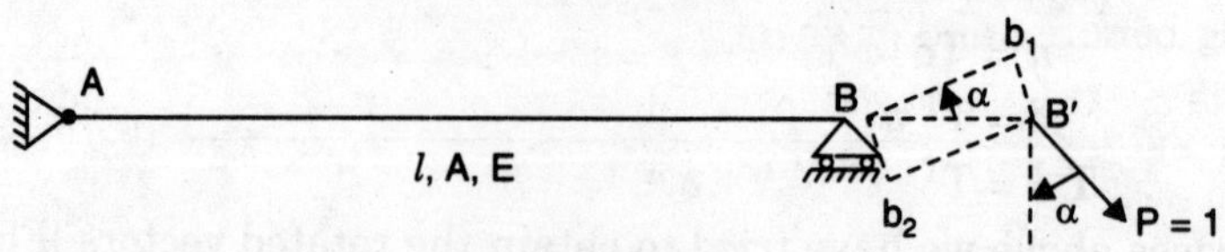

Fig. 4.10

$$BB' = 1 \,.\, \sin \alpha \left(\frac{l}{AE}\right)$$

$$Bb_2 = BB' \sin \alpha = \left(\frac{l}{AE}\right) \sin^2 \alpha$$

$$Bb_1 = BB' \cos \alpha = \left(\frac{l}{AE}\right) \sin \alpha \cos \alpha$$

Hence the flexibility matrix is given by

$$[\alpha] = \frac{l}{AE}\begin{bmatrix} \cos^2 \alpha & \sin \alpha \cos \alpha \\ \sin \alpha \cos \alpha & \sin^2 \alpha \end{bmatrix} \quad \text{...(4.3)}$$

In Figs. 4.7 and 4.8 the size and orientation of the members are the same. Only the orientation of the axes of the co-ordinates is different. We will explore how this rotation of co-ordinates can be systematised better.

4.2.7 Rotation of Forces and Displacements

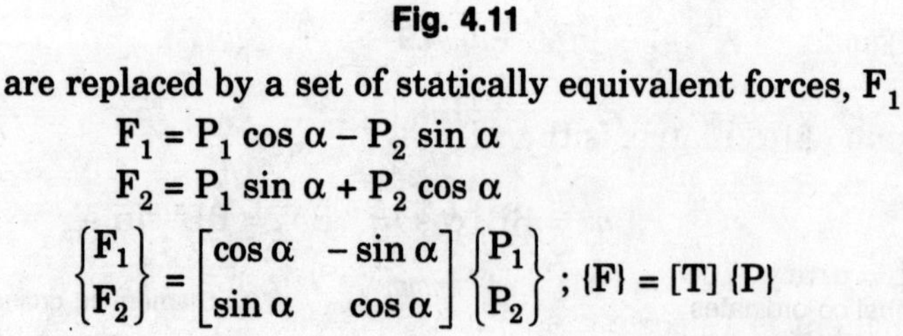

Fig. 4.11

If P_1 and P_2 are replaced by a set of statically equivalent forces, F_1 and F_2 (Fig. 4.11)

$$F_1 = P_1 \cos \alpha - P_2 \sin \alpha$$
$$F_2 = P_1 \sin \alpha + P_2 \cos \alpha$$

$$\begin{Bmatrix} F_1 \\ F_2 \end{Bmatrix} = \begin{bmatrix} \cos \alpha & -\sin \alpha \\ \sin \alpha & \cos \alpha \end{bmatrix} \begin{Bmatrix} P_1 \\ P_2 \end{Bmatrix} \; ; \{F\} = [T]\{P\} \quad \text{...(4.4)}$$

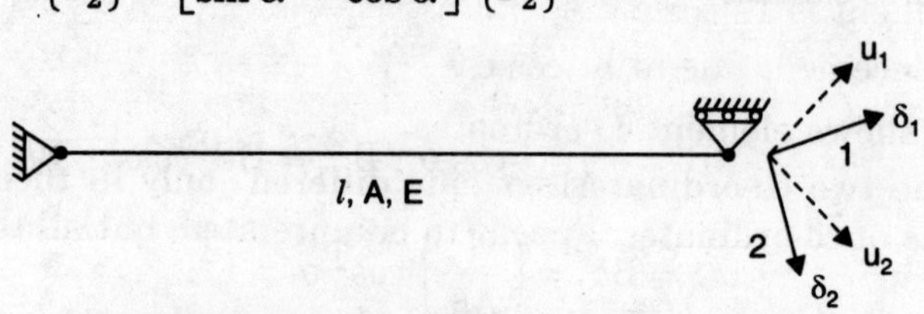

$$u_1 = \delta_1 \cos \alpha - \delta_2 \sin \alpha$$

$$u_2 = \delta_1 \sin\alpha + \delta_2 \cos\alpha$$

$$\begin{Bmatrix} u_1 \\ u_2 \end{Bmatrix} = \begin{bmatrix} \cos\alpha & -\sin\alpha \\ \sin\alpha & \cos\alpha \end{bmatrix} \begin{Bmatrix} \delta_1 \\ \delta_2 \end{Bmatrix} \quad \text{...(4.5)}$$

$$u = \mathrm{T}\delta \quad \text{...(4.6)}$$

$$\delta = \mathrm{T}^{\mathrm{T}} u \quad \text{...(4.7)}$$

$$\mathrm{T}^{-1} = \mathrm{T}^{\mathrm{T}} \quad \text{...(4.8)}$$

In both the cases above we have tried to obtain the rotated vectors {F} and {u} in terms of the original vectors {P} and {δ}.

The converse relations represented in equations (4.6) and (4.7) is termed the **principle of contragradience.**

The matrix $\begin{bmatrix} \cos\alpha & -\sin\alpha \\ \sin\alpha & \cos\alpha \end{bmatrix}$ can be called as transformation matrix [T] and more specifically a **rotation matrix.**

4.2.8 Transformation of Co-ordinates

For one and the same structure we can choose the co-ordinate system most suited to aid the solution.

In problem solving, it is usual to define a co-ordinate system dealing with the entire structure. Co-ordinates are assigned to locations, especially nodes (or joints) where loads are likely to act.

This kind of co-ordinates are called system co-ordinates or **global co-ordinates.** Fig. 4.12 (a) shows global co-ordinates.

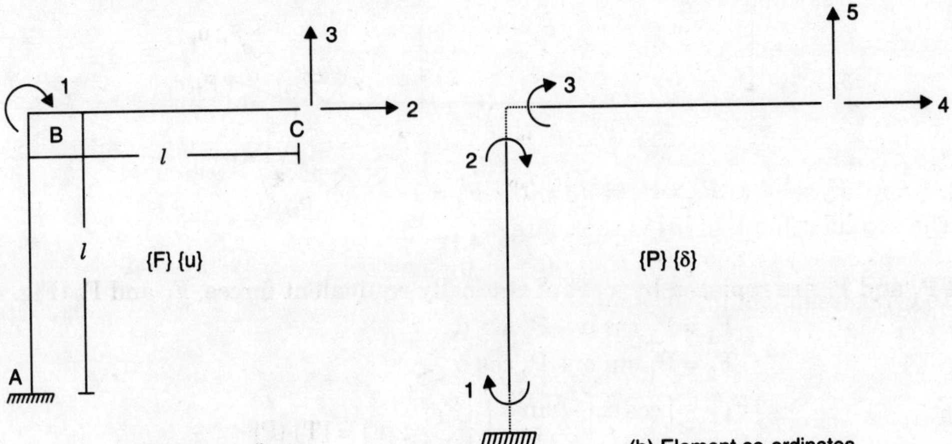

(a) Global co-ordinates (b) Element co-ordinates

Fig. 4.12

Figure 4.12 (b) shows element co-ordinates.

In Art. 4.2.7 the two co-ordinates systems differed only in their orientation. In the present case the 2 sets of co-ordinates, appear to be unrelated, but all the same relate to the same structure.

Let us derive a transformation matrix [b] such that {P} = [b] {F}.

Obviously [b] will be of size 5 × 3.

To get the first column of the [b] matrix, we have to apply a unit force at 1 on the {F} system and find the resulting moments and forces (at the nodes of elements) namely P_1, P_2, P_3, P_4 and P_5.

When $F_1 = 1$ (and $F_2 = F_3 = 0$),

$P_1 = -1$, $P_2 = 1$, $P_3 = P_4 = P_5 = 0$. Hence the first column of [b] is $[-1 \quad 1 \quad 0 \quad 0 \quad 0]^T$.

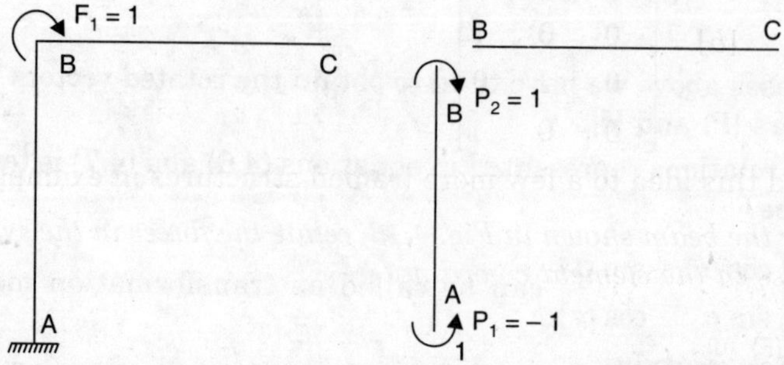

We have in fact solved the structure for $F_1 = 1$ and determined the moments at A, B and C.

We can extend this idea to get the second column of [b].

i.e., we apply $F_2 = 1$ and $F_1 = F_3 = 0$.

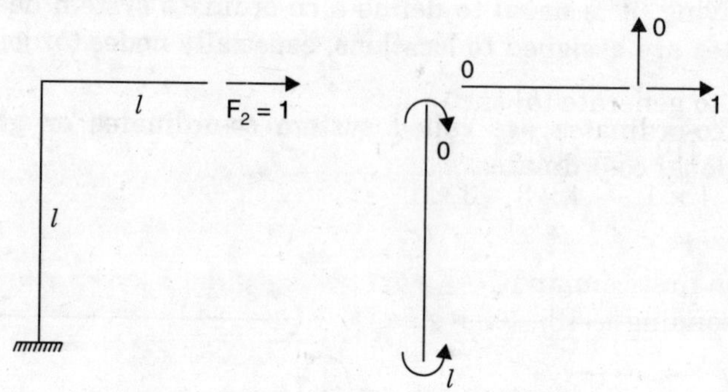

we get $P_1 = -l$; $P_2 = P_3 = P_5 = 0$; $P_4 = 1$

The second column of [b] is therefore

$$[-l \quad 0 \quad 0 \quad 1 \quad 0]^T$$

To get column 3 of [b] let apply

$$F_3 = 1 \quad \text{and} \quad F_1 = F_2 = 0$$

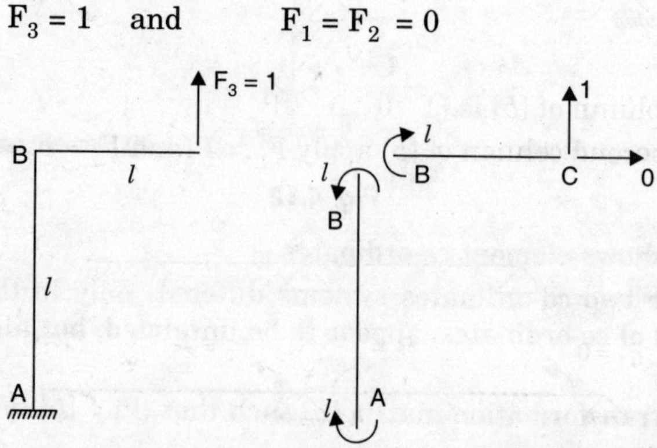

This would produce bending moment of l at A and B. Hence

$$F_1 = l \quad F_2 = -l \quad F_3 = l \quad F_4 = 0 \text{ and } F_5 = 1.$$

The third column of $[b]$ is thus $[l \;\; -l \;\; l \;\; 0 \;\; 1]^T$

Hence,
$$[b] = \begin{bmatrix} -1 & -l & l \\ 1 & 0 & -l \\ 0 & 0 & l \\ 0 & 1 & 0 \\ 0 & 0 & 1 \end{bmatrix}$$

We can extend this idea to a few more framed structures in examples 4.1, 4.3 and 4.4.

Example 4.1 *For the beam shown in Fig. 4.13, relate the forces in the system co-ordinates to those in the element co-ordinates.*

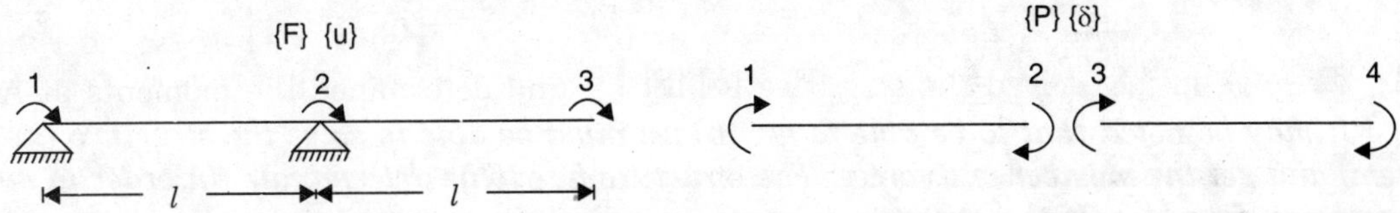

(a) System co-ordinates (b) Element co-ordinates

Fig. 4.13

Solution We have to generate $[b]$ such that

$$\underset{4\times 1}{[P]} = \underset{4\times 3}{[b]} \; \underset{3\times 1}{\{F\}}$$

$[b]$ will be of size 4×3.

To generate the first column of $[b]$, apply $F_1 = 1$ (and $F_2 = F_3 = 0$) and get the member end moments corresponding to $\{P\}$, $\{\delta\}$ system.

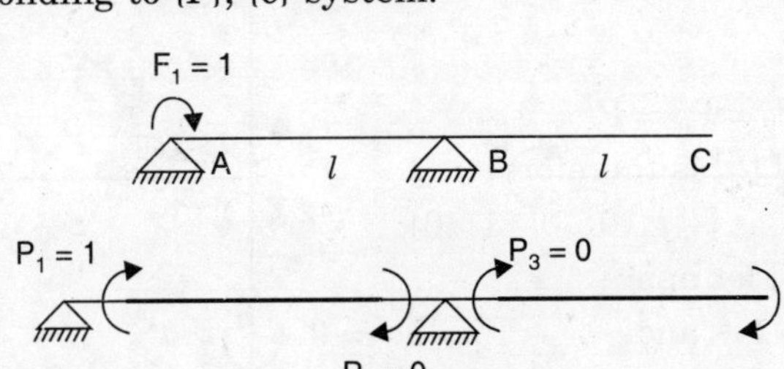

Hence the first column of $[b]$ is $[1 \;\; 0 \;\; 0 \;\; 0]^T$

To generate the second column of $[b]$ apply $F_2 = 1$ (and $F_1 = F_3 = 0$)

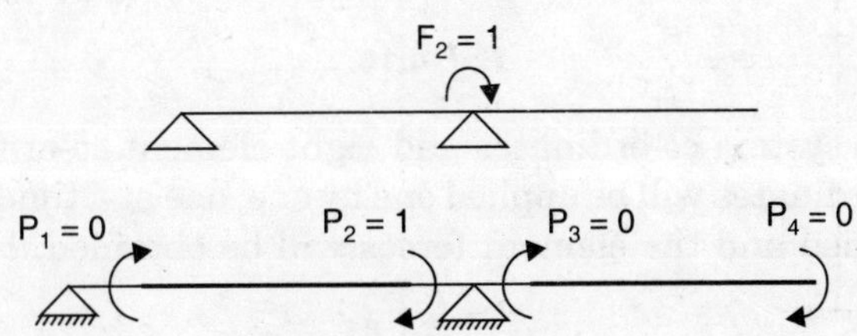

That makes the second column of $[b]$ as $[0 \quad 1 \quad 0 \quad 0]^T$

The third column of $[b]$ is derived by applying $F_3 = 1$ (and $F_1 = F_2 = 0$)

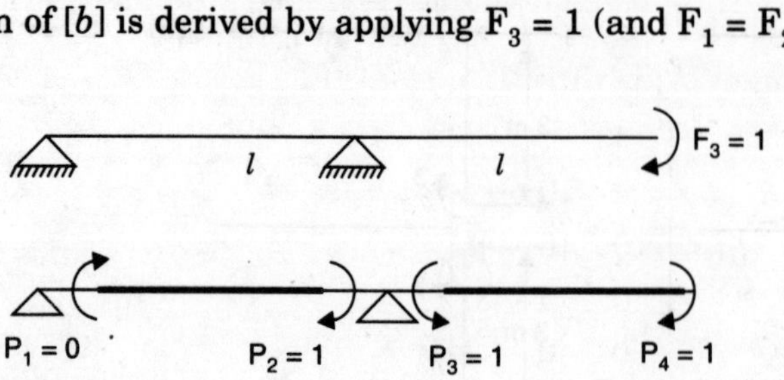

Thus the third column of $[b]$ is $[0 \quad 1 \quad -1 \quad 1]^T$ and $[b] = \begin{bmatrix} 1 & 0 & 0 \\ 0 & 1 & 1 \\ 0 & 0 & -1 \\ 0 & 0 & 1 \end{bmatrix}$

and $$\{P\} = [b]\{F\} \qquad ...(4.9)$$

It may be noted that to be able to get [b] we must be able to solve the structure (in the system) and get the member end forces. The structure has to be determinate (in order to make the solution feasible). If the structure given is not initially determinate, it has to be ***made*** *determinate, by introducing sufficient releases such as hinges and cuts. The structure so reduced (to a determinate state) is called the* ***primary*** *structure.*

It may be noted that in this case also the principle of contragradience applies, and

$$\{u\} = [b]^T \{\delta\}. \qquad ...(4.10)$$

Example 4.2 *Generate the [b] matrix for the structure shown in Fig. 4.14.*

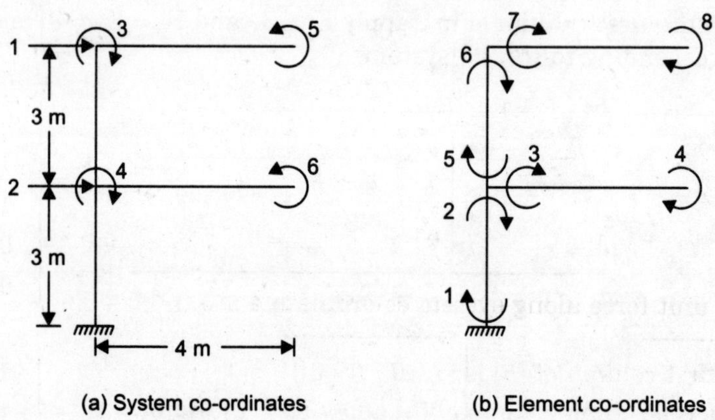

Fig. 4.14

Solution There are 6 system co-ordinates and eight element co-ordinates. Unit force along system co-ordinates will be applied one by one, one at a time. Equilibrium conditions will be applied and the element forces will be obtained.

Applying unit force along system co-ordinate 1 First column of $[b]$ matrix

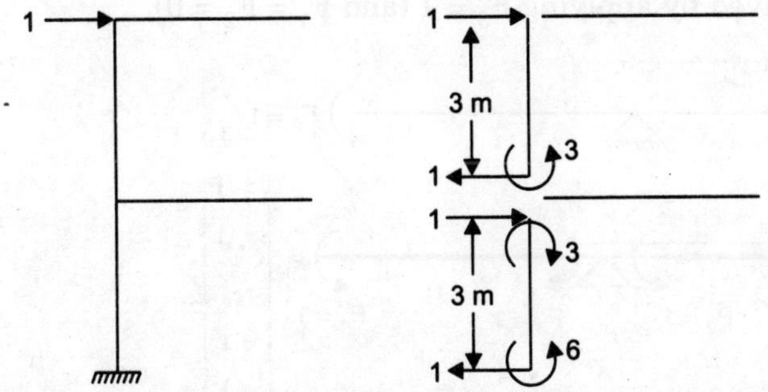

$$\begin{bmatrix} -6 \\ 3 \\ 0 \\ 0 \\ -3 \\ 0 \\ 0 \\ 0 \end{bmatrix}$$

Applying unit force along system co-ordinate 2

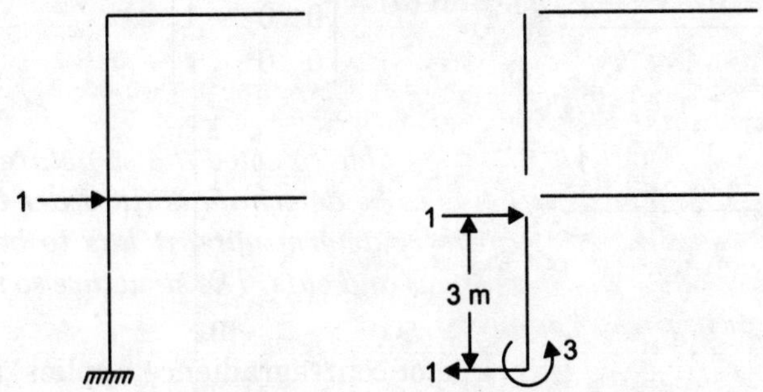

$$\begin{bmatrix} -3 \\ 0 \\ 0 \\ 0 \\ 0 \\ 0 \\ 0 \\ 0 \end{bmatrix}$$

Applying unit force along system co-ordinate 3

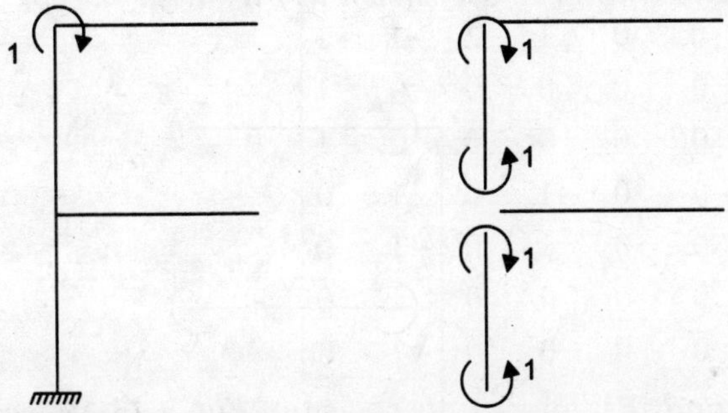

$$\begin{bmatrix} -1 \\ 1 \\ 0 \\ 0 \\ -1 \\ 1 \\ 0 \\ 0 \end{bmatrix}$$

Applying unit force along system co-ordinate 4

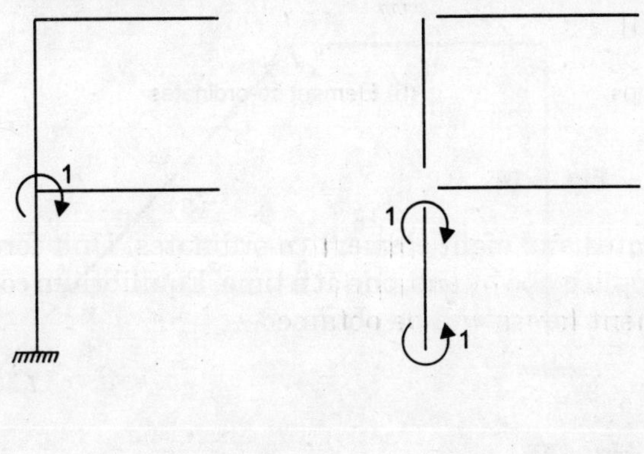

$$\begin{bmatrix} -1 \\ 1 \\ 0 \\ 0 \\ 0 \\ 0 \\ 0 \\ 0 \end{bmatrix}$$

Applying unit force along system co-ordinate 5

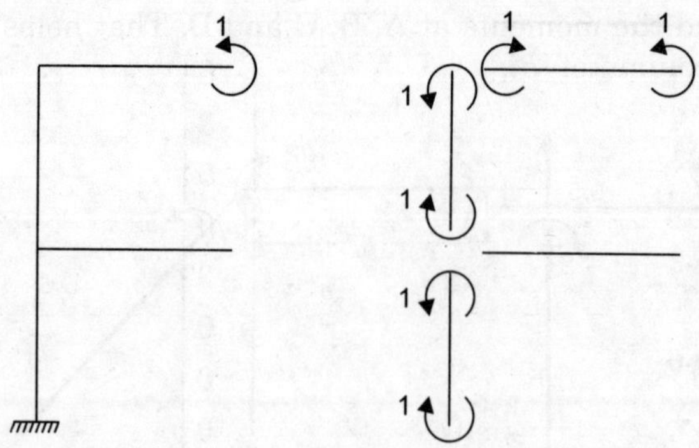

$$\begin{bmatrix} 1 \\ -1 \\ 0 \\ 0 \\ 1 \\ -1 \\ 1 \\ -1 \end{bmatrix}$$

Applying unit force along system co-ordinate 6

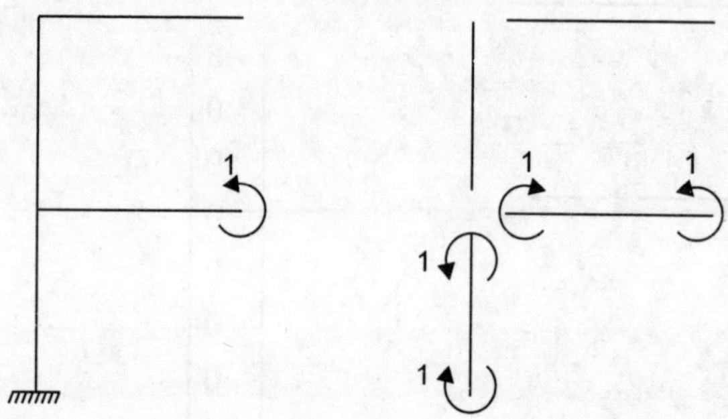

$$\begin{bmatrix} 1 \\ -1 \\ 1 \\ -1 \\ 0 \\ 0 \\ 0 \\ 0 \end{bmatrix}$$

Hence the [*b*] matrix is, $[b] = \begin{bmatrix} -6 & -3 & -1 & -1 & 1 & 1 \\ 3 & 0 & 1 & 1 & -1 & -1 \\ 0 & 0 & 0 & 0 & 0 & 1 \\ 0 & 0 & 0 & 0 & 0 & -1 \\ -3 & 0 & -1 & 0 & 1 & 0 \\ 0 & 0 & 1 & 0 & -1 & 0 \\ 0 & 0 & 0 & 0 & 1 & 0 \\ 0 & 0 & 0 & 0 & -1 & 0 \end{bmatrix}$

Example 4.3 *In respect of the* $\{F\}$, $\{u\}$ *and* $\{P\}$, $\{\delta\}$ *systems shown in Fig. 4.15, generate the* $[b]$ *matrix such that* $\{P\} = [b]\{F\}$.

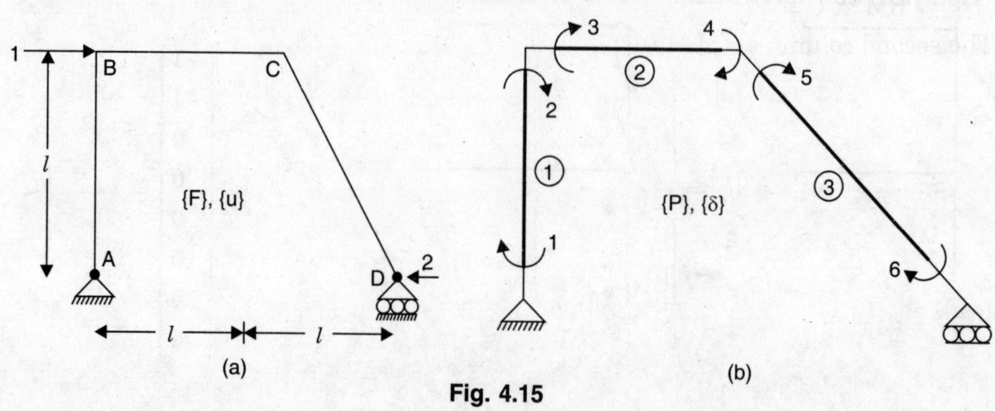

Fig. 4.15

Solution To start with, we satisfy ourself that the structure in Fig. 4.15 (a) is determinate. We then apply $F_1 = 1$ and find the moments at A, B, C and D. That helps us to determine P_1 to P_6, the first column of $[b]$.

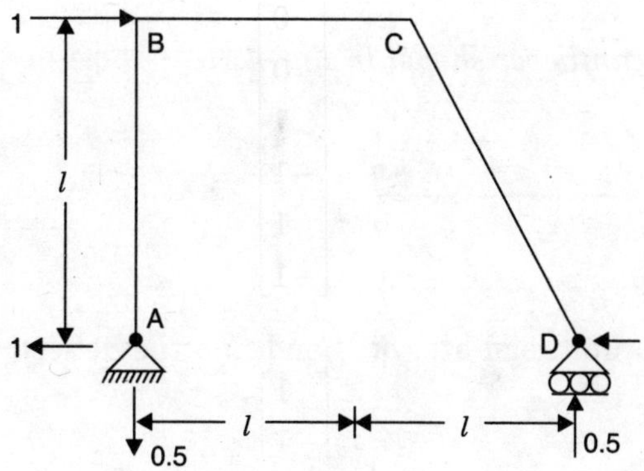

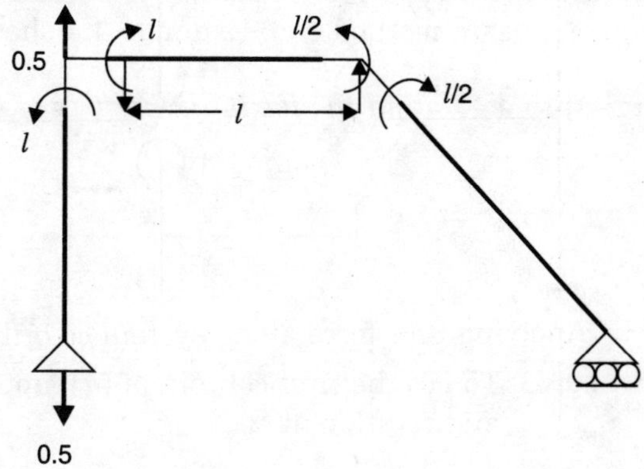

The reactions at A and D are 1.0 and 0.5 as marked. Hence BM at B is $-l$, and BM at C is $-l/2$. Hence forces at 1 to 6 are $\left\{0 \;\; -l \;\; +l \;\; \frac{-l}{2} \;\; \frac{l}{2} \;\; 0\right\}^T$.

To generate the second column of $[b]$ apply a unit force at 2. This would create a reaction of 1 at A as shown.

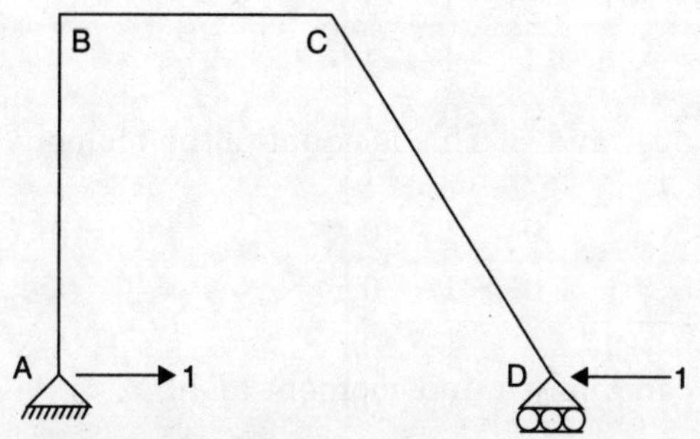

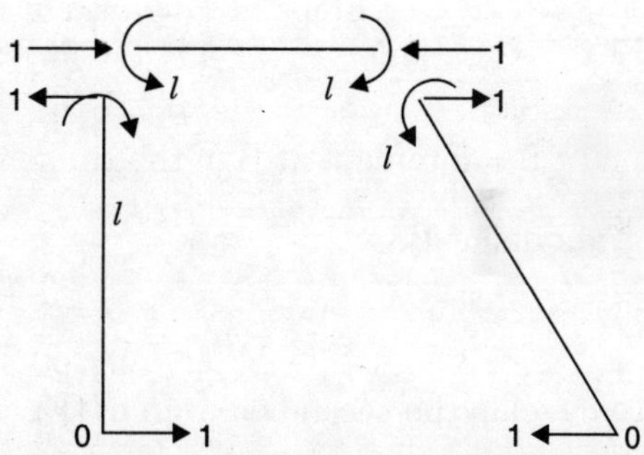

Hence, BM at B = l

BM at C = $-l$

The second column of $[b] = [0 \;\; l \;\; -l \;\; l \;\; -l \;\; 0]^T$

$$[b] = \begin{bmatrix} 0 & 0 \\ -l & +l \\ l & -l \\ -l/2 & l \\ l/2 & -l \\ 0 & 0 \end{bmatrix}.$$

4.3 DEVELOPING FLEXIBILITY MATRICES OF DETERMINATE STRUCTURES

For simple, determinate flexural members, developing the {F} matrix is only a process of finding deflections (and/ or rotations) due to a set of given forces (including moments). We may have to invoke some of the known methods of computing deflections (like Mohr's theorems, conjugate beam method etc). Example 4.4 shows how.

Example 4.4 *Find the flexibility matrix of the cantilever shown in Fig. 4.16.*

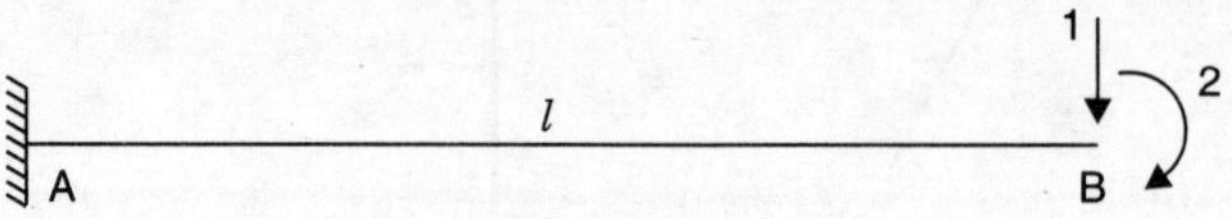

Fig. 4.16

Solution To get the first column of $[a]$, apply a unit load at 1 of B and find the deflection and rotation at B.

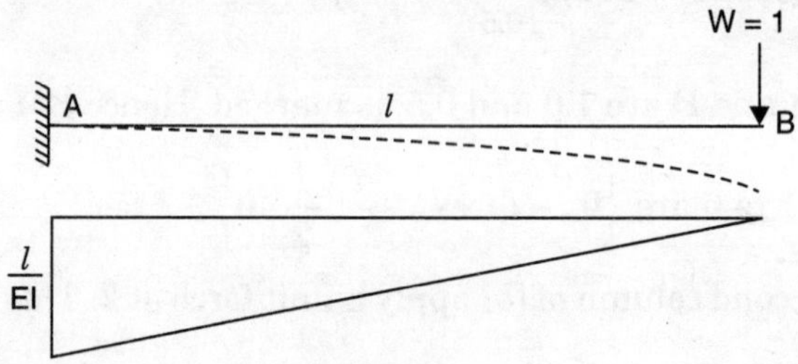

Slope at B = area of BMD ÷ EI

$$a_{21} = \frac{l \times l}{2EI} = \frac{l^2}{2EI}$$

deflection at B = intercept at B of the tangents at A and B. This is equal to the moment of the $\frac{M}{EI}$ diagram about B.

$$a_{11} = \frac{l}{EI}\frac{l}{2} \times \frac{2l}{3} = \frac{l^3}{3EI}$$

To develop the second column of [F], we can apply a unit moment along 2.

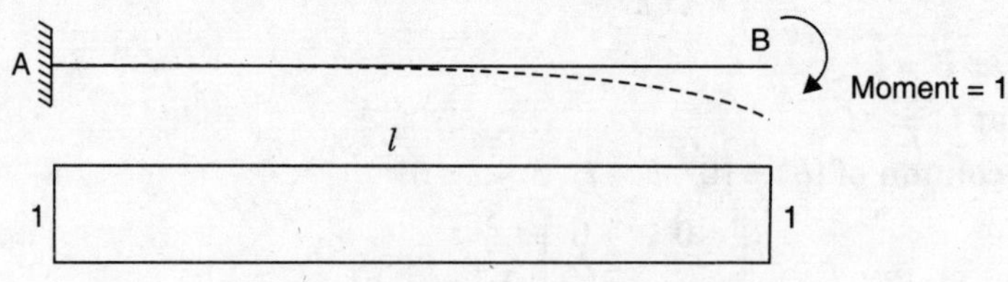

$$a_{12} = \text{deflection at B} = 1\ \frac{l}{EI} \times \frac{l}{2} = \frac{l^2}{2EI}$$

$$a_{22} = \text{slope at B} = 1\ \frac{l}{EI} = \frac{l}{EI}$$

Hence,
$$[a] = \frac{1}{EI}\begin{bmatrix} \frac{l^3}{3} & \frac{l^2}{2} \\ \frac{l^2}{2} & l \end{bmatrix}.$$

Example 4.5 *Develop the flexibility matrix for the cantilever with co-ordinates as shown in Fig. 4.17.*

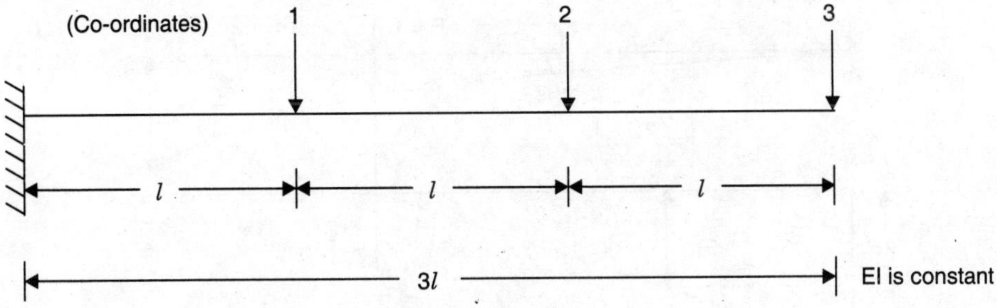

Fig. 4.17

Solution **Flexibility matrix :** $[a] = \begin{bmatrix} a_{11} & a_{12} & a_{13} \\ a_{21} & a_{22} & a_{23} \\ a_{31} & a_{32} & a_{33} \end{bmatrix}$

(*a*) **Applying a unit force along co-ordinate 1:**

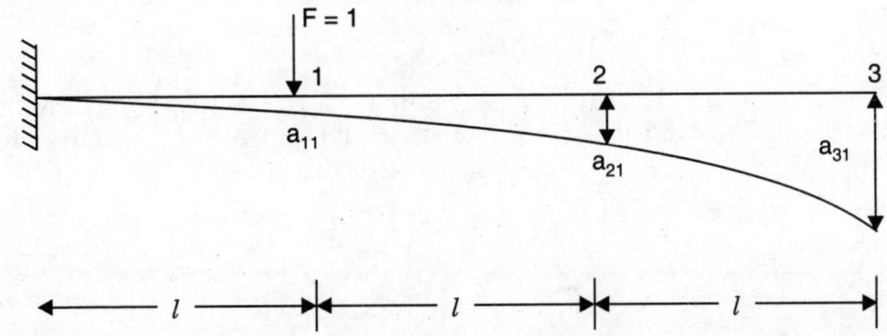

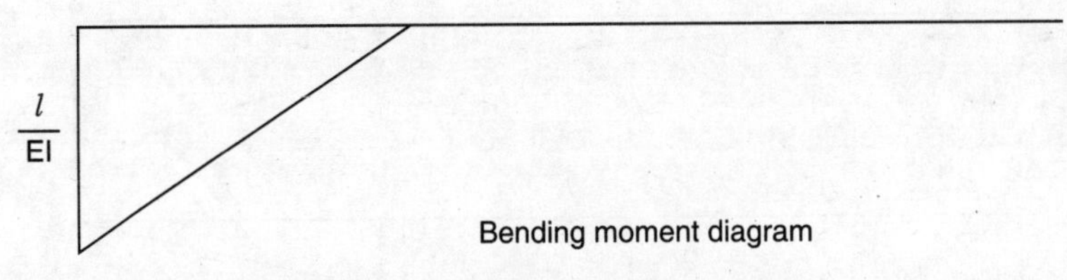

Bending moment diagram

$$a_{11} = \left(\frac{1}{2}\times l\times\frac{l}{EI}\right)\times\left(\frac{2l}{3}\right) = \frac{l^3}{3EI}$$

$$a_{21} = \left(\frac{1}{2}\times l\times\frac{l}{EI}\right)\times\left(\frac{2}{3}l+l\right) = \frac{l^2}{2EI}\times\frac{5l}{3} = \frac{5l^3}{6EI}$$

$$a_{31} = \left(\frac{1}{2}\times l\times\frac{l}{EI}\right)\times\left(\frac{2}{3}l+l+l\right) = \frac{l^2}{2EI}\times\frac{8l}{3} = \frac{4l^3}{3EI}$$

(*b*) **Applying a unit force along co-ordinate 2:**

$$a_{12} = \left(\frac{l}{EI}\times l\right)\frac{l}{2}+\left(\frac{1}{2}\times l\times\frac{l}{EI}\right)\frac{2}{3}l = \frac{l^3}{2EI}+\frac{l^3}{3EI} = \frac{5l^3}{6EI}$$

$$a_{22} = \left(\frac{1}{2}\times\frac{2l}{EI}\times 2l\right)\times\frac{2}{3}\times 2l = \frac{8l^3}{3EI}$$

$$a_{32} = \left(\frac{1}{2} \times \frac{2l}{EI} \times 2l\right) \times \left(\frac{2}{3} \times 2l + l\right) = \frac{2l^2}{EI} \times \frac{7}{3} l = \frac{14l^3}{3EI}$$

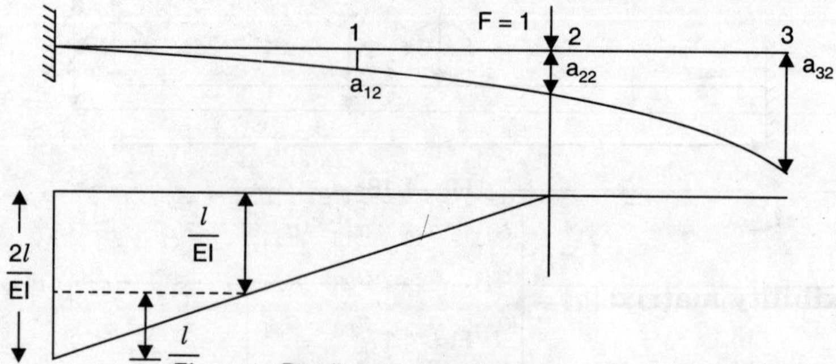

Bending moment diagram

(*c*) **Applying a unit force along co-ordinate 3:**

$$a_{13} = \left(\frac{2l}{EI} \times l\right) \times \frac{l}{2} + \left(\frac{1}{2} \times l \times \frac{l}{EI}\right) \times \frac{2l}{3}$$

$$a_{13} = \frac{l^3}{EI} + \frac{l^3}{3EI} = \frac{4l^3}{3EI}$$

$$a_{23} = \left(\frac{l}{EI} \times 2l\right) \times l + \left(\frac{1}{2} \times 2l \times \frac{2l}{EI}\right) \times \left(\frac{2}{3} \times 2l\right) = \frac{2l^3}{EI} + \frac{2l^2}{EI} \times \frac{4}{3} l = \frac{14l^3}{3EI}$$

$$a_{33} = \left(\frac{1}{2} \times 3l \times \frac{3l}{EI}\right)\left(\frac{2}{3} \times 3l\right) = \frac{9l^3}{EI}$$

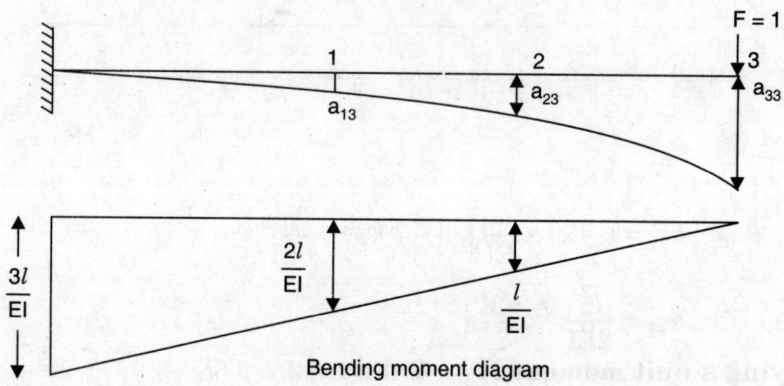

Bending moment diagram

$$\therefore \quad \text{Flexibility matrix, } [a] = \begin{bmatrix} \frac{l^3}{3EI} & \frac{5l^3}{6EI} & \frac{4l^3}{3EI} \\ \frac{5l^3}{6EI} & \frac{8l^3}{3EI} & \frac{14l^3}{3EI} \\ \frac{4l^3}{3EI} & \frac{14l^3}{3EI} & \frac{9l^3}{EI} \end{bmatrix}$$

$$[a] = \frac{l^3}{6EI} = \begin{bmatrix} 2 & 5 & 8 \\ 5 & 16 & 28 \\ 8 & 28 & 54 \end{bmatrix}$$

Example 4.6 *Develop the flexibility matrix for the beam in Fig. 4.18.*

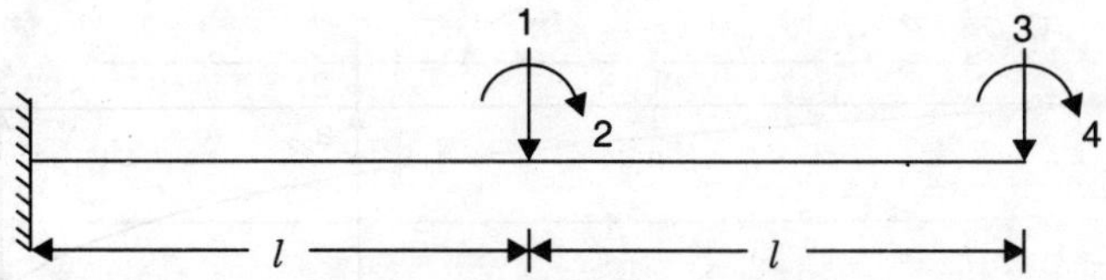

Fig. 4.18

Solution **Flexibility matrix:** $[a] = \begin{bmatrix} a_{11} & a_{12} & a_{13} & a_{14} \\ a_{21} & a_{22} & a_{23} & a_{24} \\ a_{31} & a_{32} & a_{33} & a_{34} \\ a_{41} & a_{42} & a_{43} & a_{44} \end{bmatrix}$

(*a*) **Applying a unit force along co-ordinate 1:**

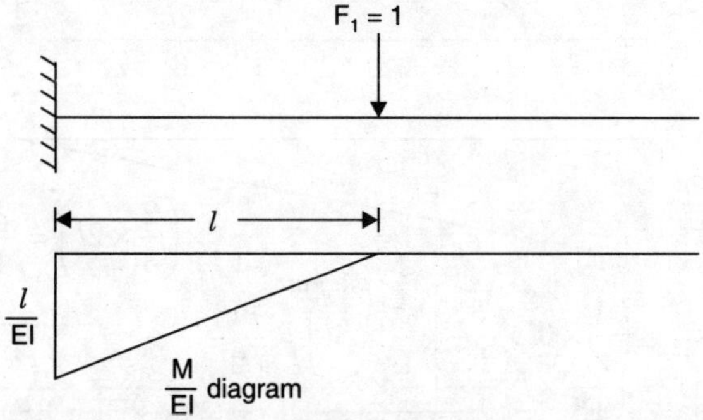

$$a_{11} = \left(\frac{1}{2} \times l \times \frac{l}{\text{EI}}\right) \times \left(\frac{2l}{3}\right) = \frac{l^3}{3\text{EI}}$$

$$a_{21} = \left(\frac{1}{2} \times l \times \frac{l}{\text{EI}}\right) = \frac{l^2}{2\text{EI}}$$

$$a_{31} = \left(\frac{1}{2} \times l \times \frac{l}{\text{EI}}\right) \times \left(\frac{2}{3}l + l\right) = \frac{5l^3}{6\text{EI}}$$

$$a_{41} = \frac{l^2}{2\text{EI}}$$

(*b*) **Applying a unit moment at co-ordinate 2:**

$$a_{12} = \frac{1}{\text{EI}} \times l \times \frac{l}{2} = \frac{l^2}{2\text{EI}}$$

$$a_{22} = \frac{1}{\text{EI}} \times l = \frac{l}{\text{EI}}$$

$$a_{32} = \left(\frac{1}{\text{EI}} \times l\right) \times \left(\frac{l}{2} + l\right) = \frac{3l^2}{2\text{EI}}$$

$$a_{42} = \frac{l}{\text{EI}}$$

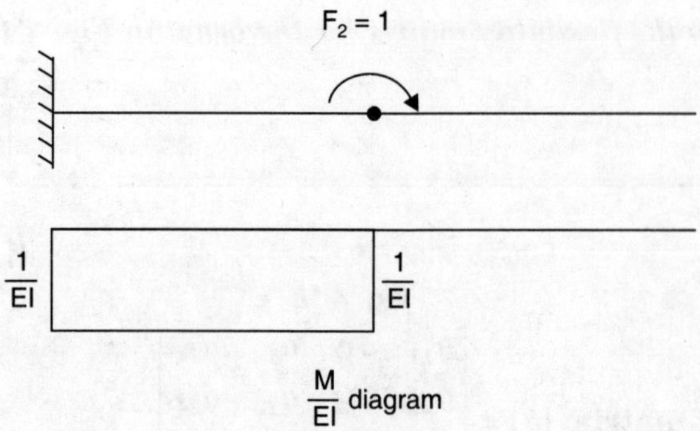

(*c*) **Applying a unit force along co-ordinate 3:**

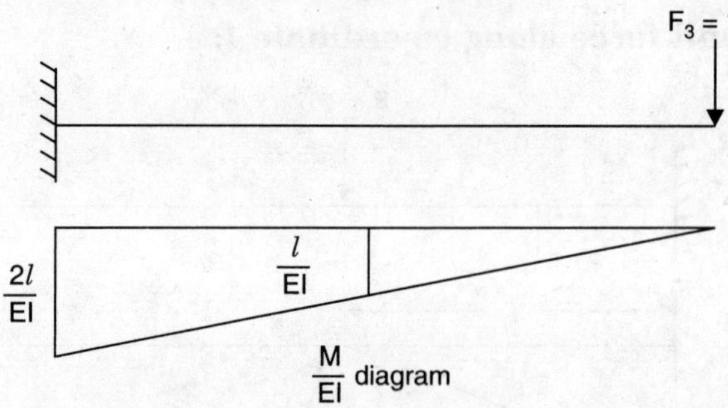

$$a_{13} = \left(\frac{l}{EI} \times l\right)\frac{l}{2} + \left(\frac{1}{2} \times l \times \frac{l}{EI}\right)\frac{2}{3}l = \frac{l^3}{2EI} + \frac{l^3}{3EI} = \frac{5l^3}{6EI}$$

$$a_{23} = \left(\frac{l}{EI} \times l\right) + \left(\frac{1}{2} \times \frac{l}{EI} \times l\right) = \frac{l^2}{EI} + \frac{l^2}{2EI} = \frac{3l^2}{2EI}$$

$$a_{33} = \left(\frac{1}{2} \times \frac{2l}{EI} \times 2l\right) \times \frac{2}{3} \times 2l = \frac{8l^3}{3EI}$$

$$a_{43} = \left(\frac{1}{2} \times 2l \times \frac{2l}{EI}\right) = \frac{2l^2}{EI}$$

(*d*) **Applying a unit force along co-ordinate 4:**

$$a_{14} = \frac{1}{EI} \times l \times \frac{l}{2} = \frac{l^2}{2EI}$$

$$a_{24} = \frac{1}{EI} \times l = \frac{l}{EI}$$

$$a_{34} = \frac{1}{EI} \times 2l \times l = \frac{2l^2}{EI}$$

$$a_{44} = \frac{1}{EI} \times 2l = \frac{2l}{EI}$$

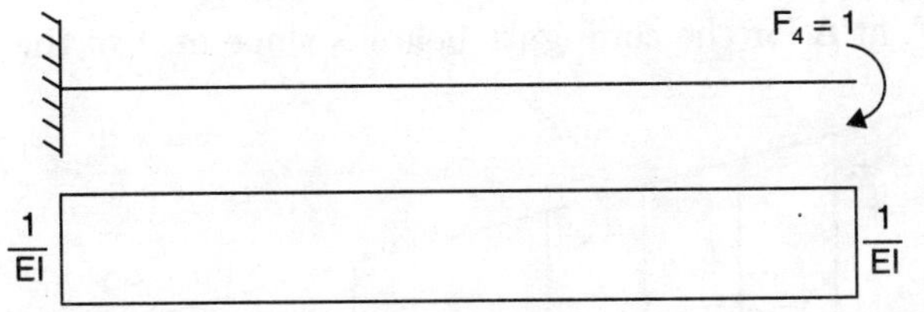

$\frac{M}{EI}$ diagram

$$\therefore \quad \text{Flexibility matrix,} \quad [a] = \begin{bmatrix} \frac{l^3}{3EI} & \frac{l^2}{2EI} & \frac{5l^3}{6EI} & \frac{l^2}{2EI} \\ \frac{l^2}{2EI} & \frac{l}{EI} & \frac{3l^2}{2EI} & \frac{l}{EI} \\ \frac{5l^3}{6EI} & \frac{3l^2}{2EI} & \frac{8l^3}{3EI} & \frac{2l^2}{EI} \\ \frac{l^2}{2EI} & \frac{l}{EI} & \frac{2l^2}{EI} & \frac{2l}{EI} \end{bmatrix}$$

Example 4.7 *Generate the flexibility matrix of the beam with the co-ordinates shown in Fig. 4.19.*

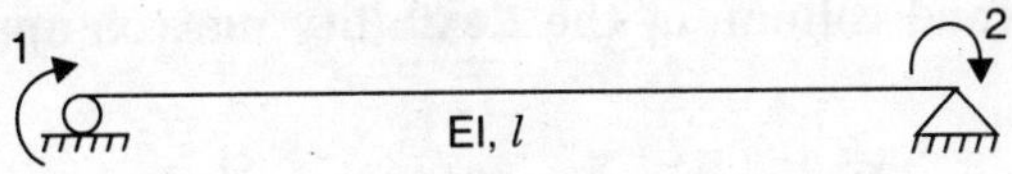

Fig. 4.19

Solution The flexibility matrix for the structure shown will be of size 2 × 2, since there are two co-ordinares. The flexibility co-efficients will be obtained by applying unit forces along each co-ordinate, one at a time and abtaining the displacements at each co-ordinate.

To generate the first column of the flexibility matrix, apply a unit force along co-ordinate 1 only.

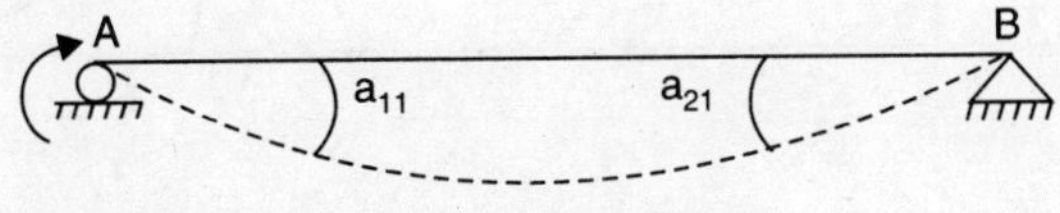

Taking moments about B (for the conjugate beam)

$$R_A \times l - \frac{1}{2} \times l \times \frac{1}{EI} \times \frac{2}{3} \times l = 0$$

$$R_A \times l = \frac{l^2}{3EI}$$

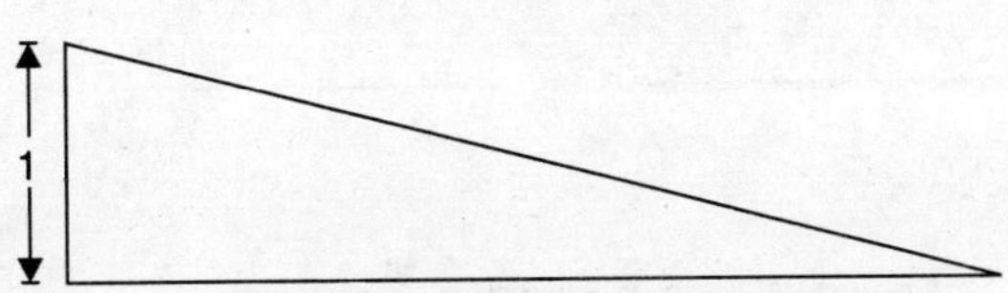

M diagram

$R_A = \dfrac{l}{3EI}$ = S.F. at A for the conjugate beam = slope at A of the given beam.

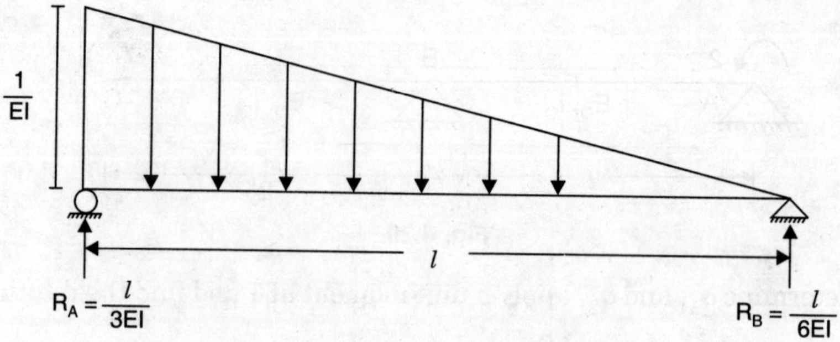

Conjugate beam loaded with $\dfrac{M}{EI}$ diagram

i.e., $$a_{11} = \frac{l}{3EI}$$

$$R_B = \left(\frac{1}{2} \times l \times \frac{1}{EI} - \frac{l}{3EI}\right) = \frac{l}{6EI}.$$

i.e., $$a_{21} = \frac{-l}{6EI} \text{ (right up –ve)}$$

To generate the second column of the flexibility matrix, apply a unit force along co-ordinate 2 only.

$$R_A' = \frac{-l}{6EI} = a_{12} \text{ (left down –ve)}$$

$$R_B' = \frac{l}{3EI} = a_{22}$$

Hence the flexibility matrix is $[a] = \begin{bmatrix} \dfrac{l}{3EI} & \dfrac{-l}{6EI} \\ -\dfrac{l}{6EI} & \dfrac{l}{3EI} \end{bmatrix} = \dfrac{l}{6EI}\begin{bmatrix} 2 & -1 \\ -1 & 2 \end{bmatrix}$

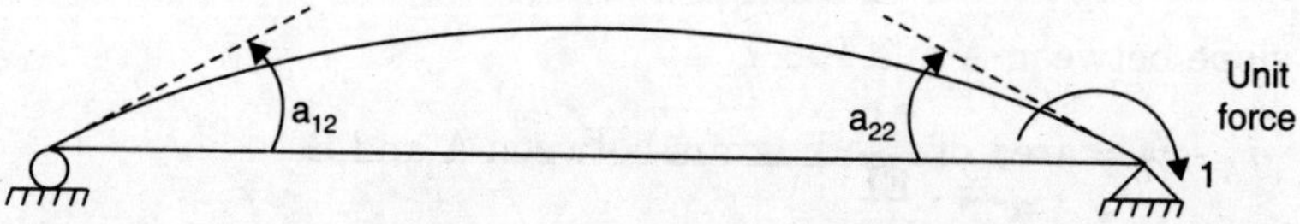

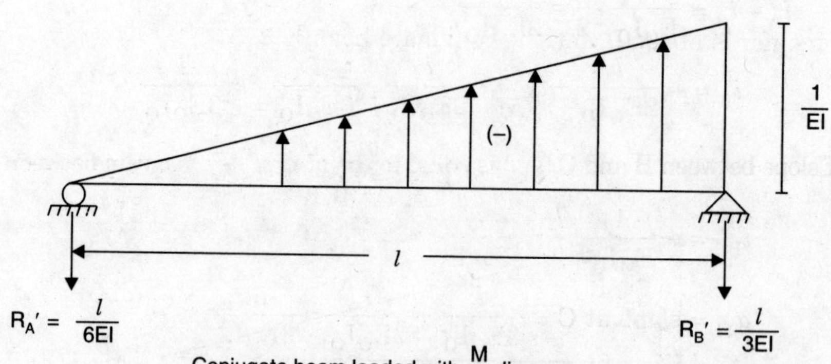

Conjugate beam loaded with $\dfrac{M}{EI}$ diagram

Example 4.8 *Generate the flexibility matrix [a] of the beam shown in Fig. 4.20, for the co-ordinates indicated.*

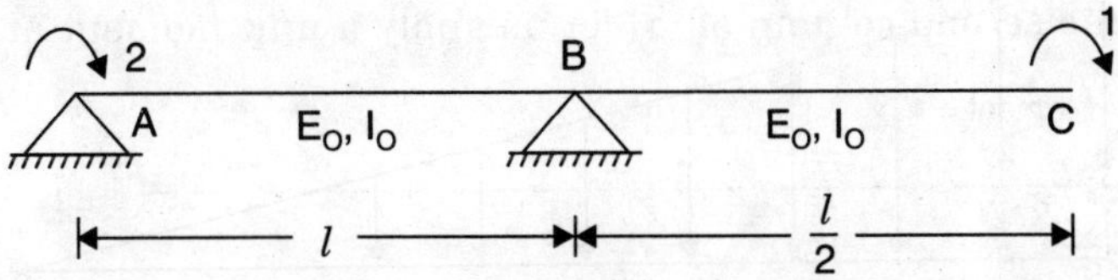

Fig. 4.20

Solution To determine a_{11} and a_{21} apply a unit moment at 1 and find the rotations at 1 and 2.

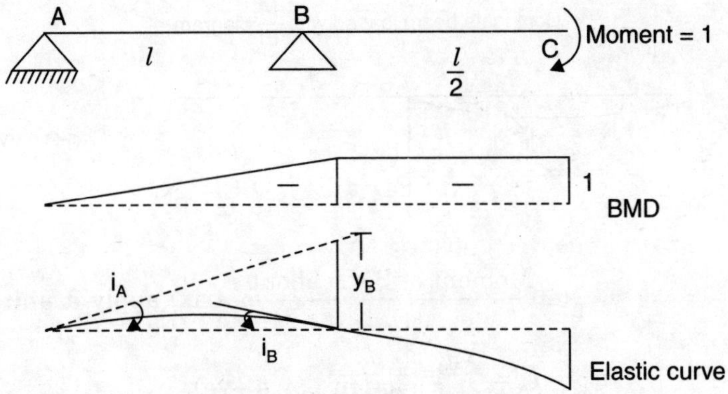

By Mohr's theorem 2, y_B = moment of $\frac{M}{EI}$ diagram about B.

$$y_B = \frac{-1}{E_O I_O} \cdot \frac{l}{2} \cdot \frac{l}{3} = \frac{-l^2}{6E_O I_O}$$

$$a_{21} = i_A = \frac{y_B}{l} = \frac{-l}{6E_O I_O}$$

Change of slope between A and B is $i_A - i_B$.

$$i_A - i_B = \text{area of } \frac{M}{EI} \text{ diagram between A and B}$$

$$i_A - i_B = \frac{-1}{E_O I_O} \cdot \frac{l}{2} = \frac{-l}{2E_O I_O}$$

$$i_B = +\frac{l}{2E_O I_O} + i_A = +\frac{l}{2E_O I_O} - \frac{l}{6E_O I_O} = +\frac{l}{3E_O I_O}$$

Change of slope between B and C (i_{BC}) is equal to the area of $\frac{M}{EI}$ diagram between B and C.

$$i_{BC} = -\frac{1}{E_O I_O} \cdot \frac{l}{2} = \frac{-l}{2E_O I_O}$$

$$a_{11} = \text{slope at C} = \frac{l}{3E_O I_O} + \frac{l}{2E_O I_O} = \frac{5l}{6E_O I_O}$$

The first column of $[a]$ is $\dfrac{l}{6E_OI_O}\begin{bmatrix}5\\-1\end{bmatrix}$

To generate the second column of $[a]$ let us apply a unit moment at 2.

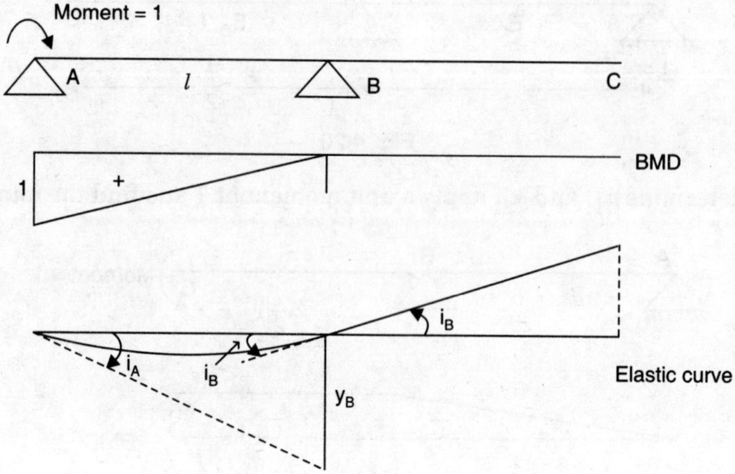

$$y_B = \frac{\text{Moment of BMD about B}}{E_OI_O} = \frac{1}{E_OI_O}\cdot\frac{l}{2}\cdot\frac{2l}{3} = \frac{+l^2}{3E_OI_O}$$

$$a_{22} = i_A = \frac{Y_B}{l} = \frac{l}{3E_OI_O}$$

$$i_a - i_b = \text{area of BMD between A and B}$$

$$i_a - i_b = \frac{l}{2E_OI_O}$$

$$i_b = i_a - \frac{l}{2E_OI_O} = \frac{l}{E_OI_O}\left(\frac{1}{3}-\frac{1}{2}\right) = \frac{-l}{6E_OI_O}$$

$$i_c = i_b = \frac{-l}{6E_OI_O} = a_{12}$$

Hence
$$[a] = \frac{l}{6E_OI_O}\begin{bmatrix}5 & -1\\-1 & 2\end{bmatrix}.$$

Example 4.9 *Find the flexibility matrix for the structure with co-ordinates shown in Fig. 4.21.*

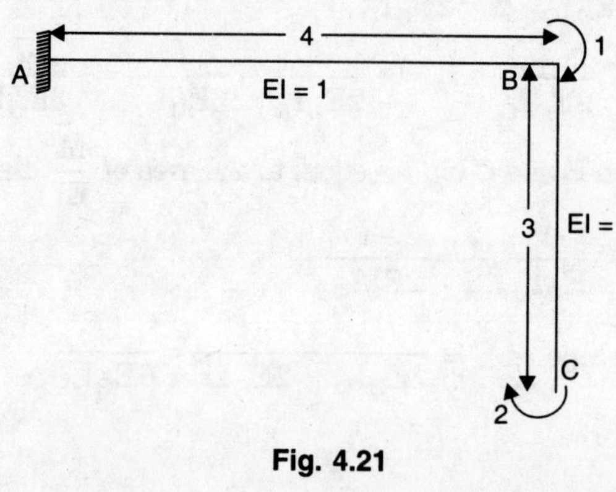

Fig. 4.21

Solution There are two co-ordinates given, in the structure. Hence the size of the flexibility matrix will be 2 × 2. The flexibility matrix will be generated by applying unit force at each co-ordinate, one at time along each co-ordinate.

To generate the first column of the flexibility matrix, apply unit force along co-ordinate 1 only.

$$\theta_B = \alpha_{11} = \frac{Ml}{EI} = \frac{1 \times 4}{1} = 4$$

$$\theta_C = \theta_B = 4 \quad i.e., \quad \alpha_{21} = 4$$

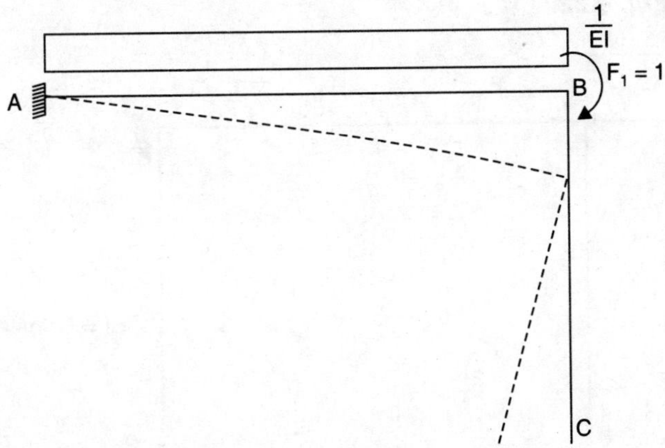

To generate the second column of the flexibility matrix, apply unit force along co-ordinate 2 only.

$$\theta_B = \frac{Ml}{EI} = \frac{1 \times 4}{1} = 4 \quad i.e., \quad \alpha_{12} = 4$$

$$\theta_C = \theta_B + \frac{Ml}{EI} = 4 + \frac{1 \times 3}{0.5} = 10$$

i.e., $\alpha_{22} = 10$

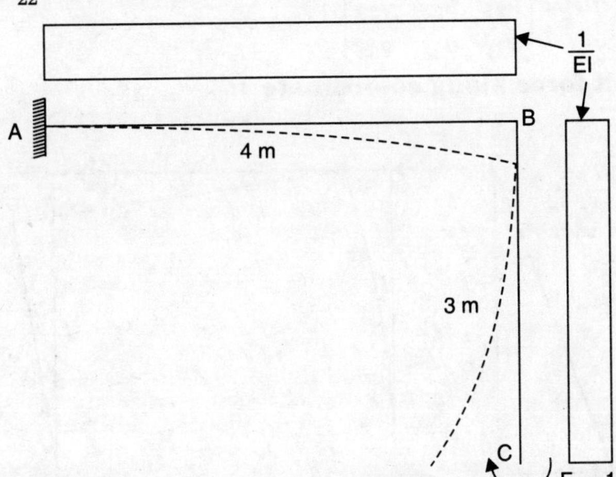

Hence the flexibility matrix is $[\alpha] = \begin{bmatrix} \alpha_{11} & \alpha_{12} \\ \alpha_{21} & \alpha_{22} \end{bmatrix} = \begin{bmatrix} 4 & 4 \\ 4 & 10 \end{bmatrix}$

In the above problem, find the [k] matrix

$$[k] = [\alpha]^{-1} = \begin{bmatrix} 4 & 4 \\ 4 & 10 \end{bmatrix}^{-1} = \frac{1}{\text{Det}} \times \text{Adjoint of the matrix}$$

Determinant $= 40 - 16 = 24$

$$\therefore \qquad [k] = \frac{1}{24}\begin{bmatrix} 10 & -4 \\ -4 & 4 \end{bmatrix}.$$

Example 4.10 *Develop the flexibility matrix for the structure with co-ordinates shown in Fig. 4.22.*

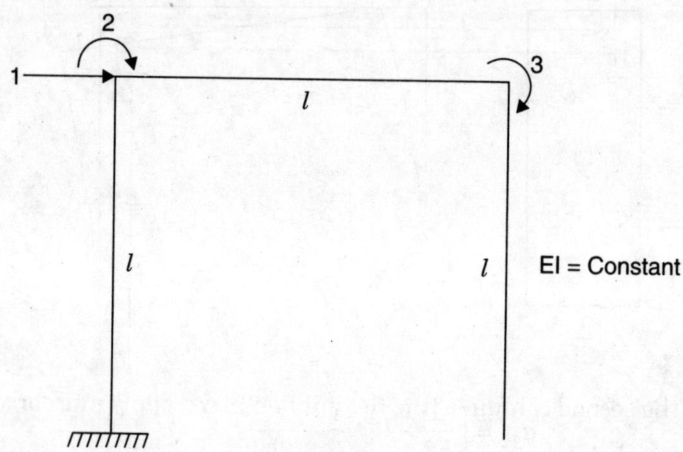

Fig. 4.22

Solution **Flexibility Matrix:**

$$[a] = \begin{bmatrix} a_{11} & a_{12} & a_{13} \\ a_{21} & a_{22} & a_{23} \\ a_{31} & a_{32} & a_{33} \end{bmatrix}$$

(a) **Apply a unit force along co-ordinate 1:**

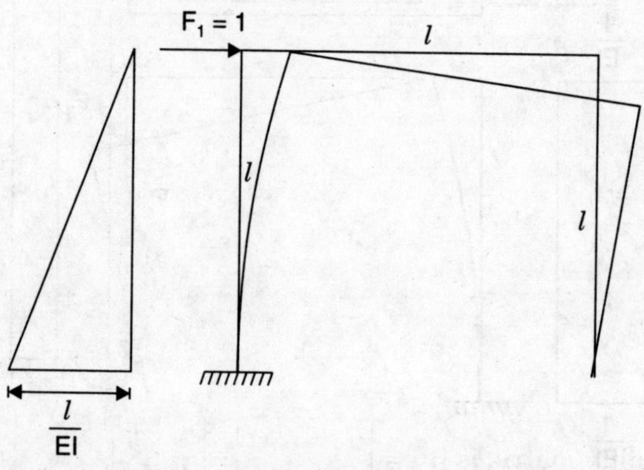

$$a_{11} = \left(\frac{1}{2} \times l \times \frac{l}{EI}\right) \times \frac{2}{3} l = \frac{l^3}{3EI}$$

$$a_{21} = \left(\frac{1}{2} \times l \times \frac{l}{EI}\right) = \frac{l^2}{2EI}$$

$$a_{31} = \frac{l^2}{2EI}$$

(*b*) **Apply a unit moment along co-ordinate 2:**

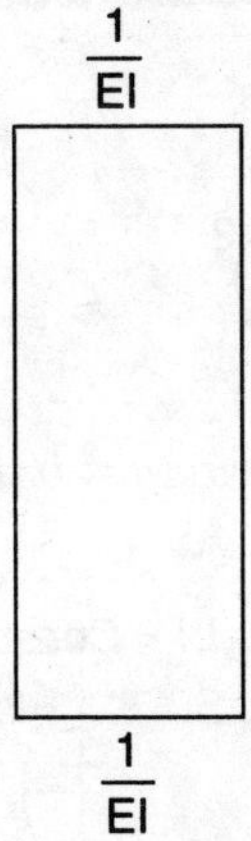

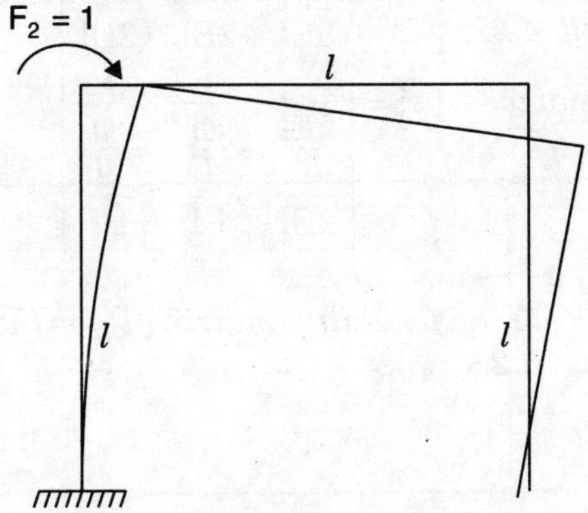

$$a_{12} = \left(\frac{1}{EI} \times l\right) \times \frac{l}{2} = \frac{l^2}{2EI}$$

$$a_{22} = \left(\frac{1}{EI} \times l\right) = \frac{l}{EI}$$

$$a_{32} = \frac{l}{EI}$$

(*c*) **Apply a unit force along co-ordinate 3:**

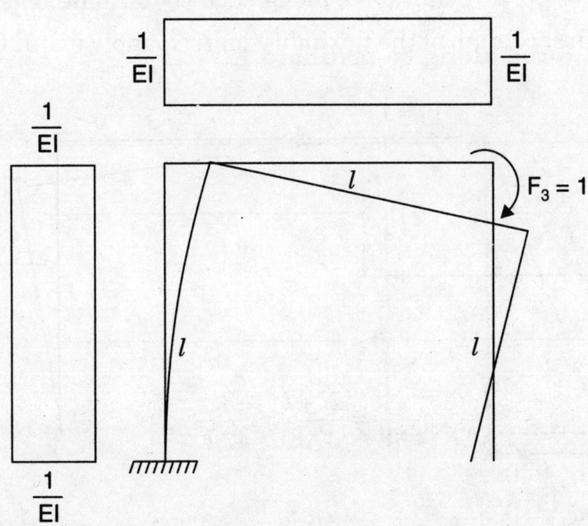

$$a_{13} = \frac{1}{EI} \times l \times \frac{l}{2} = \frac{l^2}{2EI}$$

$$a_{23} = \frac{1}{EI} \times l = \frac{l}{EI}$$

$$a_{33} = \frac{1}{EI} \times 2l = \frac{2l}{EI}$$

$$\therefore \quad \text{Flexibility matrix,} \quad [a] = \begin{bmatrix} \frac{l^3}{3EI} & \frac{l^2}{2EI} & \frac{l^2}{2EI} \\ \frac{l^2}{2EI} & \frac{l}{EI} & \frac{l}{EI} \\ \frac{l^2}{2EI} & \frac{l}{EI} & \frac{2l}{EI} \end{bmatrix}.$$

Example 4.11 *Generate the flexibility matrix [a] for the structure with co-ordinates shown in Fig. 4.23.*

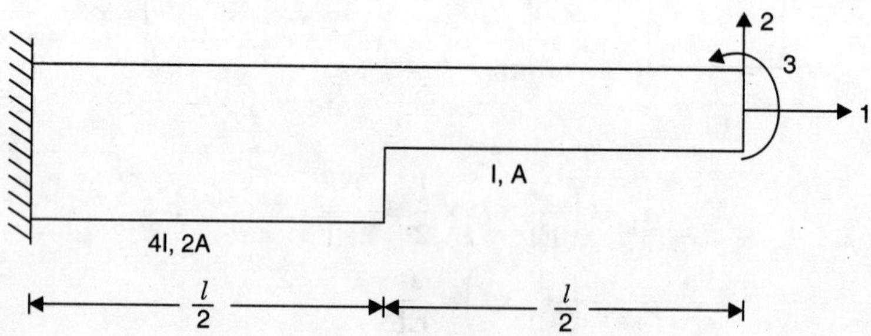

Fig. 4.23

Solution There are three co-ordinates in the given structure. The flexibility matrix is obtained by applying unit force along each co-ordinate one at a time.

To generate the first column of the flexibility matrix, apply a unit force along co-ordinate 1 only.

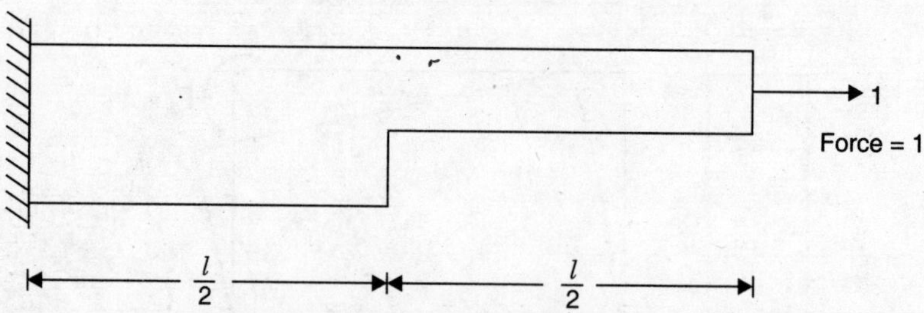

$$a_{11} = \frac{l}{2AE} + \frac{l}{4AE} = \frac{2l + l}{4AE} = \frac{3l}{4AE}$$

$$a_{12} = 0$$

$$a_{13} = 0$$

$$\therefore \begin{bmatrix} \frac{3l}{4AE} \\ 0 \\ 0 \end{bmatrix} \text{ is the first column of the flexibility matrix.}$$

To generate the second column of the flexibility matrix, apply unit force along co-ordinate 2 only.

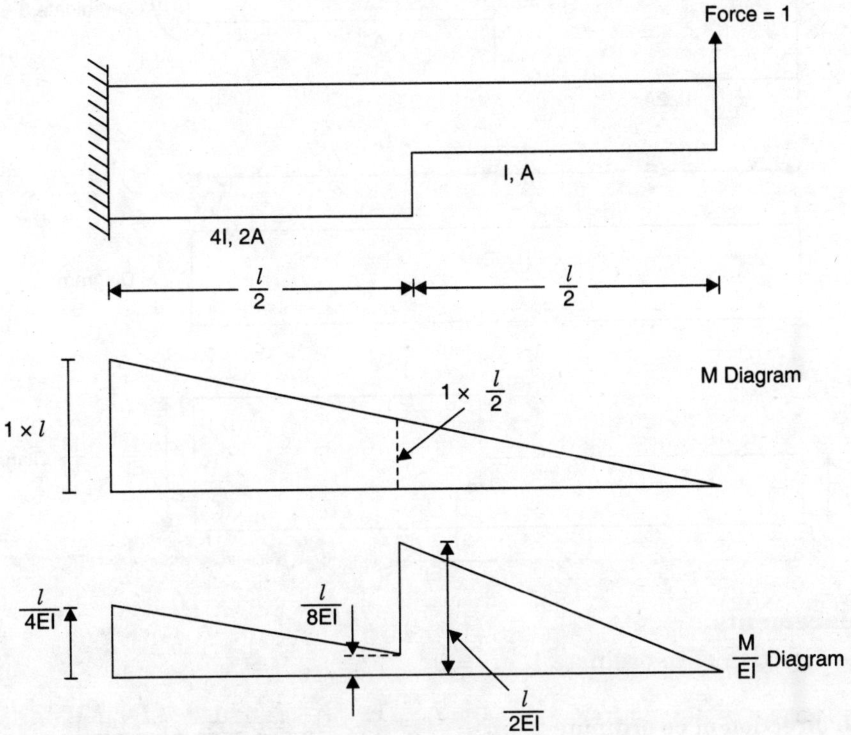

Displacement in the direction of co-ordinate 1

$$a_{12} = 0$$

Displacement of direction of co-ordinate 2.

$$a_{22} = \frac{1}{2} \times \frac{l}{2} \times \frac{l}{2EI} \times \frac{2}{3} \times \frac{l}{2} + \frac{l}{8EI} \times \frac{l}{2} \times \left(\frac{l}{2} + \frac{l}{4}\right) + \frac{1}{2} \times \frac{l}{2} \times \frac{l}{8EI} \times \left(\frac{l}{2} + \frac{2}{3} \times \frac{l}{2}\right)$$

$$= \frac{l^3}{24EI} + \frac{l^2}{16EI} \times \frac{3}{4}l + \frac{l^2}{32EI} \times \frac{5l}{6} = \frac{l^3}{24EI} + \frac{3l^3}{64EI} + \frac{5l^3}{192EI}$$

$$= \frac{8l^3 + 9l^3 + 5l^3}{192EI} = \frac{22l^3}{192EI} = \frac{11l^3}{96EI}$$

Displacement in the direction of co-ordinate 3,

$$a_{32} = \frac{1}{2} \times \frac{l}{2} \times \frac{l}{2EI} + \frac{1}{2} \times \frac{l}{2} \times \frac{l}{8EI} + \frac{l}{8EI} \times \frac{l}{2}$$

$$= \frac{l^2}{8EI} + \frac{l^2}{32EI} + \frac{l^2}{16EI} = \frac{4l^2 + l^2 + 2l^2}{32EI} = \frac{7l^2}{32EI}$$

To get the third column of the flexibility matrix, apply a unit force along co-ordinate 3 only.

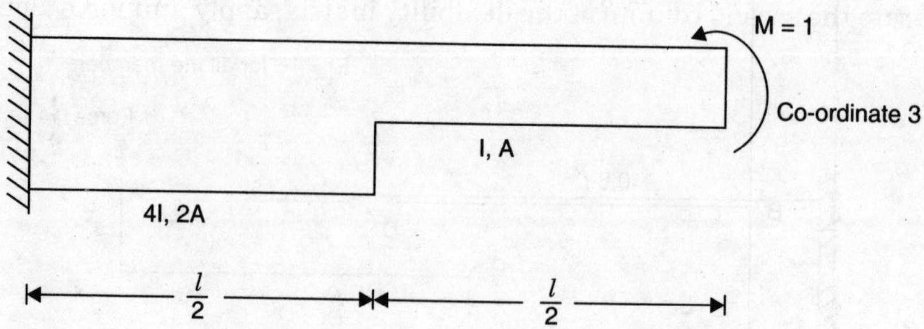

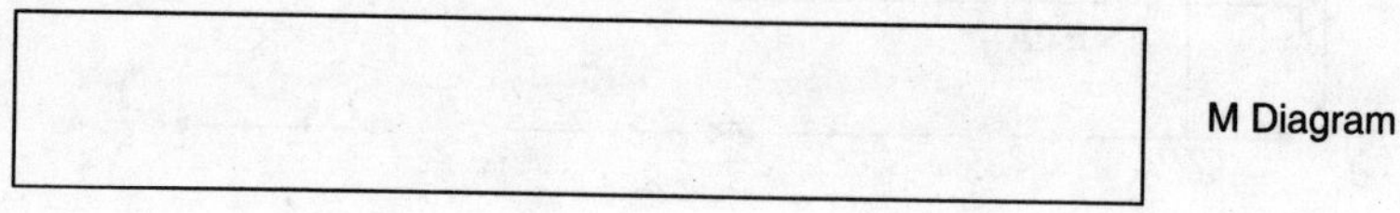

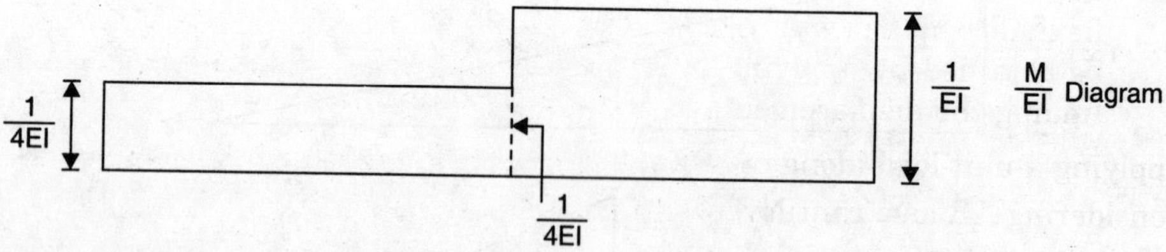

Displacements :

In the direction of co-ordinate 1, $a_{13} = 0$

In the direction of co-ordinate 2, $a_{23} = \frac{l}{2} \times \frac{1}{EI} \times \frac{l}{4} + \frac{l}{2} \times \frac{1}{4EI} \times \left(\frac{l}{2} + \frac{l}{4}\right)$

$$= \frac{l^2}{8EI} + \frac{l}{8EI} \times \frac{3}{4} l = \frac{l^2}{8EI} + \frac{3l^2}{32EI} = \frac{7l^2}{32EI}$$

In the direction of co-ordinate 3, $a_{33} = \frac{1}{EI} \times \frac{l}{2} + \frac{1}{4EI} \times \frac{l}{2} = \frac{l}{2EI} + \frac{l}{8EI} = \frac{5l}{8EI}$

Hence the flexibility matrix of the given structure,

$$[a] = \begin{bmatrix} \frac{3l}{4AE} & 0 & 0 \\ 0 & \frac{11l^3}{96EI} & \frac{7l^2}{32EI} \\ 0 & \frac{7l^2}{32EI} & \frac{5l}{8EI} \end{bmatrix}.$$

Example 4.12 *Derive the flexibility matrix [a] of the structure with two co-ordinates shown in Fig. 4.24.*

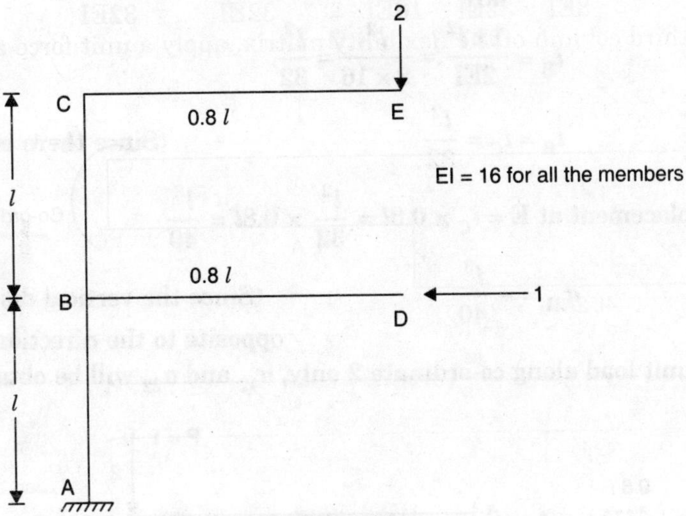

Fig. 4.24

Solution The given structure is statically determinate. There are two co-ordinates. Hence flexibility matrix will be of size 2 × 2. The co-efficients of the flexibility matrix will be obtained by applying unit forces along each co-ordinate, one at a time, and finding the displacements at 1 and 2.

Applying a unit load along co-ordinate 1 only, a_{11} and a_{21} will be obtained as below :
Considering AB as a cantilever with unit lateral load at B,

$$BB' = \delta_B = \frac{1 \times l^3}{3EI} = \frac{l^3}{3 \times 16} = \frac{l^3}{48}$$

$$DD' = BB' = \frac{l^3}{48}$$

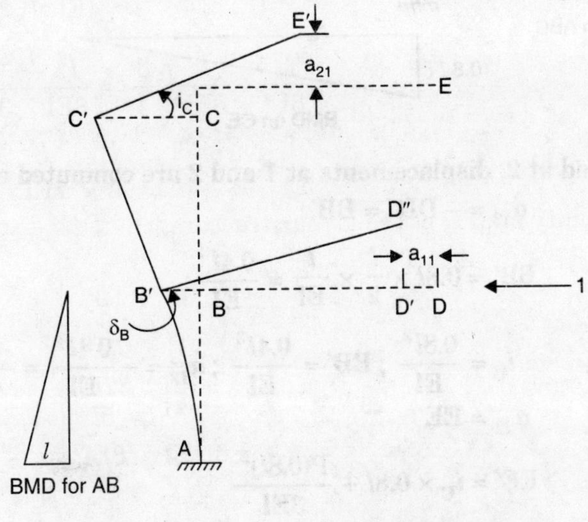

Hence, $a_{11} = \dfrac{l^3}{48}$

$$i_B = \frac{1 \times l^2}{2EI} = \frac{l^2}{2 \times 16} = \frac{l^2}{32}$$

$$i_B = i_C = \frac{l^2}{32}$$ (Since there is no load above B)

Vertical displacement at E $= i_C \times 0.8l = \dfrac{l^2}{32} \times 0.8l = \dfrac{l^3}{40}$

$a_{21} = -\dfrac{l^3}{40}$ (Since the vertical displacement at E is opposite to the direction of co-ordinate 2)

Applying a unit load along co-ordinate 2 only, a_{12} and a_{22} will be obtained as belolw :

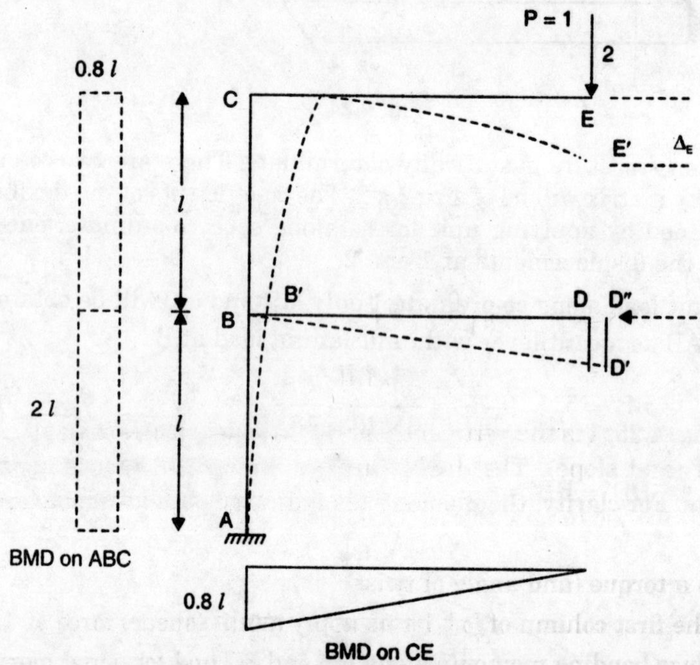

Due to a unit load at 2, displacements at 1 and 2 are computed as below :

$$a_{12} = -DD'' = BB'$$

$$BB' = 0.8l \times \frac{l}{2} \times \frac{l}{EI} = \frac{0.4l^3}{EI}$$

$$i_B = \frac{0.8l^2}{EI}\ ;\ BB' = \frac{0.4l^3}{EI}\ ;\ a_{12} = -\frac{0.4l^3}{EI} = \frac{-l^3}{40}$$

$$a_{22} = EE'$$

$$EE' = i_C \times 0.8l + \frac{P(0.8l)^3}{3EI}$$

$$i_C = \frac{2l \times 0.8l}{EI} = 0.1l^2$$

$$a_{22} = 0.1l^2 \times 0.8l + \frac{(0.8l)^3}{3EI} = 0.09067l^3$$

$$[a] = l^3 \begin{bmatrix} 0.0208 & -0.025 \\ -0.025 & 0.09067 \end{bmatrix}.$$

Example 4.13 *For the frame shown in Fig. 4.25 determine the flexibility matrix in respect of the co-ordinates indicated. Assume EI = 10 for both members and GJ = 4.*

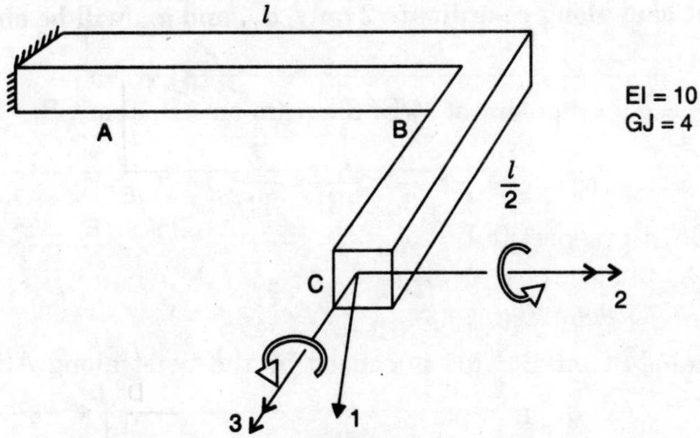

Fig. 4.25

Solution In the Fig. 4.25, 1 is the vertical shear (and vertical displacement). 2 is the bending moment (and slope). The double arrows indicate moments including torsional moments. For clarity, the moment 2 is indicated both as double arrows and as an arc.

Likewise, 3 is a torque (and angle of twist)

To generate the first column of $[a]$, let us apply a unit (shear) force at 1.

This will produce bending moments along AB and BC and torsional moments along AB.

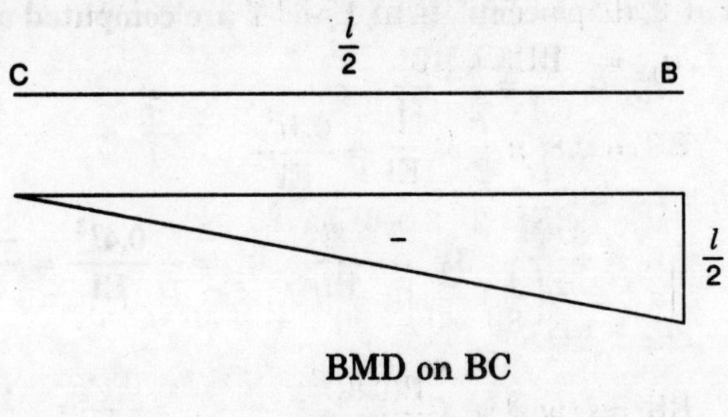

BMD on BC

BMD on AB

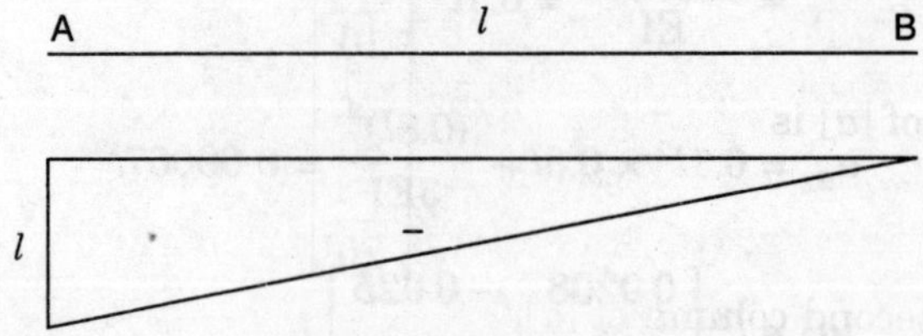

Torsion on AB

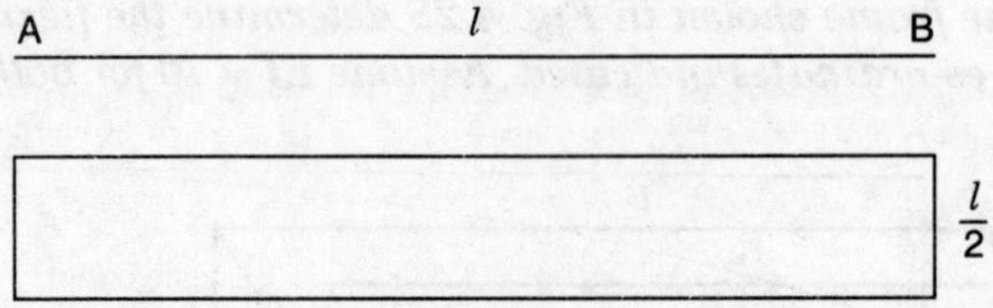

Deflection at B = d_B = moment of M/EI diagram on AB, about B.

$$\delta_B = \frac{l}{EI} \cdot \frac{l}{2} \times \frac{2l}{3} = \frac{l^3}{3EI} = \frac{l^3}{30}$$

Deflection of C with respect to B = δ_{CB}

$$\delta_{CB} = i_{BC}\,\frac{l}{2} + \frac{(l/2)^3}{3EI}$$

i_{BC} is the slope of BC at B. This is caused by the twist along AB due to the torsion diagram on AB

$$i_{BC} = \frac{Tl}{GJ} = \frac{l}{2} \cdot \frac{l}{GJ} = \frac{l^2}{8}$$

$$\delta_{CB} = \frac{l^2}{8} \cdot \frac{l}{2} + \frac{l^3}{24EI} = \frac{l^3}{16} + \frac{l^3}{240}$$

$$\delta_C = \delta_B + \delta_{BC} = \frac{l^3}{30} + \frac{l^3}{16} + \frac{l^3}{240} = \frac{l^3}{10}$$

$$a_{11} = \delta_C = \frac{l^3}{10}$$

a_{21} = slope at C due to this loading

$a_{21} = i_{BC}$ + twist angle at B of AB (i_{BA})

$$i_{BA} = \frac{Tl}{GJ} = \frac{l}{2} \cdot \frac{l}{GJ} = \frac{l^2}{8}$$

$$i_{BC} = \frac{l}{2EI} \cdot \frac{l}{2} \cdot \frac{1}{2} = \frac{l^2}{8EI} = \frac{l^2}{80}$$

$$a_{21} = l^2\left(\frac{1}{8} + \frac{1}{80}\right) = \frac{11}{80}\,l^2$$

$$a_{31} = \text{twist at C} = i_{BA} = -\frac{l}{EI} \cdot \frac{l}{2} = \frac{-l^2}{2EI} = \frac{-l^2}{20}$$

The first column of $[a]$ is $l^2 \begin{bmatrix} \frac{l}{10} \\ \frac{11}{80} \\ \frac{-1}{20} \end{bmatrix}$

To generate the second column of $[a]$,
We shall apply a unit moment at 2. Then
BMD on BC

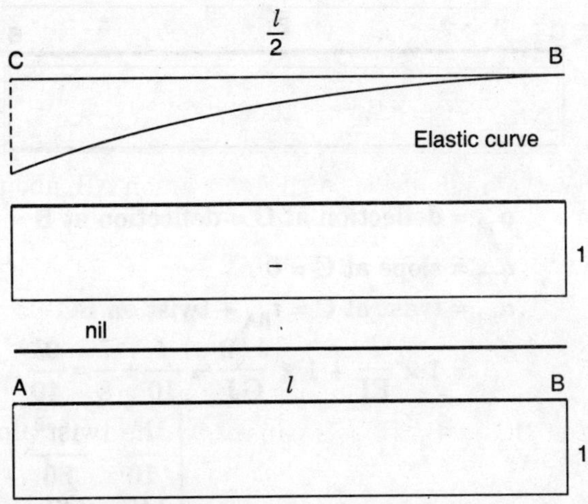

BMD on BC

BMD on BA

Torsion on BA

$i_{BA} = 0$, since there is no BM on AB.
i_{BC} = twist at B due to torsion on BA.

$$= \frac{Tl}{GJ} = \frac{l}{4}$$

$$a_{22} = i_C = i_{BC} + i_{CB} = \frac{l}{4} + \frac{l}{2EI} = \frac{l}{4} + \frac{l}{20} = \frac{6l}{20}$$

a_{12} = deflection at C = δ_C

$\delta_B = 0$ since BM on AB is zero.

$$\delta_C = i_{BC} \cdot \frac{l}{2} + \text{moment of area of BMD on BC about C}$$

$$= \frac{l}{4} \cdot \frac{l}{2} + \frac{\left(\frac{l}{2}\right)^2}{2EI} = \frac{l^2}{8} + \frac{l^2}{80} = \frac{11l^2}{80}$$

a_{32} = twist at C = 0

Hence the second column of $[a]$ is, $\begin{bmatrix} \frac{11l^2}{80} \\ \frac{24l}{80} \\ 0 \end{bmatrix}$

To generate the third column of $[a]$, let us apply a unit (twisting) moment at 3. Then

BMD on AB

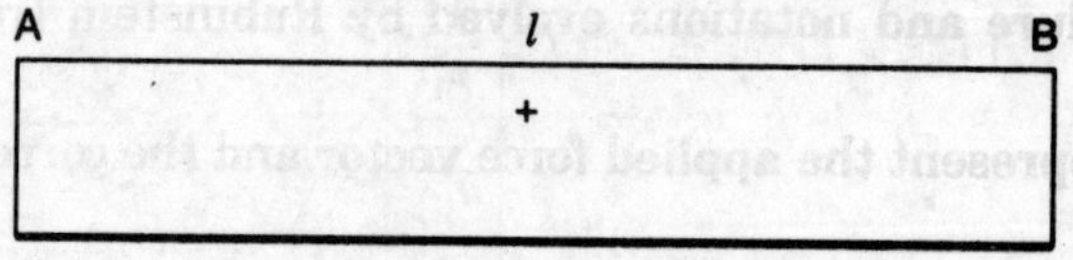

Torsion on AB nil

BM on BC nil

Torsion on BC

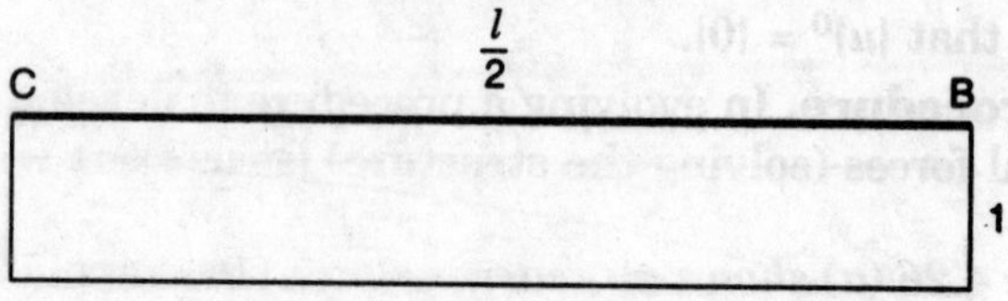

then

$$a_{13} = \text{deflection at C} = \text{deflection at B} = \frac{-1 \times l^2}{2EI} = \frac{-l^2}{20}$$

$$a_{23} = \text{slope at C} = 0$$

$$a_{33} = \text{twist at C} = i_{BA} + \text{twist on BC}$$

$$= 1 \times \frac{l}{EI} + 1 \times \frac{l/2}{GJ} = \frac{l}{10} + \frac{l}{8} = \frac{9l}{40}$$

Thus, the flexibility matrix works out as

$$[a] = \begin{bmatrix} \frac{l^3}{10} & \frac{11l^2}{80} & \frac{-l^2}{20} \\ \frac{11l^2}{80} & \frac{6l}{20} & 0 \\ \frac{-l^2}{20} & 0 & \frac{9l}{40} \end{bmatrix}.$$

4.4 FLEXIBILITY MATRICES OF INDETERMINATE STRUCTURES

We have seen that developing a flexibility matrix involves :

(*i*) Applying a force on the structure

(*ii*) Working out the internal forces and moments, and

(*iii*) Computing displacements (and rotations) at specific locations making use of the values in (*ii*) above.

In the case of determinate structures step (*ii*) does not present any problems. When it comes to indeterminate structures, solving for forces is not easy. Actually we have to employ matrix flexibility method only to solve for internal forces. Hence it becomes necessary to evolve a method for tackling indeterminate structures (including flexural frames and trusses).

4.4.1 Indeterminate Flexural Frames

(*a*) **Introduction.** Any indeterminate structure for which we seek to develop a flexibility matrix has to be first converted as a determinate structure (the primary structure). The indeterminate forces (reactions) have to be identified.

Now we have a determinate structure with two classes of loads, namely the applied forces {F}*, and the indeterminate (or redundant) forces $\{F\}^0$.

[Herein the procedure and notations evolved by Rubinstein (ref. 9) are adopted with necessary modifications]

So, {F}* and {*u*}* represent the applied force vector and the corresponding displacement vector.

$\{F\}^0$ and $\{u\}^0$ represent the indeterminate force vector and the corresponding displacement vector. It is obvious that $\{F\}^0$ are reactions and not active forces. Hence $\{u\}^0$ vector will have all elements as zero (unless there is a displacement at a support). To start with we will work on the assumption that $\{u\}^0 = \{0\}$.

(*b*) **Evolving a procedure.** In evolving a procedure to develop a flexibility matrix and to determine the internal forces (solving the structure) let us start with a simple example.

Example 4.14 *Figure 4.26 (a) shows an indeterminate structure with co-ordinates specified at load points. The applied loads are F_1^* and F_2^**

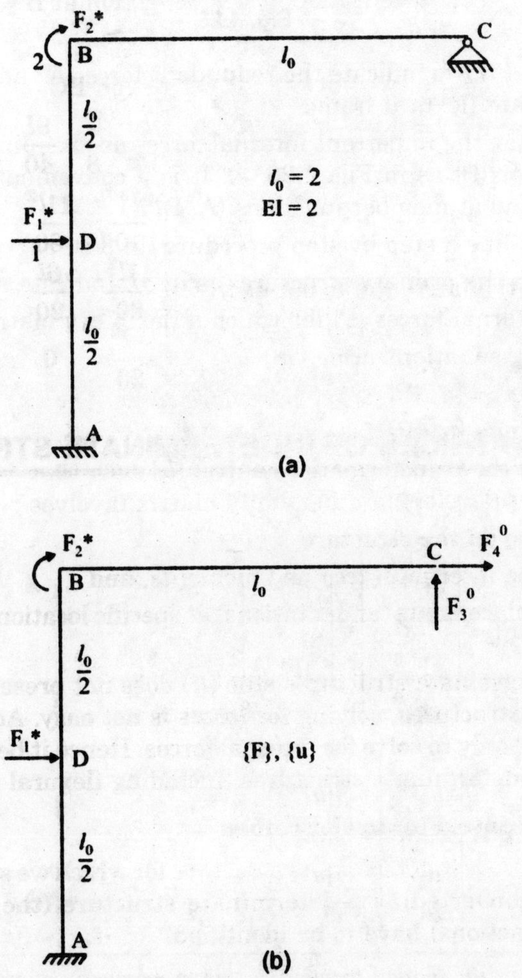

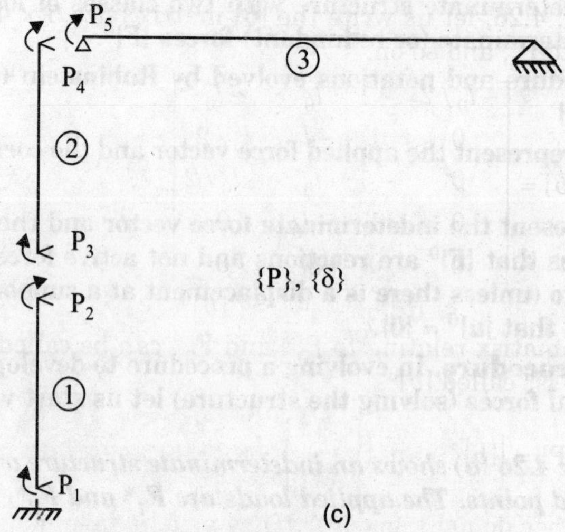

(c)

Fig. 4.26

Solution In Fig. 4.26 (*b*), we indicate the redundant forces F_3^0 and F_4^0, this being a 2 degree indeterminate flexural frame.

In a frame like this, the important internal forces are bending moments*. So we choose the internal forces vector {P} as in Fig. 4.26 (*c*). It is a convenient method to choose a node where a load acts (D), and at member junctions (A, B, C).

Now we shall out-line a step by step procedure (in 8 steps) to tackle the problem.

Step 1. Decide on the primary structure (as in *b*). Indicate redundants (F_3^0, F_4^0)

Step 2. Select internal forces {P} for which a flexibility matrix exists.

Now the following equations arise.

$$\{u\} = [a]\,\{F\}$$

$$\{\delta\} = [\alpha]\,\{P\}$$

Step 3. Generate the transformation matrix [*b*] such that

$$\{P\} = [b]\,\{F\}$$

Then,
$$\{F\} = \begin{bmatrix} F_1{}^* \\ F_2{}^* \\ \cdots\cdots \\ F_3{}^0 \\ F_4{}^0 \end{bmatrix} = \begin{bmatrix} \{F\}^* \\ \cdots\cdots \\ \{F\}^0 \end{bmatrix}$$

and
$$\{u\} = \begin{bmatrix} u_1{}^* \\ u_2{}^* \\ \cdots\cdots \\ u_3{}^0 \\ u_4{}^0 \end{bmatrix} = \begin{bmatrix} \{u\}^* \\ \cdots\cdots \\ \{u\}^0 \end{bmatrix}$$

**The energy stored in the loaded frame like this is primarily due to bending. Shear and direct thrust are there. But their strain energy contribution is much smaller and can justifiably be ignored. Hence we call these flexural frames.*

Going back to Fig. 4.26, let us write the [*b*] matrix (size 5 × 4). For column one of [*b*] apply $F_1 = 1$ and get P_1 to P_5, and so on.

$$[b] = \left[\begin{array}{cc:cc} -l_0/2 & -1 & l_0 & -l_0 \\ 0 & 1 & -l_0 & l_0/2 \\ 0 & -1 & l_0 & -l_0/2 \\ 0 & 1 & -l_0 & 0 \\ 0 & 0 & l_0 & 0 \end{array}\right]$$
$$\qquad [b]^* \qquad\qquad [b]^0$$

The part of the *b* matrix relating to F_1^* and F_2^* can be called b^* and the other part *c* corresponding to F_3^0 and F_4^0 called $[b]^0$.

Hence,
$$\{P\} = \left[[b]^* \,\vdots\, [b]^0\right]\begin{bmatrix} \{F\}^* \\ \cdots \\ \{F\}^0 \end{bmatrix} = [b]^*\{F\}^* + [b]^0\{F\}^0$$

Please remember, we do not know $\{F\}^0$. They are the redundants we have to determine.

Step 4. Synthesise the element flexibility matrix [α] from the 3 element flexibility matrices $[\alpha]_1$, $[\alpha]_2$ and $[\alpha]_3$.

For the 3 elements in Fig. 4.26 (*c*), the [α] matrices are given by

$$[\alpha]_1 = \frac{l_0}{6EI}\begin{bmatrix} 2 & -1 \\ -1 & 2 \end{bmatrix} = \frac{1}{12}\begin{bmatrix} 2 & -1 \\ -1 & 2 \end{bmatrix}$$

$$[\alpha]_2 = \frac{l_0}{6EI}\begin{bmatrix} 2 & -1 \\ -1 & 2 \end{bmatrix} = \frac{1}{12}\begin{bmatrix} 2 & -1 \\ -1 & 2 \end{bmatrix}$$

$$[\alpha]_3 = \frac{l_0}{3EI} = \frac{1}{6}$$

$$[\alpha] = \frac{1}{12}\begin{bmatrix} 2 & -1 & 0 & 0 & 0 \\ -1 & 2 & 0 & 0 & 0 \\ 0 & 0 & 2 & -1 & 0 \\ 0 & 0 & -1 & 2 & 0 \\ 0 & 0 & 0 & 0 & 2 \end{bmatrix}$$

Step 5. Compute [*a*] from the relation,

$$[a] = [b]^T[\alpha][b]$$

$$\left[\begin{array}{c:c} [a]_{11} & [a]_{12} \\ \hdashline [a]_{21} & [a]_{22} \end{array}\right] = \begin{bmatrix} [b]^{*T} \\ \cdots \\ [b]^{0T} \end{bmatrix}[\alpha]\left[[b]^* \,\vdots\, [b]^0\right]$$

Hence,
$$[a]_{11} = [b]^{*T}[\alpha][b]^*$$
$$[a]_{12} = [b]^{*T}[\alpha][b]^0$$
$$[a]_{21} = [b]^{0T}[\alpha][b]^*$$
$$[a]_{22} = [b]^{0T}[\alpha][b]^0$$

Step 6. Isolate $\{F\}^0$ from the known condition that $\{u\}^0 = \{0\}$ (no displacements at the location of redundant forces) as under.

$$\begin{Bmatrix} \{u\}^* \\ \{u\}^0 \end{Bmatrix} = \begin{bmatrix} [a]_{11} & [a]_{12} \\ [a]_{21} & [a]_{22} \end{bmatrix}\begin{Bmatrix} \{F\}^* \\ \{F\}^0 \end{Bmatrix}$$

$$\{u\}^* = [a]_{11}\ \{F\}^* + [a]_{12}\ \{F\}^0$$
$$\{u\}^0 = \{0\} = [a]_{21}\ \{F\}^* + [a]_{22}\ \{F\}^0$$
$$[a]_{22}\ \{F\}^0 = -[a]_{21}\ \{F\}^*$$
$$\{F\}^0 = -[a]_{22}^{-1}\ [a]_{21}\ \{F\}^*$$

Here we have used the condition that $\{u\}^0 = \{0\}$ and isolated the unknown redundants $\{F\}^0$ in terms of $[a]_{22}$, $[a]_{21}$, both known from step 5 and $\{F\}^*$, the given applied forces. The process involves the inversion of the $[a]_{22}$ matrix (in this case of size 2 × 2).

Step 7. Isolate $[a]^*$. We would like to express the force displacement relation for the system as $\{u\}^* = [a]^*\ \{F\}^*$

Then we must derive expressions for $[a]^*$

$\{u\}^*$ has been already defined in step 6

From step 6, $\{F\}^0 = -[a]_{22}^{-1}\ [a]_{21}\ \{F\}^*$

Hence, $\{u\}^* = [a]_{11}\ \{F\}^* + [a]_{12}\ (-[a]_{22}^{-1}\ [a]_{21}\ \{F\}^*)$

$\{u\}^* = ([a]_{11} - [a]_{12}\ [a]_{22}^{-1}\ [a]_{21})\ \{F\}^*$

So, if $\{u\}^* = [a]^*\ \{F\}^*$,

Then $[a]^* = [a]_{11} - [a]_{12}\ [a]_{22}^{-1}\ [a]_{21}$

Step 8. Get the 4 parts of $[a]$ from step 5.

Get the internal forces $\{P\}$ from step 3, (wherein we know $\{F\}^0$ now) using $\{P\} = [b]\ \{F\}$ and displacements from $\{\delta\} = [a]\ \{P\}$.

For the structure in Example 4.14, the rest of the numerical work is given under.

$$[b] = \left[\begin{array}{cc:cc} -1 & -1 & 2 & -2 \\ 0 & 1 & -2 & 1 \\ 0 & -1 & 2 & -1 \\ 0 & 1 & -2 & 0 \\ 0 & 0 & 2 & 0 \end{array}\right]$$
$$\qquad [b]^* \qquad [b]^0$$

$$[a] = b^T \propto b$$

$$[a] = \frac{1}{12}\left[\begin{array}{cc:cc} 0.5 & 1.5 & -3 & 2.5 \\ 1.5 & 12 & -24 & 12 \\ \hdashline -3 & -24 & 56 & -24 \\ 2.5 & 12 & -24 & 16 \end{array}\right]$$

$$[a]_{11} = \frac{1}{12}\begin{bmatrix} 0.5 & 1.5 \\ 1.5 & 0.5 \end{bmatrix};\ [a]_{12} = \frac{1}{12}\begin{bmatrix} -3 & 2.5 \\ -24 & 12 \end{bmatrix}$$

$$[a]_{21} = \frac{1}{12}\begin{bmatrix} -3 & -24 \\ 2.5 & 12 \end{bmatrix};\ [a]_{22} = \frac{1}{12}\begin{bmatrix} 56 & -24 \\ -24 & 16 \end{bmatrix}$$

$$[a]_{22}^{-1} = \begin{bmatrix} 0.6 & 0.9 \\ 0.9 & 2.1 \end{bmatrix}$$

$$[a]^* = [a]_{11} - [a]_{12}\ [a]_{22}^{-1}\ [a]_{21}$$

$$[a]^* = \frac{1}{12}\begin{bmatrix} 0.081 & -0.15 \\ -0.15 & 1.20 \end{bmatrix}$$

Computing Redundants

The redundant forces {F}^{0} are given by

$$\{F\}^0 = -[a]_{22}^{-1}[a]_{21}\{F\}^*$$

$$\{F\}^0 = -\frac{1}{12}\begin{bmatrix} 0.45 & -3.6 \\ 2.55 & 3.6 \end{bmatrix}\{F\}^*$$

If we assume $F_1^* = 12$, $F_2^* = 12$

$$\{F\}^0 = +\begin{Bmatrix} 3.15 \\ -6.15 \end{Bmatrix}$$

Computing the bending moments

$$\{P\} = [b]^*\{F\}^* + [b]^0\{F\}^0$$

$$= \begin{bmatrix} -1 & -1 \\ 0 & +1 \\ 0 & -1 \\ 0 & +1 \\ 0 & 0 \end{bmatrix}\begin{Bmatrix} 12 \\ 12 \end{Bmatrix} + \begin{bmatrix} 2 & -2 \\ -2 & 1 \\ 2 & -1 \\ -2 & 0 \\ 2 & 0 \end{bmatrix}\begin{Bmatrix} 3.15 \\ -6.15 \end{Bmatrix}$$

$$\{P\} = \begin{Bmatrix} -5.4 \\ -0.45 \\ 0.45 \\ 5.70 \\ 6.30 \end{Bmatrix}$$

{P}, it may be recalled, represents the member end moments from which the bending moment diagram can be drawn.

4.5 ANALYSING STRUCTURES BY FLEXIBILITY METHOD

Using Flexibility Method for analyzing simple indeterminate structures is like using a helicoptor to go to the next street. The real value of the method can be realized only when we deal with structures with a large number of redundancies. This method is specifically meant for use with computers.

However, the following examples (of manually solving simple structures) are aimed at reinforcing the fundamental principles and procedures involved.

Example 4.15 *For the beam shown in Fig. 4.27, find the deflection at mid span.*

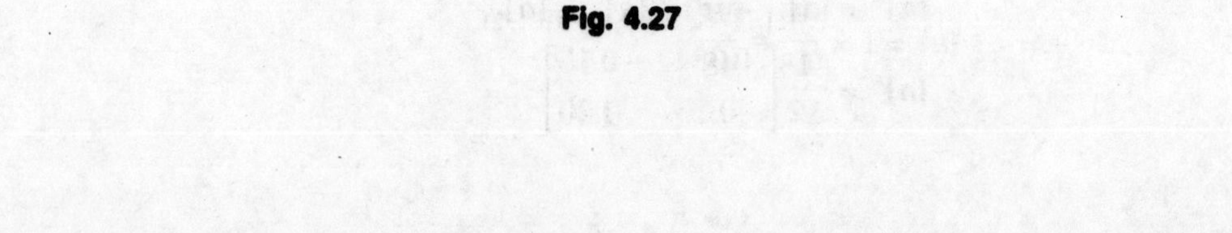

Fig. 4.27

Solution **Step 1.** The given structure, being determinate, is also the primary structure. Hence the system co-ordinate is as under

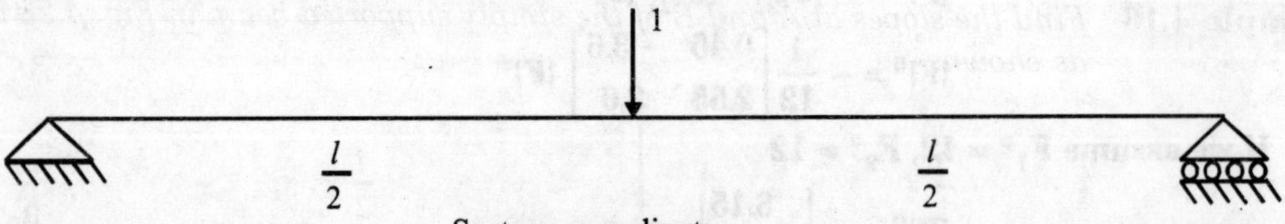

System co-ordinate

Step 2. Let us choose AC and CB as elements and define element co-ordinates as under

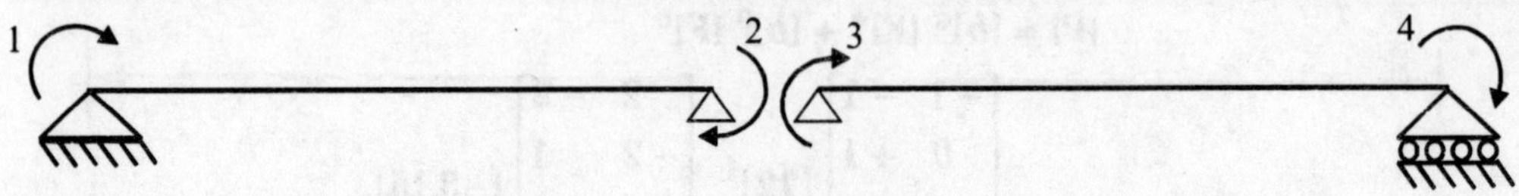

Step 3. To generate the $[b]$ matrix relating system force vector, {F} and element force vector {P}, Let us get the equilibrium conditions for the 2 elements, when $F_1 = 1$

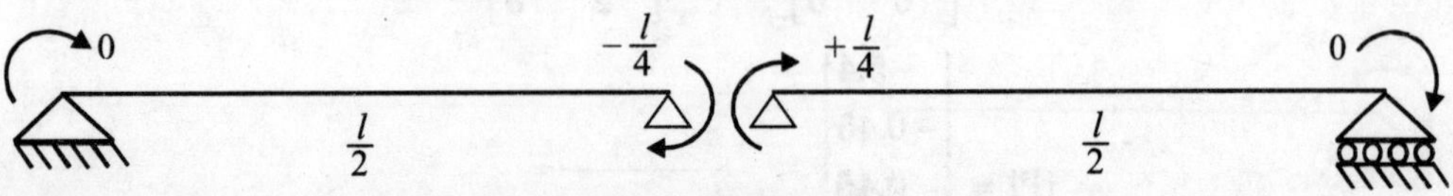

$$[b] = [0 \quad -l/4 \quad +l/4 \quad 0]^T$$
$$= [0 \quad -1/2 \quad +1/2 \quad 0]^T$$

This equation simply means that the bending moment at mid span is Wl/4, where W = 1.

Step 4. Element flexibility matrix $[\alpha]$ is given by

$$\frac{l/2}{6EI}\begin{bmatrix} 2 & -1 & 0 & 0 \\ -1 & 2 & 0 & 0 \\ 0 & 0 & 2 & -1 \\ 0 & 0 & -1 & 2 \end{bmatrix} = \frac{1}{48}\begin{bmatrix} 2 & -1 & 0 & 0 \\ -1 & 2 & 0 & 0 \\ 0 & 0 & 2 & -1 \\ 0 & 0 & -1 & 2 \end{bmatrix}$$

Step 5. Compute $[a]$

$$[a] = [b]^T \quad [\alpha] \quad [b]$$

$$= \begin{bmatrix} 0 & -\frac{1}{2} & +\frac{1}{2} & 0 \end{bmatrix} \frac{1}{48}\begin{bmatrix} 2 & -1 & 0 & 0 \\ -1 & 2 & 0 & 0 \\ 0 & 0 & 2 & -1 \\ 0 & 0 & -1 & 2 \end{bmatrix}\begin{bmatrix} 0 \\ -\frac{1}{2} \\ +\frac{1}{2} \\ 0 \end{bmatrix}$$

$$[a] = \frac{1}{48}\,[1]$$

Step 6. Find the from $\{u\} = [a] \quad \{F\}$

$$\{u\} = 1 \times \frac{W}{48} = \frac{W}{48}$$

The ludicrous fact is that we have used a lot of elaborate steps to arrive at the conclusion that the mid span deflection is $\frac{Wl^3}{48EI}$. Any way we at least landed at the correct solution.

Example 4.16 *Find the slopes at A and B of the simply supported beam in Fig. 4.28 loaded as shown.*

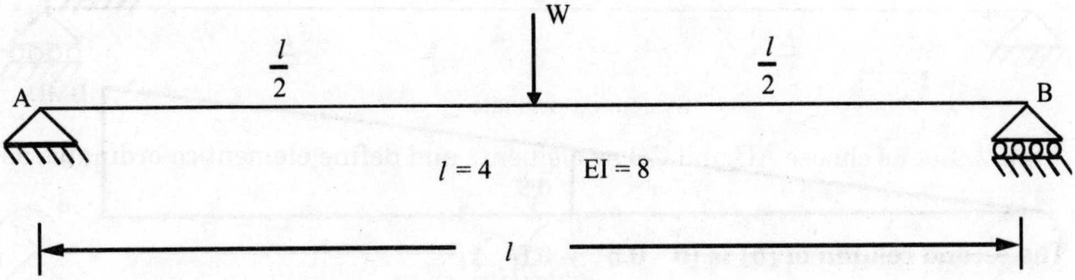

Fig. 4.28

Solution Define system co-ordinates as below :

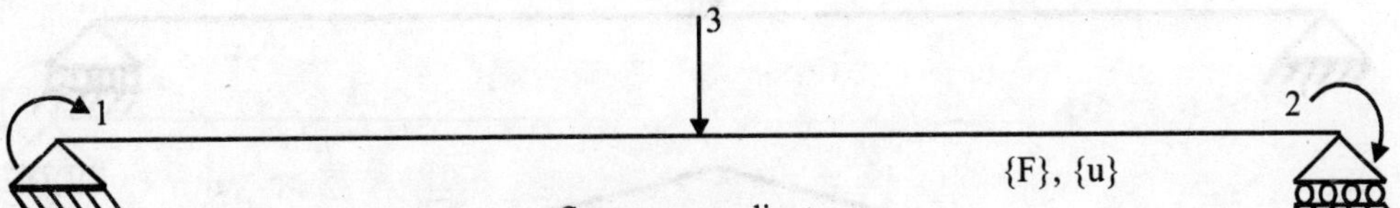

System co-ordinates

Define element co-ordinates as below :

Element co-ordinates

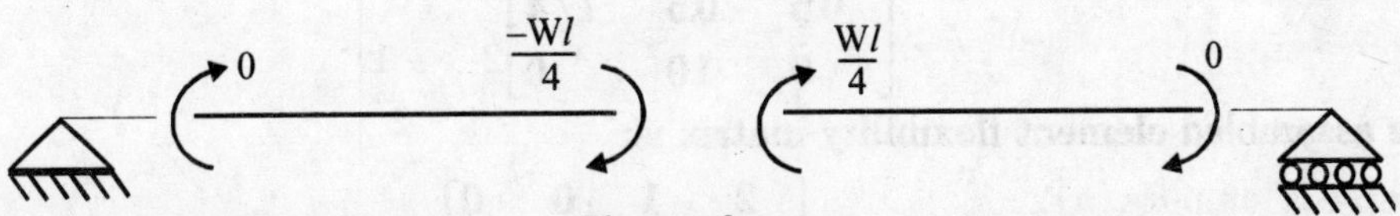

Element forces

Now we have to relate the system forces {F} and the element forces {P}. We generate the [b] matrix which will be of size 4 × 3.

Apply $F_1 = 1$

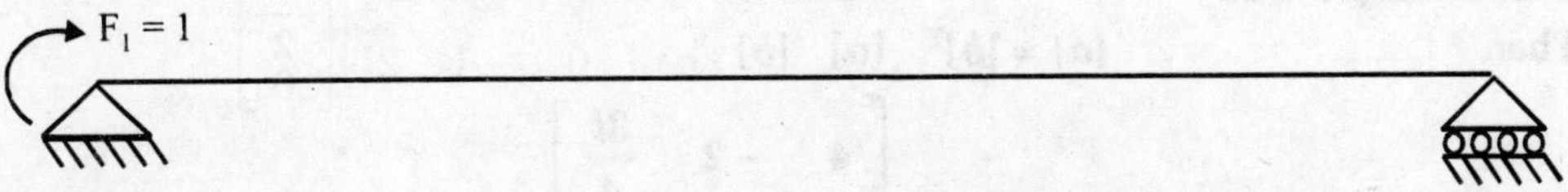

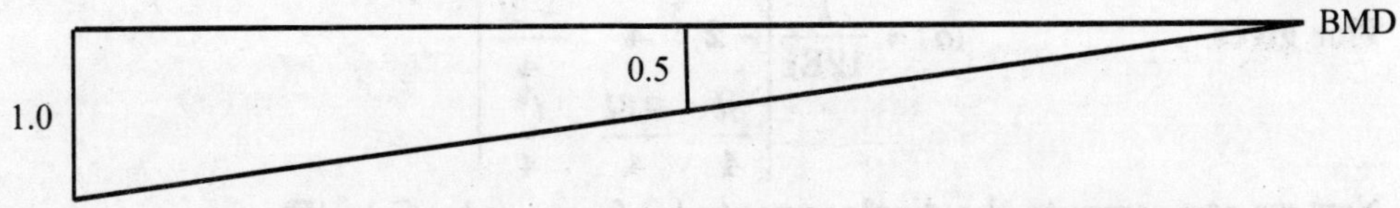

The first column of $[b]$ is therefore $[1 \quad -0.5 \quad 0.5 \quad 0]^T$

Apply $F_2 = 1$

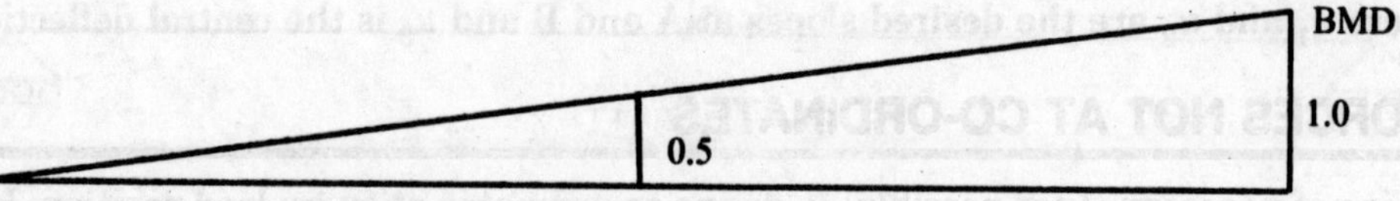

The second column of $[b]$ is $[0 \quad 0.5 \quad -0.5 \quad 1]^T$

Apply $F_3 = 1$

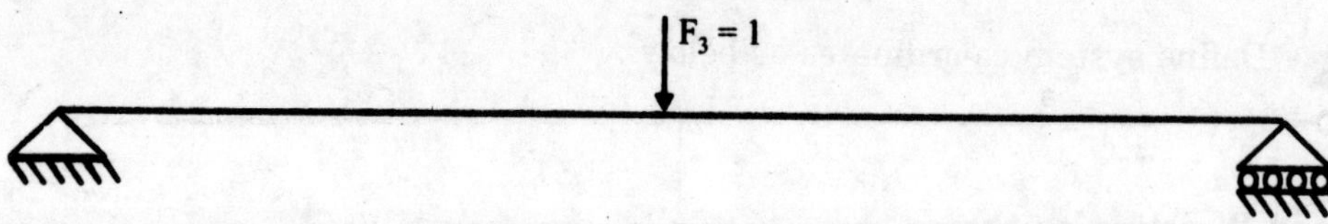

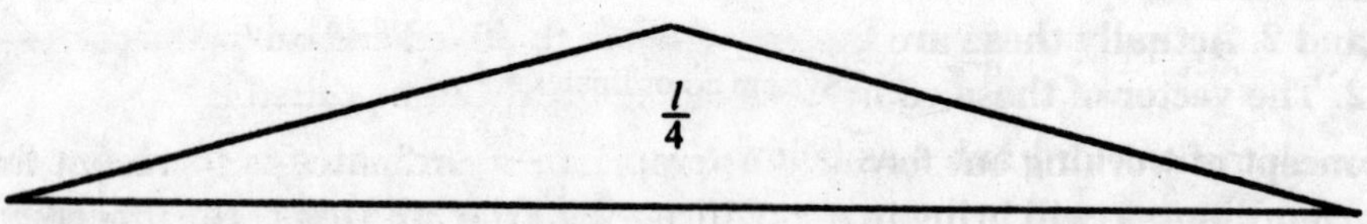

The third column of $[b]$ is $[0 \quad -l/4 \quad l/4 \quad 0]^T$

$$[b] = \begin{bmatrix} 1.0 & 0 & 0 \\ -0.5 & 0.5 & -l/4 \\ 0.5 & -0.5 & l/4 \\ 0 & 1.0 & 0 \end{bmatrix}$$

The assembled element flexibility matrix is:

$$[\alpha] = \frac{l/2}{6EI}\begin{bmatrix} 2 & -1 & 0 & 0 \\ -1 & 2 & 0 & 0 \\ 0 & 0 & 2 & -1 \\ 0 & 0 & -1 & 2 \end{bmatrix}$$

just like in Example 4.13

Then $[a] = [b]^T \; [\alpha] \; [b]$

This gives
$$[a] = \frac{l}{12EI}\begin{bmatrix} 4 & -2 & \frac{3l}{4} \\ -2 & 4 & \frac{-3l}{4} \\ \frac{3l}{4} & \frac{-3l}{4} & \frac{l^2}{4} \end{bmatrix}$$

Now we can compute the displacements $\{u\}$ from $\{u\} = [a] \; \{F\}$

Since $\{F\} = \begin{Bmatrix} 0 \\ 0 \\ W \end{Bmatrix}, \quad \{u\} = \begin{Bmatrix} \dfrac{Wl^2}{16EI} \\ \dfrac{-Wl^2}{16EI} \\ \dfrac{Wl^3}{48EI} \end{Bmatrix}$

Here u_1 and u_2 are the desired slopes at A and B and u_3 is the central deflection.

4.6 FORCES NOT AT CO-ORDINATES

It is not necessary (nor possible) to define co-ordinates at every load position. Hence we must develop a method of getting the effect of off co-ordinate loads at the joints (where co-ordinates are necessarily defined).

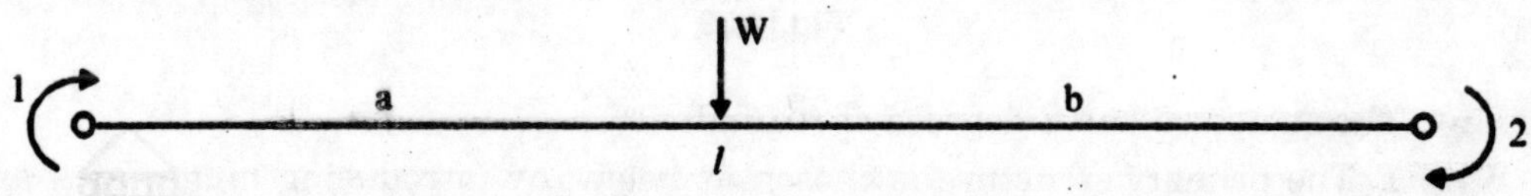

A load W acting as shown above will induce moments Wab^2/l^2 and $- Wa^2b/l^2$ at co-ordinates 1 and 2. Actually these are the opposites of the fixed end moments corresponding to W at 1 and 2. The vector of these equivalent moments would be called $\{P\}^e$.

The concept of working out forces at appropriate co-ordinates to represent forces (such as W) not at co-ordinates, will bring in a modification to the 8th step of the procedure outlined in subarticle 4.4.1(*b*). In step 3, $\{F\}^*$ would include the above forces also. The steps are rewritten in a compact form in subarticle 4.6.1.

4.6.1 8 Step Procedure for Flexibility Method

1. Decide on the primary structure; indicate redundants $\{F\}^0$.
2. Select {P} co-ordinates for internal forces.
3. Compile external forces $\{F\}^*$. This would include the effects of off node forces.

Then $\{F\} = \begin{Bmatrix} \{F\}^* \\ \cdots\cdots \\ \{F\}^0 \end{Bmatrix}$ and $\{u\} = \begin{bmatrix} \{u\}^* \\ \cdots\cdots \\ \{u\}^0 \end{bmatrix}, \; \{u\}^0 = \{0\}$

Generate $[b]$ matrix such that $\{P\} = [b] \; \{F\}$; $[b] = \left[[b]^* \; \vdots \; [b]^0\right]$

4. Synthesize element flexibility matrix $[\alpha]$ (vide Table 5.1 for standard cases)

5. Compute $[a]$ using $[a] = [b]^T \; [\alpha] \; [b]$; $[a] = \begin{bmatrix} [a]_{11} & [a]_{12} \\ [a]_{21} & [a]_{22} \end{bmatrix}$

6. Isolate $\{F\}^0$, the redundant forces, from the condition

$$\{u\}^0 = \{0\}$$

$$\{F\}^0 = -[a]_{22}^{-1} \; [a]_{21} \; \{F\}^*$$

7. Isolate $[a]^*$ $\quad [a]^* = [a]_{11} - [a]_{12} \ [a]_{22}^{-1} \ [a]_{21}$

8. Get {P} from $\quad \{P\} = [b] \ \{F\}$ and $\{P\}^f = \{P\} - \{P\}^e$

$\{P\}^f$ are the final member forces.

$\{P\}^f$ are also the support moments at all joints.

Example 4.17 *Analyse the continuous beam in Fig. 4.29 by flexibility method.*

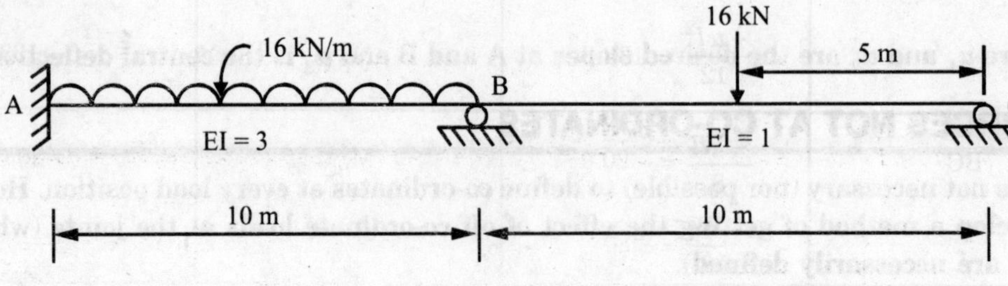

Fig. 4.29

Solution The structure has 2 degrees of redundancy.

Step 1. The primary structure is chosen as below, by introducing hinges at A and B. The moments at A and B are thus redundants.

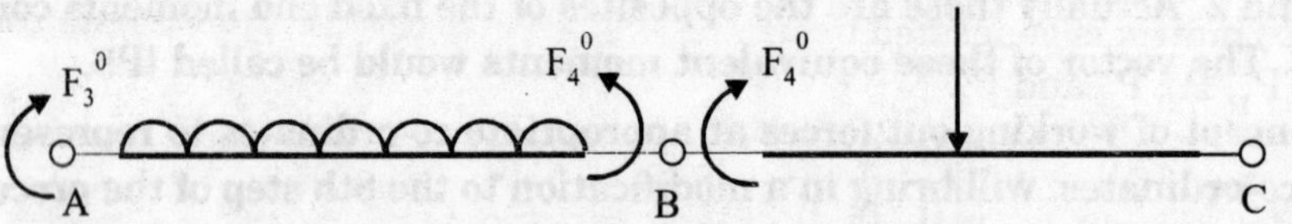

Please note that $F_4^{\,0}$ is a pair of moments making a bending moment (sagging BM in this case).

Step 2. The {P} system is as under.

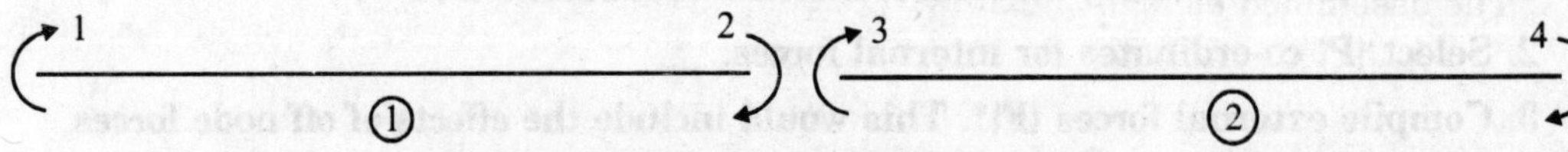

Element co-ordinates

Step 3. The {F} system is in 2 parts. $\{F\}^*$ and $\{F\}^0$.

F^* is primarily made up of the effects of the lateral loads on AB and BC. These are the opposites of fixed end moments.

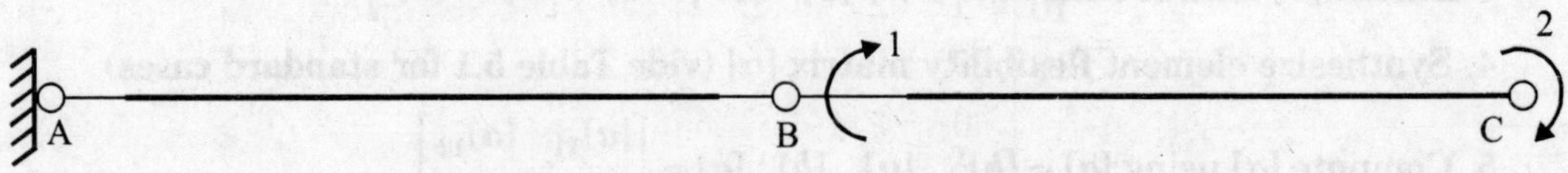

System co-ordinates

Table 4.1

SPAN	*FEM*	*Name of moment*	*Equivalent joint loads*
AB	$\frac{-wl^2}{12} = -133.3$	M_{AB}	133.3
	$\frac{+wl^2}{12} = 133.3$	M_{BA}	– 133.3
BC	$\frac{-Wl}{8} = -20.0$	M_{BC}	20.0
	$\frac{Wl}{8} = 20.0$	M_{CB}	– 20.0

Among the equivalent joint loads, the moment M_{AB} can be ignored since A is a fixed end. The moment will not get transmitted to the span. At system co-ordinate 1, the moments on BA and BC will combine to give – 113.3 kNm.

Hence, $F_1^* = -113.3$

$F_2^* = -20.0$

The [*b*] matrix is generated by introducing $F_1^* = 1$, $F_2^* = 1$, $F_3^0 = 1$ and $F_4^0 = 1$ in 4 steps and finding P_1, P_2, P_3 and P_4.

$$\text{Hence,} \quad [b] = \begin{bmatrix} 0 & 0 & \vdots & 1 & 0 \\ 0 & 0 & \vdots & 0 & -1 \\ 1 & 0 & \vdots & 0 & 1 \\ 0 & 1 & \vdots & 0 & 0 \end{bmatrix}$$

$$\qquad\qquad [b]^* \qquad [b]^0$$

The assembled element flexibility matrix [α] is:

$$[\alpha] = \begin{bmatrix} [\alpha]_1 & [0] \\ [0] & [\alpha]_2 \end{bmatrix} = \begin{bmatrix} \frac{10}{18}\begin{bmatrix} 2 & -1 \\ -1 & 2 \end{bmatrix} & 0 & 0 \\ 0 & 0 & 0 & 0 \\ 0 & 0 & \frac{10}{6}\begin{bmatrix} 2 & -1 \\ -1 & 2 \end{bmatrix} \end{bmatrix}$$

$$= \frac{10}{18}\begin{bmatrix} 2 & -1 & 0 & 0 \\ -1 & 2 & 0 & 0 \\ 0 & 0 & 6 & -3 \\ 0 & 0 & -3 & 6 \end{bmatrix}$$

$$a = [b]^T [\alpha] [b]$$

$$\begin{bmatrix} a_{11} & a_{12} \\ a_{21} & a_{22} \end{bmatrix} = \begin{bmatrix} b^{*T} \\ b^{0T} \end{bmatrix} [\alpha] \begin{bmatrix} b^* & b^0 \end{bmatrix}$$

$$a_{22} = a_{00} = b^{0T} \quad [\alpha] \quad b^0$$

$$= \begin{bmatrix} 1 & 0 & 0 & 0 \\ 0 & -1 & 1 & 0 \end{bmatrix} \frac{10}{18} \begin{bmatrix} 2 & -1 & 0 & 0 \\ -1 & 2 & 0 & 0 \\ 0 & 0 & 6 & -3 \\ 0 & 0 & -3 & 6 \end{bmatrix} \begin{bmatrix} 1 & 0 \\ 0 & -1 \\ 0 & 1 \\ 0 & 0 \end{bmatrix}$$

$$[a_{00}] = \frac{10}{18} \begin{bmatrix} 2 & 1 \\ 1 & 8 \end{bmatrix}$$

$$[a_{00}]^{-1} = 0.12 \begin{bmatrix} 8 & -1 \\ -1 & 2 \end{bmatrix}$$

$$[a_{0*}] = [b_0]^T \quad [\alpha] \quad [b_*]$$

$$= \begin{bmatrix} 1 & 0 & 0 & 0 \\ 0 & -1 & 1 & 0 \end{bmatrix} \frac{10}{18} \begin{bmatrix} 2 & -1 & 0 & 0 \\ -1 & 2 & 0 & 0 \\ 0 & 0 & 6 & -3 \\ 0 & 0 & -3 & 6 \end{bmatrix} \begin{bmatrix} 0 & 0 \\ 0 & 0 \\ 1 & 0 \\ 0 & 1 \end{bmatrix}$$

$$= \frac{10}{18} \begin{bmatrix} 0 & 0 \\ +6 & -3 \end{bmatrix}$$

$$\{F\}^0 = -[a_{00}]^{-1} \quad [a_{0*}] \quad \{F\}^*$$

$$= -0.12 \begin{bmatrix} 8 & -1 \\ -1 & 2 \end{bmatrix} \frac{10}{18} \begin{bmatrix} 0 & 0 \\ +6 & -3 \end{bmatrix} \begin{Bmatrix} -113.3 \\ -20.0 \end{Bmatrix}$$

$$\{F\}^0 = \begin{Bmatrix} -41.3 \\ 82.7 \end{Bmatrix}$$

$$\{F\}^T = [-113.3 \quad -20 \quad -41.3 \quad 82.7]$$

The element forces {P} are given by $\{P\} = [b] \quad \{F\}$

$$= \begin{bmatrix} 0 & 0 & 1 & 0 \\ 0 & 0 & 0 & -1 \\ 1 & 0 & 0 & 1 \\ 0 & 1 & 0 & 0 \end{bmatrix} \begin{Bmatrix} -113.3 \\ -20.0 \\ -41.3 \\ 82.7 \end{Bmatrix} = \begin{Bmatrix} -41.3 \\ -82.7 \\ -30.6 \\ -20.0 \end{Bmatrix}$$

The final forces $\{P\}^f$ are given by $\{P\}^f = \{P\} - \{P\}^e$, where $\{P\}^e$ is the vector of equivalent joint forces

$$\{P\}^f = \begin{Bmatrix} -41.3 \\ -82.7 \\ -30.6 \\ -20.0 \end{Bmatrix} - \begin{Bmatrix} 133.3 \\ -133.3 \\ 20 \\ -20 \end{Bmatrix} = \begin{Bmatrix} -174.6 \\ 50.6 \\ -50.6 \\ 0 \end{Bmatrix}$$

Free BM:

$$M_{max} \text{ in span } AB = \frac{wl^2}{8} = \frac{16 \times 100}{8} = 200 \text{ kNm}$$

$$M_{max} \text{ in span } BC = \frac{Wl}{4} = \frac{16 \times 10}{4} = 40 \text{ kNm}$$

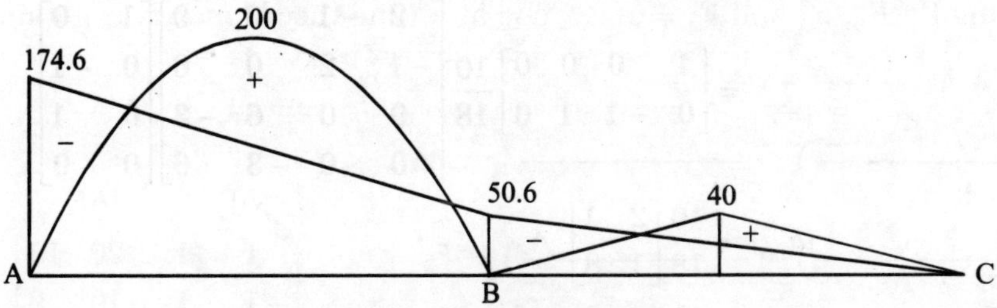

Bending moment diagram

Example 4.18 *Analyse the beam in the previous example by treating the reactions at B and C as redundants. Obtain the support moments.*

Solution The structure is statically in determinate by 2 degrees. By treating the support reactions R_B and R_C as redundants and removing them, the primary structure is shown below.

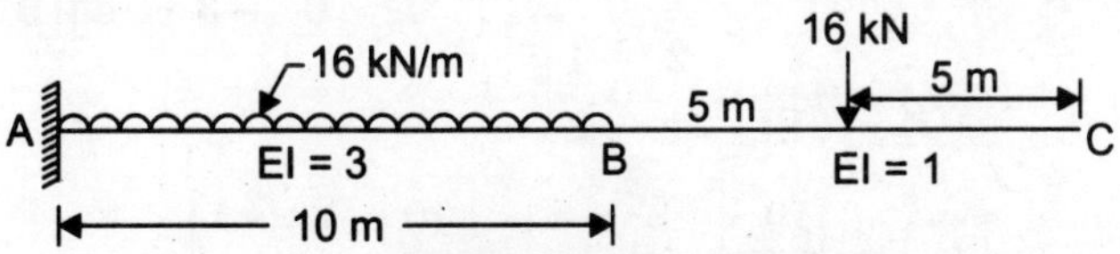

Primary structure

Fig. 4.30

$$M_{FAB} = -\frac{wl^2}{12} = -\frac{16\times10^2}{12} = -133.33 \text{ kNm}$$

$$M_{FBA} = +\frac{wl^2}{12} = +\frac{16\times10^2}{12} = 133.33 \text{ kNm}$$

$$M_{FBC} = -\frac{wl}{8} = -\frac{16\times10}{8} = -20 \text{ kNm}$$

$$M_{FCB} = +\frac{wl}{8} = +\frac{16\times10}{8} = 20 \text{ kNm}$$

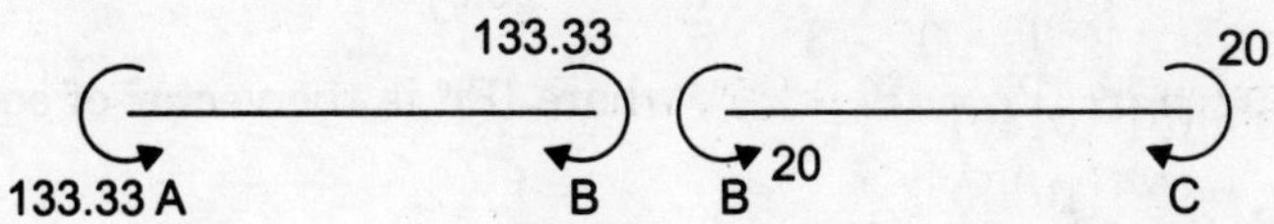

Fixed end moments

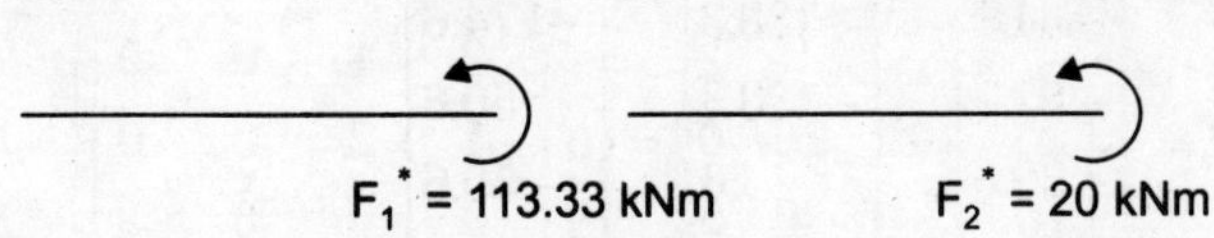

System co-ordinates

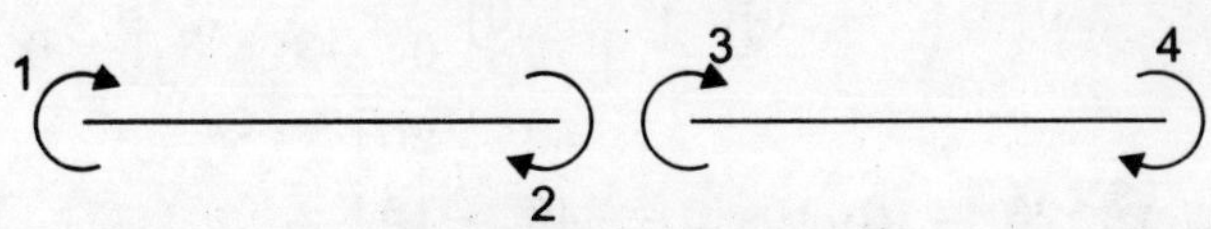

Element co-ordinates

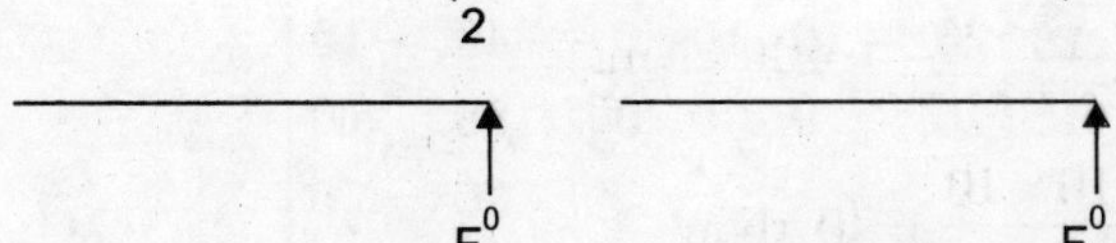

Applying F_1^*, F_2^*, F_3^0 and F_4^0 = unity, one at a time the $[b]$ matrix is obtained and given below.

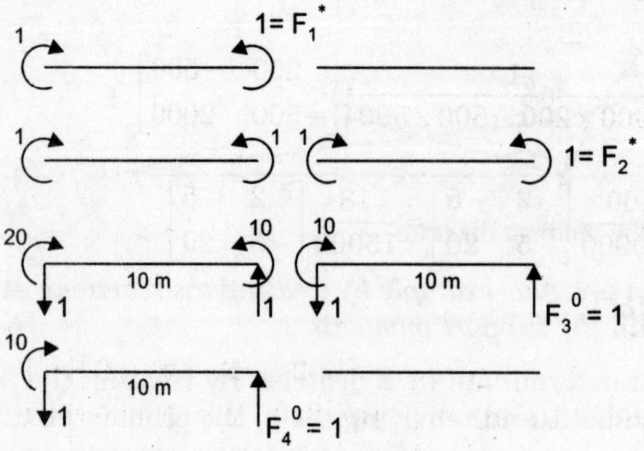

$$[b] = \left[\begin{array}{cc|cc} & [b]^* & [b]^0 & \\ 1 & 1 & 20 & 10 \\ -1 & -1 & -10 & 0 \\ 0 & 1 & 10 & 0 \\ 0 & -1 & 0 & 0 \end{array}\right]$$

Element flexibility matrix :

$$[\alpha_i] = \frac{l}{6EI}\begin{bmatrix} 2 & -1 \\ -1 & 2 \end{bmatrix}$$

$$[\alpha_1] = \frac{10}{6\times 3}\begin{bmatrix} 2 & -1 \\ -1 & 2 \end{bmatrix} = \frac{10}{18}\begin{bmatrix} 2 & -1 \\ -1 & 2 \end{bmatrix}$$

$$[\alpha_2] = \frac{10}{6\times 1}\begin{bmatrix} 2 & -1 \\ -1 & 2 \end{bmatrix} = \frac{10}{6}\times\frac{3}{3}\begin{bmatrix} 2 & -1 \\ -1 & 2 \end{bmatrix}$$

$$[\alpha_2] = \frac{10}{18}\begin{bmatrix} 6 & -3 \\ -3 & 6 \end{bmatrix}$$

$$[\alpha] = \begin{bmatrix} [\alpha_1] & 0 \\ 0 & [\alpha_2] \end{bmatrix}$$

$$[\alpha] = \frac{10}{18}\begin{bmatrix} 2 & -1 & 0 & 0 \\ -1 & 2 & 0 & 0 \\ 0 & 0 & 6 & -3 \\ 0 & 0 & -3 & 6 \end{bmatrix}$$

$$[a] = [b]^T [\alpha] [b]$$

$$a_{00} = [b^0]^T [\alpha] [b^0]$$

$$= \begin{bmatrix} 20 & -10 & 10 & 0 \\ 10 & 0 & 0 & 0 \end{bmatrix} \times \frac{10}{18}\begin{bmatrix} 2 & -1 & 0 & 0 \\ -1 & 2 & 0 & 0 \\ 0 & 0 & 6 & -3 \\ 0 & 0 & -3 & 6 \end{bmatrix}\begin{bmatrix} 20 & 10 \\ -10 & 0 \\ 10 & 0 \\ 0 & 0 \end{bmatrix}$$

$$= \frac{10}{18}\begin{bmatrix} 20 & -10 & 10 & 0 \\ 10 & 0 & 0 & 0 \end{bmatrix}\begin{bmatrix} 50 & 20 \\ -40 & -10 \\ 60 & 0 \\ -30 & 0 \end{bmatrix}$$

$$[a_{00}] = \frac{10}{18}\begin{bmatrix} 2000 & 500 \\ 500 & 200 \end{bmatrix}$$

$$[a_{00}]^{-1} = \frac{18}{10} \times \frac{1}{(2000 \times 200 - 500 \times 500)}\begin{bmatrix} 200 & -500 \\ -500 & 2000 \end{bmatrix}$$

$$= \frac{18 \times 100}{10 \times 150000}\begin{bmatrix} 2 & -5 \\ -5 & 20 \end{bmatrix} = \frac{18}{15000}\begin{bmatrix} 2 & -5 \\ -5 & 20 \end{bmatrix}$$

$$[a_{0*}] = [b^0]^T[\alpha][b^*]$$

$$= \begin{bmatrix} 20 & -10 & 10 & 0 \\ 10 & 0 & 0 & 0 \end{bmatrix} \times \frac{10}{18}\begin{bmatrix} 2 & -1 & 0 & 0 \\ -1 & 2 & 0 & 0 \\ 0 & 0 & 6 & -3 \\ 0 & 0 & -3 & 6 \end{bmatrix}\begin{bmatrix} 1 & 1 \\ -1 & -1 \\ 0 & 1 \\ 0 & -1 \end{bmatrix}$$

$$= \frac{10}{18}\begin{bmatrix} 20 & -10 & 10 & 0 \\ 10 & 0 & 0 & 0 \end{bmatrix}\begin{bmatrix} 3 & 3 \\ -3 & -3 \\ 0 & 9 \\ 0 & -9 \end{bmatrix}$$

$$= \frac{10}{18}\begin{bmatrix} 90 & 180 \\ 30 & 30 \end{bmatrix}$$

Redundant forces are given by

$$\{F^0\} = -[a_{00}]^{-1}[a_{0*}]\{F^*\}$$

$$= -\frac{18}{15000}\begin{bmatrix} 2 & -5 \\ -5 & 20 \end{bmatrix} \times \frac{10}{18}\begin{bmatrix} 90 & 180 \\ 30 & 30 \end{bmatrix}\begin{Bmatrix} 113.33 \\ 20 \end{Bmatrix}$$

$$= -\frac{1}{1500}\begin{bmatrix} 30 & 210 \\ 150 & -300 \end{bmatrix}\begin{Bmatrix} 113.33 \\ 20 \end{Bmatrix}$$

$$= -\frac{1}{1500}\begin{Bmatrix} 7599.9 \\ 10999.5 \end{Bmatrix}$$

$$\{F^0\} = \begin{Bmatrix} -5.067 \\ -7.333 \end{Bmatrix}$$

$$\{P\} = [b]\{F\}$$

$$\{P\} = \begin{bmatrix} 1 & 1 & 20 & 10 \\ -1 & -1 & -10 & 0 \\ 0 & 1 & 10 & 0 \\ 0 & -1 & 0 & 0 \end{bmatrix} \begin{Bmatrix} 113.33 \\ 20 \\ -5.067 \\ -7.333 \end{Bmatrix} = \begin{Bmatrix} -41.34 \\ -82.66 \\ -30.67 \\ -20 \end{Bmatrix}$$

Final force $$\{P^f\} = \begin{Bmatrix} -133.33 \\ 133.33 \\ -20.00 \\ +20.00 \end{Bmatrix} + \begin{Bmatrix} -41.34 \\ -82.66 \\ -30.67 \\ -20.00 \end{Bmatrix} = \begin{Bmatrix} -174.67 \\ 50.67 \\ -50.67 \\ 0 \end{Bmatrix}$$

Example 4.19 *Analyse the continuous beam shown in Fig. 4.31 by the flexibility method and draw the bending moment diagram.*

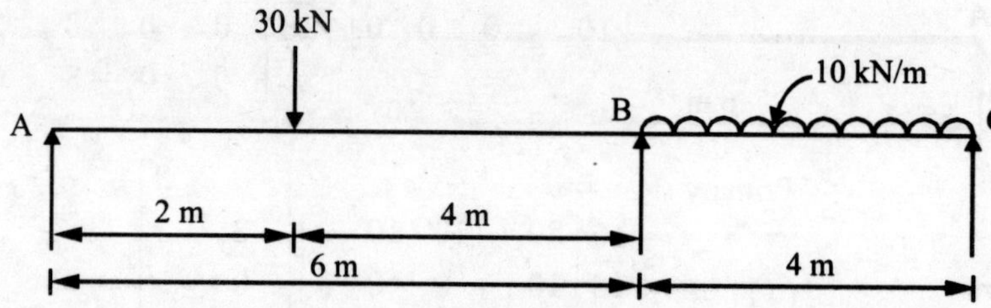

Fig. 4.31

Solution The degree of redundancy is 1.

To make the primary structure, we introduce a hinge at B. The bending moment at B will be treated as the redundant F_3^0. The system co-ordinates and element co-ordinates are shown below.

The {F}* forces are made up of equivalent joint forces arising out of transverse loads on spans AB and BC as tabulated below.

Table 4.2

SPAN	*Fixed and moments*	*Name of moment*	*Equivalent joint force (kNm)*	
AB	$-\frac{Wab^2}{l^2} = -\frac{30 \times 2 \times 4 \times 4}{6 \times 6} = -26.67$	M_{AB}	26.67	
	$\frac{Wa^2b}{l^2} = \frac{30 \times 2 \times 2 \times 4}{6 \times 6} = 13.33$	M_{BA}	− 13.33	0
BC	$-\frac{wl^2}{12} = -\frac{10 \times 4 \times 4}{12} = -13.33$	M_{BC}	13.33	
	$\frac{wl^2}{12} = \frac{10 \times 4 \times 4}{12} = 13.33$	M_{CB}	− 13.33	

since the net force transmitted to joint B happens to be zero, no co-ordinate is assigned to the same. Hence we have

$$\{F\}^* = \begin{Bmatrix} 26.67 \\ -13.33 \end{Bmatrix} \quad \text{and} \quad \{F\} = \begin{Bmatrix} \{F\}^* \\ \cdots\cdots \\ \{F\}^0 \end{Bmatrix}$$

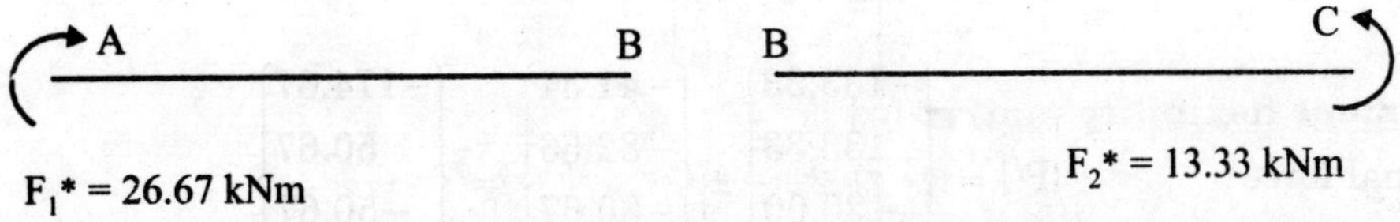

System co-ordinates with equivalent joint loads (F_1^* and F_2^*)

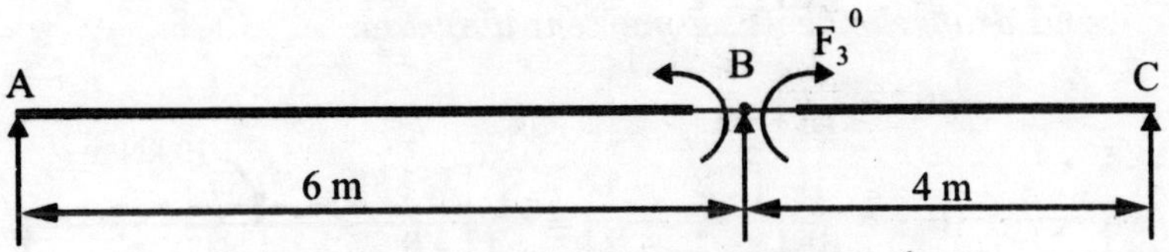

Primary structure with redundant force F_3^0

Element co-ordinates

Formation of [*b*] matrix :

Applying $F_1^* = 1$

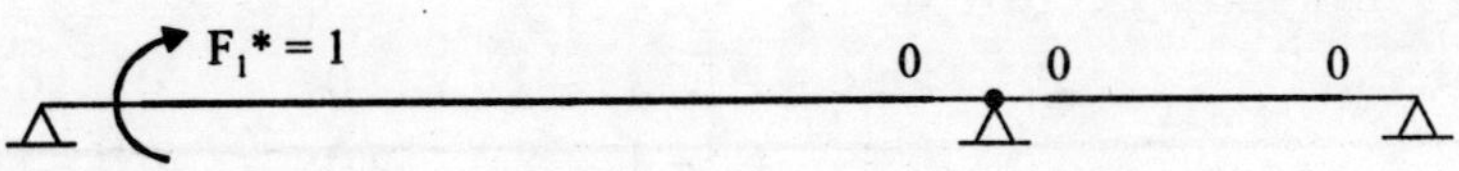

$$\begin{bmatrix} 1 \\ 0 \\ 0 \\ 0 \end{bmatrix}$$

Applying $F_2^* = 1$

$$\begin{bmatrix} 0 \\ 0 \\ 0 \\ 1 \end{bmatrix}$$

Applying $F_3^* = 1$

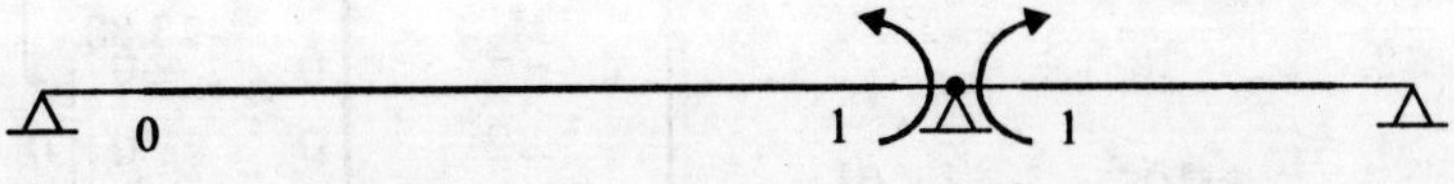

$$\begin{bmatrix} 0 \\ -1 \\ 1 \\ 0 \end{bmatrix}$$

Hence the [*b*] matrix :

$$[b] = \begin{bmatrix} 1 & 0 & 0 \\ 0 & 0 & -1 \\ 0 & 0 & +1 \\ 0 & 1 & 0 \end{bmatrix} \quad (\text{columns: } * \mid 0)$$

Element flexibility matrix [α] :

$$[\alpha_i] = \frac{l}{6EI}\begin{bmatrix} 2 & -1 \\ -1 & 2 \end{bmatrix}$$

$$[\alpha_1] = \frac{6}{6EI}\begin{bmatrix} 2 & -1 \\ -1 & 2 \end{bmatrix}$$

$$[\alpha_1] = \frac{1}{EI}\begin{bmatrix} 2 & -1 \\ -1 & 2 \end{bmatrix}$$

$$[\alpha_2] = \frac{4}{6EI}\begin{bmatrix} 2 & -1 \\ -1 & 2 \end{bmatrix} = \frac{1}{EI}\begin{bmatrix} 1.33 & -0.67 \\ -0.67 & 1.33 \end{bmatrix}$$

Assembled element flexibility matrix

$$[\alpha] = \frac{1}{EI}\begin{bmatrix} 2 & -1 & 0 & 0 \\ -1 & 2 & 0 & 0 \\ 0 & 0 & 1.33 & -0.67 \\ 0 & 0 & -0.67 & 1.33 \end{bmatrix}$$

To find redundant forces

$$\{F\}^0 = -[a^{00}]^{-1} \quad [a^{0*}] \quad \{F\}^* \qquad \{F\}^* = \begin{Bmatrix} 26.67 \\ -13.33 \end{Bmatrix}$$

$$[a_{00}] = [b^0]^T \quad [\alpha] \quad [b^0]$$

$$[a_{0*}] = [b^0]^T \quad [\alpha] \quad [b^*]$$

$$[a_{00}] = [0 \ \ -1 \ \ 1 \ \ 0] \times \frac{1}{EI}\begin{bmatrix} 2 & -1 & 0 & 0 \\ -1 & 2 & 0 & 0 \\ 0 & 0 & 1.33 & -0.67 \\ 0 & 0 & -0.67 & 1.33 \end{bmatrix}\begin{bmatrix} 0 \\ -1 \\ 1 \\ 0 \end{bmatrix}$$

$$= \frac{1}{EI}[1 \ \ -2 \ \ 1.33 \ \ -0.67]\begin{bmatrix} 0 \\ -1 \\ 1 \\ 0 \end{bmatrix} = \frac{1}{EI}[2 + 1.33]$$

$$[a_{00}] = \frac{3.33}{EI} \ ; [a_{00}]^{-1} = \frac{EI}{3.33}$$

$$[a_{0*}] = [b^0]^T \quad [\alpha] \quad [b^*]$$

$$= [0 \ \ -1 \ \ 1 \ \ 0] \times \frac{1}{EI}\begin{bmatrix} 2 & -1 & 0 & 0 \\ -1 & 2 & 0 & 0 \\ 0 & 0 & 1.33 & -0.67 \\ 0 & 0 & -0.67 & 1.33 \end{bmatrix}\begin{bmatrix} 1 & 0 \\ 0 & 0 \\ 0 & 0 \\ 0 & 1 \end{bmatrix}$$

$$[a_{0*}] = \frac{1}{EI}[1 \quad -2 \quad 1.33 \quad -0.67]\begin{bmatrix} 1 & 0 \\ 0 & 0 \\ 0 & 0 \\ 0 & 1 \end{bmatrix}$$

$$[a_{0*}] = \frac{1}{EI}[1 \quad -0.67]$$

$$\{F\}^0 = -\left[\frac{EI}{3.33}\right] \times \frac{1}{EI}[1 \quad -0.67]\begin{Bmatrix} 26.67 \\ -13.33 \end{Bmatrix}$$

$$= -\frac{1.0}{3.33} \times 35.60 = -10.69$$

$$\{F\}^0 = -10.69 \text{ kNm}$$

Element force $\{P\} = [b] \quad \{F\}$

$$\{P\} = \begin{bmatrix} 1 & 0 & 0 \\ 0 & 0 & -1 \\ 0 & 0 & +1 \\ 0 & 1 & 0 \end{bmatrix}\begin{Bmatrix} 26.67 \\ -13.33 \\ -10.69 \end{Bmatrix} = \begin{Bmatrix} 26.67 \\ 10.69 \\ -10.69 \\ -13.33 \end{Bmatrix} \text{kNm}$$

$$\{P\}^f = \{P\} - \{P\}^e = \begin{Bmatrix} 26.67 \\ 10.69 \\ -10.69 \\ -13.33 \end{Bmatrix} - \begin{Bmatrix} 26.67 \\ -13.33 \\ +13.33 \\ -13.33 \end{Bmatrix} = \begin{Bmatrix} 0 \\ 24.02 \\ -24.02 \\ 0 \end{Bmatrix}$$

Simply supported moments :

Span AB : $\frac{Wab}{l} = \frac{30 \times 2 \times 4}{6} = 40$ kNm

Span BC : $\frac{wl^2}{8} = \frac{10 \times 4^2}{8} = 20$ kNm.

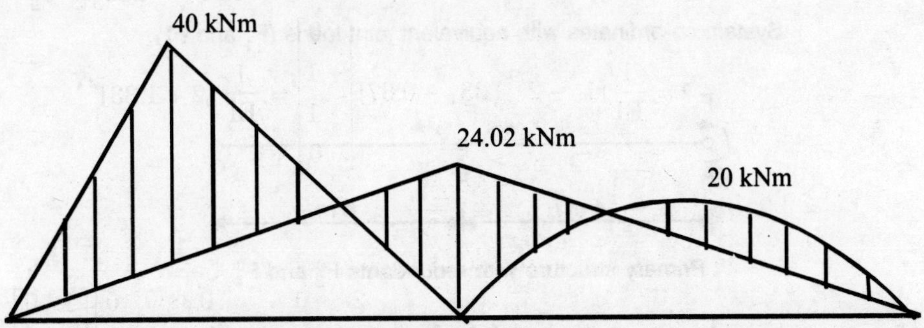

Bending moment diagram

Example 4.20 *Analyse the continuous beam loaded as shown in Fig. 4.32 by the matrix flexibility method and sketch the BMD.*

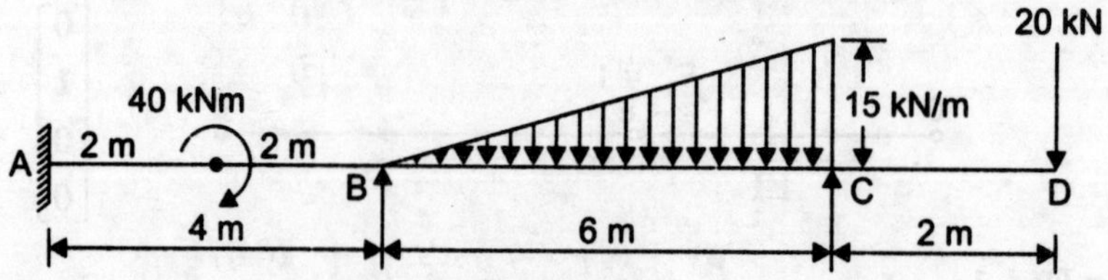

Fig. 4.32

Solution The degree of static indeterminacy is 2.

To make primary structure, we introduce hinges at A and B. The bending moments at A and B are treated as redundants. The system and element co-ordinates are shown below. Table given below summarizes the equivalent joint loads due to transverse loads on span AB, BC and CD.

SPAN	*Fixed end moments*	*Name of moment*	*Equivalent joint forces (kNm)*
AB	$\frac{M}{4} = \frac{40}{4} = 10$ kNm	M_{AB}	– 10 kNm
	$\frac{M}{4} = \frac{40}{4} = 10$ kNm	M_{BA}	– 10 kNm
BC	$-\frac{wl^2}{30} = -\frac{15 \times 6^2}{30} = -18$ kNm	M_{BC}	+ 18 kNm
	$\frac{wl^2}{20} = \frac{15 \times 6^2}{30} = 27$ kNm	M_{CB}	– 27 kNm
CD	$-20 \times 2 = -40$ kNm	M_{CD}	40 kNm

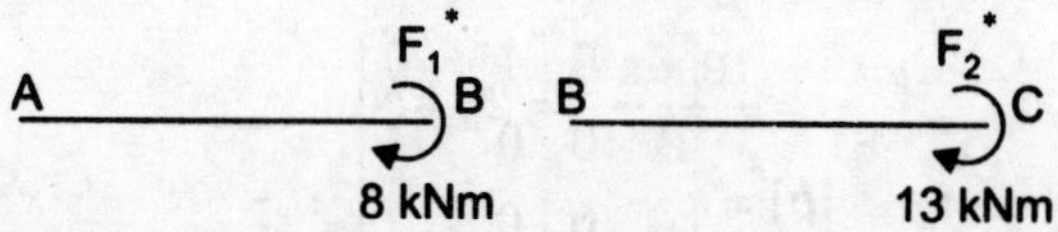

System co-ordinates with equivalent joint loads (F_1^* and F_2^*)

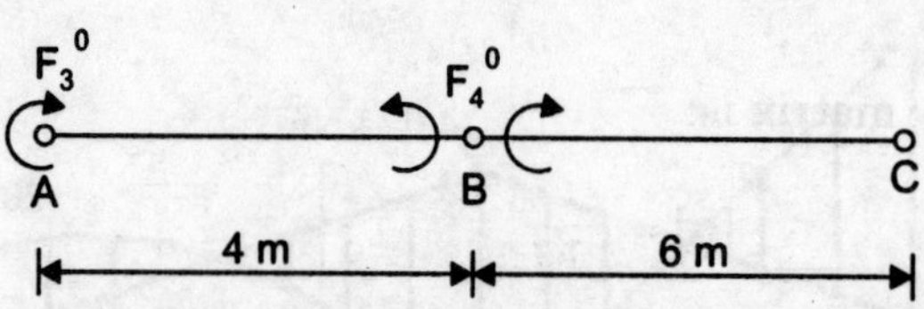

Primary structure with redundants F_3^0 and F_4^0

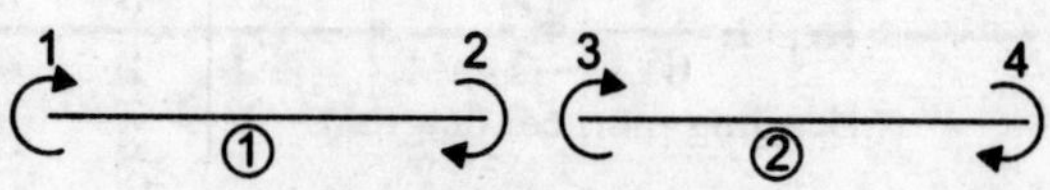

Element co-ordinates

Formation of [*b*] matrix :

Applying $F_1^* = 1$

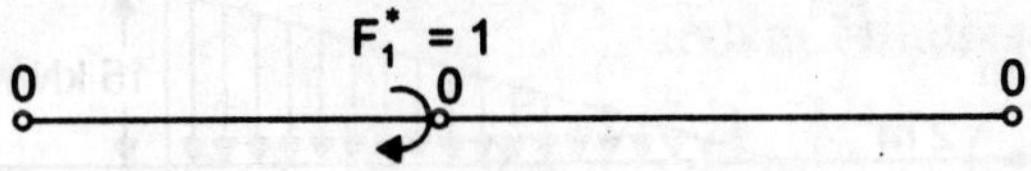

$$\begin{bmatrix} 0 \\ 1 \\ 0 \\ 0 \end{bmatrix}$$

Applying $F_2^* = 1$

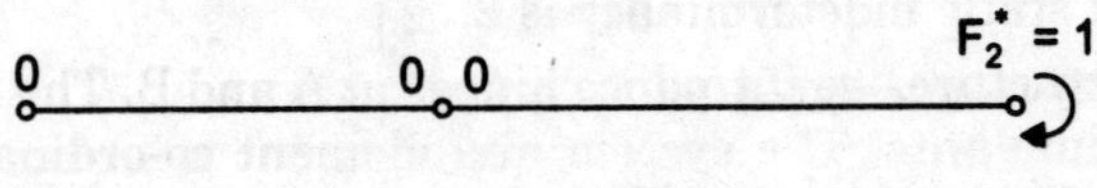

$$\begin{bmatrix} 0 \\ 0 \\ 0 \\ 1 \end{bmatrix}$$

Applying $F_3^0 = 1$

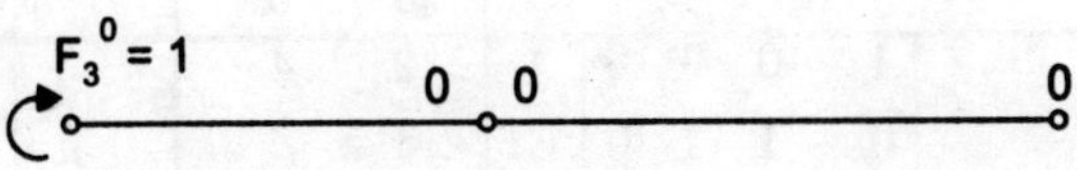

$$\begin{bmatrix} 1 \\ 0 \\ 0 \\ 0 \end{bmatrix}$$

Applying $F_4^0 = 1$

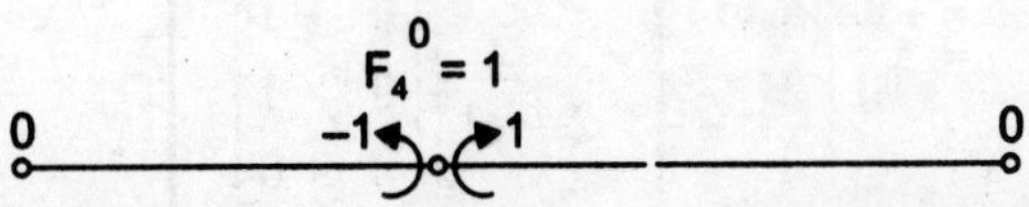

$$\begin{bmatrix} 0 \\ -1 \\ 1 \\ 0 \end{bmatrix}$$

Hence *b* matrix is given by

$$[b] = \left[\begin{array}{cc|cc} 0 & 0 & 1 & 0 \\ 1 & 0 & 0 & -1 \\ 0 & 0 & 0 & 1 \\ 0 & 1 & 0 & 0 \end{array}\right]$$

$$\qquad\qquad b^* \qquad\quad b^0$$

Element flexibility matrix is:

$$[\alpha] = \frac{l}{6EI}\begin{bmatrix} 2 & -1 \\ -1 & 2 \end{bmatrix}$$

$$[\alpha_1] = \frac{4}{6EI}\begin{bmatrix} 2 & -1 \\ -1 & 2 \end{bmatrix} = \frac{1}{EI}\begin{bmatrix} \frac{4}{3} & -\frac{2}{3} \\ -\frac{2}{3} & \frac{4}{3} \end{bmatrix}$$

$$[\alpha_2] = \frac{6}{6EI}\begin{bmatrix} 2 & -1 \\ -1 & 2 \end{bmatrix} = \frac{1}{EI}\begin{bmatrix} 2 & -1 \\ -1 & 2 \end{bmatrix}$$

Assembled element flexibility matrix

$$[\alpha] = \frac{1}{EI}\begin{bmatrix} \frac{4}{3} & -\frac{2}{3} & 0 & 0 \\ -\frac{2}{3} & \frac{4}{3} & 0 & 0 \\ 0 & 0 & 2 & -1 \\ 0 & 0 & -1 & 2 \end{bmatrix}$$

Redundant forces $\{F^0\}$ is calculated by

$$\{F^0\} = -[a_{22}^{-1}][a_{21}]\{F^*\}$$

$$[a_{22}] = [b^0]^T [\alpha] [b^0]$$

$$= \begin{bmatrix} 1 & 0 & 0 & 0 \\ 0 & -1 & 1 & 0 \end{bmatrix} \frac{1}{EI}\begin{bmatrix} \frac{4}{3} & -\frac{2}{3} & 0 & 0 \\ -\frac{2}{3} & \frac{4}{3} & 0 & 0 \\ 0 & 0 & 2 & -1 \\ 0 & 0 & -1 & 2 \end{bmatrix}\begin{bmatrix} 1 & 0 \\ 0 & -1 \\ 0 & 1 \\ 0 & 0 \end{bmatrix}$$

$$= \frac{1}{EI}\begin{bmatrix} \frac{4}{3} & -\frac{2}{3} & 0 & 0 \\ \frac{2}{3} & -\frac{4}{3} & 2 & -1 \end{bmatrix}\begin{bmatrix} 1 & 0 \\ 0 & -1 \\ 0 & 1 \\ 0 & 0 \end{bmatrix}$$

$$[a_{22}] = \frac{1}{EI}\begin{bmatrix} \frac{4}{3} & \frac{2}{3} \\ \frac{2}{3} & \frac{10}{3} \end{bmatrix}$$

$$\therefore \qquad [a_{22}]^{-1} = \frac{EI}{4}\begin{bmatrix} \frac{10}{3} & -\frac{2}{3} \\ -\frac{2}{3} & \frac{4}{3} \end{bmatrix}$$

$$[a_{21}] = [b^0]^T [\alpha] [b^*]$$

$$= \begin{bmatrix} 1 & 0 & 0 & 0 \\ 0 & -1 & 1 & 0 \end{bmatrix} \frac{1}{EI}\begin{bmatrix} \frac{4}{3} & -\frac{2}{3} & 0 & 0 \\ -\frac{2}{3} & \frac{4}{3} & 0 & 0 \\ 0 & 0 & 2 & -1 \\ 0 & 0 & -1 & 2 \end{bmatrix}\begin{bmatrix} 0 & 0 \\ 1 & 0 \\ 0 & 0 \\ 0 & 1 \end{bmatrix}$$

$$\therefore \quad [a_{21}] = \frac{1}{EI}\begin{bmatrix} \frac{4}{3} & -\frac{2}{3} & 0 & 0 \\ \frac{2}{3} & -\frac{4}{3} & 2 & -1 \end{bmatrix}\begin{bmatrix} 0 & 0 \\ 1 & 0 \\ 0 & 0 \\ 0 & 1 \end{bmatrix}$$

$$\therefore \quad [a_{21}] = \frac{1}{EI}\begin{bmatrix} -\frac{2}{3} & 0 \\ -\frac{4}{3} & -1 \end{bmatrix}$$

$$\therefore \quad \{F_0\} = -[a_{22}^{-1}][a_{21}]\{F^*\}$$

$$= -\frac{EI}{4}\begin{bmatrix} \frac{10}{3} & -\frac{2}{3} \\ -\frac{2}{3} & \frac{4}{3} \end{bmatrix} \times \frac{1}{EI}\begin{bmatrix} -\frac{2}{3} & 0 \\ -\frac{4}{3} & -1 \end{bmatrix}\begin{bmatrix} 8 \\ 13 \end{bmatrix} = \begin{bmatrix} 0.5 \\ 7.0 \end{bmatrix}$$

Element forces $\{P\} = [b]\{F\}$

$$= \begin{bmatrix} 0 & 0 & 1 & 0 \\ 1 & 0 & 0 & -1 \\ 0 & 0 & 0 & 1 \\ 0 & 1 & 0 & 0 \end{bmatrix}\begin{bmatrix} 8 \\ 13 \\ 0.5 \\ 7.0 \end{bmatrix} = \begin{bmatrix} 0.5 \\ 1 \\ 7 \\ 13 \end{bmatrix}$$

Final forces $[P]^f = [P] - [P^{(e)}]$

$$= \begin{bmatrix} 0.5 \\ 1.0 \\ 7.0 \\ 13.0 \end{bmatrix} - \begin{bmatrix} -10 \\ -10 \\ 18 \\ -27 \end{bmatrix} = \begin{bmatrix} 10.5 \\ 11 \\ -11 \\ 40 \end{bmatrix}$$

Simply supported moments: (Please Refer to example 3.10)

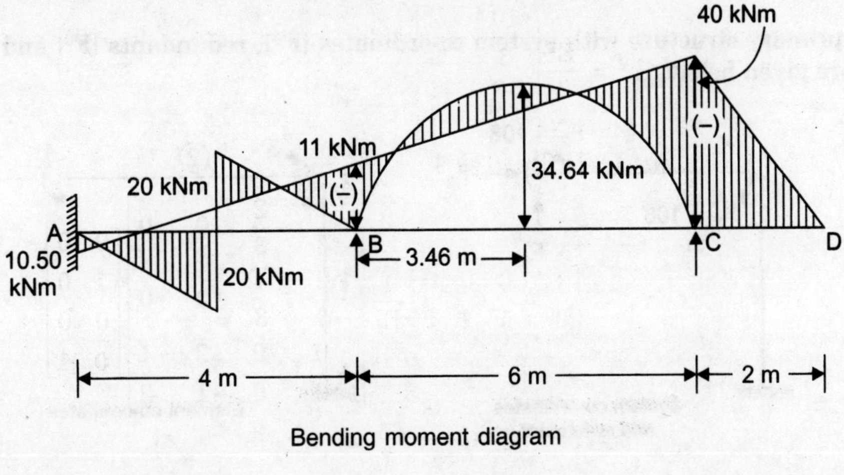

Bending moment diagram

Example 4.21 *Analyse the given frame using the matrix flexibility method.*

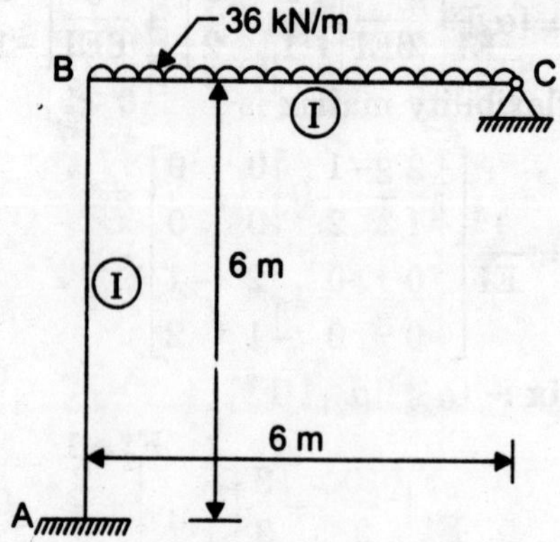

Fig. 4.33

Solution The degree of static indeterminacy = 2.

Let us make the given frame to a primary structure by removing V_C and H_C treating them as redundants.

Equivalent joint loads are calculated as follows :

$$M_{FAB} = M_{FBA} = 0$$

$$M_{FBC} = -\frac{wl^2}{12} = -\frac{36 \times 6^2}{12} = -108 \text{ kNm}$$

$$M_{FCB} = \frac{36 \times 6^2}{12} = 108 \text{ kNm}$$

$$\therefore \quad \{F\} = \begin{bmatrix} F_1^* \\ F_2^* \\ F_3^0 \\ F_4^0 \end{bmatrix}$$

The primary structure with system co-ordinates {F*}, redundants {F⁰} and element co-ordinates are given below.

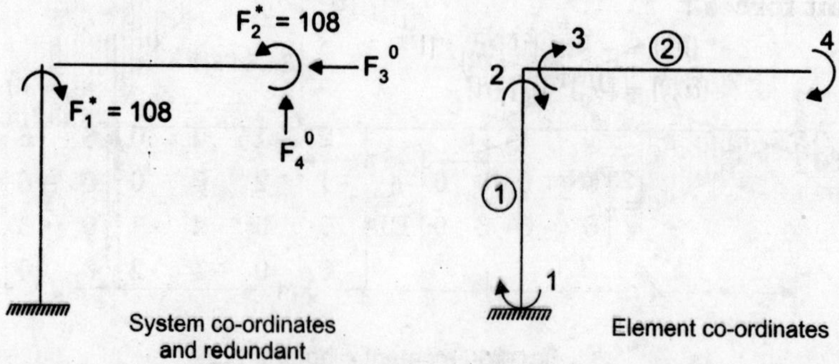

Element flexibility matrix :

$$[\alpha_1] = [\alpha_2] = \frac{l}{6EI}\begin{bmatrix} 2 & -1 \\ -1 & 2 \end{bmatrix} = \frac{6}{6EI}\begin{bmatrix} 2 & -1 \\ -1 & 2 \end{bmatrix} = \frac{1}{EI}\begin{bmatrix} 2 & -1 \\ -1 & 2 \end{bmatrix}$$

∴ **Assembled element flexibility matrix is**

$$[\alpha] = \frac{1}{EI}\begin{bmatrix} 2 & -1 & 0 & 0 \\ -1 & 2 & 0 & 0 \\ 0 & 0 & 2 & -1 \\ 0 & 0 & -1 & 2 \end{bmatrix}$$

Formation of [*b*] matrix :

$F_1^* = 1$

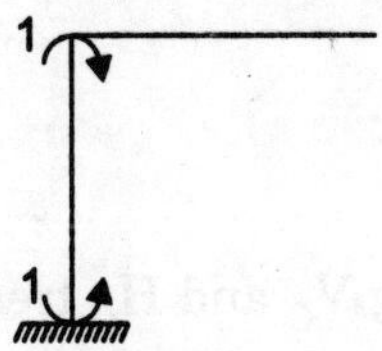

$$\begin{bmatrix} -1 \\ 1 \\ 0 \\ 0 \end{bmatrix}$$

$F_2^* = 1$

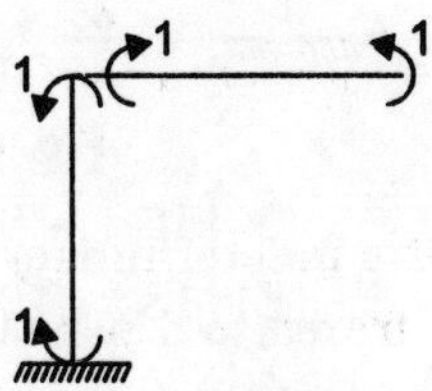

$$\begin{bmatrix} 1 \\ -1 \\ 1 \\ -1 \end{bmatrix}$$

$F_3^0 = 1$

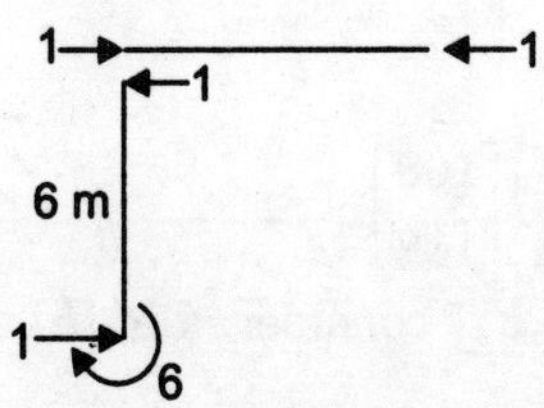

$$\begin{bmatrix} 6 \\ 0 \\ 0 \\ 0 \end{bmatrix}$$

$F_4^0 = 1$

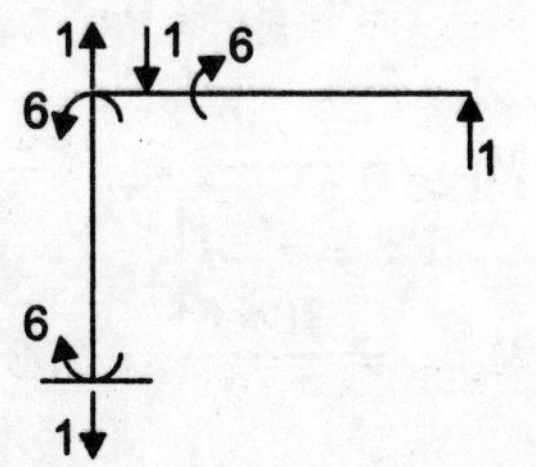

$$\begin{bmatrix} 6 \\ -6 \\ 6 \\ 0 \end{bmatrix}$$

∴ [*b*] matrix is given by

$$[b] = \left[\begin{array}{cc|cc} -1 & 1 & 6 & 6 \\ 1 & -1 & 0 & -6 \\ 0 & 1 & 0 & 6 \\ 0 & -1 & 0 & +0 \end{array}\right] \quad \begin{matrix} b^* & b^0 \end{matrix}$$

Redundant forces :

$$\{F^0\} = -[a_{22}]^{-1}[a_{21}]\{F^*\}$$

$$[a_{22}] = [b_0]^T[\alpha][b]$$

$$= \begin{bmatrix} 6 & 0 & 0 & 0 \\ 6 & -6 & 6 & 0 \end{bmatrix} \frac{1}{EI}\begin{bmatrix} 2 & -1 & 0 & 0 \\ -1 & 2 & 0 & 0 \\ 0 & 0 & 2 & -1 \\ 0 & 0 & -1 & 2 \end{bmatrix}\begin{bmatrix} 6 & 6 \\ 0 & -6 \\ 0 & 6 \\ 0 & 0 \end{bmatrix}$$

$$= \frac{1}{EI}\begin{bmatrix} 12 & -6 & 0 & 0 \\ 18 & -18 & 12 & -6 \end{bmatrix}\begin{bmatrix} 6 & 6 \\ 0 & -6 \\ 0 & 6 \\ 0 & 0 \end{bmatrix}$$

$$[a_{22}] = \frac{1}{EI}\begin{bmatrix} 72 & 108 \\ 108 & 288 \end{bmatrix}$$

$$[a_{21}] = [b_0]^T [\alpha] [b^*]$$

$$= \begin{bmatrix} 6 & 0 & 0 & 0 \\ 6 & -6 & 6 & 0 \end{bmatrix} \frac{1}{EI}\begin{bmatrix} 2 & -1 & 0 & 0 \\ -1 & 2 & 0 & 0 \\ 0 & 0 & 2 & -1 \\ 0 & 0 & -1 & 2 \end{bmatrix}\begin{bmatrix} -1 & 1 \\ 1 & -1 \\ 0 & 1 \\ 0 & -1 \end{bmatrix}$$

$$= \frac{1}{EI}\begin{bmatrix} 12 & -6 & 0 & 0 \\ 18 & -18 & 12 & -6 \end{bmatrix}\begin{bmatrix} -1 & 1 \\ 1 & -1 \\ 0 & 1 \\ 0 & -1 \end{bmatrix}$$

$$[a_{21}] = \frac{1}{EI}\begin{bmatrix} -18 & 18 \\ -36 & 54 \end{bmatrix}$$

∴ Redundant forces

$$\{F^0\} = -EI\begin{bmatrix} 72 & 108 \\ 108 & 288 \end{bmatrix}^{-1} \frac{1}{EI}\begin{bmatrix} -18 & 18 \\ -36 & 54 \end{bmatrix}\begin{bmatrix} 108 \\ 108 \end{bmatrix}$$

(Sign of F_1* and F_2* considered in [b] matrix)

$$= -\frac{1}{9072}\begin{bmatrix} 288 & -108 \\ -108 & 72 \end{bmatrix}\begin{bmatrix} 0 \\ 1944 \end{bmatrix}$$

$$\{F_0\} = \begin{Bmatrix} 23.14 \\ -15.43 \end{Bmatrix}$$

Element forces :

$$\{P\} = [b]\{F\} = [b^* \mid b^0]\begin{Bmatrix} F^* \\ F^0 \end{Bmatrix}$$

$$= \begin{bmatrix} -1 & 1 & 6 & 6 \\ 1 & -1 & 0 & -6 \\ 0 & 1 & 0 & 6 \\ 0 & -1 & 0 & 0 \end{bmatrix}\begin{bmatrix} 108 \\ 108 \\ 23.14 \\ -15.43 \end{bmatrix} = \begin{Bmatrix} 46.26 \\ 92.58 \\ 15.42 \\ -108 \end{Bmatrix}$$

Final forces

$$\{P^f\} = \{P\} + \{\text{Fixed end moments}\}$$

$$= \begin{Bmatrix} 46.26 \\ 92.58 \\ 15.42 \\ -108 \end{Bmatrix} + \begin{Bmatrix} 0 \\ 0 \\ -108 \\ 108 \end{Bmatrix}$$

$$\begin{Bmatrix} M_{AB} \\ M_{BA} \\ M_{BC} \\ M_{CB} \end{Bmatrix} = \begin{Bmatrix} 46.26 \\ 92.58 \\ -92.58 \\ 0 \end{Bmatrix}$$

Simply supported moments :

Span BC : $\dfrac{wl^2}{8} = \dfrac{36 \times 6^2}{8} = 162$ kNm

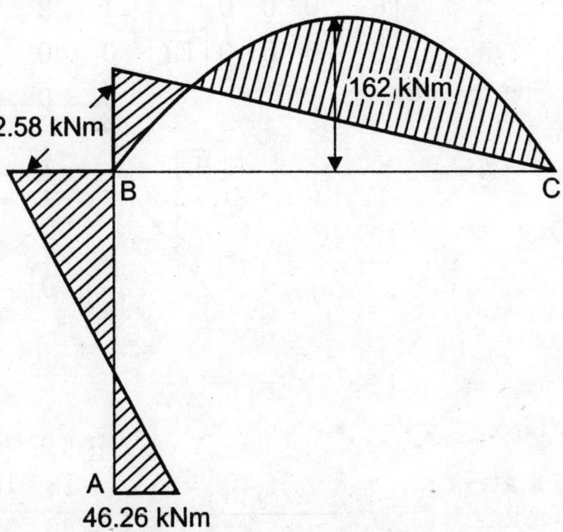

Fig. 4.34. Bending moment diagram.

Example 4.22 *Analyse the frame shown in Fig. 4.35 by the matrix flexibility method.*

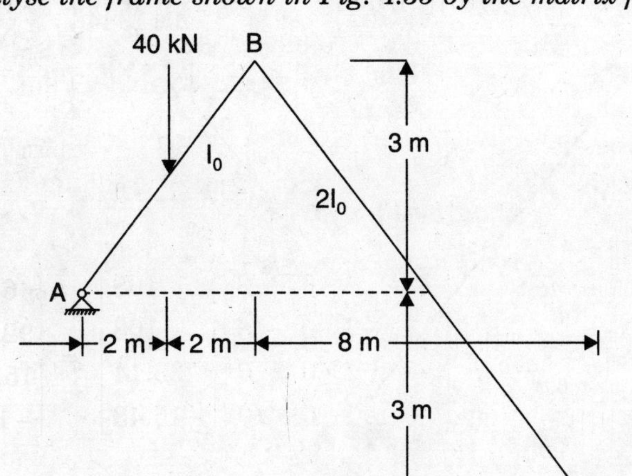

Fig. 4.35

Solution The frame has a redundancy of 1. Treating the horizontal reaction at C as redundant, the primary structure is obtained by placing C on rollers.

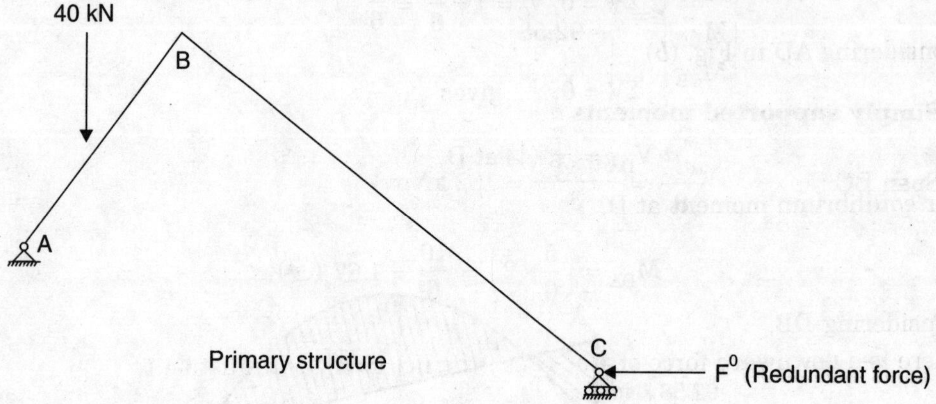

Primary structure

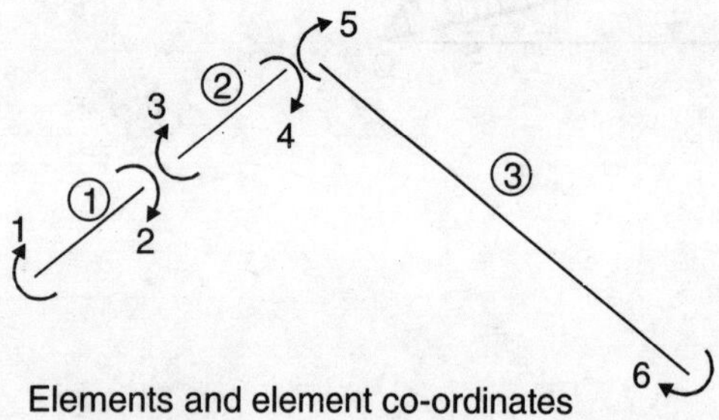

Elements and element co-ordinates

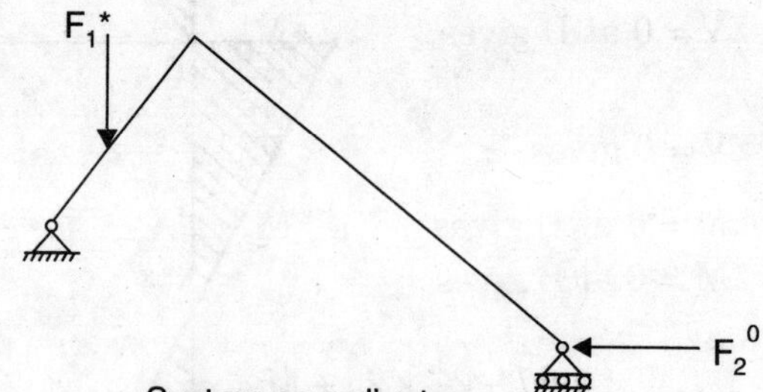

System co-ordinates

Formation of [*b*] matrix :

Applying $F_1^* = 1$

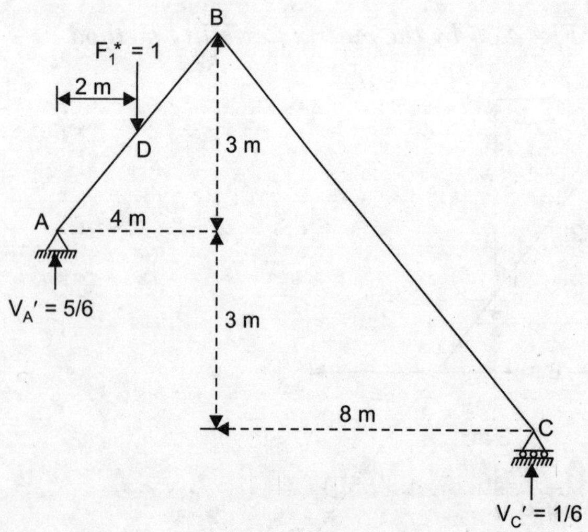

Fig. (*a*)

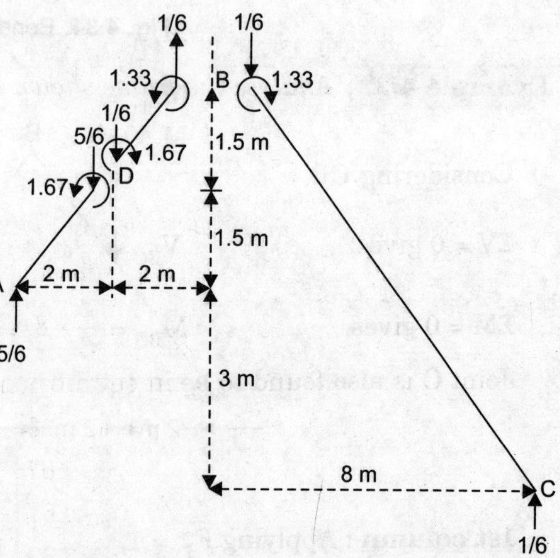

Fig. (*b*)

$$(V_A' \times 12) - (1 \times 10) = 0$$

$$V_A' = \frac{10}{12} = \frac{5}{6}$$

$$\Sigma V = 0,\ V_C' = 1 - \frac{5}{6} = \frac{1}{6}$$

Considering AD in Fig. (*b*)

$$\Sigma V = 0, \quad \text{gives}$$

$$V_{DA} = \frac{5}{6}\ (\downarrow) \text{ at D}$$

For equilibrium moment at D,

$$M_{DA} = \left(\frac{5}{6} \times 2\right) = \frac{10}{6} = 1.67\ (\circlearrowleft)$$

Considering DB,

There is a downward force of 1 at D *i.e.*, the net force at D must be 1

$\Sigma V = 0$ at D gives, $\quad V_{DB} = 1 - \frac{5}{6} = \frac{1}{6}\ (\downarrow)$

$\Sigma V = 0$ gives, $\quad V_{BD} = \frac{1}{6}\ (\uparrow)$ at B

$\Sigma M = 0$ at D gives, $\quad M_{DB} = 1.67\ (\circlearrowleft)$

$\Sigma M = 0$ at B gives

$$1.67 - \left(\frac{1}{6} \times 2\right) + M_{BD} = 0$$

$$(\circlearrowright) \qquad (\circlearrowleft) \qquad (\circlearrowright)$$

$$M_{BD} = \frac{2}{6} - 1.67$$

$$M_{BD} = -1.33$$

i.e., $\quad M_{BD} = 1.33\ (\circlearrowleft)$

Considering CB,

$\Sigma V = 0$ gives, $\quad V_{BC} = \frac{1}{6}\ (\downarrow)$

$\Sigma M = 0$ gives, $\quad M_{BC} = \frac{1}{6} \times 8 = 1.33\ (\circlearrowright)$

Joint C is also found to be in equilibrium.

1st column : Applying $F_1^* = 1$, $\begin{bmatrix} 0 \\ -1.67 \\ 1.67 \\ -1.33 \\ 1.33 \\ 0 \end{bmatrix}$ is the first column of [*b*]

2nd column: Applying $F_2^0 = 1$

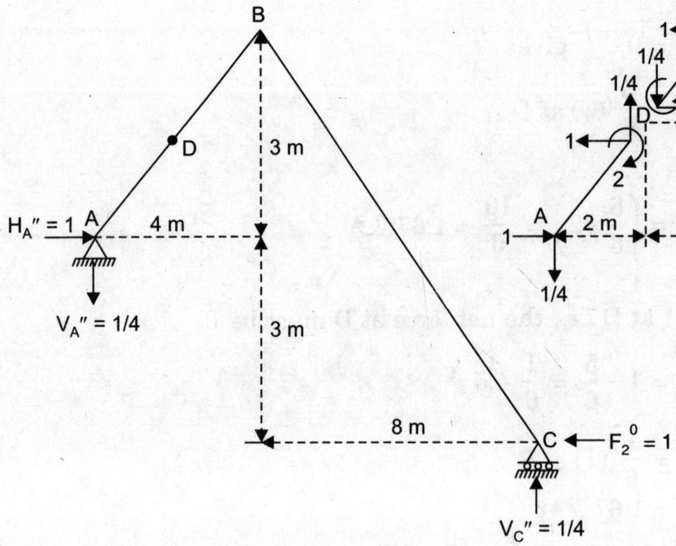

Fig. (*c*)

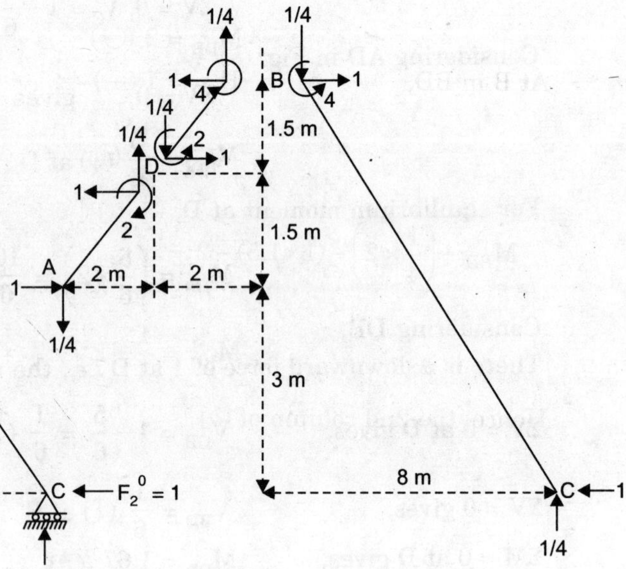

Fig. (*d*)

$\Sigma H = 0$ gives

$$H_A'' = 1\ (\rightarrow)$$

The two horizontal forces will cause a moment 1 × 3 = 3 kN. (↻)

To make $\Sigma M = 0$, there will be equal and opposite vertical reactions at A and C, whose value will be $\dfrac{1 \times 3}{12} = \dfrac{1}{4}$

Considering Fig. (*d*), on AD

$$H_{DA} = 1\ (\leftarrow)$$

$$V_{DA} = \frac{1}{4}\ (\uparrow)$$

$\Sigma M_{DA} = 0$ gives

$$\left(-\frac{1}{4} \times 2\right) - (1 \times 1.5) + M_{DA} = 0$$

$$M_{DA} = \frac{1}{2} + 1.5 = 2\ (\curvearrowright)$$

Considering DB at D, the reactions at D will be as shown in Fig. (*d*)

$$H_{DB} = 1\ (\rightarrow)$$

$$V_{DB} = \frac{1}{4}\ (\downarrow)$$

$$M_{DB} = 2\ (\circlearrowleft)$$

At B in BD, $H_{BD} = 1\ (\leftarrow)$

$$V_{BD} = \frac{1}{4}\ (\uparrow)$$

$$M_{BD} - \left(\frac{1}{4}\times 2\right) - (1\times 1.5) - 2 = 0$$

$$M_{BD} = \frac{1}{2} + 1.5 + 2 = 4\ (\circlearrowright)$$

Hence, the 2nd column of $[b]$ is $[0 \quad 2 \quad -2 \quad 4 \quad -4 \quad 0]^T$

$$\therefore \qquad [b] = \overset{[b]^* \quad [b]^0}{\begin{bmatrix} 0 & 0 \\ -1.67 & 2 \\ 1.67 & -2 \\ -1.33 & 4 \\ 1.33 & -4 \\ 0 & 0 \end{bmatrix}}$$

Element flexibility matrix [α]:

$$[\alpha_i] = \frac{l}{6EI}\begin{bmatrix} 2 & -1 \\ -1 & 2 \end{bmatrix}$$

Element 1: $$[\alpha_1] = \frac{2.5}{6EI}\begin{bmatrix} 2 & -1 \\ -1 & 2 \end{bmatrix} = \frac{1}{6EI}\begin{bmatrix} 5 & -2.5 \\ -2.5 & 5 \end{bmatrix}$$

Element 2: $$[\alpha_2] = \frac{2.5}{6EI}\begin{bmatrix} 2 & -1 \\ -1 & 2 \end{bmatrix} = \frac{1}{6EI}\begin{bmatrix} 5 & -2.5 \\ -2.5 & 5 \end{bmatrix}$$

Element 3: $$[\alpha_3] = \frac{10}{6E\times 2I}\begin{bmatrix} 2 & -1 \\ -1 & 2 \end{bmatrix} = \frac{1}{6EI}\begin{bmatrix} 10 & -5 \\ -5 & 10 \end{bmatrix}$$

Assembled element flexibility matrix :

$$[\alpha] = \begin{bmatrix} \alpha_1 & 0 & 0 \\ 0 & \alpha_2 & 0 \\ 0 & 0 & \alpha_3 \end{bmatrix}$$

$$[\alpha] = \frac{1}{6EI}\begin{bmatrix} 5 & -2.5 & 0 & 0 & 0 & 0 \\ -2.5 & 5 & 0 & 0 & 0 & 0 \\ 0 & 0 & 5 & -2.5 & 0 & 0 \\ 0 & 0 & -2.5 & 5 & 0 & 0 \\ 0 & 0 & 0 & 0 & 10 & -5 \\ 0 & 0 & 0 & 0 & -5 & 10 \end{bmatrix}$$

To find the redundant force :

$$\{F^0\} = -[a_{00}]^{-1} \quad [a_{0*}] \quad \{F\}^*$$

$$[a_{00}] = [b^0]^T \quad [\alpha] \quad [b^0]$$

$$= [0 \quad 2 \quad -2 \quad 4 \quad -4 \quad 0] \times \frac{1}{6EI}\begin{bmatrix} 5 & -2.5 & 0 & 0 & 0 & 0 \\ -2.5 & 5 & 0 & 0 & 0 & 0 \\ 0 & 0 & 5 & -2.5 & 0 & 0 \\ 0 & 0 & -2.5 & 5 & 0 & 0 \\ 0 & 0 & 0 & 0 & 10 & -5 \\ 0 & 0 & 0 & 0 & -5 & 10 \end{bmatrix}\begin{bmatrix} 0 \\ 2 \\ -2 \\ 4 \\ -4 \\ 0 \end{bmatrix}$$

$$[a_{00}] = \frac{320}{6EI}$$

$$[a_{00}]^{-1} = \frac{6EI}{320}$$

$$[a_{0*}] = [b^0]^T \quad [\alpha] \quad [b^*]$$

$$= [0 \quad 2 \quad -2 \quad 4 \quad -4 \quad 0] \times \frac{1}{6EI}\begin{bmatrix} 5 & -2.5 & 0 & 0 & 0 & 0 \\ -2.5 & 5 & 0 & 0 & 0 & 0 \\ 0 & 0 & 5 & -2.5 & 0 & 0 \\ 0 & 0 & -2.5 & 5 & 0 & 0 \\ 0 & 0 & 0 & 0 & 10 & -5 \\ 0 & 0 & 0 & 0 & -5 & 10 \end{bmatrix}\begin{bmatrix} 0 \\ -1.67 \\ 1.67 \\ -1.33 \\ 1.33 \\ 0 \end{bmatrix}$$

$$[a_{0*}] = \frac{-23.9}{EI}$$

$$\{F^0\} = -[a_{00}]^{-1} \quad [a_{0*}] \quad \{F^*\}$$

$$\{F^0\} = \frac{-EI}{53.33} \times \frac{(-23.9)}{EI} \times 40 = 17.93 \text{ kN}$$

Element forces : $\{P\} = [b] \quad \{F\}$

$$\begin{Bmatrix} P_1 \\ P_2 \\ P_3 \\ P_4 \\ P_5 \\ P_6 \end{Bmatrix} = \begin{bmatrix} 0 & 0 \\ -1.67 & 2 \\ 1.67 & -2 \\ -1.33 & 4 \\ 1.33 & -4 \\ 0 & 0 \end{bmatrix}\begin{Bmatrix} 40 \\ 17.93 \end{Bmatrix} = \begin{Bmatrix} 0 \\ -30.94 \\ 30.94 \\ 18.52 \\ -18.52 \\ 0 \end{Bmatrix}$$

Example 4.23 *Analyse the portal frame loaded as shown in Fig. 4.36 by the matrix flexibility method and sketch the bending moment diagram.*

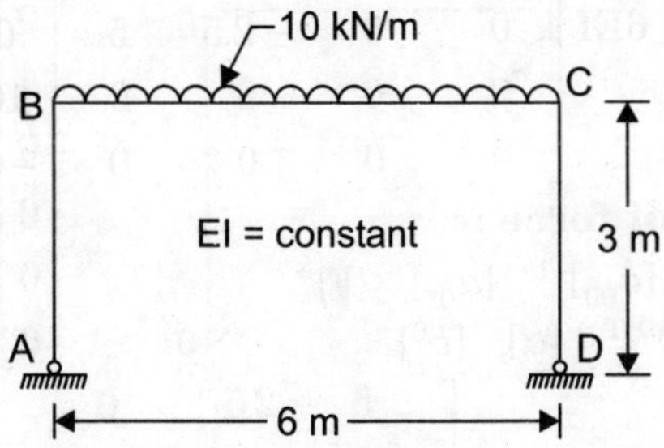

Fig. 4.36

Solution The frame is statically indeterminate to first degree. The primary structure is made by removing the horizontal reaction H_D at the right support.

∴ H_D is the redundant force.

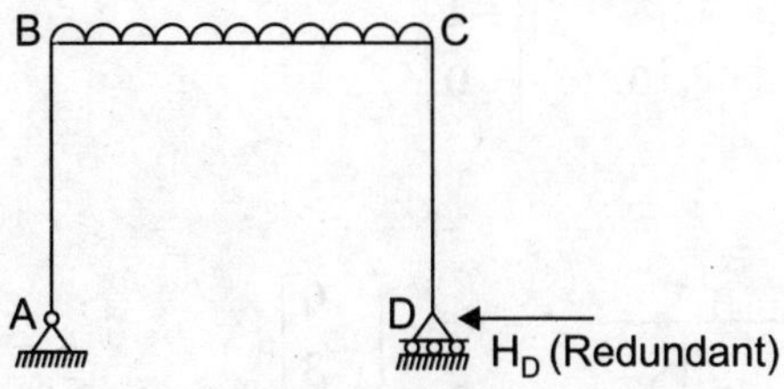

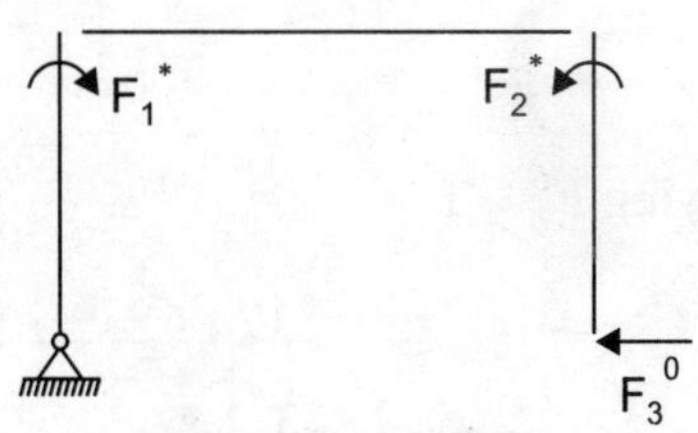

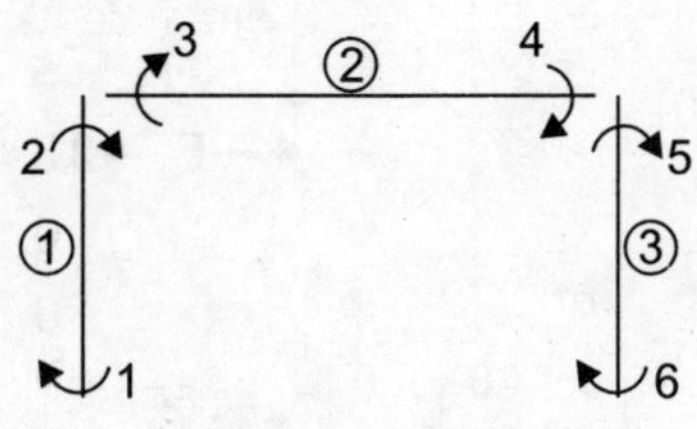

Fixed end moments :

$$M_{FAB} = M_{FBA} = M_{FCD} = M_{FDC} = 0$$

$$M_{FBC} = -\frac{10 \times 6^2}{12} = -30 \text{ kNm}$$

$$M_{FCB} = \frac{10 \times 6^2}{12} = 30 \text{ kNm}$$

Hence, equivalent joint loads $\begin{bmatrix} F_1^* = 30 \text{ kNm} \\ F_2^* = 30 \text{ kNm} \end{bmatrix}$ (Sign of F_1* and F_2* considered in [*b*] matrix)

Formation of [b] matrix :

Applying $F_1^* = 1$

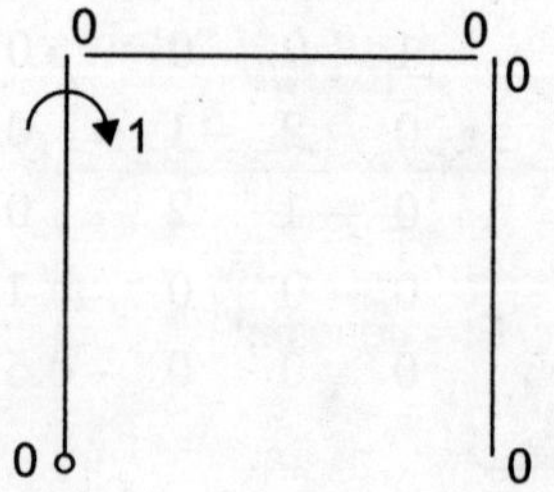

$$\begin{bmatrix} 0 \\ 1 \\ 0 \\ 0 \\ 0 \\ 0 \end{bmatrix}$$

Applying $F_2^* = 1$

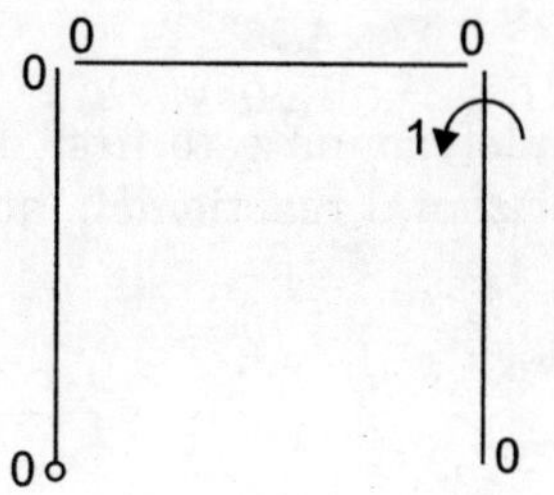

$$\begin{bmatrix} 0 \\ 0 \\ 0 \\ 0 \\ -1 \\ 0 \end{bmatrix}$$

Applying $F_3^0 = 1$

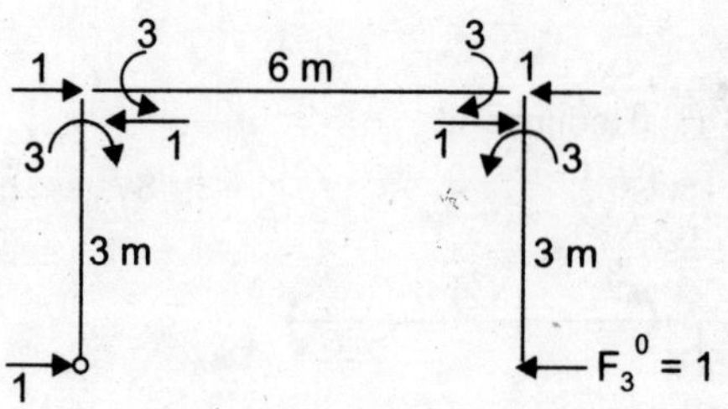

$$\begin{bmatrix} 0 \\ 3 \\ -3 \\ 3 \\ -3 \\ 0 \end{bmatrix}$$

$$\therefore \qquad [b] = \begin{array}{c} \begin{matrix} b^* & & b^0 \end{matrix} \\ \left[\begin{array}{cc|c} 0 & 0 & 0 \\ 1 & 0 & 3 \\ 0 & 0 & -3 \\ 0 & 0 & 3 \\ 0 & -1 & -3 \\ 0 & 0 & 0 \end{array}\right] \end{array}$$

Element flexibility matrix $[\alpha] = \dfrac{l}{6EI}\begin{bmatrix} 2 & -1 \\ -1 & 2 \end{bmatrix}$

$$[\alpha_1] = [\alpha_3] = \frac{3}{6EI}\begin{bmatrix} 2 & -1 \\ -1 & 2 \end{bmatrix} = \frac{1}{EI}\begin{bmatrix} 1 & -0.5 \\ -0.5 & 1 \end{bmatrix}$$

$$[\alpha_2] = \frac{6}{6EI}\begin{bmatrix} 2 & -1 \\ -1 & 2 \end{bmatrix} = \frac{1}{EI}\begin{bmatrix} 2 & -1 \\ -1 & 2 \end{bmatrix}$$

∴ Assembled element flexibility matrix is

$$[\alpha] = \frac{1}{EI}\begin{bmatrix} 1 & -0.5 & 0 & 0 & 0 & 0 \\ -0.5 & 1 & 0 & 0 & 0 & 0 \\ 0 & 0 & 2 & -1 & 0 & 0 \\ 0 & 0 & -1 & 2 & 0 & 0 \\ 0 & 0 & 0 & 0 & 1 & -0.5 \\ 0 & 0 & 0 & 0 & -0.5 & 1 \end{bmatrix}$$

Redundant forces are calculated using

$$\{F_0\} = -[a_{22}]^{-1}[a_{21}]\{F^*\}$$

$$[a_{22}] = [b^0]^T[\alpha][b^0]$$

$$= [0 \quad 3 \quad -3 \quad 3 \quad -3 \quad 0]\frac{1}{EI}\begin{bmatrix} 1 & -0.5 & 0 & 0 & 0 & 0 \\ -0.5 & 1 & 0 & 0 & 0 & 0 \\ 0 & 0 & 2 & -1 & 0 & 0 \\ 0 & 0 & -1 & 2 & 0 & 0 \\ 0 & 0 & 0 & 0 & 1 & -0.5 \\ 0 & 0 & 0 & 0 & -0.5 & 1 \end{bmatrix}\begin{bmatrix} 0 \\ 3 \\ -3 \\ 3 \\ -3 \\ 0 \end{bmatrix}$$

$$= \frac{1}{EI}[-1.5 \quad 3 \quad -9 \quad 9 \quad -3 \quad 1.5]\begin{bmatrix} 0 \\ 3 \\ -3 \\ 3 \\ -3 \\ 0 \end{bmatrix}$$

$$[a_{22}] = \left[\frac{72}{EI}\right]$$

$$[a_{22}]^{-1} = \left[\frac{EI}{72}\right]$$

$$[a_{21}] = [b^0]^T[\alpha][b^*]$$

$$= [0 \quad 3 \quad -3 \quad 3 \quad -3 \quad 0]\frac{1}{EI}\begin{bmatrix} 1 & -0.5 & 0 & 0 & 0 & 0 \\ -0.5 & 1 & 0 & 0 & 0 & 0 \\ 0 & 0 & 2 & -1 & 0 & 0 \\ 0 & 0 & -1 & 2 & 0 & 0 \\ 0 & 0 & 0 & 0 & 1 & -0.5 \\ 0 & 0 & 0 & 0 & -0.5 & 1 \end{bmatrix}\begin{bmatrix} 0 & 0 \\ 1 & 0 \\ 0 & 0 \\ 0 & 0 \\ 0 & -1 \\ 0 & 0 \end{bmatrix}$$

$$= \frac{1}{EI}[-1.5 \quad 3 \quad -9 \quad 9 \quad -3 \quad 1.5]\begin{bmatrix} 0 & 0 \\ 1 & 0 \\ 0 & 0 \\ 0 & 0 \\ 0 & -1 \\ 0 & 0 \end{bmatrix}$$

$$[a_{21}] = \frac{3}{EI}[1 \quad 1]$$

$$\therefore \quad \{F_0\} = -[a_{22}]^{-1}[a_{21}]\{F^*\} = -\left[\frac{EI}{72}\right]\left[\frac{3}{EI}(1 \quad 1)\right]\begin{Bmatrix} 30 \\ 30 \end{Bmatrix} = -2.5 \text{ kN}$$

Element forces :

$$\{P\} = [b]\{F\} = \begin{bmatrix} 0 & 0 & 0 \\ 1 & 0 & 3 \\ 0 & 0 & -3 \\ 0 & 0 & 3 \\ 0 & -1 & -3 \\ 0 & 0 & 0 \end{bmatrix}\begin{bmatrix} 30 \\ 30 \\ -2.5 \end{bmatrix} = \begin{bmatrix} 0 \\ 22.5 \\ 7.5 \\ -7.5 \\ -22.5 \\ 0 \end{bmatrix}$$

Final forces : $$\{P^f\} = [P] - \{P^e\} = \begin{bmatrix} 0 \\ 22.5 \\ 7.5 \\ -7.5 \\ -22.5 \\ 0 \end{bmatrix} - \begin{bmatrix} 0 \\ 0 \\ 30 \\ -30 \\ 0 \\ 0 \end{bmatrix}$$

$$\begin{bmatrix} M_{AB} \\ M_{BA} \\ M_{BC} \\ M_{CB} \\ M_{CD} \\ M_{DC} \end{bmatrix} = \begin{bmatrix} 0 \\ 22.5 \\ -22.5 \\ 22.5 \\ -22.5 \\ 0 \end{bmatrix}$$

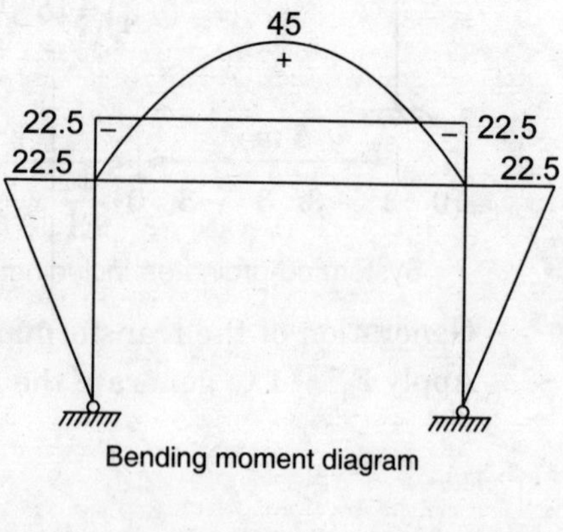

Bending moment diagram

Simply supported beam moments :

$$\text{Span BC} = \frac{wl^2}{8} = \frac{10 \times 6^2}{8}$$

$$= 45 \text{ kNm}$$

Example 4.24 *Analyse the frame shown in Fig. 4.37 by the matrix flexibility method.*

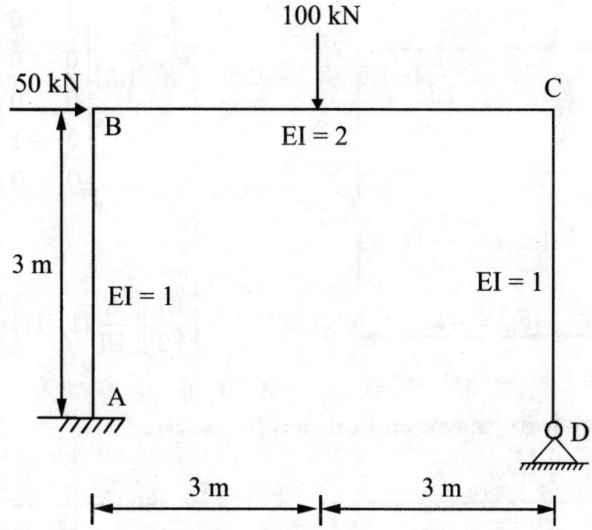

Fig. 4.37

Solution The structure is 2 degree redundant. The structure is made determinate by removing the support at D, thus removing two reaction components. That is, the two reaction components at D are treated as the redundants.

External forces F_1^* and F_2^*

Redundant forces F_3^0 and F_4^0

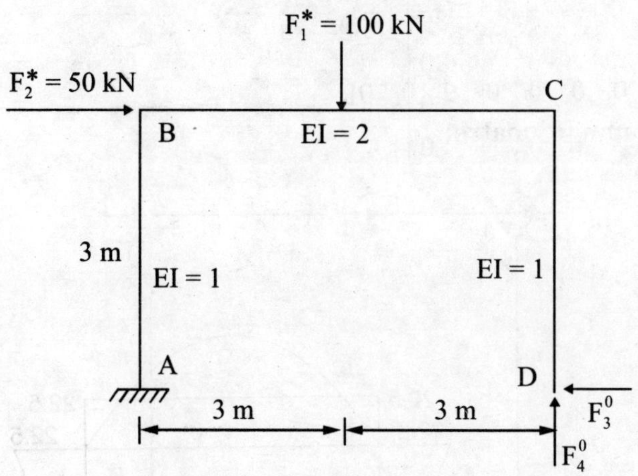

System co-ordinates including redundants

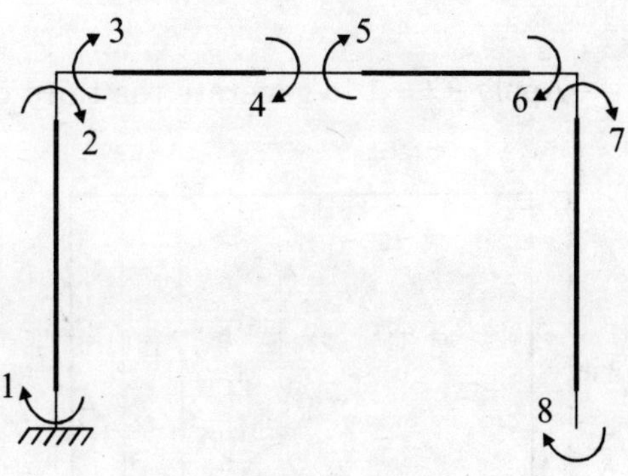

Element co-ordinates

Generation of the transformation matrix [*b*]

Apply $F_1^* = 1$ to generate the first column [*b*] matrix

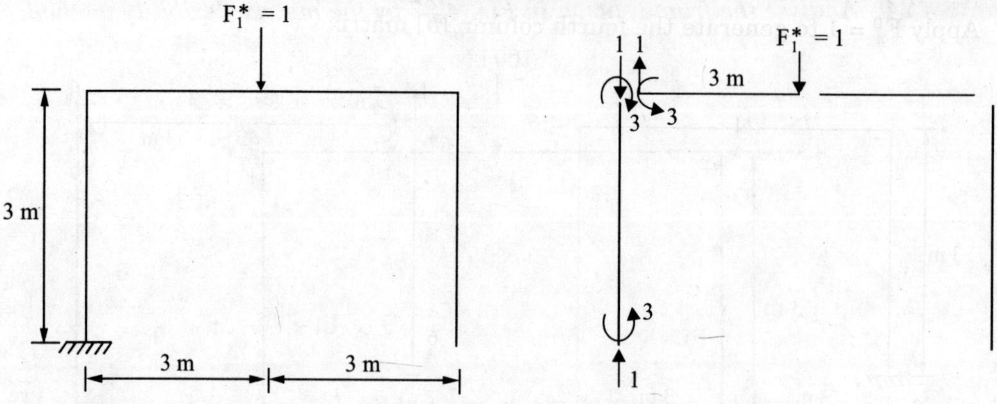

$[-3 \quad 3 \quad -3 \quad 0 \quad 0 \quad 0 \quad 0 \quad 0]^T$

Apply $F_2^* = 1$ to generate the second column [b] matrix

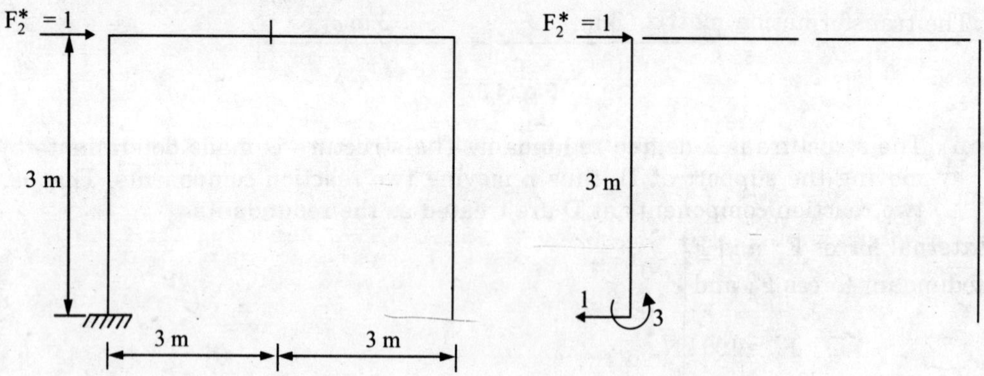

$[-3 \quad 0 \quad 0 \quad 0 \quad 0 \quad 0 \quad 0 \quad 0]^T$

Apply $F_3^0 = 1$ to generate the third column [b] matrix

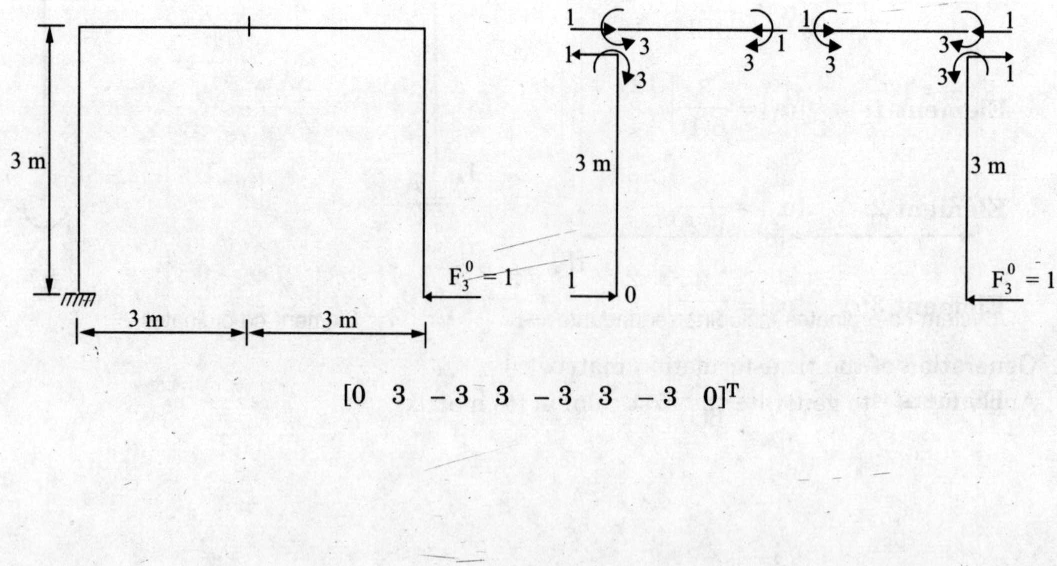

$[0 \quad 3 \quad -3 \quad 3 \quad -3 \quad 3 \quad -3 \quad 0]^T$

Apply $F_4^0 = 1$ to generate the fourth column $[b]$ matrix

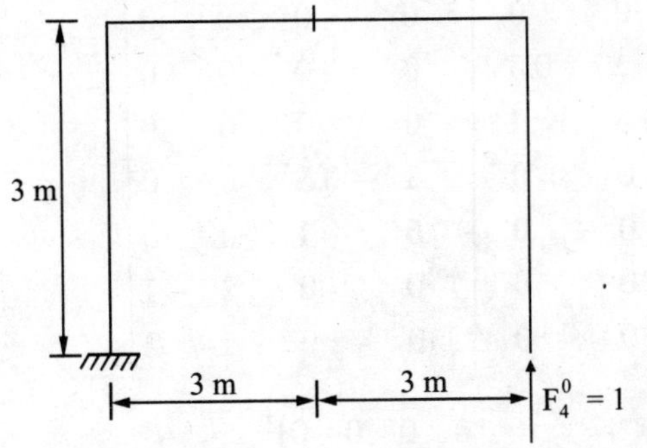

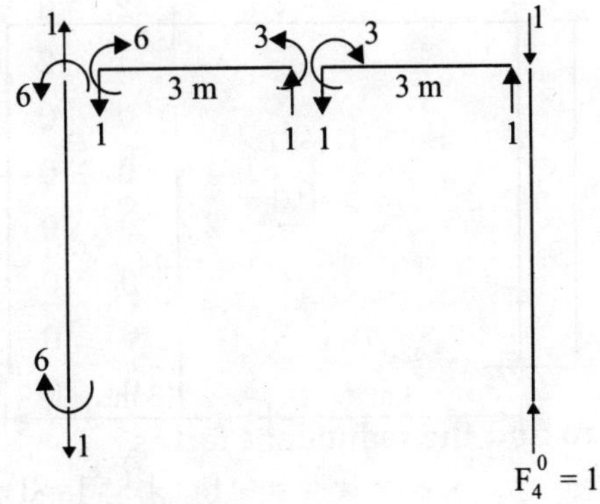

$$[6 \quad -6 \quad 6 \quad -3 \quad 3 \quad 0 \quad 0 \quad 0]^T$$

The transformation matrix

$$[b] = \begin{array}{c} \quad [b^*] \qquad\quad [b^0] \\ \left[\begin{array}{cc:cc} -3 & -3 & 0 & 6 \\ 3 & 0 & 3 & -6 \\ -3 & 0 & -3 & 6 \\ 0 & 0 & 3 & -3 \\ 0 & 0 & -3 & 3 \\ 0 & 0 & 3 & 0 \\ 0 & 0 & -3 & 0 \\ 0 & 0 & 0 & 0 \end{array}\right] \end{array}$$

Element flexibility matrix $[\alpha]$:

$$[\alpha_i] = \frac{l}{6EI}\begin{bmatrix} 2 & -1 \\ -1 & 2 \end{bmatrix}$$

Element 1: $$[\alpha_1] = \frac{3}{6(1)}\begin{bmatrix} 2 & -1 \\ -1 & 2 \end{bmatrix}$$

Element 2: $$[\alpha_2] = \frac{3}{6(2)}\begin{bmatrix} 2 & -1 \\ -1 & 2 \end{bmatrix} = \frac{3}{12}\begin{bmatrix} 2 & -1 \\ -1 & 2 \end{bmatrix} = \frac{3}{6}\begin{bmatrix} 1 & -0.5 \\ -0.5 & 1 \end{bmatrix}$$

Element 3: $$[\alpha_3] = \frac{3}{6(2)}\begin{bmatrix} 2 & -1 \\ -1 & 2 \end{bmatrix} = \frac{3}{12}\begin{bmatrix} 2 & -1 \\ -1 & 2 \end{bmatrix} = \frac{3}{6}\begin{bmatrix} 1 & -0.5 \\ -0.5 & 1 \end{bmatrix}$$

Element 4: $$[\alpha_4] = \frac{3}{6(1)}\begin{bmatrix} 2 & -1 \\ -1 & 2 \end{bmatrix}$$

Assembled element flexibility matrix :

$$[\alpha] = \frac{1}{2}\begin{bmatrix} 2 & -1 & 0 & 0 & 0 & 0 & 0 & 0 \\ -1 & 2 & 0 & 0 & 0 & 0 & 0 & 0 \\ 0 & 0 & 1 & -0.5 & 0 & 0 & 0 & 0 \\ 0 & 0 & -0.5 & 1 & 0 & 0 & 0 & 0 \\ 0 & 0 & 0 & 0 & 1 & -0.5 & 0 & 0 \\ 0 & 0 & 0 & 0 & -0.5 & 1 & 0 & 0 \\ 0 & 0 & 0 & 0 & 0 & 0 & 2 & -1 \\ 0 & 0 & 0 & 0 & 0 & 0 & -1 & 2 \end{bmatrix}$$

To find the redundant forces

$$\{F^0\} = -[a_{00}]^{-1} \quad [a_0^*] \quad \{F^*\}$$

$$[a_{00}] = [b_0]^T \quad [\alpha] \quad [b_0] = \begin{bmatrix} 0 & 3 & -3 & 3 & -3 & 3 & -3 & 0 \\ 6 & -6 & 6 & -3 & 3 & 0 & 0 & 0 \end{bmatrix}$$

$$\times \frac{1}{2}\begin{bmatrix} 2 & -1 & 0 & 0 & 0 & 0 & 0 & 0 \\ -1 & 2 & 0 & 0 & 0 & 0 & 0 & 0 \\ 0 & 0 & 1 & -0.5 & 0 & 0 & 0 & 0 \\ 0 & 0 & -0.5 & 1 & 0 & 0 & 0 & 0 \\ 0 & 0 & 0 & 0 & 1 & -0.5 & 0 & 0 \\ 0 & 0 & 0 & 0 & -0.5 & 1 & 0 & 0 \\ 0 & 0 & 0 & 0 & 0 & 0 & 2 & -1 \\ 0 & 0 & 0 & 0 & 0 & 0 & -1 & 2 \end{bmatrix}\begin{bmatrix} 0 & 6 \\ 3 & -6 \\ -3 & 6 \\ 3 & -3 \\ -3 & 3 \\ 3 & 0 \\ -3 & 0 \\ 0 & 0 \end{bmatrix}$$

$$[a_{00}] = 1/2\begin{bmatrix} 90 & -108 \\ -108 & 288 \end{bmatrix} = 9\begin{bmatrix} 5 & -6 \\ -6 & 16 \end{bmatrix}$$

$$[a_{00}]^{-1} = \frac{1}{9 \times 44}\begin{bmatrix} 16 & 6 \\ 6 & 5 \end{bmatrix} = \frac{1}{396}\begin{bmatrix} 16 & 6 \\ 6 & 5 \end{bmatrix}$$

$$[a_0^*] = [b_0]^T \quad [\alpha] \quad [b^*] = \begin{bmatrix} 0 & 3 & -3 & 3 & -3 & 3 & -3 & 0 \\ 6 & -6 & 6 & -3 & 3 & 0 & 0 & 0 \end{bmatrix}$$

$$\times \frac{1}{2}\begin{bmatrix} 2 & -1 & 0 & 0 & 0 & 0 & 0 & 0 \\ -1 & 2 & 0 & 0 & 0 & 0 & 0 & 0 \\ 0 & 0 & 1 & -0.5 & 0 & 0 & 0 & 0 \\ 0 & 0 & -0.5 & 1 & 0 & 0 & 0 & 0 \\ 0 & 0 & 0 & 0 & 1 & -0.5 & 0 & 0 \\ 0 & 0 & 0 & 0 & -0.5 & 1 & 0 & 0 \\ 0 & 0 & 0 & 0 & 0 & 0 & 2 & -1 \\ 0 & 0 & 0 & 0 & 0 & 0 & -1 & 2 \end{bmatrix}\begin{bmatrix} -3 & -3 \\ 3 & 0 \\ -3 & 0 \\ 0 & 0 \\ 0 & 0 \\ 0 & 0 \\ 0 & 0 \\ 0 & 0 \end{bmatrix}$$

$$[a_0^*] = 1/2\begin{bmatrix} 40.5 & 9 \\ -130.5 & -54 \end{bmatrix}$$

$$\{F^0\} = -[a_{00}]^{-1} \quad [a_0^*] \quad \{F^*\}$$

$$\{F\}^0 = -\frac{1}{396}\begin{bmatrix} 16 & 6 \\ 6 & 5 \end{bmatrix} 1/2 \begin{bmatrix} 40.5 & 9 \\ -130.5 & -54 \end{bmatrix}\begin{bmatrix} 100 \\ 50 \end{bmatrix}$$

$$\{F^0\} = +\begin{Bmatrix} 28.409 \\ 65.34 \end{Bmatrix}$$

Element forces $\{P\} = [b] \quad \{F\}$ where $\{F\} = \begin{Bmatrix} F^* \\ F^0 \end{Bmatrix}$

$$\begin{Bmatrix} P_1 \\ P_2 \\ P_3 \\ P_4 \\ P_5 \\ P_6 \\ P_7 \\ P_8 \end{Bmatrix} = \begin{bmatrix} -3 & -3 & 0 & 6 \\ 3 & 0 & 3 & -6 \\ -3 & 0 & -3 & 6 \\ 0 & 0 & 3 & -3 \\ 0 & 0 & -3 & 3 \\ 0 & 0 & 3 & 0 \\ 0 & 0 & -3 & 0 \\ 0 & 0 & 0 & 0 \end{bmatrix}\begin{Bmatrix} 100 \\ 50 \\ 28.409 \\ 65.34 \end{Bmatrix} = \begin{Bmatrix} -57.96 \\ -6.813 \\ 6.813 \\ -110.793 \\ 110.793 \\ 85.227 \\ -85.227 \\ 0 \end{Bmatrix}$$

Alternate method:

Here also D will be treated as redundant and removed as before. Member BC will be treated as a single element subjected to central concentrated load. Hence system forces will be F_1^*, F_2^* and F_3^*.

Redundant forces F_4^0 and F_5^0

The size of the $[\alpha]$ matrix will be 6 × 6.

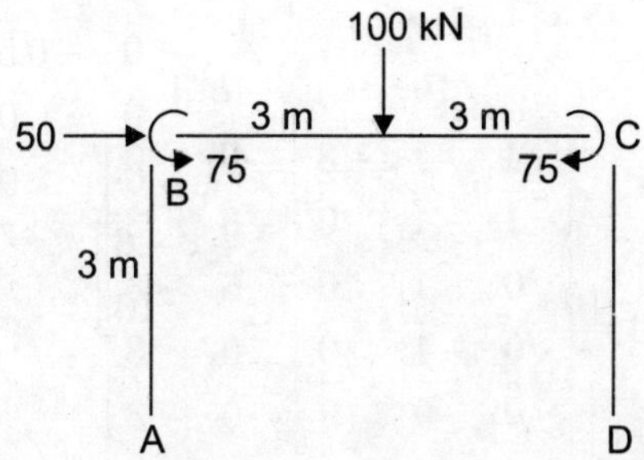

Fixed end moments :

$$M_{FBC} = -\frac{100 \times 6}{8} = -75 \text{ kNm}$$

$$M_{FCB} = \frac{100 \times 6}{8} = 75 \text{ kNm}$$

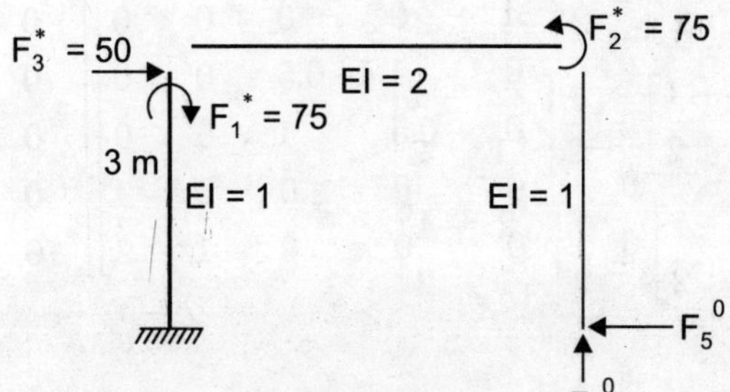

System forces and redundant forces

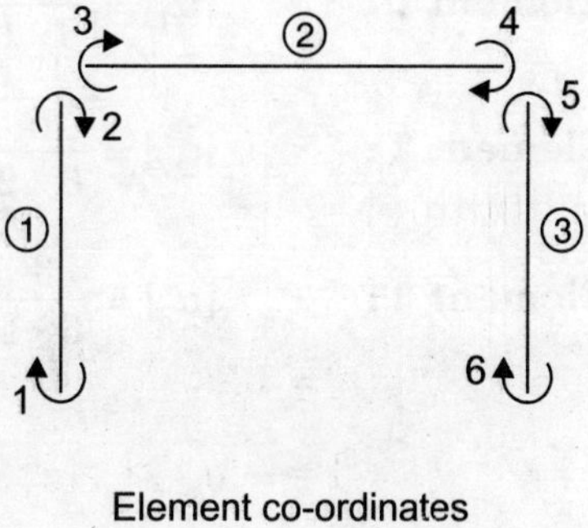

Element co-ordinates

Derivations of [*b*] matrix (Figures are self explanatory) :

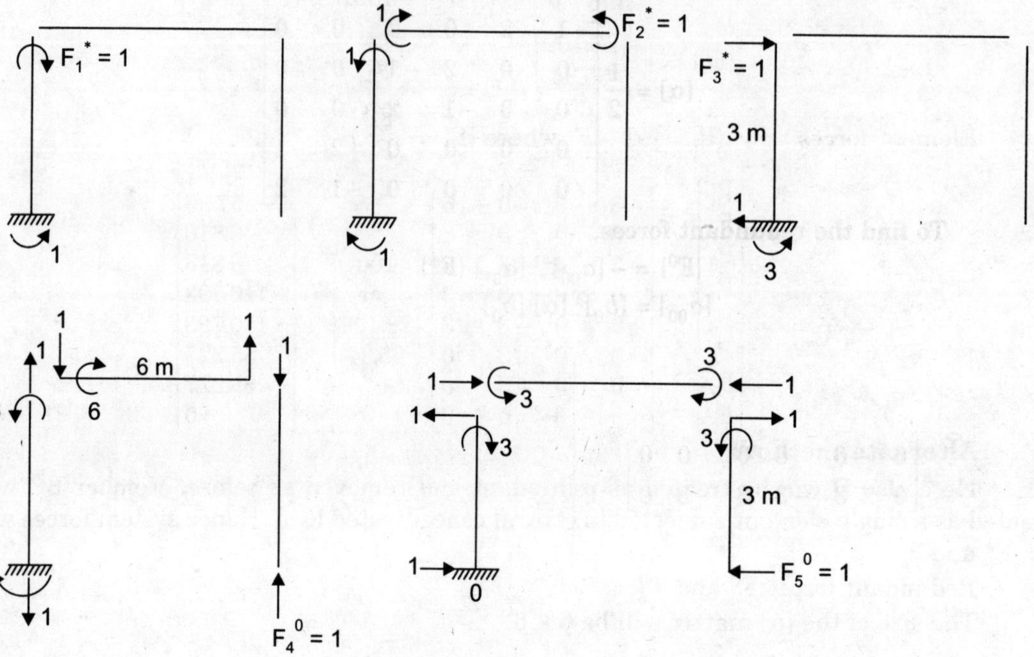

$$[b] = \begin{array}{c} \begin{array}{ccc|cc} & [b^*] & & [b^0] & \end{array} \\ \left[\begin{array}{ccc|cc} -1 & 1 & -3 & 6 & 0 \\ 1 & -1 & 0 & -6 & 3 \\ 0 & 1 & 0 & 6 & -3 \\ 0 & -1 & 0 & 0 & 3 \\ 0 & 0 & 0 & 0 & -3 \\ 0 & 0 & 0 & 0 & 0 \end{array}\right] \end{array}$$

Element flexibility matrix [α] :

$$[\alpha_i] = \frac{l}{6\text{EI}}\begin{bmatrix} 2 & -1 \\ -1 & 2 \end{bmatrix}$$

Element 1 : $$[\alpha_1] = \frac{3}{6\times1}\begin{bmatrix} 2 & -1 \\ -1 & 2 \end{bmatrix} = \frac{1}{2}\begin{bmatrix} 2 & -1 \\ -1 & 2 \end{bmatrix}$$

Element 2 : $$[\alpha_2] = \frac{6}{6\times2}\begin{bmatrix} 2 & -1 \\ -1 & 2 \end{bmatrix} = \frac{1}{2}\begin{bmatrix} 2 & -1 \\ -1 & 2 \end{bmatrix}$$

Element 3 : $$[\alpha_3] = \frac{3}{6\times1}\begin{bmatrix} 2 & -1 \\ -1 & 2 \end{bmatrix} = \frac{1}{2}\begin{bmatrix} 2 & -1 \\ -1 & 2 \end{bmatrix}$$

Assembled element flexibility matrix :

$$[\alpha] = \frac{1}{2}\begin{bmatrix} 2 & -1 & 0 & 0 & 0 & 0 \\ -1 & 2 & 0 & 0 & 0 & 0 \\ 0 & 0 & 2 & -1 & 0 & 0 \\ 0 & 0 & -1 & 2 & 0 & 0 \\ 0 & 0 & 0 & 0 & 2 & -1 \\ 0 & 0 & 0 & 0 & -1 & 2 \end{bmatrix}$$

To find the redundant forces,

$$\{F^0\} = -[a_{00}]^{-1}[a_{0*}]\{F^*\}$$

$$[a_{00}] = [b_0]^T[\alpha][b_0]$$

$$= \begin{bmatrix} 6 & -6 & 6 & 0 & 0 & 0 \\ 0 & 3 & -3 & 3 & -3 & 0 \end{bmatrix} \times \frac{1}{2}\begin{bmatrix} 2 & -1 & 0 & 0 & 0 & 0 \\ -1 & 2 & 0 & 0 & 0 & 0 \\ 0 & 0 & 2 & -1 & 0 & 0 \\ 0 & 0 & -1 & 2 & 0 & 0 \\ 0 & 0 & 0 & 0 & 2 & -1 \\ 0 & 0 & 0 & 0 & -1 & 2 \end{bmatrix}\begin{bmatrix} 6 & 0 \\ -6 & 3 \\ 6 & -3 \\ 0 & 3 \\ 0 & -3 \\ 0 & 0 \end{bmatrix}$$

$$= \frac{1}{2}\begin{bmatrix} 18 & -18 & 12 & -6 & 0 & 0 \\ -3 & 6 & -9 & 9 & -6 & 3 \end{bmatrix}\begin{bmatrix} 6 & 0 \\ -6 & 3 \\ 6 & -3 \\ 0 & 3 \\ 0 & -3 \\ 0 & 0 \end{bmatrix}$$

$$[a_{00}] = \frac{1}{2}\begin{bmatrix} 288 & -108 \\ -108 & 90 \end{bmatrix} = \begin{bmatrix} 144 & -54 \\ -54 & 45 \end{bmatrix}$$

$$[a_{00}]^{-1} = \frac{1}{(144 \times 45 - 54 \times 54)}\begin{bmatrix} 45 & 54 \\ 54 & 144 \end{bmatrix}$$

$$[a_{00}]^{-1} = \frac{1}{3564}\begin{bmatrix} 45 & 54 \\ 54 & 144 \end{bmatrix}$$

$$[a_{0*}] = [b_0]^T[\alpha][b_*]$$

$$= \begin{bmatrix} 6 & -6 & 6 & 0 & 0 & 0 \\ 0 & 3 & -3 & 3 & -3 & 0 \end{bmatrix} \times \frac{1}{2}\begin{bmatrix} 2 & -1 & 0 & 0 & 0 & 0 \\ -1 & 2 & 0 & 0 & 0 & 0 \\ 0 & 0 & 2 & -1 & 0 & 0 \\ 0 & 0 & -1 & 2 & 0 & 0 \\ 0 & 0 & 0 & 0 & 2 & -1 \\ 0 & 0 & 0 & 0 & -1 & 2 \end{bmatrix}\begin{bmatrix} -1 & 1 & -3 \\ 1 & -1 & 0 \\ 0 & 1 & 0 \\ 0 & -1 & 0 \\ 0 & 0 & 0 \\ 0 & 0 & 0 \end{bmatrix}$$

$$= \frac{1}{2}\begin{bmatrix} 18 & -18 & 12 & -6 & 0 & 0 \\ -3 & 6 & -9 & 9 & -6 & 3 \end{bmatrix}\begin{bmatrix} -1 & 1 & -3 \\ 1 & -1 & 0 \\ 0 & 1 & 0 \\ 0 & -1 & 0 \\ 0 & 0 & 0 \\ 0 & 0 & 0 \end{bmatrix}$$

$$[a_{0*}] = \frac{1}{2}\begin{bmatrix} -36 & 54 & -54 \\ 9 & -27 & 9 \end{bmatrix}$$

$$\{F^0\} = -[a_{00}]^{-1}[a_{0*}]\{F^*\}$$

$$= -\frac{1}{3564}\begin{bmatrix} 45 & 54 \\ 54 & 144 \end{bmatrix} \times \frac{1}{2}\begin{bmatrix} -36 & 54 & -54 \\ 9 & -27 & 9 \end{bmatrix}\begin{Bmatrix} 75 \\ 75 \\ 50 \end{Bmatrix}$$

$$= -\frac{1}{3564 \times 2}\begin{bmatrix} 45 & 54 \\ 54 & 144 \end{bmatrix}\begin{bmatrix} -1350 \\ -900 \end{bmatrix}$$

$$= -\frac{1}{3564 \times 2}\begin{Bmatrix} -109350 \\ -202500 \end{Bmatrix}$$

$$\{F^0\} = \begin{Bmatrix} 15.341 \\ 28.409 \end{Bmatrix}$$

Element forces : $\{P\} = [b]\{F\}$

where $\{F\} = \begin{Bmatrix} F^* \\ F^0 \end{Bmatrix}$

$$\begin{Bmatrix} P_1 \\ P_2 \\ P_3 \\ P_4 \\ P_5 \\ P_6 \end{Bmatrix} = \begin{bmatrix} -1 & 1 & -3 & 6 & 0 \\ 1 & -1 & 0 & -6 & 3 \\ 0 & 1 & 0 & 6 & -3 \\ 0 & -1 & 0 & 0 & 3 \\ 0 & 0 & 0 & 0 & -3 \\ 0 & 0 & 0 & 0 & 0 \end{bmatrix}\begin{Bmatrix} 75 \\ 75 \\ 50 \\ 15.341 \\ 28.409 \end{Bmatrix} = \begin{Bmatrix} -57.954 \\ -6.819 \\ 81.819 \\ 10.227 \\ -85.227 \\ 0 \end{Bmatrix}$$

$$\{P^f\} = \{P\} + \{FEM\}$$

$$= \begin{Bmatrix} -57.954 \\ -6.819 \\ 81.819 \\ 10.227 \\ -85.227 \\ 0 \end{Bmatrix} + \begin{Bmatrix} 0 \\ 0 \\ -75 \\ 75 \\ 0 \\ 0 \end{Bmatrix} = \begin{Bmatrix} -57.954 \\ -6.819 \\ 6.819 \\ 85.227 \\ -85.227 \\ 0 \end{Bmatrix}$$

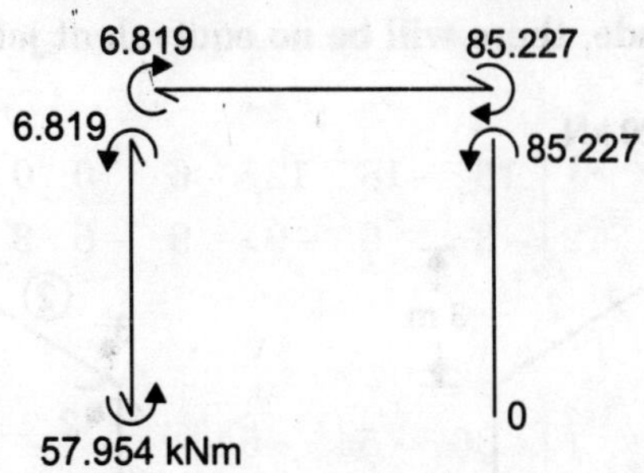

$$\text{Free BM} = \frac{Wl}{4} = \frac{100 \times 6}{4} = 150 \text{ kNm}$$

Bending moment diagram :

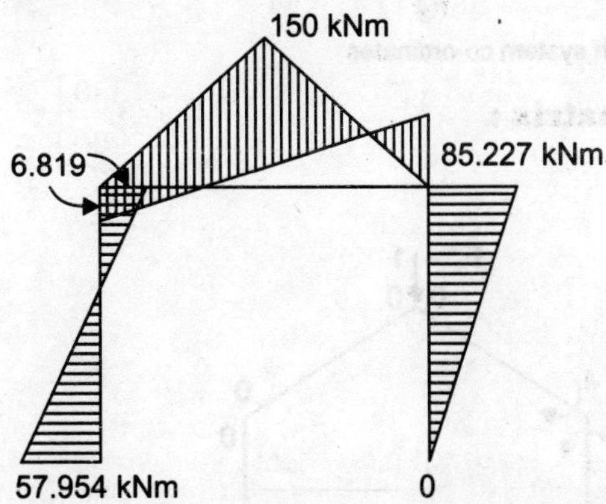

Example 4.25 *Analyse the frame shown in Fig. 4.38 by the matrix flexibility method and obtain the support moments.*

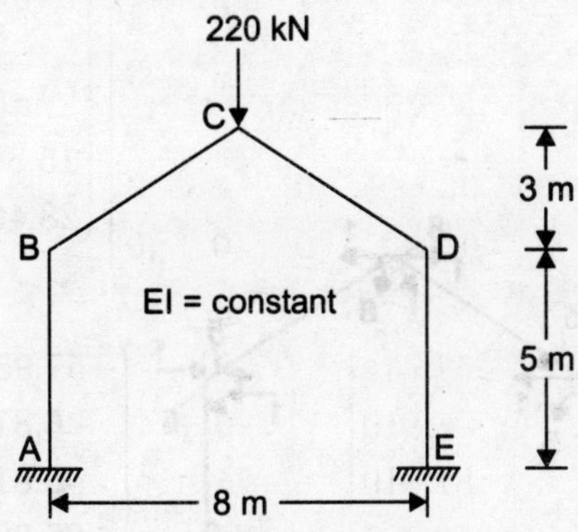

Fig. 4.38

Solution The degree of indeterminary is 3.

To make the primary structure, let us remove the horizontal, vertical reactions and moment @ right support treating H_E, V_E and M_E as redundants.

∵ **Load is acting @ node, there will be no equivalent joint load.**

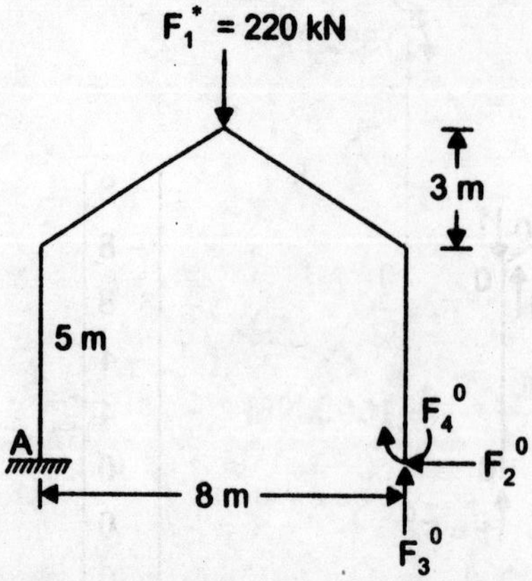

Primary structure with system co-ordinates

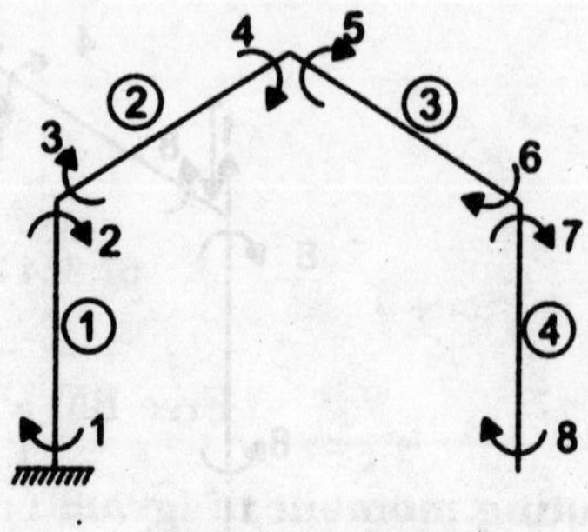

Element co-ordinates

Formation of [*b*] matrix :

Applying $\mathbf{F_1^* = 1}$

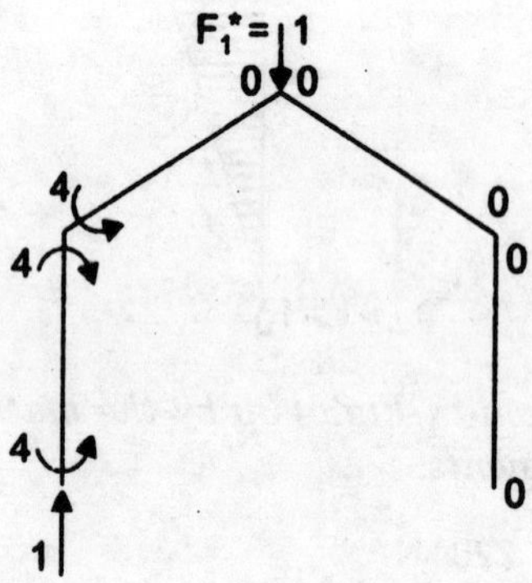

$$\begin{bmatrix} -4 \\ 4 \\ -4 \\ 0 \\ 0 \\ 0 \\ 0 \\ 0 \end{bmatrix}$$

Applying $\mathbf{F_2^0 = 1}$

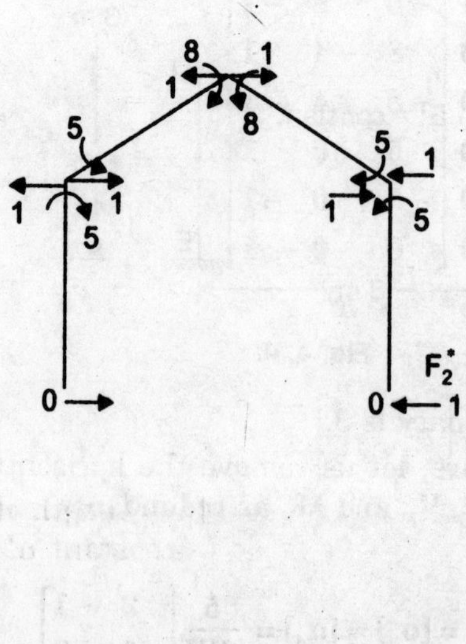

$$F_2^* = \begin{bmatrix} 0 \\ 5 \\ -5 \\ 8 \\ -8 \\ 5 \\ 5 \\ 0 \end{bmatrix}$$

Applying $F_3^0 = 1$

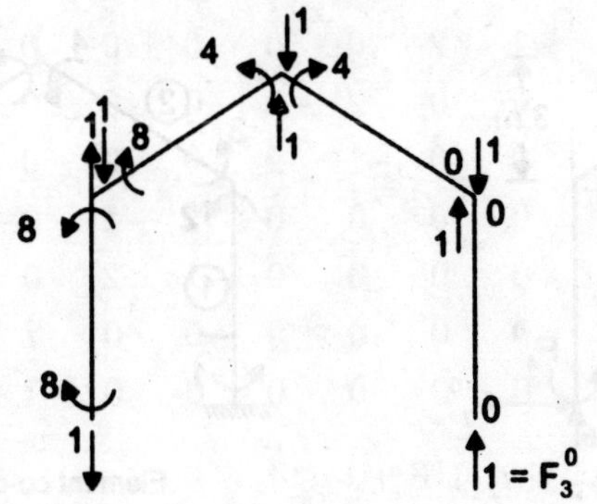

$$\begin{bmatrix} 8 \\ -8 \\ 8 \\ -4 \\ 4 \\ 0 \\ 0 \\ 0 \end{bmatrix}$$

Applying $F_4^0 = 1$

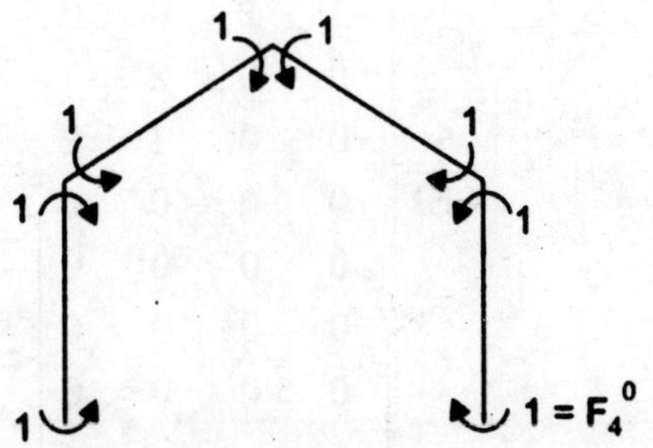

$$\begin{bmatrix} -1 \\ 1 \\ -1 \\ 1 \\ -1 \\ 1 \\ -1 \\ 1 \end{bmatrix}$$

∴ [*b*] matrix is given by

$$b = \left[\begin{array}{c|ccc} -4 & 0 & 8 & -1 \\ 4 & 5 & -8 & 1 \\ -4 & -5 & 8 & -1 \\ 0 & 8 & -4 & 1 \\ 0 & -8 & 4 & -1 \\ 0 & 5 & 0 & 1 \\ 0 & -5 & 0 & -1 \\ 0 & 0 & 0 & 1 \end{array}\right]$$

$$\qquad b^* \qquad\qquad b^0$$

Element flexibility matrix :

$$[\alpha] = \frac{l}{6EI}\begin{bmatrix} 2 & -1 \\ -1 & 2 \end{bmatrix}$$

[∵ Length of all elements and EI are constant 'α' is same for all elements]

$$\therefore \qquad [\alpha_1] = [\alpha_2] = [\alpha_3] = [\alpha_4] = \frac{5}{6EI}\begin{bmatrix} 2 & -1 \\ -1 & 2 \end{bmatrix}$$

Assembled element flexibility matrix will be

$$[\alpha] = \frac{5}{6EI}\begin{bmatrix} 2 & -1 & 0 & 0 & 0 & 0 & 0 & 0 \\ -1 & 2 & 0 & 0 & 0 & 0 & 0 & 0 \\ 0 & 0 & 2 & -1 & 0 & 0 & 0 & 0 \\ 0 & 0 & -1 & 2 & 0 & 0 & 0 & 0 \\ 0 & 0 & 0 & 0 & 2 & -1 & 0 & 0 \\ 0 & 0 & 0 & 0 & -1 & 2 & 0 & 0 \\ 0 & 0 & 0 & 0 & 0 & 0 & 2 & -1 \\ 0 & 0 & 0 & 0 & 0 & 0 & -1 & 2 \end{bmatrix}$$

Redundant forces $\{F_0\} = -[a_{22}^{-1}][a_{21}]\{F^*\}$

$$[a_{22}] = [b^{0T}]\,[\alpha]\,[b^0]$$

$$= \begin{bmatrix} 0 & 5 & -5 & 8 & -8 & 5 & -5 & 0 \\ 8 & -8 & 8 & -4 & 4 & 0 & 0 & 0 \\ -1 & 1 & -1 & 1 & -1 & 1 & -1 & 1 \end{bmatrix} \frac{5}{6EI}\begin{bmatrix} 2 & -1 & 0 & 0 & 0 & 0 & 0 & 0 \\ -1 & 2 & 0 & 0 & 0 & 0 & 0 & 0 \\ 0 & 0 & 2 & -1 & 0 & 0 & 0 & 0 \\ 0 & 0 & -1 & 2 & 0 & 0 & 0 & 0 \\ 0 & 0 & 0 & 0 & 2 & -1 & 0 & 0 \\ 0 & 0 & 0 & 0 & -1 & 2 & 0 & 0 \\ 0 & 0 & 0 & 0 & 0 & 0 & 2 & -1 \\ 0 & 0 & 0 & 0 & 0 & 0 & -1 & 2 \end{bmatrix}$$

$$\times \begin{bmatrix} 0 & 8 & -1 \\ 5 & -8 & 1 \\ -5 & 8 & -1 \\ 8 & -4 & 1 \\ -8 & 4 & -1 \\ 5 & 0 & 1 \\ -5 & 0 & -1 \\ 0 & 0 & 1 \end{bmatrix}$$

$$[a_{22}] = \frac{5}{6EI}\begin{bmatrix} -5 & 10 & -18 & 21 & -21 & 18 & -10 & 5 \\ 24 & -24 & 20 & -16 & 8 & -4 & 0 & 0 \\ -3 & 3 & -3 & 3 & -3 & 3 & -3 & 3 \end{bmatrix}\begin{bmatrix} 0 & 8 & -1 \\ 5 & -8 & 1 \\ -5 & 8 & -1 \\ 8 & -4 & 1 \\ -8 & 4 & -1 \\ 5 & 0 & 1 \\ -5 & 0 & -1 \\ 0 & 0 & 1 \end{bmatrix}$$

$$= \frac{5}{6EI}\begin{bmatrix} 616 & -432 & 108 \\ -432 & 640 & -96 \\ 108 & -96 & 24 \end{bmatrix}$$

$$[a_{21}] = [b^0]^T [\alpha] [b^*]$$

$$= \frac{5}{6EI}\begin{bmatrix} 0 & 5 & -5 & 8 & -8 & 5 & -5 & 0 \\ 8 & -8 & 8 & -4 & 4 & 0 & 0 & 0 \\ -1 & 1 & -1 & 1 & -1 & 1 & -1 & 1 \end{bmatrix}\begin{bmatrix} 2 & -1 & 0 & 0 & 0 & 0 & 0 & 0 \\ -1 & 2 & 0 & 0 & 0 & 0 & 0 & 0 \\ 0 & 0 & 2 & -1 & 0 & 0 & 0 & 0 \\ 0 & 0 & -1 & 2 & 0 & 0 & 0 & 0 \\ 0 & 0 & 0 & 0 & 2 & -1 & 0 & 0 \\ 0 & 0 & 0 & 0 & -1 & 2 & 0 & 0 \\ 0 & 0 & 0 & 0 & 0 & 0 & 2 & -1 \\ 0 & 0 & 0 & 0 & 0 & 0 & -1 & 2 \end{bmatrix}\begin{bmatrix} -4 \\ 4 \\ -4 \\ 0 \\ 0 \\ 0 \\ 0 \\ 0 \end{bmatrix}$$

$$= \frac{5}{6EI}\begin{bmatrix} 132 \\ -272 \\ 36 \end{bmatrix}$$

$$\{F_0\} = -[a_{22}^{-1}][a_{21}][F^*]$$

$$= -\frac{6EI}{5}\begin{bmatrix} 616 & -432 & 108 \\ -432 & 640 & -96 \\ 108 & -96 & 24 \end{bmatrix}^{-1} \times \frac{5}{6EI}\begin{bmatrix} 132 \\ -272 \\ 36 \end{bmatrix} \times [220]$$

$$F_0 = \begin{bmatrix} 50.76 \\ 110 \\ -118.46 \end{bmatrix}$$

$$\{P\} = [b]\,\{F\}$$

$$= \begin{bmatrix} -4 & 0 & 8 & -1 \\ 4 & 5 & -8 & 1 \\ -4 & -5 & 8 & -1 \\ 0 & 8 & -4 & 1 \\ 0 & -8 & 4 & -1 \\ 0 & 5 & 0 & 1 \\ 0 & -5 & 0 & -1 \\ 0 & 0 & 0 & 1 \end{bmatrix}\begin{bmatrix} 220 \\ 50.76 \\ 110 \\ -118.46 \end{bmatrix}$$

$$\begin{bmatrix} M_{AB} \\ M_{BA} \\ M_{BC} \\ M_{CB} \\ M_{CD} \\ M_{DC} \\ M_{DE} \\ M_{ED} \end{bmatrix} = \begin{bmatrix} 118.46 \\ 135.34 \\ -135.34 \\ -152.38 \\ 152.38 \\ 135.34 \\ -135.34 \\ -118.46 \end{bmatrix} \text{kNm}$$

Refer to example 2.18 for bending moment diagram.

Example 4.26 *A stiff jointed frame ABCD of constant section throughout has the following dimensions as shown in Fig. 4.39. Calculate the moments at the joints. Use matrix flexibility method.*

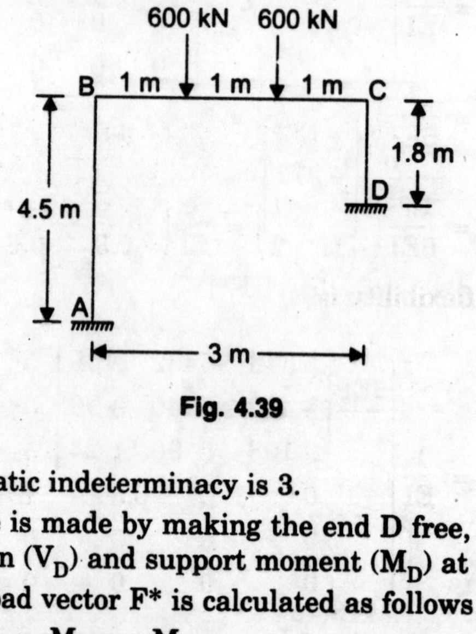

Fig. 4.39

Solution The degree of static indeterminacy is 3.

The primary structure is made by making the end D free, *i.e.*, by removing horizontal reaction (H_D) vertical reaction (V_D) and support moment (M_D) at the right support.

The equivalent joint load vector F* is calculated as follows :

$$M_{FAB} = M_{FBA} = M_{FCD}$$
$$= M_{FDC} = 0$$

$$M_{FBC} = -\frac{600 \times 1 \times 2^2}{3^2} - \frac{600 \times 2 \times 1^2}{3^2}$$
$$= -400 \text{ kNm}$$

$$M_{FCB} = \frac{600 \times 1 \times 2^2}{3^2} + \frac{600 \times 2 \times 1^2}{3^2}$$
$$= 400 \text{ kNm}$$

∴ The primary structure with system co-ordinates and element co-oridnates is

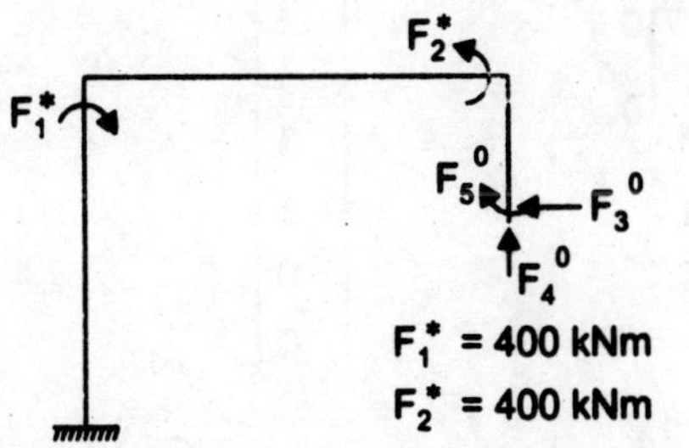

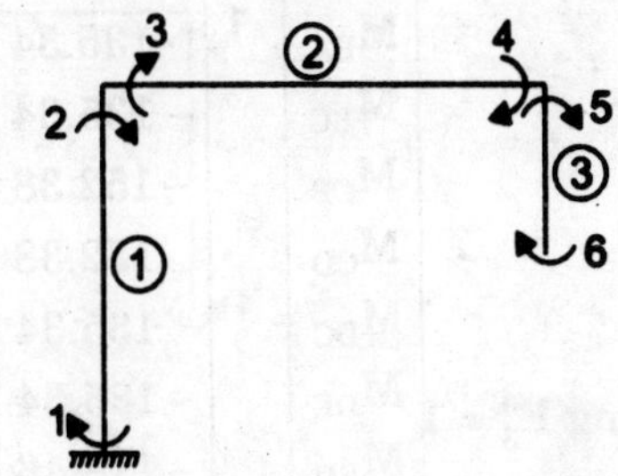

Element flexibility matrix :

$$[\alpha_1] = \frac{l_1}{6EI}\begin{bmatrix} 2 & -1 \\ -1 & 2 \end{bmatrix} = \frac{4.5}{6EI}\begin{bmatrix} 2 & -1 \\ -1 & 2 \end{bmatrix}$$

$$= \frac{1}{EI}\begin{bmatrix} 1.5 & -0.75 \\ -0.75 & 1.5 \end{bmatrix}$$

$$[\alpha_2] = \frac{3}{6EI}\begin{pmatrix} 2 & -1 \\ -1 & 2 \end{pmatrix} = \frac{1}{EI}\begin{pmatrix} 1 & -0.5 \\ -0.5 & 1 \end{pmatrix}$$

$$[\alpha_3] = \frac{1.8}{6EI}\begin{pmatrix} 2 & -1 \\ -1 & 2 \end{pmatrix} = \frac{1}{EI}\begin{pmatrix} 0.6 & -0.3 \\ -0.3 & 0.6 \end{pmatrix}$$

∴ Assembled element flexibility is

$$[\alpha] = \frac{1}{EI}\begin{pmatrix} 1.5 & -0.75 & 0 & 0 & 0 & 0 \\ -0.75 & 1.5 & 0 & 0 & 0 & 0 \\ 0 & 0 & 1 & -0.5 & 0 & 0 \\ 0 & 0 & -0.5 & 1 & 0 & 0 \\ 0 & 0 & 0 & 0 & 0.6 & -0.3 \\ 0 & 0 & 0 & 0 & -0.3 & 0.6 \end{pmatrix}$$

Formation of [*b*] matrix :

Applying $F_1^* = 1$

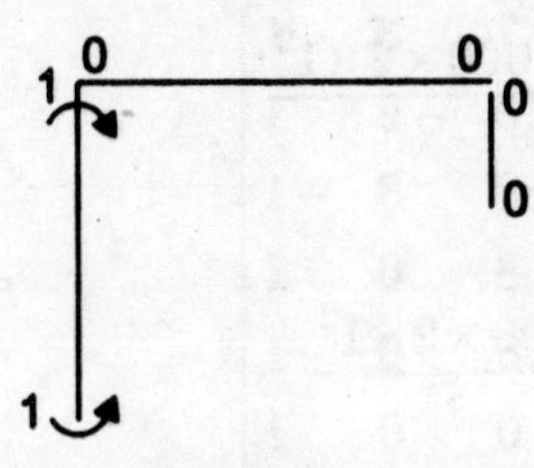

$$\begin{bmatrix} -1 \\ 1 \\ 0 \\ 0 \\ 0 \\ 0 \end{bmatrix}$$

Applying $F_2^* = 1$

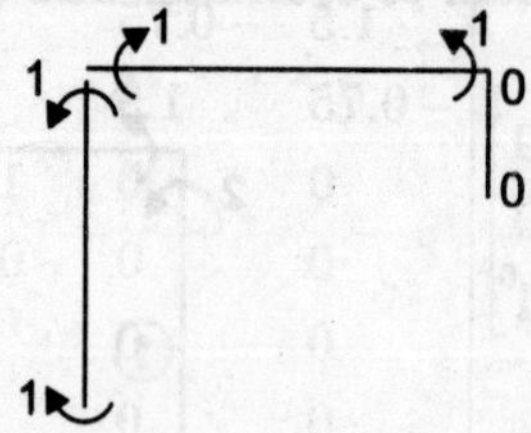

$$\begin{bmatrix} 1 \\ -1 \\ 1 \\ -1 \\ 0 \\ 0 \end{bmatrix}$$

Applying $F_3^0 = 1$

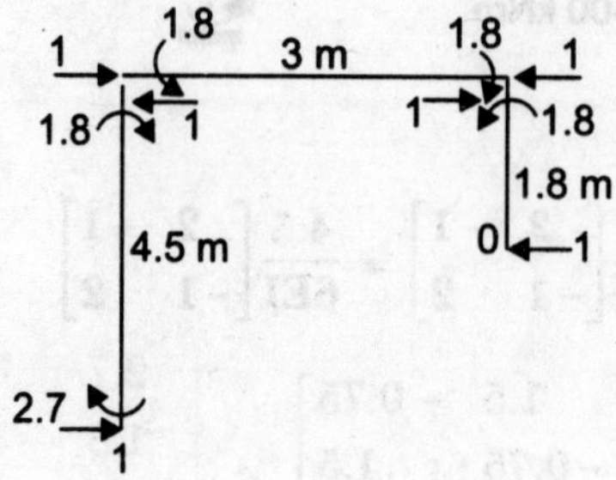

$$\begin{bmatrix} 2.7 \\ 1.8 \\ -1.8 \\ 1.8 \\ -1.8 \\ 0 \end{bmatrix}$$

Applying $F_4^0 = 1$

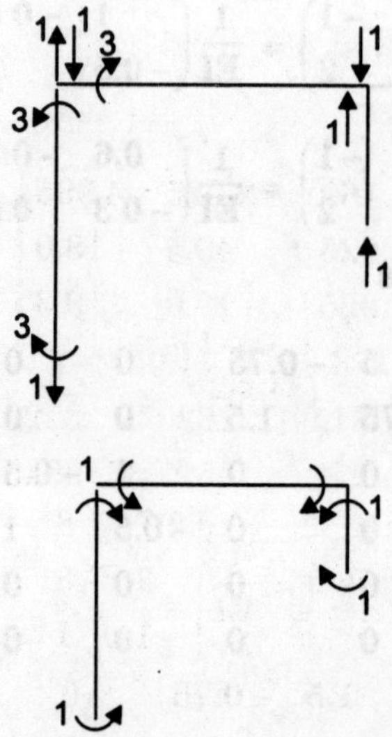

$$\begin{bmatrix} 3 \\ -3 \\ 3 \\ 0 \\ 0 \\ 0 \end{bmatrix}$$

Applying $F_5^0 = 1$

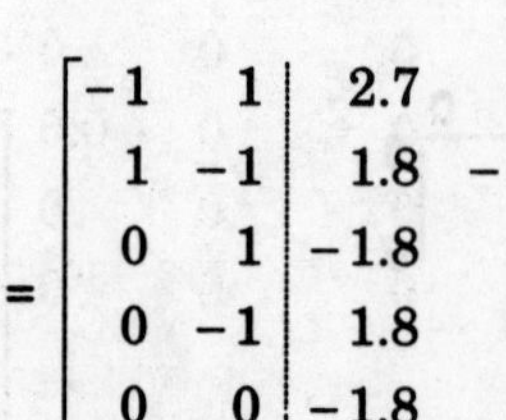

$$\begin{bmatrix} -1 \\ 1 \\ -1 \\ 1 \\ -1 \\ 1 \end{bmatrix}$$

$$\therefore \quad b = \left[\begin{array}{cc|ccc} -1 & 1 & 2.7 & 3 & -1 \\ 1 & -1 & 1.8 & -3 & 1 \\ 0 & 1 & -1.8 & 3 & -1 \\ 0 & -1 & 1.8 & 0 & 1 \\ 0 & 0 & -1.8 & 0 & -1 \\ 0 & 0 & 0 & 0 & 1 \end{array}\right]$$

$$b^* \qquad\qquad b^0$$

Redundant forces $\{F_0\} = -[a_{22}]^{-1}[a_{21}]\{F^*\}$

$$[a_{22}] = [b^0]^T [\alpha] [b^0]$$

$$= \frac{1}{EI}\begin{bmatrix} 2.7 & 1.8 & -1.8 & 1.8 & -1.8 & 0 \\ 3 & -3 & 3 & 0 & 0 & 0 \\ -1 & 1 & -1 & 1 & -1 & 1 \end{bmatrix}\begin{bmatrix} 1.5 & -0.75 & 0 & 0 & 0 & 0 \\ -0.75 & 1.5 & 0 & 0 & 0 & 0 \\ 0 & 0 & 1.0 & -0.5 & 0 & 0 \\ 0 & 0 & -0.5 & 1 & 0 & 0 \\ 0 & 0 & 0 & 0 & 0.6 & -0.3 \\ 0 & 0 & 0 & 0 & -0.3 & 0.6 \end{bmatrix} \times \begin{bmatrix} 2.7 & 3 & -1 \\ 1.8 & -3 & 1 \\ -1.8 & 3 & -1 \\ 1.8 & 0 & 1 \\ -1.8 & 0 & -1 \\ 0 & 0 & 1 \end{bmatrix}$$

$$= \frac{1}{EI}\begin{bmatrix} 2.7 & 0.675 & -2.7 & 2.7 & -1.08 & 0.54 \\ 6.75 & -6.75 & 3 & -1.5 & 0 & 0 \\ -2.25 & 2.25 & -1.5 & 1.5 & -0.9 & 0.9 \end{bmatrix}\begin{bmatrix} 2.7 & 3 & -1 \\ 1.8 & -3 & 1 \\ -1.8 & 3 & -1 \\ 1.8 & 0 & 1 \\ -1.8 & 0 & -1 \\ 0 & 0 & 1 \end{bmatrix}$$

$$[a_{22}] = \frac{1}{EI}\begin{bmatrix} 20.169 & -2.025 & 4.995 \\ -2.025 & 49.5 & -18.0 \\ 4.995 & -18.0 & 9.3 \end{bmatrix}$$

$$[a_{22}]^{-1} = \frac{EI}{1841}\begin{bmatrix} 136.35 & -71.08 & -210.8 \\ -71.08 & 162.62 & 352.93 \\ -210.8 & 352.93 & 994.26 \end{bmatrix}$$

$$[a_{21}] = [b^{0T}]\,[\alpha]\,[b^{*}] = \frac{1}{EI}\begin{bmatrix} 2.7 & 1.8 & -1.8 & 1.8 & -1.8 & 0 \\ 3 & -3 & 3 & 0 & 0 & 0 \\ -1 & 1 & -1 & 1 & -1 & 1 \end{bmatrix}$$

$$\begin{bmatrix} 1.5 & -0.75 & 0 & 0 & 0 & 0 \\ -0.75 & 1.5 & 0 & 0 & 0 & 0 \\ 0 & 0 & 1.0 & -0.5 & 0 & 0 \\ 0 & 0 & -0.5 & 1.0 & 0 & 0 \\ 0 & 0 & 0 & 0 & 0.6 & -0.3 \\ 0 & 0 & 0 & 0 & -0.3 & 0.6 \end{bmatrix} \times \begin{bmatrix} -1 & 1 \\ 1 & -1 \\ 0 & 1 \\ 0 & -1 \\ 0 & 0 \\ 0 & 0 \end{bmatrix}$$

$$= \frac{1}{EI}\begin{bmatrix} 2.7 & 0.675 & -2.7 & 2.7 & -1.08 & 0.54 \\ 6.75 & -6.75 & 3 & -1.5 & 0 & 0 \\ -2.25 & 2.25 & -1.5 & 1.5 & -0.9 & 0.9 \end{bmatrix}\begin{bmatrix} -1 & 1 \\ 1 & -1 \\ 0 & 1 \\ 0 & -1 \\ 0 & 0 \\ 0 & 0 \end{bmatrix}$$

$$[a_{21}] = \frac{1}{EI}\begin{bmatrix} -2.025 & -3.375 \\ -13.5 & 18.0 \\ 4.5 & -7.5 \end{bmatrix}$$

$$\therefore \quad [F_0] = -\frac{EI}{1841}\begin{bmatrix} 136.35 & -71.08 & -210.8 \\ -71.08 & 162.62 & 352.93 \\ -210.8 & 352.93 & 994.26 \end{bmatrix} \frac{1}{EI}\begin{bmatrix} -2.025 & -3.375 \\ -13.5 & 18 \\ 4.5 & -7.5 \end{bmatrix} \times \begin{bmatrix} 400 \\ 400 \end{bmatrix}$$

$$= \begin{bmatrix} 92.06 \\ -12.34 \\ 55.68 \end{bmatrix}$$

Element forces : $\{P\} = [b]\{F\} = [b^* \; b^0]\begin{Bmatrix} F^* \\ F^0 \end{Bmatrix}$

$$= \begin{bmatrix} -1 & 1 & 2.7 & 3 & -1 \\ 1 & -1 & 1.8 & -3 & 1 \\ 0 & 1 & -1.8 & 3 & -1 \\ 0 & -1 & 1.8 & 0 & 1 \\ 0 & 0 & -1.8 & 0 & -1 \\ 0 & 0 & 0 & 0 & 1 \end{bmatrix}\begin{bmatrix} 400 \\ 400 \\ 92.06 \\ -12.34 \\ 55.68 \end{bmatrix} = \begin{bmatrix} 155.86 \\ 258.41 \\ 141.59 \\ -178.61 \\ -221.39 \\ 55.68 \end{bmatrix}$$

Final forces : $[P^f] = [P] + [FEM]$

$$= \begin{bmatrix} 155.86 \\ 258.41 \\ 141.59 \\ -178.61 \\ -221.39 \\ 55.68 \end{bmatrix} + \begin{bmatrix} 0 \\ 0 \\ -400 \\ 400 \\ 0 \\ 0 \end{bmatrix} = \begin{bmatrix} 155.86 \\ 258.41 \\ -258.41 \\ 221.39 \\ -221.39 \\ 55.68 \end{bmatrix}$$

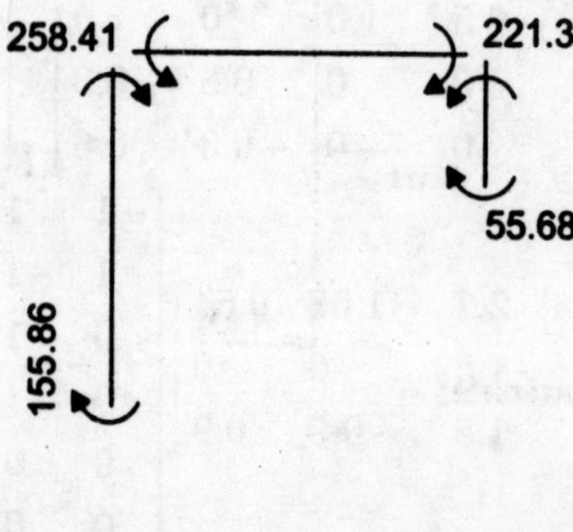

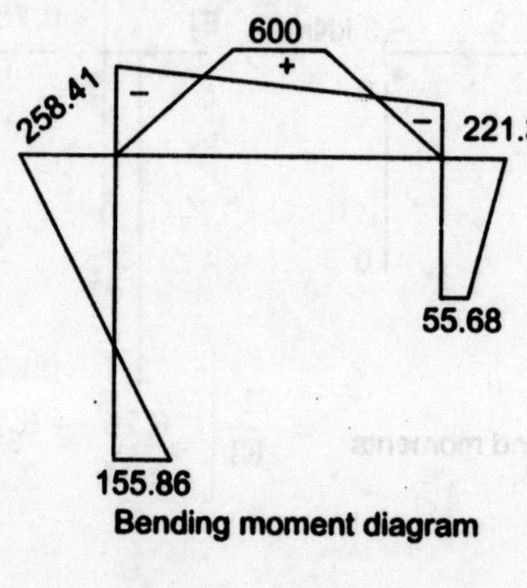

Bending moment diagram

Example 4.27 *Analyse the frame shown in Fig. 4.40 by matrix flexibility method.*

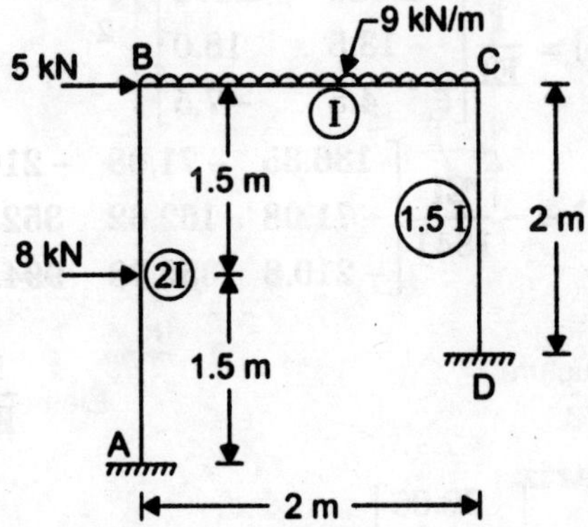

Fig. 4.40

Solution The structure is 3 degrees redundant. The structure is made determinate by removing the support at D, thus removing two reaction components (H, V) and one moment component (M). That is, the three components at D are treated as the redundants.

Fixed end moments :

$$M_{FAB} = -\frac{Wl}{8} = -\frac{8\times3}{8} = -3 \text{ kNm}$$

$$M_{FBA} = \frac{Wl}{8} = \frac{8\times3}{8} = 3 \text{ kNm}$$

$$M_{FBC} = -\frac{wl^2}{12} = -\frac{9\times2^2}{12} = -3 \text{ kNm}$$

$$M_{FCB} = \frac{wl^2}{12} = \frac{9\times2^2}{12} = 3 \text{ kNm}$$

$$M_{FCD} = M_{FDC} = 0$$

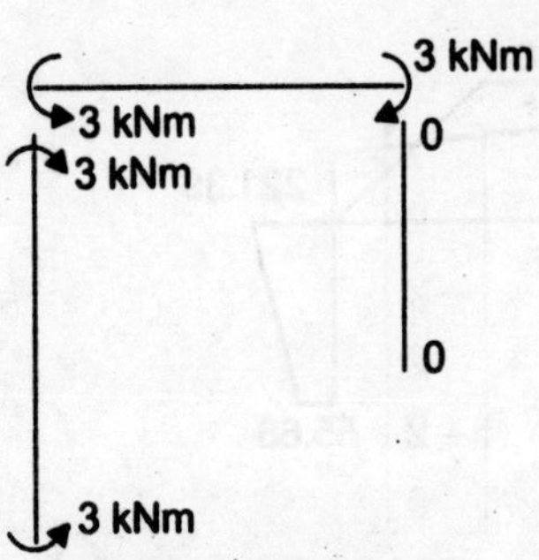

Fixed end moments

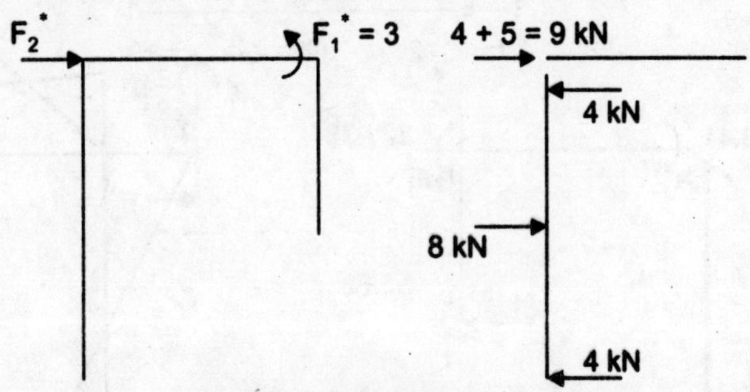

System co-ordinates

$$F^* = \begin{bmatrix} 3 \\ 9 \end{bmatrix}$$

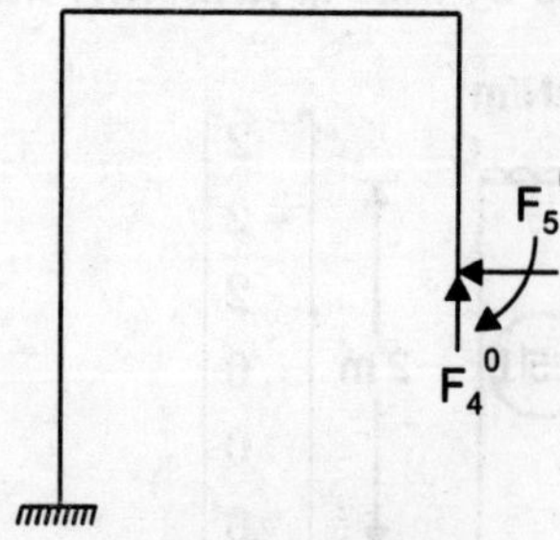

Primary structure

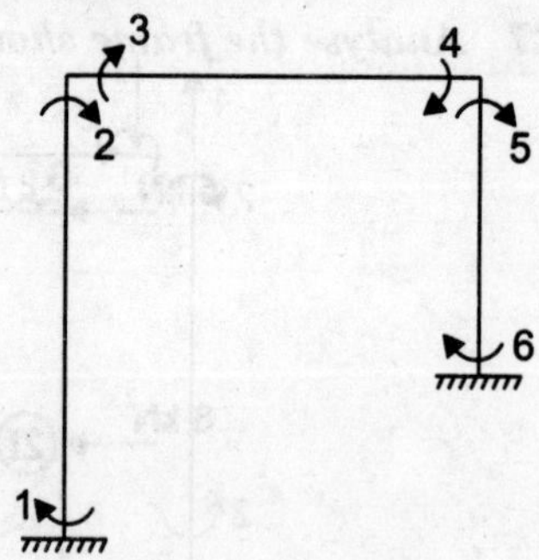

Element co-ordinates

Derivation of [*b*] matrix :

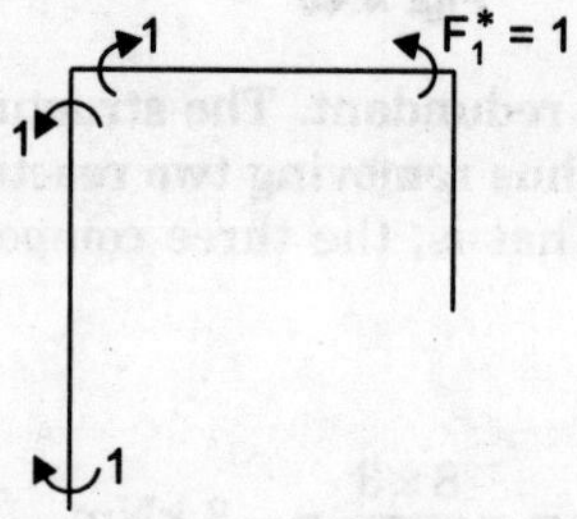

$$\begin{bmatrix} 1 \\ -1 \\ 1 \\ -1 \\ 0 \\ 0 \end{bmatrix}$$

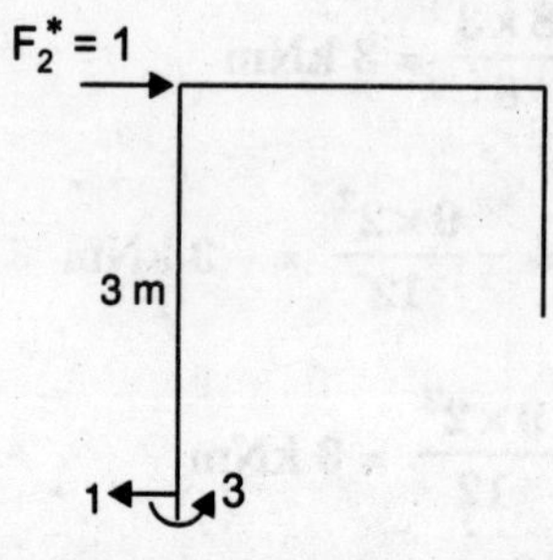

$$\begin{bmatrix} -3 \\ 0 \\ 0 \\ 0 \\ 0 \\ 0 \end{bmatrix}$$

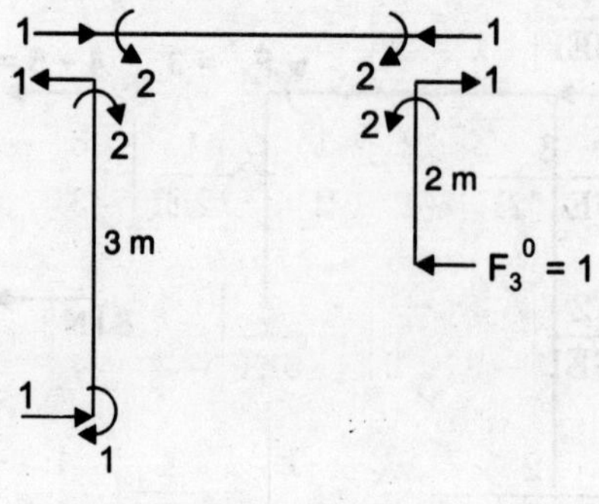

$$\begin{bmatrix} 1 \\ 2 \\ -2 \\ 2 \\ -2 \\ 0 \end{bmatrix}$$

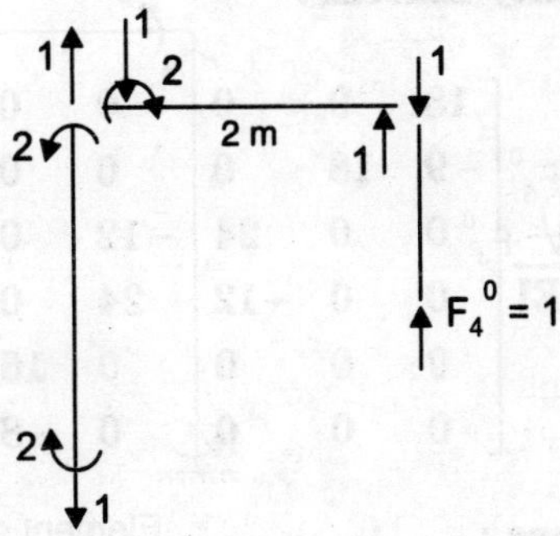

$$\begin{bmatrix} 2 \\ -2 \\ 2 \\ 0 \\ 0 \\ 0 \end{bmatrix}$$

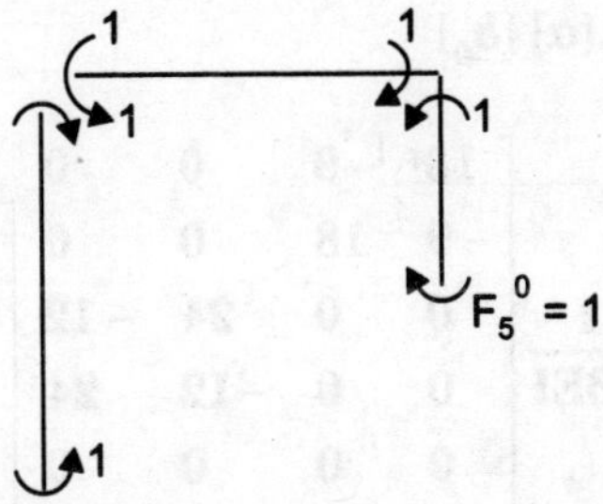

$$\begin{bmatrix} -1 \\ 1 \\ -1 \\ 1 \\ -1 \\ 1 \end{bmatrix}$$

Hence,

$$[b] = \left[\begin{array}{cc|ccc} \multicolumn{2}{c|}{b^*} & & b^0 & \\ 1 & -3 & 1 & 2 & -1 \\ -1 & 0 & 2 & -2 & 1 \\ 1 & 0 & -2 & 2 & -1 \\ -1 & 0 & 2 & 0 & 1 \\ 0 & 0 & -2 & 0 & -1 \\ 0 & 0 & 0 & 0 & 1 \end{array}\right]$$

Element flexibility matrix [α] :

$$[\alpha] = \frac{l}{6EI}\begin{bmatrix} 2 & -1 \\ -1 & 2 \end{bmatrix}$$

Element 1 : $$[\alpha_1] = \frac{3}{6E.[2I]}\begin{bmatrix} 2 & -1 \\ -1 & 2 \end{bmatrix} = \frac{1}{12EI}\begin{bmatrix} 6 & -3 \\ -3 & 6 \end{bmatrix}$$

Element 2 : $$[\alpha_2] = \frac{2}{6EI}\begin{bmatrix} 2 & -1 \\ -1 & 2 \end{bmatrix} = \frac{1}{6EI}\begin{bmatrix} 4 & -2 \\ -2 & 4 \end{bmatrix}$$

Element 3 : $$[\alpha_3] = \frac{2}{6E\cdot[1.5I]}\begin{bmatrix} 2 & -1 \\ -1 & 2 \end{bmatrix} = \frac{1}{9EI}\begin{bmatrix} 4 & -2 \\ -2 & 4 \end{bmatrix}$$

Assembled element flexibility matrix :

$$[\alpha] = \frac{1}{36EI}\begin{bmatrix} 18 & -9 & 0 & 0 & 0 & 0 \\ -9 & 18 & 0 & 0 & 0 & 0 \\ 0 & 0 & 24 & -12 & 0 & 0 \\ 0 & 0 & -12 & 24 & 0 & 0 \\ 0 & 0 & 0 & 0 & 16 & -8 \\ 0 & 0 & 0 & 0 & -8 & 16 \end{bmatrix}$$

To find the redundant forces :

$$\{F^0\} = -[a_{00}]^{-1} \cdot [a_{0*}] \cdot \{F^*\}$$

$$[a_{00}] = [b_0]^T [\alpha] [b_0]$$

$$= \begin{bmatrix} 1 & 2 & -2 & 2 & -2 & 0 \\ 2 & -2 & 2 & 0 & 0 & 0 \\ -1 & 1 & -1 & 1 & -1 & 1 \end{bmatrix} \cdot \frac{1}{36EI}\begin{bmatrix} 18 & -9 & 0 & 0 & 0 & 0 \\ -9 & 18 & 0 & 0 & 0 & 0 \\ 0 & 0 & 24 & -12 & 0 & 0 \\ 0 & 0 & -12 & 24 & 0 & 0 \\ 0 & 0 & 0 & 0 & 16 & -8 \\ 0 & 0 & 0 & 0 & -8 & 16 \end{bmatrix}\begin{bmatrix} 1 & 2 & -1 \\ 2 & -2 & 1 \\ -2 & 2 & -1 \\ 2 & 0 & 1 \\ -2 & 0 & -1 \\ 0 & 0 & 1 \end{bmatrix}$$

$$= \frac{1}{36EI}\begin{bmatrix} 0 & 27 & -72 & 72 & -32 & 16 \\ 54 & -54 & 48 & -24 & 0 & 0 \\ -27 & 27 & -36 & 36 & -24 & 24 \end{bmatrix}\begin{bmatrix} 1 & 2 & -1 \\ 2 & -2 & 1 \\ -2 & 2 & -1 \\ 2 & 0 & 1 \\ -2 & 0 & -1 \\ 0 & 0 & 1 \end{bmatrix}$$

$$[a_{00}] = \frac{1}{36EI}\begin{bmatrix} 406 & -198 & 219 \\ -198 & 312 & -180 \\ 219 & -180 & 174 \end{bmatrix}$$

$$[a_{00}]^{-1} = 36EI\begin{bmatrix} 0.00807 & -0.00183 & -0.01205 \\ -0.00183 & 0.00836 & 0.01096 \\ -0.01205 & 0.01096 & 0.03225 \end{bmatrix}$$

$$[a_{0*}] = [b_0]^T \cdot [\alpha] \cdot [b^*]$$

$$= \begin{bmatrix} 1 & 2 & -2 & 2 & -2 & 0 \\ 2 & -2 & 2 & 0 & 0 & 0 \\ -1 & 1 & -1 & 1 & -1 & 1 \end{bmatrix} \cdot \frac{1}{36EI} \begin{bmatrix} 18 & -9 & 0 & 0 & 0 & 0 \\ -9 & 18 & 0 & 0 & 0 & 0 \\ 0 & 0 & 24 & -12 & 0 & 0 \\ 0 & 0 & -12 & 24 & 0 & 0 \\ 0 & 0 & 0 & 0 & 16 & -8 \\ 0 & 0 & 0 & 0 & -8 & 16 \end{bmatrix} \cdot \begin{bmatrix} 1 & -3 \\ -1 & 0 \\ 1 & 0 \\ -1 & 0 \\ 0 & 0 \\ 0 & 0 \end{bmatrix}$$

$$= \frac{1}{36EI} \begin{bmatrix} 0 & 27 & -72 & 72 & -32 & 16 \\ 54 & -54 & 48 & -24 & 0 & 0 \\ -27 & 27 & -36 & 36 & -24 & 24 \end{bmatrix} \begin{bmatrix} 1 & -3 \\ -1 & 0 \\ 1 & 0 \\ -1 & 0 \\ 0 & 0 \\ 0 & 0 \end{bmatrix}$$

$$[a_{0*}] = \frac{1}{36EI} \begin{bmatrix} -171 & 0 \\ 180 & -162 \\ -126 & 81 \end{bmatrix}$$

$$\{F^0\} = -[a_{00}]^{-1} [a_{0*}] \{F^*\}$$

$$= -36EI \begin{bmatrix} 0.00807 & -0.00183 & -0.01205 \\ -0.00183 & 0.00836 & 0.01096 \\ -0.01205 & 0.01096 & 0.03225 \end{bmatrix} \cdot \frac{1}{36EI} \begin{bmatrix} -171 & 0 \\ 180 & -162 \\ -126 & 81 \end{bmatrix} \begin{bmatrix} 3 \\ 9 \end{bmatrix}$$

$$= - \begin{bmatrix} -0.191 & -0.679 \\ 0.438 & -0.467 \\ -0.0302 & 0.837 \end{bmatrix} \begin{bmatrix} 3 \\ 9 \end{bmatrix}$$

$$\{F^0\} = \begin{bmatrix} 6.69 \\ 2.893 \\ -7.446 \end{bmatrix}$$

Element forces $\{P\} = [b]\{F\}$, where $\{F\} = \begin{Bmatrix} F^* \\ F^0 \end{Bmatrix}$

$$\begin{bmatrix} P_1 \\ P_2 \\ P_3 \\ P_4 \\ P_5 \\ P_6 \end{bmatrix} = \begin{bmatrix} 1 & -3 & 1 & 2 & -1 \\ -1 & 0 & 2 & -2 & 1 \\ 1 & 0 & -2 & 2 & -1 \\ -1 & 0 & 2 & 0 & 1 \\ 0 & 0 & -2 & 0 & -1 \\ 0 & 0 & 0 & 0 & 1 \end{bmatrix} \begin{bmatrix} 3 \\ 9 \\ 6.69 \\ 2.893 \\ -7.446 \end{bmatrix}$$

$$\{P\} = \begin{bmatrix} -4.078 \\ -2.852 \\ 2.852 \\ 2.934 \\ -5.934 \\ -7.446 \end{bmatrix}$$

$$\{P^f\} = \{P\} + \{FEM\} = \begin{bmatrix} -4.078 \\ -2.852 \\ 2.852 \\ 2.934 \\ -5.934 \\ -7.446 \end{bmatrix} + \begin{bmatrix} -3 \\ 3 \\ -3 \\ 3 \\ 0 \\ 0 \end{bmatrix}$$

$$\{P^f\} = \begin{bmatrix} -7.078 \\ 0.148 \\ -0.148 \\ 5.934 \\ -5.934 \\ -7.446 \end{bmatrix} \text{kNm}$$

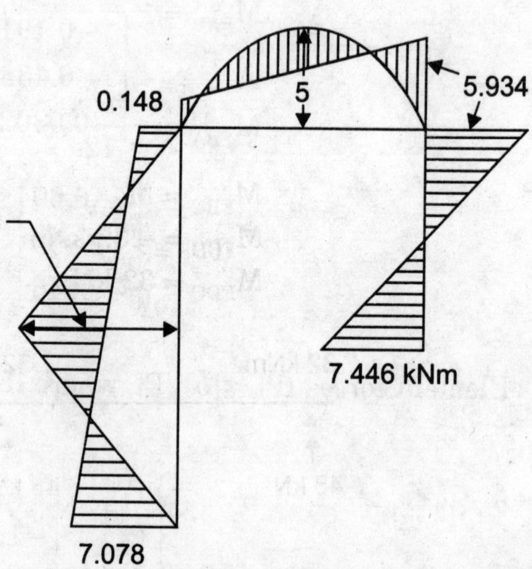

Bending moment diagram

Example 4.28 *Analyse the two level beam loaded as shown in figure below by the matrix flexibility method. Sketch the bending moment diagram.*

Solution The primary structure will be obtained by releasing the right support D.

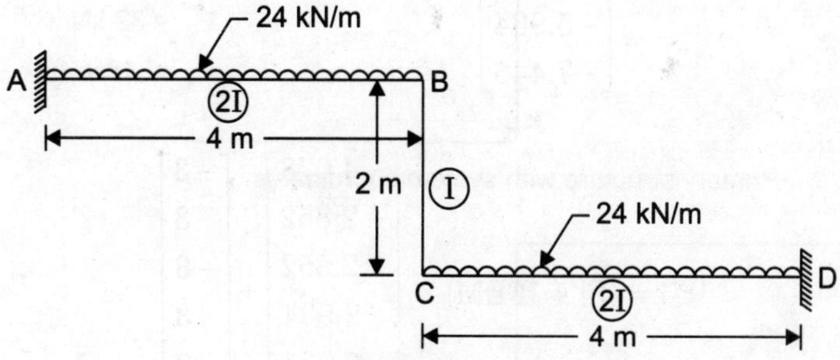

Given frame

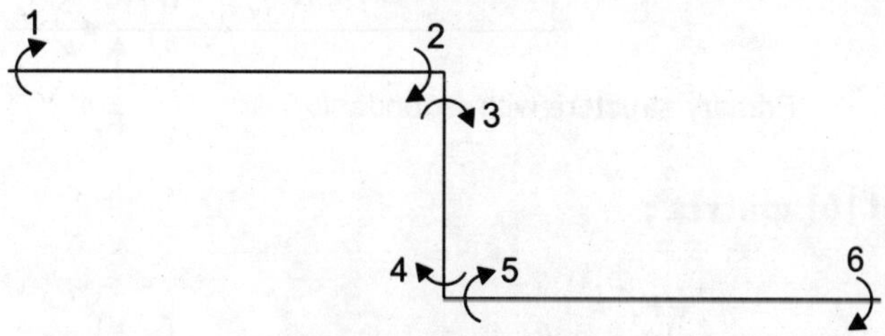

Element coordinates

Fixed end moments :

$$M_{FAB} = -\frac{wl^2}{12} = -\frac{24 \times 4^2}{12} = -32 \text{ kNm}$$

$$M_{FBA} = +\frac{wl^2}{12} = \frac{24 \times 4^2}{12} = 32 \text{ kNm}$$

$$M_{FBC} = M_{FCB} = 0$$
$$M_{FCD} = -32 \text{ kNm}$$
$$M_{FDC} = 32 \text{ kNm}$$

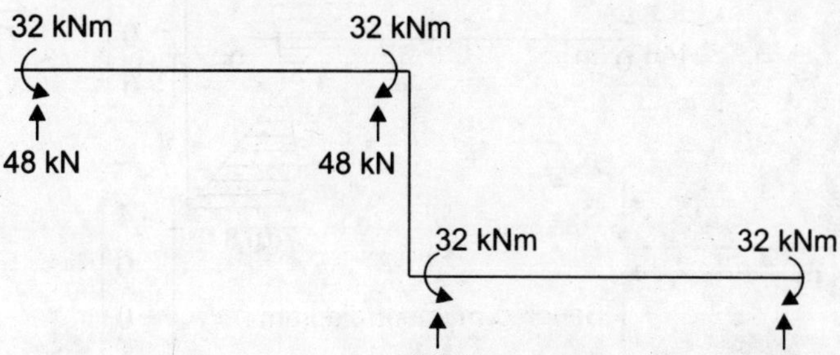

Fixed end moments

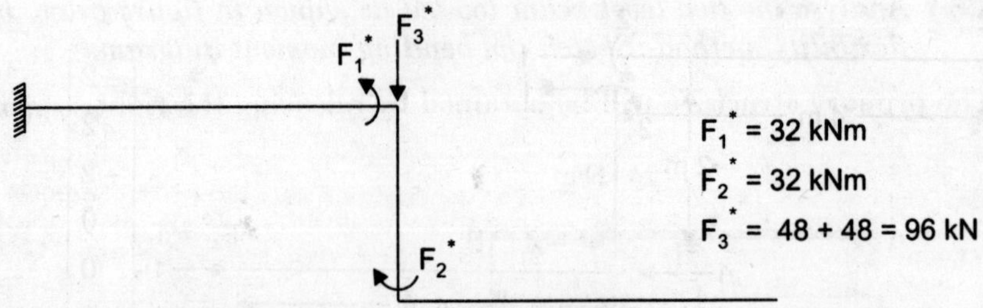

Primary structure with system coordinates

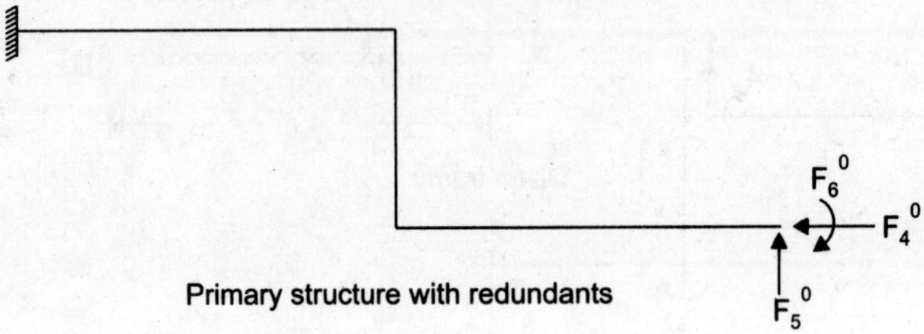

Primary structure with redundants

Derivation of [*b*] matrix :

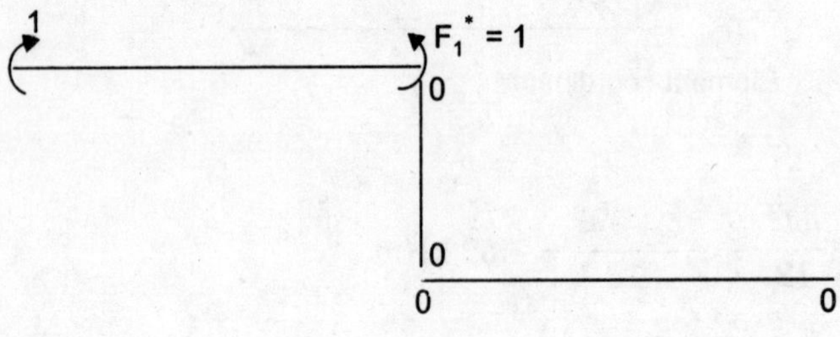

$$\begin{bmatrix} 1 \\ -1 \\ 0 \\ 0 \\ 0 \\ 0 \end{bmatrix}$$

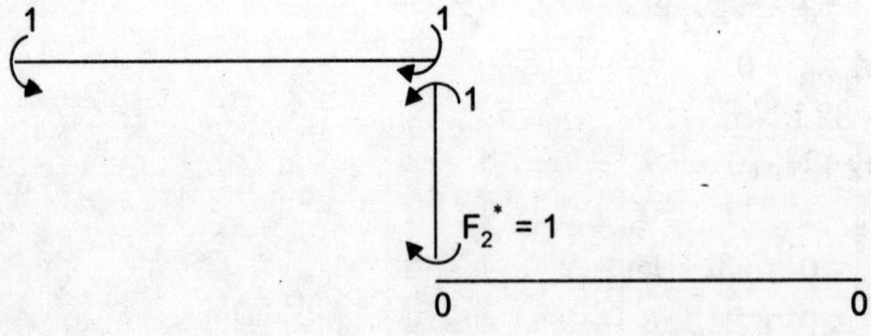

$$\begin{bmatrix} -1 \\ 1 \\ -1 \\ 1 \\ 0 \\ 0 \end{bmatrix}$$

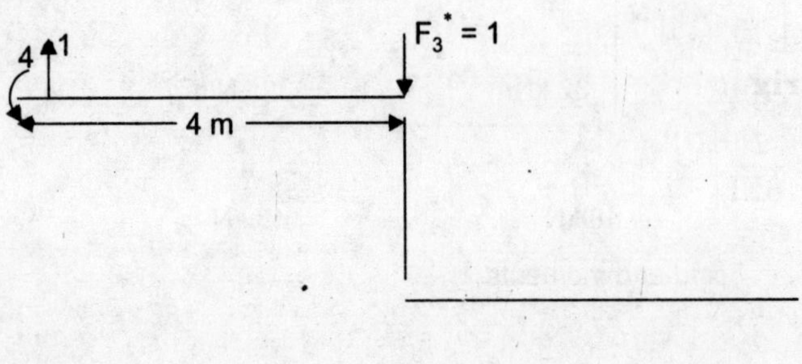

$$\begin{bmatrix} -4 \\ 0 \\ 0 \\ 0 \\ 0 \\ 0 \end{bmatrix}$$

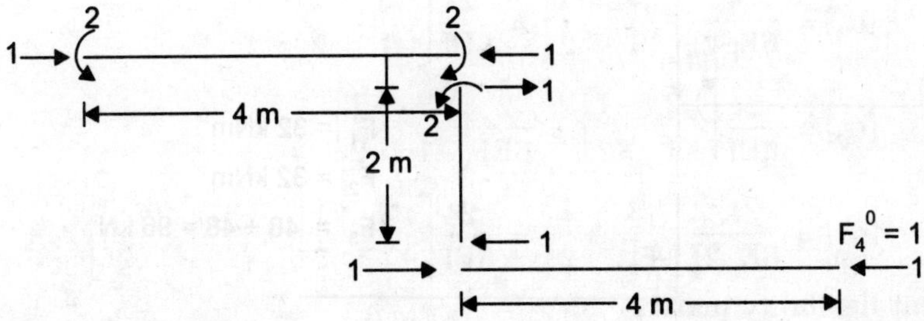

$$\begin{bmatrix} -2 \\ 2 \\ -2 \\ 0 \\ 0 \\ 0 \end{bmatrix}$$

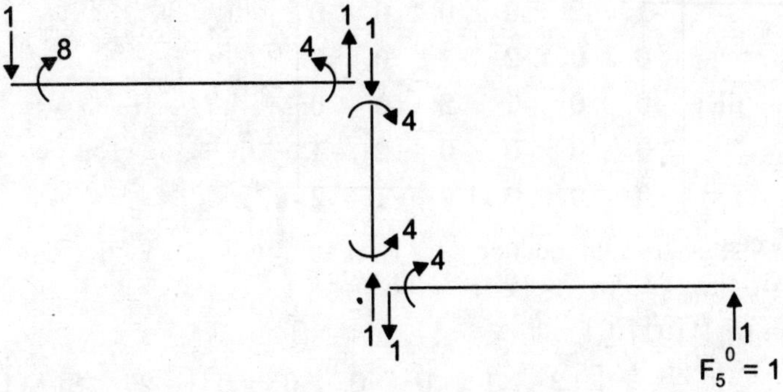

$$\begin{bmatrix} 8 \\ -4 \\ 4 \\ -4 \\ 4 \\ 0 \end{bmatrix}$$

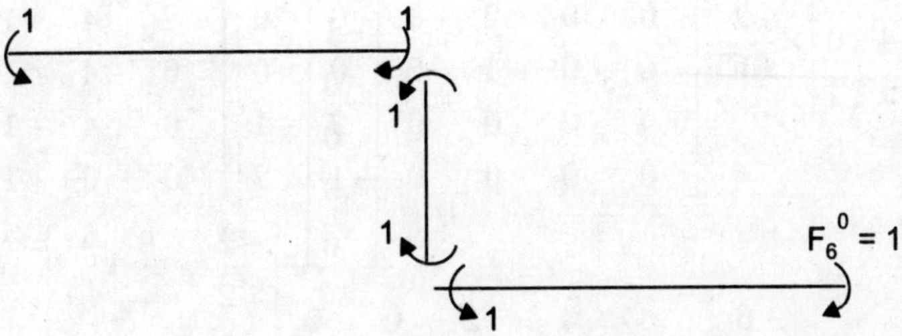

$$\begin{bmatrix} -1 \\ 1 \\ -1 \\ 1 \\ -1 \\ 1 \end{bmatrix}$$

$$[b] = \left[\begin{array}{ccc|ccc} & b^* & & & b^0 & \\ 1 & -1 & -4 & -2 & 8 & -1 \\ -1 & 1 & 0 & 2 & -4 & 1 \\ 0 & -1 & 0 & -2 & 4 & -1 \\ 0 & 1 & 0 & 0 & -4 & 1 \\ 0 & 0 & 0 & 0 & 4 & -1 \\ 0 & 0 & 0 & 0 & 0 & 1 \end{array}\right]$$

Element flexibility matrix [α] :

$$[\alpha] = \frac{l}{6EI}\begin{bmatrix} 2 & -1 \\ -1 & 2 \end{bmatrix}$$

Element 1 : $[\alpha_1] = \dfrac{4}{6E \cdot 2I}\begin{bmatrix} 2 & -1 \\ -1 & 2 \end{bmatrix} = \dfrac{2}{6EI}\begin{bmatrix} 2 & -1 \\ -1 & 2 \end{bmatrix}$

Element 2 : $[\alpha_2] = \dfrac{2}{6EI}\begin{bmatrix} 2 & -1 \\ -1 & 2 \end{bmatrix} = \dfrac{2}{6EI}\begin{bmatrix} 2 & -1 \\ -1 & 2 \end{bmatrix}$

Element 3 : $[\alpha_3] = \dfrac{4}{6E \cdot 2I}\begin{bmatrix} 2 & -1 \\ -1 & 2 \end{bmatrix} = \dfrac{2}{6EI}\begin{bmatrix} 2 & -1 \\ -1 & 2 \end{bmatrix}$

Assembled element flexibility matrix

$$[\alpha] = \frac{2}{6EI}\begin{bmatrix} 2 & -1 & 0 & 0 & 0 & 0 \\ -1 & 2 & 0 & 0 & 0 & 0 \\ 0 & 0 & 2 & -1 & 0 & 0 \\ 0 & 0 & -1 & 2 & 0 & 0 \\ 0 & 0 & 0 & 0 & 2 & -1 \\ 0 & 0 & 0 & 0 & -1 & 2 \end{bmatrix}$$

To find the redundant forces

$$\{F^0\} = -[a_{00}]^{-1} \cdot [a_{0*}] \cdot [F^*]$$

$$[a_{00}] = [b_0]^T [\alpha] [b_0]$$

$$= \begin{bmatrix} -2 & 2 & -2 & 0 & 0 & 0 \\ 8 & -4 & 4 & -4 & 4 & 0 \\ -1 & 1 & -1 & 1 & -1 & 1 \end{bmatrix} \times \frac{2}{6EI}\begin{bmatrix} 2 & -1 & 0 & 0 & 0 & 0 \\ -1 & 2 & 0 & 0 & 0 & 0 \\ 0 & 0 & 2 & -1 & 0 & 0 \\ 0 & 0 & -1 & 2 & 0 & 0 \\ 0 & 0 & 0 & 0 & 2 & -1 \\ 0 & 0 & 0 & 0 & -1 & 2 \end{bmatrix}\begin{bmatrix} -2 & 8 & -1 \\ 2 & -4 & 1 \\ -2 & 4 & -1 \\ 0 & -4 & 1 \\ 0 & 4 & -1 \\ 0 & 0 & 1 \end{bmatrix}$$

$$= \frac{2}{6EI}\begin{bmatrix} -6 & 6 & -4 & 2 & 0 & 0 \\ 20 & -16 & 12 & -12 & 8 & -4 \\ -3 & 3 & -3 & 3 & -3 & 3 \end{bmatrix}\begin{bmatrix} -2 & 8 & -1 \\ 2 & -4 & 1 \\ -2 & 4 & -1 \\ 0 & -4 & 1 \\ 0 & 4 & -1 \\ 0 & 0 & 1 \end{bmatrix}$$

$$[a_{00}] = \frac{2}{6EI}\begin{bmatrix} 32 & -96 & 18 \\ -96 & 352 & -72 \\ 18 & -72 & 18 \end{bmatrix}$$

$$[a_{00}]^{-1} = \frac{6EI}{2}\begin{bmatrix} 0.2 & 0.075 & 0.1 \\ 0.075 & 0.04375 & 0.1 \\ 0.1 & 0.1 & 0.3555 \end{bmatrix}$$

$$[a_{0*}] = [b_0]^T [\alpha] [b^*]$$

$$= \begin{bmatrix} -2 & 2 & -2 & 0 & 0 & 0 \\ 8 & -4 & 4 & -4 & 4 & 0 \\ -1 & 1 & -1 & 1 & -1 & 1 \end{bmatrix} \times \frac{2}{6EI} \begin{bmatrix} 2 & -1 & 0 & 0 & 0 & 0 \\ -1 & 2 & 0 & 0 & 0 & 0 \\ 0 & 0 & 2 & -1 & 0 & 0 \\ 0 & 0 & -1 & 2 & 0 & 0 \\ 0 & 0 & 0 & 0 & 2 & -1 \\ 0 & 0 & 0 & 0 & -1 & 2 \end{bmatrix} \begin{bmatrix} 1 & -1 & -4 \\ -1 & 1 & 0 \\ 0 & -1 & 0 \\ 0 & 1 & 0 \\ 0 & 0 & 0 \\ 0 & 0 & 0 \end{bmatrix}$$

$$= \frac{2}{6EI} \begin{bmatrix} -6 & 6 & -4 & 2 & 0 & 0 \\ 20 & -16 & 12 & -12 & 8 & -4 \\ -3 & 3 & -3 & 3 & -3 & 3 \end{bmatrix} \begin{bmatrix} 1 & -1 & -4 \\ -1 & 1 & 0 \\ 0 & -1 & 0 \\ 0 & 1 & 0 \\ 0 & 0 & 0 \\ 0 & 0 & 0 \end{bmatrix}$$

$$[a_{0*}] = \frac{2}{6EI} \begin{bmatrix} -12 & 18 & 24 \\ 36 & -60 & -80 \\ -6 & 12 & 12 \end{bmatrix}$$

$$\{F^0\} = -[a_{00}]^{-1}[a_{0*}]\{F^*\}$$

$$= -\frac{6EI}{2} \begin{bmatrix} 0.2 & 0.075 & 0.1 \\ 0.075 & 0.04375 & 0.1 \\ 0.1 & 0.1 & 0.3555 \end{bmatrix} \times \frac{2}{6EI} \begin{bmatrix} -12 & 18 & 24 \\ 36 & -60 & -80 \\ -6 & 12 & 12 \end{bmatrix} \begin{bmatrix} 32 \\ 32 \\ 96 \end{bmatrix}$$

$$= -\begin{bmatrix} -0.3 & 0.3 & 0 \\ 0.0750 & -0.075 & -0.50 \\ 0.267 & 0.066 & -1.334 \end{bmatrix} \begin{bmatrix} 32 \\ 32 \\ 96 \end{bmatrix} = -\begin{bmatrix} 0 \\ -48 \\ -117.33 \end{bmatrix}$$

$$\{F^0\} = \begin{bmatrix} 0 \\ 48 \\ 117.33 \end{bmatrix}$$

Element forces $\{P\} = [b]\{F\}$, where $\{F\} = \begin{Bmatrix} F^* \\ F^0 \end{Bmatrix}$

$$\begin{bmatrix} P_1 \\ P_2 \\ P_3 \\ P_4 \\ P_5 \\ P_6 \end{bmatrix} = \begin{bmatrix} 1 & -1 & -4 & -2 & 8 & -1 \\ -1 & 1 & 0 & 2 & -4 & 1 \\ 0 & -1 & 0 & -2 & 4 & -1 \\ 0 & 1 & 0 & 0 & -4 & 1 \\ 0 & 0 & 0 & 0 & 4 & -1 \\ 0 & 0 & 0 & 0 & 0 & 1 \end{bmatrix} \begin{bmatrix} 32 \\ 32 \\ 96 \\ 0 \\ 48 \\ 117.33 \end{bmatrix}$$

$$\{P\} = \begin{bmatrix} -117.33 \\ -74.67 \\ 42.67 \\ -42.67 \\ 74.67 \\ 117.33 \end{bmatrix}$$

$$\{P^f\} = \{P\} + \{FEM\}$$

$$= \begin{bmatrix} -117.33 \\ -74.67 \\ 42.67 \\ -42.67 \\ 74.67 \\ 117.33 \end{bmatrix} + \begin{bmatrix} -32 \\ 32 \\ 0 \\ 0 \\ -32 \\ 32 \end{bmatrix} = \begin{bmatrix} -149.33 \\ -42.67 \\ 42.67 \\ -42.67 \\ 42.67 \\ 149.33 \end{bmatrix} \text{ kNm}$$

Simply supported beam moments :

$$\mu_{AB} = \mu_{CD} = \frac{wl^2}{8} = \frac{24 \times 4^2}{8} = 48 \text{ kNm}$$

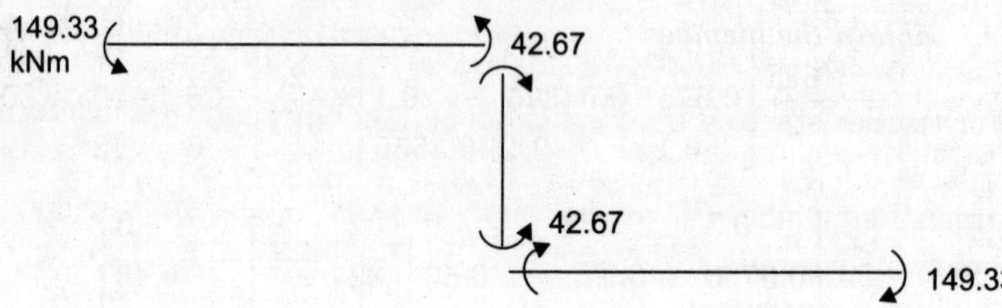

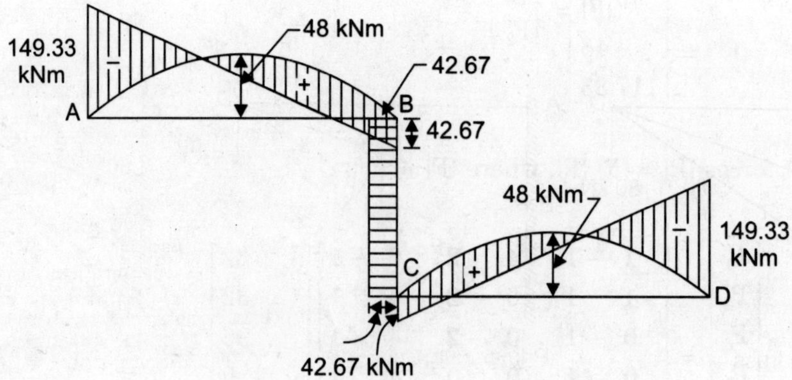

Bending moment diagram

4.7. [b] MATRIX FOR PIN JOINTED TRUSSES

Finding the [b] matrix for pin jointed trusses is just a process of finding the internal forces (in members) due to a given loading.

Here, in the relation {P} = {b} {F}, {F} stands for the system of external loads and {P} stands for the set of member forces.

If the truss is internally (or externally) indeterminate, then we shall consider all indeterminate forces as part of $\{F\}^0$ system.

In Fig. 4.41 of example 4.30, the truss is one degree indeterminate. Member AD is identified as indeterminate. Here, Fig. 4.42 (a) is the primary structure. Figure 4.42 (b) shows the {P} system consisting of member forces in the primary structure.

Thus, when we apply the {F} forces on the primary structure one by one the vectors of {P} forces give the appropriate columns of [b] matrix. When F_1^* (= 1) is applied at D, the 5 internal forces give us $[b]^*$. [Fig. 4.42 (c)]

When $F_2^0 = 1$ is applied at A and D, the corresponding 5 member forces give us $[b]^0$. [Fig. 4.42 (d)]

It may be noted that in designating the co-ordinates for {P}, we take tension as positive. The arrows indicate the forces coming *on the members* (and not the forces exerted by the members on the joints).

Example 4.29 *Analyse the truss shown in figure below by the matrix force method and obtain the member forces. Take* $\frac{AE}{l} = 1$ *for all the members.*

Solution For trusses starting from a rigid base, degree of internal indeterminacy

$$I = m - 2j = 3 - 2 \times 1 = 1$$

Let us treat the member CD as redundant.

The primary structure is shown in figure.

The force vector consists of

$$\{F\} = \begin{Bmatrix} F_1^* \\ F_2^0 \end{Bmatrix} = \begin{Bmatrix} 80 \\ F_2^0 \end{Bmatrix}$$

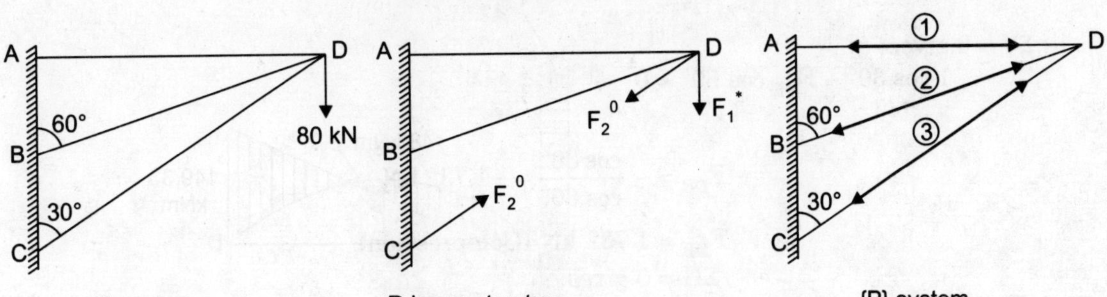

Element forces are

$$\{P\} = \begin{Bmatrix} P_1 \\ P_2 \\ P_3 \end{Bmatrix}$$

Transformation matrix [*b*] :

To obtain the 1st column of [*b*] matrix, applying $F_1^* = 1$ the forces in the elements will be obtained using method of joints.

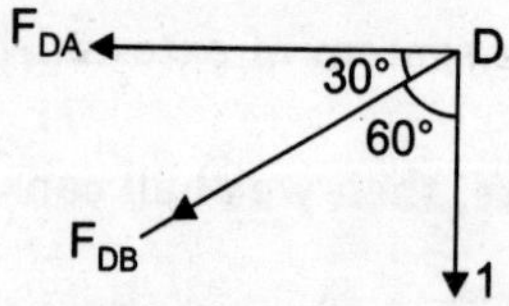

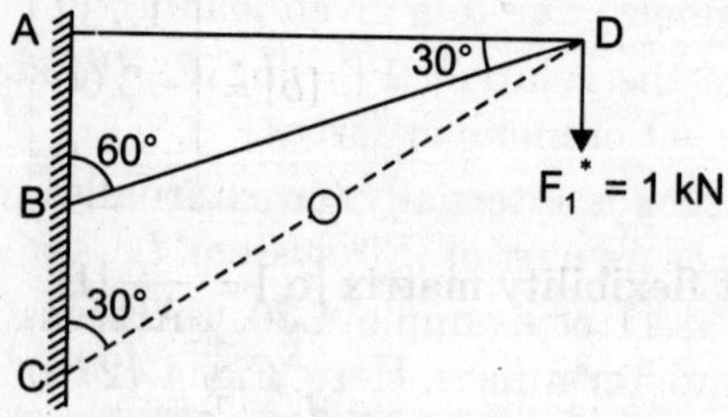

$\Sigma V = 0$ gives

$$\underset{(\downarrow)}{F_{DB} \cos 60°} + \underset{(\downarrow)}{1} = 0$$

$$F_{DB} = -\frac{1}{\cos 60°} = -2 \text{ kN}$$

$$F_{DB} = 2 \text{ kN (Compression)}$$

$\Sigma H = 0$ gives

$$\underset{(\leftarrow)}{F_{DA}} + \underset{(\leftarrow)}{F_{DB}} \cos 30° = 0$$

$$F_{DA} = -F_{DB} \cos 30° = -(-2) \cos 30° = 1.732 \text{ kN}$$

$$F_{DA} = 1.732 \text{ kN (Tension)}$$

$\therefore$ $$[b]^* = [1.732 \quad -2.0 \quad 0]^T$$

To obtain the second column of [*b*] matrix, $F_2^0 = 1$ will be applied along the redundant member and the element forces will be obtained,

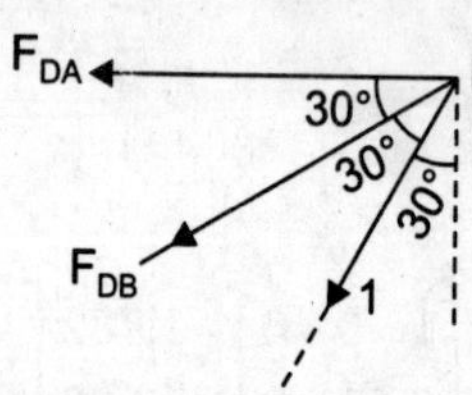

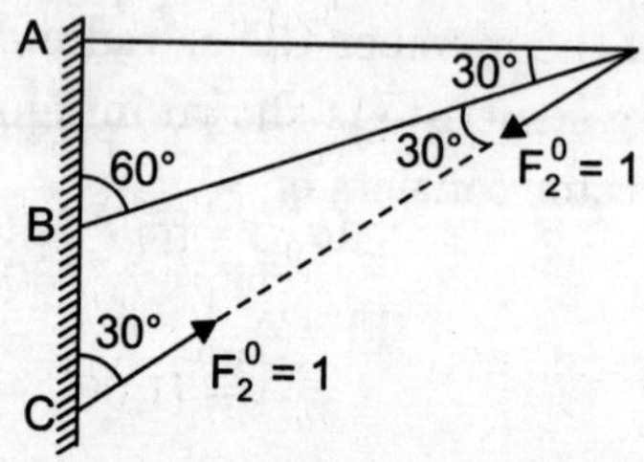

$\Sigma V = 0$ gives

$$\underset{(\downarrow)}{1 \cos 30°} + \underset{(\downarrow)}{F_{DB} \cos 60°} = 0$$

$$F_{DB} = -\frac{\cos 30°}{\cos 60°} = -1.732 \text{ kN}$$

$$F_{DB} = 1.732 \text{ kN (Compression)}$$

$\Sigma H = 0$ gives

$$\underset{(\leftarrow)}{F_{DA}} + \underset{(\leftarrow)}{F_{DB}} \cos 30° + \underset{(\leftarrow)}{1 \cos 60°} = 0$$

$$F_{DA} = -F_{DB} \cos 30° - 1 \cos 60°$$

$$= -(-1.732) \cos 30° - 1 \cos 60° = 1.500 - 0.5 = 1.00$$

$$F_{DA} = 1.00 \text{ kN (Tension)}$$

$\therefore$ $$[b]^0 = [1.00 \quad -1.732 \quad 1.00]^T$$

Hence, $$[b] = \begin{bmatrix} 1.732 & 1.00 \\ -2.00 & -1.732 \\ 0 & 1.00 \end{bmatrix}$$

Element flexibility matrix $[\alpha_i] = \dfrac{l}{AE}[1]$

$\because$ $\dfrac{l}{AE} = 1$ for all members

Assembled element flexibility matrix

$$[\alpha] = \frac{l}{AE}\begin{bmatrix} 1 & 0 & 0 \\ 0 & 1 & 0 \\ 0 & 0 & 1 \end{bmatrix} = \begin{bmatrix} 1 & 0 & 0 \\ 0 & 1 & 0 \\ 0 & 0 & 1 \end{bmatrix}$$

Redundant force, $$\{F\}^0 = -[a_{00}]^{-1}[a_{0*}]\{F^*\}$$

$$[a_{00}] = [b_0]^T [\alpha] [b_0]$$

$$= [1.00 \quad -1.732 \quad 1.00]\begin{bmatrix} 1 & 0 & 0 \\ 0 & 1 & 0 \\ 0 & 0 & 1 \end{bmatrix}\begin{bmatrix} 1.00 \\ -1.732 \\ 1.00 \end{bmatrix}$$

$$= [1.00 \quad -1.732 \quad 1.00]\begin{bmatrix} 1.00 \\ -1.732 \\ 1.00 \end{bmatrix} = [5.00]$$

$$[a_{00}]^{-1} = \frac{1}{5} = 0.20$$

$$[a_{0*}] = [b_0]^T [\alpha] [b_*]$$

$$= [1.00 \quad -1.732 \quad 1.00]\begin{bmatrix} 1 & 0 & 0 \\ 0 & 1 & 0 \\ 0 & 0 & 1 \end{bmatrix}\begin{bmatrix} 1.732 \\ -2.00 \\ 0 \end{bmatrix}$$

$$= [1.00 \quad -1.732 \quad 1.00]\begin{bmatrix} 1.732 \\ -2.00 \\ 0 \end{bmatrix} = 5.196$$

$$\{F\}^0 = -[0.20][5.196]\{80\} = -83.136 \text{ kN}$$

Member forces, $$\{P\} = [b]\{F\}$$

$$= \begin{bmatrix} 1.732 & 1.00 \\ -2.00 & -1.732 \\ 0 & 1.00 \end{bmatrix}\begin{Bmatrix} 80 \\ -83.136 \end{Bmatrix} = \begin{Bmatrix} 55.424 \\ -16.00 \\ -83.136 \end{Bmatrix}$$

$\therefore$ $F_{AD} = 55.424$ kN (Tension) ; $F_{BD} = 16.00$ kN (Compression) ;
$F_{CD} = 83.136$ kN (Compression)

Example 4.30 *A statically indeterminate frame shown in Fig. 4.41 carries a load of 80 kN. Analyse the frame by matrix flexibility method. A and E are same for all members.*

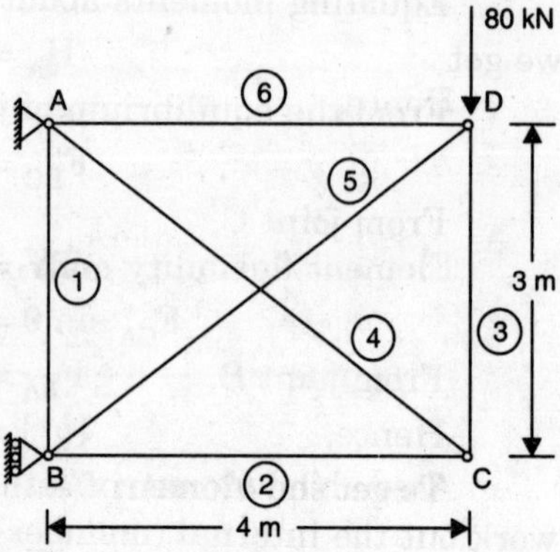

Fig. 4.41

Solution Internal indeterminacy, $I = m - (2j - 3)$

$$= 6 - (8 - 3) = 1$$

External indeterminacy $E = r - R = 3 - 3 = 0$

The frame is statically (internally) indeterminate to one degree. Treating member AD as redundant, the primary structure is as shown in Fig. 4.42.

To form the [*b*] matrix, we would treat the external load position as the global system and internal forces in the 6 members as local co-ordinates. Let us now apply $F^* = 1$. [vide Fig. 4.42 (*c*)]

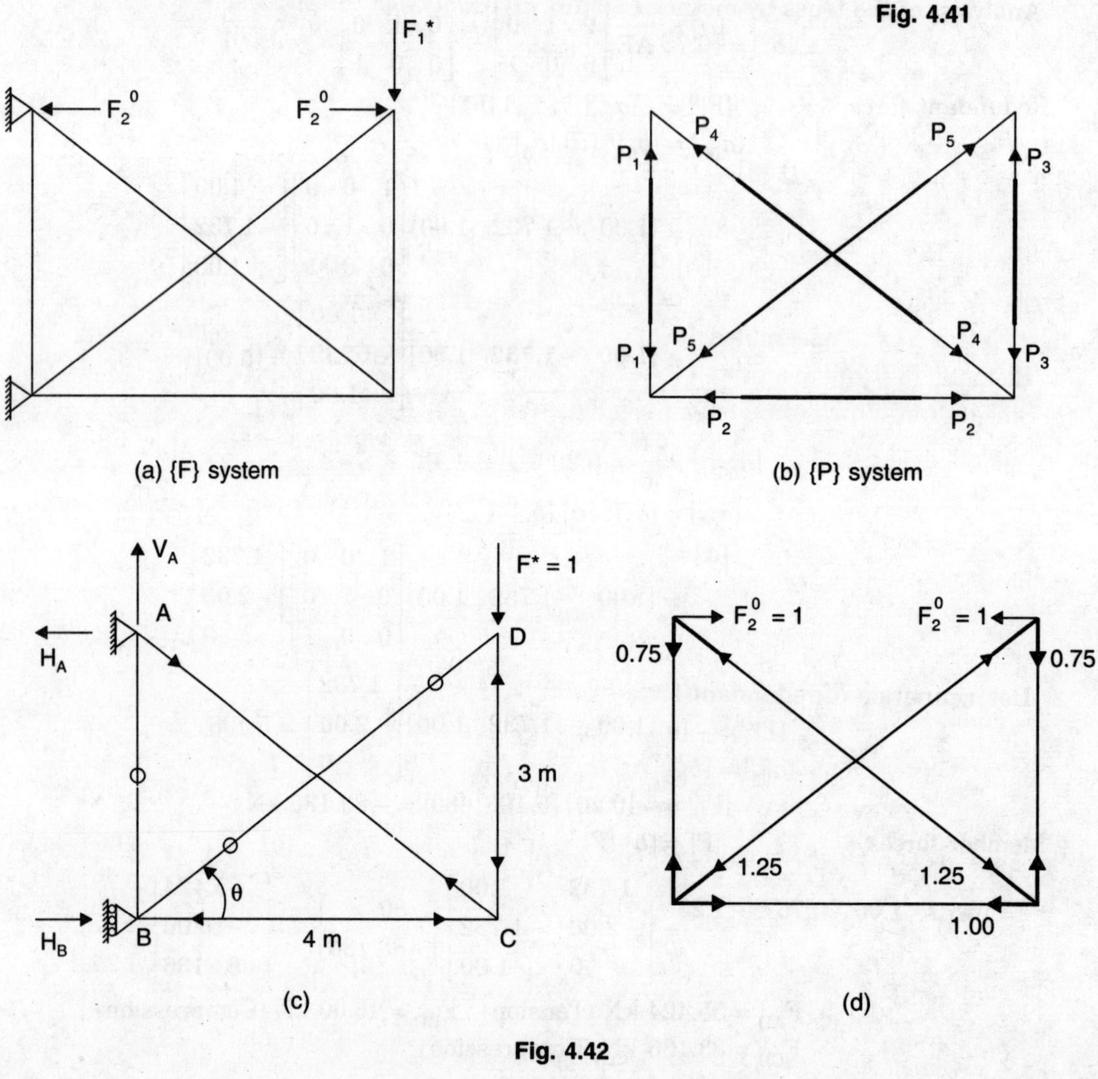

Fig. 4.42

$$\text{Tan } \theta = 3/4 = 0.75 \; ; \; \sin\theta = 0.6 \; ; \; \cos\theta = 0.8$$

Since $\Sigma V = 0, \quad V_A = 1$

Equating moments about B to zero, and ΣH to zero,

we get $H_A = 1.333$ and $H_B = 1.333$

From the equilibrium of joint D,

$$F_{DC} = 1 \text{ (Compression)} = -1$$

From joint C,

$$\Sigma V = 0 \text{ gives}$$

$$F_{CA} \sin\theta = 1 \; ; \; F_{CA} = 1.667 \; ; \; F_{CB} = 1.333$$

From joint B, $F_{BA} = 0$ and $F_{BC} = 1.33$ (as already seen)

Hence, $[b^*] = [0 \quad -1.333 \quad -1.00 \quad 1.667 \quad 0 \quad 0]^T$

To get the b^0 matrix, let us apply a unit force $F^0 = 1$ along member (redundant) AD and work out the internal (member) forces. [vide Fig. 4.33 (*d*)]

Analysis of the truss by method of joints gives member forces as above. Hence

$$[b^0] = [0.75 \quad 1.00 \quad 0.75 \quad -1.25 \quad -1.25 \quad 1.00]^T$$

$$[b] = \left[\begin{array}{c|c} 0 & 0.75 \\ -1.333 & 1.00 \\ -1.00 & 0.75 \\ +1.667 & -1.25 \\ 0 & -1.25 \\ 0 & 1.00 \end{array}\right]$$
$$\quad b^* \qquad b^0$$

Assembled Element Flexibility matrix :

Since A and E are same for all members,

$$[\alpha] = \frac{1}{AE}\begin{bmatrix} 3 & & & & & \\ & 4 & & \text{zeros} & & \\ & & 3 & & & \\ & & & 5 & & \\ & \text{zeros} & & & 5 & \\ & & & & & 4 \end{bmatrix}$$

Determination of redundant force F^0 :

$$\{F\}^0 = -[a_{00}]^{-1}[a_{0*}]\{F\}^*$$

$$[a_{00}] = [b_0]^T[\alpha][b_0]$$

$$= [0.75 \quad 1.00 \quad 0.75 \quad -1.25 \quad -1.25 \quad 1.00]\frac{1}{AE}\begin{bmatrix} 3 & & & & & \\ & 4 & & [0] & & \\ & & 3 & & & \\ & & & 5 & & \\ & [0] & & & 5 & \\ & & & & & 4 \end{bmatrix}\begin{bmatrix} 0.75 \\ 1.00 \\ 0.75 \\ -1.25 \\ -1.25 \\ 1.00 \end{bmatrix}$$

$$[a_{00}] = \frac{27}{AE}$$

$$[a_{00}]^{-1} = \frac{AE}{27}$$

$$[a_{0*}] = [b^0]^T\,[\alpha]\,[b^*]$$

$$= [0.75 \quad 1.00 \quad 0.75 \quad -1.25 \quad -1.25 \quad 1.00]\,\frac{1}{AE}\begin{bmatrix} 3 & & & & & \\ & 4 & & [0] & & \\ & & 3 & & & \\ & & & 5 & & \\ & [0] & & & 5 & \\ & & & & & 4 \end{bmatrix}\begin{bmatrix} 0 \\ -1.333 \\ -1.00 \\ 1.667 \\ 0 \\ 0 \end{bmatrix}$$

$$= \frac{-18.00}{AE}$$

$$\{F\}^0 = -[a_{00}]^{-1}\,[a_{0*}]\,\{F\}^*$$

$$\{F\}^0 = -\frac{AE}{27} \times \frac{(-18.00)}{AE} \times 80 = 53.33 \text{ kN.}$$

Element forces $\{P\}^f = [b]\,\{F\}$

$$\begin{matrix}\{P\}^f \\ \text{(Final Forces)}\end{matrix} = \begin{bmatrix} 0 & 0.75 \\ -1.333 & 1.00 \\ -1.00 & 0.75 \\ 1.667 & -1.25 \\ 0 & -1.25 \\ 0 & 1.00 \end{bmatrix}\begin{Bmatrix} 80 \\ 53.33 \end{Bmatrix} = \begin{bmatrix} 39.998 \\ -53.31 \\ -40.00 \\ 66.698 \\ -66.663 \\ 53.33 \end{bmatrix}$$

Example 4.31 *Using matrix flexibility method analyse the truss loaded as shown in Fig. 4.43 and find the member forces. A and E are the same for all members.*

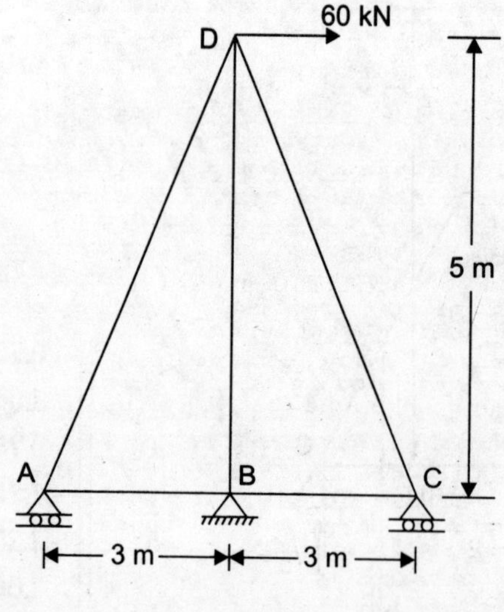

Fig. 4.43

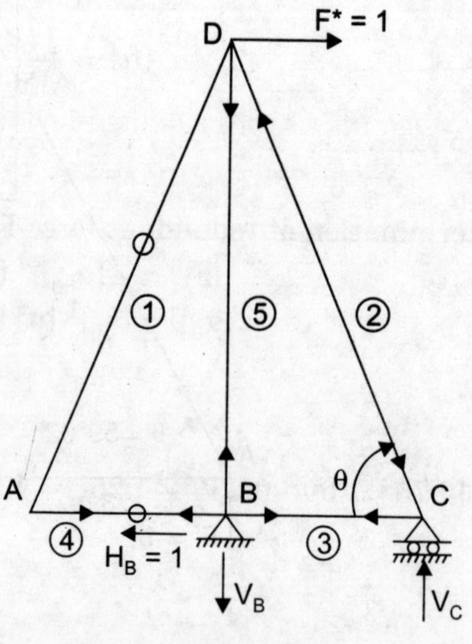

Fig. 4.44

Solution Degree of internal indeterminacy

$$I = m - (2j - 3) = 5 - (2 \times 4 - 3) = 0$$

Degree of external indeterminacy

$$E = r - R = 4 - 3 = 1$$

The structure is externally indeterminate to one degree. Treating support A as redundant, the primary structure is shown in Fig. 4.44.

To get the $[b]$ matrix :

The $[b]$ matrix would be in 2 parts, b^* and b^0. To get the b^* matrix let us apply a unit force ($F^* = 1$) at D as shown and get the member forces by method of joints.

Reactions : $H_B = 1$

Equating moments about C to zero,

$$V_B \times 3 = 1 \times 5 \; ; \; V_B = 1.667$$

Hence, $V_C = 1.667$

At joint A, $F_{AB} = F_{AD} = 0$

At joint B, $F_{BA} = V_B = 1.667$ (Tension)

At joint C, $F_{CD} \sin \theta = V_C = 1.667$

$$F_{CD} = \frac{1.667}{\sin \theta} = 1.944 \text{ (Compression)}$$

$$\tan \theta = \frac{5}{3} = 1.667 \quad \theta = 59°02' \quad \cos \theta = 0.515, \quad \sin \theta = 0.857$$

$$F_{BC} = F_{CD} \cos \theta = 1.00 \text{ (tallies)}$$

Thus, $[b^*] = [0 \quad -1.944 \quad 1.00 \quad 0 \quad 1.667]^T$

$[b]^0$ is obtained by applying a unit force at the location of the redundant F_0. (see Fig. 4.45)

Equating to zero, moments about B,

$$V_C \times 3 = 1 \times 3 \; ; \; V_C = 1$$

From $\Sigma V = 0 \; ; \; V_B = 2$

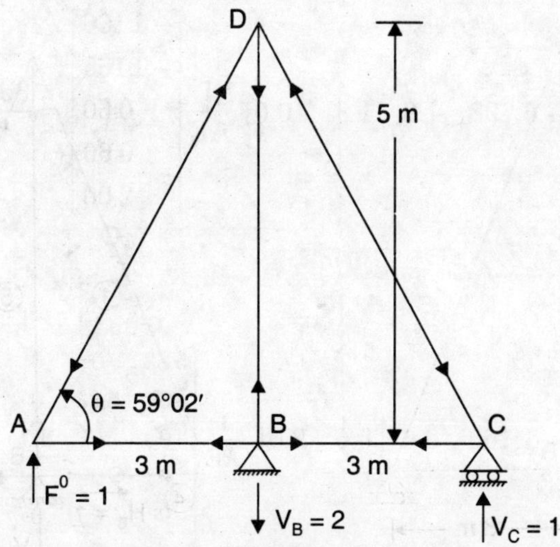

Fig. 4.45

At joint A, $F_{AD} \sin\theta = 1$; $F_{AD} = 1.167$ (Compression)

$F_{AB} = F_{AD} \cos\theta$; $F_{AB} = 0.60$ (tension)

At joint B, $F_{BC} = F_{AB} = 0.60$ (tension)

$F_{BD} = 2$ (tension)

At joint C, $F_{CD} \sin\theta = 1$; $F_{CD} = 1.167$ (Compression)

$F_{BC} = F_{CD} \cos\theta = 0.60$ (tallies.)

Hence, $[b^0] = [-1.166 \quad -1.166 \quad 0.60 \quad 0.60 \quad 2.00]^T$

$$[b] = [b^* \vdots b^0] = \begin{bmatrix} 0 & -1.166 \\ -1.944 & -1.166 \\ 1.00 & 0.60 \\ 0 & 0.60 \\ 1.667 & 2.00 \end{bmatrix}$$

Redundant force, $\{F^0\} = -[a_{00}]^{-1} [a_{0*}] \{F^*\}$

$[a_{00}] = [b^0]^T [\alpha] [b^0]$

$$[\alpha] = \frac{1}{AE} \begin{bmatrix} 5.83 & & & & \\ & 5.83 & & [0] & \\ & & 3.0 & & \\ & [0] & & 3.0 & \\ & & & & 5.0 \end{bmatrix}$$

$$[a_{00}] = [-1.166 \quad -1.166 \quad 0.6 \quad 0.6 \quad 2.0] \frac{1}{AE} \begin{bmatrix} 5.83 & 0 & 0 & 0 & 0 \\ 0 & 5.83 & 0 & 0 & 0 \\ 0 & 0 & 3.0 & 0 & 0 \\ 0 & 0 & 0 & 3.0 & 0 \\ 0 & 0 & 0 & 0 & 5.0 \end{bmatrix} \begin{bmatrix} -1.166 \\ -1.166 \\ 0.60 \\ 0.60 \\ 2.00 \end{bmatrix}$$

$$= [-6.798 \quad -6.798 \quad 1.8 \quad 1.8 \quad 10.0] \frac{1}{AE} \begin{bmatrix} -1.166 \\ -1.166 \\ 0.60 \\ 0.60 \\ 2.00 \end{bmatrix} = \frac{38.01}{AE}$$

$$[a_{00}]^{-1} = \frac{AE}{38.01}$$

$[a_{0*}] = [b^0] [\alpha] [b^*]$

$$= [-1.166 \quad -1.166 \quad 0.6 \quad 0.6 \quad 2.0] \frac{1}{AE} \begin{bmatrix} 5.83 & 0 & 0 & 0 & 0 \\ 0 & 5.83 & 0 & 0 & 0 \\ 0 & 0 & 3.0 & 0 & 0 \\ 0 & 0 & 0 & 3.0 & 0 \\ 0 & 0 & 0 & 0 & 5.0 \end{bmatrix} \begin{bmatrix} 0 \\ -1.944 \\ 1.00 \\ 0 \\ 1.667 \end{bmatrix}$$

$$= [-6.798 \quad -6.798 \quad 1.80 \quad 1.80 \quad 10.00]\ \frac{1}{AE}\begin{bmatrix} 0 \\ -1.944 \\ 1.00 \\ 0 \\ 1.667 \end{bmatrix} = \frac{31.59}{AE}$$

$$\{F^0\} = -[a_{00}]^{-1}[a_{0*}]\{F^*\} = -\frac{AE}{38.01} \times \frac{31.59}{AE} \times 60 = -49.87 \text{ kN}$$

Element forces $\{P\} = [b]\{F\}$

$$\{P\} = \begin{bmatrix} 0 & -1.667 \\ -1.944 & -1.166 \\ 1.00 & 0.6 \\ 0 & 0.6 \\ 1.667 & 2.0 \end{bmatrix} \begin{Bmatrix} 60 \\ -49.87 \end{Bmatrix} = \begin{Bmatrix} 58.15 \\ -58.25 \\ 30.08 \\ -29.92 \\ -0.26 \end{Bmatrix}$$

These forces are shown in Fig. 4.46.

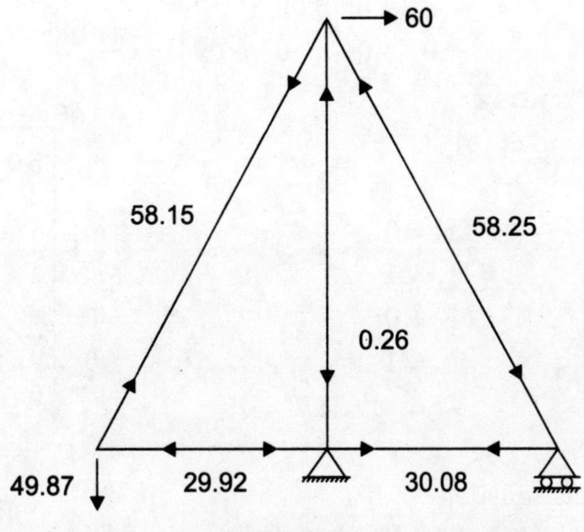

Fig. 4.46

Example 4.32 *Analyse the pin jointed truss given below using matrix flexibility method. AE is taken uniform for all the members.*

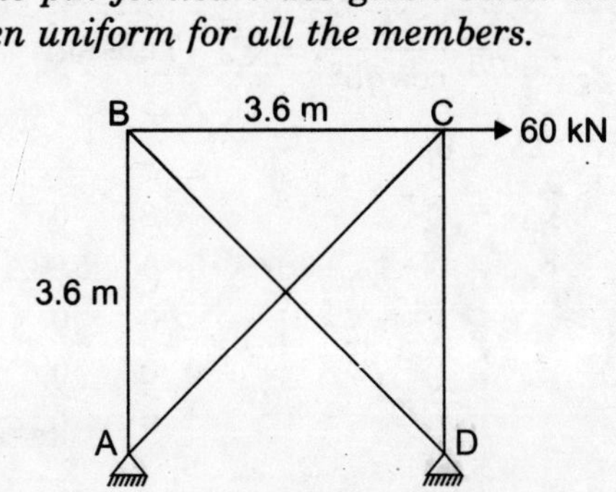

Fig. 4.47

Solution Degree of external indeterminacy = No. of reactions – 3

$$= 4 - 3 = 1$$

Degree of internal indeterminacy $= m - (2j - 3)$

$$= 5 - (2 \times 4 - 3)$$

$$= 0$$

Given truss is externally indeterminate to 1st degree.

The structure is made determinate by converting the right support (D) into roller support *i.e.,* treating H_D as redundant.

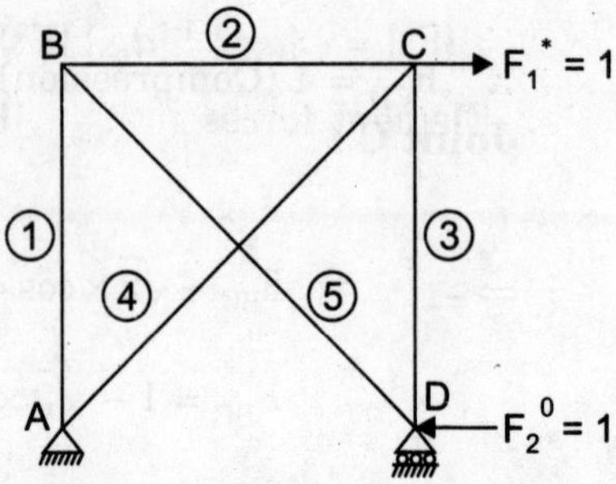

[*b*] matrix is obtained by applying unit force along co-ordinates given in following figure.

Element flexibility matrix is given by

$$[\alpha] = \frac{1}{AE}\begin{bmatrix} 3.6 & 0 & 0 & 0 & 0 \\ 0 & 3.6 & 0 & 0 & 0 \\ 0 & 0 & 3.6 & 0 & 0 \\ 0 & 0 & 0 & 5.09 & 0 \\ 0 & 0 & 0 & 0 & 5.09 \end{bmatrix}$$

Formation of [*b*] matrix :

To obtain first column of [*b*]

Applying $\mathbf{F_1^* = 1}$

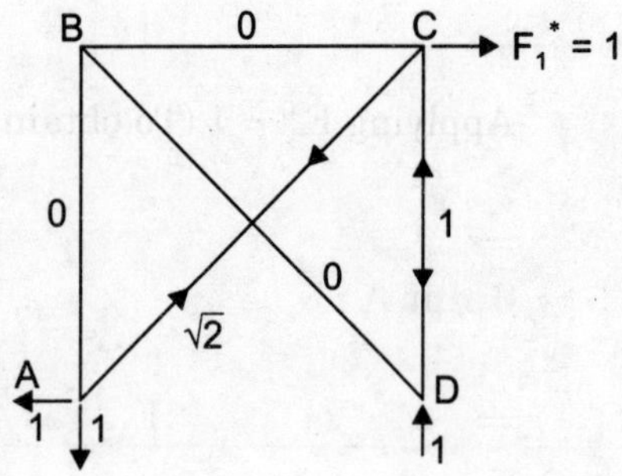

$$\Sigma H = 0$$

$$\Rightarrow \quad H_A = 1$$

$$\Sigma M_A = 0$$

$$\Rightarrow \quad V_D = 1$$

$$\therefore \quad V_A = -1 \quad [\Sigma V = 0]$$

Joint A :

Treating F_{AB}, F_{AC} as tensile,

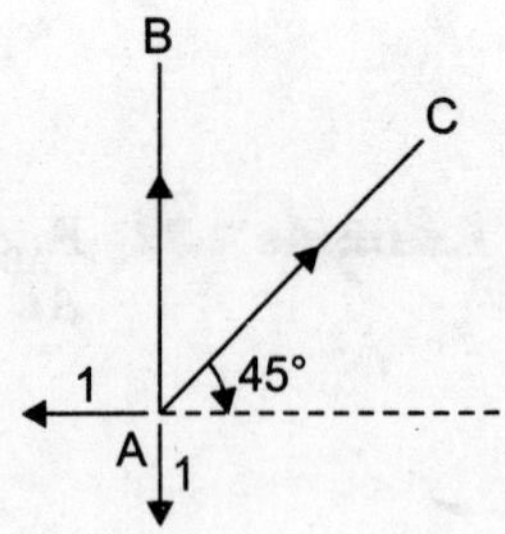

$$\Sigma H = 0 \text{ @ joint A}$$

$$\Rightarrow \quad F_{AC} \cos 45° = 1$$

$$F_{AC} = \frac{1}{\cos 45°} = \sqrt{2} \text{ (Tension)}$$

$$\Sigma V = 0$$

$$\Rightarrow \quad F_{AC} \sin 45° + F_{AB} = 1$$

$$\therefore \quad F_{AB} = 1 - F_{AC} \sin 45°$$

$$= 1 - \sqrt{2} \times \frac{1}{\sqrt{2}} = 0$$

Joint D :

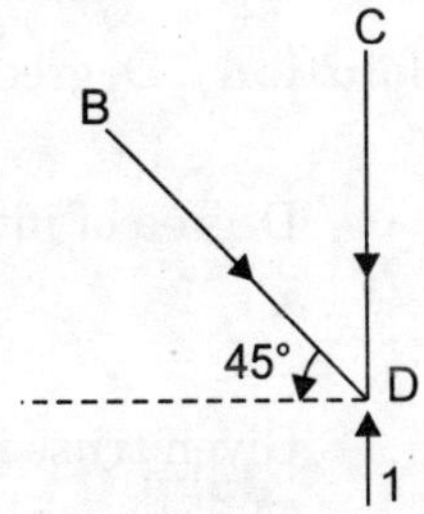

Considering F_{CD} and F_{BD} as compression

$\therefore$ $\Sigma V = 0$

$\Rightarrow$ $F_{BD} \sin 45° + F_{CD} = 1$

$\Sigma H = 0$

$\Rightarrow$ $F_{BD} \cos 45° = 0$

$\therefore$ $F_{BD} = 0$

$\therefore$ $F_{CD} = 1$ (Compression)

Joint C :

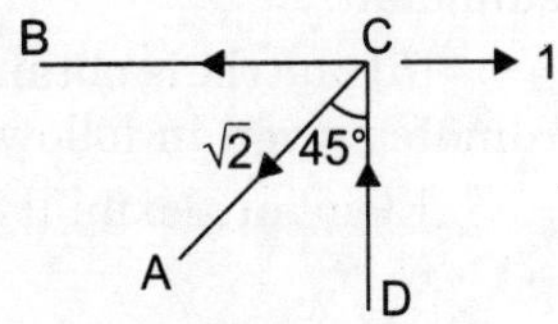

$\Sigma H = 0$

$$\Rightarrow \quad F_{BC} + \sqrt{2} \times \cos 45° = 1$$

$$F_{BC} = 1 - \sqrt{2} \times \frac{1}{\sqrt{2}} = 0$$

$$\begin{bmatrix} 0 \\ 0 \\ -1 \\ \sqrt{2} \\ 0 \end{bmatrix} \text{ is the first column of } [b] \text{ matrix}$$

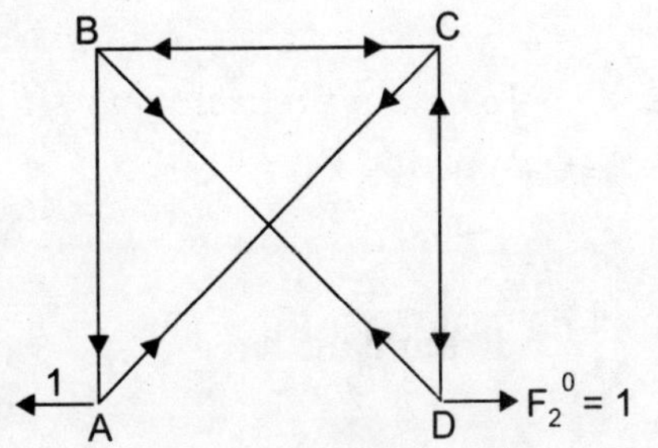

Applying $F_2^0 = 1$ (To obtain the second column of $[b]$)

$\Sigma H = 0$

$\Rightarrow$ $H_A = 1$

Joint A :

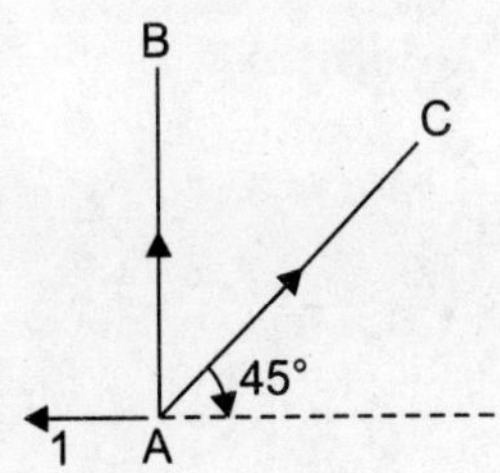

$\Sigma H = 0$

$\Rightarrow$ $F_{AC} \cos 45° = 1$

$$F_{AC} = \frac{1}{\cos 45°}$$

$F_{AC} = \sqrt{2}$ (Tension)

$\Sigma V = 0$

$\Rightarrow$ $F_{AB} + F_{AC} \sin 45° = 0$

$F_{AB} = - F_{AC} \sin 45°$

$$= - \sqrt{2} \times \frac{1}{\sqrt{2}} = -1$$

$F_{AB} = 1$ (Compression)

Joint D :

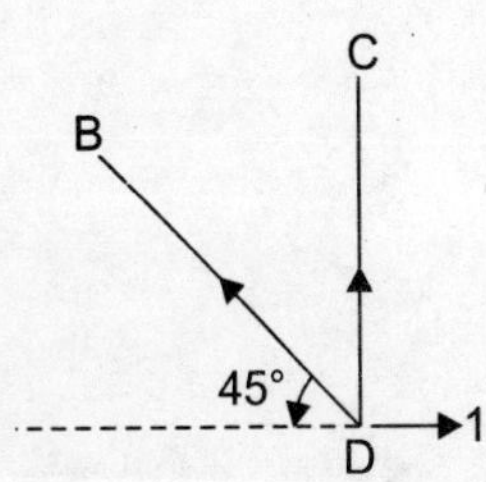

$\Sigma H = 0$

$\Rightarrow$ $F_{BD} \cos 45° = 1$

$$F_{BD} = \frac{1}{\cos 45°} = \sqrt{2} \text{ (Tension)}$$

$\Sigma V = 0$

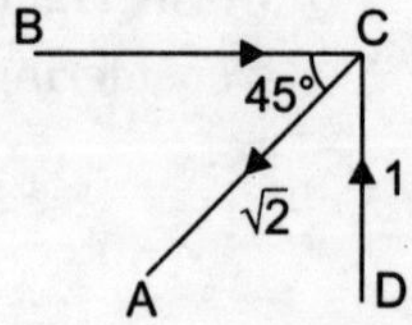

$$F_{BD} \sin 45° + F_{CD} = 0$$

$$F_{CD} = -F_{BD} \sin 45°$$

$$= -\sqrt{2} \times \frac{1}{\sqrt{2}} = -1$$

$$F_{CD} = 1 \text{ (Compression)}$$

Joint C :

$$\Sigma H = 0$$

$$F_{BC} = \sqrt{2} \times \cos 45°$$

$$= \sqrt{2} \times \frac{1}{\sqrt{2}}$$

$$F_{BC} = 1 \text{ (Compression)}$$

$\begin{bmatrix} -1 \\ -1 \\ -1 \\ \sqrt{2} \\ \sqrt{2} \end{bmatrix}$ is the second column of $[b]$ matrix

∴ $[b]$ matrix is given by

$$[b] = \begin{bmatrix} 0 & -1 \\ 0 & -1 \\ -1 & -1 \\ \sqrt{2} & \sqrt{2} \\ 0 & \sqrt{2} \end{bmatrix}$$

$$b^* \mid b^0$$

Redundant force $\{F_0\} = -[a_{22}]^{-1} [a_{21}] \{F^*\}$

$$[a_{22}] = [b_0]^* [\alpha] [b_0]$$

$$= [-1 \quad -1 \quad -1 \quad \sqrt{2} \quad \sqrt{2}] \frac{1}{AE} \begin{bmatrix} 3.6 & 0 & 0 & 0 & 0 \\ 0 & 3.6 & 0 & 0 & 0 \\ 0 & 0 & 3.6 & 0 & 0 \\ 0 & 0 & 0 & 5.09 & 0 \\ 0 & 0 & 0 & 0 & 5.09 \end{bmatrix} \begin{bmatrix} -1 \\ -1 \\ -1 \\ \sqrt{2} \\ \sqrt{2} \end{bmatrix}$$

$$= \frac{1}{AE} [-3.6 \quad -3.6 \quad -3.6 \quad 7.2 \quad 7.2] \begin{bmatrix} -1 \\ -1 \\ -1 \\ \sqrt{2} \\ \sqrt{2} \end{bmatrix}$$

$$[a_{22}] = \frac{31.16}{AE}$$

$$[a_{21}] = [b_0]^T [\alpha] [b_*]$$

$$= [-1 \ \ -1 \ \ -1 \ \ \sqrt{2} \ \ \sqrt{2}] \frac{1}{AE} \begin{bmatrix} 3.6 & 0 & 0 & 0 & 0 \\ 0 & 3.6 & 0 & 0 & 0 \\ 0 & 0 & 3.6 & 0 & 0 \\ 0 & 0 & 0 & 5.09 & 0 \\ 0 & 0 & 0 & 0 & 5.09 \end{bmatrix} \begin{bmatrix} 0 \\ 0 \\ -1 \\ \sqrt{2} \\ 0 \end{bmatrix}$$

$$= \frac{1}{AE} [-3.6 \ \ -3.6 \ \ -3.6 \ \ 7.2 \ \ 7.2] \begin{bmatrix} 0 \\ 0 \\ -1 \\ \sqrt{2} \\ 0 \end{bmatrix}$$

$$[a_{21}] = \frac{1}{AE} [13.78]$$

$$\therefore \qquad \{F_0\} = -[a_{22}]^{-1} [a_{21}] \{F^*\}$$

$$= -\left[\frac{31.16}{AE}\right]^{-1} \left[\frac{13.78}{AE}\right] \{60\}$$

$$F_0 = -26.53 \text{ kN}$$

Element forces :

$$\{P\} = [b] \{F\}$$

$$= \begin{bmatrix} 0 & -1 \\ 0 & -1 \\ -1 & -1 \\ \sqrt{2} & \sqrt{2} \\ 0 & \sqrt{2} \end{bmatrix} \begin{Bmatrix} 60 \\ -26.53 \end{Bmatrix} = \begin{bmatrix} 26.53 \\ 26.53 \\ -33.47 \\ 47.33 \\ -37.51 \end{bmatrix} \text{kN}$$

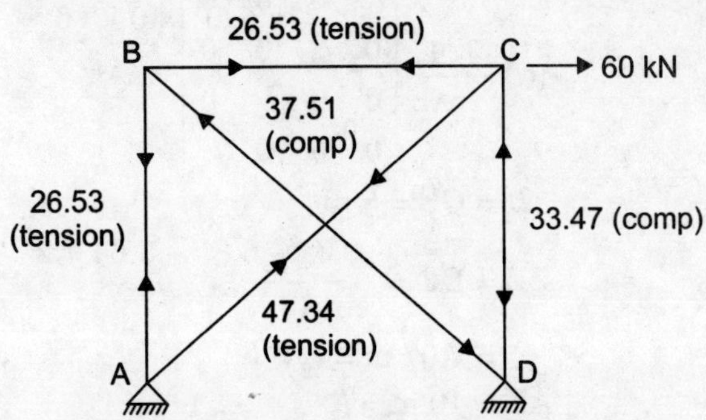

Example 4.33 *Find the forces in all the members of pin jointed framework shown in Fig. 4.48 using flexibility method. All the members have the same area of cross-section and Young's modulus.*

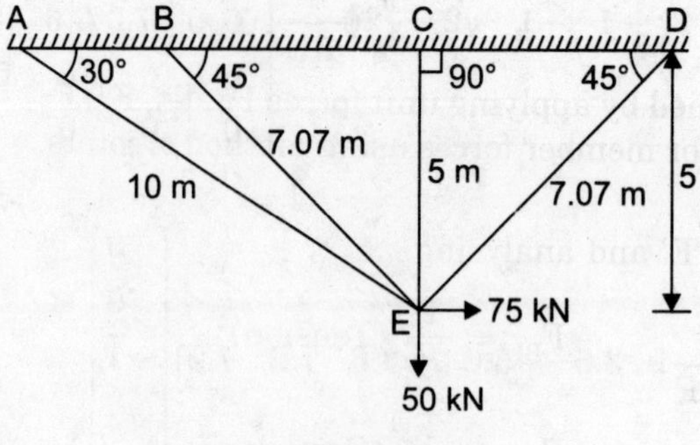

Fig. 4.48

Solution The degree of internal indeterminacy of the given truss is

$$i = m - 2j$$
$$= 4 - 2(1)$$
$$i = 2$$

Treating AE and CE as redundants, the primary structure with external and redundant forces is

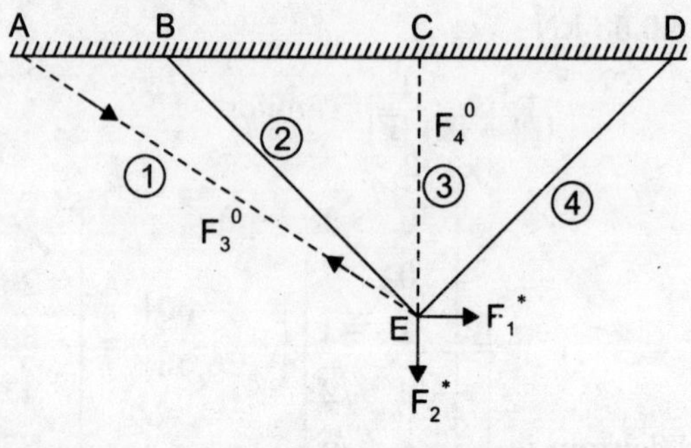

Element flexibility matrix

$$[\alpha] = \frac{1}{AE}\begin{bmatrix} l_1 & 0 & 0 & 0 \\ 0 & l_2 & 0 & 0 \\ 0 & 0 & l_3 & 0 \\ 0 & 0 & 0 & l_4 \end{bmatrix}$$

$$l_3 = CE = 5 \text{ m}$$

$$BE = DE = \frac{5}{\sin 45°}$$

$$= 7.07 \text{ m} = l_2 = l_4$$

$$AE = 10 \text{ m} = l_1$$

$$\because \qquad \alpha = \frac{1}{AE}\begin{bmatrix} 10 & 0 & 0 & 0 \\ 0 & 7.07 & 0 & 0 \\ 0 & 0 & 5 & 0 \\ 0 & 0 & 0 & 7.07 \end{bmatrix}$$

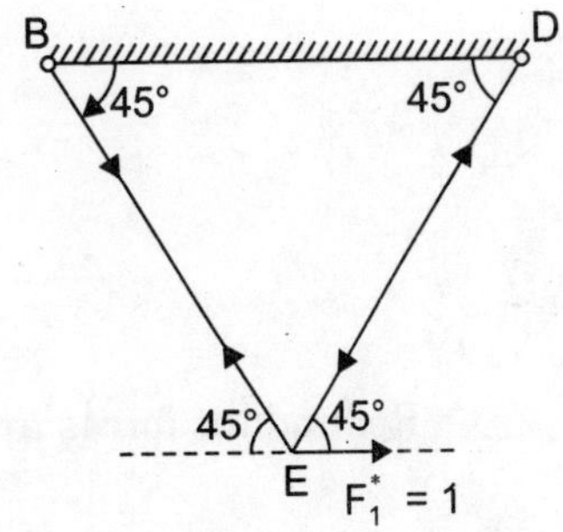

[*b*] matrix is obtained by applying unit forces F_1^*, F_2^*, F_3^0, F_4^0 and analysing the truss for member forces using method of joints.

Applying $F_1^* = 1$

Considering joint 'E' and analysing

$$F_{BE} = \frac{1}{\sqrt{2}} \text{ (Tension)}$$

$$F_{DE} = \frac{1}{\sqrt{2}} \text{ (Compression)}$$

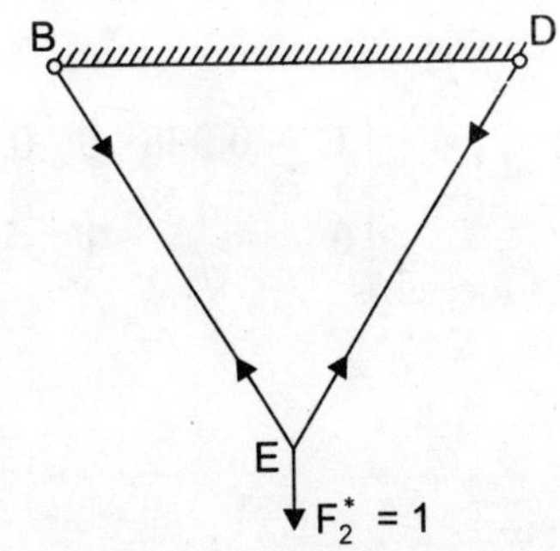

Hence the first column is $\begin{bmatrix} 0 \\ \frac{1}{\sqrt{2}} \\ 0 \\ -\frac{1}{\sqrt{2}} \end{bmatrix}$

Applying $F_2^* = 1$

$$F_{BE} = \frac{1}{\sqrt{2}} \text{ (Tension)}$$

$$F_{DE} = \frac{1}{\sqrt{2}} \text{ (Tension)}$$

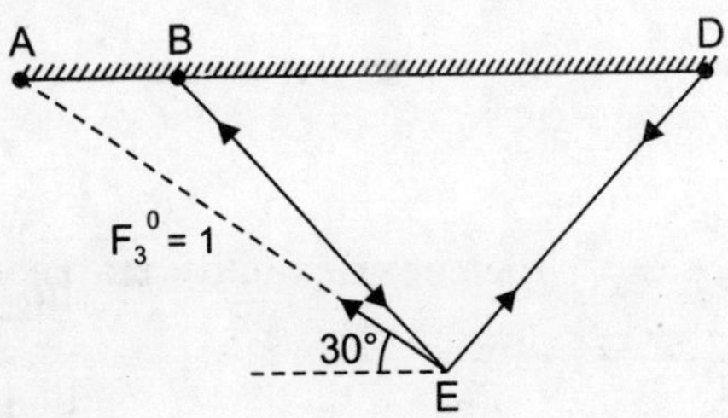

Hence the second column is $\begin{bmatrix} 0 \\ \frac{1}{\sqrt{2}} \\ 0 \\ \frac{1}{\sqrt{2}} \end{bmatrix}$

Applying $F_3^0 = 1$

$$F_{DE} = 0.259 \text{ (Tension) and } F_{BE} = 0.966 \text{ (Compression)}$$

Applying $F_4^0 = 1$

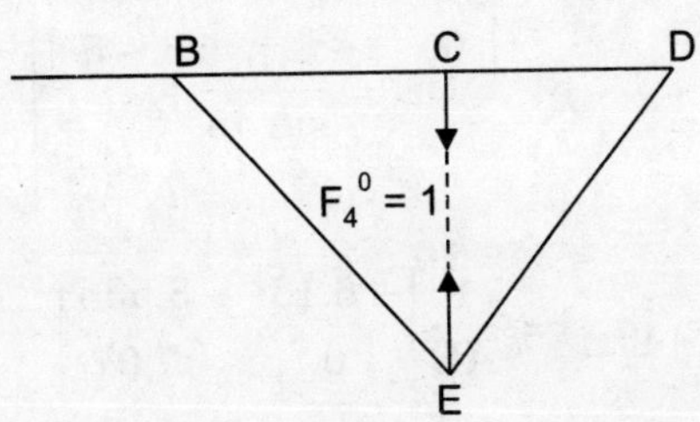

$$F_{BE} = F_{DE} = \frac{1}{\sqrt{2}} \text{ (Compression)}$$

$\therefore$ [b] matrix is given by

$$[b] = \left[\begin{array}{cc|cc} \multicolumn{2}{c|}{[b^*]} & \multicolumn{2}{c}{[b^0]} \\ 0 & 0 & 1 & 0 \\ \frac{1}{\sqrt{2}} & \frac{1}{\sqrt{2}} & -0.966 & -\frac{1}{\sqrt{2}} \\ 0 & 0 & 0 & 1 \\ -\frac{1}{\sqrt{2}} & \frac{1}{\sqrt{2}} & 0.259 & -\frac{1}{\sqrt{2}} \end{array}\right]$$

Redundant forces are given by

$$\{F_0\} = -[a_{22}^{-1}]\,[a_{21}]\,\{F^*\}$$

$$[a_{22}] = [b_0]^T\,[\alpha]\,[b^0]$$

$$= \begin{bmatrix} 1 & -0.966 & 0 & 0.259 \\ 0 & -\frac{1}{\sqrt{2}} & 1 & -\frac{1}{\sqrt{2}} \end{bmatrix} \frac{1}{AE} \begin{bmatrix} 10 & 0 & 0 & 0 \\ 0 & 7.07 & 0 & 0 \\ 0 & 0 & 5 & 0 \\ 0 & 0 & 0 & 7.07 \end{bmatrix} \begin{bmatrix} 1 & 0 \\ -0.966 & -\frac{1}{\sqrt{2}} \\ 0 & 1 \\ 0.259 & -\frac{1}{\sqrt{2}} \end{bmatrix}$$

$$= \begin{bmatrix} 10 & -6.83 & 0 & 1.83 \\ 0 & -5 & 5 & -5 \end{bmatrix} \frac{1}{AE} \begin{bmatrix} 1 & 0 \\ -0.966 & \frac{1}{\sqrt{2}} \\ 0 & 1 \\ 0.259 & -\frac{1}{\sqrt{2}} \end{bmatrix}$$

$$[a_{22}] = \frac{1}{AE}\begin{bmatrix} 17.07 & 3.535 \\ 3.535 & 12.07 \end{bmatrix}$$

$$[a_{21}] = [b^0]^T\,[\alpha]\,[b^*]$$

$$= \begin{bmatrix} 10 & -6.83 & 0 & 1.83 \\ 0 & -5 & 5 & -5 \end{bmatrix} \begin{bmatrix} 0 & 0 \\ \frac{1}{\sqrt{2}} & \frac{1}{\sqrt{2}} \\ 0 & 0 \\ -\frac{1}{\sqrt{2}} & \frac{1}{\sqrt{2}} \end{bmatrix} \frac{1}{AE}$$

$$[a_{21}] = \frac{1}{AE}\begin{bmatrix} -6.13 & -3.535 \\ 0 & -7.07 \end{bmatrix}$$

$$\{F_0\} = -[a_{22}]^{-1}[a_{21}][F^*]$$

$$= -AE\begin{bmatrix}17.07 & 3.535\\ 3.535 & 12.07\end{bmatrix}^{-1}\frac{1}{AE}\begin{bmatrix}-6.13 & -3.535\\ 0 & -7.07\end{bmatrix}\begin{bmatrix}75\\50\end{bmatrix}$$

$$= \frac{-1}{193.54}\begin{bmatrix}12.07 & -3.535\\ -3.535 & 17.07\end{bmatrix}\begin{bmatrix}-636.5\\-353.5\end{bmatrix}$$

$$= \begin{bmatrix}33.23\\19.56\end{bmatrix}$$

Element forces [P] $\quad p = [b]\,[F]$

$$= \begin{bmatrix}0 & 0 & 1 & 0\\ \frac{1}{\sqrt{2}} & \frac{1}{\sqrt{2}} & -0.966 & -\frac{1}{\sqrt{2}}\\ 0 & 0 & 0 & 1\\ -\frac{1}{\sqrt{2}} & \frac{1}{\sqrt{2}} & 0.209 & -\frac{1}{\sqrt{2}}\end{bmatrix}\begin{bmatrix}75\\50\\33.23\\19.56\end{bmatrix} = \begin{bmatrix}33.23\\42.46\\19.56\\-22.90\end{bmatrix}\text{kN}$$

Hence, $\quad F_{AE} = 33.23$ (Tension) ; $F_{BE} = 42.46$ (Tension)

$F_{CE} = 19.56$ (Tension) ; $F_{DE} = 22.90$ (Compression)

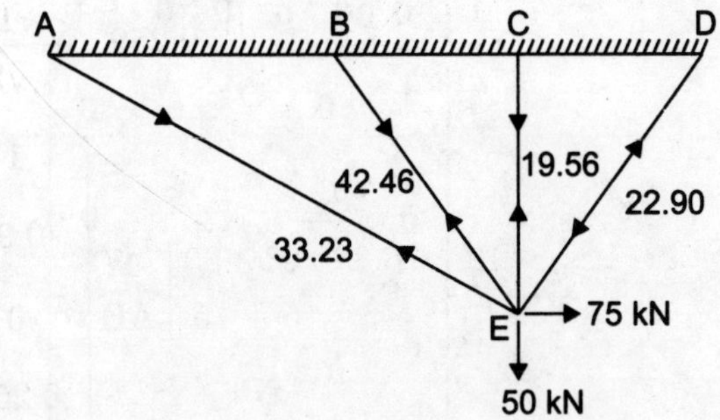

Example 4.34 *While fabricating the pin jointed frame shown in Fig. 4.49, the member AC was the last member to be fitted and was found to be 1.5 mm short of required length. Determine the forces induced in all the members when AC is forced into position. Diagonal members are having 1000 mm² area while remaining members are having an area of 2000 mm². Take E = 2 × 10⁵ N/mm².*

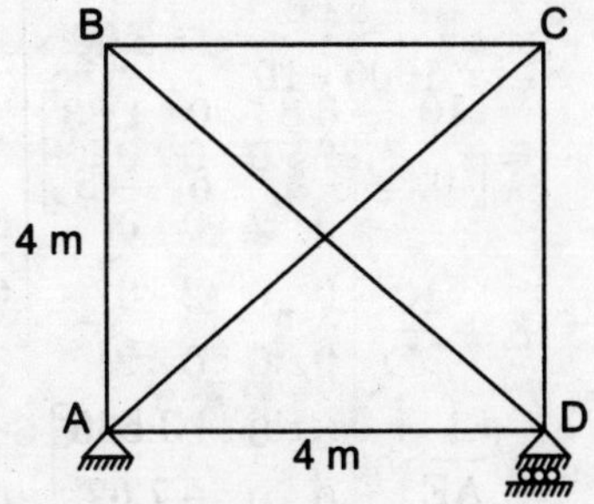

Fig. 4.49

Solution The given truss is analysed for lack of fit problem.

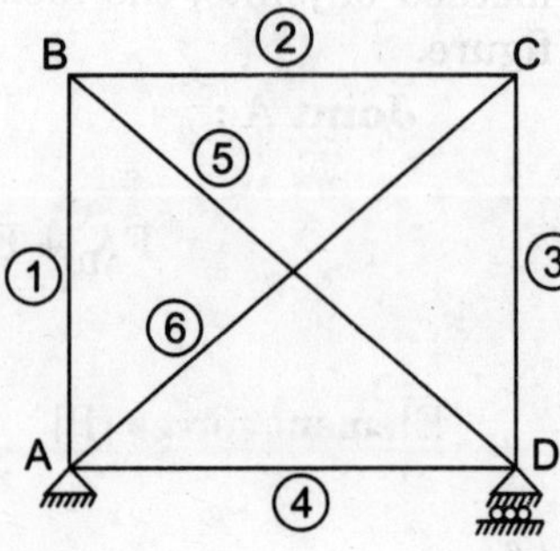

Degree of Indeterminacy :

External

No. of reactions – No. of equilibrium equations

$= 3 - 3 = 0$

Statically determinate externally

Internal $I = m - [2j - 3]$

$= 6 - [2(4) - 3]$

$= 6 - 5 = 1$

Internally indeterminate to one degree.

∴ Considering **AC** as redundant.

Remove member "**AC**".

Element flexibility matrix (α) :

$$\alpha = \frac{1}{A_i}\begin{bmatrix} l_1 & 0 & 0 & 0 & 0 & 0 \\ 0 & l_2 & 0 & 0 & 0 & 0 \\ 0 & 0 & l_3 & 0 & 0 & 0 \\ 0 & 0 & 0 & l_4 & 0 & 0 \\ 0 & 0 & 0 & 0 & l_5 & 0 \\ 0 & 0 & 0 & 0 & 0 & l_6 \end{bmatrix}$$

$$= \frac{1}{E}\begin{bmatrix} \frac{l_1}{A_1} & 0 & & 0 \\ 0 & \frac{l_2}{A_2} & & 0 \\ & & \ddots & \\ 0 & 0 & & \frac{l_6}{A_6} \end{bmatrix}$$

$$\frac{l_1}{A_1} = \frac{4}{2000 \times 10^{-6}} = 2 \times 10^3 \text{ m}$$

$$\frac{l_5}{A_5} = \frac{5.66}{1000 \times 10^{-6}} = 5.66 \times 10^3 \text{ m}$$

$$\therefore \quad \alpha = \frac{10^3}{E}\begin{bmatrix} 2 & 0 & 0 & 0 & 0 & 0 \\ 0 & 2 & 0 & 0 & 0 & 0 \\ 0 & 0 & 2 & 0 & 0 & 0 \\ 0 & 0 & 0 & 2 & 0 & 0 \\ 0 & 0 & 0 & 0 & 5.66 & 0 \\ 0 & 0 & 0 & 0 & 0 & 5.66 \end{bmatrix}$$

Formation of [b] matrix :

Applying $\mathbf{F_1^0 = 1}$ and determining member forces using method of joints, the member forces obtained are shown in figure.

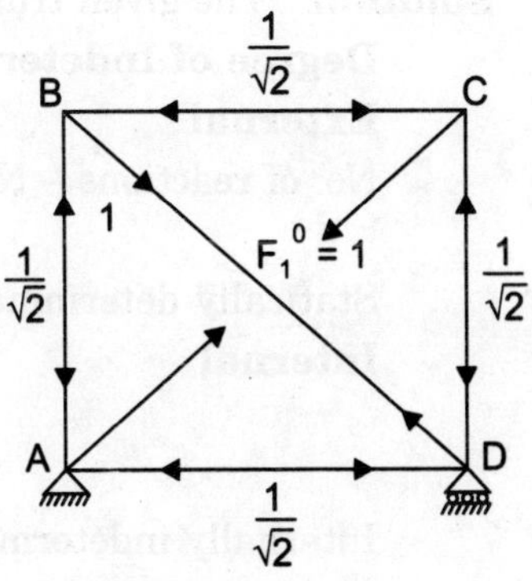

Joint A :

$$\Sigma H = 0 \text{ gives}$$

$$F_{AD} + F_{AC} \cos 45° = 0$$

$$F_{AD} = -F_{AC} \cos 45° = -\frac{1}{\sqrt{2}}$$

i.e.,

$$F_{AD} = \frac{1}{\sqrt{2}} \text{ (Compression)}$$

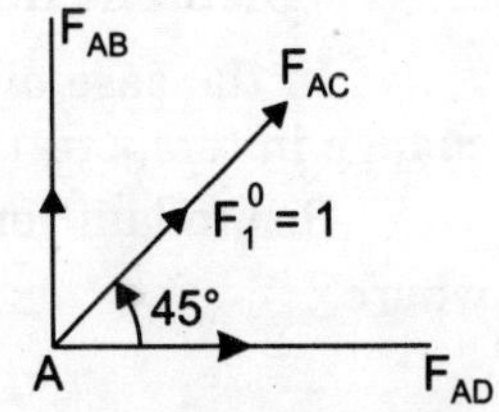

$$\Sigma V = 0 \text{ gives}$$

$$F_{AB} + F_{AC} \sin 45° = 0$$

$$F_{AB} = -F_{AC} \sin 45° = -1 \times \frac{1}{\sqrt{2}}$$

$$F_{AB} = \frac{1}{\sqrt{2}} \text{ (Compression)}$$

Joint C :

$$\Sigma H = 0 \text{ gives}$$

$$F_{CB} + F_{CA} \cos 45° = 0$$

$$F_{CB} = -F_{CA} \cos 45° = -1 \times \frac{1}{\sqrt{2}}$$

$$F_{CB} = \frac{1}{\sqrt{2}} \text{ (Compression)}$$

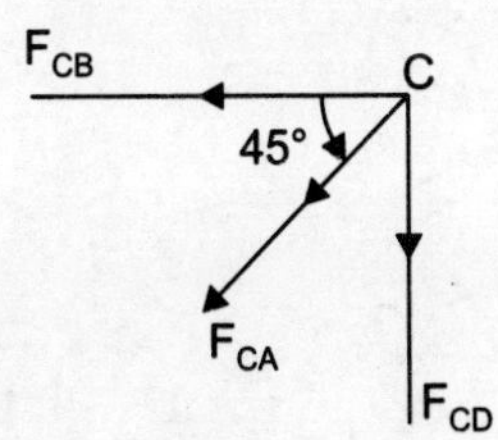

$$\Sigma V = 0 \text{ gives}$$

$$F_{CD} + F_{CA} \sin 45° = 0$$

$$F_{CD} = -F_{CA} \sin 45° = -1 \times \frac{1}{\sqrt{2}}$$

$$F_{CD} = \frac{1}{\sqrt{2}} \text{ (Compression)}$$

Joint D :

$$\Sigma H = 0 \text{ gives}$$

$$F_{DB} \cos 45° = F_{DA}$$

$$F_{DB} = \frac{F_{DA}}{\cos 45°} = \frac{\frac{1}{\sqrt{2}}}{\frac{1}{\sqrt{2}}} = 1$$

$$F_{DB} = 1 \text{ (Tension)}$$

$$\therefore \qquad [b] = \begin{bmatrix} -\frac{1}{\sqrt{2}} \\ -\frac{1}{\sqrt{2}} \\ -\frac{1}{\sqrt{2}} \\ -\frac{1}{\sqrt{2}} \\ 1 \\ 1 \end{bmatrix}$$

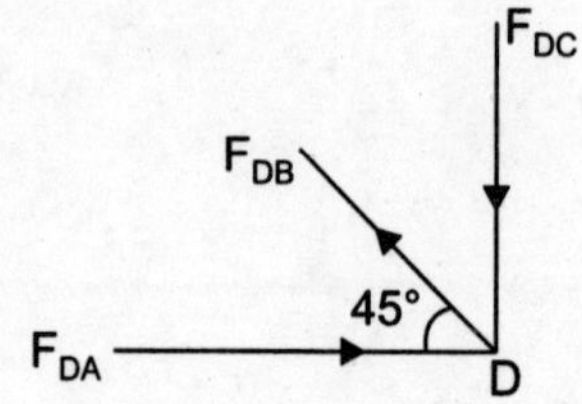

In the case of trusses with member's having lack of fit only or members subjected to change in temperature only.

Redundant force $\{F^0\} = -[a_{22}^{-1}]\,[b^{0T}]\,\{h\}$ ...(1)

where $\{h\}$ = Initial strain matrix

$$= \begin{bmatrix} 0 \\ 0 \\ 0 \\ 0 \\ 0 \\ -0.0015 \end{bmatrix}$$

$$[a_{22}] = [b^{0T}]\,[\alpha]\,[b^0]$$

$$= \begin{bmatrix} -\frac{1}{\sqrt{2}} & -\frac{1}{\sqrt{2}} & -\frac{1}{\sqrt{2}} & -\frac{1}{\sqrt{2}} & 1 & 1 \end{bmatrix} \times \frac{10^3}{E} \begin{bmatrix} 2 & 0 & 0 & 0 & 0 & 0 \\ 0 & 2 & 0 & 0 & 0 & 0 \\ 0 & 0 & 2 & 0 & 0 & 0 \\ 0 & 0 & 0 & 2 & 0 & 0 \\ 0 & 0 & 0 & 0 & 5.66 & 0 \\ 0 & 0 & 0 & 0 & 0 & 5.66 \end{bmatrix} \begin{bmatrix} -\frac{1}{\sqrt{2}} \\ -\frac{1}{\sqrt{2}} \\ -\frac{1}{\sqrt{2}} \\ -\frac{1}{\sqrt{2}} \\ 1 \\ 1 \end{bmatrix}$$

$$= \frac{10^3}{E}[-\sqrt{2} \quad -\sqrt{2} \quad -\sqrt{2} \quad -\sqrt{2} \quad 5.66 \quad 5.66]\begin{bmatrix} -\frac{1}{\sqrt{2}} \\ -\frac{1}{\sqrt{2}} \\ -\frac{1}{\sqrt{2}} \\ -\frac{1}{\sqrt{2}} \\ 1 \\ 1 \end{bmatrix}$$

$$[a_{22}] = \frac{10^3}{E}.[15.31]$$

$$\therefore \qquad [a_{22}^{-1}] = \left\{\frac{E}{10^3 \times 15.31}\right\}$$

$$\{F^0\} = -[a_{22}^{-1}][b^{0T}][h]$$

$$= -\left[\frac{E}{10^3\,(15.31)}\right]\left[-\frac{1}{\sqrt{2}} \quad -\frac{1}{\sqrt{2}} \quad -\frac{1}{\sqrt{2}} \quad -\frac{1}{\sqrt{2}} \quad 1 \quad 1\right]\begin{bmatrix} 0 \\ 0 \\ 0 \\ 0 \\ 0 \\ -0.0015 \end{bmatrix}$$

$$= +\frac{E}{10^3 \times 15.31}(+0.0015)$$

$$\left[E = 2 \times 10^5 \text{ N/mm}^2 = \frac{2 \times 10^5 \times 10^6}{10^3} \text{ kN/m}^2 = 2 \times 10^8 \text{ kN/m}^2\right]$$

$$\{F^0\} = \frac{2 \times 10^8}{15.31 \times 10^3} \times 0.0015 = 19.59 \text{ kN}$$

$\therefore$ **Element forces :**

$$\{P\} = [b]\{F^0\} = \begin{bmatrix} -\frac{1}{\sqrt{2}} \\ -\frac{1}{\sqrt{2}} \\ -\frac{1}{\sqrt{2}} \\ -\frac{1}{\sqrt{2}} \\ 1 \\ 1 \end{bmatrix}[19.59] = \begin{bmatrix} -13.85 \\ -13.85 \\ -13.85 \\ -13.85 \\ 19.59 \\ 19.59 \end{bmatrix}$$

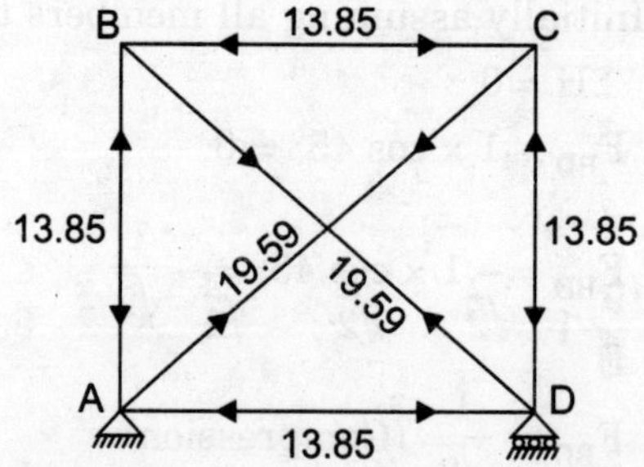

–ve ⇒ Compression
+ve ⇒ Tension

Example 4.35 *Determine the forces in all the members of frame as shown in Fig. 4.50 due to a fall in temperature in member BC by 28°C. All the members have c/s area of 1300 mm². Take $\alpha = 0.000012/°C$, $E = 2 \times 10^5\ N/mm^2$.*

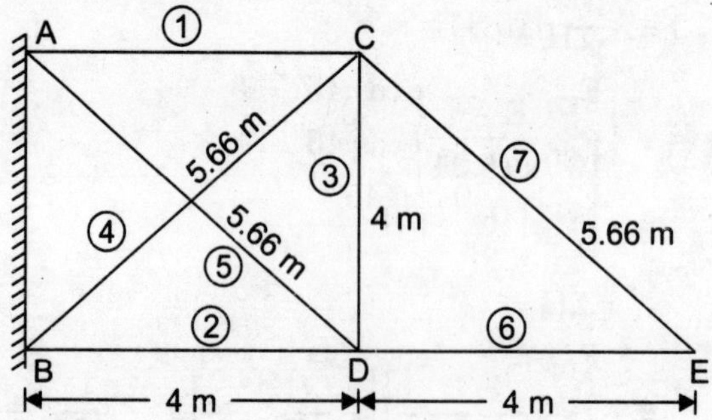

Fig. 4.50

Solution For the truss given above

Degree of redundancy = $m - 2j = 7 - 2(3) = 1$

Let us consider member BC as redundant.

∴ Removing member BC and applying unit force along its direction, we can arrive at the [b] matrix.

∵ There is no external force. The member forces due to fall in temperature of member BC alone is calculated by

$$\{P\} = [b]\{F^0\}$$
$$\{F^0\} = -[a_{22}]^{-1}[b_0]\{h\}$$

where h ⇒ Initial strain matrix

Assembled element flexibility matrix :

$$[\alpha] = \frac{1}{AE}\begin{bmatrix} 4 & 0 & 0 & 0 & 0 & 0 & 0 \\ 0 & 4 & 0 & 0 & 0 & 0 & 0 \\ 0 & 0 & 4 & 0 & 0 & 0 & 0 \\ 0 & 0 & 0 & 5.66 & 0 & 0 & 0 \\ 0 & 0 & 0 & 0 & 5.66 & 0 & 0 \\ 0 & 0 & 0 & 0 & 0 & 4 & 0 \\ 0 & 0 & 0 & 0 & 0 & 0 & 5.66 \end{bmatrix}$$

Considering joint 'B' : Initially assuming all members to be under tension

$$\Sigma H = 0$$

$\Rightarrow$ $$F_{BD} + 1 \times \cos 45° = 0$$

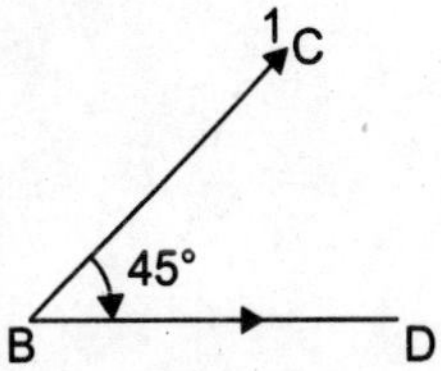

$$F_{BD} = -1 \times \cos 45° = -\frac{1}{\sqrt{2}}$$

i.e., $$F_{BD} = \frac{1}{\sqrt{2}} \text{ (Compression)}$$

Joint E : $$\Sigma V = 0$$

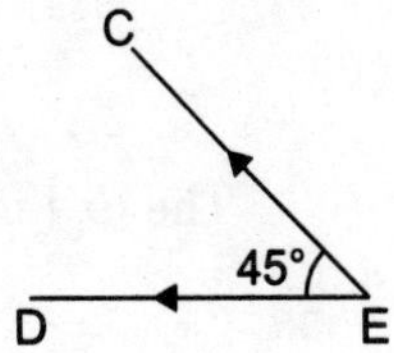

$$F_{EC} \sin 45° = 0$$

$\therefore$ $$F_{EC} = 0$$

$$\Sigma H = 0$$

$\Rightarrow$ $$F_{ED} + F_{EC} \times \cos 45° = 0$$

or $$F_{ED} = -F_{EC} \cos 45°$$

$$F_{ED} = -0 \cos 45°$$

$$F_{ED} = 0$$

Joint D : $$\Sigma H = 0$$

$\Rightarrow$ $$F_{DB} = F_{DA} \cos 45°$$

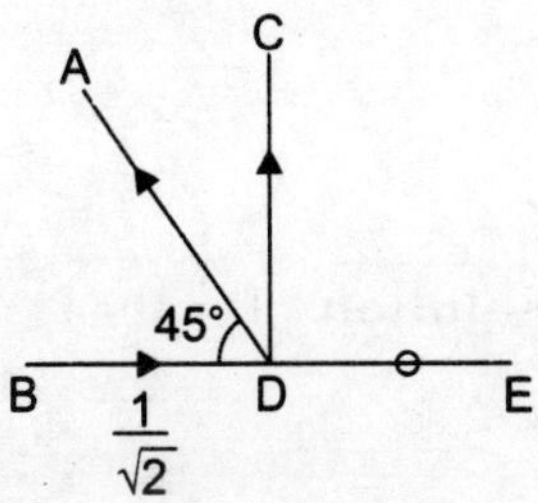

$$F_{DA} = \frac{F_{DB}}{\cos 45°}$$

$$= \frac{\frac{1}{\sqrt{2}}}{\frac{1}{\sqrt{2}}}$$

$$F_{DA} = 1 \text{ (Tension)}$$

$$\Sigma V = 0 \quad \text{gives}$$

$$F_{DA} + F_{DC} = 0$$

$$F_{DC} = -F_{DA}$$

$$= -(1)$$

i.e., $$F_{DC} = 1 \text{ (Compression)}$$

Joint C : $$\Sigma H = 0$$

$\Rightarrow$ $$F_{CA} + F_{CB} \times \cos 45° = 0$$

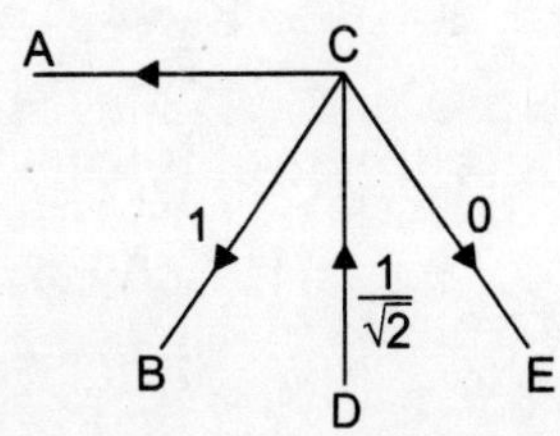

$$F_{CA} = -F_{CB} \cos 45°$$

$$= -(1) \times \cos 45°$$

$$= -\frac{1}{\sqrt{2}}$$

i.e., $$F_{CA} = \frac{1}{\sqrt{2}} \text{ (Compression)}$$

Formation of $[b_0]$ matrix :

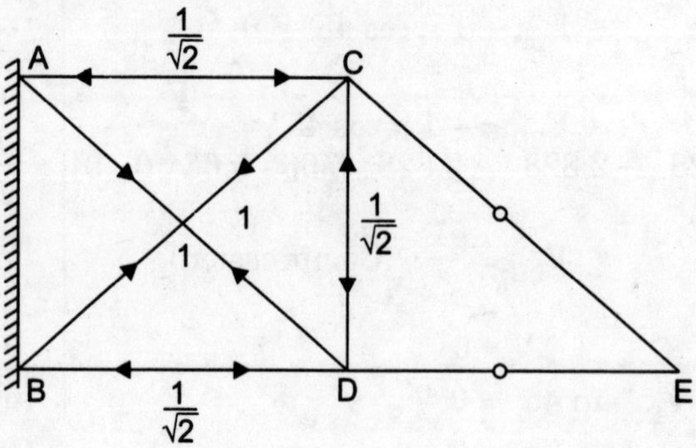

The $[b_0]$ matrix is given by

$$[b_0] = \begin{bmatrix} -\frac{1}{\sqrt{2}} \\ -\frac{1}{\sqrt{2}} \\ -\frac{1}{\sqrt{2}} \\ 1 \\ 1 \\ 0 \\ 0 \end{bmatrix}$$

$$[a_{22}] = [b^{0T}]\,[\alpha]\,[b^0]$$

$$= \begin{bmatrix} -\frac{1}{\sqrt{2}} & -\frac{1}{\sqrt{2}} & -\frac{1}{\sqrt{2}} & 1 & 1 & 0 & 0 \end{bmatrix} \frac{1}{AE} \begin{bmatrix} 4 & 0 & 0 & 0 & 0 & 0 & 0 \\ 0 & 4 & 0 & 0 & 0 & 0 & 0 \\ 0 & 0 & 4 & 0 & 0 & 0 & 0 \\ 0 & 0 & 0 & 5.66 & 0 & 0 & 0 \\ 0 & 0 & 0 & 0 & 5.66 & 0 & 0 \\ 0 & 0 & 0 & 0 & 0 & 4 & 0 \\ 0 & 0 & 0 & 0 & 0 & 0 & 5.66 \end{bmatrix} \begin{bmatrix} -\frac{1}{\sqrt{2}} \\ -\frac{1}{\sqrt{2}} \\ -\frac{1}{\sqrt{2}} \\ 1 \\ 1 \\ 0 \\ 0 \end{bmatrix}$$

$$= \frac{1}{AE}[-2.828 \quad -2.828 \quad -2.828 \quad 5.66 \quad 5.66 \quad 0 \quad 0]\begin{bmatrix} -\frac{1}{\sqrt{2}} \\ -\frac{1}{\sqrt{2}} \\ -\frac{1}{\sqrt{2}} \\ 1 \\ 1 \\ 0 \\ 0 \end{bmatrix}$$

$$= \frac{1}{AE}[2 + 2 + 2 + 5.66 + 5.66]$$

$$= \frac{17.32}{AE}$$

$$[a_{22}^{-1}] = \frac{AE}{17.32}$$

Redundant force $\{F_0\} = -[a_{22}^{-1}]\,[b^{0T}]\,\{h\}$

$$= -\frac{AE}{(17.32)}\begin{bmatrix} -\frac{1}{\sqrt{2}} & -\frac{1}{\sqrt{2}} & -\frac{1}{\sqrt{2}} & 1 & 1 & 0 & 0 \end{bmatrix}\{h\}$$

$$h \Rightarrow \text{initial strain matrix} = \begin{bmatrix} 0 \\ 0 \\ 0 \\ l \propto T \\ 0 \\ 0 \\ 0 \end{bmatrix}$$

$$l \propto T = 5.66 \times 0.000012 \times 28$$
$$= -1.90 \times 10^{-3} \text{ m}$$

$$AE = \frac{1300}{10^6} \times \frac{2 \times 10^5 \times 10^6}{10^3}$$
$$= 2.6 \times 10^5 \text{ kN}$$

$$\therefore \quad \{F_0\} = -\left(\frac{AE}{17.32}\right)\left(-\frac{1}{\sqrt{2}} \quad -\frac{1}{\sqrt{2}} \quad -\frac{1}{\sqrt{2}} \quad 1 \quad 1 \quad 0 \quad 0\right)\begin{bmatrix} 0 \\ 0 \\ 0 \\ -1.90 \times 10^3 \\ 0 \\ 0 \\ 0 \end{bmatrix}$$

$$= -\left(\frac{2.6 \times 10^5}{17.32}\right)(-1.90 \times 10^{-3})$$

$$\{F_0\} = +\ 28.52 \text{ kN (Tension)}$$

$\therefore$ **Element forces :**

$$[P] = [b]\,\{F_0\}$$

$$\begin{bmatrix} F_{AC} \\ F_{BD} \\ F_{CD} \\ F_{BC} \\ F_{AD} \\ F_{DE} \\ F_{CE} \end{bmatrix} = \begin{bmatrix} -\frac{1}{\sqrt{2}} \\ -\frac{1}{\sqrt{2}} \\ -\frac{1}{\sqrt{2}} \\ 1 \\ 1 \\ 0 \\ 0 \end{bmatrix} [28.52] = \begin{bmatrix} -20.16 \\ -20.16 \\ -20.16 \\ 28.52 \\ 28.52 \\ 0 \\ 0 \end{bmatrix} \text{kN}$$

The member forces are given by

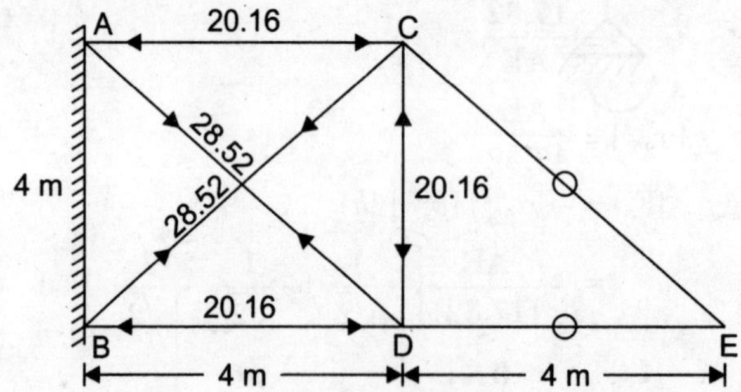

PROBLEMS

1. Generate the flexibility matrix for the simply supported beam AB with the co-ordinates shown in Fig. 4.51.

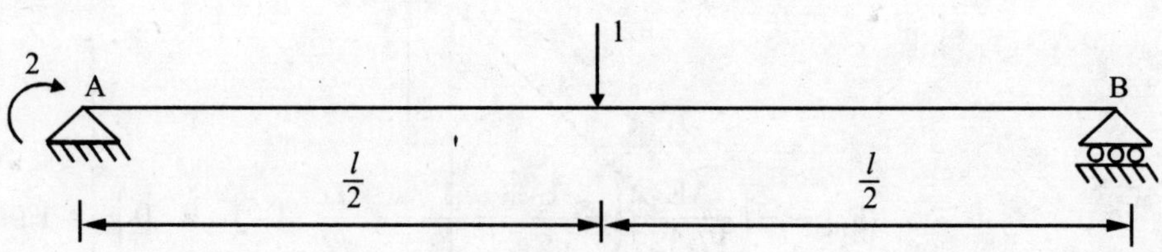

Fig. 4.51

$$\left[\textbf{Ans. } (a) = \frac{l}{48EI} \begin{bmatrix} l & 3l \\ 3l & 16 \end{bmatrix} \right].$$

2. Derive the flexibility matrix for the rigid jointed structure shown in Fig. 4.52

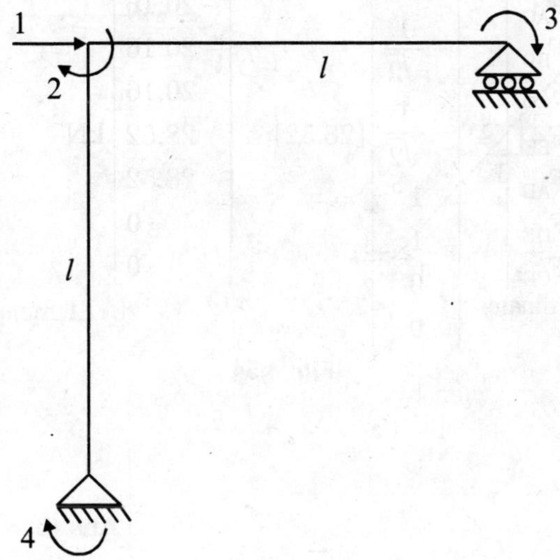

Fig. 4.52

$$\textbf{Ans.}\ (a) = \left[\frac{l}{6\text{EI}}\begin{bmatrix} 4l^2 & 2l & -l & 5l \\ 2l & 2 & -1 & 2 \\ -l & -1 & 2 & -1 \\ 5l & 2 & -1 & 8 \end{bmatrix}\right].$$

3. Generate the flexibility matrix for the pin jointed frame with reference to the co-ordinates shown in Fig. 4.53 AE is same for all the members.

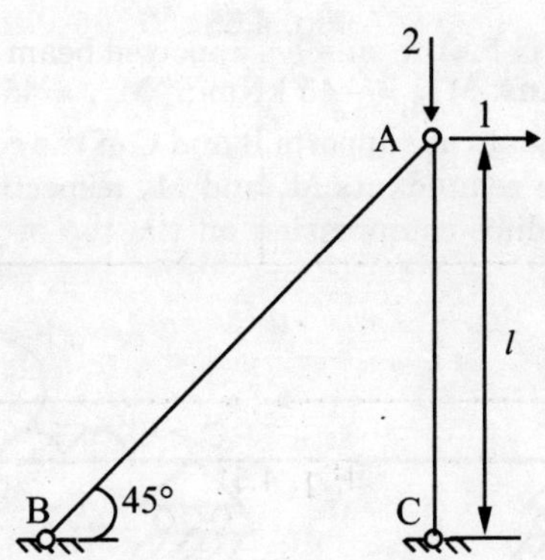

Fig. 4.53

4. Generate the (b) matrix for the structure in Fig. 4.54.

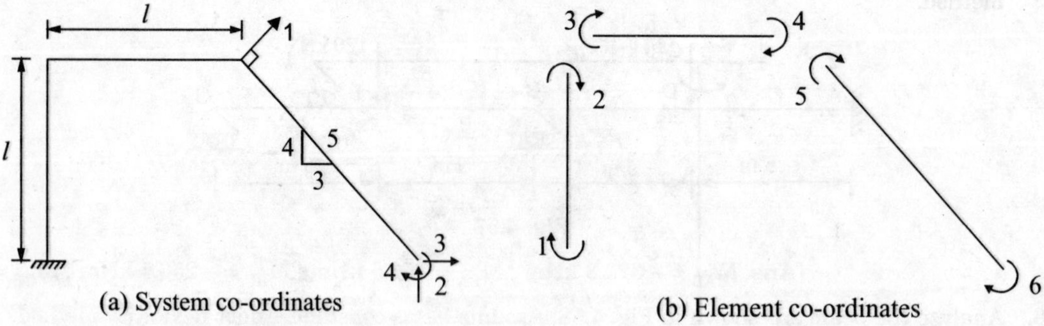

Fig. 4.54

$$\textbf{Ans.}\ (b) = \begin{bmatrix} -0.2l & 1.75l & 0 & -1 \\ -0.6l & -1.75l & -l & 1 \\ +0.6l & 1.75l & l & -1 \\ 0 & -0.175l & l & -1 \\ 0 & 0.175l & l & -1 \\ 0 & 0 & 0 & 1 \end{bmatrix}$$

5. Analyse the continuous beam shown in Fig. 4.55, using the matrix flexibility method.

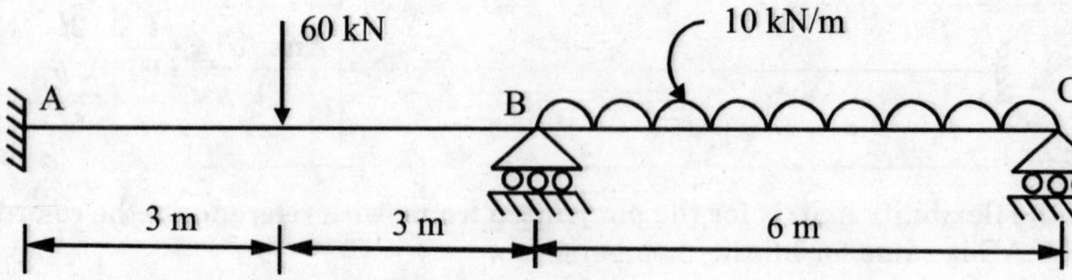

Fig. 4.55

[**Ans.** $M_{AB} = -45$ kNm ; $M_{BA} = 45$ kNm ; $M_{BC} = -45$ kNm ; $M_{CB} = 0$]

6. Determine the bending moments at supports B and C of the continuous beam shown in Fig. 4.56, using these moments as the redundants M_1 and M_2 respectively. Assume that the redundants are positive when they produce compression on the top of the beam. The beam has uniform flexural rigidity EI.

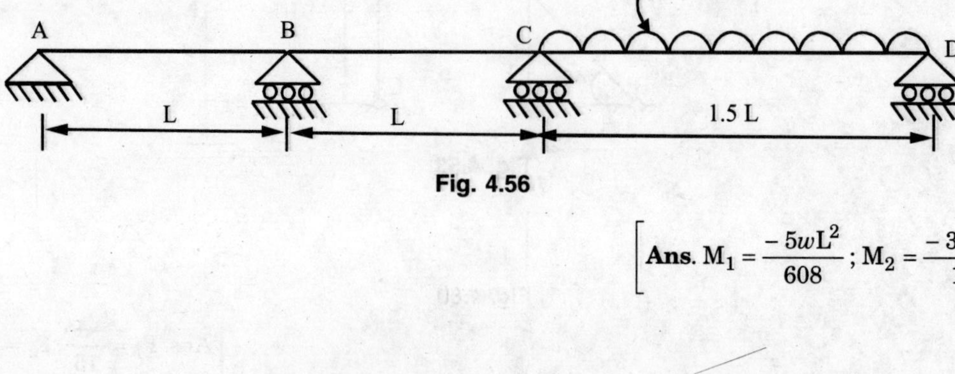

Fig. 4.56

$$\left[\textbf{Ans.}\ M_1 = \frac{-5wL^2}{608}\ ;\ M_2 = \frac{-33wL^2}{152}\right].$$

7. Analyse the continuous beam shown in Fig. 4.57. Assume EI as uniform. Use matrix flexibility method.

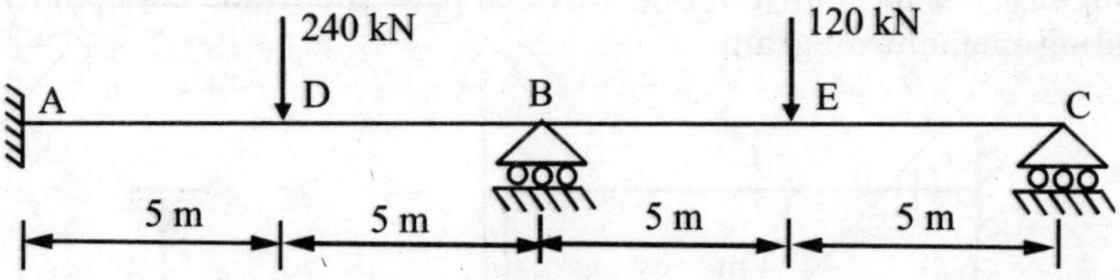

Fig. 4.57

[**Ans.** M_{AB} = – 321.3 kNm ; M_{BA} = 256.5 kNm ; M_{BC} = – 256.5 kNm ; M_{CB} = 0]

8. Analyse the beam AB shown in Fig. 4.58. Assume EI as constant. Adopt flexibility method.

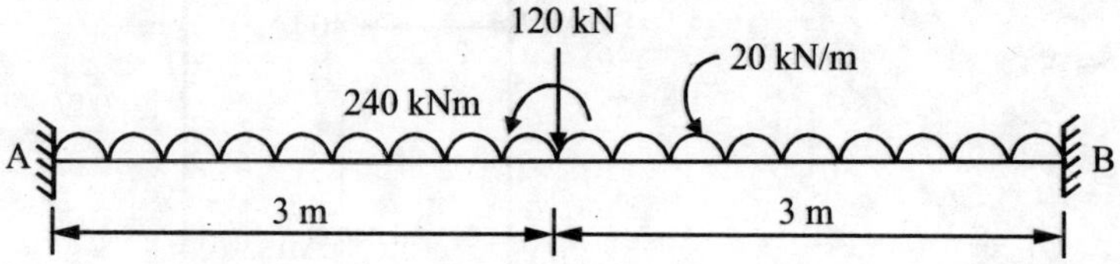

Fig. 4.58

[**Ans.** M_A = – 210 kNm ; M_{BA} = 390 kNm]

9. Using matrix flexibility method, analyse the continuous beam loaded as shown in Fig. 4.59.

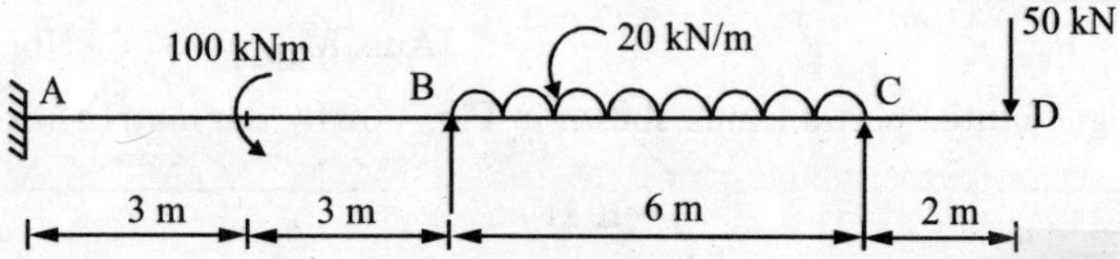

Fig. 4.59

[**Ans.** M_A = – 6.43 kNm ; M_{BA} = 12.14 kNm]

10. Find the redundant F_1 and F_2 for the plane frame shown in Fig. 4.60 considering only flexural deformations. The flexural rigidity EI is the same for all members.

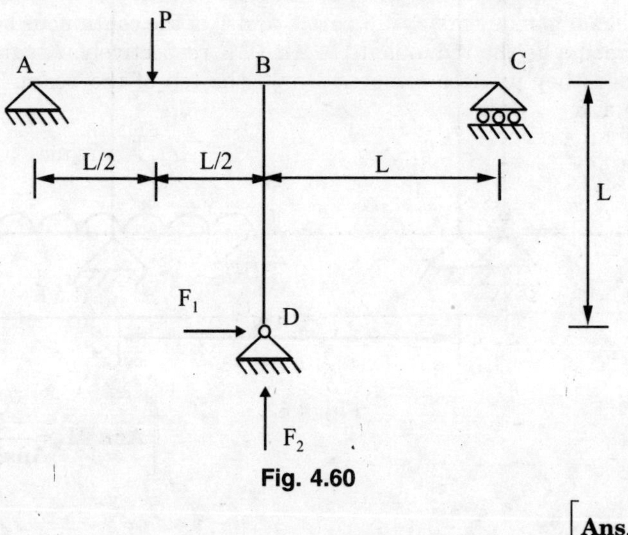

Fig. 4.60

$$\left[\textbf{Ans. } F_1 = \frac{-P}{16}, F_2 = \frac{11P}{16}\right].$$

11. A rigid frame ABC is fixed at A and hinged at C. AB is horizontal and 4 m long, BC is vertical and 6 m high. Moment of Inertia of member AB is I_0 and that of BC is $2I_0$. A concerted load of 40 kN acts at mid height of BC as shown in Fig. 4.61. Analyse the frame using matrix force method and draw the bending moment diagram.

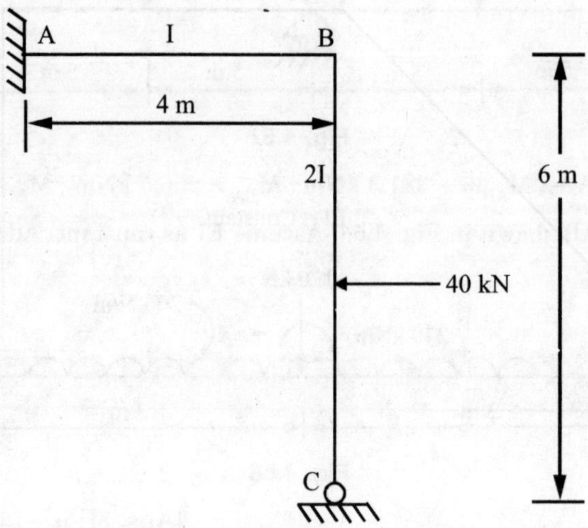

Fig. 4.61

[**Ans.** M_{AB} = – 97.6 kNm ; M_{BA} = 240 kNm]

12. Analyse the rigid jointed portal frame shown in Fig. 4.62 by the matrix flexibility method.

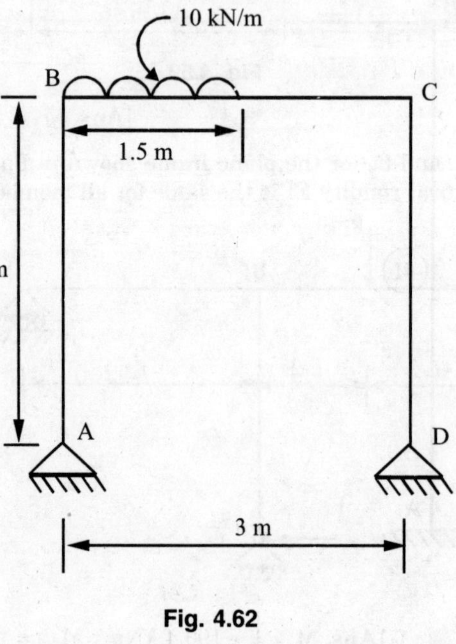

Fig. 4.62

[**Ans.** $M_B = M_C$ = 2.25 kNm]

13. Analyse the portal frame with inclined leg shown in Fig. 4.63 using the matrix flexibility method.

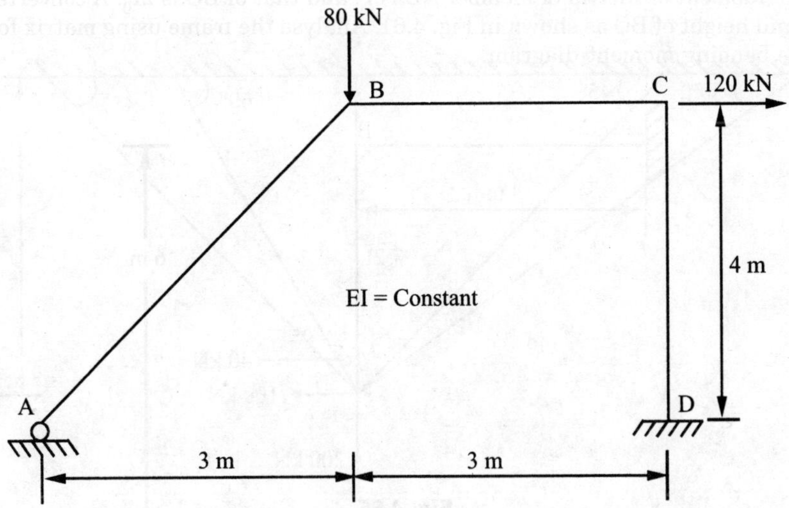

Fig. 4.63

[**Ans.** M_{AB} = 0 ; M_{BA} = – 112.91 kNm ; M_{BC} = 112.91 kNm ; M_{CB} = 164.84 kNm ; M_{CD} = – 164.86 kNm ; M_{DC} = – 164.46 kNm]

14. Analyse the portal frame shown in Fig. 4.64 using matrix flexibility method.

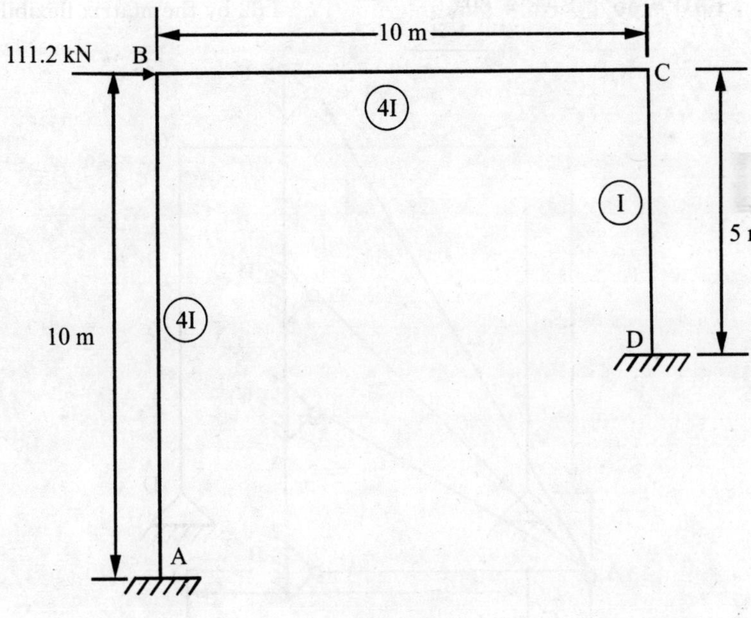

Fig. 4.64

[**Ans.** M_{AB} = – 196.4 kNm ; M_{BA} = 152.6 kNm ; M_{BC} = – 152.6 kNm ; M_{CB} = 164.84 kNm ; M_{CD} = – 207.1 kNm]

15. Analyse the pin jointed structure shown in Fig. 4.65. Using matrix flexibility method. The cross-sectional area of each member is 2000 mm^2. Assume E = 2 × 10^5 N/mm^2.

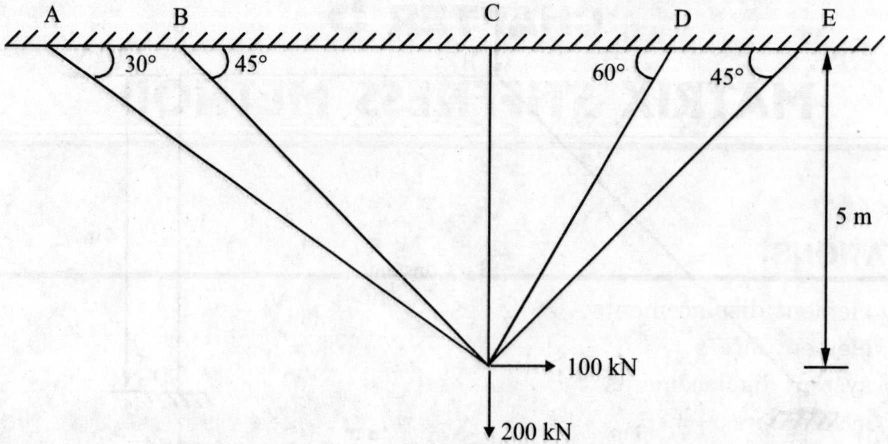

Fig. 4.65

[**Ans.** F_{0A} = 60 kN ; F_{0B} = 87 kN ; F_{0C} = 86 kN ; F_{0D} = 27 kN ; F_{0E} = – 1 kN]

16. Figure 4.66 shows the members of a truss structure, subjected to a vertical load of 100 kN as shown. Calculate the displacements of node A using matrix flexibility method. The cross-section of each member is 0.001 m^2 and modulus of elasticity of the material (E) is 2.0 × 10^8 kN/m^2. Lengths AB = 3.0 m, AC = 3.46 m, AD = 4.243 m, AE = 6.0 m. Angles BAC = 30° ; BAD = 45° ; BAE = 60°.

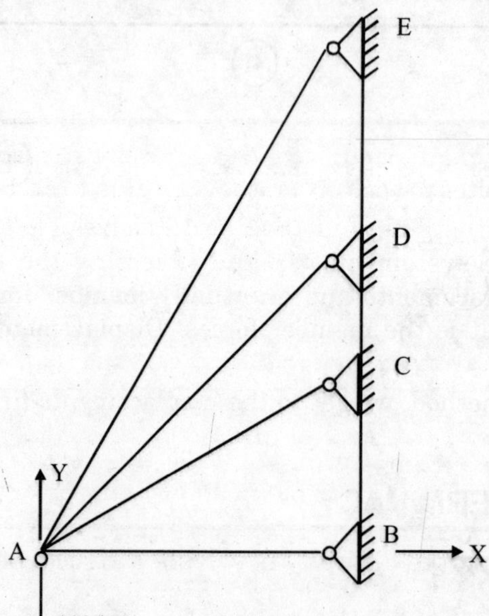

Fig. 4.66

[**Ans.** Δ_{HA} = 1.721 mm (←) ; Δ_{VA} = 3.246 mm (↓)]

CHAPTER 5
MATRIX STIFFNESS METHOD

5.1 NOTATIONS

{δ}	element displacements
{P}	element forces
{u}	system displacements
{F}	system forces
[k_n]	stiffness matrix of element n
[K]	system stiffness matrix
[k]	assembled element stiffness matrix
[β]	transformation matrix
{P^0}	forces in element co-ordinates in the fixed state
{F^f}	forces applied at system co-ordinates
{F^0}	forces at fixed co-ordinates
{P′}	element forces consistent with displacements
{P^f}	final forces

5.2 INTRODUCTION

Stiffness method is the more popular younger brother of Flexibility method. Although the two methods are the opposites to each other, they are akin to each other in several respects.

Like Flexibility method, stiffness method also involves generating element matrices, assembling them to get the system matrix and inverting the same to solve for nodal displacements, member displacements and eventually member forces. In the solution of a structure the important result is the member forces. Displacements are generally of lesser importance.

However in Stiffness method, we get to the displacements first and thence to member forces.

5.3 KINEMATIC INDETERMINACY

In the Flexibility method the difficulty of solving a structure increases with the static indeterminacy of a structure.

In Stiffness method the difficulty increases with its kinematic indeterminacy. Thus, structures with more constraints (supports, fixities etc.) are more easily solved than structures with more freedom. Strangely, the structure in Fig. 5.1 (a) is easier to tackle than the structure in Fig. 5.1 (b).

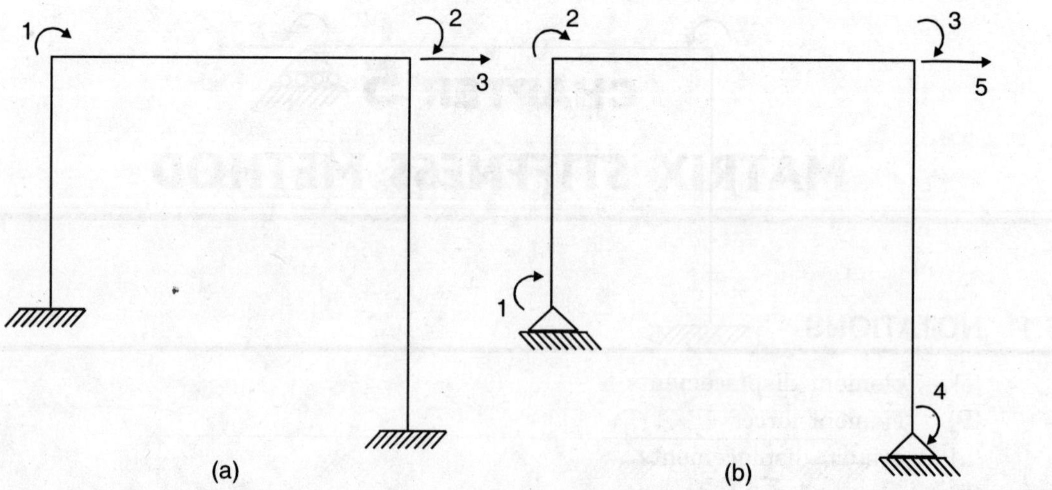

Fig. 5.1

We have to get familiar with kinematic indeterminacies (or freedoms). Figure 5.1 (*a*) has three freedoms as shown and Fig. 5.1 (*b*) has five and thus more complex for Stiffness methods.

By flexibility method the frame in Fig. 5.1 (*a*) would have been more complex (indeterminacy of three) than that in Fig. 5.1 (*b*) (indeterminacy of one).

Example 5.1 *Determine the kinematic indeterminacies of the frame and beams in Fig. 5.2.*

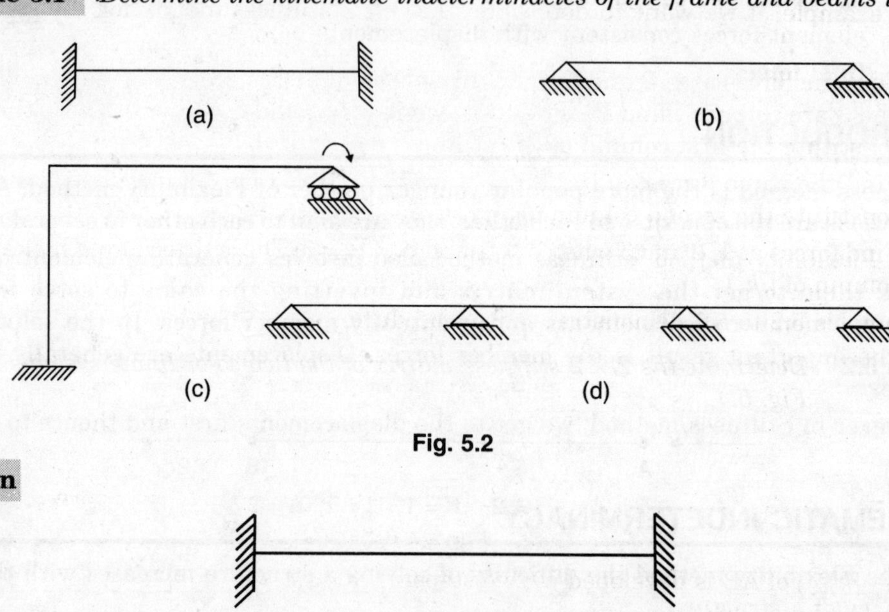

Fig. 5.2

Solution

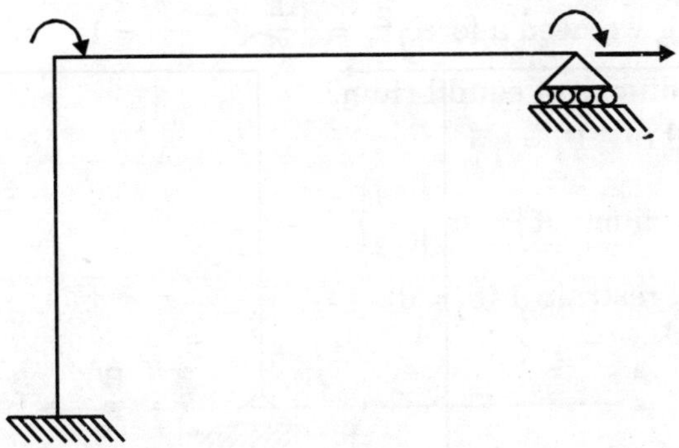

KI = 3

(c)

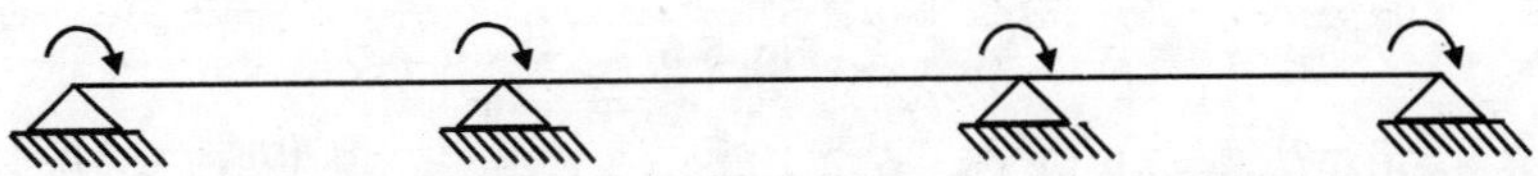

KI = 4

(d)

Fig. 5.3. (*a*) to (*d*).

5.4 GENERATING STIFFNESS MATRICES

The $n \times n$ stiffness matrix of a structure with a specified set of n co-ordinates is determined by applying one unit displacement at a time and determining the forces at each co-ordinate to sustain that displacement.

For example, if we want to determine the 3×3 stiffness matrix for the structure in Fig. 5.1 (*a*).

(*i*) Find the forces at 1, 2 and 3 when displacement at 1 is unity and displacements at 2 and 3 are zero, *i.e.*, find P_1, P_2 and P_3 when $\delta_1 = 1$ and $\delta_2 = \delta_3 = 0$. These three forces constitute the first column of the stiffness matrix $[k_1]$.

(*ii*) Find the three forces at 1, 2 and 3 when $\delta_2 = 1$ and $\delta_1 = \delta_3 = 0$. These three forces constitute the second column of the stiffness matrix $[k_1]$.

(*iii*) Find forces at 1, 2 and 3 when $\delta_3 = 1$ and $\delta_1 = \delta_2 = 0$. These three forces make the third column of $[\ k_1]$.

Examples 5.2 to 5.5 demonstrate this procedure.

Example 5.2 *Determine the 2×2 stiffness matrix of the two co-ordinate systems shown in Fig. 5.4.*

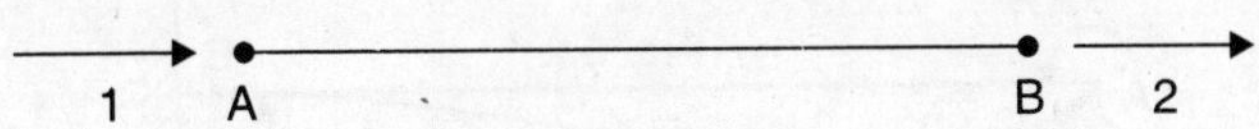

$l = 2,\ \ E = 1,\ \ I = 1,\ \ A = 2$

Fig. 5.4

Solution $[k]_{2\times 2}$ of AB is developed in two steps:

(*i*) Apply $\delta_1 = 1$ and restrain 2 ($\delta_2 = 0$)

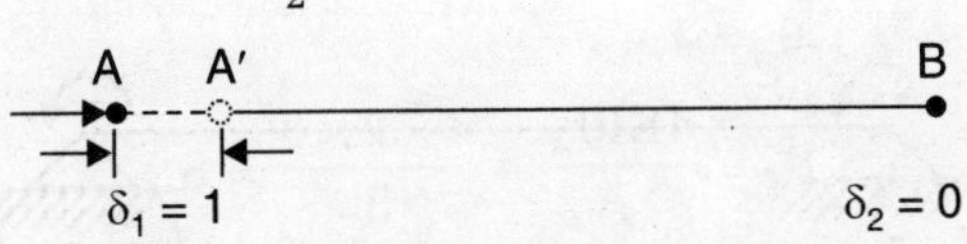

Fig. 5.5

To sustain $\delta_1 = 1$, we need a force $P_1 = \frac{AE}{l} = \frac{2 \times 1}{2} = 1$

Then , from conditions of equilibrium,

P_2 shall be equal to $- P_1 = - 1$

Hence the first column of [k] is $\begin{Bmatrix} 1 \\ -1 \end{Bmatrix}$

(*ii*) Apply $\delta_2 = 1$ and restrain 1 ($\delta_1 = 0$)

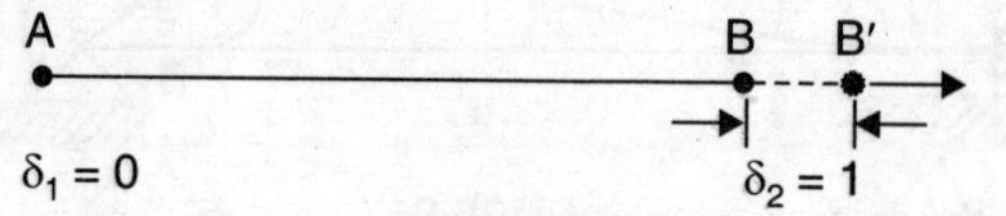

Fig. 5.6

To cause a displacement of 1 at B, we need a force at 2 equal to $\frac{AE}{l}$ (= 1)

To keep this force in equilibrium we need an opposite force at A, *i.e.*, $P_1 = - 1$

Hence the second column of [k_1] is $\begin{Bmatrix} -1 \\ 1 \end{Bmatrix}$

Hence, $$[k] = \begin{bmatrix} 1 & -1 \\ -1 & 1 \end{bmatrix}.$$

Example 5.3 *Determine the 2 × 2 stiffness matrix of the beam system shown in Fig. 5.7.*

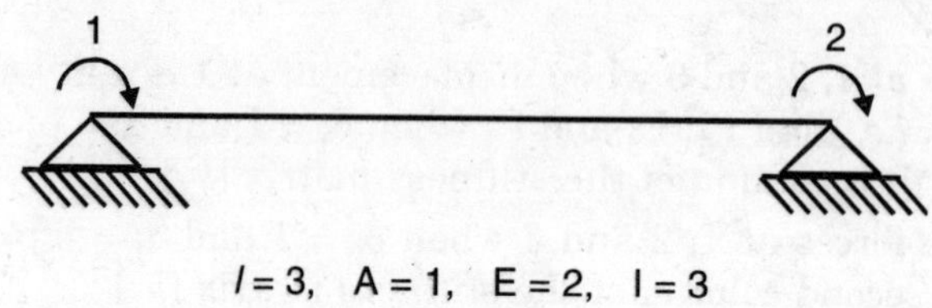

Fig. 5.7

Solution **Step 1.** To find the first column of [k] apply a unit displacement at 1 only and restrain 2 from rotating

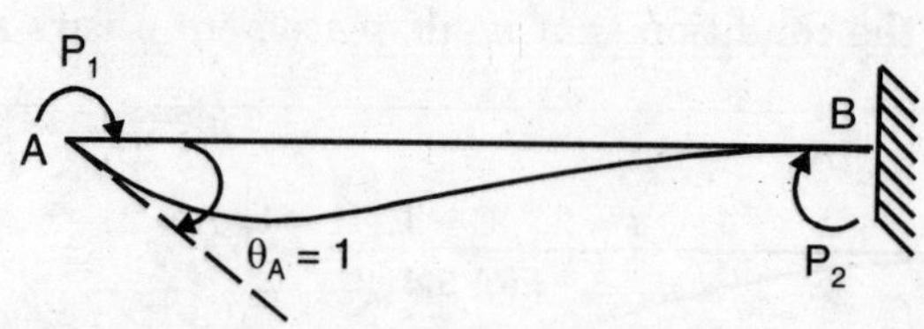

Fig. 5.8

If $$\theta_A = 1,$$

$$P_1 = \frac{4EI\theta_A}{l} = \frac{4 \times 2 \times 3}{3} = 8$$

$$P_2 = \frac{2EI\theta_A}{l} = \frac{2 \times 2 \times 3}{3} = 4$$

Hence, $$\begin{Bmatrix} k_{11} \\ k_{21} \end{Bmatrix} = \begin{Bmatrix} 8 \\ 4 \end{Bmatrix}$$

Step 2. To get the second column of [*k*] apply a unit rotation at B and restrain A

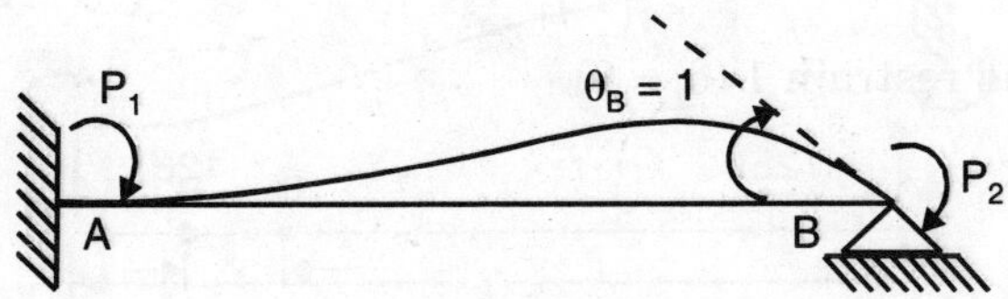

Fig. 5.9

$$P_2 = \frac{4EI\theta_B}{l} = \frac{4 \times 2 \times 3}{3} = 8$$

$$P_1 = \frac{2EI\theta_B}{l} = \frac{2 \times 2 \times 3}{3} = 4$$

Hence, $$\begin{Bmatrix} k_{12} \\ k_{22} \end{Bmatrix} = \begin{Bmatrix} 4 \\ 8 \end{Bmatrix} \text{ and } [k] = \begin{bmatrix} 8 & 4 \\ 4 & 8 \end{bmatrix}.$$

Example 5.4 *Develop the stiffness matrix for the beam element with respect to the two co-ordinates shown in Fig. 5.10.*

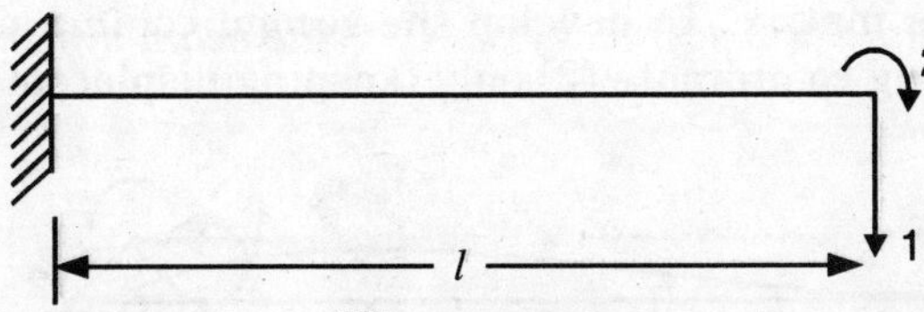

Fig. 5.10

Solution To develop the stiffness matrix, unit displacements are introduced along co-ordinates (1) and (2), one at a time and the forces developed in the co-ordinates given are calculated.

To develop the first column of the stiffness matrix, apply unit displacement along co-ordinate (1) only, with the condition that no displacement occurs at co-ordinate (2).

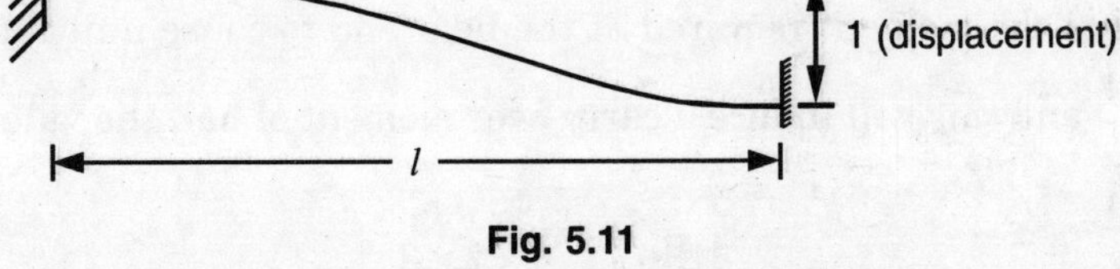

Fig. 5.11

Here the right end settles down by unity (*i.e.*, $\delta = 1$). The moment developed at either end will be $\frac{6EI}{l^2}$ (anti-clockwise).

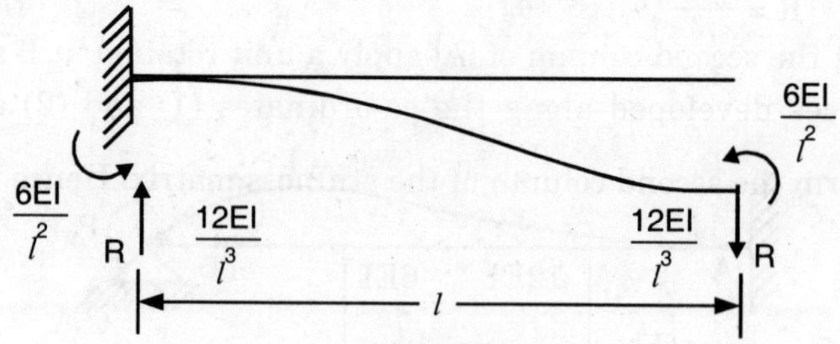

Fig. 5.12

For equilibrium, the reactions will be such that $\frac{6EI}{l^2} + \frac{6EI}{l^2} = R \times l$

(↺) (↺) (↻)

$\therefore$ $R = \frac{12EI}{l^3}$ (Left up and right down, since this has to develop a clockwise moment.)

Hence the forces along co-ordinates (1) and (2) are $\frac{12EI}{l^3}$ and $\frac{-6EI}{l^2}$ which form the first column of the stiffness matrix. To develop the second column of the stiffness matrix, apply unit displacement along co-ordinate (2) only (keeping displacement along 1 as zero.)

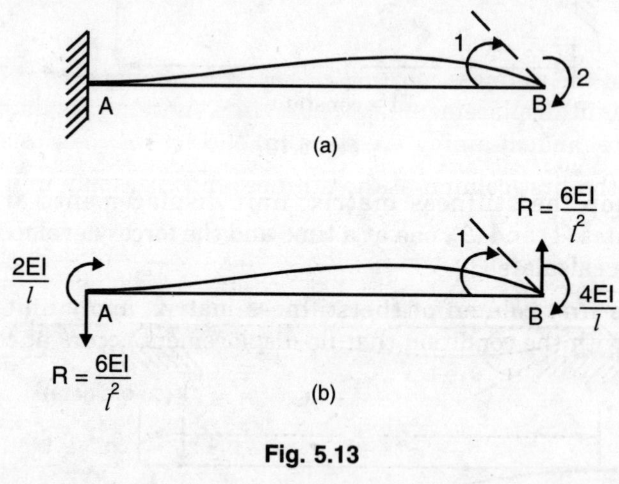

Fig. 5.13

We know that the moment required at the near end to cause unit rotation, the far end being fixed, is $\frac{4EI}{l}$ and this will induce a carry over moment of half the value and in the same direction, *i.e.*, $\frac{2EI}{l}$.

Reactions : $\frac{2EI}{l} + \frac{4EI}{l} - R \times l = 0$

(↶) (↶) (↶)

$$R = \frac{6EI}{l^2}$$ (Left down, right up)

Hence the forces developed along the co-ordinates (1) and (2) are $-\frac{6EI}{l^2}$ and $\frac{4EI}{l}$ respectively which form the second column of the stiffness matrix. Hence the stiffness matrix is

$$[k] = \begin{bmatrix} \frac{12EI}{l^3} & -\frac{6EI}{l^2} \\ -\frac{6EI}{l^2} & \frac{4EI}{l} \end{bmatrix}$$

$$[k] = \frac{EI}{l^3}\begin{bmatrix} 12 & -6l \\ -6l & 4l^2 \end{bmatrix}.$$

Example 5.5 *Develop the stiffness matrix for the beam element shown in Fig. 5.14 with respect to the degrees of freedom shown.*

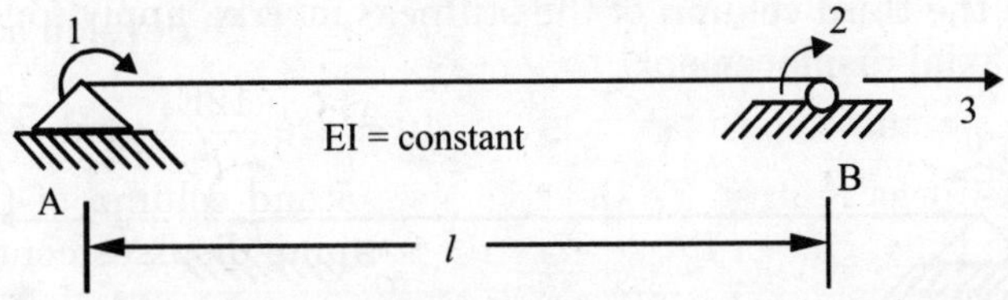

Fig. 5.14

Solution The size of the stiffness matrix is 3 × 3. The stiffness matrix will be obtained by applying unit displacements one at a time, (with no displacements at any other co-ordinate) and obtaining the forces in each of the co-ordinates.

(*i*) To generate the first column of the stiffness matrix, apply unit displacement along co-ordinate (1) only.

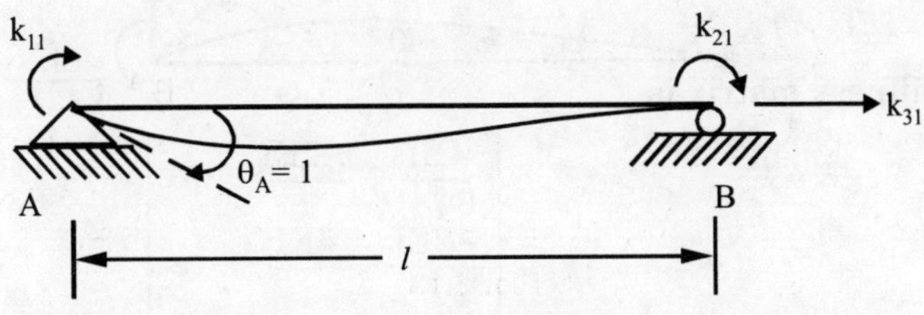

Fig. 5.15. Unit displacement applied along co-ordinate (1) only.

k_{11} = Force developed along co-ordinate (1) due to unit displacement along co-ordinate (1) only = $\frac{4EI}{l}$.

This is the moment required to cause unit rotation at the near end without translation, the far end being fixed.

$$k_{21} = \frac{2EI}{l}$$

$$k_{31} = 0^*$$

(*ii*) To generate the second column of stiffness matrix, apply unit displacement along co-ordinate (2) only.

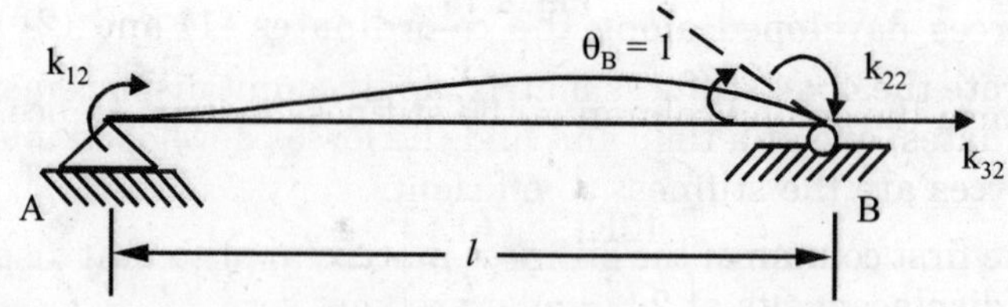

Fig. 5.16. Unit displacement applied along co-ordinate (2) only.

$$k_{22} = \frac{4EI}{l}$$

$$k_{12} = \frac{2EI}{l}$$

$$k_{32} = 0^*$$

(*iii*) To generate the third column of the stiffness matrix, apply unit displacement along co-ordinate (3) only (axial displacement).

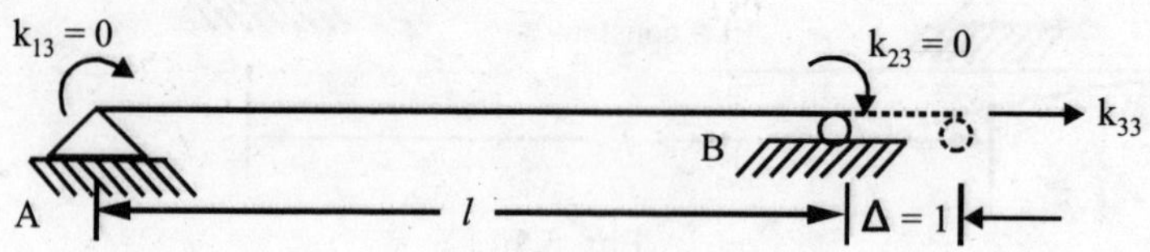

Fig. 5.17

$$\frac{k_{33} \times l}{AE} = 1$$

$$k_{33} = \frac{AE}{l}$$

$$k_{13} = k_{23} = 0$$

Hence the stiffness matrix is

$$[k] = \begin{bmatrix} \frac{4EI}{l} & +\frac{2EI}{l} & 0 \\ +\frac{2EI}{l} & \frac{4EI}{l} & 0 \\ 0 & 0 & \frac{AE}{l} \end{bmatrix}.$$

Example 5.6 *Develop the stiffness matrix of the beam shown in Fig. 5.18 with respect to the 4 degrees of freedom given.*

*In either case, displacement along AB is zero both at A and at B. Hence there is no axial strain and no axial force.

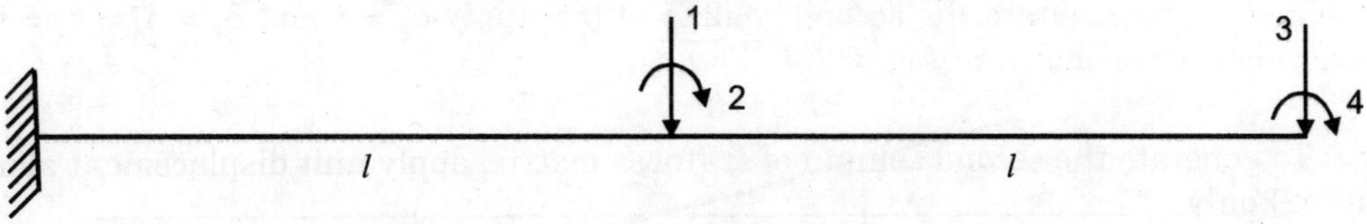

Fig. 5.18

Solution To generate the 4 × 4 stiffness matrix, apply a unit displacement along each of the 4 co-ordinates, one at a time and find the forces developed in all the co-ordinates. These forces are the stiffness co-efficients.

To generate the first column of the stiffness matrix, apply a unit displacement at 1 only, taking care that the displacements at 2, 3 and 4 are kept zero.

i.e., $\delta_1 = 1$ and $\delta_2 = \delta_3 = \delta_4 = 0.$

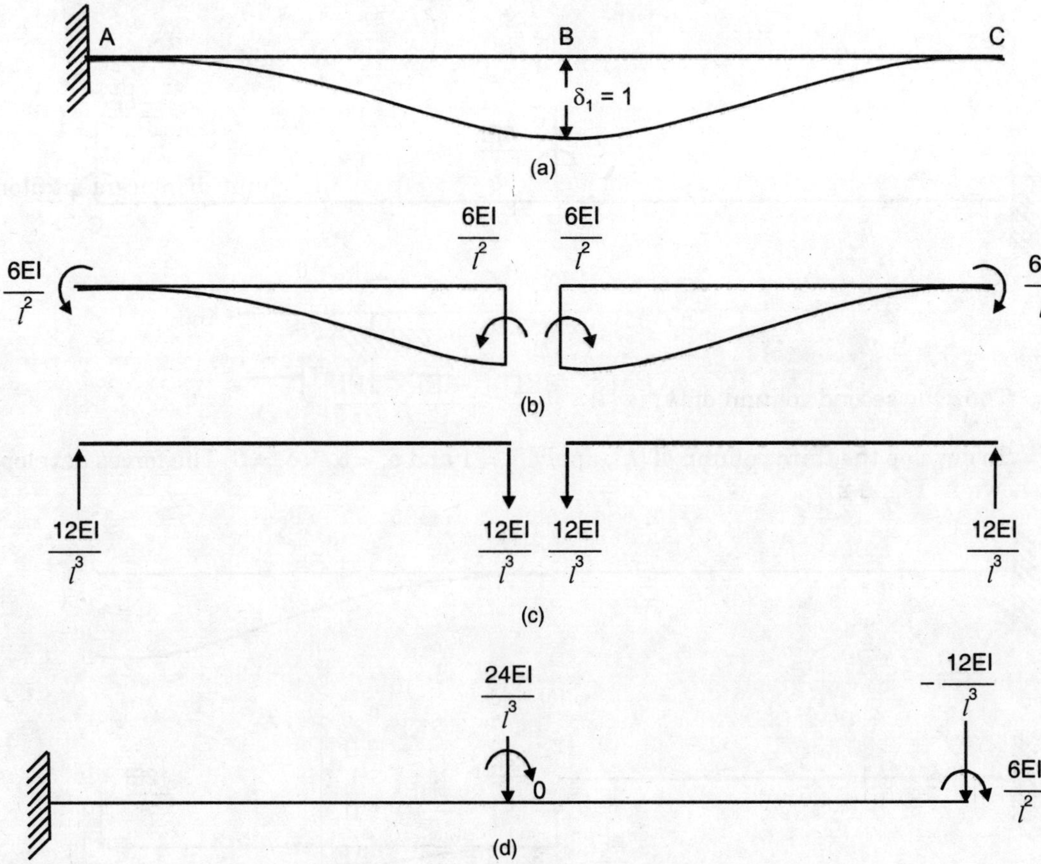

Fig. 5.19

The forces consistent with $\delta_1 = 1$ are shown in Fig 5.19 (*b*) and (*c*). These are summed up and shown in Fig. 5.19 (*d*). Hence the first column of [*k*] matrix is,

$$\begin{bmatrix} \dfrac{24EI}{l^3} & 0 & -\dfrac{12EI}{l^3} & \dfrac{6EI}{l^2} \end{bmatrix}^T$$

Likewise, to generate the second column of $[k]$, apply $\delta_2 = 1$ and $\delta_1 = \delta_3 = \delta_4 = 0$. The forces developed are shown in Fig. 5.20.

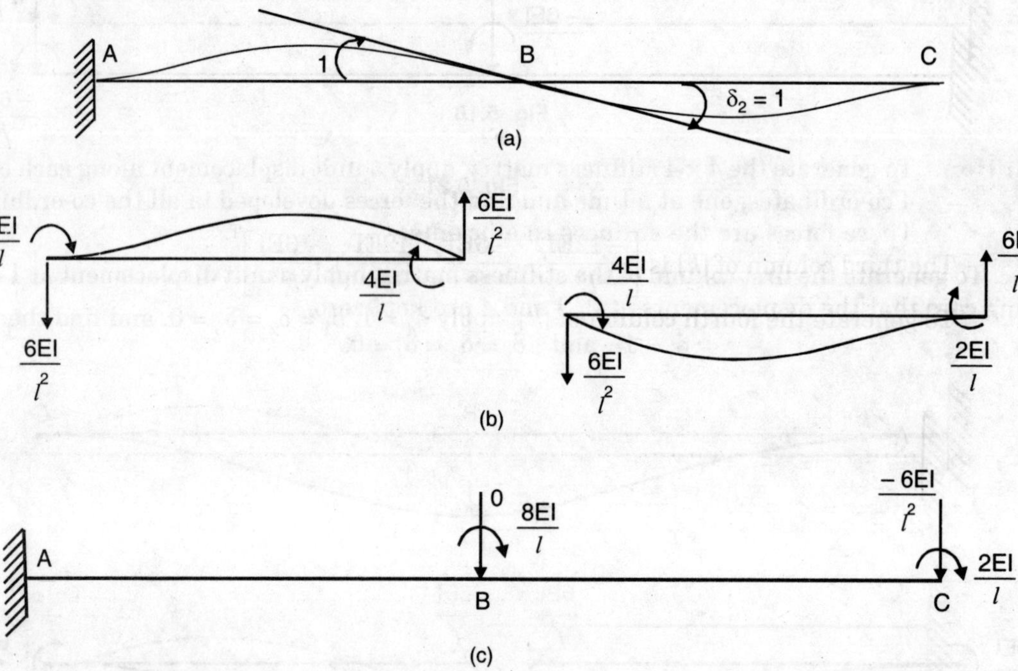

Fig. 5.20

Thus the second column of $[k]$ is $\begin{bmatrix} 0 & \dfrac{8EI}{l} & -\dfrac{6EI}{l^2} & \dfrac{2EI}{l} \end{bmatrix}^T$

To develop the third column of $[k]$, apply $\delta_3 = 1$ and $\delta_1 = \delta_2 = \delta_4 = 0$. The forces developed are shown in Fig. 5.21.

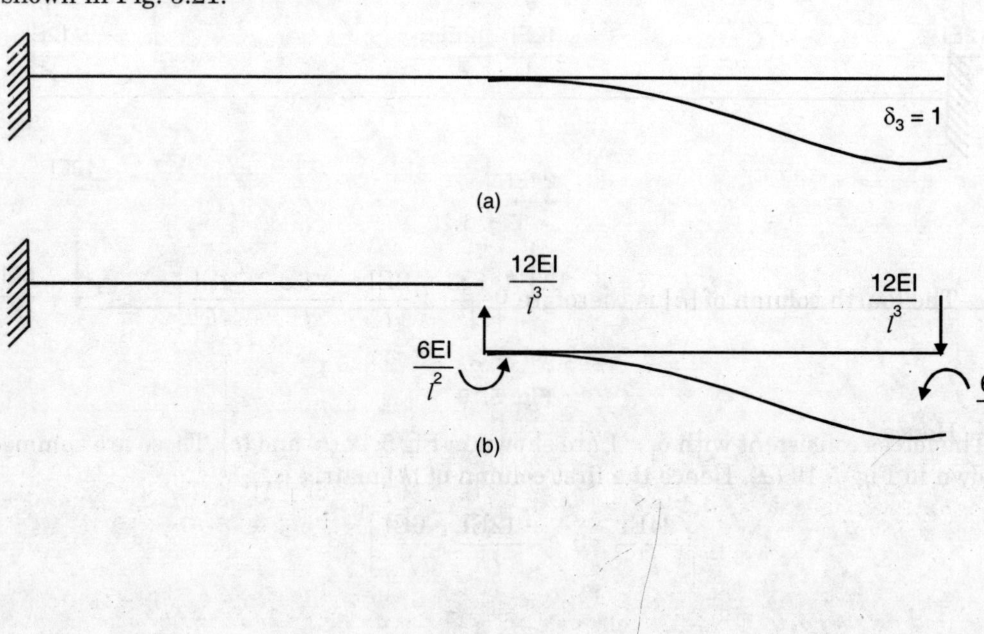

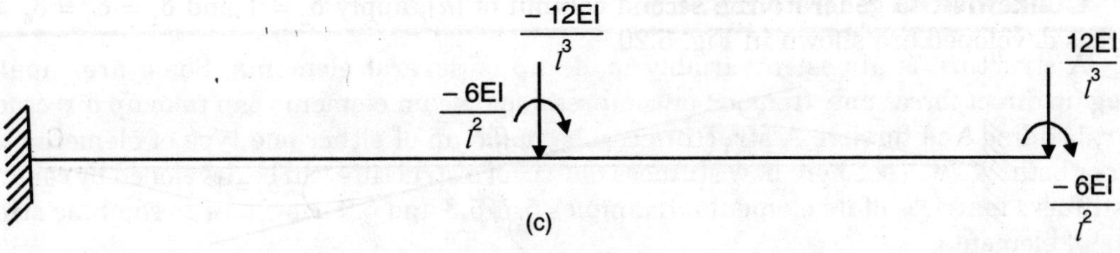

Fig. 5.21

The third column of $[k]$ is $\begin{bmatrix} \dfrac{-12EI}{l^3} & \dfrac{-6EI}{l^2} & \dfrac{12EI}{l^3} & \dfrac{-6EI}{l^2} \end{bmatrix}^T$

To generate the fourth column of $[k]$, apply $\delta_4 = 1$, $\delta_1 = \delta_2 = \delta_3 = 0$, and find the forces.

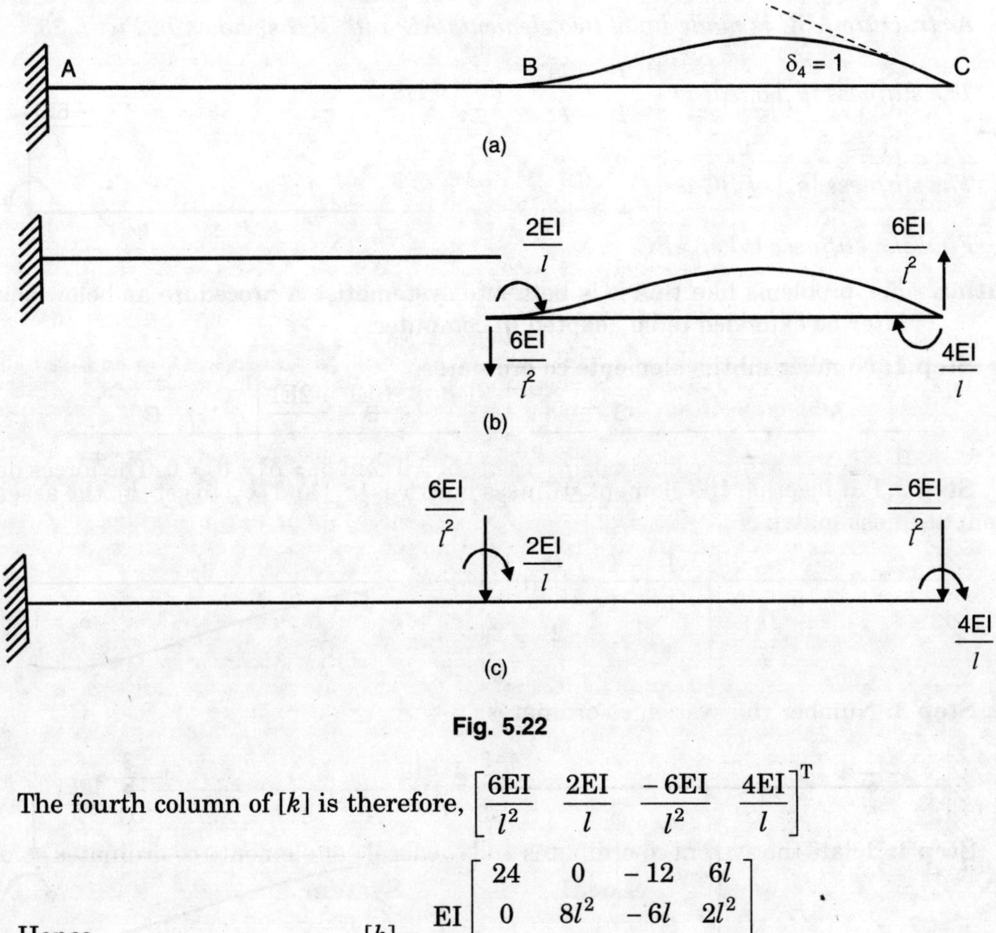

Fig. 5.22

The fourth column of $[k]$ is therefore, $\begin{bmatrix} \dfrac{6EI}{l^2} & \dfrac{2EI}{l} & \dfrac{-6EI}{l^2} & \dfrac{4EI}{l} \end{bmatrix}^T$

Hence,
$$[k] = \frac{EI}{l^3}\begin{bmatrix} 24 & 0 & -12 & 6l \\ 0 & 8l^2 & -6l & 2l^2 \\ -12 & -6l & 12 & -6l \\ 6l & 2l^2 & -6l & 4l^2 \end{bmatrix}$$

5.5 COMBINING STIFFNESSES

A structure is almost invariably made up of several elements. Some are capable of taking up direct forces only (tension or compression). Beam elements can take up direct forces, shear, bending and torsion. A structure can be made up of either one type of elements or the other or both. We will see here how stiffness matrix of a structure can be developed by combining the stiffness matrices of its elements. Examples 5.7, 5.8 and 5.9 show how to combine stiffness terms of elements.

Example 5.7

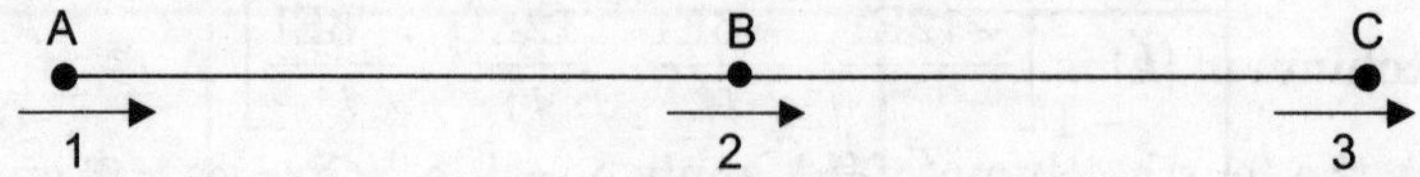

Fig. 5.23

A structure ABC is made up of two elements AB and BC as shown in Fig. 5.23.

The stiffness $[k_1]$ of AB is $\begin{bmatrix} 1 & -1 \\ -1 & 1 \end{bmatrix}$

The stiffness $[k_2]$ of BC is $\begin{bmatrix} 2 & -2 \\ -2 & 2 \end{bmatrix}$

Find the stiffness [k] of ABC.

Solution For problems like this it is better to systematise a procedure as below that can later be extended to be adopted in computers.

Step 1: Number all the elements co-ordinates.

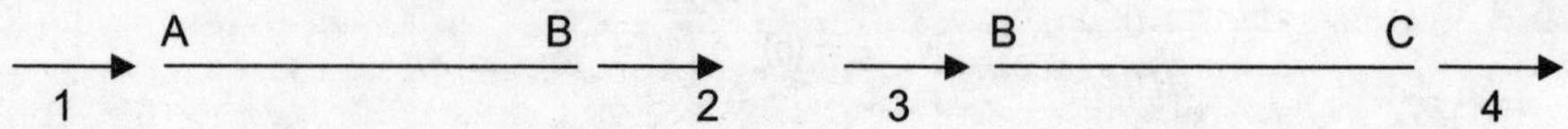

Step 2: Put together the element stiffness matrices $[k_1]$ and $[k_2]$ to get $[k]$, the assembled element stiffness matrix.

$$[k] = \begin{bmatrix} \begin{matrix} 1 & -1 \\ -1 & 1 \end{matrix} & [0] \\ [0] & \begin{matrix} 2 & -2 \\ -2 & 2 \end{matrix} \end{bmatrix}$$

Step 3: Number the system co-ordinates.

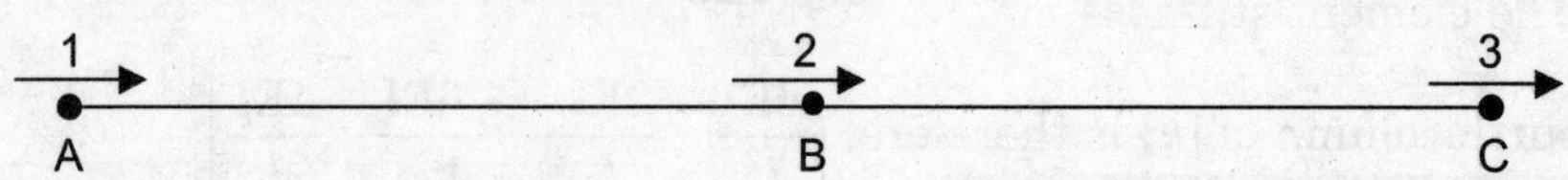

Step 4: Relate the system co-ordinates and the local (or element) co-ordinates as under.

Local	System
1	1
2	2
3	2
4	3

Step 5: Rewrite the elements of the [k] matrix on a 3 × 3 framework using the relationships in step 4 above, to from the system matrix [K]

Element →		1	2	3	4
	System →	1	2		3
1	1	1	− 1	0	0
2	2	− 1	1	0	0
3		0	0	2	− 2
4	3	0	0	− 2	2

$$[K] = \begin{bmatrix} 1 & -1 & 0 \\ -1 & 3 & -2 \\ 0 & -2 & 2 \end{bmatrix}$$

This method can be called Rubinstein's method.

Example 5.8 *Two springs of stiffnesses 20 N/mm and 30 N/mm are connected in series. Find the stiffness matrix of the assembly for the three co-ordinates shown in Fig. 5.24.*

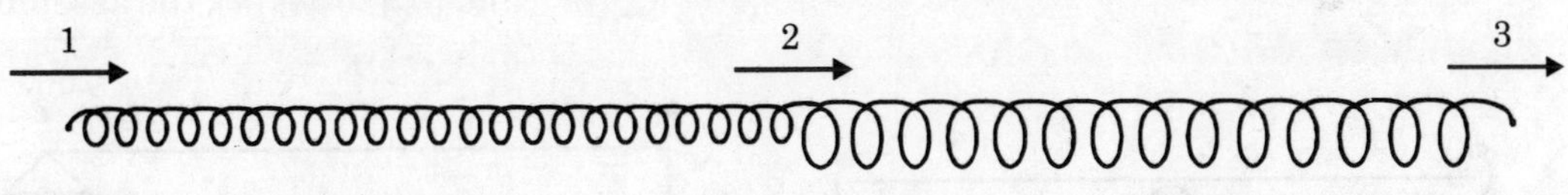

Fig. 5.24

Solution The element stiffness matrices are as under:

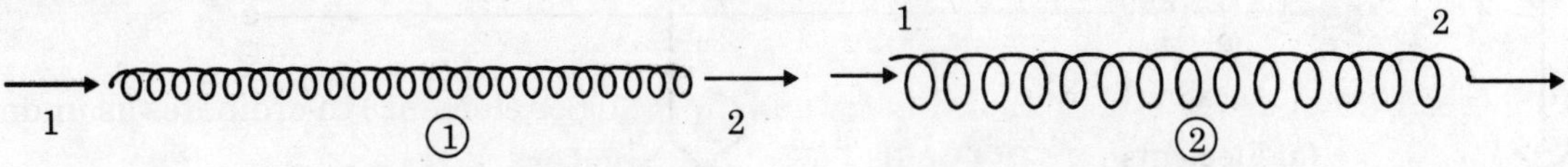

$$[k_1] = \begin{bmatrix} 20 & -20 \\ -20 & 20 \end{bmatrix}; \qquad [k_2] = \begin{bmatrix} 30 & -30 \\ -30 & 30 \end{bmatrix}$$

The assembled element stiffness matrix

$$[k] \text{ is } \begin{bmatrix} 20 & -20 & & \\ -20 & 20 & \multicolumn{2}{c}{[0]} \\ & [0] & 30 & -30 \\ & & -30 & 30 \end{bmatrix}$$

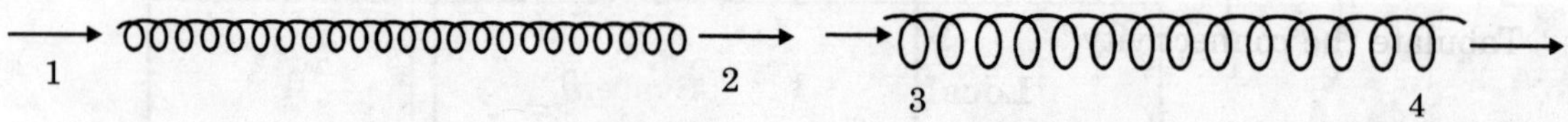

The relationship between element and system co-ordinates (connectivity) is as below

Local	**System**
1	1
2	2
3	2
4	3

$$\text{Hence, [K] becomes} \left[\begin{array}{c|c|cc} 20 & -20 & 0 & 0 \\ \hline -20 & 20 & 0 & 0 \\ 0 & 0 & 30 & -30 \\ \hline 0 & 0 & -30 & 30 \end{array}\right]$$

$$[K] = \begin{bmatrix} 20 & -20 & 0 \\ -20 & 50 & -30 \\ 0 & -30 & 30 \end{bmatrix}.$$

Example 5.9 *Assemble $[k_1]$ and $[k_2]$ of the two element frame in Fig. 5.25 to generate $[K]$ of the system.*

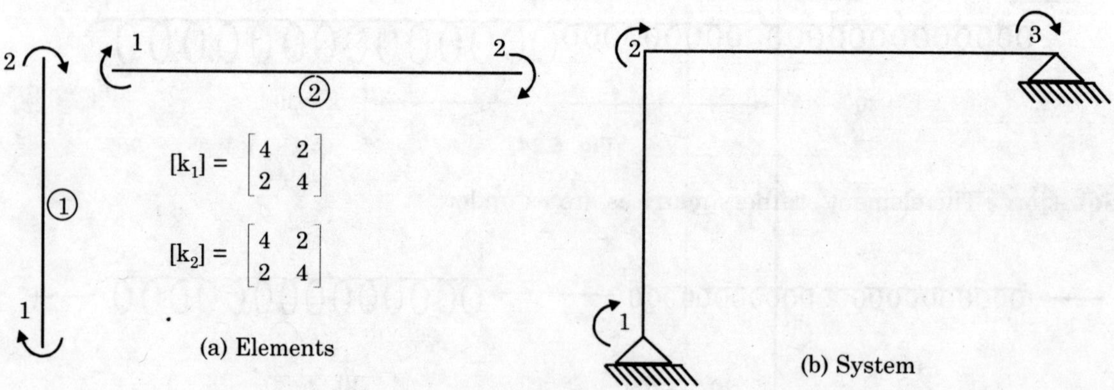

Fig. 5.25

Solution Re-number the co-ordinates of the elements as under. Write [k], the assembled element stiffness matrix.

$$[k] = \begin{bmatrix} \begin{matrix} 4 & 2 \\ 2 & 4 \end{matrix} & [0] \\ [0] & \begin{matrix} 4 & 2 \\ 2 & 4 \end{matrix} \end{bmatrix}$$

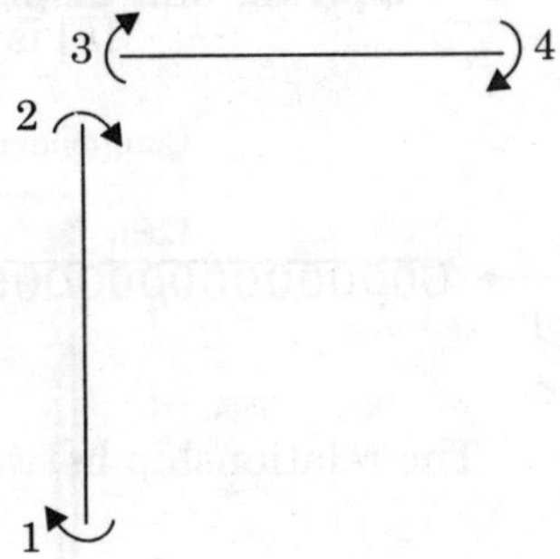

Tabulate the connectivity

Local	**System**
1	1
2	2
3	2
4	3

Generate [K]

$$[K] = \begin{bmatrix} 4 & 2 & 0 & 0 \\ 2 & 4 & 0 & 0 \\ 0 & 0 & 4 & 2 \\ 0 & 0 & 2 & 4 \end{bmatrix} = \begin{bmatrix} 4 & 2 & 0 \\ 2 & 8 & 2 \\ 0 & 2 & 4 \end{bmatrix}.$$

5.6 GENERATING STIFFNESS MATRICES DIRECTLY

The [k] matrix of a structure for a given set of co-ordinates can also be generated directly without taking recourse to the assembling techniques adopted in the earlier examples. The procedure is exemplified in examples 5.10 to 5.14.

Example 5.10 *Generate the stiffness matrix for the structure with co-ordinates as shown in Fig. 5.26.*

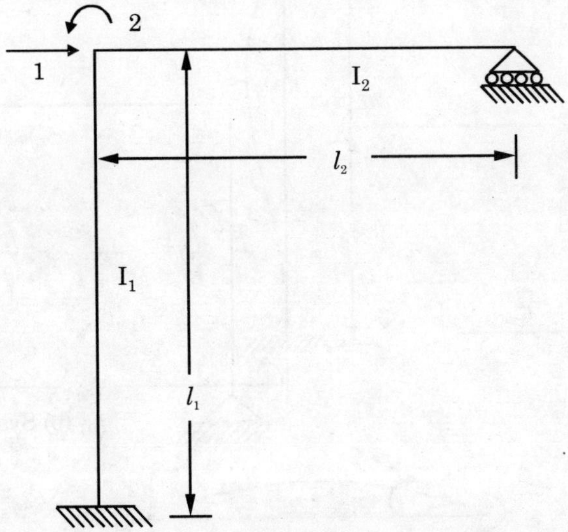

Fig. 5.26

Solution There are two co-ordinates as shown. Hence the size of the stiffness matrix will be 2×2.

Applying unit displacement along co-ordinate (1) only.

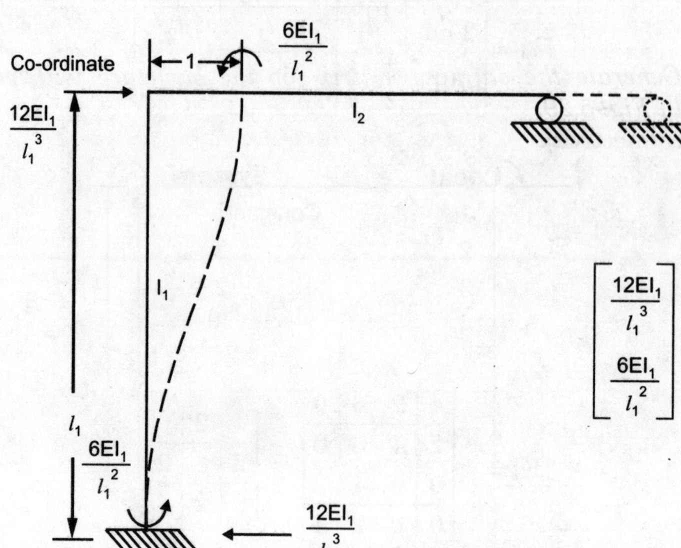

Fig. 5.27

Applying unit displacement along co-ordinate (2) only.

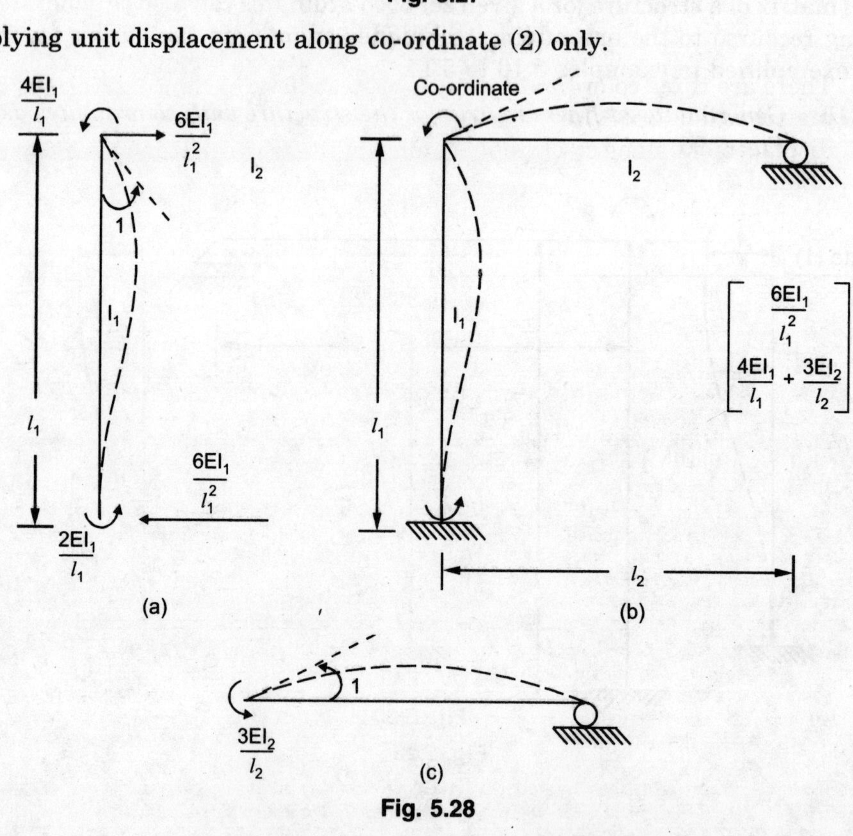

Fig. 5.28

Hence the stiffness matrix , $[K] = \begin{bmatrix} \dfrac{12EI_1}{l_1^3} & \dfrac{6EI_1}{l_1^2} \\ \dfrac{6EI_1}{l_1^2} & \dfrac{4EI_1}{l_1} + \dfrac{3EI_2}{l_2} \end{bmatrix}$.

Example 5.11 *Generate the stiffness matrix for the structure with co-ordinates shown in Fig. 5.29.*

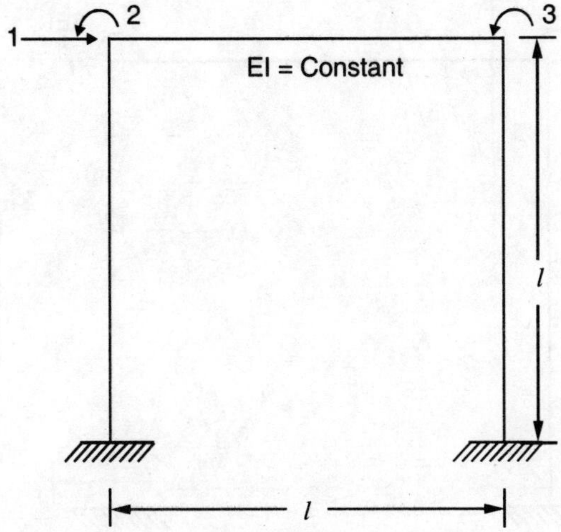

Fig. 5.29

Solution There are three co-ordinates in the given structure. Hence the size of the stiffness matrix will be 3 × 3. To generate the first column of the stiffness matrix apply unit displacement along co-ordinate (1) only with zero displacements along co-ordinates (2) and (3).

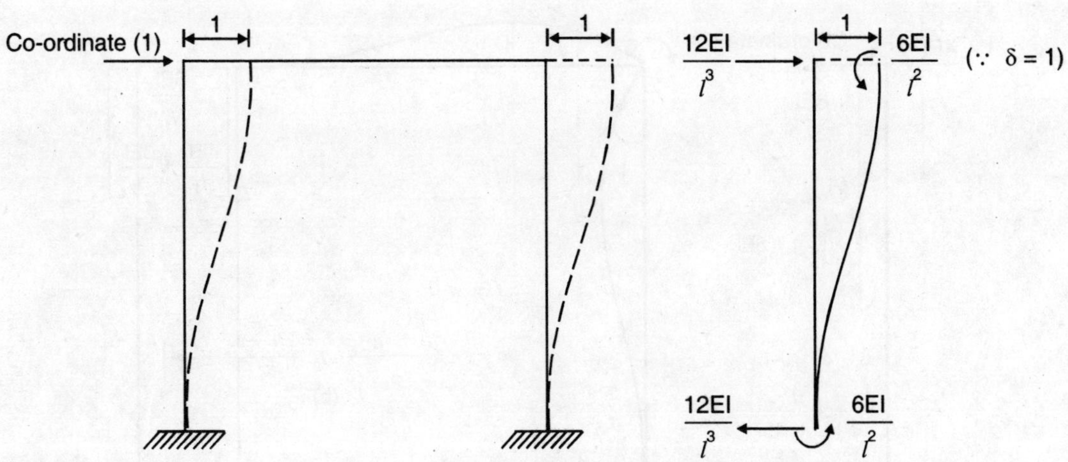

Fig. 5.30

Reactions : $$\frac{6EI}{l^2} + \frac{6EI}{l^2} - R \times l = 0$$

$$(\circlearrowleft) \quad (\circlearrowleft) \quad (\circlearrowright)$$

$$R = \frac{12EI}{l^3} \qquad \text{(Top–right ; Bottom–left)}$$

Combining the terms along the given co-ordinates only

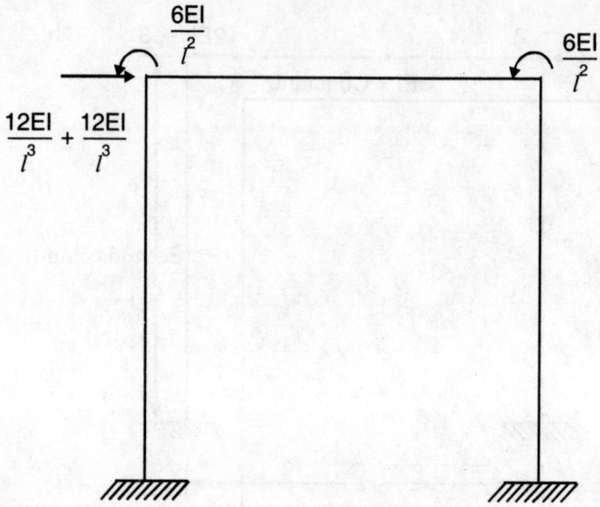

First column

$$\begin{bmatrix} \dfrac{24EI}{l^3} \\ \dfrac{6EI}{l^2} \\ \dfrac{6EI}{l^2} \end{bmatrix}$$

Fig. 5.31

To generate the second column of the stiffness matrix, apply unit displacement along co-ordinate (2) only. (Fig. 5.32)

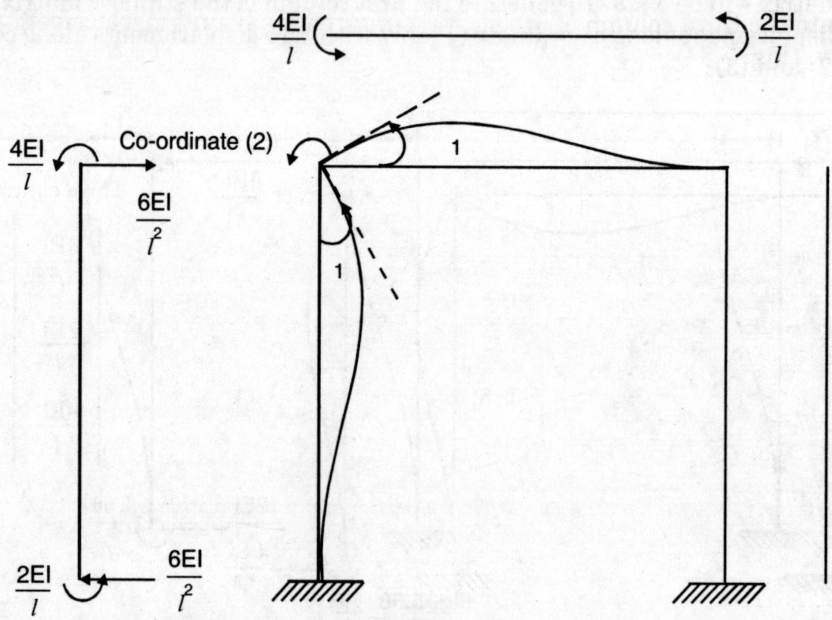

Fig. 5.32

Column reactions : $\frac{4EI}{l} + \frac{2EI}{l} - R \times l = 0$

(↺) (↺) (↻)

$$R = \frac{6EI}{l^2} \qquad \text{(Top – right ; Bottom – left)}$$

Combining the terms along the co-ordinates, we get the second column.

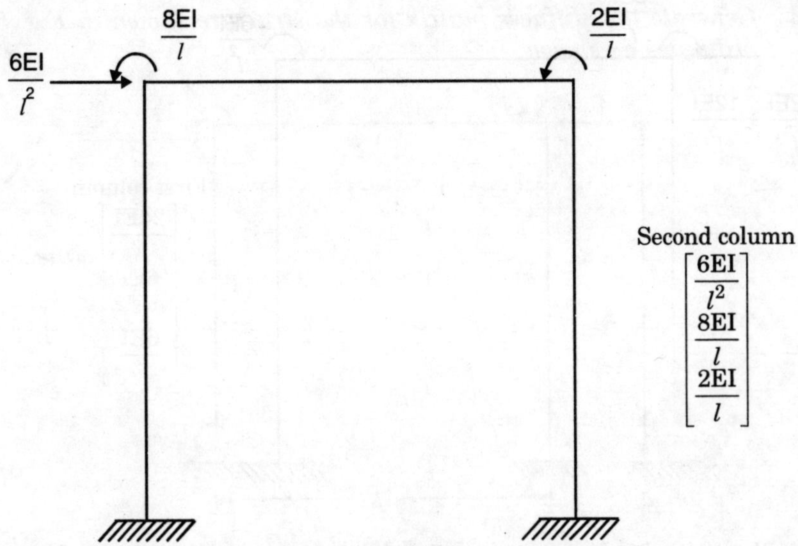

Fig. 5.33

To generate the third column of the stiffness matrix, apply unit displacement along co-ordinate (3) only. (Fig. 5.34)

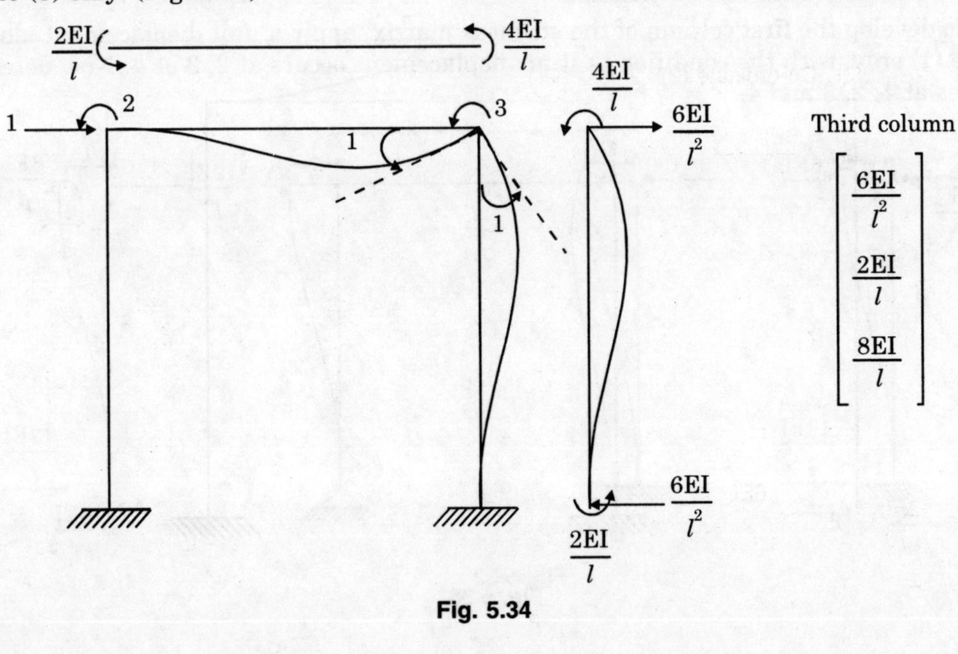

Fig. 5.34

Hence the stiffness matrix,

$$[K] = \begin{bmatrix} \frac{24EI}{l^3} & \frac{6EI}{l^2} & \frac{6EI}{l^2} \\ \frac{6EI}{l^2} & \frac{8EI}{l} & \frac{2EI}{l} \\ \frac{6EI}{l^2} & \frac{2EI}{l} & \frac{8EI}{l} \end{bmatrix} = \frac{EI}{l^3}\begin{bmatrix} 24 & 6l & 6l \\ 6l & 8l^2 & 2l^2 \\ 6l & 2l^2 & 8l^2 \end{bmatrix}.$$

Example 5.12 *Generate the stiffness matrix for the structure shown in Fig. 5.35 with co-ordinates as shown.*

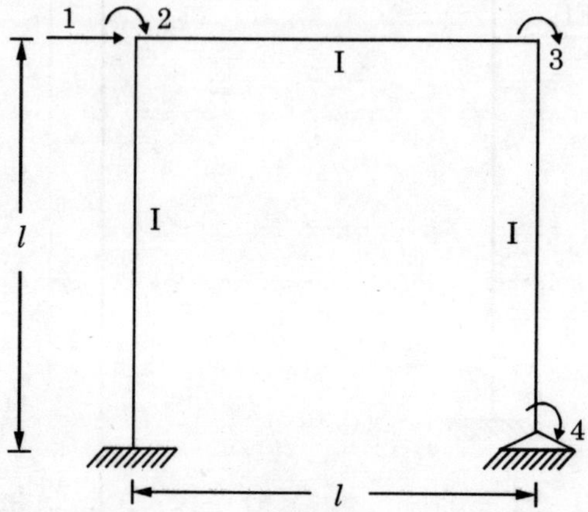

Fig. 5.35

Solution To develop the stiffness matrix for the structure, unit displacements are applied along each of the four co-ordinates, one at a time and the forces along the given co-ordinates are determined.

To develop the first column of the stiffness matrix, apply a unit displacement along co-ordinate (1) only, with the condition that no displacement occurs at 2, 3 or 4. Now determine the forces at 1, 2, 3 and 4.

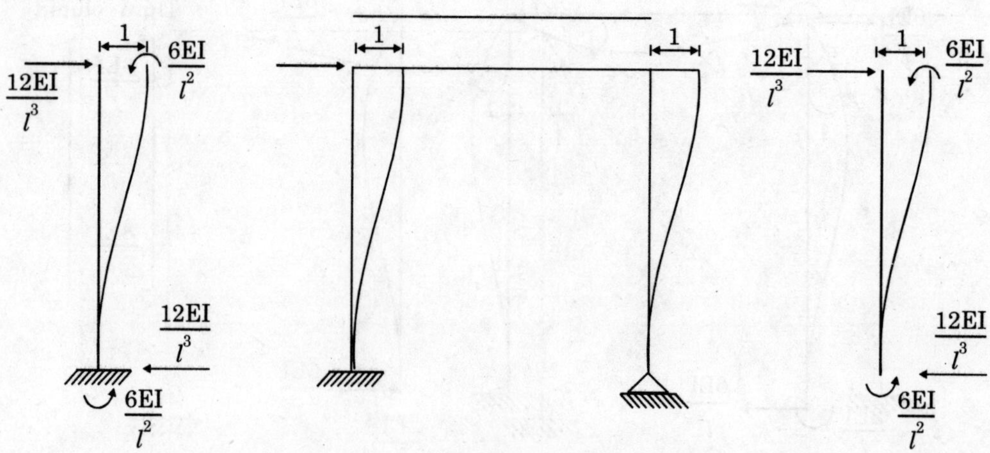

Fig. 5.36

Reactions : $\dfrac{6EI}{l^2} + \dfrac{6EI}{l^2} = R \times l$

(↺) (↺) (↻)

$$R = \frac{12EI}{l^3} \qquad \text{(Top – right ; Bottom – left)}$$

Combining, and showing values only along the co-ordinates

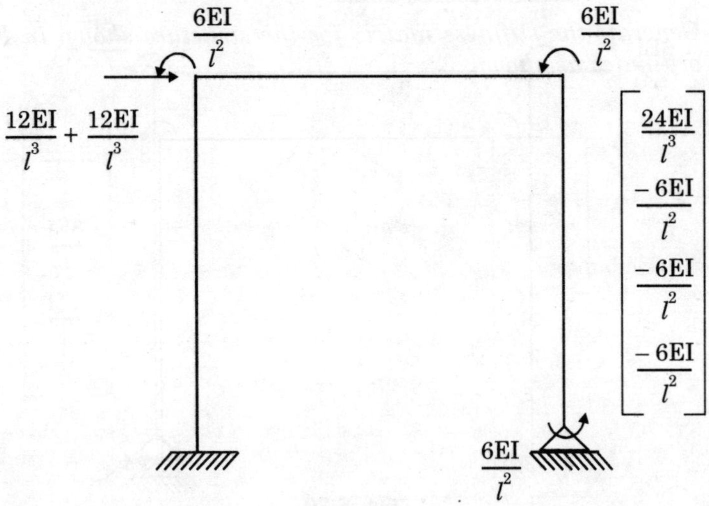

Fig. 5.37

To develop the second column of the stiffness matrix, apply a unit displacement along co-ordinate (2) only.

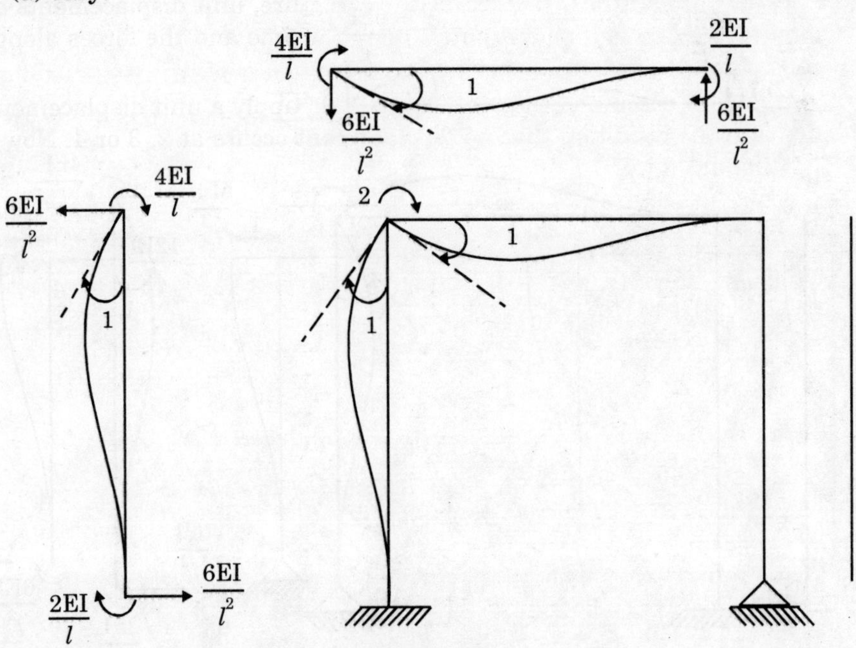

Fig. 5.38

Reactions : $\frac{4EI}{l} + \frac{2EI}{l} = R \times l$

(↻) (↻) (↺)

$R = \frac{6EI}{l^2}$ (Top – left ; Bottom – Right)

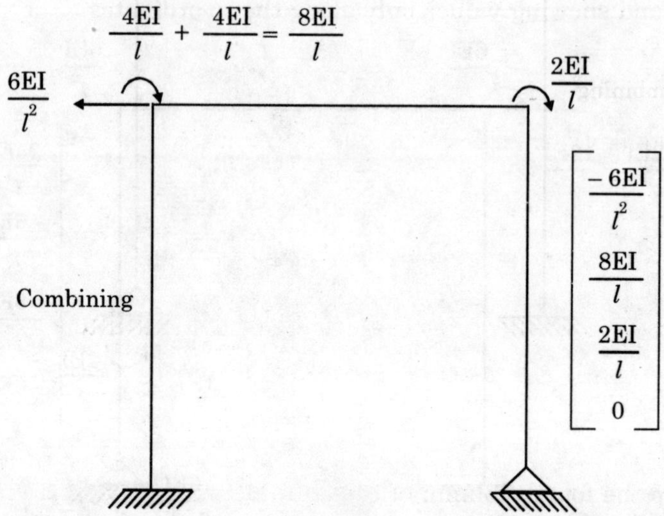

Fig. 5.39

To develop the third column of the stiffness matrix, apply unit displacement along co-ordinate (3) only.

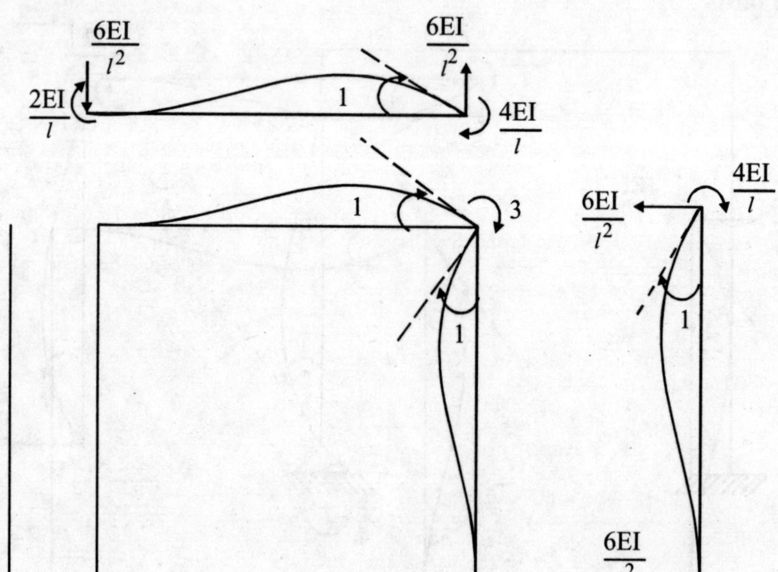

Fig. 5.40

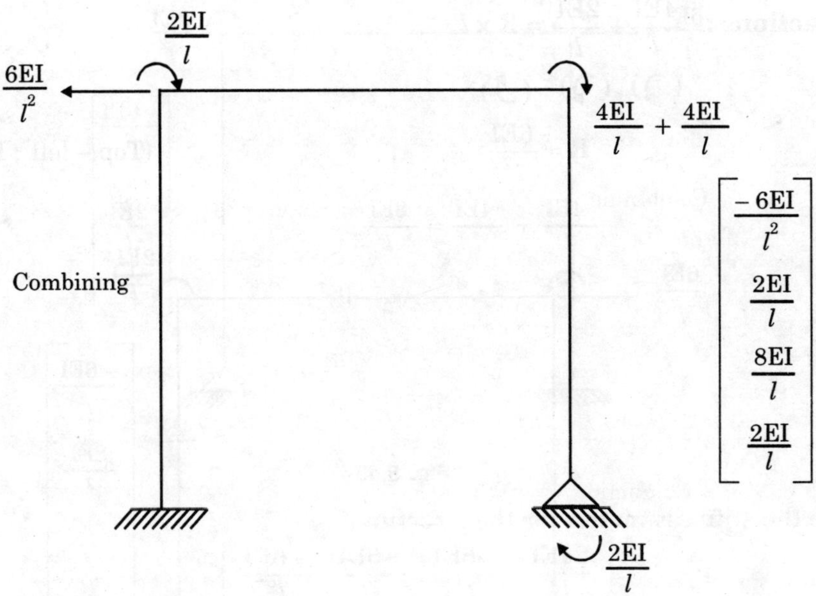

Fig. 5.41

To develop the fourth column of the stiffness matrix, apply a unit displacement along co-ordinate (4) only. (Fig. 5.42)

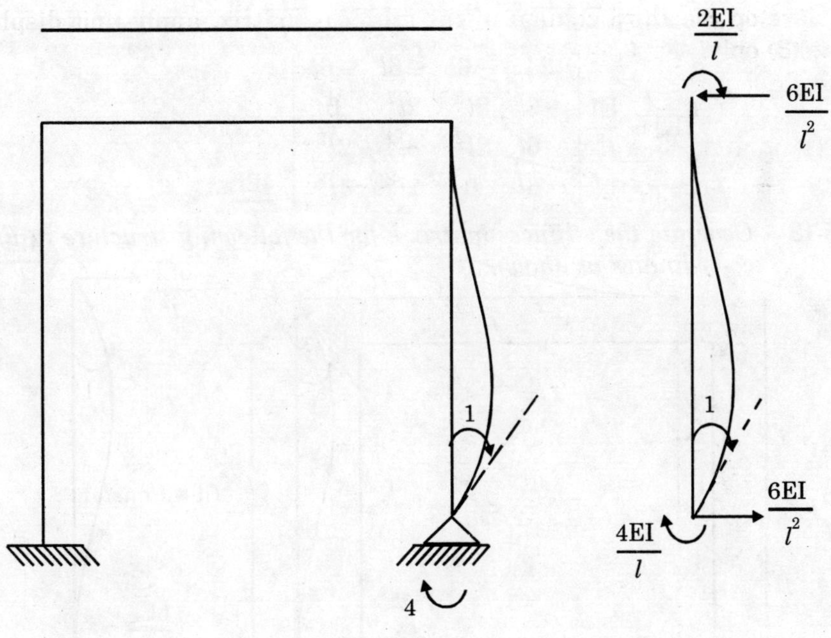

Fig. 5.42

Combining

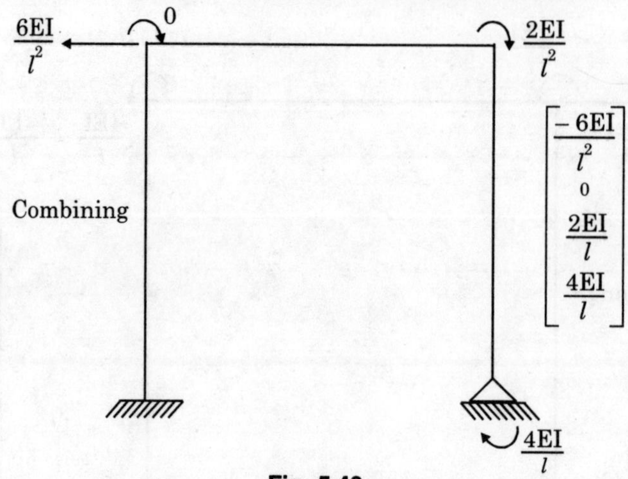

Fig. 5.43

Hence the stiffness matrix for the structure,

$$[K] = \begin{bmatrix} \frac{24EI}{l^3} & \frac{-6EI}{l^2} & \frac{-6EI}{l^2} & \frac{-6EI}{l^2} \\ \frac{-6EI}{l^2} & \frac{8EI}{l} & \frac{2EI}{l} & 0 \\ \frac{-6EI}{l^2} & \frac{2EI}{l} & \frac{8EI}{l} & \frac{2EI}{l} \\ \frac{-6EI}{l^2} & 0 & \frac{2EI}{l} & \frac{4EI}{l} \end{bmatrix}$$

$$[K] = \frac{EI}{l^3} \begin{bmatrix} 24 & -6l & -6l & -6l \\ -6l & 8l^2 & 2l^2 & 0 \\ -6l & 2l^2 & 8l^2 & 2l^2 \\ -6l & 0 & 2l^2 & 4l^2 \end{bmatrix}.$$

Example 5.13 *Generate the stiffness matrix k for the following structure in Fig. 5.44 with co-ordinates as shown.*

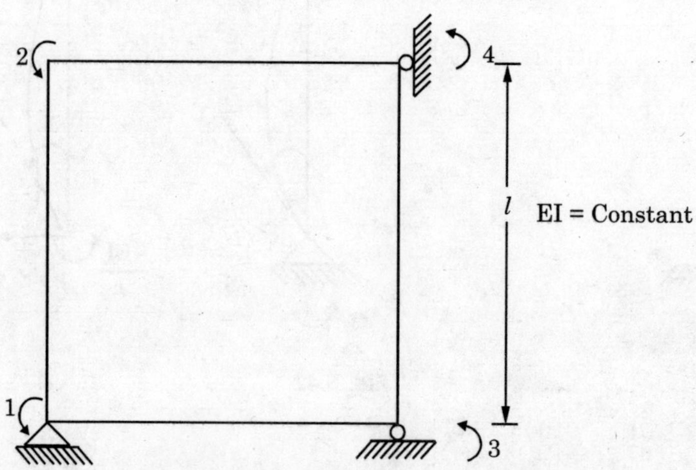

Fig. 5.44

Solution There are four co-ordinates. Hence the size of the stiffness matrix will be 4×4. The stiffness matrix will be developed by applying unit displacements along each of the co-ordinate, one at a time and the forces developed 1, 2, 3 and 4 will be obtained.

To generate the first column of the stiffness matrix, apply unit displacement along co-ordinate (1) only. (Fig. 5.45)

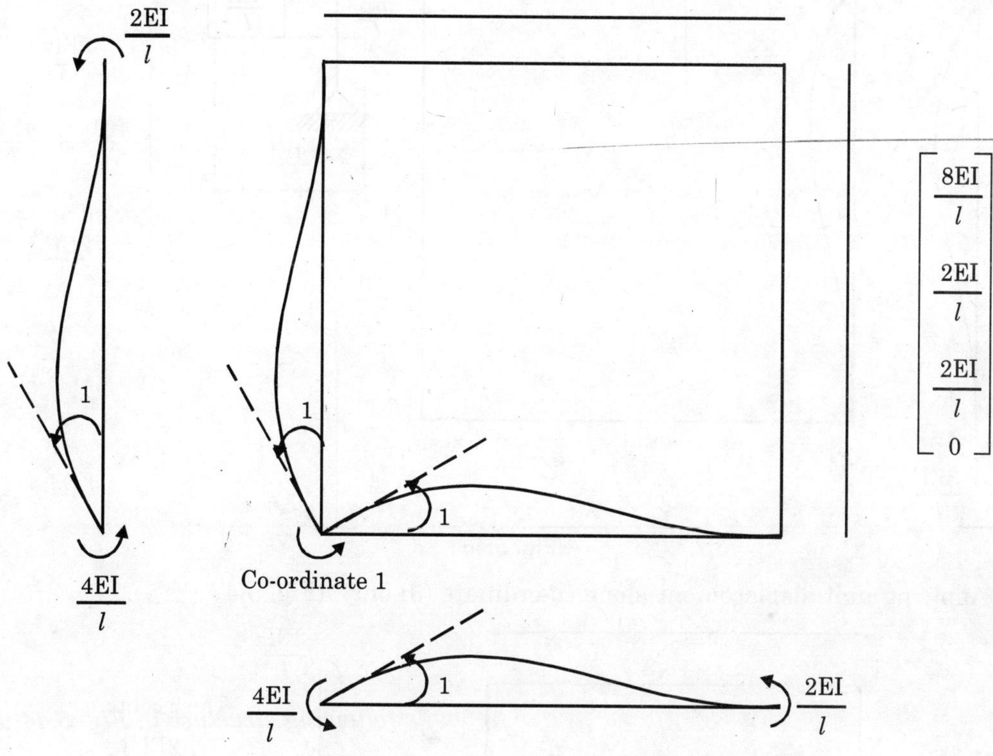

Fig. 5.45

$$\text{Force along co-ordinate (1)} = \frac{4EI}{l} + \frac{4EI}{l} = \frac{8EI}{l}$$

(↻) (↻) (↻)

$$\text{Force along co-ordinate (2)} = \frac{2EI}{l}$$ (↺) and so on.

Applying unit displacement along co-ordinate (2) only. (Fig. 5.46)

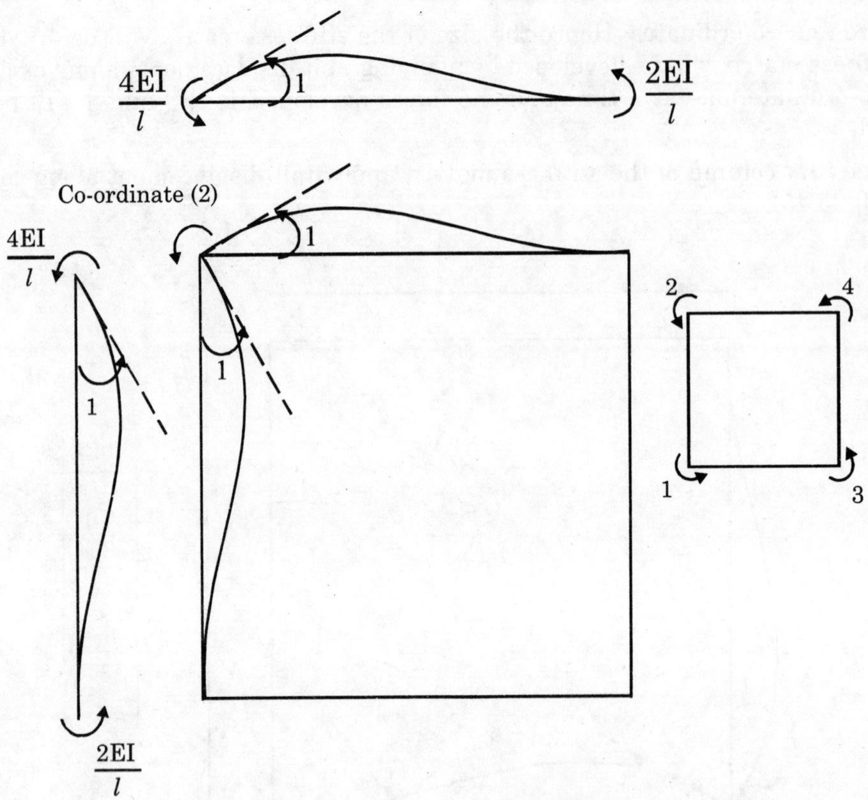

Second column

$$\begin{bmatrix} \frac{2EI}{l} \\ \frac{4EI}{l} + \frac{4EI}{l} \\ 0 \\ \frac{2EI}{l} \end{bmatrix}$$

Fig. 5.46

Applying unit displacement along co-ordinate (3) only. (Fig. 5.47)

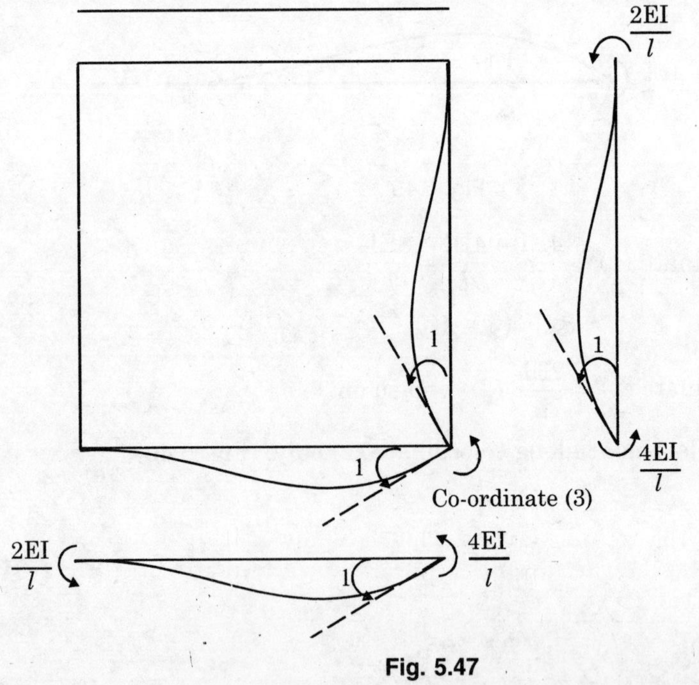

Third column

$$\begin{bmatrix} \frac{2EI}{l} \\ 0 \\ \frac{8EI}{l} \\ \frac{2EI}{l} \end{bmatrix}$$

Fig. 5.47

Applying unit displacement along co-ordinate (4) only. (Fig. 5.48)

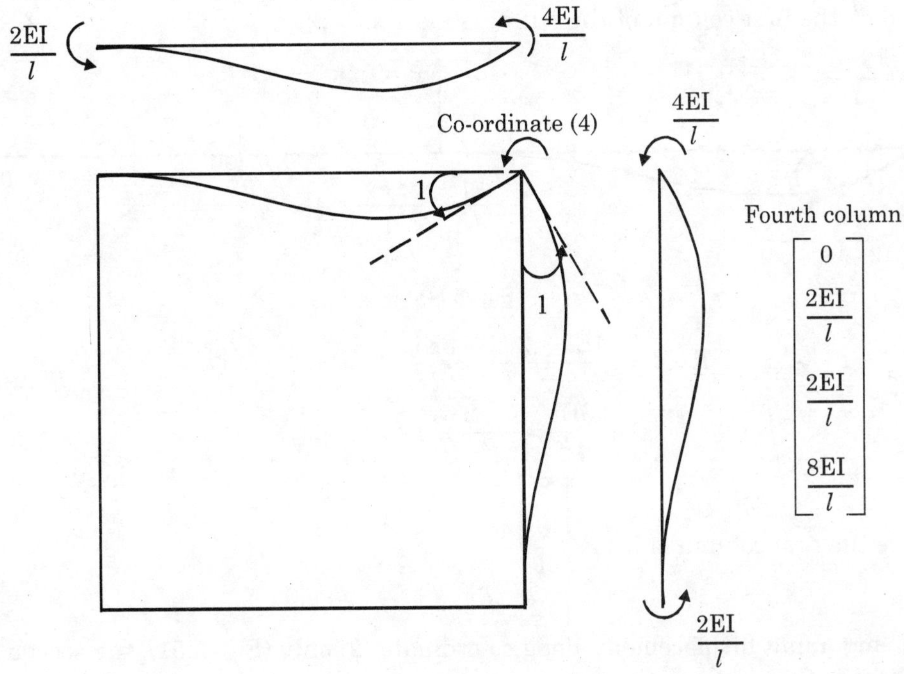

Fig. 5.48

Hence the stiffness matrix,

$$[K] = \begin{bmatrix} \frac{8EI}{l} & \frac{2EI}{l} & \frac{2EI}{l} & 0 \\ \frac{2EI}{l} & \frac{8EI}{l} & 0 & \frac{2EI}{l} \\ \frac{2EI}{l} & 0 & \frac{8EI}{l} & \frac{2EI}{l} \\ 0 & \frac{2EI}{l} & \frac{2EI}{l} & \frac{8EI}{l} \end{bmatrix}.$$

Example 5.14 *Generate the stiffness matrix of the structure.*

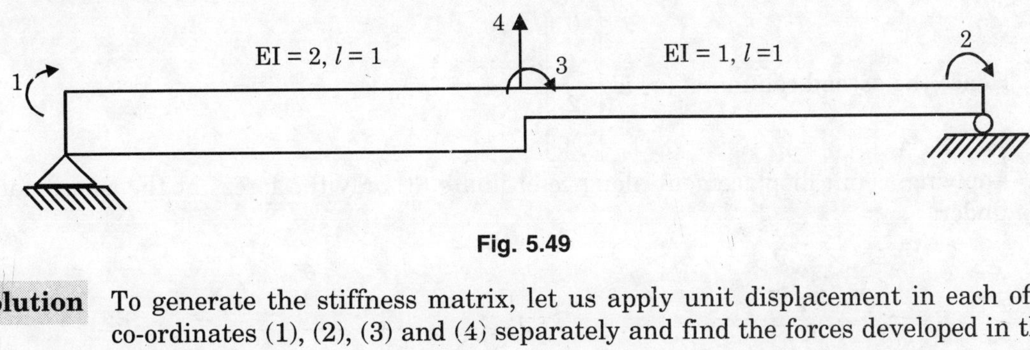

Fig. 5.49

Solution To generate the stiffness matrix, let us apply unit displacement in each of the co-ordinates (1), (2), (3) and (4) separately and find the forces developed in the 4 co-ordinates.

Applying a unit displacement along co-ordinate (1) only and not allowing displacement in any other co-ordinate, the forces developed in the co-ordinates are determined and represented as the first column of the stiffness matrix.

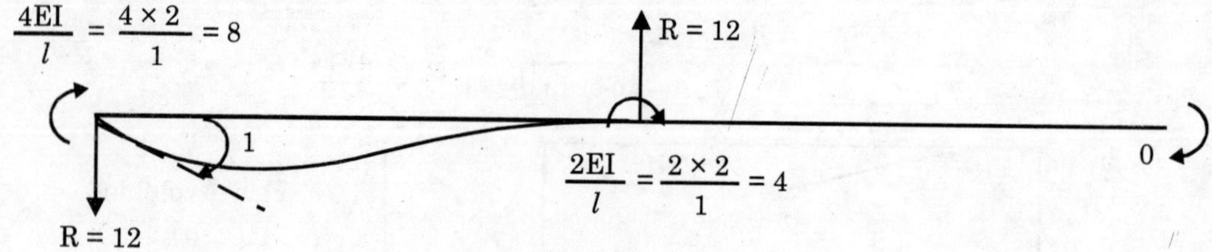

Fig. 5.50

$$R \times l = \frac{4EI}{l} + \frac{2EI}{l} = \frac{6EI}{l}$$

$$R = \frac{6EI}{l^2} = \frac{6 \times 2}{1^2} = 12$$

Hence the first column of $[k]$ is $\begin{bmatrix} 8 \\ 0 \\ 4 \\ 12 \end{bmatrix}$

Applying a unit displacement along co-ordinate (2) only (Fig. 5.51), the second column of the stiffness matrix is obtained.

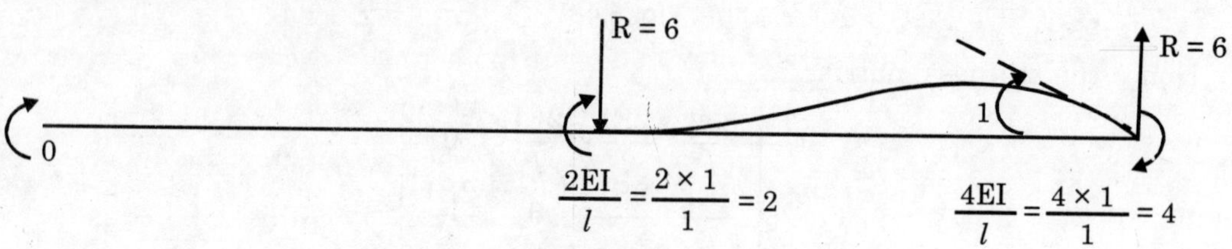

Fig. 5.51

$$R \times l = \frac{2EI}{l} + \frac{4EI}{l} = \frac{6EI}{l}$$

$$R = \frac{6EI}{l} = \frac{6 \times 1}{1} = 6$$

Hence the second column of $[k]$ is $\begin{bmatrix} 0 \\ 4 \\ 2 \\ -6 \end{bmatrix}$

Applying a unit displacement along co-ordinate (3) only the forces at the co-ordinates are as under.

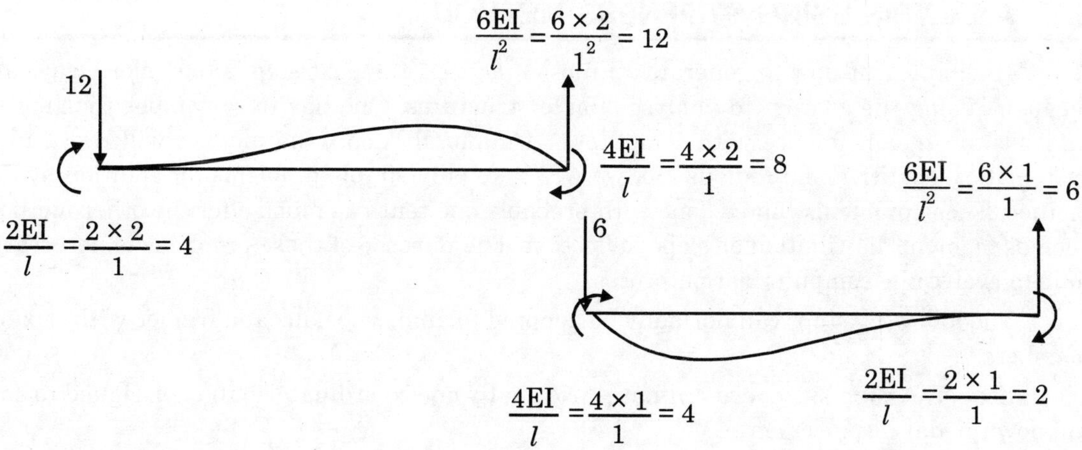

Fig. 5.52

Hence the third column of $[k]$ is $\begin{bmatrix} 4 \\ 2 \\ 12 \\ 6 \end{bmatrix}$

Applying a unit displacement along co-ordinate (4) only the forces are as below

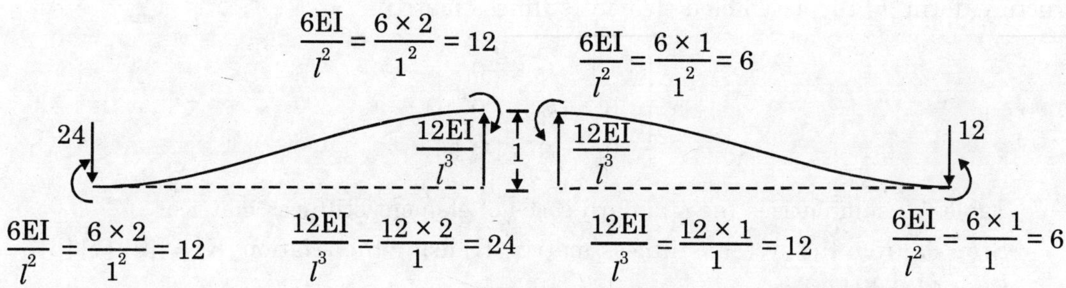

Fig. 5.53

Hence the fourth column of $[k]$ is $\begin{bmatrix} 12 \\ -6 \\ 6 \\ 36 \end{bmatrix}$

$$\text{Hence the stiffness matrix is } [k] = \begin{bmatrix} 8 & 0 & 4 & +12 \\ 0 & 4 & 2 & -6 \\ 4 & 2 & 12 & 6 \\ 12 & -6 & 6 & 36 \end{bmatrix}.$$

5.7 ANALYSIS USING STIFFNESS METHOD

Having learnt how to generate stiffness matrices, the next step is to explore ways and means of using the concept to analyse simple structures. One has to remember that matrix methods are meant for use with computers. If manually done, the method will prove to be systematic, repetitive and tedious. So, we will take only simple problems for solution by this method. These problems can be done with probably one tenth as much effort by other methods such as moment distribution or slope deflection. The purpose of these problems is necessarily to help evolve the computerisation process.

The following steps can normally be adopted to analyse beams and frames with flexural members.

Step 1. Assign system co-ordinates. Normally one co-ordinate shall be assigned to each unknown nodal displacement.

Step 2. Split the structure into component units of elements. Assign nodal co-ordinates for the extremities of each element.

Step 3. Evaluate element forces. For flexure members, these forces are usually support moments and reactions due to forces not acting at the co-ordinates.

Step 4. From the [β] matrix. This is similar to the [b] matrix defined in Chapter 4, Flexibility method. Examples 5.15, 5.16 and 5.17 show how β matrices are computed.

[β] connects the system displacements {u} and the element displacements {δ}.

Step 5. From the element stiffness matrices [k_1], [k_2] ... [k_n] for all the elements in the structure, form [k] the assembled element stiffness matrix.

$$[k] = \begin{bmatrix} [k_1] & 0 & 0 \\ 0 & [k_2] & 0 \\ 0 & 0 & [k_n] \end{bmatrix}$$

Table 5.2 summarises the standard cases of element stiffness matrices.

Step 6. From the system stiffness matrix [K] using the relation $[K] = [\beta]^T [k] [\beta]$

Step 7. Get $[K]^{-1}$.

Step 8. Compute the system displacement vector {u}. Using the relationship,

$$\{u\} = [K]^{-1} \{\{F^f\} - \{F^0\}\}.$$

Step 9. Get the element displacements {δ} using the equation {δ} = [β] {u}.

Step 10. Get the corresponding element forces {P′} using {P′} = [k] {δ}.

Step 11. The final forces {P^f} are given by $\{P^f\} = \{P^0\} + \{P'\}$.

Step 12. Sketch the end moments diagram and the free bending moment diagram superposed on it.

It may be noted that equations in steps 5, 6, 8, 9, 10 and 11 are given without a valid proof. For proof the reader is referred to Rubinstein (ref 9). The steps and concepts listed above are borrowed from the same (ref 9). Each of the fifteen examples (5.21 through 5.35) meticulously follows this step by step method of solving structures.

Example 5.15 *For the element co-ordinates and system co-ordinates described in Fig. 5.54, determine the β matrix such that*

$$[K] = [\beta]^T [k] [\beta]$$

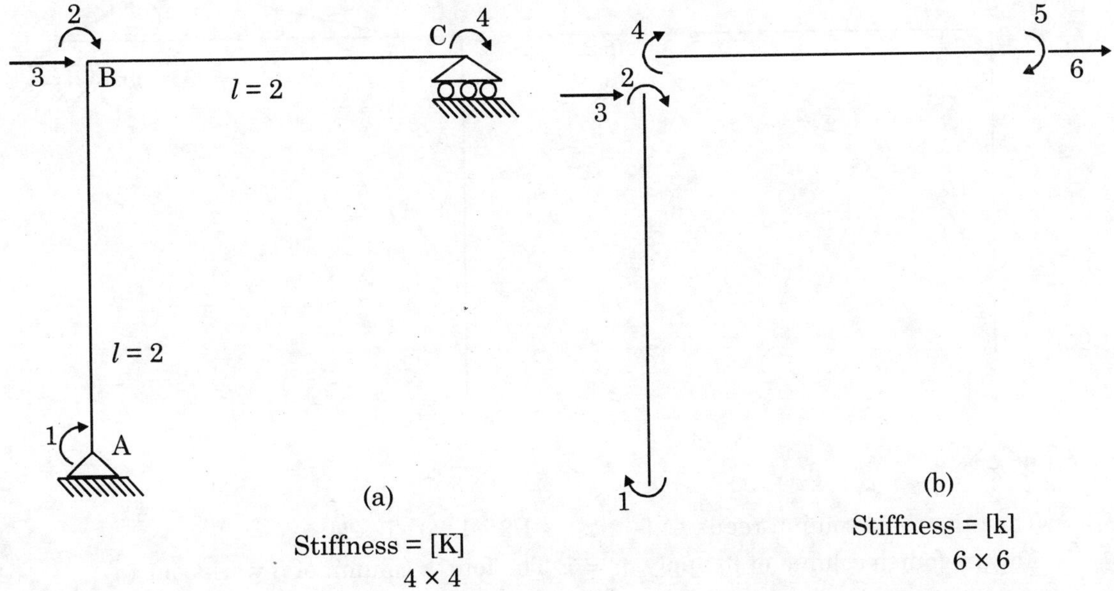

Fig. 5.54

Solution [β] matrix is developed column by column. There will be 4 columns corresponding to the 4 co-ordinates in Fig. 5.54 (*a*). To develop the first column of [β], apply a unit displacement at 1 only.

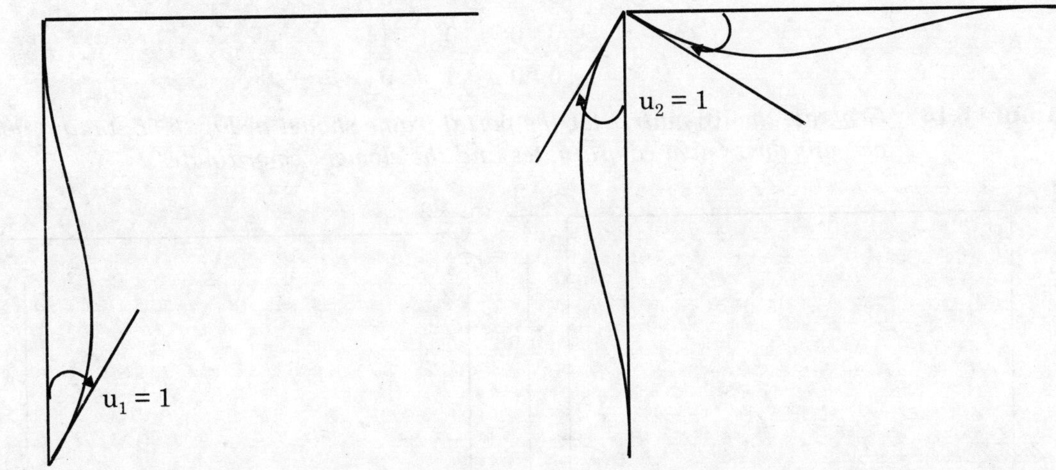

This corresponds to displacements $\delta_1 = 1$ in the element co-ordinates. So the first column of [β] is $[1\ 0\ 0\ 0\ 0\ 0]^T$. The second column is got by applying $u_2 = 1$.

The second column of β is obviously $[0\ 1\ 0\ 1\ 0\ 0]^T$. The third column of β is obtained by applying $u_3 = 1$. While applying a horizontal displacement of 1 along BB′, member AB gets tilted to the position AB′. Since slopes at A and B will be zero, tangents at A and B′ will be vertical. So there will be effective anti-clockwise rotations at A and B [element co-ordinates at (1) and (2)] of magnitude 1/2.

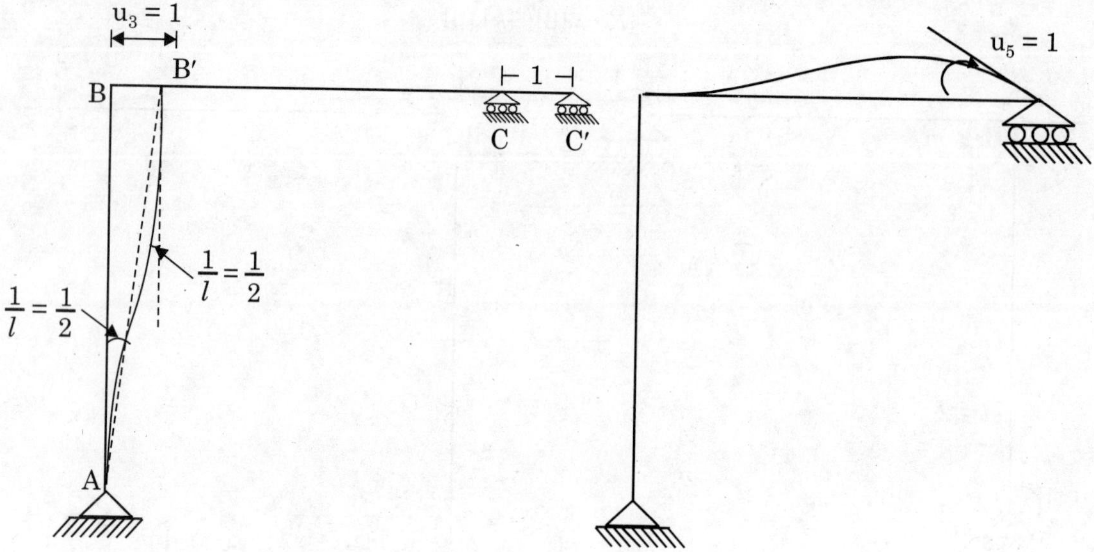

Thus the third column reads as $[-1/2 \quad -1/2 \quad 1 \quad 0 \quad 0 \quad 1]^T$

For the fourth column of β, apply $u_4 = 1$. The fourth column of β works out to

$$[\,0\,0\,0\,0\,1\,0\,]^T$$

Hence,

$$[\beta] = \begin{bmatrix} 1 & 0 & -1/2 & 0 \\ 0 & 1 & -1/2 & 0 \\ 0 & 0 & 1 & 0 \\ 0 & 1 & 0 & 0 \\ 0 & 0 & 0 & 1 \\ 0 & 0 & 1 & 0 \end{bmatrix}.$$

Example 5.16 *Generate the* [β] *matrix for the portal frame shown in Fig. 5.55, taking into account the system co-ordinates and the element co-ordinates.*

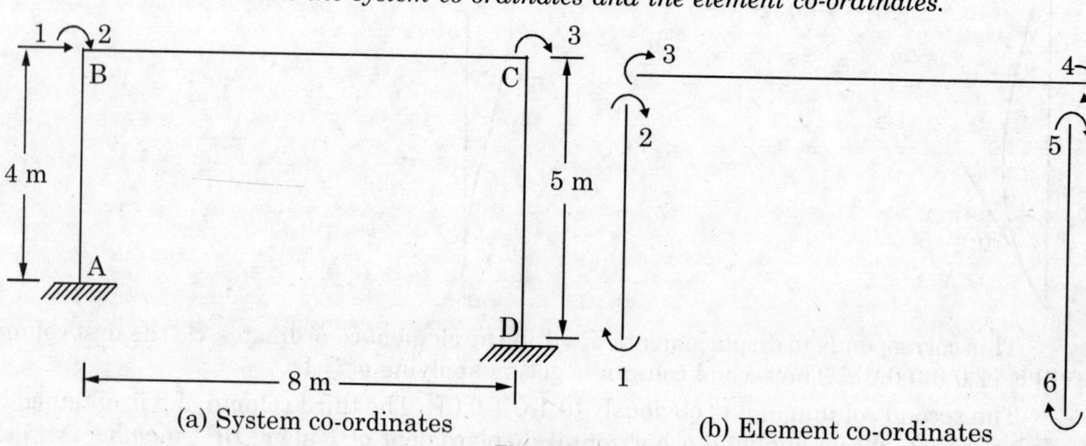

Fig. 5.55

Solution Number of system co-ordinates = 3

Number of element co-ordinates = 6

Hence the size of the [β] matrix is 6 × 3. To generate the first column of (β), apply a unit displacement along co-ordinate (1) of the system. This would effectively shift the centre lines of columns AB and DC to tilted positions AB′ and DC′, with respect to which the rotations at A, B, C and D have to be reckoned. (vide Fig. 5.56)

Fig. 5.56

Hence the displacements at the six-element co-ordinates are – 1/4, – 1/4, 0, 0, – 1/5 and – 1/5. To generate column 2 of [β], apply a unit rotation at 2 only and to generate column 3, apply a unit rotation at 3 only as below.

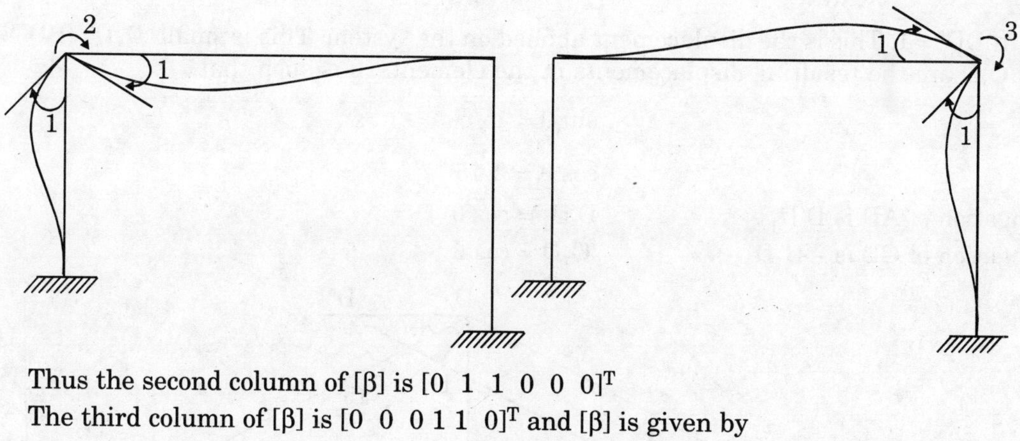

Thus the second column of [β] is $[0\ \ 1\ \ 1\ \ 0\ \ 0\ \ 0]^T$

The third column of [β] is $[0\ \ 0\ \ 0\ 1\ 1\ \ 0]^T$ and [β] is given by

$$[\beta] = \begin{bmatrix} -0.25 & 0 & 0 \\ -0.25 & 1 & 0 \\ 0 & 1 & 0 \\ 0 & 0 & 1 \\ -0.20 & 0 & 1 \\ -0.20 & 0 & 0 \end{bmatrix}.$$

Example 5.17 *For the element co-ordinates and system co-ordinates shown in Figs. 5.57 (a) and (b), derive the* [β] *matrix such that* $[K] = [\beta]^T[k][\beta]$.

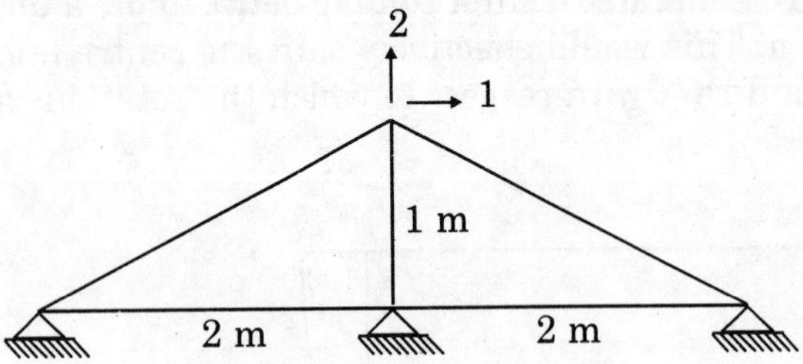

(a) System stiffness = $\underset{2 \times 2}{[K]}$

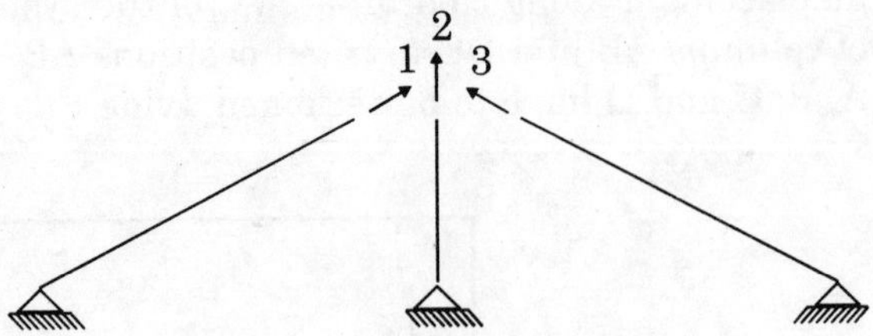

(b) Element stiffness = $\underset{3 \times 3}{[k]}$

Fig. 5.57

Solution [β] matrix is of size 3 × 2. The two columns of [β] are generated by

(*i*) Applying a unit displacement along 1 in the system and determining the corresponding displacements along 1, 2 and 3 in the elements.

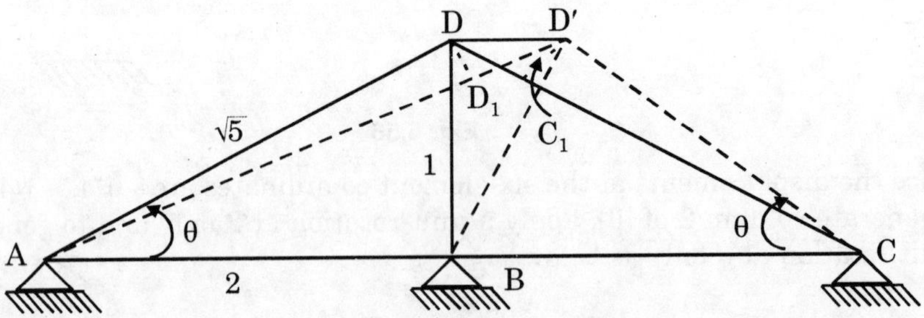

DD′ = 1. This is the displacement applied on the system. This is small. D_1D', (BD′ – BD) and C_1D are the resulting displacements in the elements. θ is such that

$$\sin\theta = 1/\sqrt{5}$$

$$\cos\theta = 2/\sqrt{5}$$

elongation of AD is $D'D_1$ $\qquad D'D_1 = \cos\theta$

elongation of CD is $-C_1D$ $\qquad C_1D = \cos\theta$

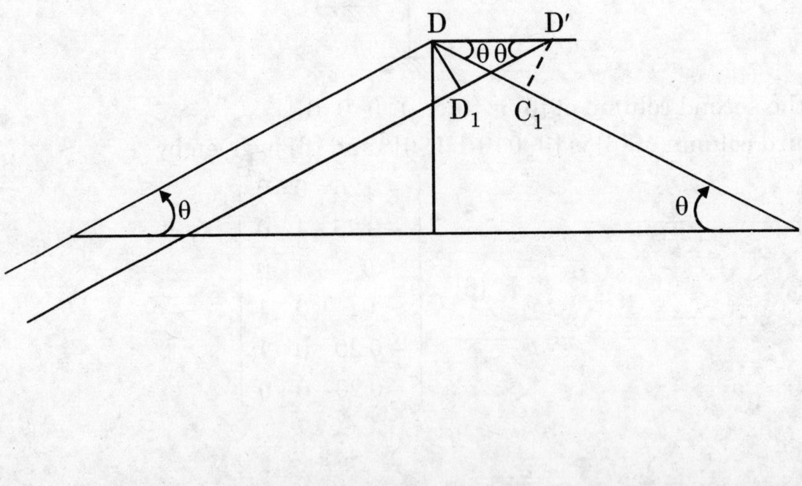

BD does not change its length since DD′ is small. So elongation of BD is zero. Thus the first column of [β] is [cos θ   0 – cos θ]^T.

(*ii*) The second column of β is generated by applying a unit displacement along 2 in the system, and determining the resulting displacements along 1, 2 and 3 of the elements.

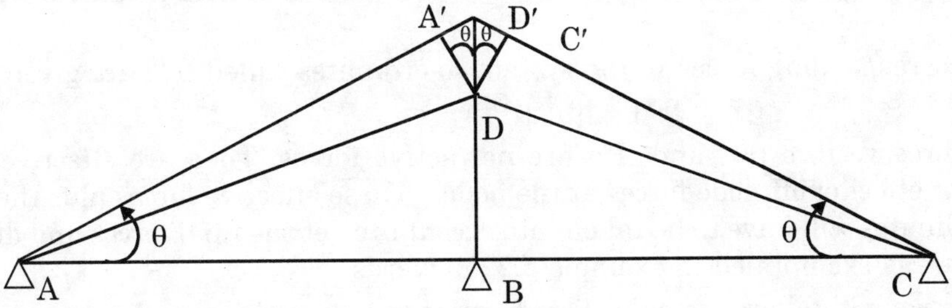

$\sin\theta = 1/\sqrt{5}$ $\qquad$ $\cos\theta = 2/\sqrt{5}$

Displacement along AD′ = A′D′ A′D′ = sin θ

Displacement along BD′ = 1

Displacement along CD′ = 1

Displacement along BD = 1

Hence the second column of [β] is [sin θ   0 sin θ]^T

The [β] matrix is then given by

$$[\beta] = \begin{bmatrix} \cos\theta & \sin\theta \\ 0 & 1 \\ -\cos\theta & \sin\theta \end{bmatrix} = \frac{1}{\sqrt{5}}\begin{bmatrix} 2 & 1 \\ 0 & \sqrt{5} \\ -2 & 1 \end{bmatrix}.$$

5.8 HOW TO ASSIGN CO-ORDINATES

5.8.1 Frames in Flexure

Before we start analysing structural frames, we shall learn how best to assign co-ordinates.

Matrix methods deal with forces and displacements at assigned co-ordinates only.

For the analysis to be useful, we would need the bending moment at all joints and supports. Hence all joints and supports need to be assigned co-ordinate numbers in the element co-ordinates.

However, loads can act anywhere on a structure (and uniform loads act everywhere on a structure). Hence we shall evolve a technique of handling loads not at joints (hence not at co-ordinates).

In flexure members we solve the problem by our earlier concept of fixed end moments,

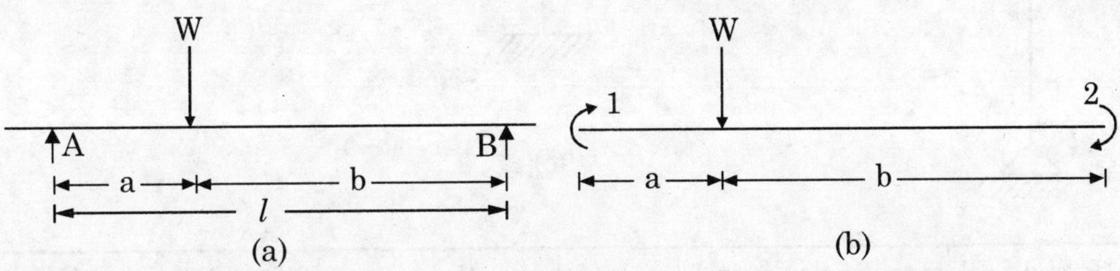

Fig. 5.58

Figure 5.58 shows a span in a structure in which a load W acts as shown. We can invoke moments $-\dfrac{Wab^2}{l^2}$ and $\dfrac{Wa^2b}{l^2}$ (the fixed end moments) at co-ordinates (1) and (2) that would maintain fixity (or zero displacement) at 1 and 2. These fixed end moments would be called $\{P^0\}$.

The corresponding forces in the system co-ordinates called $\{F^0\}$ are given by

$$\{F^0\} = [\beta]^T \{P^0\}^*$$

The force vectors $\{F^0\}$ and $\{P^0\}$ are not active forces. They are effective forces. They represent the effect of off-node forces at the nodes. These effective forces plus the active forces at the co-ordinates will have to be taken into account to get the final forces and displacements. The procedure is exemplified in example 5.22 onwards.

Hence we will follow the rules listed under to assign the co-ordinates in a frame to be analysed.

(*i*) Assign system co-ordinate number to every nodal degree of freedom-rotational or translational. A few examples are given below.

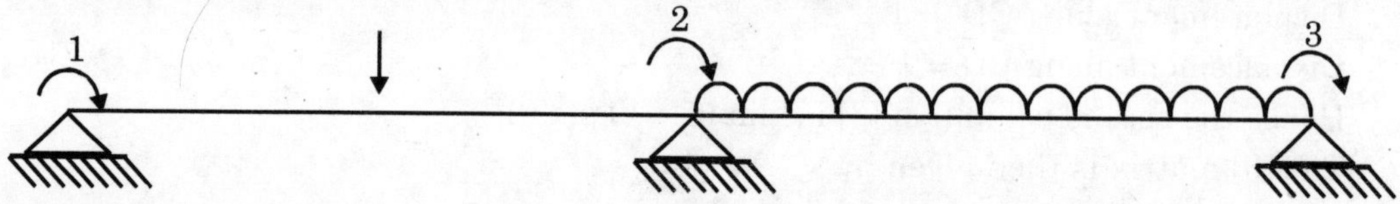

(a)

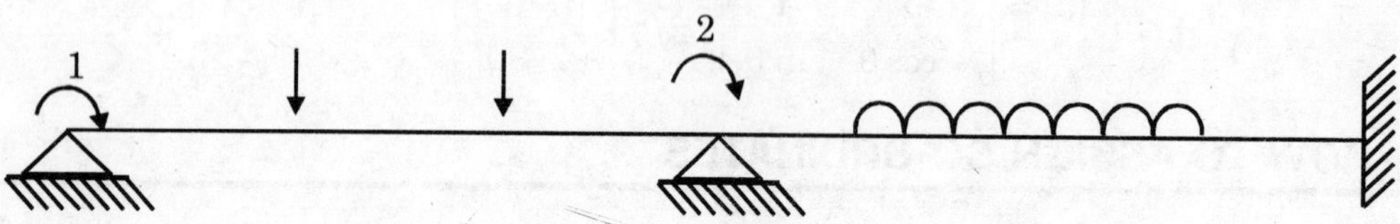

(b)

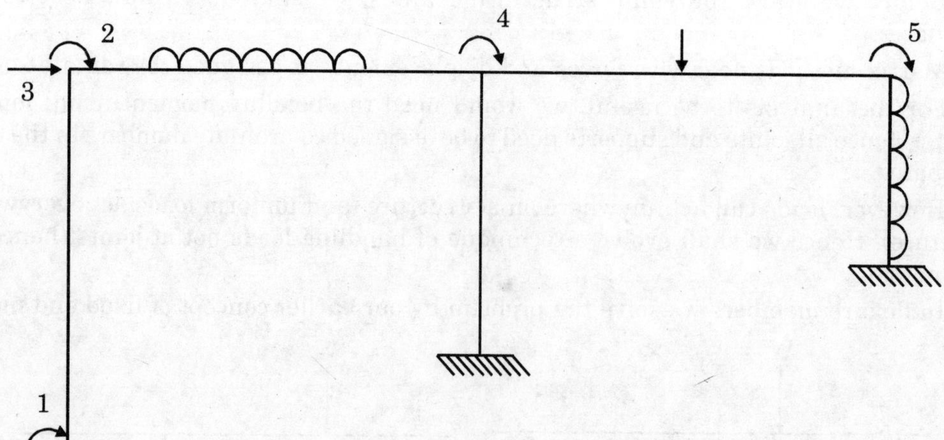

(c)

*P^0 and δ relate to the elements, F^0 and u relate to the system. If $\{\delta\} = [\beta]\{u\}$, then $\{F^0\} = [\beta]^T\{P^0\}$. This property is called the principle of contragradience.

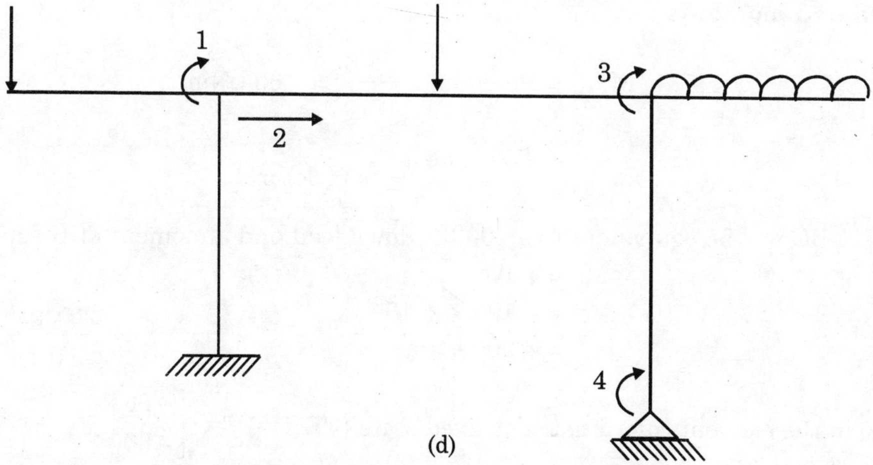

(d)

Free ends need not be considered as nodes.

In all the four examples above the effect of the external loads at the 'spans' on the nodes will constitute effective forces. This does away with the necessity of defining co-ordinates at the load positions.

(*ii*) Separate the frame into component elements and assign rotation co-ordinates to the extremities. This would ensure that we can eventually evaluate the corresponding rotations and moments.

[If additional information such as shear, thrust, torsion is needed to be directly evaluated, co-ordinates corresponding to the same shall be assigned to the extremities. No such examples are worked out in this volume.]

5.8.2 Pin Jointed Frames

In the case of pin jointed plane frames, we have to assign 2 degrees of freedom to each node.

The elements in trusses are very distinct. Each element shall have one degree of freedom for each end except the ends that are restrained. Normally the question of forces not at co-ordinates will not arise in trusses.

Example 5.18 *Analyse the beam loaded as shown in Fig. 5.59 by matrix stiffness method and sketch the BMD.*

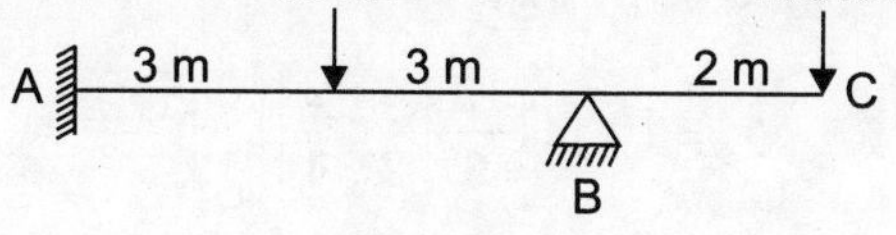

Fig. 5.59

Solution Since the only possible independent joint displacement happens to be θ_B, the kinematic indeterminacy of the given structure is 1.

System co-ordinates:

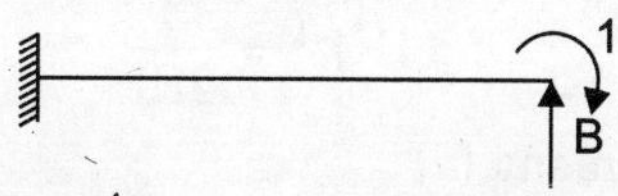

Element co-ordinates:

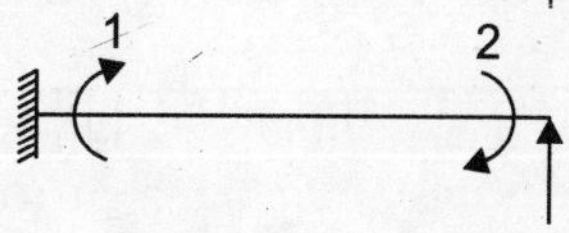

Fixed end moments :

$$M_{FAB} = -\frac{Wl}{8} = -\frac{80 \times 6}{8} = -60 \text{ kNm}$$

$$M_{FBA} = \frac{Wl}{8} = \frac{80 \times 6}{8} = 60 \text{ kNm}$$

Load in BC will be equivalent to a 100 kN down load and a moment at B equal to

$$M_B = +Wl$$
$$= +100 \times 2$$
$$= +200 \text{ kNm}$$

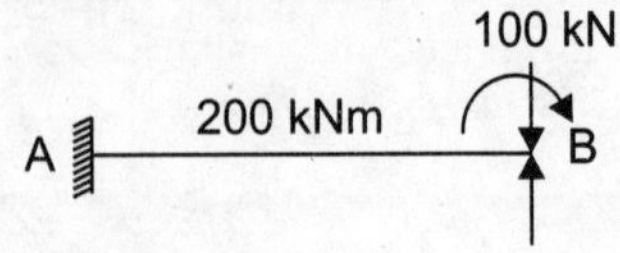

Force in the element co-ordinates at fixed state (FEM) $\{P^0\} = \begin{bmatrix} -60 \\ 60 \end{bmatrix}$

Transformation matrix [β]

Apply unit displacement at system co-ordinate and the values arise in element co-ordinates as [β]

$$[\beta] = \begin{bmatrix} 0 \\ 1 \end{bmatrix}$$

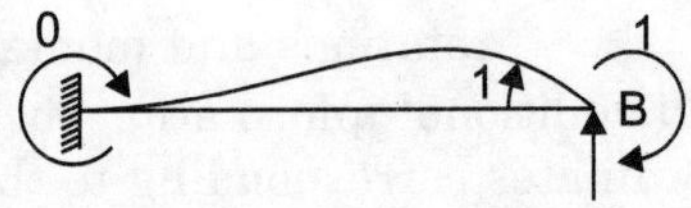

Fixed co-ordinate forces $\{F^0\} = [\beta]^T - \{P^0\}$

$$= [0 \quad 1]\begin{Bmatrix} -60 \\ 60 \end{Bmatrix}$$

$$= \{60\}$$

Force applied at the system co-ordinates $[F^F] = \{200\}$

$$[F] = [F^F] - [F^0]$$
$$= \{200\} - \{60\}$$
$$= [140]$$

Element stiffness matrix [*k*]

$$[k] = \frac{EI}{l}\begin{bmatrix} 4 & 2 \\ 2 & 4 \end{bmatrix} = \frac{EI}{6}\begin{bmatrix} 4 & 2 \\ 2 & 4 \end{bmatrix}$$

System stiffness matrix [K]

$$[K] = [\beta]^T [k] [\beta]$$

$$= [0 \quad 1]\frac{EI}{6}\begin{bmatrix} 4 & 2 \\ 2 & 4 \end{bmatrix}\begin{bmatrix} 0 \\ 1 \end{bmatrix}$$

$$= \frac{EI}{6}[2 \quad 4]\begin{bmatrix} 0 \\ 1 \end{bmatrix} = \frac{2EI}{3}$$

$$[K]^{-1} = \frac{3}{2EI}$$

System displacement $\{u\}$

$$\{u\} = [K^{-1}]\{F\} = \left[\frac{3}{2EI}\right][140] = \left[\frac{210}{EI}\right]$$

Element forces [P′]

$$\{P'\} = [k]\,[\beta]\,\{u\}$$

$$= \frac{EI}{6}\begin{bmatrix}4 & 2\\ 2 & 4\end{bmatrix}\begin{bmatrix}0\\ 1\end{bmatrix}\begin{bmatrix}210\\ \hline EI\end{bmatrix}$$

$$= \begin{bmatrix}70\\ 140\end{bmatrix}$$

Final forces, $$[P^F] = [P^0] + [P']$$

$$= \begin{bmatrix}-60\\ 60\end{bmatrix} + \begin{bmatrix}70\\ 140\end{bmatrix} = \begin{bmatrix}10\\ 200\end{bmatrix}$$

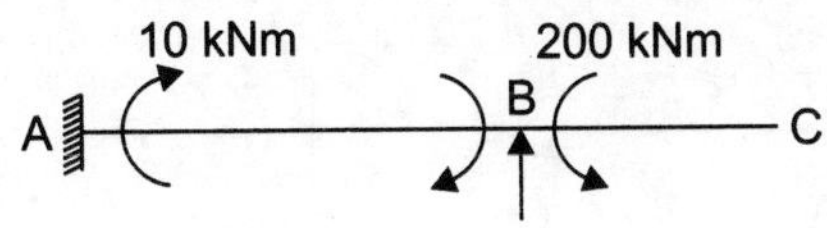

Free BM, $$M'_{AB} = \frac{Wab}{l} = \frac{80\times 3\times 3}{6} = 120 \text{ kNm}$$

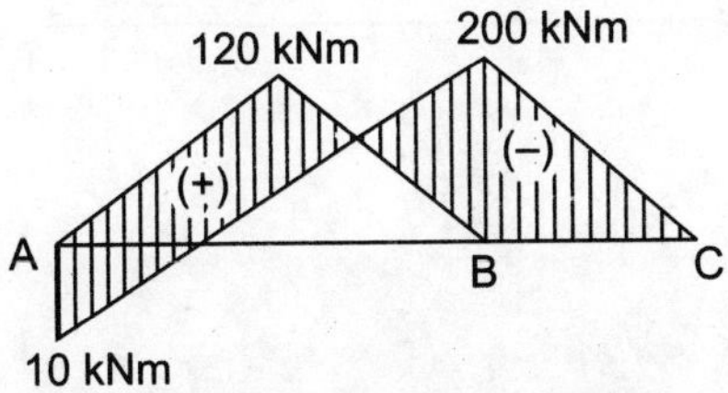

Bending moment diagram

Example 5.19 *Analyse the beam loaded as shown in Fig. 5.60 by stiffness method.*

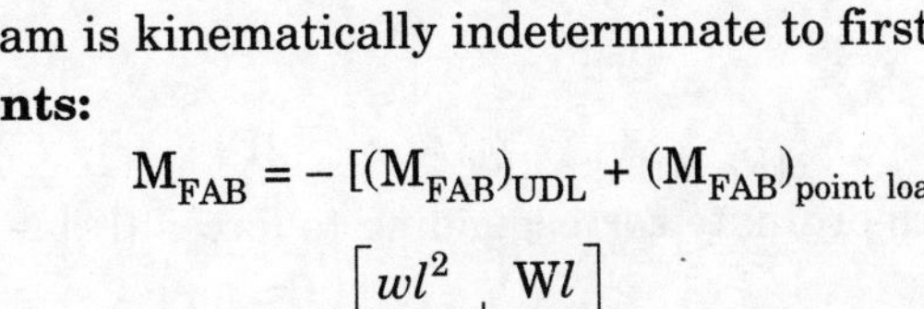

Fig. 5.60

Solution The given beam is kinematically indeterminate to first degree, *i.e.*, rotation at B.

Fixed end moments:

$$M_{FAB} = -\left[(M_{FAB})_{UDL} + (M_{FAB})_{\text{point load}}\right]$$

$$= -\left[\frac{wl^2}{12} + \frac{Wl}{8}\right]$$

$$M_{FAB} = -\left[\frac{12\times 6^2}{12} + \frac{24\times 6}{8}\right] = -54 \text{ kNm}$$

$$M_{FBA} = +\left[(M_{FBA})_{UDL} + (M_{FBA})_{point\ load}\right]$$

$$= \left[\frac{wl^2}{12} + \frac{Wl}{8}\right]$$

$$M_{FBA} = \left[\frac{12 \times 6^2}{12} + \frac{24 \times 6}{8}\right] = 54 \text{ kNm}$$

$$\{P^0\} = \begin{Bmatrix} -54 \\ 54 \end{Bmatrix}$$

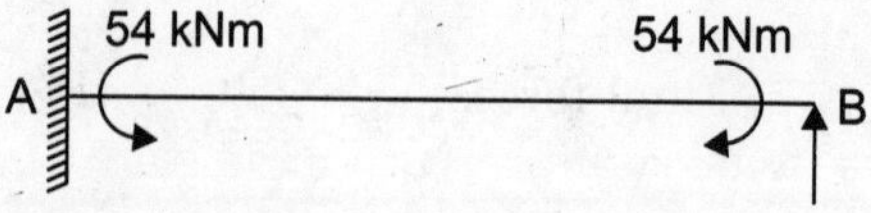

Transformation matrix (β) :

Apply a unit rotation at B in system co-ordinate and find the element co-ordinate forces.

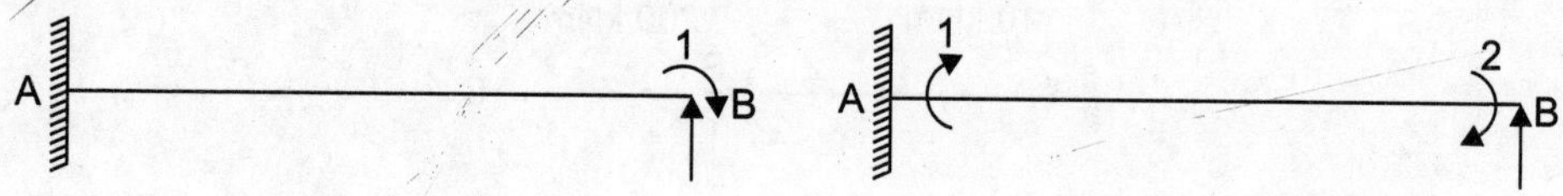

System co-ordinate   Element co-ordinate

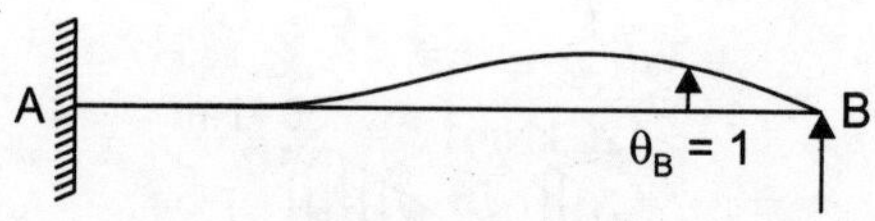

$$[\beta] = \begin{bmatrix} 0 \\ 1 \end{bmatrix}$$

Element stiffness matrix (k)

$$[k] = \frac{EI}{l}\begin{bmatrix} 4 & 2 \\ 2 & 4 \end{bmatrix}, \quad [k] = \frac{EI}{6}\begin{bmatrix} 4 & 2 \\ 2 & 4 \end{bmatrix}$$

System stiffness matrix (K)

$$[K] = [\beta^T]\,[k]\,[\beta] = [0 \quad 1]\frac{EI}{6}\begin{bmatrix} 4 & 2 \\ 2 & 4 \end{bmatrix}\begin{bmatrix} 0 \\ 1 \end{bmatrix}$$

$$[K] = [0 \quad 1]\frac{EI}{6}\begin{bmatrix} 2 \\ 4 \end{bmatrix}, \quad [K] = \frac{EI}{6}[4] = \frac{2}{3}EI$$

$$[K]^{-1} = \frac{3}{2EI}$$

Forces (F)

$$\{F\} = \{F^f\} - \{F^0\}$$

Force in the system co-ordinate corresponding to forces $\{P^0\}$

$$\{F^0\} = [\beta^T]\,[P^0] = [0 \quad 1]\begin{bmatrix} -54 \\ 54 \end{bmatrix} = [54]$$

Force applied at system co-ordinate

$$\{F^f\} = [10]$$

$$\{F\} = [10] - [54] = [-44].$$

System displacement (u)

$$\{u\} = [K^{-1}]\{F\} = \frac{3}{2EI}[-44] = \left[-\frac{66}{EI}\right]$$

Element forces {P′}

$$\{P'\} = [k]\{\delta\} = [k][\beta]\{u\} = \frac{EI}{6}\begin{bmatrix}4 & 2\\2 & 4\end{bmatrix}\begin{bmatrix}0\\1\end{bmatrix}\left[-\frac{66}{EI}\right]$$

$$\{P'\} = -11\begin{bmatrix}2\\4\end{bmatrix} = \begin{bmatrix}-22\\-44\end{bmatrix}$$

Final forces $$\{P^f\} = \{P'\} + \{P^0\} = \begin{bmatrix}-22\\-44\end{bmatrix} + \begin{bmatrix}-54\\54\end{bmatrix} = \begin{bmatrix}-76\\10\end{bmatrix}$$

$M_{AB} = -76$ kNm

$M_{BA} = 10$ kNm.

Example 5.20 *Analyse the fixed beam loaded as shown by the matrix stiffness method and obtain the support moments.*

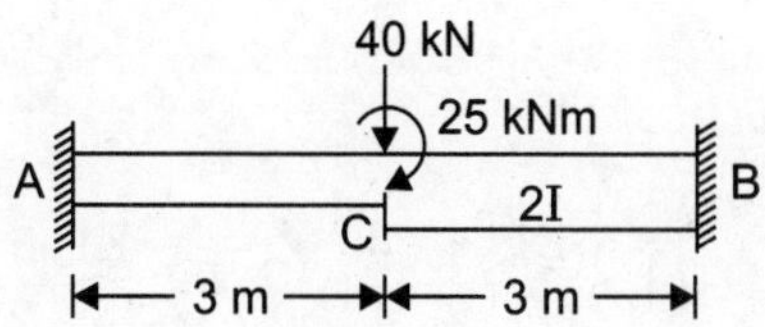

Solution The given structure is kinematically indeterminate to second degree, *i.e.*, displacement and rotation at 'C'

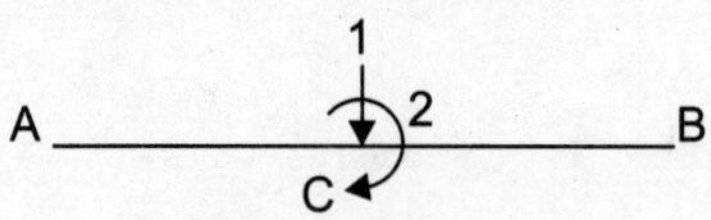

System co-ordinates

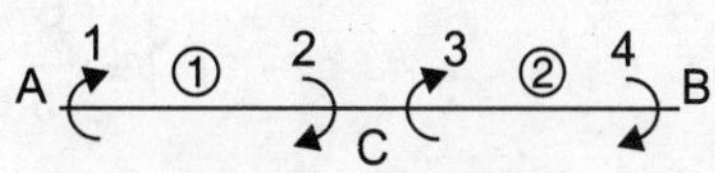

Element co-ordinates

$$\{F\} = \begin{bmatrix}40\\25\end{bmatrix}$$

Transformation matrix (β):

Apply a unit displacement and unit rotation at 'C' and find the corresponding co-ordinate displacements to form transformation matrix.

Apply a unit displacement at C along system co-ordinate (1)

$$\delta_C = 1$$

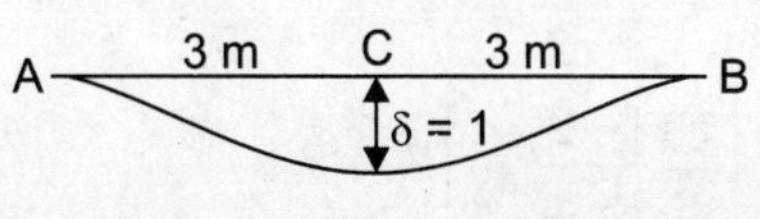

$$\begin{bmatrix}-\frac{1}{3}\\-\frac{1}{3}\\\frac{1}{3}\\\frac{1}{3}\end{bmatrix}$$

Apply a unit rotation at C along system co-ordinate (2)

$$\theta_C = 1$$

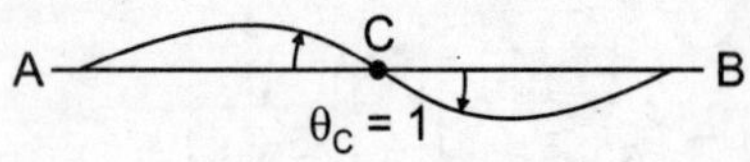

$$\begin{bmatrix} 0 \\ 1 \\ 1 \\ 0 \end{bmatrix}$$

Transformation matrix $[\beta] = \begin{bmatrix} -\frac{1}{3} & 0 \\ -\frac{1}{3} & 1 \\ \frac{1}{3} & 1 \\ \frac{1}{3} & 0 \end{bmatrix}$

Element stiffness matrix (k)

$$[k] = \frac{EI}{l}\begin{bmatrix} 4 & 2 \\ 2 & 4 \end{bmatrix}, \; [k_1] = \frac{EI}{3}\begin{bmatrix} 4 & 2 \\ 2 & 4 \end{bmatrix}, \quad [k_2] = \frac{2EI}{3}\begin{bmatrix} 4 & 2 \\ 2 & 4 \end{bmatrix}$$

Assembled element stiffness matrix

$$[k] = \frac{EI}{3}\begin{bmatrix} 4 & 2 & 0 & 0 \\ 2 & 4 & 0 & 0 \\ 0 & 0 & 8 & 4 \\ 0 & 0 & 4 & 8 \end{bmatrix}$$

System stiffness matrix (K)

$$[K] = [\beta^T]\,[k]\,[\beta]$$

$$= \begin{bmatrix} -\frac{1}{3} & -\frac{1}{3} & \frac{1}{3} & \frac{1}{3} \\ 0 & 1 & 1 & 0 \end{bmatrix} \frac{EI}{3} \begin{bmatrix} 4 & 2 & 0 & 0 \\ 2 & 4 & 0 & 0 \\ 0 & 0 & 8 & 4 \\ 0 & 0 & 4 & 8 \end{bmatrix} \begin{bmatrix} -\frac{1}{3} & 0 \\ -\frac{1}{3} & 1 \\ \frac{1}{3} & 1 \\ \frac{1}{3} & 0 \end{bmatrix}$$

$$= \begin{bmatrix} -\frac{1}{3} & -\frac{1}{3} & \frac{1}{3} & \frac{1}{3} \\ 0 & 1 & 1 & 0 \end{bmatrix} \frac{EI}{3} \begin{bmatrix} -2 & 2 \\ -2 & 4 \\ 4 & 8 \\ 4 & 4 \end{bmatrix}$$

$$[K] = \frac{EI}{3}\begin{bmatrix} 4 & 2 \\ 2 & 12 \end{bmatrix}$$

$$[K^{-1}] = \frac{3}{EI}\begin{bmatrix} 12 & -2 \\ -2 & 4 \end{bmatrix}\frac{1}{44} = \frac{3}{44EI}\begin{bmatrix} 12 & -2 \\ -2 & 4 \end{bmatrix}$$

System displacement (u)

$$\{u\} = [K^{-1}]\,[F]$$

$$\{u\} = \frac{3}{44EI}\begin{bmatrix} 12 & -2 \\ -2 & 4 \end{bmatrix}\begin{bmatrix} 40 \\ 25 \end{bmatrix} = \frac{3}{EI}\begin{bmatrix} 9.772 \\ 0.454 \end{bmatrix}$$

Element forces

$$\{P\} = [k]\,[\beta]\,[u]$$

$$= \frac{EI}{3}\begin{bmatrix} 4 & 2 & 0 & 0 \\ 2 & 4 & 0 & 0 \\ 0 & 0 & 8 & 4 \\ 0 & 0 & 4 & 8 \end{bmatrix}\begin{bmatrix} -\frac{1}{3} & 0 \\ -\frac{1}{3} & 1 \\ \frac{1}{3} & 1 \\ \frac{1}{3} & 0 \end{bmatrix}\frac{3}{EI}\begin{bmatrix} 9.772 \\ 0.454 \end{bmatrix}$$

$$\{P\} = \begin{bmatrix} -2 & 2 \\ -2 & 4 \\ 4 & 8 \\ 4 & 4 \end{bmatrix}\begin{bmatrix} 9.772 \\ 0.454 \end{bmatrix} = \begin{bmatrix} -18.636 \\ -17.728 \\ 42.72 \\ 40.904 \end{bmatrix}$$

$M_{AB} = -\ 18.638$ kNm
$M_{BA} = -\ 17.728$ kNm
$M_{BC} = 42.72$ kNm
$M_{CB} = 40.904$ kNm.

Example 5.21 *Analyse the continuous beam shown in Fig. 5.61 by stiffness method. Draw the bending moment diagram.*

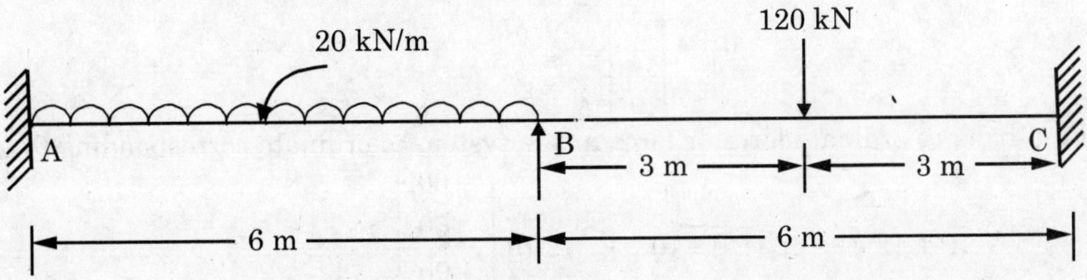

Fig. 5.61

Solution The structure is kinematically indeterminate to first degree, *i.e.*, θ_B is an unknown independent displacement (system co-ordinate)

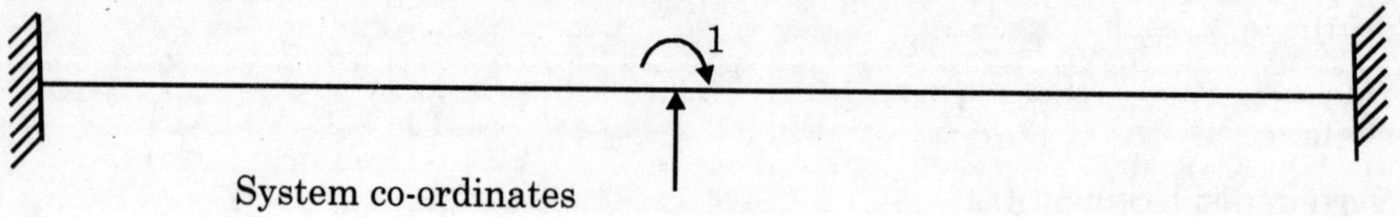

System co-ordinates

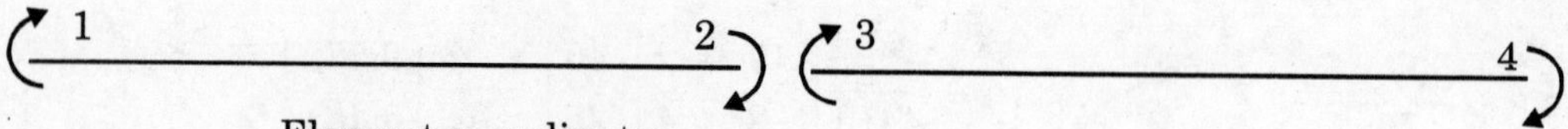

Element co-ordinates

Joint loads (Fixed end moments) :

$$M_{FAB} = \frac{-wl^2}{12} = \frac{-20 \times 6^2}{12} = -60 \text{ kNm}$$

$$M_{FBA} = \frac{wl^2}{12} = \frac{20 \times 6^2}{12} = 60 \text{ kNm}$$

$$M_{FBC} = \frac{-Wl}{8} = \frac{-120 \times 6}{8} = -90 \text{ kNm}$$

$$M_{FCB} = \frac{Wl}{8} = \frac{120 \times 6}{8} = 90 \text{ kNm}$$

Forces in element co-ordinates :

i.e., Fixed end moments $[P^0] = \begin{bmatrix} -60 \\ 60 \\ -90 \\ 90 \end{bmatrix}$.

To form [β] : Applying unit displacement at the system co-ordinate, the resulting displacements at element co-ordinates are given by {β}

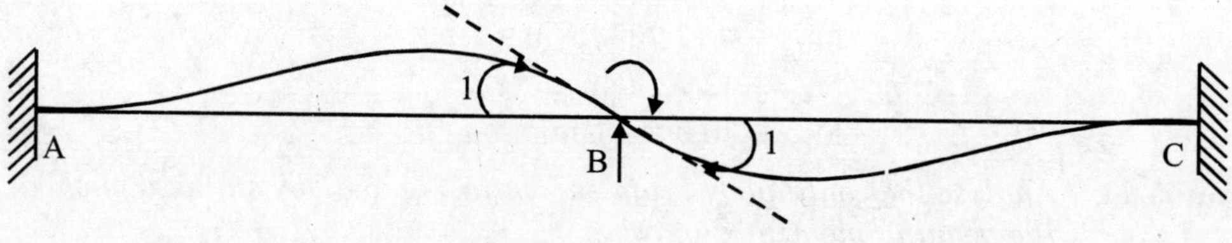

$$[\beta] = \begin{bmatrix} 0 \\ 1 \\ 1 \\ 0 \end{bmatrix}$$

Find fixed co-ordinate force or force in the system co-ordinate corresponding to forces

$$\{P^0\} \; [F^0] = [\beta]^T [P^0] = [0 \quad 1 \quad 1 \quad 0] \begin{bmatrix} -60 \\ 60 \\ -90 \\ 90 \end{bmatrix}$$

Fixed co-ordinate force, $[F^0] = [-30]$

Forces applied at system co-ordinate, $[F^f] = [0]$

($\because$ There is no external force at the system co-ordinate)

Element stiffness matrix $[K_n]$:

$$[k_1] = \frac{EI}{l}\begin{bmatrix} 4 & 2 \\ 2 & 4 \end{bmatrix} = \frac{EI}{6}\begin{bmatrix} 4 & 2 \\ 2 & 4 \end{bmatrix} = EI\begin{bmatrix} 0.67 & 0.33 \\ 0.33 & 0.67 \end{bmatrix}$$

$$[k_1] = [k_2] = EI\begin{bmatrix} 0.67 & 0.33 \\ 0.33 & 0.67 \end{bmatrix}$$

Assembled element stiffness matrix $[k]$:

$$[k] = EI\begin{bmatrix} 0.67 & 0.33 & 0 & 0 \\ 0.33 & 0.67 & 0 & 0 \\ 0 & 0 & 0.67 & 0.33 \\ 0 & 0 & 0.33 & 0.67 \end{bmatrix}$$

System stiffness matrix [K] :

System stiffness matrix $[K] = [\beta]^T [k] [\beta]$

$$= [0 \quad 1 \quad 1 \quad 0]\, EI\begin{bmatrix} 0.67 & 0.33 & 0 & 0 \\ 0.33 & 0.67 & 0 & 0 \\ 0 & 0 & 0.67 & 0.33 \\ 0 & 0 & 0.33 & 0.67 \end{bmatrix}\begin{bmatrix} 0 \\ 1 \\ 1 \\ 0 \end{bmatrix}$$

$$= EI\,[0.33 \quad 0.67 \quad 0.67 \quad 0.33]\begin{bmatrix} 0 \\ 1 \\ 1 \\ 0 \end{bmatrix} = EI\,[0.67 + 0.67]$$

$$[K] = 1.34\,EI$$

$$[K]^{-1} = \frac{1}{1.34EI}$$

System displacement $\{u\}$

$$\{u\} = [K]^{-1}\{[F^f] - [F^0]\}$$

$$[K^{-1}] = \frac{1}{1.34EI}$$

$$\{u\} = \frac{1}{1.34EI}[0 - [-30]]$$

$$\{u\} = \frac{30}{1.34EI} = \frac{22.39}{EI}$$

Element displacement $[\delta]$:

$$[\delta] = [\beta]\,[u] = \begin{bmatrix} 0 \\ 1 \\ 1 \\ 0 \end{bmatrix} \times \frac{22.39}{EI} = \frac{1}{EI}\begin{bmatrix} 0 \\ 22.39 \\ 22.39 \\ 0 \end{bmatrix}$$

$$[P'] = [k]\,[\delta] = EI \begin{bmatrix} 0.67 & 0.33 & 0 & 0 \\ 0.33 & 0.67 & 0 & 0 \\ 0 & 0 & 0.67 & 0.33 \\ 0 & 0 & 0.33 & 0.67 \end{bmatrix} \frac{1}{EI} \begin{bmatrix} 0 \\ 22.39 \\ 22.39 \\ 0 \end{bmatrix}$$

$$[P'] = \begin{bmatrix} 7.39 \\ 15.00 \\ 15.00 \\ 7.39 \end{bmatrix}$$

Final forces, $[P^f] = [P^0] + [P']$

$$= \begin{bmatrix} -60 \\ 60 \\ -90 \\ 90 \end{bmatrix} + \begin{bmatrix} 7.39 \\ 15.00 \\ 15.00 \\ 7.39 \end{bmatrix} = \begin{bmatrix} -52.61 \\ 75.00 \\ -75.00 \\ 97.39 \end{bmatrix} \text{kNm}$$

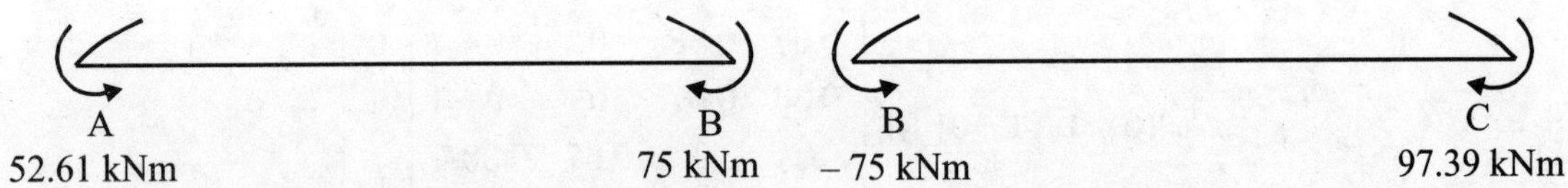

Simply supported beam moments :

$$M_{FAB} = \frac{wl^2}{8} = \frac{20 \times 6^2}{8} = 90 \text{ kNm}$$

$$M_{FBC} = \frac{Wl}{4} = \frac{120 \times 6}{4} = 180 \text{ kNm}$$

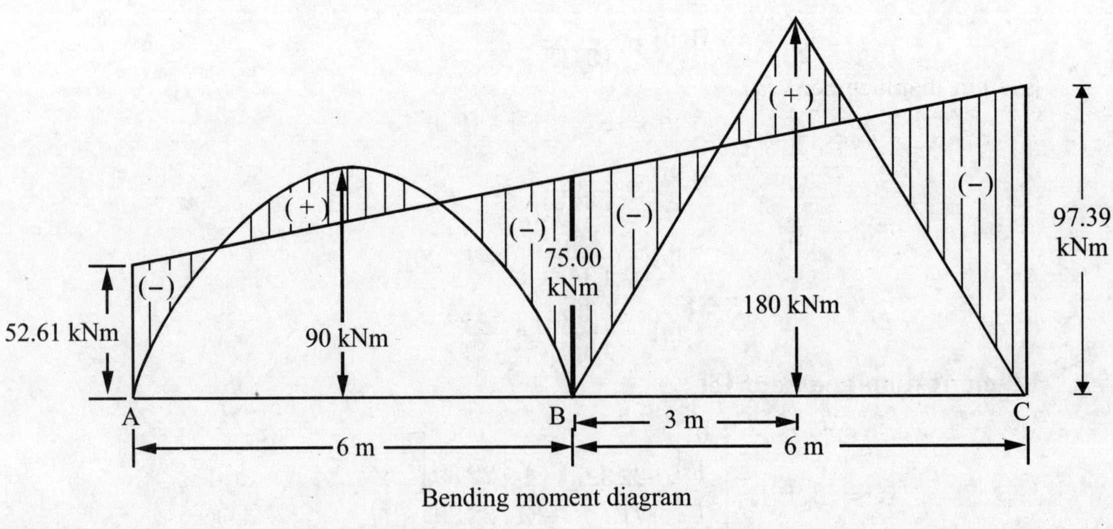

Bending moment diagram

Example 5.22 *A two span continuous beam ABC is fixed at A and simply supported over the supports B and C. AB = 6 m and BC = 6 m. The moments of inertia is constant throughout. It is loaded as shown in Fig. 5.62. Analyse the beam by matrix stiffness method.*

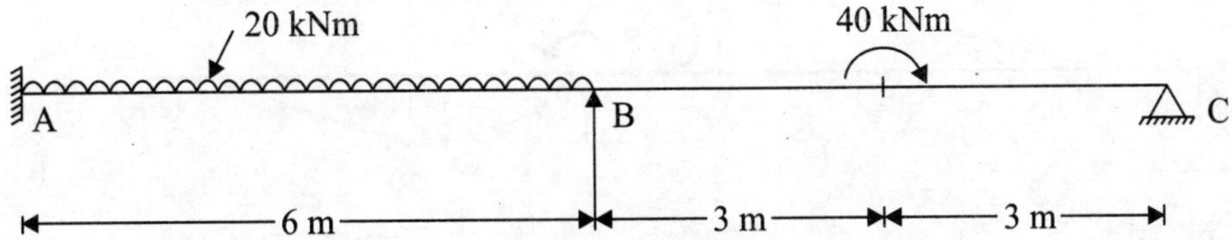

Fig. 5.62

Solution θ_B and θ_C are the unknown displacements, *i.e.*, the structure kinematically indeterminate to the second degree. Hence the stiffness matrix k will be of the order 2 × 2.

Fixed joint forces :

$$M_{FAB} = \frac{-wl^2}{12} = \frac{-20 \times 6^2}{12} = -60 \text{ kNm}$$

$$M_{FBA} = \frac{wl^2}{12} = \frac{20 \times 6^2}{12} = 60 \text{ kNm}$$

$$M_{FBC} = \frac{M}{4} = \frac{40}{4} = 10 \text{ kNm}$$

$$M_{FCB} = \frac{M}{4} = \frac{40}{4} = 10 \text{ kNm}$$

Joint forces :

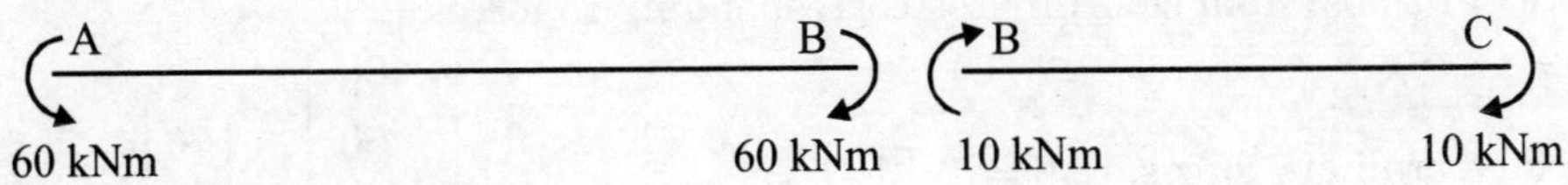

Equivalent joint loads :

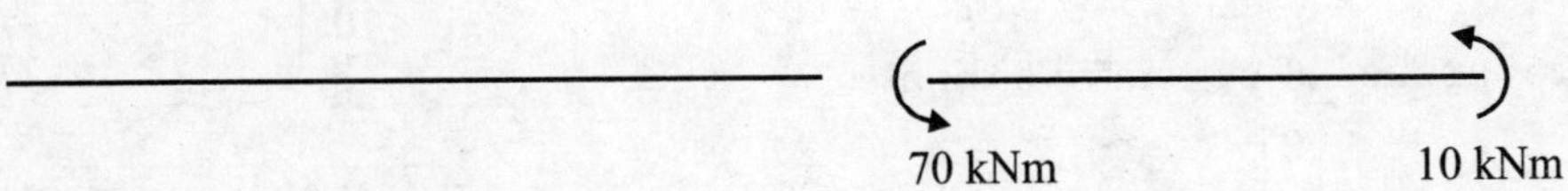

Element co-ordinates :

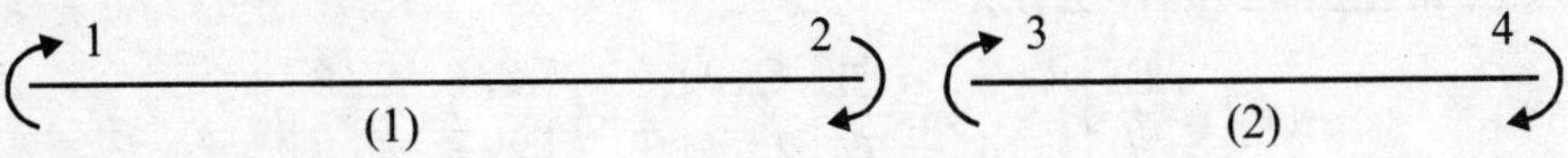

System co-ordinates :

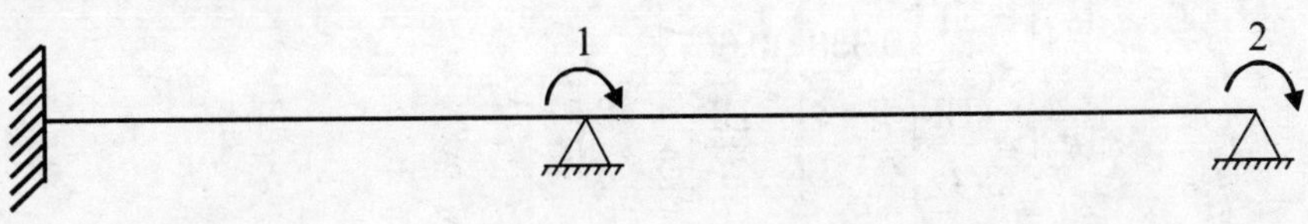

Formation of [β] matrix :

1st Column. Apply a unit displacement at system co-ordinate (1) only. The consequent element co-ordinate displacements are :

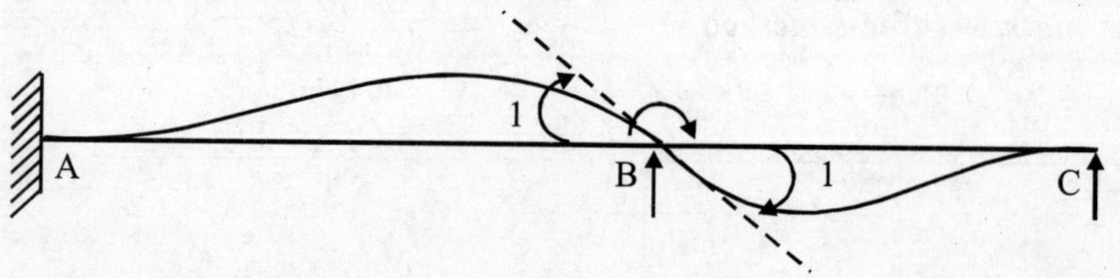

$$\begin{bmatrix} 0 \\ 1 \\ 1 \\ 0 \end{bmatrix}$$

2nd Column. Apply unit displacement at system co-ordinate (2) only.

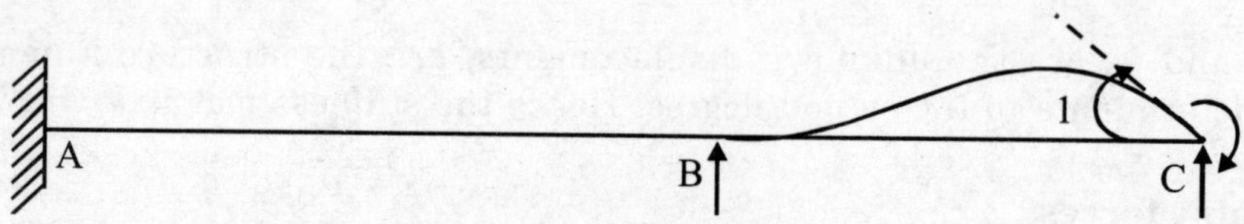

$$\begin{bmatrix} 0 \\ 0 \\ 0 \\ 1 \end{bmatrix}$$

$$\therefore \qquad \beta = \begin{bmatrix} 0 & 0 \\ 1 & 0 \\ 1 & 0 \\ 0 & 1 \end{bmatrix}$$

[P⁰] Values. Forces at the element co-ordinates in the fixed state.

$$[P^0] = \begin{bmatrix} -60 \\ 60 \\ 10 \\ 10 \end{bmatrix} \qquad \text{[i.e., fixed end moments]}$$

Forces in the system co-ordinates corresponding to forces $\{P^0\}$

i.e., Fixed co-ordinate forces, $[F^0] = [\beta]^T [P^0] = \begin{bmatrix} 0 & 1 & 1 & 0 \\ 0 & 0 & 0 & 1 \end{bmatrix} \begin{bmatrix} -60 \\ 60 \\ 10 \\ 10 \end{bmatrix} = \begin{bmatrix} 70 \\ 10 \end{bmatrix}$

Forces applied at system co-ordinates,

$$[F^f] = \begin{bmatrix} 0 \\ 0 \end{bmatrix} \qquad (\because \text{ No forces in the given loading at the system co-ordinates})$$

Element stiffness matrix $[k_i]$:

$$[k_1] = \frac{EI}{l}\begin{bmatrix} 4 & 2 \\ 2 & 4 \end{bmatrix} = \frac{EI}{6}\begin{bmatrix} 4 & 2 \\ 2 & 4 \end{bmatrix} = EI\begin{bmatrix} 4/6 & 2/6 \\ 2/6 & 4/6 \end{bmatrix}$$

$$[k_1] = EI\begin{bmatrix} 0.667 & 0.333 \\ 0.333 & 0.667 \end{bmatrix}$$

$$[k_2] = \frac{EI}{l}\begin{bmatrix} 4 & 2 \\ 2 & 4 \end{bmatrix} = \frac{EI}{6}\begin{bmatrix} 4 & 2 \\ 2 & 4 \end{bmatrix}$$

$$[k_2] = EI\begin{bmatrix} 0.667 & 0.333 \\ 0.333 & 0.667 \end{bmatrix}$$

Assembled element stiffness matrix is

$$[k] = EI\begin{bmatrix} 0.667 & 0.333 & 0 & 0 \\ 0.333 & 0.667 & 0 & 0 \\ 0 & 0 & 0.667 & 0.333 \\ 0 & 0 & 0.333 & 0.667 \end{bmatrix}$$

System stiffness matrix is

$[K] = [\beta]^T [k] [\beta]$, where $[k]$ is given above

and
$$[\beta]^T = \begin{bmatrix} 0 & 1 & 1 & 0 \\ 0 & 0 & 0 & 1 \end{bmatrix}$$

$$[K] = EI\begin{bmatrix} 1.333 & 0.333 \\ 0.333 & 0.667 \end{bmatrix}; [K]^{-1} = \frac{1}{EI}\begin{bmatrix} 0.857 & -0.429 \\ -0.429 & 1.714 \end{bmatrix}$$

$\{u\}$, the system displacements are given by,

$$\{u\} = [K]^{-1} \{ [F^f] - [F^0]\}$$

$$= \frac{1}{EI}\begin{bmatrix} 0.857 & -0.429 \\ -0.429 & 1.714 \end{bmatrix}\left\{\begin{Bmatrix} 0 \\ 0 \end{Bmatrix} - \begin{Bmatrix} 70 \\ 10 \end{Bmatrix}\right\}$$

$$[u] = \frac{1}{EI}\begin{bmatrix} -55.71 \\ -12.86 \end{bmatrix}$$

Element displacements $[\delta]$ are

$$\{\delta\} = [\beta] [u] = \begin{bmatrix} 0 & 0 \\ 1 & 0 \\ 1 & 0 \\ 0 & 1 \end{bmatrix} \frac{1}{EI}\begin{bmatrix} -55.71 \\ -12.86 \end{bmatrix} = \frac{1}{EI}\begin{bmatrix} 0 \\ -55.71 \\ -55.71 \\ 12.86 \end{bmatrix}$$

The element forces $\{P'\} = [k] \{\delta\}$. Substituting,

$$\{P'\} = [-18.57 \quad -37.14 \quad -32.86 \quad -10]^T$$

$$\text{The find forces } \{P^f\} = \{P^0\} + [P'] = \begin{bmatrix} -60 \\ 60 \\ 10 \\ 10 \end{bmatrix} + \begin{bmatrix} -18.57 \\ -37.14 \\ -32.86 \\ -10 \end{bmatrix} = \begin{bmatrix} -78.57 \\ 22.86 \\ -22.86 \\ 0 \end{bmatrix}$$

Simply supported beam moments :

Moment in AB = $\frac{wl^2}{8} = \frac{20 \times 6^2}{8}$ = 90 kNm

Moment in BC, 40 kNm concentrated at midspan.

The bending moment diagram is drawn below on the tension sides.

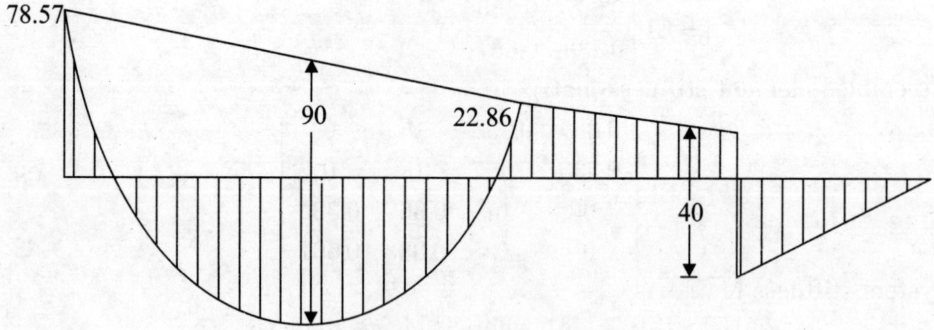

Bending moment diagram

Example 5.23 *Analyse the continuous beam by matrix stiffness method.*

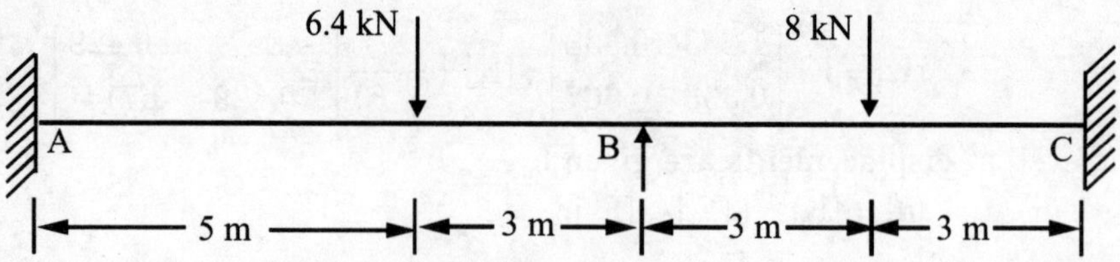

Fig. 5.63

Solution Since the only unknown independent joint displacement happens at B, kinematic indeterminacy of the given structure is 1.

System co-ordinates :

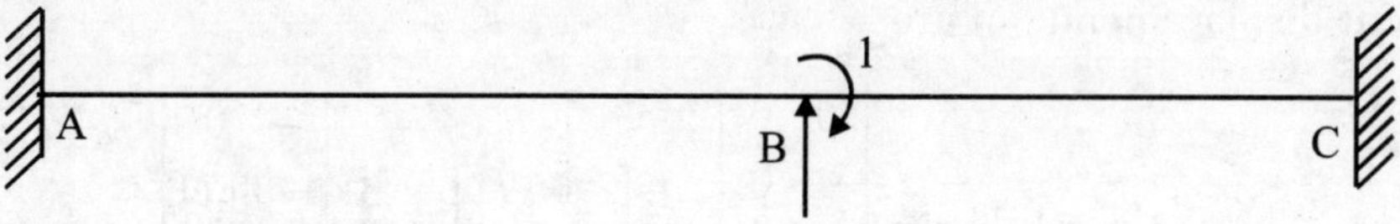

Element co-ordinates :

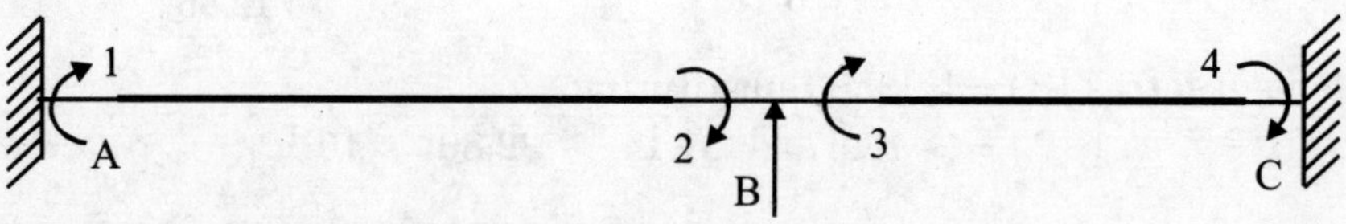

Fixed end moments :

$$M_{FAB} = \frac{-Wab^2}{l^2} = \frac{-6.4 \times 5 \times 3^2}{8^2} = -4.5 \text{ kNm}$$

$$M_{FBA} = \frac{Wa^2b}{l^2} = \frac{6.4 \times 5^2 \times 3}{8^2} = 7.5 \text{ kNm}$$

$$M_{FBC} = \frac{-Wl}{8} = \frac{-8 \times 6}{8} = -6 \text{ kNm}$$

$$M_{FCB} = \frac{Wl}{8} = \frac{8 \times 6}{8} = 6 \text{ kNm}$$

Forces in element co-ordinates at fixed state [*i.e.*, Fixed end moments], $[P^0]$

$$\{P^0\} = \begin{bmatrix} -4.5 \\ 7.5 \\ -6.0 \\ 6.0 \end{bmatrix}$$

To form transformation matrix [β] :

Apply unit displacement at the system co-ordinate.

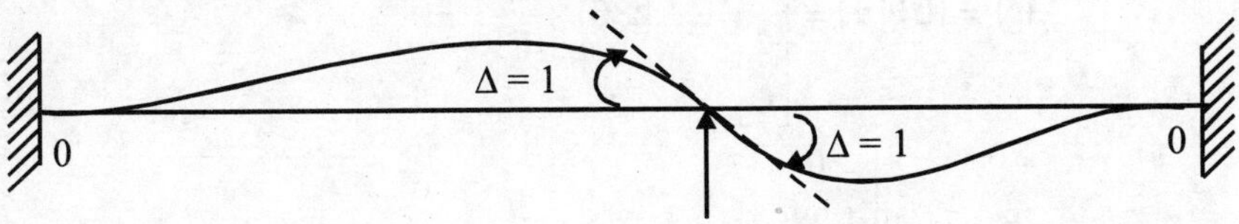

$$\text{Transformation matrix } [\beta] = \begin{bmatrix} 0 \\ 1 \\ 1 \\ 0 \end{bmatrix}$$

$$\text{Fixed co-ordinate forces, } [F^0] = [\beta]^T [P^0] = [\,0\ 1\ 1\ 0\,] \begin{bmatrix} -4.5 \\ 7.5 \\ -6.0 \\ 6.0 \end{bmatrix}$$

$$[F^0] = [1.50]$$

Forces applied at system co-ordinates, $[F^f] = [0]$

[This is because there are no external forces at the system co-ordinate]

Element stiffness matrix :

$$\text{Element stiffness matrix, } [k_n] = \frac{EI}{l}\begin{bmatrix} 4 & 2 \\ 2 & 4 \end{bmatrix}$$

For first element, $[k_1] = \frac{EI}{8}\begin{bmatrix} 4 & 2 \\ 2 & 4 \end{bmatrix} = EI \begin{bmatrix} 0.50 & 0.25 \\ 0.25 & 0.50 \end{bmatrix}$

For second element, $[k_2] = \frac{EI}{6}\begin{bmatrix} 4 & 2 \\ 2 & 4 \end{bmatrix} = EI \begin{bmatrix} 4/6 & 2/6 \\ 2/6 & 4/6 \end{bmatrix} = EI \begin{bmatrix} 0.667 & 0.333 \\ 0.333 & 0.667 \end{bmatrix}$

$$\text{Assembled element stiffness matrix, } [k] = EI \begin{bmatrix} 0.50 & 0.25 & 0 & 0 \\ 0.25 & 0.50 & 0 & 0 \\ 0 & 0 & 0.667 & 0.333 \\ 0 & 0 & 0.333 & 0.667 \end{bmatrix}$$

System stiffness matrix :

System stiffness matrix, $[K] = [\beta]^T [k] [\beta]$

$$[K] = 1.167\ EI\ ;\ [K]^{-1} = \frac{0.857}{EI}$$

System displacements {*u*} :

$$\{u\} = [K]^{-1} \{[F^f] - [F^0]\} = \frac{0.857}{EI} \times (-1.50)$$

$$\{u\} = -\frac{1.286}{EI}$$

Element displacements {δ} :

$$\{\delta\} = [\beta]\,[u] = \begin{bmatrix} 0 \\ 1 \\ 1 \\ 0 \end{bmatrix} [-1.286] \times \frac{1}{EI}$$

$$\{\delta\} = \frac{-1}{EI} \begin{bmatrix} 0 \\ 1.286 \\ 1.286 \\ 0 \end{bmatrix}$$

$$[P'] = [k]\,[\delta] = \begin{bmatrix} -0.322 \\ -0.643 \\ -0.857 \\ -0.448 \end{bmatrix}$$

Final forces, $$[P^f] = [P^0] + [P'] = \begin{bmatrix} -4.5 \\ 7.5 \\ -6.0 \\ 6.0 \end{bmatrix} + \begin{bmatrix} -0.322 \\ -0.643 \\ -0.857 \\ -0.448 \end{bmatrix}$$

Final forces, $$[P^f] = \begin{bmatrix} -4.822 \\ 6.857 \\ -6.857 \\ 5.572 \end{bmatrix}$$

Span bending moments :

Moment in span AB $= \dfrac{Wab}{l} = \dfrac{6.4 \times 5 \times 3}{8} = 12.0$ kNm

Moment in span BC $= \dfrac{Wl}{4} = \dfrac{8 \times 6}{4} = 12.0$ kNm

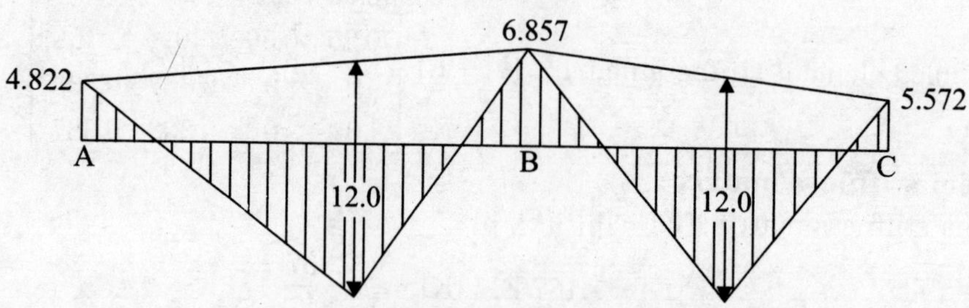

Bending moment diagram
(Drawn on the tension sides)

5.9 ANALYSIS OF RIGID JOINTED FRAMES

Example 5.24 *Analyse the structure shown in Fig. 5.64 by stiffness method.*

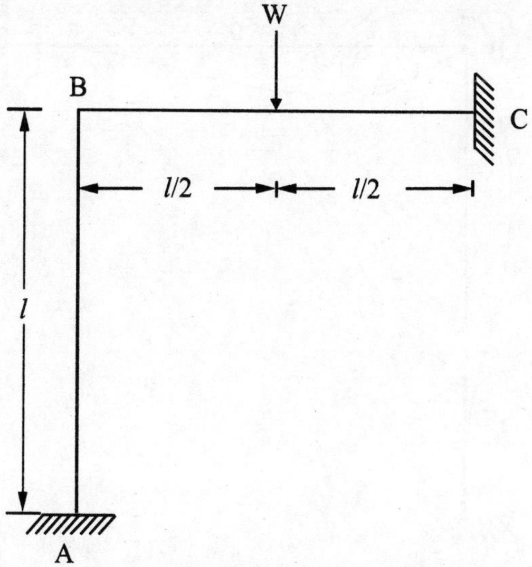

Fig. 5.64

Solution The structure is kinematically indeterminate to first degree, *i.e.*, θ_B is the only unknown joint displacement. Hence the size of the structure stiffness matrix will be 1 × 1.

Fixed end moments :

$$M_{FAB} = M_{FBA} = 0$$

$$M_{FBC} = -\frac{Wl}{8}$$

$$M_{FCB} = \frac{Wl}{8}$$

θ_B is the structure co-ordinate

Element co-ordinates

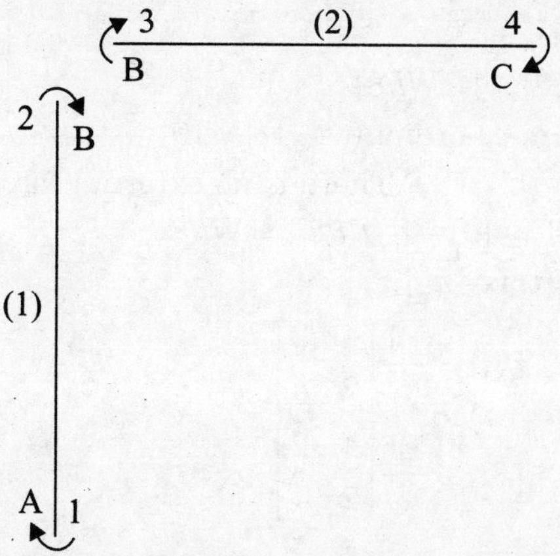

The displacement transformation matrix β will be generated by imposing unit rotation at B and keeping the far ends fixed.

Applying $\theta_B = 1$;

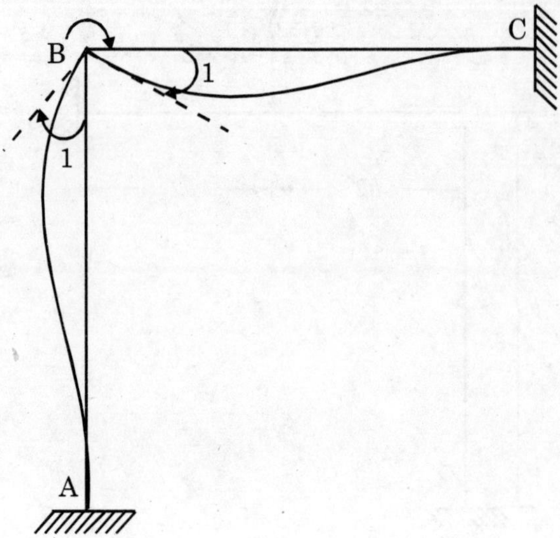

$$[\beta] = \begin{bmatrix} 0 \\ 1 \\ 1 \\ 0 \end{bmatrix}$$

Forces in element co-ordinates, *i.e.*, fixed end moments

$$[P^0] = \begin{bmatrix} 0 \\ 0 \\ -Wl/8 \\ Wl/8 \end{bmatrix}$$

Fixed end co-ordinate forces, $[F^0] = [\beta]^T [P^0] = [0 \;\; 1 \;\; 1 \;\; 0] \begin{bmatrix} 0 \\ 0 \\ -Wl/8 \\ Wl/8 \end{bmatrix}$

$$[F^0] = [-Wl/8]$$

Forces applied at system co-ordinates, $[F^f] = [0]$

(∵ There is no external force at the system co-ordinates)

$$\{[F^f] - [F^0]\} = \{[0] - [-Wl/8]\} = Wl/8$$

Element stiffness matrix $[k_n]$:

$$[k_1] = \frac{EI}{l}\begin{bmatrix} 4 & 2 \\ 2 & 4 \end{bmatrix}$$

$$[k_2] = \frac{EI}{l}\begin{bmatrix} 4 & 2 \\ 2 & 4 \end{bmatrix}$$

Assembled element stiffness matrix [k] :

$$[k] = \begin{bmatrix} k_1 & 0 \\ 0 & k_2 \end{bmatrix} = \frac{EI}{l}\begin{bmatrix} 4 & 2 & 0 & 0 \\ 2 & 4 & 0 & 0 \\ 0 & 0 & 4 & 2 \\ 0 & 0 & 2 & 4 \end{bmatrix}$$

System stiffness matrix [K] :

$$[K] = [\beta]^T [k] [\beta] = [0 \ \ 1 \ \ 1 \ \ 0] \times \frac{EI}{l}\begin{bmatrix} 4 & 2 & 0 & 0 \\ 2 & 4 & 0 & 0 \\ 0 & 0 & 4 & 2 \\ 0 & 0 & 2 & 4 \end{bmatrix}\begin{bmatrix} 0 \\ 1 \\ 1 \\ 0 \end{bmatrix} = \frac{EI}{l}[2 \ \ 4 \ \ 4 \ \ 2]\begin{bmatrix} 0 \\ 1 \\ 1 \\ 0 \end{bmatrix}$$

$$[K] = \frac{EI}{l}[8]$$

System displacements {u} :

$$\{u\} = [K]^{-1} \{ [F^f] - [F^0]\} = \frac{l}{8EI} \times \frac{Wl}{8} = \frac{Wl^2}{64EI}$$

Element displacements {δ} :

$$\{\delta\} = [\beta] [u] = \begin{bmatrix} 0 \\ 1 \\ 1 \\ 0 \end{bmatrix} \times \frac{Wl^2}{64EI}[1]$$

$$\{\delta\} = \frac{Wl^2}{64EI}\begin{bmatrix} 0 \\ 1 \\ 1 \\ 0 \end{bmatrix}$$

$$[P'] = [k] [\delta] = \frac{EI}{l}\begin{bmatrix} 4 & 2 & 0 & 0 \\ 2 & 4 & 0 & 0 \\ 0 & 0 & 4 & 2 \\ 0 & 0 & 2 & 4 \end{bmatrix} \times \frac{Wl^2}{64EI}\begin{bmatrix} 0 \\ 1 \\ 1 \\ 0 \end{bmatrix}$$

$$[P'] = \frac{Wl}{64}\begin{bmatrix} 2 \\ 4 \\ 4 \\ 2 \end{bmatrix}$$

Final forces, $[P^f] = [P'] + [P^0]$

$$= \frac{Wl}{64}\begin{bmatrix} 2 \\ 4 \\ 4 \\ 2 \end{bmatrix} + \begin{bmatrix} 0 \\ 0 \\ -Wl/8 \\ Wl/8 \end{bmatrix} = Wl\begin{bmatrix} 0.03125 \\ 0.0625 \\ 0.0625 \\ 0.03125 \end{bmatrix} + Wl\begin{bmatrix} 0 \\ 0 \\ -0.125 \\ 0.125 \end{bmatrix}$$

$$[P^f] = Wl \begin{bmatrix} 0.03125 \\ 0.0625 \\ -0.0625 \\ 0.15625 \end{bmatrix}$$

Simply supported bending moments :

$$M_{AB} = 0$$

$$M_{BC} = \frac{Wl}{4}$$

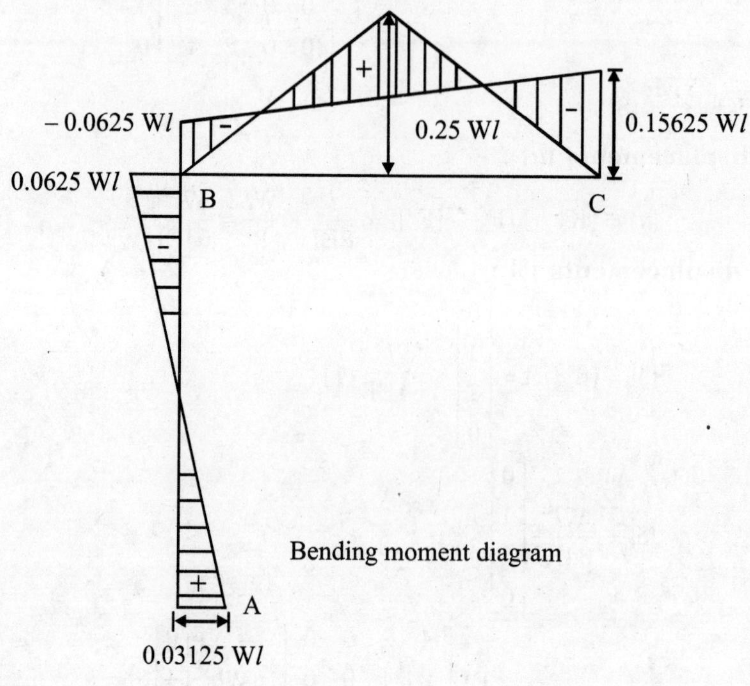

Bending moment diagram

Example 5.25 *Analyse the structure shown in Fig. 5.65 by the matrix stiffness method and sketch the bending moment diagram.*

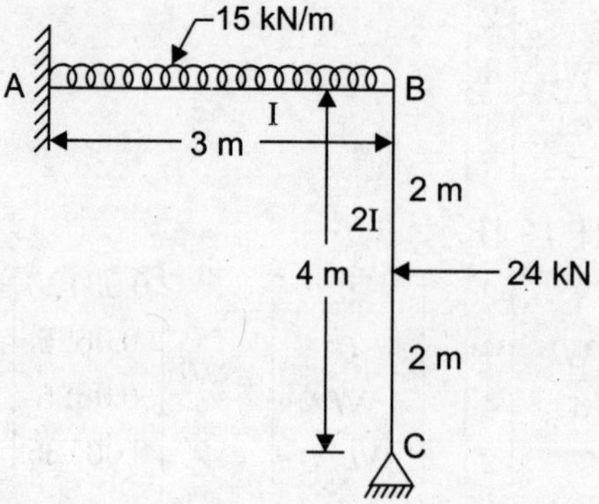

Fig. 5.65

Solution Structure is kinematically indeterminate to the second degree and the rotations at B and C are the kinematic indeterminacies.

Fixed end moments :

$$M_{FAB} = -\frac{wl^2}{12} = -\frac{15 \times 3^2}{12} = -\ 11.25 \text{ kNm}$$

$$M_{FBA} = \frac{wl^2}{12} = \frac{15 \times 3^2}{12} = 11.25 \text{ kNm}$$

$$M_{FBC} = -\frac{Wl}{8} = -\frac{24 \times 4}{8} = -\ 12 \text{ kNm}$$

$$M_{FCB} = \frac{Wl}{8} = \frac{24 \times 4}{8} = 12 \text{ kNm}$$

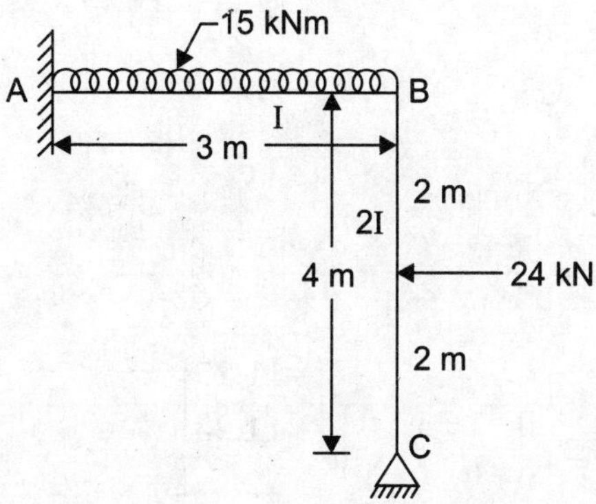

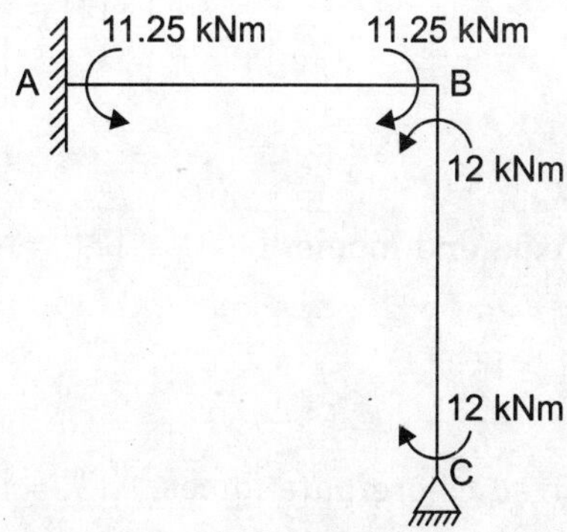

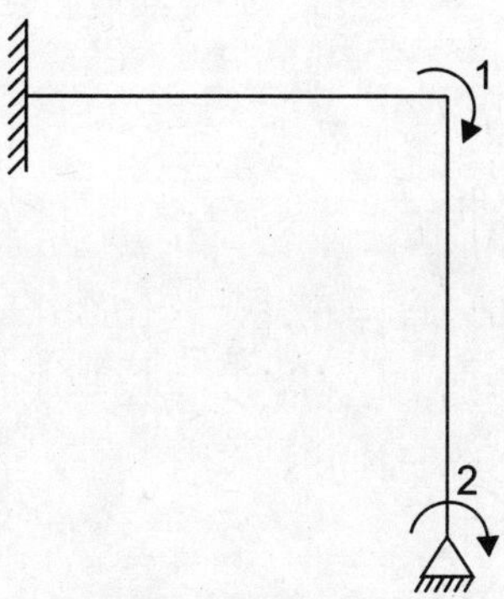

System co-ordinates

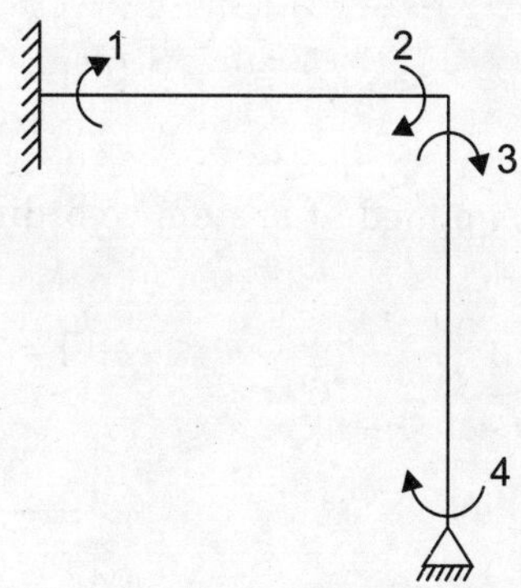

Element co-ordinates

To form transformation matrix (β)

(*i*) Apply unit rotation at joint 'B' (*ii*) Apply unit rotation at joint 'C'

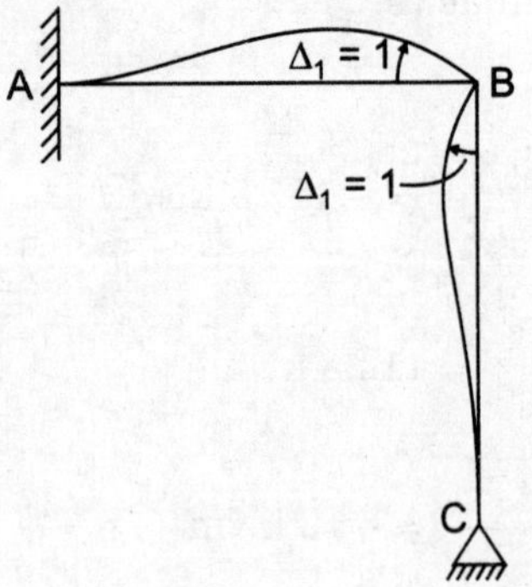

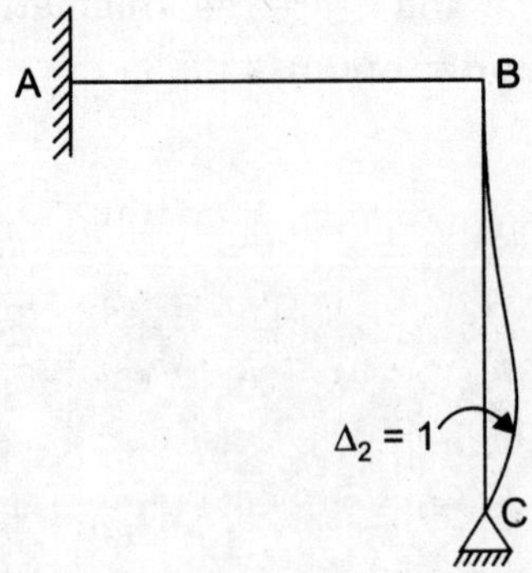

$$[\beta] = \begin{array}{c} \Delta_1 = 1 \quad \Delta_2 = 1 \\ \begin{bmatrix} 0 & 0 \\ 1 & 0 \\ 1 & 0 \\ 0 & 1 \end{bmatrix} \end{array}$$

Fixed end moments, $\{P^0\} = \begin{Bmatrix} -11.25 \\ 11.25 \\ -12 \\ 12 \end{Bmatrix}$

Fixed co-ordinate forces, $\{F^0\} = [\beta]^T \{P^0\} = \begin{bmatrix} 0 & 1 & 1 & 0 \\ 0 & 0 & 0 & 1 \end{bmatrix} \begin{Bmatrix} -11.25 \\ 11.25 \\ -12 \\ 12 \end{Bmatrix}$

$$\{F^0\} = \begin{Bmatrix} -0.75 \\ 12 \end{Bmatrix}$$

Forces applied at system co-ordinates $\{F^f\} = \begin{Bmatrix} 0 \\ 0 \end{Bmatrix}$

$$\{F\} = \{F^f\} - \{F^0\} = \begin{Bmatrix} 0 \\ 0 \end{Bmatrix} - \begin{Bmatrix} -0.75 \\ 12 \end{Bmatrix} = \begin{Bmatrix} 0.75 \\ -12 \end{Bmatrix}$$

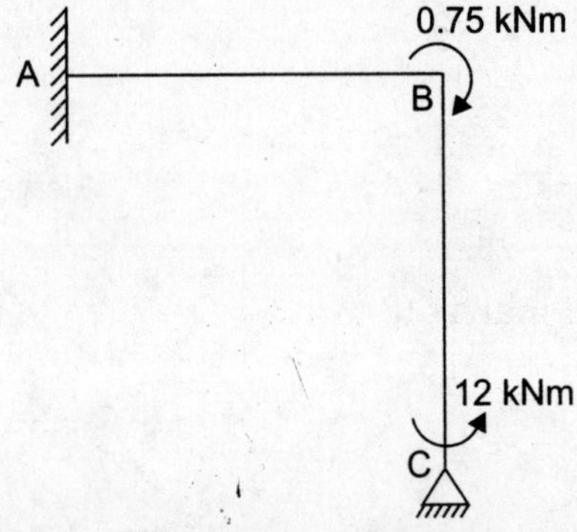

Element stiffness matrix :

$$[k] = \frac{EI}{l}\begin{bmatrix}4 & 2\\ 2 & 4\end{bmatrix},\ [k_1] = \frac{EI}{3}\begin{bmatrix}4 & 2\\ 2 & 4\end{bmatrix},\ [k_2] = \frac{2EI}{4}\begin{bmatrix}4 & 2\\ 2 & 4\end{bmatrix} = \frac{EI}{3}\begin{bmatrix}6 & 3\\ 3 & 6\end{bmatrix}$$

Assembled element

Stiffness matrix
$$[k] = \frac{EI}{3}\begin{bmatrix}4 & 2 & 0 & 0\\ 2 & 4 & 0 & 0\\ 0 & 0 & 6 & 3\\ 0 & 0 & 3 & 6\end{bmatrix}$$

System stiffness matrix :

$$[K] = [\beta^T]\,[k][\beta] = \begin{bmatrix}0 & 1 & 1 & 0\\ 0 & 0 & 0 & 1\end{bmatrix}\frac{EI}{3}\begin{bmatrix}4 & 2 & 0 & 0\\ 2 & 4 & 0 & 0\\ 0 & 0 & 6 & 3\\ 0 & 0 & 3 & 6\end{bmatrix}\begin{bmatrix}0 & 0\\ 1 & 0\\ 1 & 0\\ 0 & 1\end{bmatrix}$$

$$[K] = \frac{EI}{3}\begin{bmatrix}2 & 4 & 6 & 3\\ 0 & 0 & 3 & 6\end{bmatrix}\begin{bmatrix}0 & 0\\ 1 & 0\\ 1 & 0\\ 0 & 1\end{bmatrix} = \frac{EI}{3}\begin{bmatrix}10 & 3\\ 3 & 6\end{bmatrix}$$

$$[K^{-1}] = \frac{3}{EI}\times\frac{\begin{bmatrix}6 & -3\\ -3 & 10\end{bmatrix}}{51} = \frac{3}{EI}\times\frac{1}{(60-9)}\begin{bmatrix}6 & -3\\ -3 & 10\end{bmatrix}$$

$$[K^{-1}] = \frac{3}{51EI}\begin{bmatrix}6 & -3\\ -3 & 10\end{bmatrix} = \frac{1}{17EI}\begin{bmatrix}6 & -3\\ -3 & 10\end{bmatrix}$$

System displacements :

$$\{u\} = [K^{-1}]\,\{F\}$$

$$= \frac{1}{17EI}\begin{bmatrix}6 & -3\\ -3 & 10\end{bmatrix}\begin{bmatrix}0.75\\ -12\end{bmatrix} = \frac{1}{17EI}\begin{bmatrix}40.5\\ -122.25\end{bmatrix}$$

$$\{P'\} = [k]\,\{\delta\} = [k]\,[\beta]\,\{u\}$$

$$= \frac{EI}{3}\begin{bmatrix}4 & 2 & 0 & 0\\ 2 & 4 & 0 & 0\\ 0 & 0 & 6 & 3\\ 0 & 0 & 3 & 6\end{bmatrix}\begin{bmatrix}0 & 0\\ 1 & 0\\ 1 & 0\\ 0 & 1\end{bmatrix}\frac{1}{17EI}\begin{bmatrix}40.5\\ -122.25\end{bmatrix}$$

$$= \frac{1}{51}\begin{bmatrix}2 & 0\\ 4 & 0\\ 6 & 3\\ 3 & 6\end{bmatrix}\begin{bmatrix}40.5\\ -122.25\end{bmatrix} = \frac{1}{51}\begin{bmatrix}81\\ 162\\ -123.75\\ -612\end{bmatrix}$$

$$\{P'\} = \begin{bmatrix}1.59\\ 3.17\\ -2.42\\ -12\end{bmatrix}$$

Final forces :

$$\{P^f\} = [P'] + [P^0]$$

$$\{P^f\} = \begin{bmatrix} 1.59 \\ 3.17 \\ -2.42 \\ -12 \end{bmatrix} + \begin{bmatrix} -11.25 \\ +11.25 \\ -12 \\ +12 \end{bmatrix} = \begin{bmatrix} -9.66 \\ 14.42 \\ -14.42 \\ 0 \end{bmatrix}$$

S.S. Beam bending moments:

$$\text{Bending moment in span AB} = \frac{wl^2}{8} = \frac{15 \times 3^2}{8} = 16.875 \text{ kNm}$$

$$\text{Bending moment in span BC} = \frac{Wl}{4} = \frac{24 \times 4}{4} = 24 \text{ kNm}$$

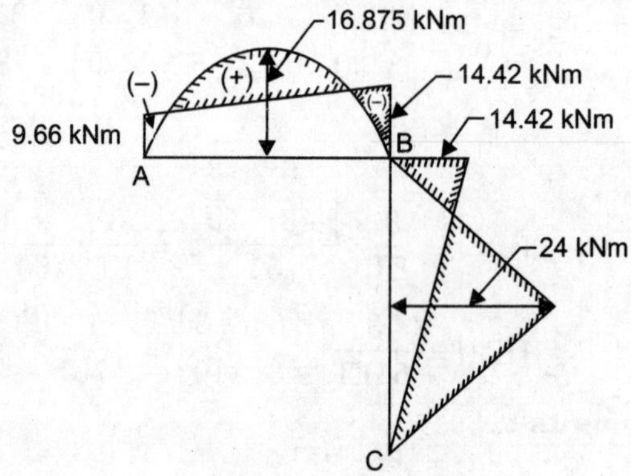

Bending moment diagram

Example 5.26 *Analyse the structure shown in Fig. 5.66 by the matrix stiffness method. Sketch the bending moment diagram.*

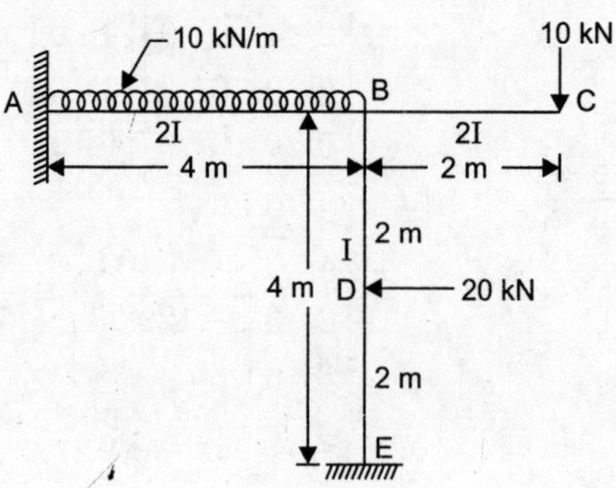

Fig. 5.66

Solution Degree of kinematic indeterminacy is one, *i.e.,* rotation at B

Fixed end moments :

$$M_{FAB} = -\frac{wl^2}{12} = -\frac{10 \times 4^2}{12} = -13.33 \text{ kNm}$$

$$M_{FBA} = \frac{wl^2}{12} = \frac{10 \times 4^2}{12} = 13.33 \text{ kNm}$$

$$M_B = 10 \times 2 = 20 \text{ kNm}$$

(The force at C will be equivalent to a downward force of 10 kN and a moment of 20 kNm at B)

$$M_{FBE} = -\frac{Wl}{8} = -\frac{20 \times 4}{8} = -10 \text{ kNm}$$

$$M_{FEB} = \frac{Wl}{8} = \frac{20 \times 4}{8} = 10 \text{ kNm}$$

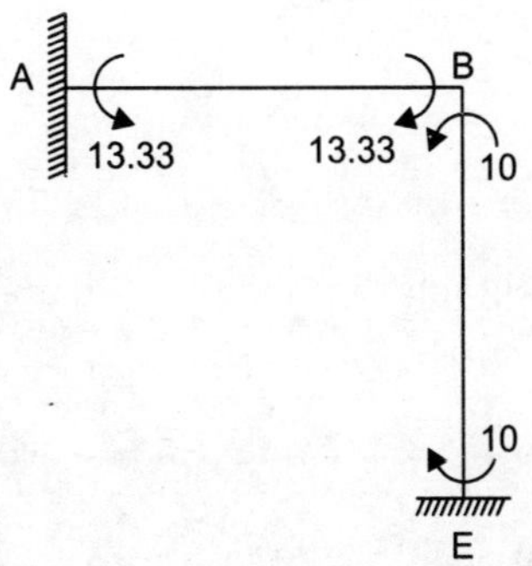

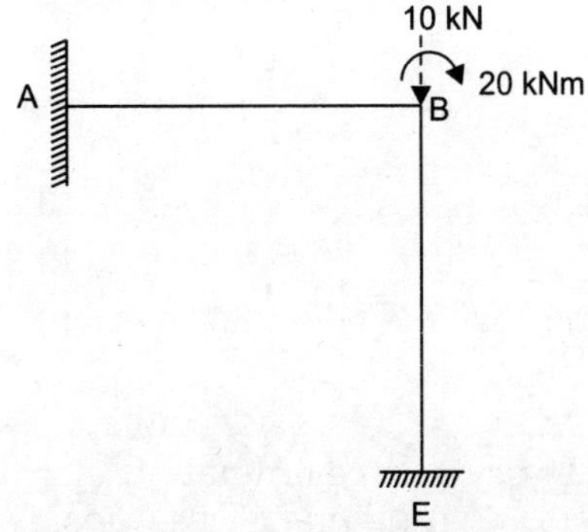

Equivalent forces at B due to the force in BC

$$\text{Fixed end moments } \{P^0\} = \begin{bmatrix} -13.33 \\ 13.33 \\ -10 \\ 10 \end{bmatrix}$$

Transformation matrix [β]:

The displacement transformation matrix [β] would be obtained by applying a unit displacement at B only and finding the resulting displacements at element co-ordinates.

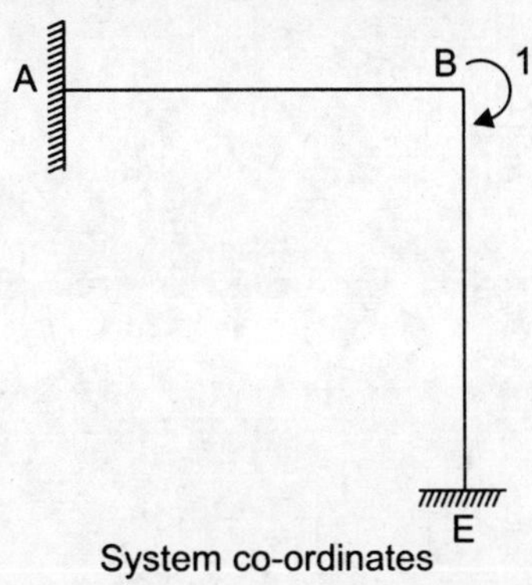

System co-ordinates

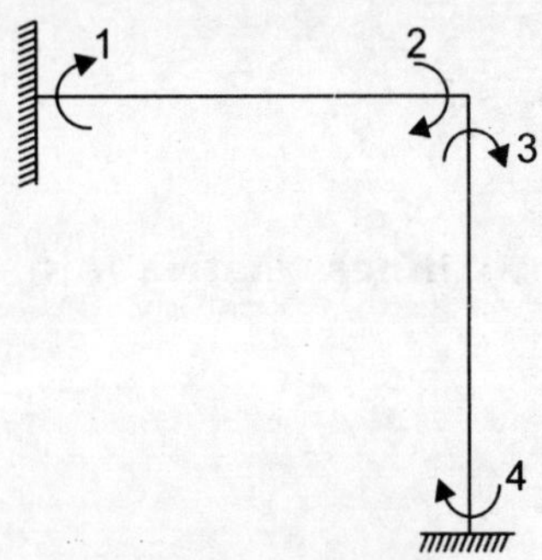

Element co-ordinates

Applying unit displacement along system co-ordinate

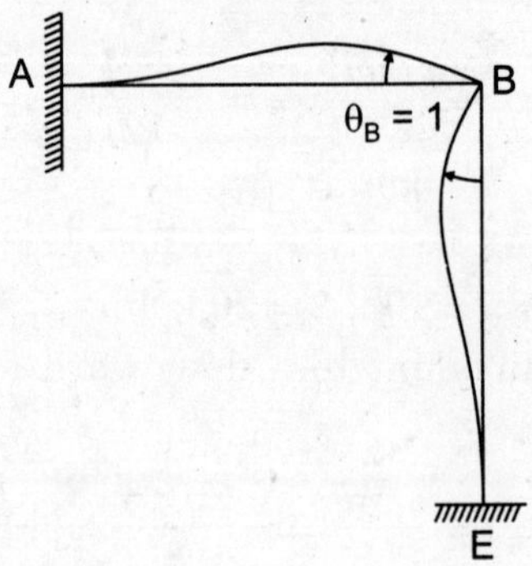

Transformation matrix $[\beta] = \begin{bmatrix} 0 \\ 1 \\ 1 \\ 0 \end{bmatrix}$

Fixed co-ordinate forces $\{F^0\} = [\beta]^T\{P^0\}$

$$= \begin{bmatrix} 0 & 1 & 1 & 0 \end{bmatrix} \begin{bmatrix} -13.33 \\ 13.33 \\ -10 \\ 10 \end{bmatrix} = \{3.33\}$$

Force applied at system co-ordinate $\{F^f\} = \{20\}$

$$\{F\} = \{F^f\} - \{F^0\} = \{20\} - \{3.33\}$$

$$\{F\} = \{16.67\}$$

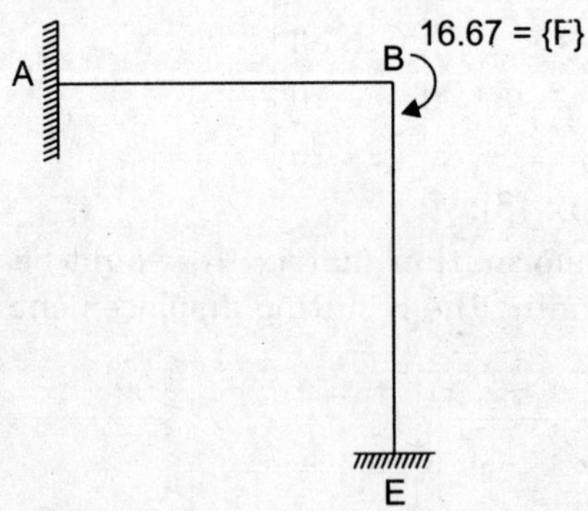

Element stiffness matrix [*k*]:

$$[k_i] = \frac{EI}{l}\begin{bmatrix} 4 & 2 \\ 2 & 4 \end{bmatrix}$$

$$[k_1] = \frac{2EI}{4}\begin{bmatrix} 4 & 2 \\ 2 & 4 \end{bmatrix} = EI\begin{bmatrix} 2 & 1 \\ 1 & 2 \end{bmatrix}$$

$$[k_2] = \frac{EI}{4}\begin{bmatrix} 4 & 2 \\ 2 & 4 \end{bmatrix} = EI\begin{bmatrix} 1 & 0.5 \\ 0.5 & 1 \end{bmatrix}$$

Assembled element stiffness matrix:

$$[k] = \begin{bmatrix} k_1 & 0 \\ 0 & k_2 \end{bmatrix} = EI\begin{bmatrix} 2 & 1 & 0 & 0 \\ 1 & 2 & 0 & 0 \\ 0 & 0 & 1 & 0.5 \\ 0 & 0 & 0.5 & 1 \end{bmatrix}$$

System stiffness matrix:

$$[K] = [\beta^T]\,[k]\,[\beta]$$

$$= [0 \quad 1 \quad 1 \quad 0]\,EI\begin{bmatrix} 2 & 1 & 0 & 0 \\ 1 & 2 & 0 & 0 \\ 0 & 0 & 1 & 0.5 \\ 0 & 0 & 0.5 & 1 \end{bmatrix}\begin{bmatrix} 0 \\ 1 \\ 1 \\ 0 \end{bmatrix}$$

$$[K] = [0 \quad 1 \quad 1 \quad 0]\,EI\begin{bmatrix} 1 \\ 2 \\ 1 \\ 0.5 \end{bmatrix} = EI\,[3] = [3EI]$$

$$[K^{-1}] = \left[\frac{1}{3EI}\right]$$

System displacement vector is

$$\{u\} = [K^{-1}]\,\{F\}$$

$$\{u\} = \frac{1}{3EI} \times [+16.67] = \frac{5.55}{EI}$$

Element forces

$$\{P'\} = [k]\,[\beta]\,[u]$$

$$\{P'\} = EI\begin{bmatrix} 2 & 1 & 0 & 0 \\ 1 & 2 & 0 & 0 \\ 0 & 0 & 1 & 0.5 \\ 0 & 0 & 0.5 & 1 \end{bmatrix}\begin{bmatrix} 0 \\ 1 \\ 1 \\ 0 \end{bmatrix}\frac{1}{EI}[5.55]$$

$$\{P'\} = \begin{bmatrix} 1 \\ 2 \\ 1 \\ 0.5 \end{bmatrix}[5.55] = \begin{bmatrix} 5.55 \\ 11.11 \\ 5.55 \\ 2.79 \end{bmatrix}$$

Final forces (P^f)

$$\{P^f\} = \{P'\} + \{P^0\} = \begin{bmatrix} 5.55 \\ 11.11 \\ 5.55 \\ 2.77 \end{bmatrix} + \begin{bmatrix} -13.33 \\ 13.33 \\ -10 \\ 10 \end{bmatrix} = \begin{bmatrix} -7.77 \\ 24.44 \\ -4.44 \\ 12.77 \end{bmatrix}$$

S.S. Span bending moment :

$$\text{Bending moment in span AB} = \frac{wl^2}{8} = \frac{10 \times 4^2}{8} = 20 \text{ kNm}$$

$$\text{Bending moment in span BC} = -Wl = -10 \times 2 = -20 \text{ kNm}$$

$$\text{Bending moment in span BE} = \frac{Wl}{4} = \frac{20 \times 4}{4} = 20 \text{ kNm}$$

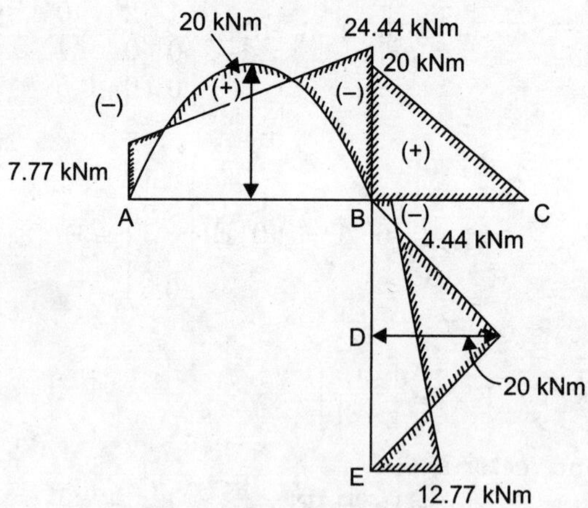

Bending moment diagram

Example 5.27 *Analyse the frame shown in Fig. 5.67 by the matrix stiffness method.*

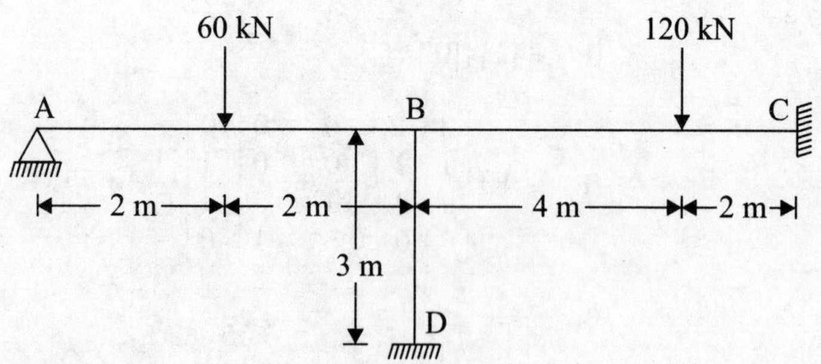

Fig. 5.67

Solution Degree of kinematic indeterminacy = 2 (one rotation each at A and B)

Fixed end moments :

$$M_{FAB} = \frac{-Wl}{8} = \frac{-60 \times 4}{8} = -30 \text{ kNm}$$

$$M_{FBA} = 30 \text{ kNm}$$

$$M_{FBC} = \frac{-Wab^2}{l^2} = \frac{-120 \times 4 \times 2^2}{6^2} = -53.33 \text{ kNm}$$

$$M_{FCB} = \frac{Wa^2b}{l^2} = \frac{120 \times 4^2 \times 2}{6^2} = 106.67 \text{ kNm}$$

Transformation matrix [β] :

The displacement transformation matrix [β] would be obtained by applying a displacement at B (only) at first and at C only next, and finding the resulting displacements at element co-ordinates.

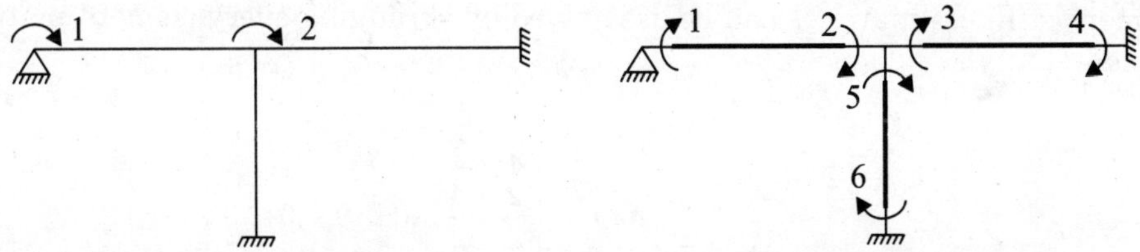

System co-ordinates Element co-ordinates

Applying $\theta_A = 1$, the first column of β is obtained as $[1 \;\; 0 \;\; 0 \;\; 0 \;\; 0 \;\; 0]^T$

Applying $\theta_B = 1$,

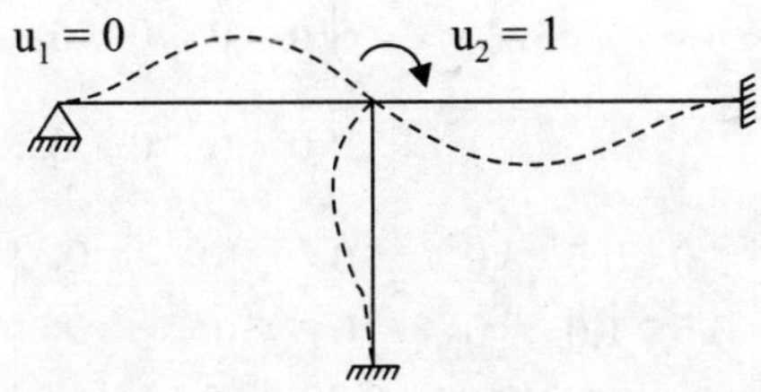

the second column of [β] is obtained as $[0 \;\; 1 \;\; 1 \;\; 0 \;\; 1 \;\; 0]^T$

Thus,
$$[\beta] = \begin{bmatrix} 1 & 0 \\ 0 & 1 \\ 0 & 1 \\ 0 & 0 \\ 0 & 1 \\ 0 & 0 \end{bmatrix}$$

Forces in the element co-ordinates (fixed end moments) are given by

$$[P^0] = [-30 \;\; 30 \;\; -53.33 \;\; 106.67 \;\; 0 \;\; 0]^T$$

Forces at the fixed co-ordinates of the system are

$$\{F^0\} = [\beta]^T \{P^0\}$$

$$= \begin{bmatrix} 1 & 0 & 0 & 0 & 0 & 0 \\ 0 & 1 & 1 & 0 & 1 & 0 \end{bmatrix} \begin{bmatrix} -30 \\ 30 \\ -53.33 \\ 106.67 \\ 0 \\ 0 \end{bmatrix}$$

$$\{F^0\} = \begin{bmatrix} -30 \\ -23.33 \end{bmatrix}$$

$$\{F^f\} = \{0\} \ ; \{F^f\} - \{F^0\} = \begin{bmatrix} 30 \\ 23.33 \end{bmatrix}$$

Element stiffness matrices can be assembled by arranging every element matrix diagonally as under:

$$[k] = \begin{bmatrix} k_1 & & [0] \\ & k_2 & \\ [0] & & k_3 \end{bmatrix} = \text{EI} \begin{bmatrix} \frac{4}{4} & \frac{2}{4} & 0 & 0 & 0 & 0 \\ \frac{2}{4} & \frac{4}{4} & 0 & 0 & 0 & 0 \\ 0 & 0 & \frac{4}{6} & \frac{2}{6} & 0 & 0 \\ 0 & 0 & \frac{2}{6} & \frac{4}{6} & 0 & 0 \\ 0 & 0 & 0 & 0 & \frac{4}{3} & \frac{2}{3} \\ 0 & 0 & 0 & 0 & \frac{2}{3} & \frac{4}{3} \end{bmatrix}$$

$$= \begin{bmatrix} 1.0 & 0.5 & 0 & 0 & 0 & 0 \\ 0.5 & 1.0 & 0 & 0 & 0 & 0 \\ 0 & 0 & 0.67 & 0.33 & 0 & 0 \\ 0 & 0 & 0.33 & 0.67 & 0 & 0 \\ 0 & 0 & 0 & 0 & 1.33 & 0.67 \\ 0 & 0 & 0 & 0 & 0.67 & 1.33 \end{bmatrix} \text{EI}$$

System stiffness matrix is given by

$$[K] = [\beta]^T [k] [\beta]$$

$$= \begin{bmatrix} 1 & 0 & 0 & 0 & 0 & 0 \\ 0 & 1 & 1 & 0 & 1 & 0 \end{bmatrix} \text{EI} \begin{bmatrix} 1.0 & 0.5 & 0 & 0 & 0 & 0 \\ 0.5 & 1.0 & 0 & 0 & 0 & 0 \\ 0 & 0 & 0.67 & 0.33 & 0 & 0 \\ 0 & 0 & 0.33 & 0.67 & 0 & 0 \\ 0 & 0 & 0 & 0 & 1.33 & 0.67 \\ 0 & 0 & 0 & 0 & 0.67 & 1.33 \end{bmatrix} \begin{bmatrix} 1 & 0 \\ 0 & 1 \\ 0 & 1 \\ 0 & 0 \\ 0 & 1 \\ 0 & 0 \end{bmatrix}$$

$$[K] = EI\begin{bmatrix}1.0 & 0.5\\ 0.5 & 3.0\end{bmatrix}$$

$$[K]^{-1} = \frac{1}{EI}\frac{1}{2.75}\begin{bmatrix}3.0 & -0.5\\ -0.5 & 1.0\end{bmatrix}$$

System displacement vector is

$$\{u\} = [K]^{-1}\,\{\{F^f\} - \{F^0\}\} = \frac{1}{2.75}\times\frac{1}{EI}\begin{bmatrix}3 & -0.5\\ -0.5 & 1.0\end{bmatrix}\begin{Bmatrix}30\\ 23.3\end{Bmatrix}$$

$$\{u\} = \frac{1}{EI}\begin{Bmatrix}28.49\\ 3.02\end{Bmatrix}$$

The element displacements $\{\delta\}$ are given by

$$\{\delta\} = [\beta]\,\{u\} = \begin{bmatrix}1 & 0\\ 0 & 1\\ 0 & 1\\ 0 & 0\\ 0 & 1\\ 0 & 0\end{bmatrix}\frac{1}{EI}\begin{Bmatrix}28.49\\ 3.02\end{Bmatrix} = \frac{1}{EI}\begin{Bmatrix}28.49\\ 3.02\\ 3.02\\ 0\\ 3.02\\ 0\end{Bmatrix}$$

The element forces corresponding to these displacements are

$$\{P'\} = [k]\,\{\delta\} = EI\begin{bmatrix}1.0 & 0.5 & 0 & 0 & 0 & 0\\ 0.5 & 1.0 & 0 & 0 & 0 & 0\\ 0 & 0 & 0.67 & 0.33 & 0 & 0\\ 0 & 0 & 0.33 & 0.67 & 0 & 0\\ 0 & 0 & 0 & 0 & 1.33 & 0.67\\ 0 & 0 & 0 & 0 & 0.67 & 1.33\end{bmatrix}\begin{bmatrix}28.49\\ 3.02\\ 3.02\\ 0\\ 3.02\\ 0\end{bmatrix}\frac{1}{EI}$$

$$\{P'\} = [30.0 \quad 17.27 \quad 2.02 \quad 1.00 \quad 4.02 \quad 2.02]^T$$

$$\{P^0\} = [-30.0 \quad 30.0 \quad -53.33 \quad 106.67 \quad 0 \quad 0]^T$$

P^f, the find forces are = $\{P^0\} + \{P'\}$

$$\{P^f\} = [0 \quad 47.27 \quad -51.31 \quad 107.67 \quad 4.02 \quad 2.02]^T$$

The span moments in AB and BC are as under

$$\text{Moment in AB} = \frac{Wl}{4} = \frac{60\times 4}{4} = 60 \text{ kNm}$$

$$\text{Moment in BC} = \frac{Wab}{l} = \frac{120\times 4\times 2}{6} = 160 \text{ kNm}$$

The bending moments are plotted on the following figure.

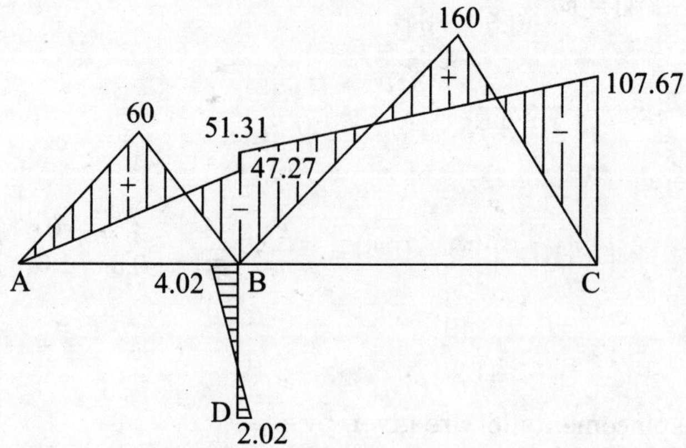

Bending moment diagram

Example 5.28 *Analyse the continuous beam ABC supported on an elastic column BD and loaded as shown in Fig. 5.68 by the matrix stiffness method. Sketch the bending moment diagram.*

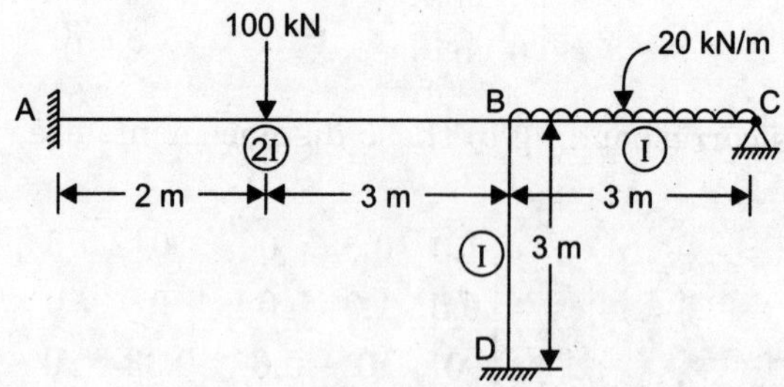

Fig. 5.68

Solution Degree of kinematic indeterminacy = 2 (rotation at B and C)

Fixed end moments :

$$M_{FAB} = -\frac{Wab^2}{l^2} = -\frac{100 \times 2 \times 3^2}{5^2} = -72 \text{ kNm}$$

$$M_{FBA} = \frac{Wa^2b}{l^2} = \frac{100 \times 2^2 \times 3}{5^2} = 48 \text{ kNm}$$

$$M_{FBC} = -\frac{wl^2}{12} = -\frac{20 \times 3^2}{12} = -15 \text{ kNm}$$

$$M_{FCB} = \frac{wl^2}{12} = \frac{20 \times 3^2}{12} = 15 \text{ kNm}$$

Transformation matrix [β] :

The displacement transformation matrix [β] would be obtained by applying a displacement at B and C respectively one at a time and finding the resulting displacements at element co-ordinates.

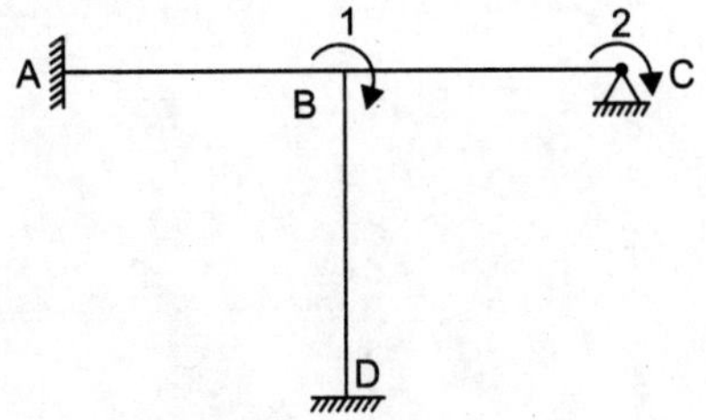

System co-ordinates

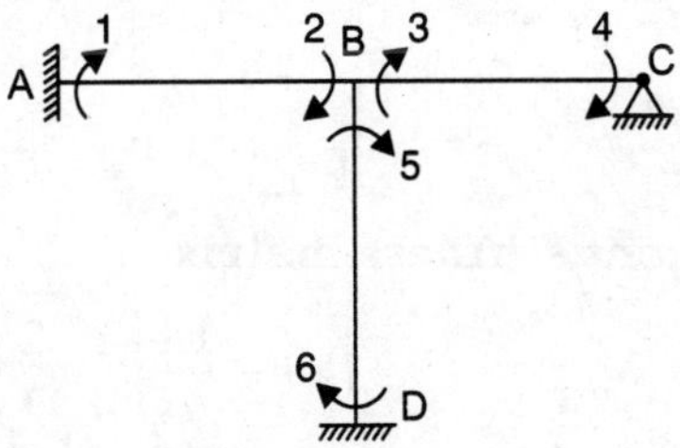

Element co-ordinates

$$\text{Transformation matrix } [\beta] = \begin{bmatrix} \Delta_1 = 1 & \Delta_2 = 1 \\ 0 & 0 \\ 1 & 0 \\ 1 & 0 \\ 0 & 1 \\ 1 & 0 \\ 0 & 0 \end{bmatrix}$$

Fixed end moments :

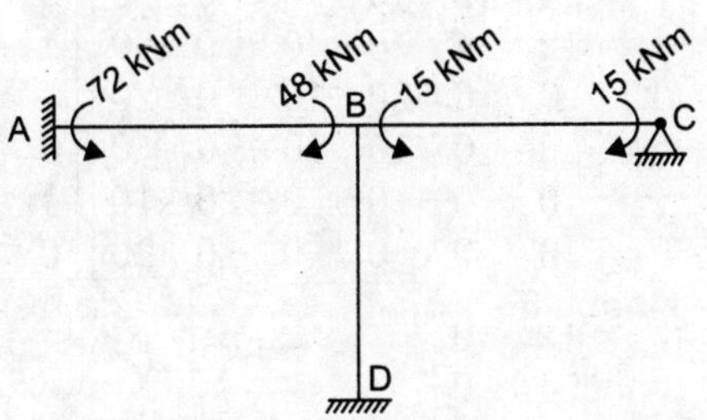

$$\{P^0\} = \begin{bmatrix} -72 \\ 48 \\ -15 \\ 15 \\ 0 \\ 0 \end{bmatrix}$$

Fixed co-ordinate forces $\{F^0\} = \{\beta\}^T\{P^0\}$

$$= \begin{bmatrix} 0 & 1 & 1 & 0 & 1 & 0 \\ 0 & 0 & 0 & 1 & 0 & 0 \end{bmatrix} \begin{Bmatrix} -72 \\ 48 \\ -15 \\ 15 \\ 0 \\ 0 \end{Bmatrix} = \begin{Bmatrix} 33 \\ 15 \end{Bmatrix}$$

Forces applied at system co-ordinate $\{F^f\} = \begin{Bmatrix} 0 \\ 0 \end{Bmatrix}$

$$\{F\} = \{F^f\} - \{F^0\} = \begin{Bmatrix} 0 \\ 0 \end{Bmatrix} - \begin{Bmatrix} 33 \\ 15 \end{Bmatrix}$$

Forces at co-ordinates

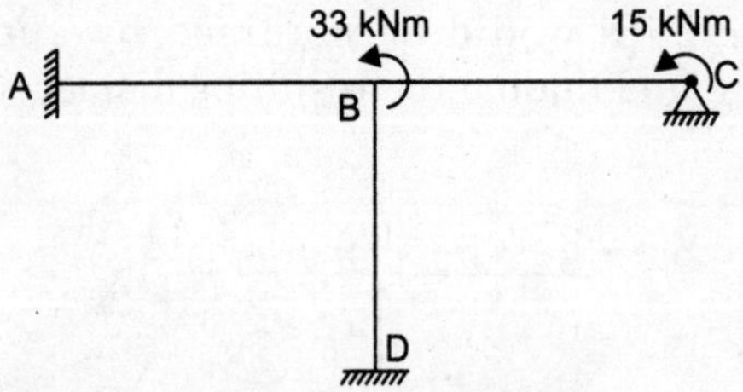

$$F = \begin{bmatrix} -33 \\ -15 \end{bmatrix}$$

Element stiffness matrix :

$$[k_i] = \frac{EI}{l}\begin{bmatrix} 4 & 2 \\ 2 & 4 \end{bmatrix}$$

$$[k_1] = \frac{2EI}{5}\begin{bmatrix} 4 & 2 \\ 2 & 4 \end{bmatrix};\ [k_2] = \frac{EI}{3}\begin{bmatrix} 4 & 2 \\ 2 & 4 \end{bmatrix};\ [k_3] = \frac{EI}{3}\begin{bmatrix} 4 & 2 \\ 2 & 4 \end{bmatrix}$$

$$[k] = \frac{EI}{15}\begin{bmatrix} 24 & 12 & 0 & 0 & 0 & 0 \\ 12 & 24 & 0 & 0 & 0 & 0 \\ 0 & 0 & 20 & 10 & 0 & 0 \\ 0 & 0 & 10 & 20 & 0 & 0 \\ 0 & 0 & 0 & 0 & 20 & 10 \\ 0 & 0 & 0 & 0 & 10 & 20 \end{bmatrix}$$

System stiffness matrix:

$$[K] = [\beta]\,[k]\,[\beta]$$

$$[K] = \begin{bmatrix} 0 & 1 & 1 & 0 & 1 & 0 \\ 0 & 0 & 0 & 1 & 0 & 0 \end{bmatrix}\frac{EI}{15}\begin{bmatrix} 24 & 12 & 0 & 0 & 0 & 0 \\ 12 & 24 & 0 & 0 & 0 & 0 \\ 0 & 0 & 20 & 10 & 0 & 0 \\ 0 & 0 & 10 & 20 & 0 & 0 \\ 0 & 0 & 0 & 0 & 20 & 10 \\ 0 & 0 & 0 & 0 & 10 & 20 \end{bmatrix}\begin{bmatrix} 0 & 0 \\ 1 & 0 \\ 1 & 0 \\ 0 & 1 \\ 1 & 0 \\ 0 & 0 \end{bmatrix}$$

$$[K] = \begin{bmatrix} 0 & 1 & 1 & 0 & 1 & 0 \\ 0 & 0 & 0 & 1 & 0 & 0 \end{bmatrix}\frac{EI}{15}\begin{bmatrix} 12 & 0 \\ 24 & 0 \\ 20 & 10 \\ 10 & 20 \\ 20 & 0 \\ 10 & 0 \end{bmatrix} = \frac{EI}{15}\begin{bmatrix} 64 & 10 \\ 10 & 20 \end{bmatrix}$$

$$[K]^{-1} = \frac{15}{EI}\begin{bmatrix} \frac{1}{59} & -\frac{1}{118} \\ -\frac{1}{118} & \frac{16}{295} \end{bmatrix}$$

System displacements

$$\{u\} = [K]^{-1}\{F\} = \frac{15}{EI}\begin{bmatrix} \frac{1}{59} & -\frac{1}{118} \\ -\frac{1}{118} & \frac{16}{295} \end{bmatrix}\begin{bmatrix} -33 \\ -15 \end{bmatrix} = \frac{15}{EI}\begin{bmatrix} -\frac{51}{118} \\ -\frac{63}{118} \end{bmatrix}$$

Element forces

$$\{P'\} = [k]\,[\beta]\,\{u\}$$

$$= \frac{EI}{15}\begin{bmatrix} 24 & 12 & 0 & 0 & 0 & 0 \\ 12 & 24 & 0 & 0 & 0 & 0 \\ 0 & 0 & 20 & 10 & 0 & 0 \\ 0 & 0 & 10 & 20 & 0 & 0 \\ 0 & 0 & 0 & 0 & 20 & 10 \\ 0 & 0 & 0 & 0 & 10 & 20 \end{bmatrix}\begin{bmatrix} 0 & 0 \\ 1 & 0 \\ 1 & 0 \\ 0 & 1 \\ 1 & 0 \\ 0 & 0 \end{bmatrix}\frac{15}{EI}\begin{bmatrix} -\dfrac{51}{118} \\ -\dfrac{63}{118} \end{bmatrix}$$

$$\{P'\} = \begin{bmatrix} -5.186 \\ -10.37 \\ -13.986 \\ -15 \\ -8.644 \\ -4.32 \end{bmatrix}$$

$$\{P^f\} = \{P'\} + \{P^0\}$$

$$= \begin{bmatrix} -5.186 \\ -10.37 \\ -13.986 \\ -15 \\ -8.644 \\ -4.32 \end{bmatrix} + \begin{bmatrix} -72 \\ 48 \\ -15 \\ 15 \\ 0 \\ 0 \end{bmatrix} = \begin{bmatrix} -77.186 \\ 37.627 \\ -28.983 \\ 0 \\ -8.644 \\ -4.322 \end{bmatrix}$$

S.S. Span bending moment :

Bending moment in span AB = $\dfrac{Wab}{l} = \dfrac{100 \times 2 \times 3}{5}$ = 120 kNm

Bending moment in span BC = $\dfrac{wl^2}{8} = \dfrac{20 \times 3^2}{8}$ = 22.5 kNm

Bending moment in span BD = 0 (No external load)

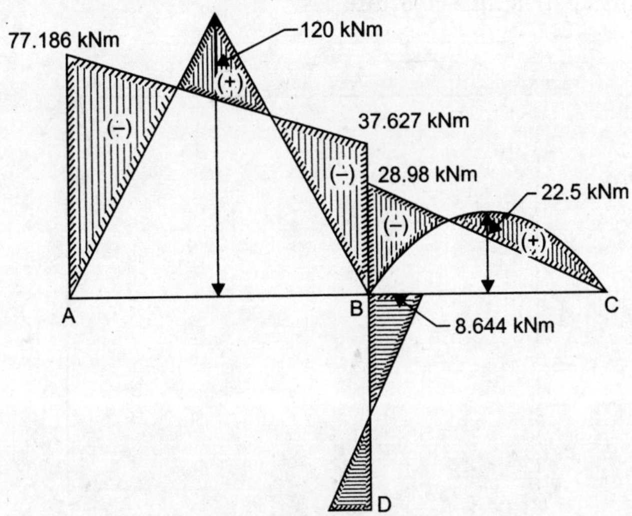

Bending moment diagram

Example 5.29 *Analyse the frame shown in Fig. 5.69 by stiffness method.*

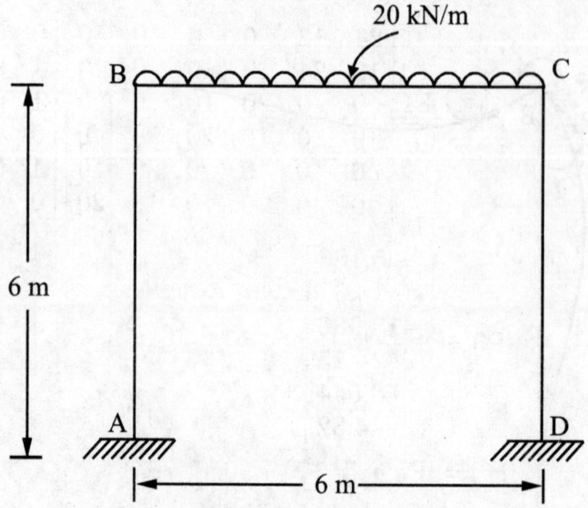

Fig. 5.69

Solution The above frame is symmetrical. ∴ There is no sway in the structure. The structure is kinematically indeterminate to second degree, *i.e.*, θ_B and θ_C are the unknown joint displacements. Hence the size of the stiffness matrix will be 2 × 2.

Fixed end moments :

$$M_{FAB} = M_{FBA} = 0$$

$$M_{FBC} = \frac{-wl^2}{12} = \frac{-20 \times 6^2}{12} = -60 \text{ kNm}$$

$$M_{FCB} = \frac{wl^2}{12} = \frac{20 \times 6^2}{12} = 60 \text{ kNm}$$

$$M_{FCD} = M_{FDC} = 0$$

θ_B and θ_C are the structure co-ordinates.

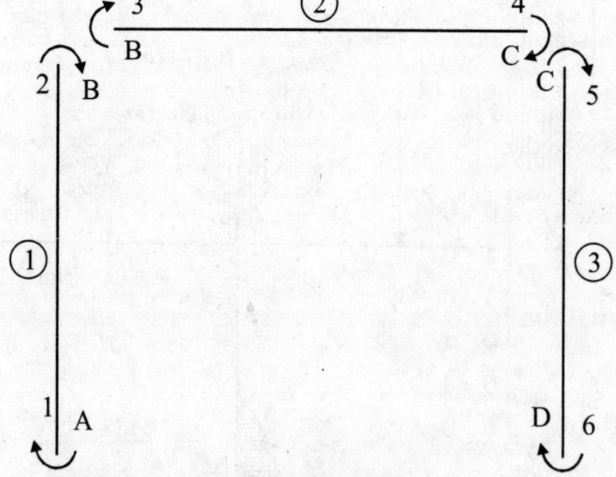

Element co-ordinates

The displacement transformation matrix β will be generated by imposing unit rotation at B and C separately and successively and keeping the far ends fixed.

Applying $\theta_B = 1$

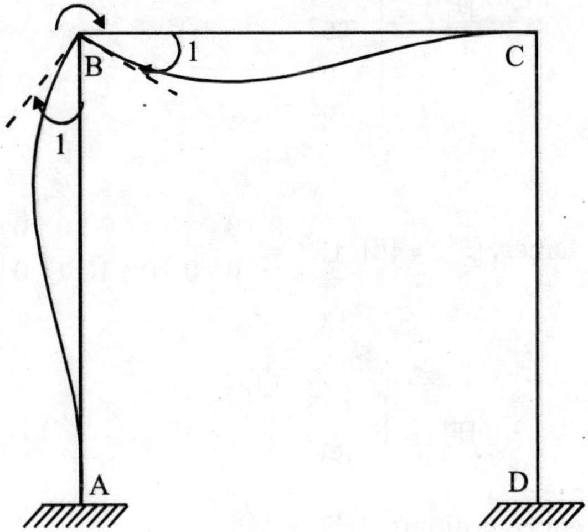

$$\begin{bmatrix} 0 \\ 1 \\ 1 \\ 0 \\ 0 \\ 0 \end{bmatrix}$$

Applying $\theta_C = 1$

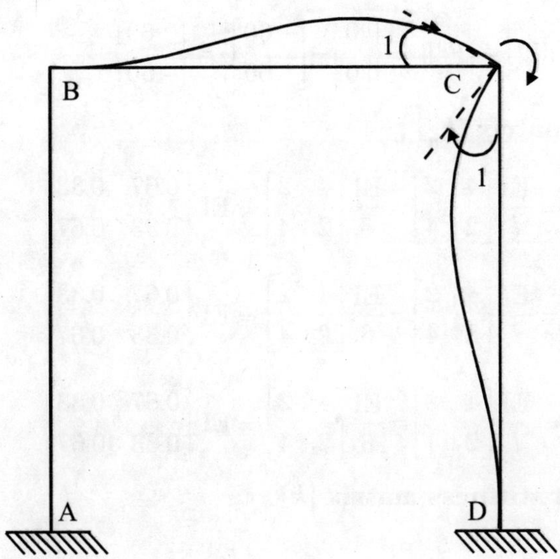

$$\begin{bmatrix} 0 \\ 0 \\ 0 \\ 1 \\ 1 \\ 0 \end{bmatrix}$$

Hence [β] matrix will be $\begin{bmatrix} 0 & 0 \\ 1 & 0 \\ 1 & 0 \\ 0 & 1 \\ 0 & 1 \\ 0 & 0 \end{bmatrix}$

Forces in element co-ordinates *i.e.*, fixed end moments

$$[P^0] = \begin{bmatrix} 0 \\ 0 \\ -60 \\ 60 \\ 0 \\ 0 \end{bmatrix}$$

Fixed end co-ordinate forces, $[F^0] = [\beta]^T [P^0] = \begin{bmatrix} 0 & 1 & 1 & 0 & 0 & 0 \\ 0 & 0 & 0 & 1 & 1 & 0 \end{bmatrix} \begin{bmatrix} 0 \\ 0 \\ -60 \\ 60 \\ 0 \\ 0 \end{bmatrix}$

$$[F^0] = \begin{bmatrix} -60 \\ 60 \end{bmatrix}$$

Forces applied at system co-ordinates, $[F^f] = \begin{bmatrix} 0 \\ 0 \end{bmatrix}$

($\because$ There is no external force at the system co-ordinates)

$$\{[F^f] - [F^0]\} = \left\{ \begin{bmatrix} 0 \\ 0 \end{bmatrix} - \begin{bmatrix} -60 \\ 60 \end{bmatrix} \right\} = \begin{Bmatrix} 60 \\ -60 \end{Bmatrix}$$

Element stiffness matrix $[k_n]$:

$$[k_1] = \frac{EI}{l}\begin{bmatrix} 4 & 2 \\ 2 & 4 \end{bmatrix} = \frac{EI}{6}\begin{bmatrix} 4 & 2 \\ 2 & 4 \end{bmatrix} = EI\begin{bmatrix} 0.67 & 0.33 \\ 0.33 & 0.67 \end{bmatrix}$$

$$[k_2] = \frac{EI}{l}\begin{bmatrix} 4 & 2 \\ 2 & 4 \end{bmatrix} = \frac{EI}{6}\begin{bmatrix} 4 & 2 \\ 2 & 4 \end{bmatrix} = EI\begin{bmatrix} 0.67 & 0.33 \\ 0.33 & 0.67 \end{bmatrix}$$

$$[k_3] = \frac{EI}{l}\begin{bmatrix} 4 & 2 \\ 2 & 4 \end{bmatrix} = \frac{EI}{6}\begin{bmatrix} 4 & 2 \\ 2 & 4 \end{bmatrix} = EI\begin{bmatrix} 0.67 & 0.33 \\ 0.33 & 0.67 \end{bmatrix}$$

Assembled element stiffness matrix $[k]$:

$$[k] = \begin{bmatrix} k_1 & 0 & 0 \\ 0 & k_2 & 0 \\ 0 & 0 & k_3 \end{bmatrix} = EI\begin{bmatrix} 0.67 & 0.33 & 0 & 0 & 0 & 0 \\ 0.33 & 0.67 & 0 & 0 & 0 & 0 \\ 0 & 0 & 0.67 & 0.33 & 0 & 0 \\ 0 & 0 & 0.33 & 0.67 & 0 & 0 \\ 0 & 0 & 0 & 0 & 0.67 & 0.33 \\ 0 & 0 & 0 & 0 & 0.33 & 0.67 \end{bmatrix}$$

System stiffness matrix [K] :

$$[K] = [\beta]^T [k] [\beta]$$

$$= \begin{bmatrix} 0 & 1 & 1 & 0 & 0 & 0 \\ 0 & 0 & 0 & 1 & 1 & 0 \end{bmatrix} EI \begin{bmatrix} 0.67 & 0.33 & 0 & 0 & 0 & 0 \\ 0.33 & 0.67 & 0 & 0 & 0 & 0 \\ 0 & 0 & 0.67 & 0.33 & 0 & 0 \\ 0 & 0 & 0.33 & 0.67 & 0 & 0 \\ 0 & 0 & 0 & 0 & 0.67 & 0.33 \\ 0 & 0 & 0 & 0 & 0.33 & 0.67 \end{bmatrix} \times \begin{bmatrix} 0 & 0 \\ 1 & 0 \\ 1 & 0 \\ 0 & 1 \\ 0 & 1 \\ 0 & 0 \end{bmatrix}$$

$$= EI \begin{bmatrix} 0.33 & 0.67 & 0.67 & 0.33 & 0 & 0 \\ 0 & 0 & 0.33 & 0.67 & 0.67 & 0.33 \end{bmatrix} \begin{bmatrix} 0 & 0 \\ 1 & 0 \\ 1 & 0 \\ 0 & 1 \\ 0 & 1 \\ 0 & 0 \end{bmatrix}$$

$$[K] = EI \begin{bmatrix} 1.34 & 0.33 \\ 0.33 & 1.34 \end{bmatrix}$$

System displacements {*u*} :

$$\{u\} = [K]^{-1} \{[F^f] - [F^0]\} = \frac{1}{EI\,[1.34 \times 1.34 - 0.33 \times 0.33]} \begin{bmatrix} 1.34 & -0.33 \\ -0.33 & 1.34 \end{bmatrix} \begin{bmatrix} 60 \\ -60 \end{bmatrix}$$

$$= \frac{1}{EI\,[1.6867]} \begin{bmatrix} 100.2 \\ -100.2 \end{bmatrix}$$

$$\{u\} = \frac{1}{EI} \begin{bmatrix} 59.406 \\ -59.406 \end{bmatrix}$$

Element displacements [δ] :

$$\{\delta\} = [\beta]\,[u]$$

$$= \begin{bmatrix} 0 & 0 \\ 1 & 0 \\ 1 & 0 \\ 0 & 1 \\ 0 & 1 \\ 0 & 0 \end{bmatrix} \times \frac{1}{EI} \begin{bmatrix} 59.406 \\ -59.406 \end{bmatrix} = \frac{1}{EI} \begin{bmatrix} 0 \\ 59.406 \\ 59.406 \\ -59.406 \\ -59.406 \\ 0 \end{bmatrix}$$

Now,

$$[P'] = [k]\,[\delta]$$

$$= EI \begin{bmatrix} 0.67 & 0.33 & 0 & 0 & 0 & 0 \\ 0.33 & 0.67 & 0 & 0 & 0 & 0 \\ 0 & 0 & 0.67 & 0.33 & 0 & 0 \\ 0 & 0 & 0.33 & 0.67 & 0 & 0 \\ 0 & 0 & 0 & 0 & 0.67 & 0.33 \\ 0 & 0 & 0 & 0 & 0.33 & 0.67 \end{bmatrix} \times \frac{1}{EI} \begin{bmatrix} 0 \\ 59.406 \\ 59.406 \\ -59.406 \\ -59.406 \\ 0 \end{bmatrix}$$

$$[P'] = \begin{bmatrix} 19.604 \\ 39.802 \\ 20.198 \\ -20.198 \\ -39.802 \\ -19.604 \end{bmatrix}$$

$$\text{Final forces, } [P^f] = [P^0] + [P'] = \begin{bmatrix} 0 \\ 0 \\ -60 \\ 60 \\ 0 \\ 0 \end{bmatrix} + \begin{bmatrix} 19.604 \\ 39.802 \\ 20.198 \\ -20.198 \\ -39.802 \\ -19.604 \end{bmatrix} = \begin{bmatrix} 19.604 \\ 39.802 \\ -39.802 \\ 39.802 \\ -39.802 \\ -19.604 \end{bmatrix}$$

Simply supported bending moments :

$$M_{AB} = M_{CD} = 0$$

$$M_{BC} = \frac{wl^2}{8} = \frac{20 \times 6^2}{8} = 90 \text{ kNm}$$

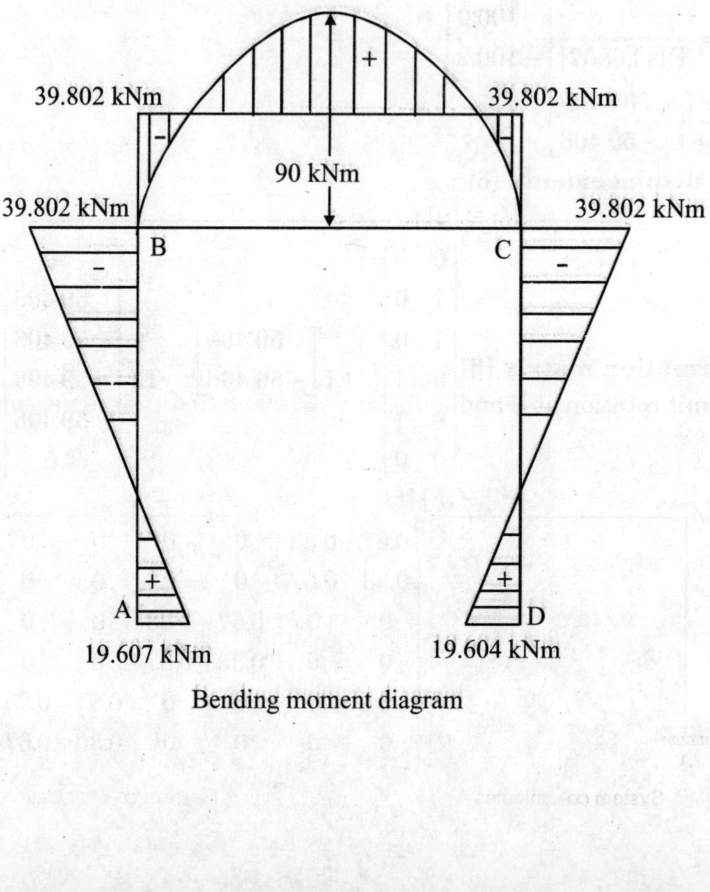

Bending moment diagram

Example 5.30 *Analyse the following structure by matrix stiffness method and sketch the Bending moment diagram.*

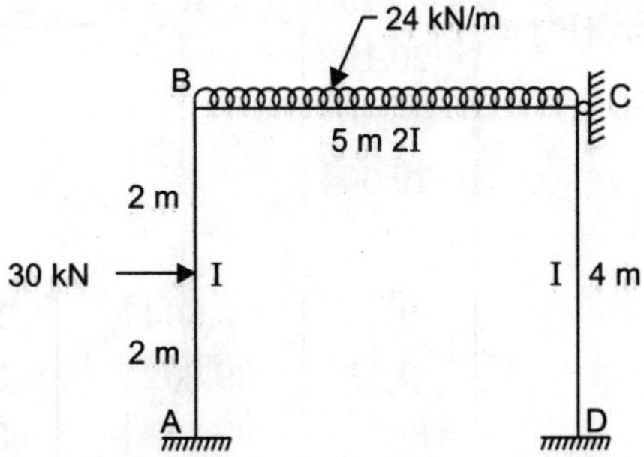

Solution This structure is kinematically indeterminate to the second degree, *i.e.*, rotation at B and C (sway arrested by the support at C).

Fixed end moments :

$$M_{FAB} = -\frac{Wl}{8} = -\frac{30 \times 4}{8} = -15 \text{ kNm}$$

$$M_{FAB} = \frac{Wl}{8} = \frac{30 \times 4}{8} = 15 \text{ kNm}$$

$$M_{FBC} = -\frac{wl^2}{12} = -\frac{24 \times 5^2}{12} = -50 \text{ kNm}$$

$$M_{FCB} = \frac{wl^2}{12} = \frac{24 \times 5^2}{12} = 50 \text{ kNm}$$

$$\{P^0\} = \begin{bmatrix} -15 \\ 15 \\ -50 \\ 50 \\ 0 \\ 0 \end{bmatrix}$$

Transformation matrix [β] :

Apply a unit rotation at B and C one at a time and find the displacements in the element co-ordinates.

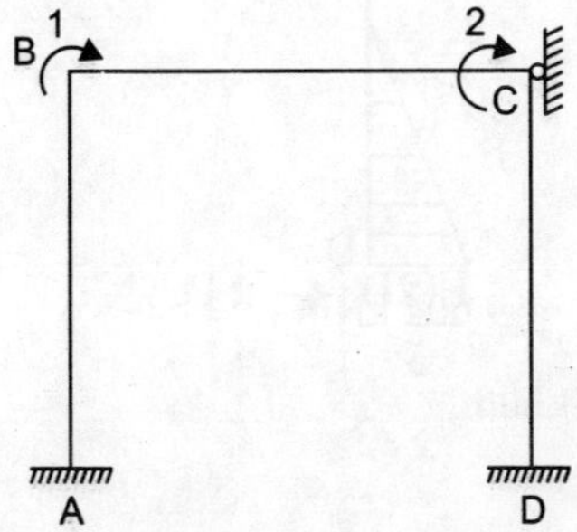

System co-ordinates

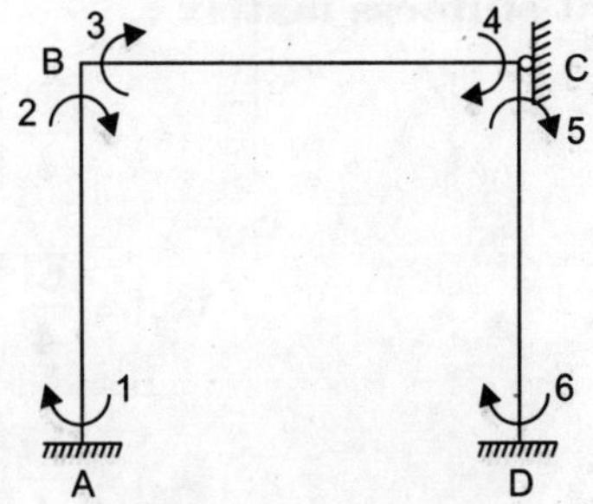

Element co-ordinates

Apply unit rotation at B

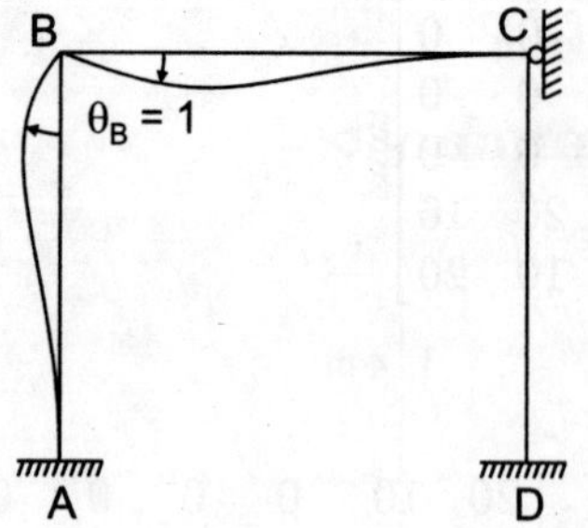

Apply unit rotation at C

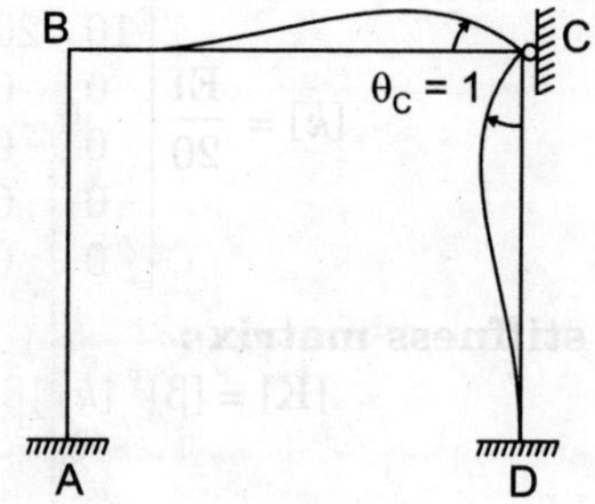

$$\text{Transformation matrix } [\beta] = \begin{array}{c} \begin{matrix} \theta_B = 1 & \theta_C = 1 \end{matrix} \\ \begin{bmatrix} 0 & 0 \\ 1 & 0 \\ 1 & 0 \\ 0 & 1 \\ 0 & 1 \\ 0 & 0 \end{bmatrix} \end{array}$$

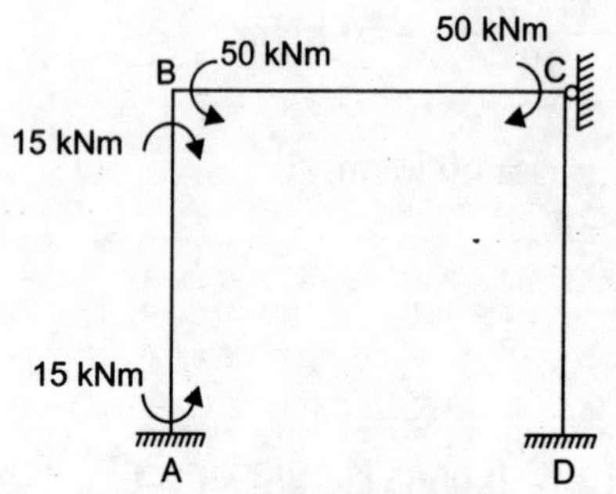

Fixed end moments

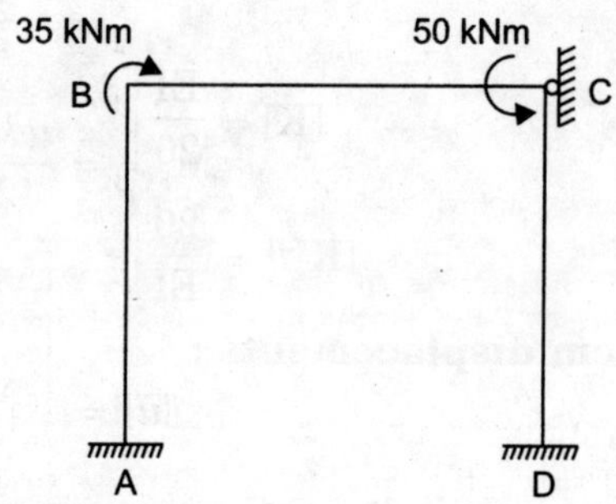

Moment at system co-ordinates

By usual calculations, $\{F\} = \begin{bmatrix} 35 \\ -50 \end{bmatrix}$

Element stiffness matrix :

$$[k] = \frac{EI}{l}\begin{bmatrix} 4 & 2 \\ 2 & 4 \end{bmatrix}$$

$$[k_1] = \frac{EI}{4}\begin{bmatrix} 4 & 2 \\ 2 & 4 \end{bmatrix} ; [k_2] = \frac{E(2I)}{5}\begin{bmatrix} 4 & 2 \\ 2 & 4 \end{bmatrix}$$

$$[k_3] = \frac{EI}{4}\begin{bmatrix} 4 & 2 \\ 2 & 4 \end{bmatrix}$$

Assembled element stiffness matrix :

$$[k] = \frac{EI}{20}\begin{bmatrix} 20 & 10 & 0 & 0 & 0 & 0 \\ 10 & 20 & 0 & 0 & 0 & 0 \\ 0 & 0 & 32 & 16 & 0 & 0 \\ 0 & 0 & 16 & 32 & 0 & 0 \\ 0 & 0 & 0 & 0 & 20 & 10 \\ 0 & 0 & 0 & 0 & 10 & 20 \end{bmatrix}$$

System stiffness matrix :

$$[K] = [\beta]^T [k] [\beta]$$

$$= \begin{bmatrix} 0 & 1 & 1 & 0 & 0 & 0 \\ 0 & 0 & 0 & 1 & 1 & 0 \end{bmatrix} \frac{EI}{20}\begin{bmatrix} 20 & 10 & 0 & 0 & 0 & 0 \\ 10 & 20 & 0 & 0 & 0 & 0 \\ 0 & 0 & 32 & 16 & 0 & 0 \\ 0 & 0 & 16 & 32 & 0 & 0 \\ 0 & 0 & 0 & 0 & 20 & 10 \\ 0 & 0 & 0 & 0 & 10 & 20 \end{bmatrix}\begin{bmatrix} 0 & 0 \\ 1 & 0 \\ 1 & 0 \\ 0 & 1 \\ 0 & 1 \\ 0 & 0 \end{bmatrix}$$

$$= \begin{bmatrix} 0 & 1 & 1 & 0 & 0 & 0 \\ 0 & 0 & 0 & 1 & 1 & 0 \end{bmatrix} \frac{EI}{20}\begin{bmatrix} 10 & 0 \\ 20 & 0 \\ 32 & 16 \\ 16 & 32 \\ 0 & 20 \\ 0 & 10 \end{bmatrix}$$

$$[K] = \frac{EI}{20}\begin{bmatrix} 52 & 16 \\ 16 & 52 \end{bmatrix}$$

$$[K]^{-1} = \frac{20}{EI}\begin{bmatrix} 0.0212 & -0.0065 \\ -0.0065 & 0.0212 \end{bmatrix}$$

System displacements :

$$\{u\} = [K]^{-1} \{F\}$$

$$= \frac{20}{EI}\begin{bmatrix} 0.0212 & -0.0065 \\ -0.0065 & 0.0212 \end{bmatrix}\begin{Bmatrix} 35 \\ -50 \end{Bmatrix}$$

$$= \frac{20}{EI}\begin{Bmatrix} 1.067 \\ -1.2875 \end{Bmatrix}$$

Element forces :

$$\{P'\} = [k] [\beta] \{u\}$$

$$= \frac{EI}{20}\begin{bmatrix} 20 & 10 & 0 & 0 & 0 & 0 \\ 10 & 20 & 0 & 0 & 0 & 0 \\ 0 & 0 & 32 & 16 & 0 & 0 \\ 0 & 0 & 16 & 32 & 0 & 0 \\ 0 & 0 & 0 & 0 & 20 & 10 \\ 0 & 0 & 0 & 0 & 10 & 20 \end{bmatrix}\begin{bmatrix} 0 & 0 \\ 1 & 0 \\ 1 & 0 \\ 0 & 1 \\ 0 & 1 \\ 0 & 0 \end{bmatrix}\frac{20}{EI}\begin{Bmatrix} 1.067 \\ -1.2875 \end{Bmatrix}$$

$$\{P'\} = \begin{bmatrix} 10 & 0 \\ 20 & 0 \\ 32 & 16 \\ 16 & 32 \\ 0 & 20 \\ 0 & 10 \end{bmatrix} \begin{bmatrix} 1.067 \\ -1.287 \end{bmatrix} = \begin{bmatrix} 10.67 \\ 21.34 \\ +13.55 \\ -24.2 \\ -25.8 \\ -12.9 \end{bmatrix}$$

$$\{P^f\} = \{P'\} + \{P^0\}$$

$$= \begin{bmatrix} 10.67 \\ 21.34 \\ 13.55 \\ -24.2 \\ -25.8 \\ -12.9 \end{bmatrix} + \begin{bmatrix} -15 \\ 15 \\ -50 \\ 50 \\ 0 \\ 0 \end{bmatrix} = \begin{bmatrix} -4.33 \\ 36.45 \\ -36.45 \\ 25.8 \\ -25.8 \\ -12.9 \end{bmatrix}$$

S.S. Span bending moment :

$$\text{Bending moment in span AB} = \frac{Wl}{4} = \frac{30 \times 4}{4}$$

$$= 30 \text{ kNm}$$

$$\text{Bending moment in span BC} = \frac{wl^2}{8} = \frac{24 \times 5^2}{8}$$

$$= 75 \text{ kNm}$$

Bending moment in span CD = 0 (No external load)

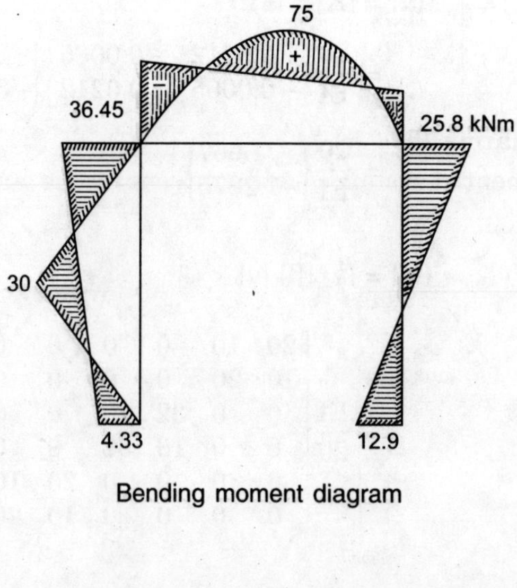

Bending moment diagram

Example 5.31 *Analyse the frame shown in Fig. 5.70 by the matrix stiffness method.*

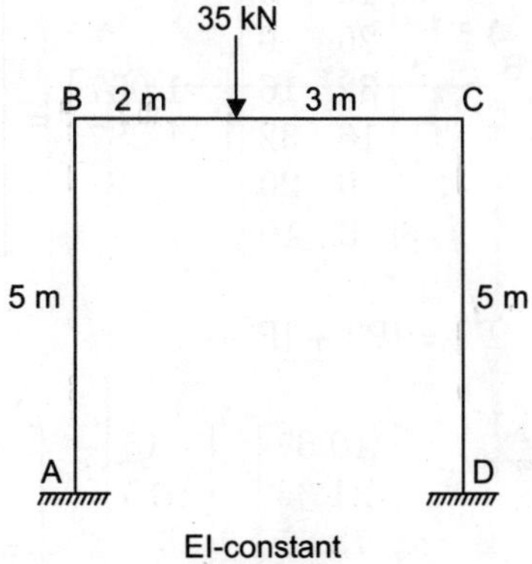

Fig. 5.70

Solution The given frame is kinematically indeterminate to third degree, *i.e.,* rotation at B and C and sway of frame due to eccentric loading on frame.

Fixed end moments :

$M_{FAB} = M_{FBA} = 0$ (No lateral load on AB)

$$M_{FBC} = -\frac{Wab^2}{l^2} = -\frac{35 \times 2 \times 3^2}{5^2} = -25.2 \text{ kNm}$$

$$M_{FCB} = +\frac{Wa^2b}{l^2} = \frac{35 \times 2^2 \times 3}{5^2} = 16.8 \text{ kNm}$$

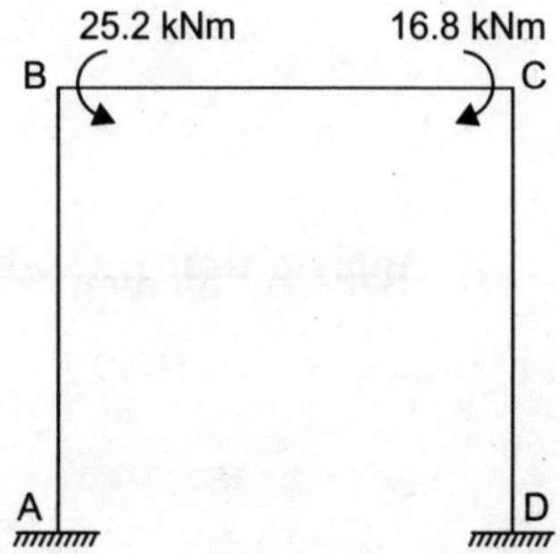

$M_{FCD} = M_{FDC} = 0$ (No lateral load on CD)

$$\text{FEM} = [P^0] = \begin{bmatrix} 0 \\ 0 \\ -25.2 \\ 16.8 \\ 0 \\ 0 \end{bmatrix}$$

Transformation matrix (β) :

Apply unit displacement at system co-ordinates one by one and obtain the displacements at element.

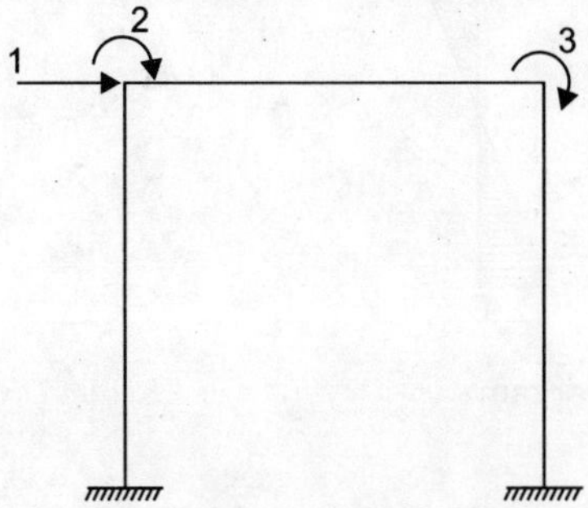

System co-ordinates

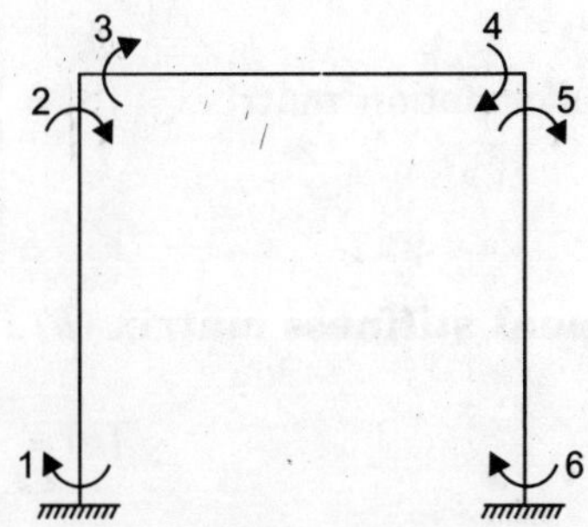

Element co-ordinates

Apply a unit displacement at system co-ordinate 1 ($\Delta = 1$)

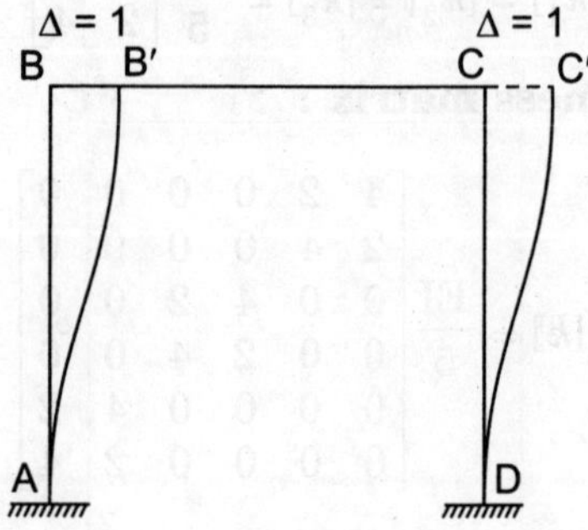

$$\beta_1 = \begin{bmatrix} -\frac{1}{5} \\ -\frac{1}{5} \\ 0 \\ 0 \\ -\frac{1}{5} \\ \frac{1}{5} \end{bmatrix} = \begin{bmatrix} -0.2 \\ -0.2 \\ 0 \\ 0 \\ -0.2 \\ -0.2 \end{bmatrix}$$

Apply a unit rotation at B ($\theta_B = 1$)

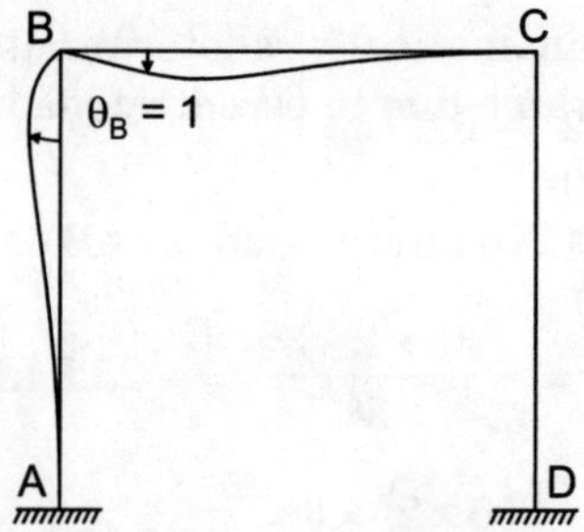

$$\beta_2 = \begin{bmatrix} 0 \\ 1 \\ 1 \\ 0 \\ 0 \\ 0 \end{bmatrix}$$

Apply a unit rotation at C ($\theta_C = 1$)

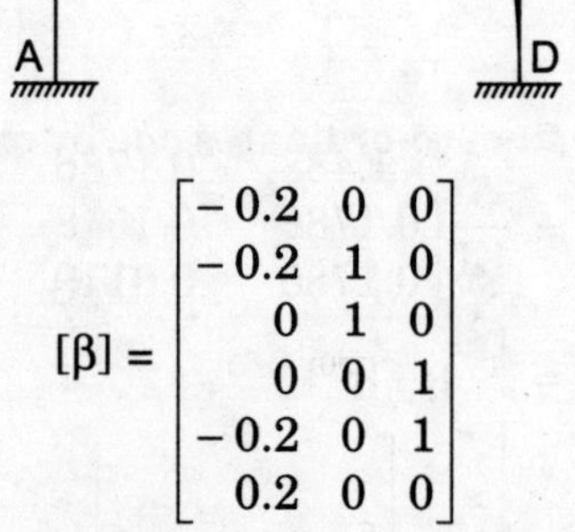

$$\beta_3 = \begin{bmatrix} 0 \\ 0 \\ 0 \\ 1 \\ 1 \\ 0 \end{bmatrix}$$

Transformation matrix, $[\beta] = \begin{bmatrix} -0.2 & 0 & 0 \\ -0.2 & 1 & 0 \\ 0 & 1 & 0 \\ 0 & 0 & 1 \\ -0.2 & 0 & 1 \\ 0.2 & 0 & 0 \end{bmatrix}$

Element stiffness matrix (*k*) :

$$[k] = \frac{EI}{l}\begin{bmatrix} 4 & 2 \\ 2 & 4 \end{bmatrix}$$

$$[k_1] = [k_2] = [k_3] = \frac{EI}{5}\begin{bmatrix} 4 & 2 \\ 2 & 4 \end{bmatrix}$$

Assembled element stiffness matrix :

$$[k] = \frac{EI}{5}\begin{bmatrix} 4 & 2 & 0 & 0 & 0 & 0 \\ 2 & 4 & 0 & 0 & 0 & 0 \\ 0 & 0 & 4 & 2 & 0 & 0 \\ 0 & 0 & 2 & 4 & 0 & 0 \\ 0 & 0 & 0 & 0 & 4 & 2 \\ 0 & 0 & 0 & 0 & 2 & 4 \end{bmatrix}$$

System stiffness matrix (K) :

$$[K] = [\beta]^T [k] [\beta]$$

$$= \begin{bmatrix} -0.2 & -0.2 & 0 & 0 & -0.2 & -0.2 \\ 0 & 1 & 1 & 0 & 0 & 0 \\ 0 & 0 & 0 & 1 & 1 & 0 \end{bmatrix} \frac{EI}{5}\begin{bmatrix} 4 & 2 & 0 & 0 & 0 & 0 \\ 2 & 4 & 0 & 0 & 0 & 0 \\ 0 & 0 & 4 & 2 & 0 & 0 \\ 0 & 0 & 2 & 4 & 0 & 0 \\ 0 & 0 & 0 & 0 & 4 & 2 \\ 0 & 0 & 0 & 0 & 2 & 4 \end{bmatrix}\begin{bmatrix} -0.2 & 0 & 0 \\ -0.2 & 1 & 0 \\ 0 & 1 & 0 \\ 0 & 0 & 1 \\ -0.2 & 0 & 1 \\ -0.2 & 0 & 0 \end{bmatrix}$$

$$[K] = \begin{bmatrix} -0.2 & -0.2 & 0 & 0 & -0.2 & -0.2 \\ 0 & 1 & 1 & 0 & 0 & 0 \\ 0 & 0 & 0 & 1 & 1 & 0 \end{bmatrix} \frac{EI}{5}\begin{bmatrix} -1.2 & 2 & 0 \\ -1.2 & 4 & 0 \\ 0 & 4 & 2 \\ 0 & 2 & 4 \\ -1.2 & 0 & 4 \\ -1.2 & 0 & 2 \end{bmatrix}$$

$$= \frac{EI}{5}\begin{bmatrix} 0.96 & -1.2 & -1.2 \\ -1.2 & 8 & 2 \\ -1.2 & 2 & 8 \end{bmatrix}$$

$$[K^{-1}] = \frac{5}{EI}\begin{bmatrix} 1.488 & 0.1786 & 0.1786 \\ 0.1786 & 0.1548 & -0.0119 \\ 0.1786 & -0.0119 & 0.1548 \end{bmatrix}$$

$$\{F\} = \{F^f\} - \{F^0\}$$

$$= \begin{bmatrix} 0 \\ 0 \\ 0 \end{bmatrix} - \left[\begin{bmatrix} -0.2 & -0.2 & 0 & 0 & -0.2 & -0.2 \\ 0 & 1 & 1 & 0 & 0 & 0 \\ 0 & 0 & 0 & 1 & 1 & 0 \end{bmatrix}\begin{bmatrix} 0 \\ 0 \\ -25.2 \\ +16.8 \\ 0 \\ 0 \end{bmatrix}\right]$$

$$(\because \quad F^0 = \beta^T P^0)$$

$$\{F\} = \begin{bmatrix} 0 \\ 0 \\ 0 \end{bmatrix} - \begin{bmatrix} -0 \\ -25.2 \\ +16.8 \end{bmatrix} = \begin{bmatrix} 0 \\ 25.2 \\ -16.8 \end{bmatrix}$$

System displacement (u)

$$\{u\} = [K]^{-1}\{F\}$$

$$\{u\} = \frac{5}{EI}\begin{bmatrix} 1.488 & 0.1786 & 0.1786 \\ 0.1786 & 0.1548 & -0.0119 \\ 0.1786 & -0.0119 & 0.1548 \end{bmatrix}\begin{bmatrix} 0 \\ 25.2 \\ -16.8 \end{bmatrix}$$

$$\{u\} = \frac{5}{EI}\begin{bmatrix} 1.5 \\ 4.1 \\ -2.9 \end{bmatrix}$$

Element forces (P′)

$$\{P'\} = [k]\,[\beta]\,\{u\}$$

$$= \frac{EI}{5}\begin{bmatrix} 4 & 2 & 0 & 0 & 0 & 0 \\ 2 & 4 & 0 & 0 & 0 & 0 \\ 0 & 0 & 4 & 2 & 0 & 0 \\ 0 & 0 & 2 & 4 & 0 & 0 \\ 0 & 0 & 0 & 0 & 4 & 2 \\ 0 & 0 & 0 & 0 & 2 & 4 \end{bmatrix}\begin{bmatrix} -0.2 & 0 & 0 \\ -0.2 & 1 & 0 \\ 0 & 1 & 0 \\ 0 & 0 & 1 \\ -0.2 & 0 & 1 \\ -0.2 & 0 & 0 \end{bmatrix}\frac{5}{EI}\begin{bmatrix} 1.5 \\ 4.1 \\ -2.9 \end{bmatrix}$$

$$= \begin{bmatrix} -1.2 & 2 & 0 \\ -1.2 & 4 & 0 \\ 0 & 4 & 2 \\ 0 & 2 & 4 \\ -1.2 & 0 & 4 \\ -1.2 & 0 & 2 \end{bmatrix}\begin{bmatrix} 1.5 \\ 4.1 \\ -2.9 \end{bmatrix} = \begin{bmatrix} 6.4 \\ 14.6 \\ 10.6 \\ -3.4 \\ -13.4 \\ -7.6 \end{bmatrix}$$

$$\{P^f\} = \begin{bmatrix} 6.4 \\ 14.6 \\ 10.6 \\ -3.4 \\ -13.4 \\ -7.6 \end{bmatrix} + \begin{bmatrix} 0 \\ 0 \\ -25.2 \\ 16.8 \\ 0 \\ 0 \end{bmatrix} = \begin{bmatrix} 6.4 \\ 14.6 \\ -14.6 \\ 13.4 \\ -13.4 \\ -7.6 \end{bmatrix}$$

$M_{AB} = 6.4$ kNm $\quad$ $M_{BA} = 14.6$ kNm

$M_{BC} = -14.6$ kNm $\quad$ $M_{CB} = 13.4$ kNm

$M_{CD} = -13.4$ kNm $\quad$ $M_{DC} = -7.6$ kNm

Example 5.32 *Analyse the following frame by matrix stiffness method and obtain the support/joint moments.*

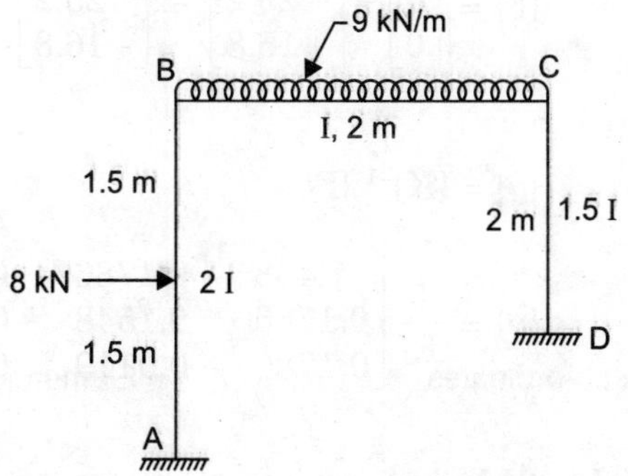

Solution The given frame is kinematically indeterminate to third degree, *i.e.*, rotation at B and C and sway of frame.

Fixed end moments :

$$M_{FAB} = -\frac{Wl}{8} = -\frac{8 \times 3}{8} = -3 \text{ kNm}$$

$$M_{FBA} = \frac{Wl}{8} = \frac{8 \times 3}{8} = 3 \text{ kNm}$$

$$M_{FBC} = -\frac{wl^2}{12} = -\frac{9 \times 2^2}{12} = -3 \text{ kNm}$$

$$M_{FCB} = \frac{wl^2}{12} = \frac{9 \times 2^2}{12} = 3 \text{ kNm}$$

$$M_{FCD} = M_{FDC} = 0 \text{ (No lateral load on CD)}$$

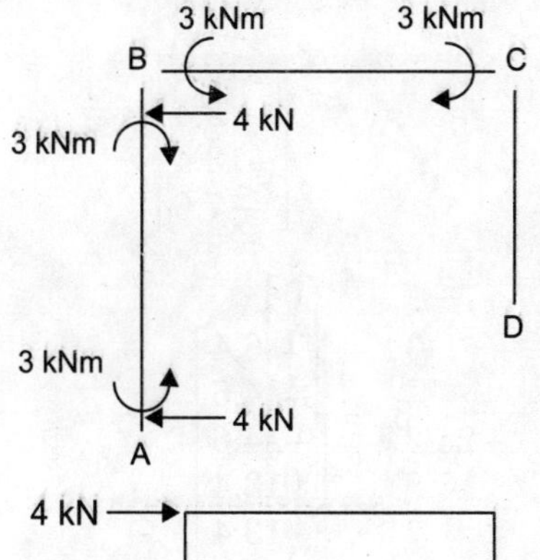

$$\{P^0\} = \begin{bmatrix} -3 \\ 3 \\ -3 \\ 3 \\ 0 \\ 0 \end{bmatrix}$$

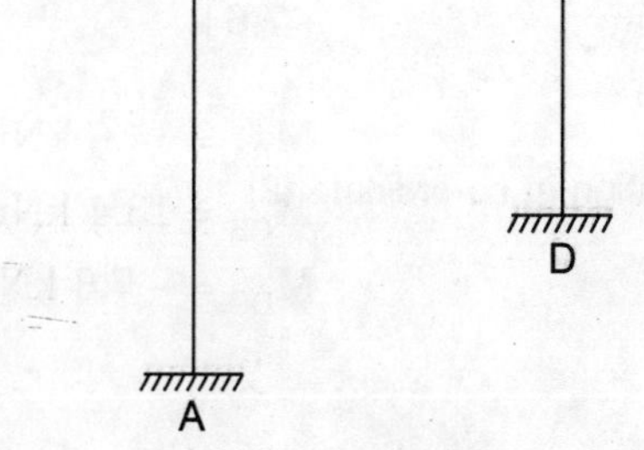

$$\{F^f\} = \begin{bmatrix} 4 \\ 0 \\ 0 \end{bmatrix}$$

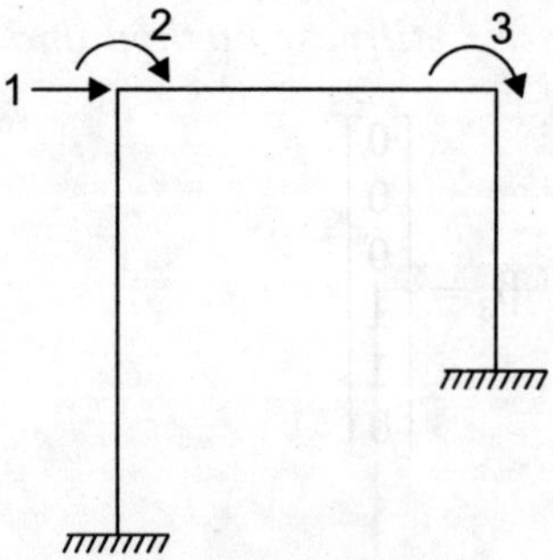

System co-ordinates

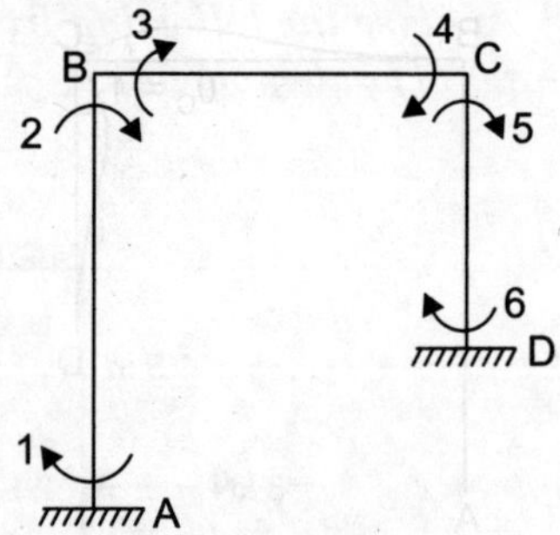

Element co-ordinates

Transformation matrix (β) :

Apply a unit displacement in co-ordinate (1) (Δ = 1) and unit rotation in co-ordinates (2) and (3) and find the element co-ordinate displacements.

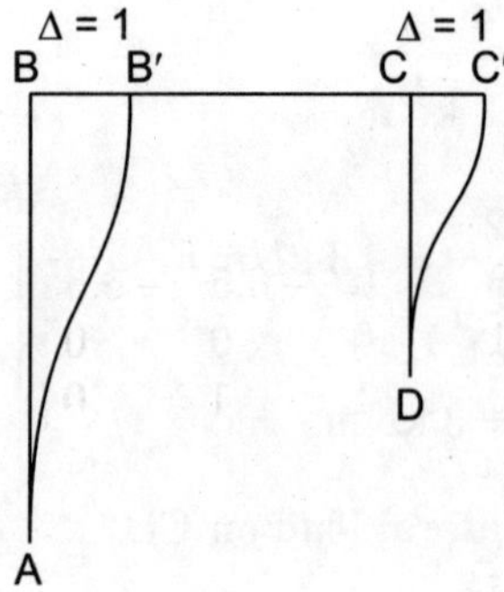

$$\beta_1 = \begin{bmatrix} -\dfrac{1}{l_{AB}} \\ -\dfrac{1}{l_{AB}} \\ 0 \\ 0 \\ -\dfrac{1}{l_{CD}} \\ -\dfrac{1}{l_{CD}} \end{bmatrix} = \begin{bmatrix} -\dfrac{1}{3} \\ -\dfrac{1}{3} \\ 0 \\ 0 \\ -\dfrac{1}{2} \\ -\dfrac{1}{2} \end{bmatrix} = \begin{bmatrix} -0.33 \\ -0.33 \\ 0 \\ 0 \\ -0.5 \\ -0.5 \end{bmatrix}$$

Unit displacement, (Δ = 1)

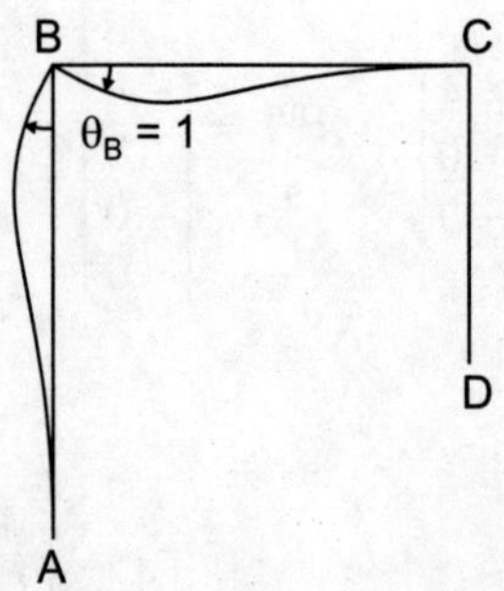

$$\beta_2 = \begin{bmatrix} 0 \\ 1 \\ 1 \\ 0 \\ 0 \\ 0 \end{bmatrix}$$

Unit rotation at co-ordinate (2)

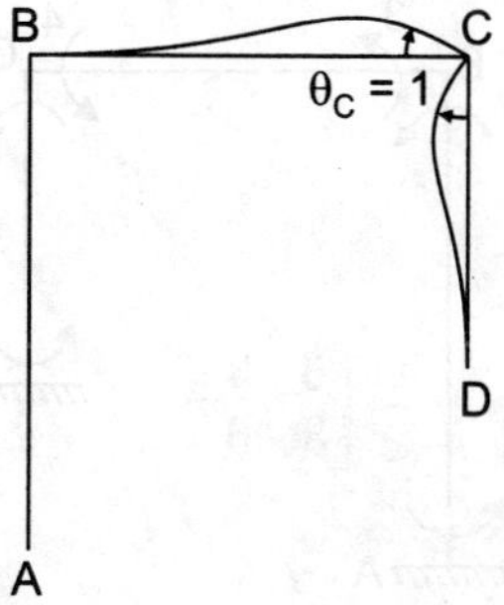

$$\beta_3 = \begin{bmatrix} 0 \\ 0 \\ 0 \\ 1 \\ 1 \\ 0 \end{bmatrix}$$

Unit rotation at co-ordinate (3)

$$\text{Transformation matrix } (\beta) = \begin{bmatrix} -0.33 & 0 & 0 \\ -0.33 & 1 & 0 \\ 0 & 1 & 0 \\ 0 & 0 & 1 \\ -0.5 & 0 & 1 \\ -0.5 & 0 & 0 \end{bmatrix}$$

Fixed co-ordinate forces, $\{F^0\} = [\beta]^T \{P^0\}$

$$= \begin{bmatrix} -0.33 & -0.33 & 0 & 0 & -0.5 & -0.5 \\ 0 & 1 & 1 & 0 & 0 & 0 \\ 0 & 0 & 0 & 1 & 1 & 0 \end{bmatrix} \begin{Bmatrix} -3 \\ 3 \\ -3 \\ 3 \\ 0 \\ 0 \end{Bmatrix}$$

$$\{F^0\} = \begin{Bmatrix} 0 \\ 0 \\ 3 \end{Bmatrix}$$

$$\{F\} = \{F^f\} - \{F^0\}$$

$$= \begin{Bmatrix} 4 \\ 0 \\ 0 \end{Bmatrix} - \begin{Bmatrix} 0 \\ 0 \\ 3 \end{Bmatrix} = \begin{Bmatrix} 4 \\ 0 \\ -3 \end{Bmatrix}$$

Element stiffness matrix (*k*) :

$$[k] = \frac{EI}{l} \begin{bmatrix} 4 & 2 \\ 2 & 4 \end{bmatrix}$$

For member AB

$$[k_1] = \frac{E(2I)}{3} \begin{bmatrix} 4 & 2 \\ 2 & 4 \end{bmatrix} = \frac{2EI}{3} \begin{bmatrix} 4 & 2 \\ 2 & 4 \end{bmatrix} = \frac{EI}{2} \begin{bmatrix} 5.33 & 2.67 \\ 2.67 & 5.33 \end{bmatrix}$$

For member BC

$$[k_2] = \frac{EI}{2}\begin{bmatrix} 4 & 2 \\ 2 & 4 \end{bmatrix}$$

For member CD

$$[k_3] = \frac{E(1.5I)}{2}\begin{bmatrix} 4 & 2 \\ 2 & 4 \end{bmatrix} = \frac{EI}{2}\begin{bmatrix} 6 & 3 \\ 3 & 6 \end{bmatrix}$$

Assembled element stiffness matrix (*k*) :

$$[k] = \frac{EI}{2}\begin{bmatrix} 5.33 & 2.67 & 0 & 0 & 0 & 0 \\ 2.67 & 5.33 & 0 & 0 & 0 & 0 \\ 0 & 0 & 4 & 2 & 0 & 0 \\ 0 & 0 & 2 & 4 & 0 & 0 \\ 0 & 0 & 0 & 0 & 6 & 3 \\ 0 & 0 & 0 & 0 & 3 & 6 \end{bmatrix}$$

System stiffness matrix [K] :

$$[K] = [\beta^T]\,[k]\,[\beta]$$

$$= \begin{bmatrix} -0.33 & -0.33 & 0 & 0 & -0.5 & -0.5 \\ 0 & 1 & 1 & 0 & 0 & 0 \\ 0 & 0 & 0 & 1 & 1 & 0 \end{bmatrix} \frac{EI}{2}\begin{bmatrix} 5.33 & 2.67 & 0 & 0 & 0 & 0 \\ 2.67 & 5.33 & 0 & 0 & 0 & 0 \\ 0 & 0 & 4 & 2 & 0 & 0 \\ 0 & 0 & 2 & 4 & 0 & 0 \\ 0 & 0 & 0 & 0 & 6 & 3 \\ 0 & 0 & 0 & 0 & 3 & 6 \end{bmatrix}\begin{bmatrix} -0.33 & 0 & 0 \\ -0.33 & 1 & 0 \\ 0 & 1 & 0 \\ 0 & 0 & 1 \\ -0.5 & 0 & 1 \\ -0.5 & 0 & 0 \end{bmatrix}$$

$$= \begin{bmatrix} -0.33 & -0.33 & 0 & 0 & -0.5 & -0.5 \\ 0 & 1 & 1 & 0 & 0 & 0 \\ 0 & 0 & 0 & 1 & 1 & 0 \end{bmatrix} \frac{EI}{2}\begin{bmatrix} -2.64 & 2.67 & 0 \\ -2.64 & 5.33 & 0 \\ 0 & 4 & 2 \\ 0 & 2 & 4 \\ -4.5 & 0 & 6 \\ -4.5 & 0 & 3 \end{bmatrix}$$

$$[K] = \frac{EI}{2}\begin{bmatrix} 6.242 & -2.64 & -4.50 \\ -2.64 & 9.33 & 2 \\ -4.50 & 2 & 10 \end{bmatrix}$$

$$[K]^{-1} = \frac{2}{EI}\begin{bmatrix} 0.2578 & 0.0502 & 0.10598 \\ 0.0502 & 0.1218 & -0.00175 \\ 0.10598 & -0.00175 & 0.1480 \end{bmatrix}$$

System displacement (u)

$$\{u\} = [K]^{-1}\,\{F\}$$

$$= \frac{2}{EI}\begin{bmatrix} 0.2578 & 0.0502 & 0.10598 \\ 0.0502 & 0.1218 & -0.00175 \\ 0.10598 & -0.00175 & 0.1480 \end{bmatrix}\begin{bmatrix} 4 \\ 0 \\ -3 \end{bmatrix}$$

$$\{u\} = \frac{2}{EI}\begin{bmatrix} 0.7134 \\ 0.2062 \\ -0.02 \end{bmatrix}$$

Element forces, $\{P'\} = [k]\,[\beta]\,\{u\}$

$$= \frac{EI}{2}\begin{bmatrix} 5.33 & 2.67 & 0 & 0 & 0 & 0 \\ 2.67 & 5.33 & 0 & 0 & 0 & 0 \\ 0 & 0 & 4 & 2 & 0 & 0 \\ 0 & 0 & 2 & 4 & 0 & 0 \\ 0 & 0 & 0 & 0 & 6 & 3 \\ 0 & 0 & 0 & 0 & 3 & 6 \end{bmatrix}\begin{bmatrix} -0.33 & 0 & 0 \\ -0.33 & 1 & 0 \\ 0 & 1 & 0 \\ 0 & 0 & 1 \\ -0.5 & 0 & 1 \\ -0.5 & 0 & 0 \end{bmatrix}\frac{2}{EI}\begin{bmatrix} 0.7134 \\ 0.2062 \\ -0.02 \end{bmatrix}$$

$$\{P'\} = \begin{bmatrix} -2.64 & 2.67 & 0 \\ -2.64 & 5.33 & 0 \\ 0 & 4 & 2 \\ 0 & 2 & 4 \\ -4.5 & 0 & 6 \\ -4.5 & 0 & 3 \end{bmatrix}\begin{bmatrix} 0.7134 \\ 0.2062 \\ -0.02 \end{bmatrix} = \begin{bmatrix} -1.33 \\ -0.7843 \\ 0.7848 \\ 0.33 \\ -3.33 \\ -3.27 \end{bmatrix}$$

Final forces at supports/joints :

$$\{P^f\} = [P'] + \{P^0\} = \begin{bmatrix} -1.33 \\ -0.7843 \\ 0.7848 \\ 0.33 \\ -3.33 \\ -3.27 \end{bmatrix} + \begin{bmatrix} -3 \\ 3 \\ -3 \\ 3 \\ 0 \\ 0 \end{bmatrix} = \begin{bmatrix} -4.33 \\ 2.2157 \\ -2.2152 \\ 3.33 \\ -3.33 \\ -3.27 \end{bmatrix}$$

$M_{AB} = -4.33$ kNm, $M_{BA} = 2.2157$ kNm

$M_{BC} = -2.2152$ kNm, $M_{CB} = 3.33$ kNm

$M_{CD} = -3.33$ kNm, $M_{DC} = -3.27$ kNm

Example 5.33 *Analyse the frame shown in Fig. 5.71 by matrix stiffness method.*

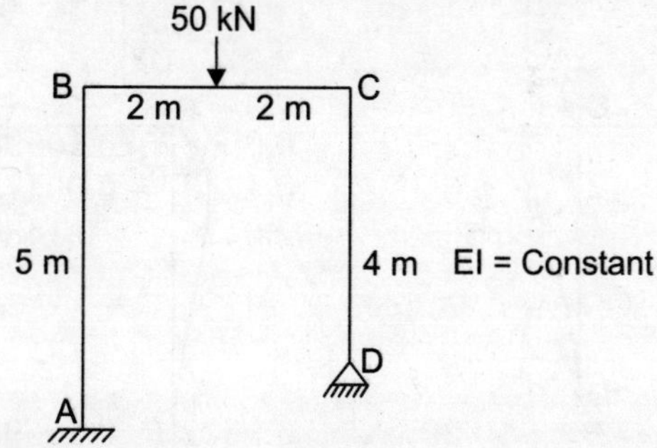

Fig. 5.71

Solution The given frame is kinematically indeterminate to fourth degree, *i.e.*, rotation at B, C, D and sway of frame due to unsymmetry in geometry.

Fixed end moments :

$$M_{FAB} = M_{FBA} = 0 \text{ (No lateral load on AB)}$$

$$M_{FBC} = -\frac{Wl}{8} = -\frac{50 \times 4}{8} = -25 \text{ kNm}$$

$$M_{FCB} = \frac{Wl}{8} = \frac{50 \times 4}{8} = 25 \text{ kNm}$$

$$M_{FCD} = M_{FDC} = 0 \text{ (No lateral load on CD)}$$

Fixed end moments (P^0)

$$P^0 = \begin{bmatrix} 0 \\ 0 \\ -25 \\ 25 \\ 0 \\ 0 \end{bmatrix}$$

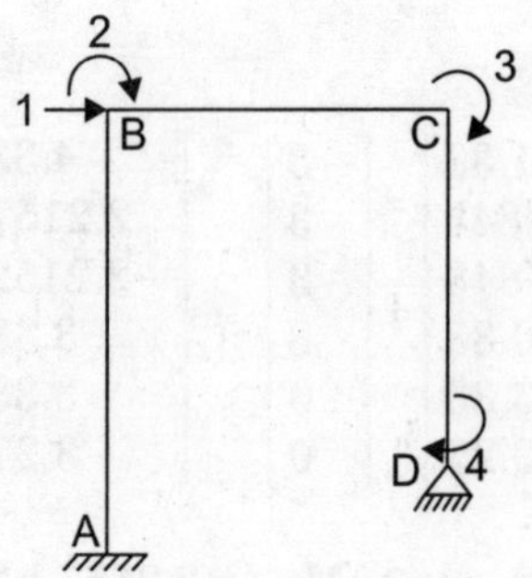

System co-ordinates

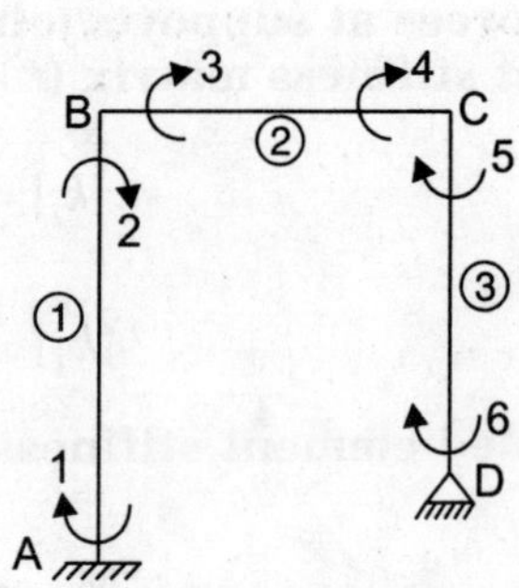

Element co-ordinates

Transformation matrix (β) :

Apply unit rotation in system co-ordinates (2), (3), (4) and unit displacement 'Δ' along 1 one at a time and find the displacements corresponding to element co-ordinates to form transformation matrix (β).

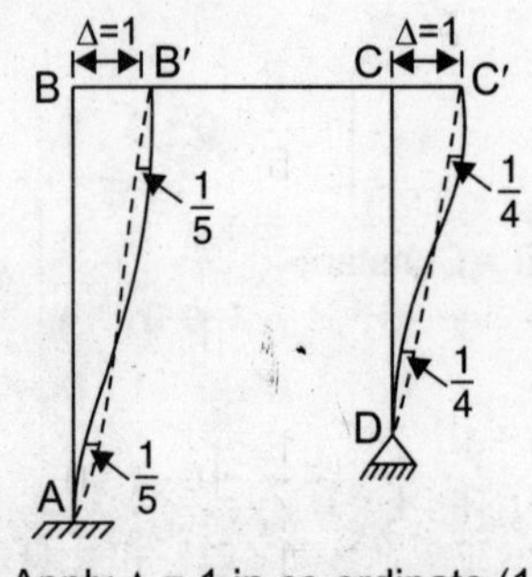

Apply Δ = 1 in co-ordinate (1)

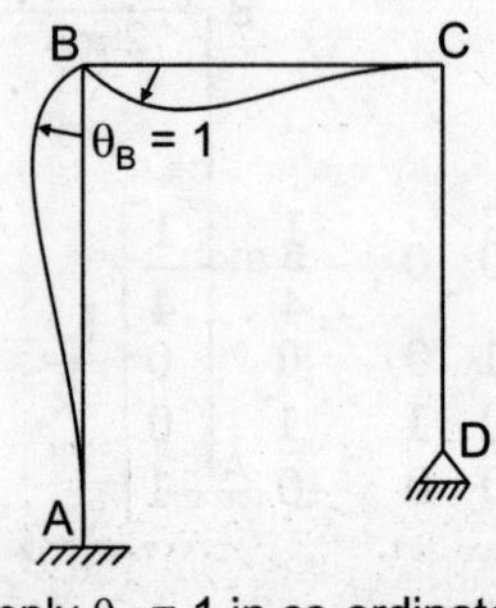

Apply θ_B = 1 in co-ordinate (2)

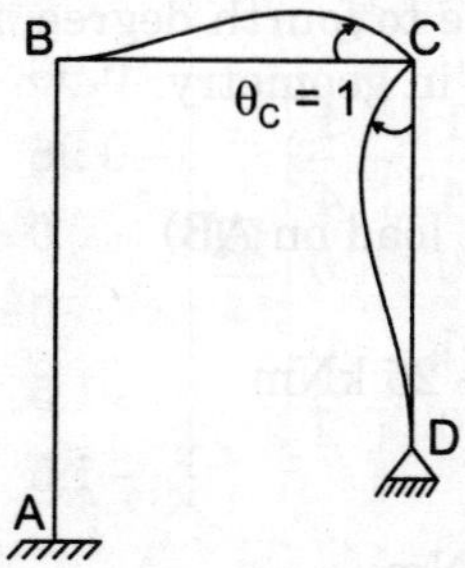

Apply $\theta_C = 1$ in co-ordinate (3)

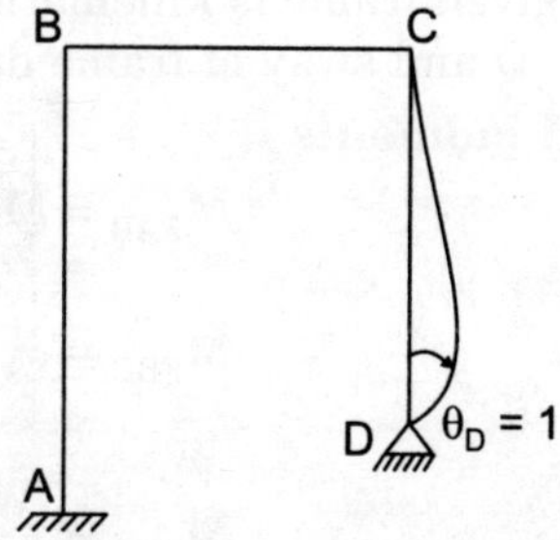

Apply $\theta_D = 1$ in co-ordinate (4)

$$[\beta] = \begin{array}{c} \begin{array}{cccc} \delta = 1 & \theta_B = 1 & \theta_C = 1 & \theta_D = 1 \end{array} \\ \begin{bmatrix} -\frac{1}{5} & 0 & 0 & 0 \\ -\frac{1}{5} & 1 & 0 & 0 \\ 0 & 1 & 0 & 0 \\ 0 & 0 & 1 & 0 \\ -\frac{1}{4} & 0 & 1 & 0 \\ -\frac{1}{4} & 0 & 0 & 1 \end{bmatrix} \end{array}$$

Element stiffness matrix (k) :

$$[k_i] = \frac{EI}{l}\begin{bmatrix} 4 & 2 \\ 2 & 4 \end{bmatrix}$$

$$[k_1] = \frac{EI}{5}\begin{bmatrix} 4 & 2 \\ 2 & 4 \end{bmatrix}, \quad [k_2] = \frac{EI}{4}\begin{bmatrix} 4 & 2 \\ 2 & 4 \end{bmatrix}, \quad [k_3] = \frac{EI}{4}\begin{bmatrix} 4 & 2 \\ 2 & 4 \end{bmatrix}$$

Assembled element stiffness matrix :

$$[k] = \frac{EI}{4}\begin{bmatrix} 3.2 & 1.6 & 0 & 0 & 0 & 0 \\ 1.6 & 3.2 & 0 & 0 & 0 & 0 \\ 0 & 0 & 4 & 2 & 0 & 0 \\ 0 & 0 & 2 & 4 & 0 & 0 \\ 0 & 0 & 0 & 0 & 4 & 2 \\ 0 & 0 & 0 & 0 & 2 & 4 \end{bmatrix}$$

System stiffness matrix (K) :

$$[K] = [\beta^T]\,[k]\,[\beta]$$

$$= \begin{bmatrix} -\frac{1}{5} & -\frac{1}{5} & 0 & 0 & -\frac{1}{4} & -\frac{1}{4} \\ 0 & 1 & 1 & 0 & 0 & 0 \\ 0 & 0 & 0 & 1 & 1 & 0 \\ 0 & 0 & 0 & 0 & 0 & 1 \end{bmatrix} \frac{EI}{4}\begin{bmatrix} 3.2 & 1.6 & 0 & 0 & 0 & 0 \\ 1.6 & 3.2 & 0 & 0 & 0 & 0 \\ 0 & 0 & 4 & 2 & 0 & 0 \\ 0 & 0 & 2 & 4 & 0 & 0 \\ 0 & 0 & 0 & 0 & 4 & 2 \\ 0 & 0 & 0 & 0 & 2 & 4 \end{bmatrix} \begin{bmatrix} -\frac{1}{5} & 0 & 0 & 0 \\ -\frac{1}{5} & 1 & 0 & 0 \\ 0 & 1 & 0 & 0 \\ 0 & 0 & 1 & 0 \\ -\frac{1}{4} & 0 & 1 & 0 \\ -\frac{1}{4} & 0 & 0 & 1 \end{bmatrix}.$$

$$= \begin{bmatrix} -\frac{1}{5} & -\frac{1}{5} & 0 & 0 & -\frac{1}{4} & -\frac{1}{4} \\ 0 & 1 & 1 & 0 & 0 & 0 \\ 0 & 0 & 0 & 1 & 1 & 0 \\ 0 & 0 & 0 & 0 & 0 & 1 \end{bmatrix} \frac{EI}{4} \begin{bmatrix} -0.96 & 1.6 & 0 & 0 \\ -0.96 & 3.2 & 0 & 0 \\ 0 & 4 & 2 & 0 \\ 0 & 2 & 4 & 0 \\ -1.5 & 0 & 4 & 2 \\ -1.5 & 0 & 2 & 4 \end{bmatrix}$$

$$[K] = \frac{EI}{4} \begin{bmatrix} 1.134 & -0.96 & -1.5 & -1.5 \\ -0.96 & 7.2 & 2 & 0 \\ -1.5 & 2 & 8 & 2 \\ -1.5 & 0 & 2 & 4 \end{bmatrix}$$

$$[K^{-1}] = \frac{\text{Adjoint of K}}{|K|}$$

$$\text{Adjoint of K} = \frac{4}{EI} \begin{bmatrix} K_{11} & K_{12} & K_{13} & K_{14} \\ K_{21} & K_{22} & K_{23} & K_{24} \\ K_{31} & K_{32} & K_{33} & K_{34} \\ K_{41} & K_{42} & K_{43} & K_{44} \end{bmatrix}$$

$$K_{11} = + \begin{vmatrix} 7.2 & 2 & 0 \\ 2 & 8 & 2 \\ 0 & 2 & 4 \end{vmatrix} = 185.6$$

$$K_{12} = - \begin{vmatrix} -0.96 & 2 & 0 \\ -1.5 & 8 & 2 \\ -1.5 & 2 & 4 \end{vmatrix} = -(-20.88) = 20.88$$

$$K_{13} = + \begin{vmatrix} -0.96 & 7.2 & 0 \\ -1.5 & 2 & 2 \\ -1.5 & 0 & 4 \end{vmatrix} = 13.92$$

$$K_{14} = - \begin{vmatrix} -0.96 & 7.2 & 2 \\ -1.5 & 2 & 8 \\ -1.5 & 0 & 2 \end{vmatrix} = -(-62.64) = 62.64$$

Similarly

$K_{21} = 20.86$ $\quad K_{22} = 13.752$ $\quad K_{23} = -1.692$ $\quad K_{24} = 8.676$

$K_{31} = 13.92$ $\quad K_{32} = -1.692$ $\quad K_{33} = 12.7728$ $\quad K_{34} = -1.1664$

$K_{41} = 62.64$ $\quad K_{42} = 8.676$ $\quad K_{43} = -1.1664$ $\quad K_{44} = 42.9696$

$$\text{Adjoint of K} = \frac{4}{EI}\begin{bmatrix} 185.6 & 20.88 & 13.92 & 62.64 \\ 20.88 & 13.752 & -1.692 & 8.676 \\ 13.92 & -1.692 & 12.7728 & -1.1664 \\ 62.64 & 8.676 & -1.1664 & 42.9696 \end{bmatrix}$$

$$|K| = (1.134 \times |K_{11}|) - (-0.96 \times |K_{12}|) + (-1.5 \times |K_{13}|) - (-1.5 \times |K_{14}|)$$

$$= (1.134 \times 185.6) - (-0.96 \times -20.88) + (-1.5 \times 13.92) - (-1.5 \times -62.64)$$

$$|K| = 75.5856$$

$$[K^{-1}] = \frac{\text{Adjoint K}}{|K|} = \frac{\frac{4}{EI}\begin{bmatrix} 185.6 & 20.88 & 13.92 & 62.64 \\ 20.88 & 13.752 & -1.692 & 8.676 \\ 13.92 & -1.692 & 12.7728 & -1.1664 \\ 62.64 & 8.676 & -1.1664 & 42.9696 \end{bmatrix}}{75.5856}$$

$$[K]^{-1} = \frac{4}{EI}\begin{bmatrix} 2.45549 & 0.27624 & 0.18416 & 0.82873 \\ 0.27624 & 0.181939417 & -0.02239 & 0.11478 \\ 0.18416 & -0.02239 & 0.16898 & -0.01543 \\ 0.82873 & 0.11478 & -0.01543 & 0.56849 \end{bmatrix}$$

System co-ordinate forces (F)

$$\{F\} = \{F^f\} - \{F^0\}$$

$$\{F^0\} = [\beta]^T\{P^0\} = \begin{bmatrix} -\frac{1}{5} & -\frac{1}{5} & 0 & 0 & -\frac{1}{4} & -\frac{1}{4} \\ 0 & 1 & 1 & 0 & 0 & 0 \\ 0 & 0 & 0 & 1 & 1 & 0 \\ 0 & 0 & 0 & 0 & 0 & 1 \end{bmatrix}\begin{bmatrix} 0 \\ 0 \\ -25 \\ 25 \\ 0 \\ 0 \end{bmatrix} = \begin{bmatrix} 0 \\ -25 \\ 25 \\ 0 \end{bmatrix}$$

$$\{F\}^f = \begin{bmatrix} 0 \\ 0 \\ 0 \\ 0 \end{bmatrix} = \text{external forces at co-ordinates}$$

$$\{F\} = \{F\}^f - \{F^0\} = \begin{bmatrix} 0 \\ 0 \\ 0 \\ 0 \end{bmatrix} - \begin{bmatrix} 0 \\ -25 \\ 25 \\ 0 \end{bmatrix} = \begin{bmatrix} 0 \\ 25 \\ -25 \\ 0 \end{bmatrix}$$

System displacement (u)

$$\{u\} = [K]^{-1}\{F\}$$

$$= \frac{4}{EI}\begin{bmatrix} 2.45549 & 0.27624 & 0.18416 & 0.82873 \\ 0.27624 & 0.181939417 & -0.02239 & 0.11478 \\ 0.18416 & -0.02239 & 0.16898 & -0.01543 \\ 0.82873 & 0.11478 & -0.01543 & 0.56849 \end{bmatrix}\begin{bmatrix} 0 \\ 25 \\ -25 \\ 0 \end{bmatrix}$$

$$\{u\} = \frac{4}{EI}\begin{bmatrix} 2.302 \\ 5.10824 \\ -4.78425 \\ 3.25525 \end{bmatrix}$$

Element forces (P′)

$$\{P'\} = [k][\beta][u]$$

$$= \frac{EI}{4}\begin{bmatrix} 3.2 & 1.6 & 0 & 0 & 0 & 0 \\ 1.6 & 3.2 & 0 & 0 & 0 & 0 \\ 0 & 0 & 4 & 2 & 0 & 0 \\ 0 & 0 & 2 & 4 & 0 & 0 \\ 0 & 0 & 0 & 0 & 4 & 2 \\ 0 & 0 & 0 & 0 & 2 & 4 \end{bmatrix}\begin{bmatrix} -\frac{1}{5} & 0 & 0 & 0 \\ -\frac{1}{5} & 1 & 0 & 0 \\ 0 & 1 & 0 & 0 \\ 0 & 0 & 1 & 0 \\ -\frac{1}{4} & 0 & 1 & 0 \\ -\frac{1}{4} & 0 & 0 & 1 \end{bmatrix}\frac{4}{EI}\begin{bmatrix} 2.302 \\ 5.10824 \\ -4.78425 \\ 3.25525 \end{bmatrix}$$

$$\{P'\} = \begin{bmatrix} -0.96 & 1.6 & 0 & 0 \\ -0.96 & 3.2 & 0 & 0 \\ 0 & 4 & 2 & 0 \\ 0 & 2 & 4 & 0 \\ -1.5 & 0 & 4 & 2 \\ -1.5 & 0 & 2 & 4 \end{bmatrix}\begin{bmatrix} 2.302 \\ 5.10824 \\ -4.78425 \\ 3.25525 \end{bmatrix} = \begin{bmatrix} 5.96326 \\ 14.13645 \\ 10.86446 \\ -8.92052 \\ -16.0795 \\ 0 \end{bmatrix}$$

$$\{P^f\} = \{P'\} + \{P^0\} = \begin{bmatrix} 5.96326 \\ 14.13645 \\ 10.86446 \\ -8.92052 \\ -16.0795 \\ 0 \end{bmatrix} + \begin{bmatrix} 0 \\ 0 \\ -25 \\ 25 \\ 0 \\ 0 \end{bmatrix} = \begin{bmatrix} 5.963 \\ 14.14 \\ -14.14 \\ 16.0795 \\ -16.0795 \\ 0 \end{bmatrix}$$

$M_{AB} = 5.963$ kNm $\quad M_{BA} = 14.14$ kNm

$M_{BC} = -14.14$ kNm $\quad M_{CB} = 16.0795$ kNm

$M_{CD} = -16.0795$ kNm $\quad M_{DC} = 0$

Example 5.34 *Analyse the frame shown in Fig. 5.72 by the matrix stiffness method. Sketch the bending moment diagram.*

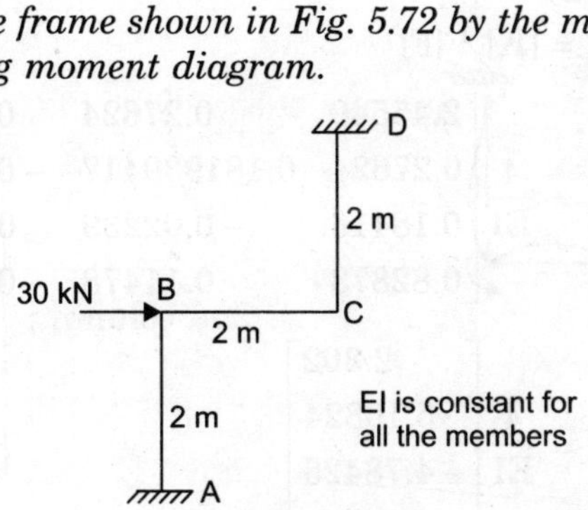

Fig. 5.72

Solution The frame given is kinematically indeterminate to third degree *i.e.*, θ_B, θ_C and Δ are the unknown joint displacements. Hence the size of the stiffness matrix will be 3 × 3.

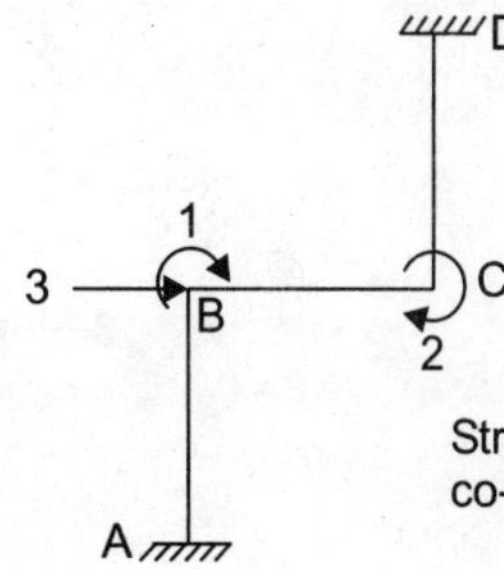

Structure co-ordinate

Fixed end moments :

Since there is no lateral loading on the spans,

$$M_{FAB} = M_{FBA} = M_{FBC} = M_{FCB} = M_{FCD} = M_{FDC} = 0$$

θ_B, θ_C and Δ are the structure co-ordinates.

Element co-ordinates :

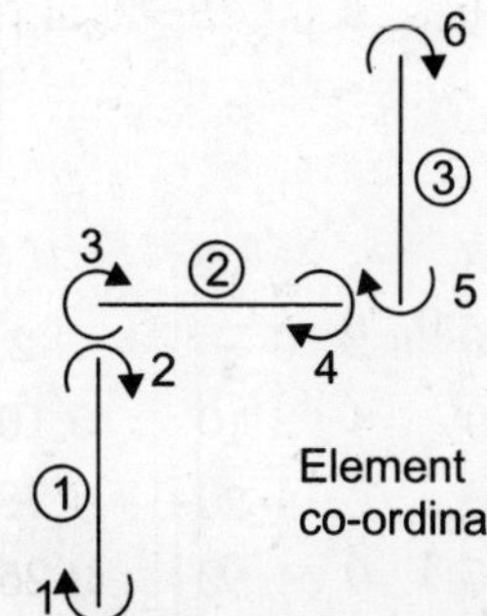

Element co-ordinates

Displacement transformation matrix, β : β matrix will be generated by imposing unit rotation at B and C and unit displacement along BC allowing only one at a time and getting the displacements along the element co-ordinates.

Apply $\theta_B = 1$

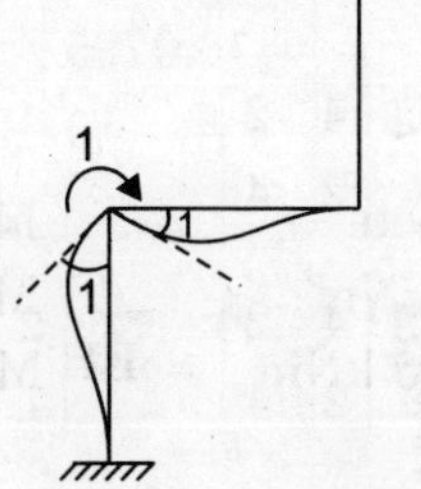

1st column : $\begin{bmatrix} 0 \\ 1 \\ 1 \\ 0 \\ 0 \\ 0 \end{bmatrix}$

Applying $\theta_C = 1$

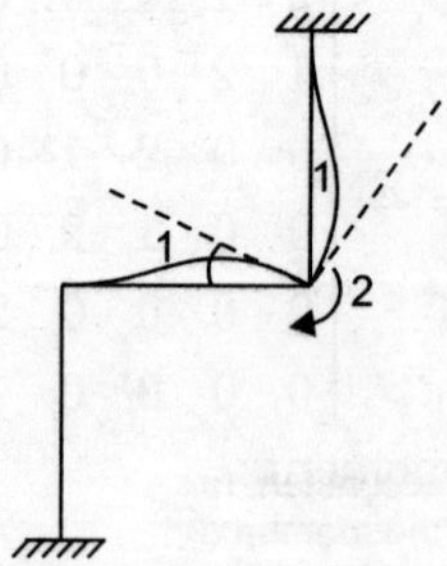

$$\text{2nd column : } \begin{bmatrix} 0 \\ 0 \\ 0 \\ 1 \\ 1 \\ 0 \end{bmatrix}$$

Applying $\Delta = 1$

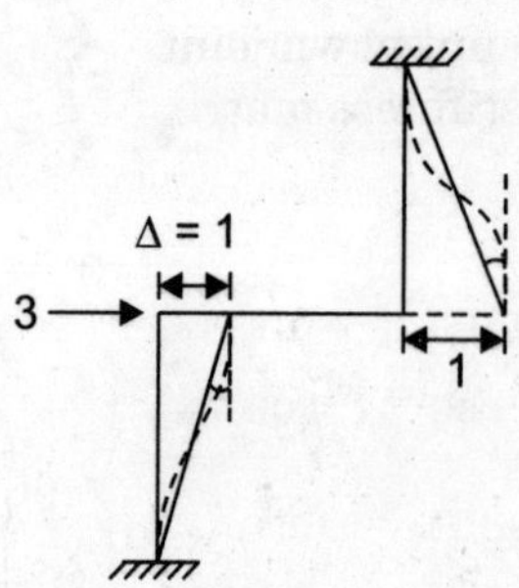

$$\text{3rd column : } \begin{bmatrix} -\frac{1}{2} \\ -\frac{1}{2} \\ 0 \\ 0 \\ \frac{1}{2} \\ \frac{1}{2} \end{bmatrix}$$

Hence, the [β] matrix will be,

$$[\beta] = \begin{bmatrix} 0 & 0 & -\frac{1}{2} \\ 1 & 0 & -\frac{1}{2} \\ 1 & 0 & 0 \\ 0 & 1 & 0 \\ 0 & 1 & \frac{1}{2} \\ 0 & 0 & \frac{1}{2} \end{bmatrix}$$

Element stiffness matrix $[k_i]$:

$$[k_1] = \frac{EI}{l}\begin{bmatrix} 4 & 2 \\ 2 & 4 \end{bmatrix}$$

$$= \frac{EI}{2}\begin{bmatrix} 4 & 2 \\ 2 & 4 \end{bmatrix} = EI\begin{bmatrix} 2 & 1 \\ 1 & 2 \end{bmatrix} = [k_2] = [k_3]$$

Assembled element stiffness matrix [*k*] :

$$[k] = \begin{bmatrix} k_1 & 0 & 0 \\ 0 & k_2 & 0 \\ 0 & 0 & k_3 \end{bmatrix} = EI \begin{bmatrix} 2 & 1 & 0 & 0 & 0 & 0 \\ 1 & 2 & 0 & 0 & 0 & 0 \\ 0 & 0 & 2 & 1 & 0 & 0 \\ 0 & 0 & 1 & 2 & 0 & 0 \\ 0 & 0 & 0 & 0 & 2 & 1 \\ 0 & 0 & 0 & 0 & 1 & 2 \end{bmatrix}$$

Forces in element co-ordinates, *i.e.,* fixed end moments,

$$[P^0] = \begin{bmatrix} 0 \\ 0 \\ 0 \\ 0 \\ 0 \\ 0 \end{bmatrix}$$

Fixed end co-ordinate forces, $\{F^0\} = [\beta^T]\{P^0\}$

$$= \begin{bmatrix} 0 & 1 & 1 & 0 & 0 & 0 \\ 0 & 0 & 0 & 1 & 1 & 0 \\ -\frac{1}{2} & -\frac{1}{2} & 0 & 0 & \frac{1}{2} & \frac{1}{2} \end{bmatrix} \begin{Bmatrix} 0 \\ 0 \\ 0 \\ 0 \\ 0 \\ 0 \end{Bmatrix} = \begin{Bmatrix} 0 \\ 0 \\ 0 \end{Bmatrix}$$

Forces applied at system co-ordinates, $\{F^f\} = \begin{Bmatrix} 0 \\ 0 \\ 30 \end{Bmatrix}$

[∵ 30 kN force applied along system co-ordinate (3)]

$$[\{F^f\} - \{F^0\}] = \left[\begin{Bmatrix} 0 \\ 0 \\ 30 \end{Bmatrix} - \begin{Bmatrix} 0 \\ 0 \\ 0 \end{Bmatrix}\right] = \begin{Bmatrix} 0 \\ 0 \\ 30 \end{Bmatrix}$$

System stiffness matrix [K] :

$$[K] = [\beta^T][k][\beta]$$

$$= \begin{bmatrix} 0 & 1 & 1 & 0 & 0 & 0 \\ 0 & 0 & 0 & 1 & 1 & 0 \\ -\frac{1}{2} & -\frac{1}{2} & 0 & 0 & \frac{1}{2} & \frac{1}{2} \end{bmatrix} EI \begin{bmatrix} 2 & 1 & 0 & 0 & 0 & 0 \\ 1 & 2 & 0 & 0 & 0 & 0 \\ 0 & 0 & 2 & 1 & 0 & 0 \\ 0 & 0 & 1 & 2 & 0 & 0 \\ 0 & 0 & 0 & 0 & 2 & 1 \\ 0 & 0 & 0 & 0 & 1 & 2 \end{bmatrix} \begin{bmatrix} 0 & 0 & -\frac{1}{2} \\ 1 & 0 & -\frac{1}{2} \\ 1 & 0 & 0 \\ 0 & 1 & 0 \\ 0 & 1 & \frac{1}{2} \\ 0 & 0 & \frac{1}{2} \end{bmatrix}$$

$$= \begin{bmatrix} 0 & 1 & 1 & 0 & 0 & 0 \\ 0 & 0 & 0 & 1 & 1 & 0 \\ -\frac{1}{2} & -\frac{1}{2} & 0 & 0 & \frac{1}{2} & \frac{1}{2} \end{bmatrix} \text{EI} \begin{bmatrix} 1 & 0 & -\frac{3}{2} \\ 2 & 0 & -\frac{3}{2} \\ 2 & 1 & 0 \\ 1 & 2 & 0 \\ 0 & 2 & \frac{3}{2} \\ 0 & 1 & \frac{3}{2} \end{bmatrix}$$

$$= \text{EI} \begin{bmatrix} 4 & 1 & -\frac{3}{2} \\ 1 & 4 & \frac{3}{2} \\ -\frac{3}{2} & \frac{3}{2} & 3 \end{bmatrix}$$

System displacements {*u*} :

$$\{u\} = [\text{K}]^{-1} \{[\text{F}^f] - [\text{F}^0]\}$$

$$[\text{K}]^{-1} = \frac{1}{|\text{K}|} \text{Adj}\,[\text{K}]$$

$$[\text{K}]^{-1} = \frac{1}{\text{EI}} \begin{bmatrix} 0.433 & -0.233 & 0.333 \\ -0.233 & 0.433 & -0.333 \\ 0.333 & -0.333 & 0.666 \end{bmatrix}$$

$$\{u\} = \frac{1}{\text{EI}} \begin{bmatrix} 0.433 & -0.233 & 0.333 \\ -0.233 & 0.433 & -0.333 \\ 0.333 & -0.333 & 0.666 \end{bmatrix} \begin{Bmatrix} 0 \\ 0 \\ 30 \end{Bmatrix} = \frac{1}{\text{EI}} \begin{Bmatrix} 10 \\ -10 \\ 20 \end{Bmatrix}$$

Element displacements {δ} :

$$\{\delta\} = [\beta]\,\{u\}$$

$$= \begin{bmatrix} 0 & 0 & -\frac{1}{2} \\ 1 & 0 & -\frac{1}{2} \\ 1 & 0 & 0 \\ 0 & 1 & 0 \\ 0 & 1 & \frac{1}{2} \\ 0 & 0 & \frac{1}{2} \end{bmatrix} \times \frac{1}{\text{EI}} \begin{Bmatrix} 10 \\ -10 \\ 20 \end{Bmatrix} = \frac{1}{\text{EI}} \begin{Bmatrix} -10 \\ 0 \\ 10 \\ -10 \\ 0 \\ 10 \end{Bmatrix}$$

$$[P'] = [k]\{\delta\} = EI\begin{bmatrix}2&1&0&0&0&0\\1&2&0&0&0&0\\0&0&2&1&0&0\\0&0&1&2&0&0\\0&0&0&0&2&1\\0&0&0&0&1&2\end{bmatrix}\times\frac{1}{EI}\begin{Bmatrix}-10\\0\\10\\-10\\0\\10\end{Bmatrix}$$

$$\{P'\} = \begin{Bmatrix}-20\\-10\\10\\-10\\10\\20\end{Bmatrix}$$

$$\{P^f\} = \{P^0\} + \{P'\} = \begin{Bmatrix}0\\0\\0\\0\\0\\0\end{Bmatrix} + \begin{Bmatrix}-20\\-10\\10\\-10\\10\\20\end{Bmatrix}$$

$$\{P^f\} = \begin{Bmatrix}-20\\-10\\10\\-10\\10\\20\end{Bmatrix}\text{ kNm}$$

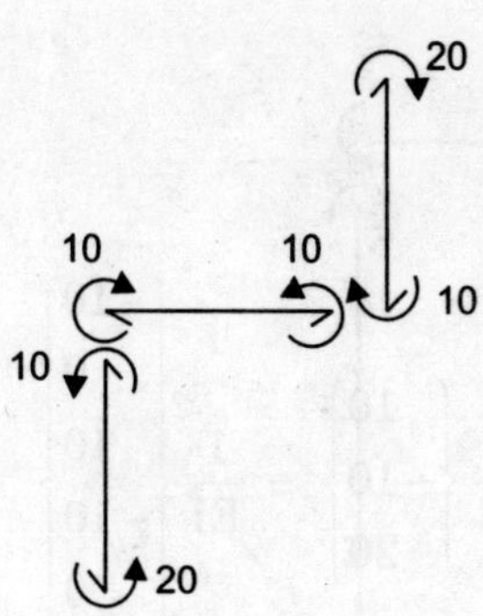

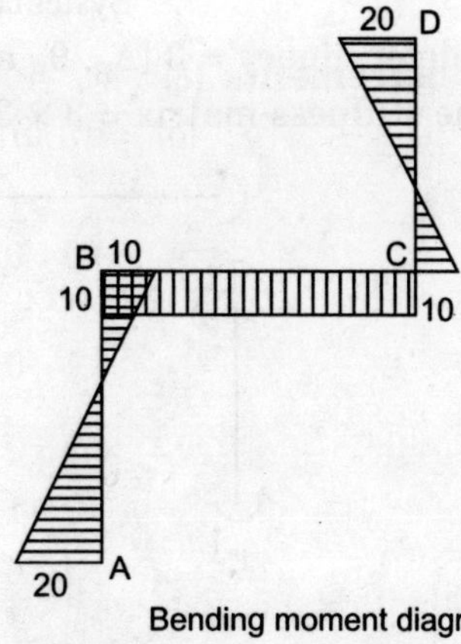

Bending moment diagram

Example 5.35 *Analyse the frame shown in Fig. 5.73 by matrix stiffness method.*

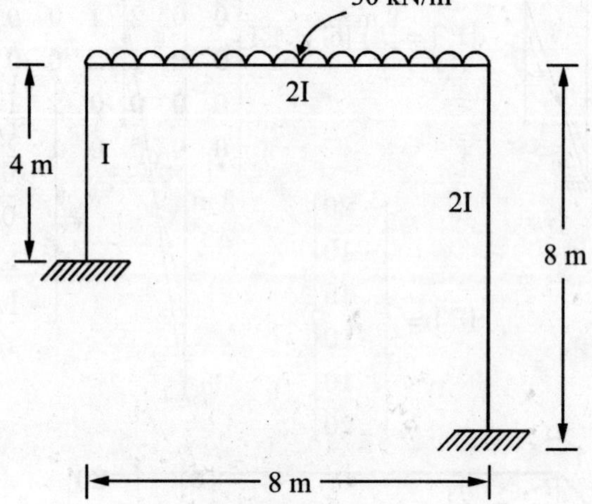

Fig. 5.73

Solution

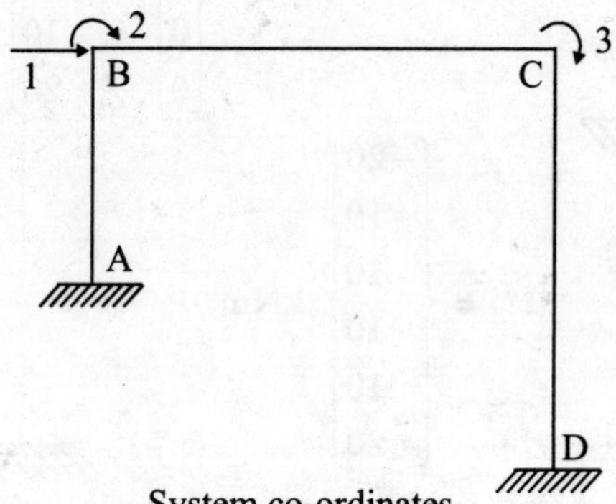

System co-ordinates

Kinematic indeterminacy = 3 [Δ_B, θ_B and θ_C]

$\therefore$ Size of the stiffness matrix = 3 × 3

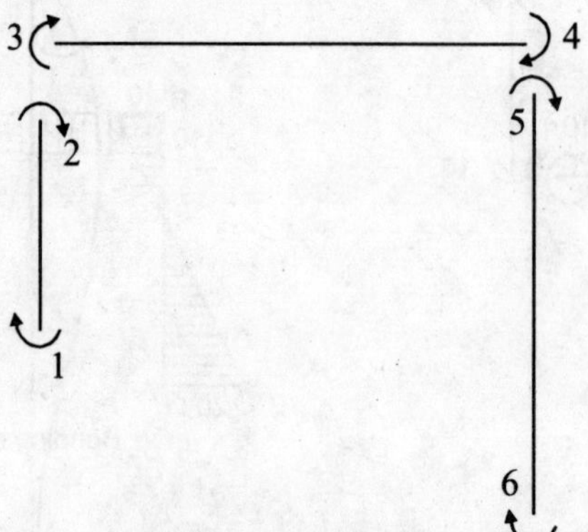

Element co-ordinates

[β] matrix :

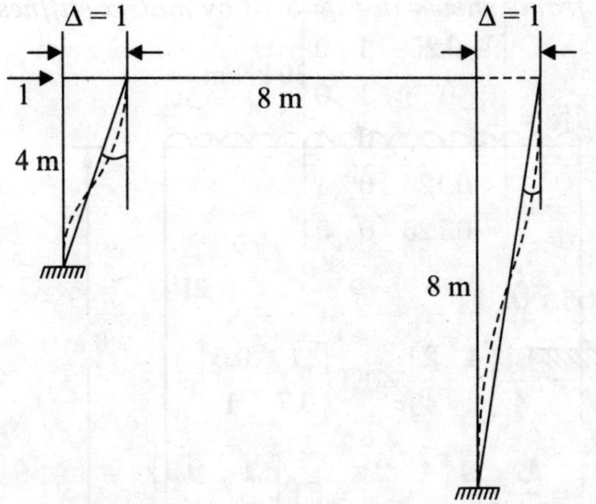

$$\begin{bmatrix} -1/4 \\ -1/4 \\ 0 \\ 0 \\ -1/8 \\ -1/8 \end{bmatrix}$$

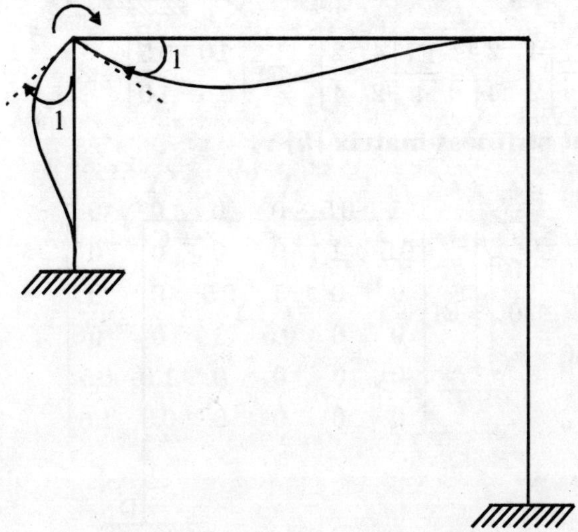

$$\begin{bmatrix} 0 \\ 1 \\ 1 \\ 0 \\ 0 \\ 0 \end{bmatrix}$$

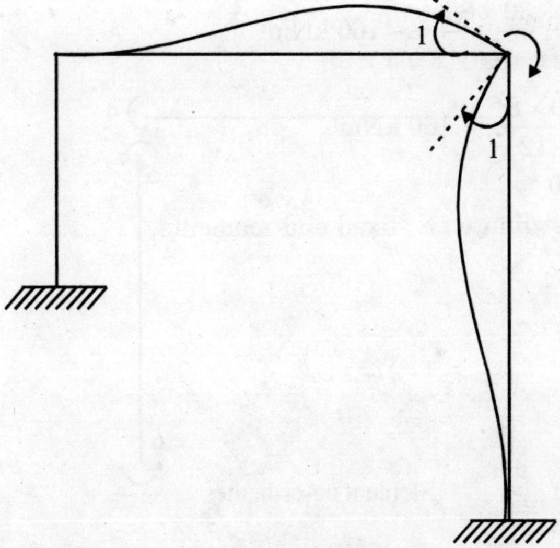

$$\begin{bmatrix} 0 \\ 0 \\ 0 \\ 1 \\ 1 \\ 0 \end{bmatrix}$$

$$[\beta] = \begin{bmatrix} -0.25 & 0 & 0 \\ -0.25 & 1 & 0 \\ 0 & 1 & 0 \\ 0 & 0 & 1 \\ -0.125 & 0 & 1 \\ -0.125 & 0 & 0 \end{bmatrix}$$

Element stiffness matrix $[k_n]$:

$$[k_1] = \frac{EI}{l}\begin{bmatrix} 4 & 2 \\ 2 & 4 \end{bmatrix} = \frac{EI}{4}\begin{bmatrix} 4 & 2 \\ 2 & 4 \end{bmatrix} = EI\begin{bmatrix} 1 & 0.5 \\ 0.5 & 1 \end{bmatrix}$$

$$[k_2] = \frac{EI}{l}\begin{bmatrix} 4 & 2 \\ 2 & 4 \end{bmatrix} = \frac{E \times 2I}{8}\begin{bmatrix} 4 & 2 \\ 2 & 4 \end{bmatrix} = EI\begin{bmatrix} 1 & 0.5 \\ 0.5 & 1 \end{bmatrix}$$

$$[k_3] = \frac{E \times 2I}{l}\begin{bmatrix} 4 & 2 \\ 2 & 4 \end{bmatrix} = \frac{EI}{4}\begin{bmatrix} 4 & 2 \\ 2 & 4 \end{bmatrix} = EI\begin{bmatrix} 1.0 & 0.5 \\ 0.5 & 1.0 \end{bmatrix}$$

Assembled element stiffness matrix $[k]$:

$$[k] = \begin{bmatrix} k_1 & 0 & 0 \\ 0 & k_2 & 0 \\ 0 & 0 & k_3 \end{bmatrix} = EI\begin{bmatrix} 1 & 0.5 & 0 & 0 & 0 & 0 \\ 0.5 & 1 & 0 & 0 & 0 & 0 \\ 0 & 0 & 1 & 0.5 & 0 & 0 \\ 0 & 0 & 0.5 & 1 & 0 & 0 \\ 0 & 0 & 0 & 0 & 1.0 & 0.5 \\ 0 & 0 & 0 & 0 & 0.5 & 1.0 \end{bmatrix}$$

Fixed end moments :

$$M_{FAB} = M_{FBA} = 0$$

$$M_{FBC} = -\frac{wl^2}{12} = -\frac{30 \times 8^2}{12} = -160 \text{ kNm}$$

$$M_{FCB} = \frac{wl^2}{12} = \frac{30 \times 8^2}{12} = 160 \text{ kNm}$$

$$M_{FCD} = M_{FDC} = 0$$

Forces in element co-ordinate *i.e.*, fixed end moments,

$$[P^0] = \begin{bmatrix} 0 \\ 0 \\ -160 \\ 160 \\ 0 \\ 0 \end{bmatrix}$$

Fixed end co-ordinate forces, $[F^0] = [\beta]^T [P^0]$

$$= \begin{bmatrix} -0.25 & -0.25 & 0 & 0 & -0.125 & -0.125 \\ 0 & 1 & 1 & 0 & 0 & 0 \\ 0 & 0 & 0 & 1 & 1 & 0 \end{bmatrix} \times \begin{bmatrix} 0 \\ 0 \\ -160 \\ 160 \\ 0 \\ 0 \end{bmatrix}$$

$$[F^0] = \begin{bmatrix} 0 \\ -160 \\ 160 \end{bmatrix}$$

Forces applied at system co-ordinates, $[F^f] = \begin{bmatrix} 0 \\ 0 \\ 0 \end{bmatrix}$

[$\therefore$ There are no external forces at the system co-ordinates]

$$\{[F^f] - [F^0]\} = \left\{ \begin{bmatrix} 0 \\ 0 \\ 0 \end{bmatrix} - \begin{bmatrix} 0 \\ -160 \\ 160 \end{bmatrix} \right\} = \begin{bmatrix} 0 \\ 160 \\ -160 \end{bmatrix}$$

System stiffness matrix [K] :

$$[K] = [\beta]^T [k] [\beta]$$

$$= \begin{bmatrix} -0.25 & -0.25 & 0 & 0 & -0.125 & -0.125 \\ 0 & 1 & 1 & 0 & 0 & 0 \\ 0 & 0 & 0 & 1 & 1 & 0 \end{bmatrix}$$

$$\times EI \begin{bmatrix} 1 & 0.5 & 0 & 0 & 0 & 0 \\ 0.5 & 1 & 0 & 0 & 0 & 0 \\ 0 & 0 & 1 & 0.5 & 0 & 0 \\ 0 & 0 & 0.5 & 1 & 0 & 0 \\ 0 & 0 & 0 & 0 & 1.0 & 0.5 \\ 0 & 0 & 0 & 0 & 0.5 & 1.0 \end{bmatrix} \times \begin{bmatrix} -0.25 & 0 & 0 \\ -0.25 & 1 & 0 \\ 0 & 1 & 0 \\ 0 & 0 & 1 \\ -0.125 & 0 & 1 \\ -0.125 & 0 & 0 \end{bmatrix}$$

$$[K] = EI \begin{bmatrix} 0.235 & -0.375 & -0.188 \\ -0.375 & 2.00 & 0.50 \\ -0.188 & 0.50 & 2.00 \end{bmatrix}$$

System displacements {*u*} :

$$\{u\} = [K]^{-1} \{[F^f] - [F^0]\}, \quad [K]^{-1} = \frac{1}{|K|} \text{Adj } [K]$$

$$[K]^{-1} = \frac{1}{EI} \begin{bmatrix} 6.25 & 1.093 & 0.314 \\ 1.093 & 0.725 & -0.078 \\ 0.314 & -0.078 & 0.548 \end{bmatrix}$$

$$\therefore \qquad [u] = \frac{1}{EI}\begin{bmatrix} 124.64 \\ 128.48 \\ -100.16 \end{bmatrix}$$

Element displacements {δ} :

$$\{\delta\} = [\beta]\,[u] = \begin{bmatrix} -0.25 & 0 & 0 \\ -0.25 & 1 & 0 \\ 0 & 1 & 0 \\ 0 & 0 & 1 \\ -0.125 & 0 & 1 \\ -0.125 & 0 & 0 \end{bmatrix} \times \frac{1}{EI}\begin{bmatrix} 124.64 \\ 128.48 \\ -100.16 \end{bmatrix} = \begin{bmatrix} -31.16 \\ 97.32 \\ 128.48 \\ -100.16 \\ -115.74 \\ -15.58 \end{bmatrix}$$

$$[P'] = [k]\,\{\delta\}$$

$$EI = \begin{bmatrix} 1 & 0.5 & 0 & 0 & 0 & 0 \\ 0.5 & 1 & 0 & 0 & 0 & 0 \\ 0 & 0 & 1 & 0.5 & 0 & 0 \\ 0 & 0 & 0.5 & 1 & 0 & 0 \\ 0 & 0 & 0 & 0 & 1.0 & 0.5 \\ 0 & 0 & 0 & 0 & 0.5 & 1.0 \end{bmatrix}\begin{bmatrix} -31.16 \\ 97.32 \\ 128.48 \\ -100.16 \\ -115.74 \\ -15.58 \end{bmatrix} \quad [P'] = \begin{bmatrix} 17.5 \\ 81.7 \\ 78.4 \\ -35.9 \\ -123.5 \\ -73.5 \end{bmatrix}$$

$$[P^f] = [P^0] + [P'] = \begin{bmatrix} 0 \\ 0 \\ -160 \\ 160 \\ 0 \\ 0 \end{bmatrix} + \begin{bmatrix} 17.5 \\ 81.7 \\ 78.4 \\ -35.9 \\ -123.5 \\ -73.5 \end{bmatrix} = \begin{bmatrix} 17.5 \\ 81.7 \\ -81.6 \\ 124.1 \\ -123.5 \\ -73.5 \end{bmatrix}$$

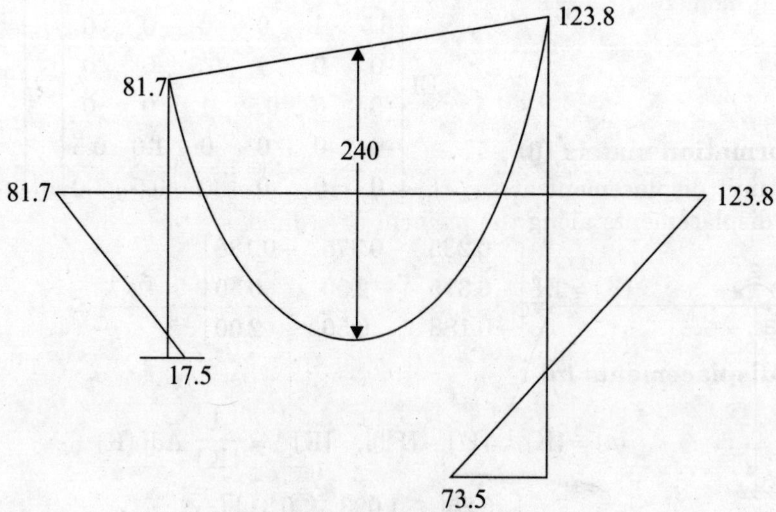

Bending moment diagram
(drawn on tension sides)

Fig. 5.74

Example 5.36 *Analyse the following frame by stiffness method and obtain the support moments.*

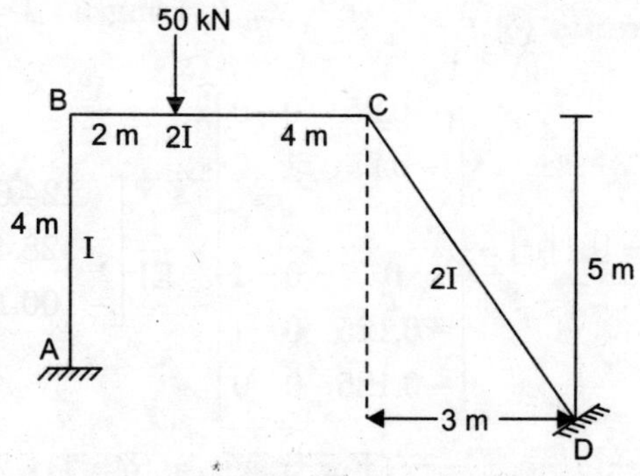

Solution The given structure is kinematically indeterminate to third degree, *i.e.,* rotation at B and C and horizontal sway due to unsymmetry in geometry and loading.

Fixed end moments :

$$M_{FBC} = -\frac{Wab^2}{l^2} = -\frac{50 \times 2 \times 4^2}{6^2} = -44.44 \text{ kNm}$$

$$M_{FCB} = \frac{Wa^2b}{l^2} = \frac{50 \times 2^2 \times 4}{6^2} = 22.22 \text{ kNm}$$

$$M_{FAB} = M_{FBA} = M_{FCD} = M_{FDC} = 0 \text{ (No external load)}$$

Fixed end moments, $(P^0) = \begin{bmatrix} 0 \\ 0 \\ -44.44 \\ 22.22 \\ 0 \\ 0 \end{bmatrix}$

Transformation matrix (β) :

Apply a unit displacement at the three system co-ordinates one by one and find the corresponding displacements along the element co-ordinates.

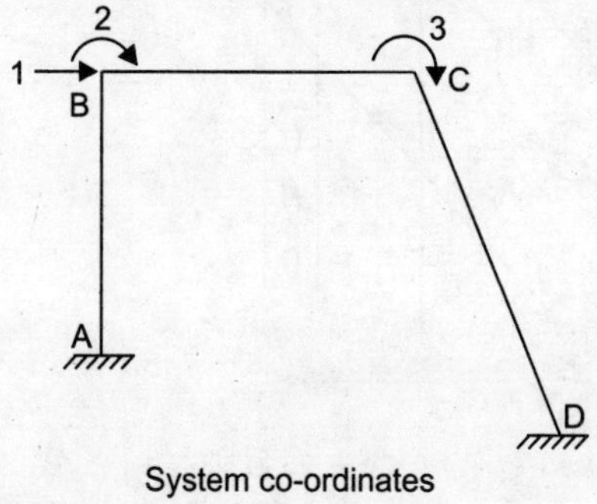

System co-ordinates

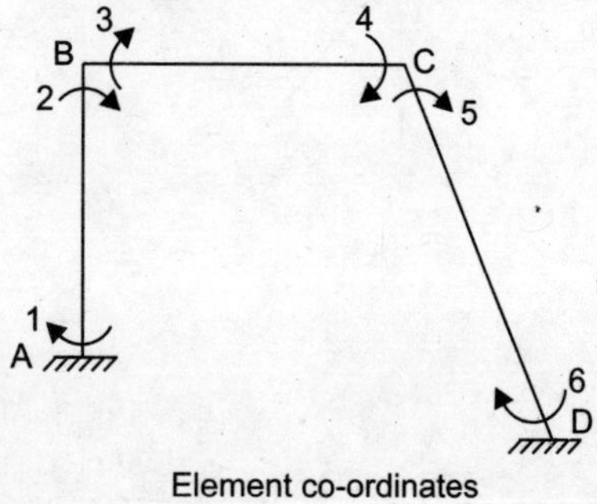

Element co-ordinates

Apply a unit displacement at co-ordinate (1) ($\delta = 1$)

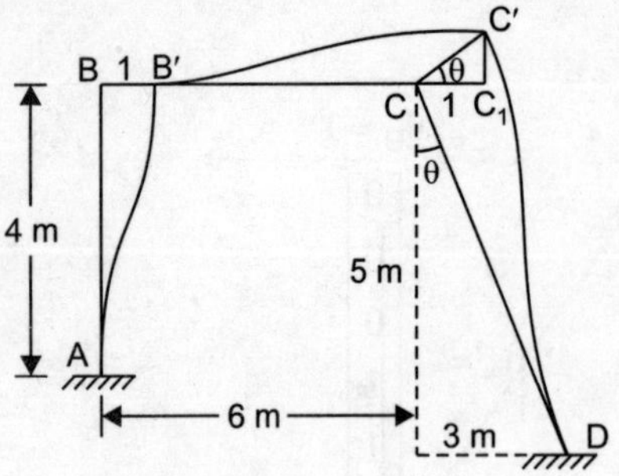

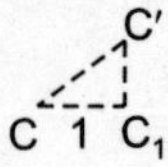

$$\text{Length CD} = \sqrt{5^2 + 3^2}$$

$$= 5.83 \text{ m}$$

$$\theta = \tan^{-1}\frac{3}{5}$$

$$= 30° \; 58'$$

Consider $\Delta CC_1C'$:

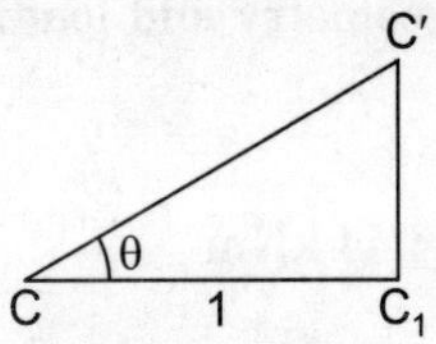

$$\frac{C_1C'}{CC_1} = \tan\theta$$

$$C_1C' = (\tan 30° \; 58') \times 1$$

$$C_1C' = 0.6$$

$$CC' = \sqrt{1^2 + 0.6^2}$$

$$CC' = 1.167$$

For this case ($\delta = 1$)

$$\beta_1 = \begin{bmatrix} -\dfrac{BB'}{l_{AB}} \\ -\dfrac{BB'}{l_{AB}} \\ \dfrac{C_1C'}{l_{BC}} \\ \dfrac{C_1C'}{l_{BC}} \\ -\dfrac{CC'}{l_{CD}} \\ -\dfrac{CC'}{l_{CD}} \end{bmatrix} = \begin{bmatrix} -\dfrac{1}{4} \\ -\dfrac{1}{4} \\ \dfrac{0.6}{6} \\ \dfrac{0.6}{6} \\ -\dfrac{1.167}{5.83} \\ -\dfrac{1.167}{5.83} \end{bmatrix} = \begin{bmatrix} -0.25 \\ -0.25 \\ 0.1 \\ 0.1 \\ -0.2 \\ -0.2 \end{bmatrix}$$

Apply a unit rotation at co-ordinate (2)

$$\theta_B = 1$$

$$\beta_2 = \begin{bmatrix} 0 \\ 1 \\ 1 \\ 0 \\ 0 \\ 0 \end{bmatrix}$$

Apply a unit rotation at co-ordinate (3)

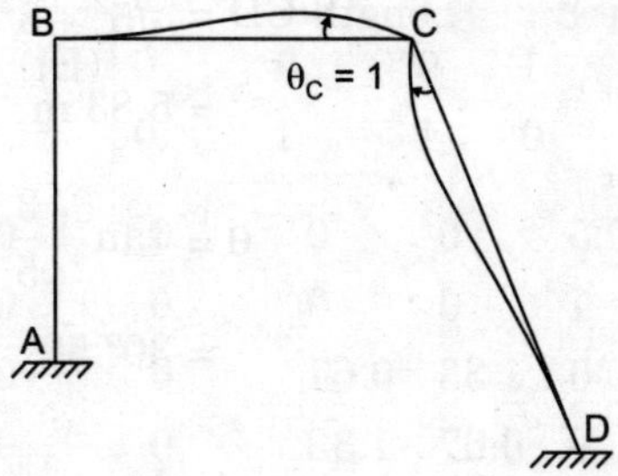

$$\theta_C = 1 \qquad \beta_3 = \begin{bmatrix} 0 \\ 0 \\ 0 \\ 1 \\ 1 \\ 0 \end{bmatrix}$$

Transformation matrix (β) :

$$[\beta] = \begin{bmatrix} -0.25 & 0 & 0 \\ -0.25 & 1 & 0 \\ 0.1 & 1 & 0 \\ 0.1 & 0 & 1 \\ -0.2 & 0 & 1 \\ -0.2 & 0 & 0 \end{bmatrix}$$

Element stiffness matrix (*k*) :

$$[k_i] = \frac{EI}{l}\begin{bmatrix} 4 & 2 \\ 2 & 4 \end{bmatrix}$$

Element AB : $$[k_1] = \frac{EI}{4}\begin{bmatrix} 4 & 2 \\ 2 & 4 \end{bmatrix} = EI\begin{bmatrix} 1 & 0.5 \\ 0.5 & 1 \end{bmatrix}$$

Element BC : $$[k_2] = \frac{E(2I)}{6}\begin{bmatrix} 4 & 2 \\ 2 & 4 \end{bmatrix} = EI\begin{bmatrix} 1.33 & 0.67 \\ 0.67 & 1.33 \end{bmatrix}$$

Element CD : $$[k_3] = \frac{E(2I)}{5.83}\begin{bmatrix} 4 & 2 \\ 2 & 4 \end{bmatrix} = EI\begin{bmatrix} 1.372 & 0.686 \\ 0.686 & 1.372 \end{bmatrix}$$

Assembled element stiffness matrix (*k*) :

$$[k] = EI\begin{bmatrix} 1 & 0.5 & 0 & 0 & 0 & 0 \\ 0.5 & 1 & 0 & 0 & 0 & 0 \\ 0 & 0 & 1.33 & 0.67 & 0 & 0 \\ 0 & 0 & 0.67 & 1.33 & 0 & 0 \\ 0 & 0 & 0 & 0 & 1.372 & 0.686 \\ 0 & 0 & 0 & 0 & 0.686 & 1.372 \end{bmatrix}$$

System stiffness matrix :

$$[K] = [\beta]^T [k] [\beta]$$

$$[K] = \begin{bmatrix} -0.25 & -0.25 & 0.1 & 0.1 & -0.2 & -0.2 \\ 0 & 1 & 1 & 0 & 0 & 0 \\ 0 & 0 & 0 & 1 & 1 & 0 \end{bmatrix} (EI)$$

$$\begin{bmatrix} 1 & 0.5 & 0 & 0 & 0 & 0 \\ 0.5 & 1 & 0 & 0 & 0 & 0 \\ 0 & 0 & 1.33 & 0.67 & 0 & 0 \\ 0 & 0 & 0.67 & 1.33 & 0 & 0 \\ 0 & 0 & 0 & 0 & 1.372 & 0.686 \\ 0 & 0 & 0 & 0 & 0.686 & 1.372 \end{bmatrix} \begin{bmatrix} -0.25 & 0 & 0 \\ -0.25 & 1 & 0 \\ 0.1 & 1 & 0 \\ 0.1 & 0 & 1 \\ -0.2 & 0 & 1 \\ -0.2 & 0 & 0 \end{bmatrix}$$

$$[K] = \begin{bmatrix} -0.25 & -0.25 & 0.1 & 0.1 & -0.2 & -0.2 \\ 0 & 1 & 1 & 0 & 0 & 0 \\ 0 & 0 & 0 & 1 & 1 & 0 \end{bmatrix} (EI) \begin{bmatrix} -0.375 & 0.5 & 0 \\ -0.375 & 1 & 0 \\ 0.2 & 1.33 & 0.67 \\ 0.2 & 0.67 & 1.33 \\ -0.4116 & 0 & 1.372 \\ -0.4116 & 0 & 0.686 \end{bmatrix}$$

$$[K] = EI \begin{bmatrix} 0.3921 & -0.175 & -0.2116 \\ -0.175 & 2.33 & 0.67 \\ -0.2116 & 0.67 & 2.702 \end{bmatrix}$$

$$[K^{-1}] = \frac{\text{Adjoint of } |K|}{|K|}$$

$$= \frac{1}{EI} \begin{bmatrix} 2.71 & 0.15 & 0.1744 \\ 0.15 & 0.4708 & -0.1047 \\ 0.1744 & -0.1047 & 0.4097 \end{bmatrix}$$

$$\{F^0\} = \begin{bmatrix} -0.25 & -0.25 & 0.1 & 0.1 & -0.2 & -0.2 \\ 0 & 1 & 1 & 0 & 0 & 0 \\ 0 & 0 & 0 & 1 & 1 & 0 \end{bmatrix} \begin{bmatrix} 0 \\ 0 \\ -44.44 \\ 22.22 \\ 0 \\ 0 \end{bmatrix}$$

$$(\because \ F^0 = \beta^T P^0)$$

$$\{F^0\} = \begin{bmatrix} -2.22 \\ -44.44 \\ 22.22 \end{bmatrix}$$

System displacement, (u)

$$\{u\} = [K^{-1}]\{F\}$$

$$\{F\} = (F^f - F^0)$$

$$= \begin{bmatrix} 0 \\ 0 \\ 0 \end{bmatrix} - \begin{bmatrix} -2.22 \\ -44.44 \\ 22.22 \end{bmatrix} = \begin{bmatrix} 2.22 \\ 44.44 \\ -22.22 \end{bmatrix}$$

$$\{u\} = \frac{1}{EI}\begin{bmatrix} 2.71 & 0.1536 & 0.1744 \\ 0.1536 & 0.4708 & -0.1047 \\ 0.1744 & -0.1047 & 0.4097 \end{bmatrix}\begin{bmatrix} 2.22 \\ 44.44 \\ -22.22 \end{bmatrix}$$

$$\{u\} = \frac{1}{EI}\begin{bmatrix} 8.968 \\ 23.58 \\ -13.369 \end{bmatrix}$$

Element co-ordinate forces (P′)

$$\{P'\} = [k]\,[\beta]\,\{u\}$$

$$\{P'\} = EI\begin{bmatrix} 1 & 0.5 & 0 & 0 & 0 & 0 \\ 0.5 & 1 & 0 & 0 & 0 & 0 \\ 0 & 0 & 1.33 & 0.67 & 0 & 0 \\ 0 & 0 & 0.67 & 1.33 & 0 & 0 \\ 0 & 0 & 0 & 0 & 1.372 & 0.686 \\ 0 & 0 & 0 & 0 & 0.686 & 1.372 \end{bmatrix}\begin{bmatrix} -0.25 & 0 & 0 \\ -0.25 & 1 & 0 \\ 0.1 & 1 & 0 \\ 0.1 & 0 & 1 \\ -0.2 & 0 & 1 \\ -0.2 & 0 & 0 \end{bmatrix}\frac{1}{EI}\begin{bmatrix} 8.968 \\ 23.58 \\ -13.369 \end{bmatrix}$$

$$\{P'\} = \begin{bmatrix} -0.375 & 0.5 & 0 \\ -0.375 & 1 & 0 \\ 0.2 & 1.33 & 0.67 \\ 0.2 & 0.67 & 1.33 \\ -0.4116 & 0 & 1.372 \\ -0.4116 & 0 & 0.686 \end{bmatrix}\begin{bmatrix} 8.968 \\ 23.58 \\ -13.369 \end{bmatrix} = \begin{bmatrix} 8.427 \\ 20.217 \\ 24.1977 \\ -0.1885 \\ -22.0335 \\ -12.86 \end{bmatrix}$$

Final forces :

$$\{P^f\} = \{P'\} + \{P^0\} = \begin{bmatrix} 8.427 \\ 20.217 \\ 24.1977 \\ -0.1885 \\ -22.0335 \\ -12.86 \end{bmatrix} + \begin{bmatrix} 0 \\ 0 \\ -44.44 \\ 22.22 \\ 0 \\ 0 \end{bmatrix} = \begin{bmatrix} 8.427 \\ 20.217 \\ -20.242 \\ 22.031 \\ -22.0335 \\ -12.86 \end{bmatrix} \text{kNm}$$

5.10 ANALYSIS OF PIN JOINTED FRAMES

Members in a pin jointed frame are predominantly in tension or compression. These members seldom develop bending or shear. The member ends can undergo linear displacements in the co-ordinate directions (X, Y or Z).

Hence the procedure for analysing these structures is easier than what we adopted for flexural frames and are illustrated in the examples below.

Example 5.37 *Using matrix stiffness method, analyse the truss for the member forces in the truss loaded as shown in Fig. 5.75. AE and l are tabulated below for all the three members.*

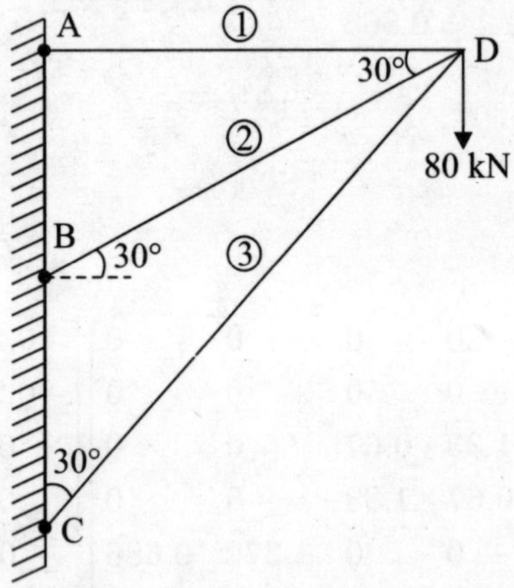

Member	AE	l
AD	400	400
BD	461.9	461.9
CD	800	800

Fig. 5.75

Solution **Global system :** There are two possible displacements at D. Those are the vertical and horizontal displacements.

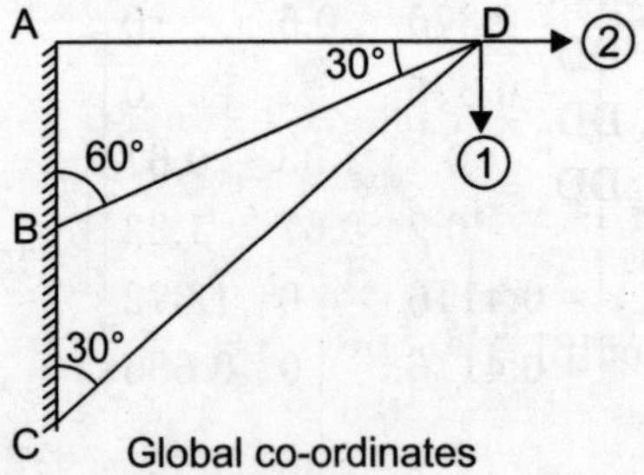

Global co-ordinates

∴ Degree of kinematic indeterminancy-2

System forces :

$$\{F\} = \begin{Bmatrix} 80 \\ 0 \end{Bmatrix}$$

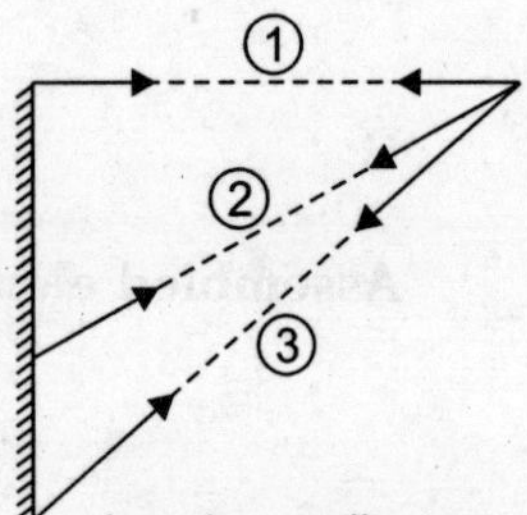

Local co-ordinates

Element co-ordinate forces {P}

$$\{P\} = \begin{Bmatrix} P_1 \\ P_2 \\ P_3 \end{Bmatrix}$$

Transformation matrix [β] :

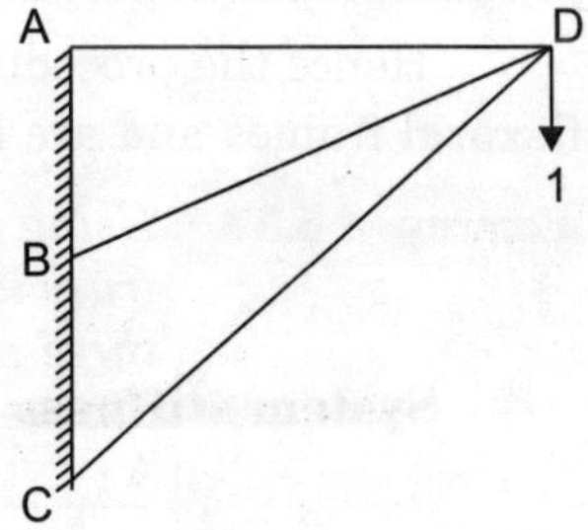

(*i*) To obtain the first column of [β] matrix unit displacement is applied along (1) in global system and corresponding values in element system are obtained.

$DD_1 = 0$

$DD_2 = \cos 60° \times 1 = 0.5$

$DD_3 = \cos 30° \times 1 = 0.866$

$$\therefore \quad \text{1st column of } [\beta] = \begin{bmatrix} 0 \\ -0.5 \\ -0.866 \end{bmatrix}$$

–ve sign due to shortening of length.

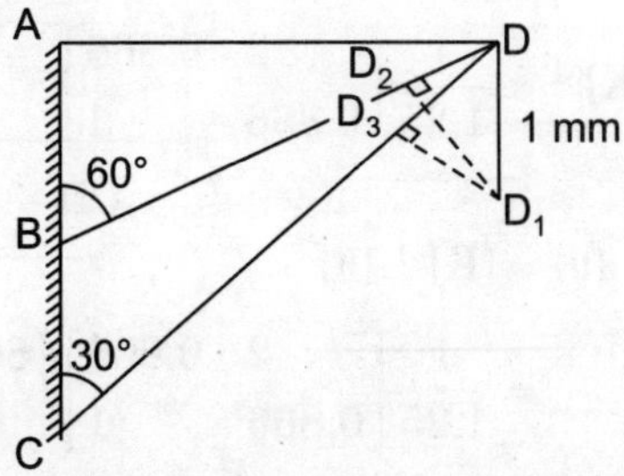

(*ii*) To obtain second column of β matrix unit displacement is applied along (2) in global system and corresponding values in element system are obtained.

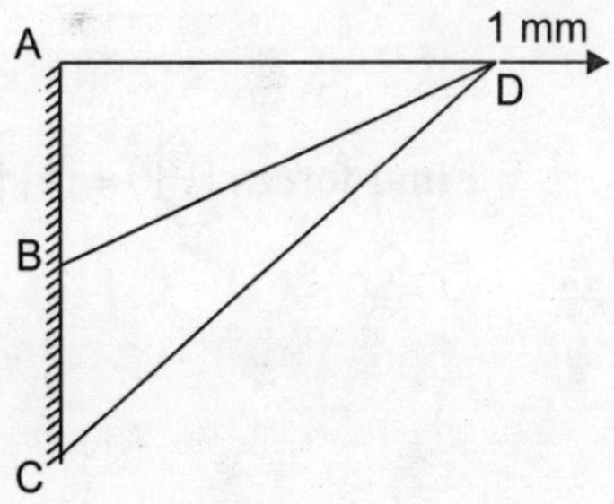

$DD_1 = 1$

$DD_2 = \cos 30° \times 1 = 0.866$

$DD_3 = \cos 60° \times 1 = 0.5$

$$\therefore \quad \text{2nd column of } [\beta] = \begin{bmatrix} 1 \\ 0.866 \\ 0.5 \end{bmatrix}$$

$$\therefore \quad [\beta] = \begin{bmatrix} 0 & 1 \\ -0.5 & 0.866 \\ -0.866 & 0.5 \end{bmatrix}$$

Assembled element stiffness matrix [*k*] :

$$[k] = \frac{AE}{l}\begin{bmatrix} 1 & 0 & 0 \\ 0 & 1 & 0 \\ 0 & 0 & 1 \end{bmatrix} = \begin{bmatrix} 1 & 0 & 0 \\ 0 & 1 & 0 \\ 0 & 0 & 1 \end{bmatrix} \qquad \left(\because \frac{AE}{l} = 1\right)$$

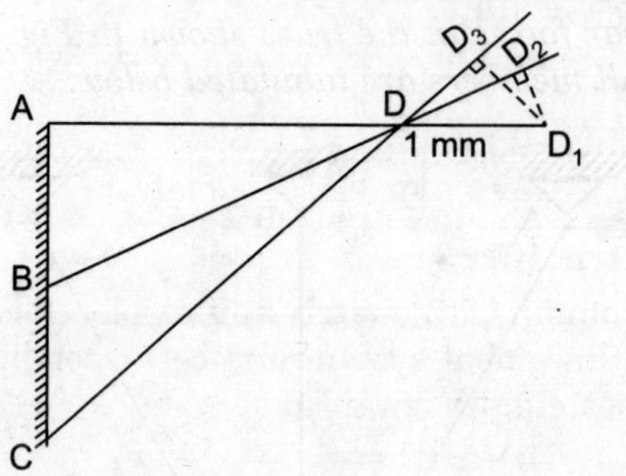

System stiffness matrix : $[K] = [\beta]^T [k] [\beta]$

$$[K] = \begin{bmatrix} 0 & -0.5 & -0.866 \\ 1 & 0.866 & 0.5 \end{bmatrix} \begin{bmatrix} 1 & 0 & 0 \\ 0 & 1 & 0 \\ 0 & 0 & 1 \end{bmatrix} \begin{bmatrix} 0 & 1 \\ -0.5 & 0.866 \\ -0.866 & 0.5 \end{bmatrix}$$

$$[K] = \begin{bmatrix} 1 & -0.866 \\ -0.866 & 2 \end{bmatrix}$$

$$[K]^{-1} = \frac{1}{1.25} \begin{bmatrix} 2 & 0.866 \\ 0.866 & 1 \end{bmatrix}$$

System displacements :

$$\{u\} = [K]^{-1} \{F\}$$

$$= \frac{1}{1.25} \begin{bmatrix} 2 & 0.866 \\ 0.866 & 1 \end{bmatrix} \begin{Bmatrix} 80 \\ 0 \end{Bmatrix}$$

$$\{u\} = \begin{Bmatrix} 128 \\ 55.424 \end{Bmatrix}$$

Final forces $[P]^F = [k] [\beta] \{u\}$

$$[P]^F = \begin{bmatrix} 1 & 0 & 0 \\ 0 & 1 & 0 \\ 0 & 0 & 1 \end{bmatrix} \begin{bmatrix} 0 & 1 \\ -0.5 & 0.866 \\ -0.866 & 0.5 \end{bmatrix} \begin{Bmatrix} 128 \\ 55.424 \end{Bmatrix}$$

$$= \begin{bmatrix} 0 & 1 \\ -0.5 & 0.866 \\ -0.866 & 0.5 \end{bmatrix} \begin{Bmatrix} 128 \\ 55.424 \end{Bmatrix}$$

$$= \begin{bmatrix} 55.424 \\ -16.0 \\ -83.136 \end{bmatrix}$$

F_{AD} = 55.424 kN (Tensile)

F_{BD} = 16.0 kN (Compressive)

F_{CD} = 83.136 kN (Compressive)

Example 5.38 *Find the bar forces in the truss shown in Fig. 5.76 by stiffness method. AE and l for all members are tabulated below.*

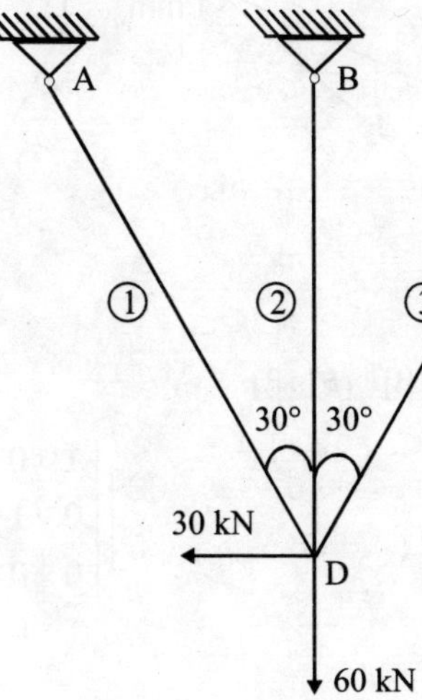

Fig. 5.76

Member	AE (MN)	*l* (cm)
AD, CD	300	300
BD	259.8	259.8

Solution We identify the local and global co-ordinates as under.

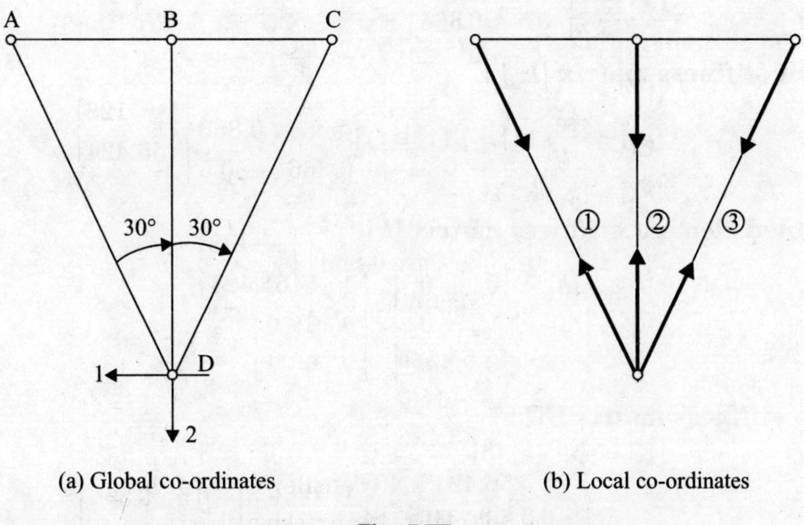

(a) Global co-ordinates (b) Local co-ordinates

Fig. 5.77

We give unit displacement along Global (1) and find the elongations in local (1) (2) and (3). These make the first column of [β] matrix.

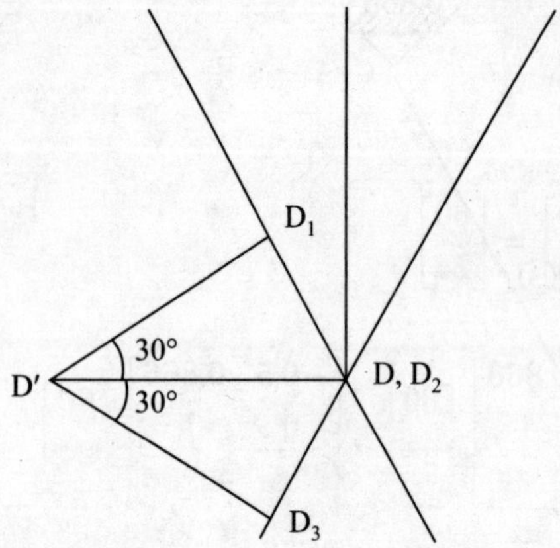

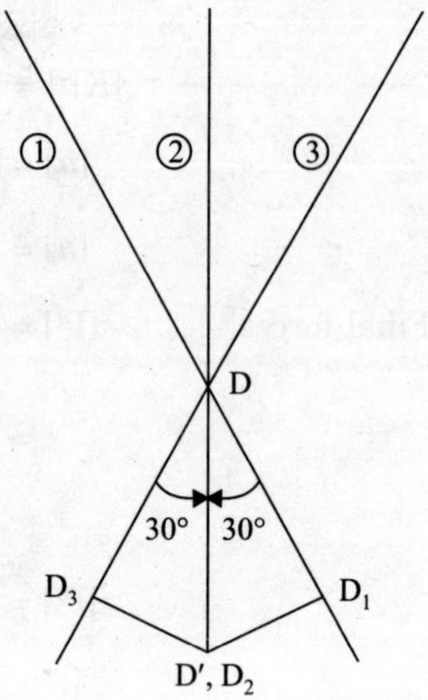

$DD' = 1 \qquad DD' = 1 = DD_2$

$DD_1 = \sin 30° = 0.5 \quad DD_3 = \cos 30° = 0.866$

$DD_2 = 0 \qquad DD_1 = \cos 30° = 0.866$

$DD_3 = \sin 30° = 0.5 \quad [\beta_2] = [0.866 \quad 1 \quad 0.866]^T$

$[\beta_1] = [-0.5 \quad 0 \quad 0.5]^T$

Now,

$$\text{Displacement matrix, } \beta = \begin{bmatrix} -0.5 & 0.866 \\ 0 & 1 \\ 0.5 & 0.866 \end{bmatrix}$$

Element stiffness matrix $[k_n]$:

$$[k_1] = \frac{AE}{l}[1] = 1\,[1] = [1]$$

$$[k_2] = [k_3] = [1]$$

Assembled element stiffness matrix $[k]$:

$$[k] = \begin{bmatrix} [k_1] & 0 & 0 \\ 0 & [k_2] & 0 \\ 0 & 0 & [k_3] \end{bmatrix} = \begin{bmatrix} 1 & 0 & 0 \\ 0 & 1 & 0 \\ 0 & 0 & 1 \end{bmatrix}$$

System stiffness matrix [K] :

$$[K] = [\beta]^T [k] [\beta]$$

$$= \begin{bmatrix} -0.5 & 0 & 0.5 \\ 0.866 & 1 & 0.866 \end{bmatrix} \begin{bmatrix} 1 & 0 & 0 \\ 0 & 1 & 0 \\ 0 & 0 & 1 \end{bmatrix} \begin{bmatrix} -0.5 & 0.866 \\ 0 & 1 \\ 0.5 & 0.866 \end{bmatrix}$$

$$= \begin{bmatrix} -0.5 & 0 & 0.5 \\ 0.866 & 1 & 0.866 \end{bmatrix} \begin{bmatrix} -0.5 & 0.866 \\ 0 & 1 \\ 0.5 & 0.866 \end{bmatrix} = \begin{bmatrix} 0.5 & 0 \\ 0 & 2.5 \end{bmatrix}$$

$$[K]^{-1} = \frac{1}{1.25} \begin{bmatrix} 2.5 & 0 \\ 0 & 0.5 \end{bmatrix}$$

$$[u] = [K]^{-1} [F]$$

$$[u] = \frac{1}{1.25} \begin{bmatrix} 2.5 & 0 \\ 0 & 0.5 \end{bmatrix} \begin{bmatrix} 30 \\ 60 \end{bmatrix} = \begin{bmatrix} 60 \\ 24 \end{bmatrix}$$

Final forces, $[P^f] = [k] [\beta] \{u\}$

$$= \begin{bmatrix} 1 & 0 & 0 \\ 0 & 1 & 0 \\ 0 & 0 & 1 \end{bmatrix} \begin{bmatrix} -0.5 & 0.866 \\ 0 & 1 \\ 0.5 & 0.866 \end{bmatrix} \begin{bmatrix} 60 \\ 24 \end{bmatrix} = \begin{bmatrix} -0.5 & 0.866 \\ 0 & 1 \\ 0.5 & 0.866 \end{bmatrix} \begin{bmatrix} 60 \\ 24 \end{bmatrix}$$

$$[P^f] = \begin{bmatrix} -9.216 \\ 24 \\ 50.784 \end{bmatrix} \text{kN.}$$

Example 5.39 *Analyse the following truss by matrix stiffness method.*

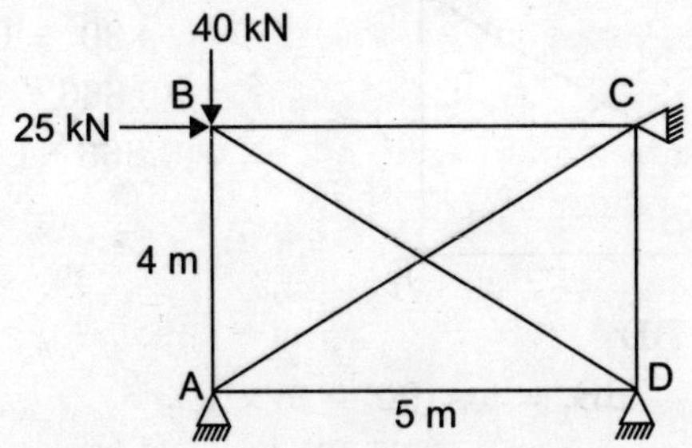

Solution The given truss is kinematically indeterminate to the second degree, *i.e.*, horizontal and vertical movement of joint B.

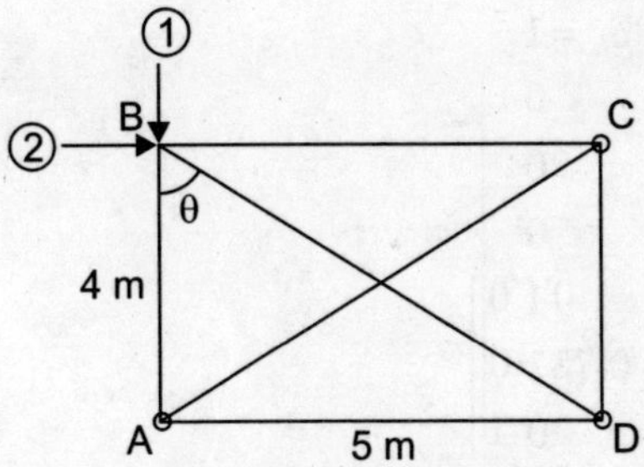

System co-ordinates

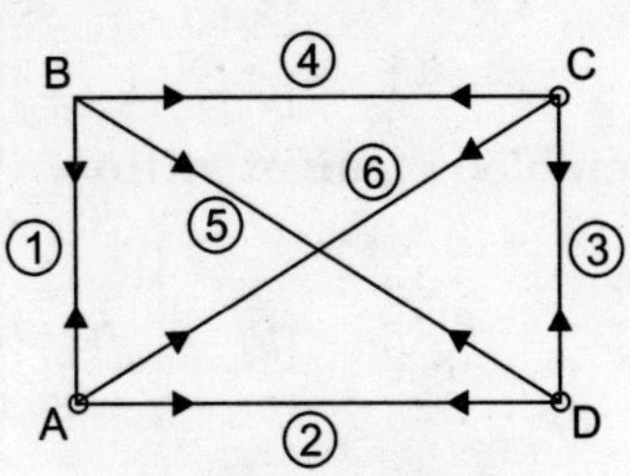

Element co-ordinates

$$\theta = \tan^{-1}\left(\frac{5}{4}\right) = 51° \, 20'$$

$$\text{Length BD} = \text{AC} = \sqrt{5^2 + 4^2} = 6.403 \text{ m}$$

Transformation matrix [β] :

Applying a unit deformation in force direction at co-ordinate (1) and (2) of system, one at a time, the transformation matrix will be obtained.

Applying $\delta_v = 1$, at B :

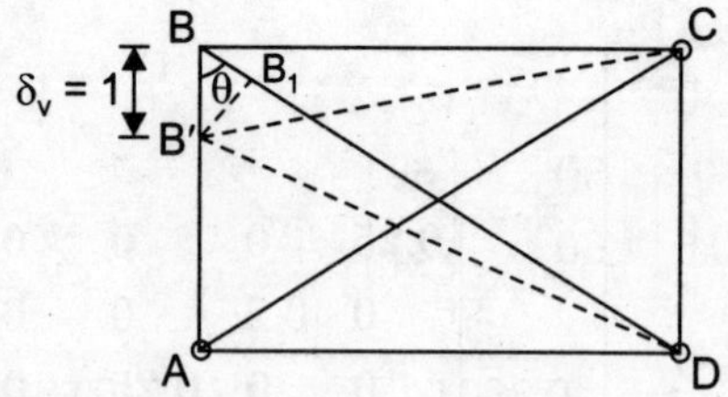

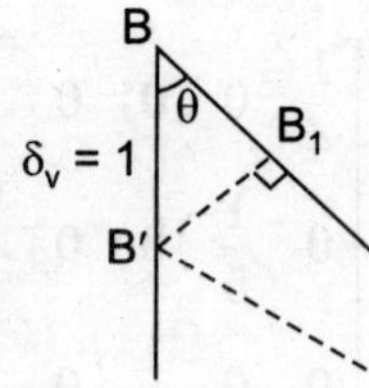

Shortening of length in BD

$$BB_1 = \cos\theta \times 1$$
$$= \cos 51° \, 20' = 0.625$$

Shortening in member AB = 1 (contraction) all other members remain the same.

Apply $\delta_h = 1$.

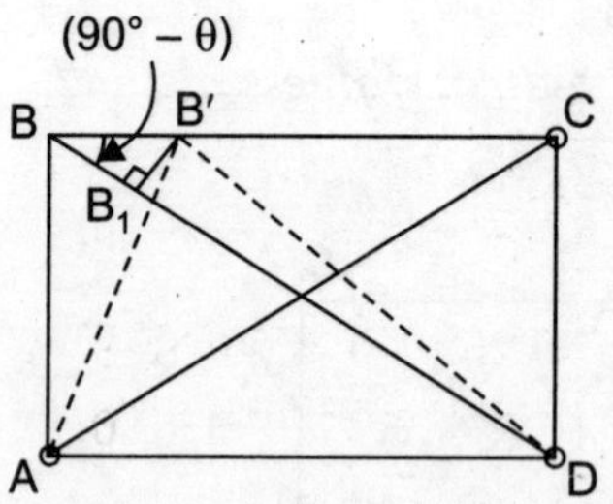

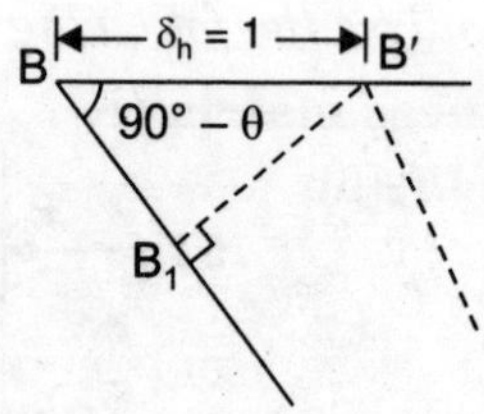

Shortening in member BD

$$BB_1 = \cos(90° - \theta) \times 1$$
$$= \cos 38° \, 40'$$
$$= 0.781 \text{ (Contraction)}$$

Shortening in member BC = 1 (Contraction) all other members remain the same.

$$[\beta] = \begin{array}{c} \begin{array}{cc} \delta_v = 1 & \delta_h = 1 \end{array} \\ \begin{bmatrix} -1 & 0 \\ 0 & 0 \\ 0 & 0 \\ 0 & -1 \\ -0.625 & -0.781 \\ 0 & 0 \end{bmatrix} \end{array}$$

Element stiffness matrix (*k*) :

$$[k_i] = \frac{AE}{l}[1] \text{ for truss}$$

$$[k_1] = \frac{AE}{4}[1], \quad [k_2] = \frac{AE}{5}[1], \quad [k_3] = \frac{AE}{4}[1]$$

$$[k_4] = \frac{AE}{5}[1], \quad [k_5] = \frac{AE}{6.403}[1], \quad [k_6] = \frac{AE}{6.403}[1]$$

Assembled element stiffness matrix (*k*) :

$$k = AE\begin{bmatrix} \frac{1}{4} & 0 & 0 & 0 & 0 & 0 \\ 0 & \frac{1}{5} & 0 & 0 & 0 & 0 \\ 0 & 0 & \frac{1}{4} & 0 & 0 & 0 \\ 0 & 0 & 0 & \frac{1}{5} & 0 & 0 \\ 0 & 0 & 0 & 0 & \frac{1}{6.403} & 0 \\ 0 & 0 & 0 & 0 & 0 & \frac{1}{6.403} \end{bmatrix} = AE\begin{bmatrix} 0.25 & 0 & 0 & 0 & 0 & 0 \\ 0 & 0.2 & 0 & 0 & 0 & 0 \\ 0 & 0 & 0.25 & 0 & 0 & 0 \\ 0 & 0 & 0 & 0.2 & 0 & 0 \\ 0 & 0 & 0 & 0 & 0.156 & 0 \\ 0 & 0 & 0 & 0 & 0 & 0.156 \end{bmatrix}$$

System stiffness matrix (K) :

$$[K] = [\beta]^T [k] [\beta]$$

$$[K] = \begin{bmatrix} -1 & 0 & 0 & 0 & -0.625 & 0 \\ 0 & 0 & 0 & -1 & -0.781 & 0 \end{bmatrix} AE \begin{bmatrix} 0.25 & 0 & 0 & 0 & 0 & 0 \\ 0 & 0.2 & 0 & 0 & 0 & 0 \\ 0 & 0 & 0.25 & 0 & 0 & 0 \\ 0 & 0 & 0 & 0.2 & 0 & 0 \\ 0 & 0 & 0 & 0 & 0.156 & 0 \\ 0 & 0 & 0 & 0 & 0 & 0.156 \end{bmatrix} \begin{bmatrix} -1 & 0 \\ 0 & 0 \\ 0 & 0 \\ 0 & -1 \\ -0.625 & -0.781 \\ 0 & 0 \end{bmatrix}$$

$$\Rightarrow [K] = \begin{bmatrix} -1 & 0 & 0 & 0 & -0.625 & 0 \\ 0 & 0 & 0 & -1 & -0.781 & 0 \end{bmatrix} AE \begin{bmatrix} -0.25 & 0 \\ 0 & 0 \\ 0 & 0 \\ 0 & -0.2 \\ -0.0976 & -0.1219 \\ 0 & 0 \end{bmatrix}$$

$$[K] = AE\begin{bmatrix} 0.311 & 0.0762 \\ 0.0762 & 0.295 \end{bmatrix}$$

$$[K]^{-1} = \frac{\text{Adjoint K}}{|K|} = \frac{1}{AE}\begin{bmatrix} 3.43 & -0.886 \\ -0.886 & 3.633 \end{bmatrix}$$

System displacement (u) :

$$\{u\} = [K]^{-1}\{F\}$$

$$= \frac{1}{AE}\begin{bmatrix} 3.43 & -0.886 \\ -0.886 & 3.633 \end{bmatrix}\begin{bmatrix} 40 \\ 25 \end{bmatrix}$$

$$\{u\} = \frac{1}{AE}\begin{bmatrix} 115.2 \\ 55.385 \end{bmatrix}$$

Forces in elements :

$$\{P\} = [k]\{\delta\} = [k][\beta]\{u\}$$

$$= AE\begin{bmatrix} 0.25 & 0 & 0 & 0 & 0 & 0 \\ 0 & 0.2 & 0 & 0 & 0 & 0 \\ 0 & 0 & 0.25 & 0 & 0 & 0 \\ 0 & 0 & 0 & 0.2 & 0 & 0 \\ 0 & 0 & 0 & 0 & 0.156 & 0 \\ 0 & 0 & 0 & 0 & 0 & 0.156 \end{bmatrix}\begin{bmatrix} -1 & 0 \\ 0 & 0 \\ 0 & 0 \\ 0 & -1 \\ -0.625 & -0.781 \\ 0 & 0 \end{bmatrix}\frac{1}{AE}\begin{bmatrix} 115.2 \\ 55.385 \end{bmatrix}$$

$$= \begin{bmatrix} -0.25 & 0 \\ 0 & 0 \\ 0 & 0 \\ 0 & -0.2 \\ -0.0976 & -0.1219 \\ 0 & 0 \end{bmatrix}\begin{bmatrix} 115.2 \\ 55.385 \end{bmatrix}$$

$$\{P\} = \begin{bmatrix} -28.8 \\ 0 \\ 0 \\ -11.080 \\ -17.99 \\ 0 \end{bmatrix}$$

Forces in members :

F_{AB} = 28.8 kN (Compression)

$F_{AD} = 0$

$F_{DC} = 0$

F_{BC} = 11.080 kN (Compression)

F_{BD} = 17.99 kN (Compression)

$F_{AC} = 0$

PROBLEMS

1. Using stiffness matrix method find the end moments at A and B for the given beam shown in Fig. 5.78.

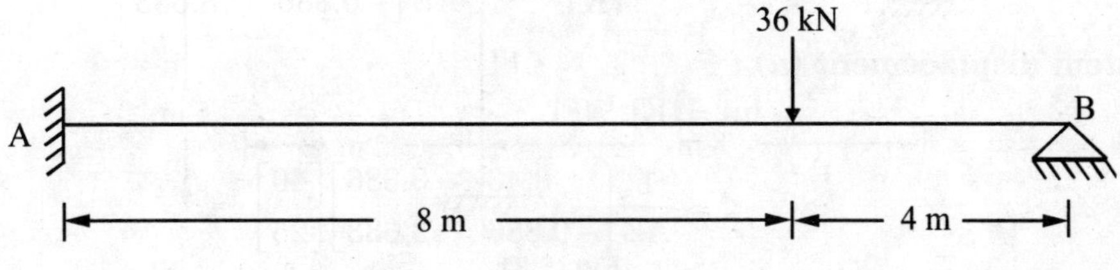

Fig. 5.78

[**Ans.** $M_A = -64$ kNm ; $M_B = 0$]

2. A continuous beam ABC shown in Fig. 5.79 has two equal spans of 10 m each with A as fixed and support C as hinged. Spans AB and BC carry central concentrated loads of 240 kN and 200 kN respectively. Supports B and C settle by $\frac{2000}{EI}$ and $\frac{1000}{EI}$ respectively. Calculate the slopes at B and C in terms of EI and hence find the end moment at B using displacement method of analysis.

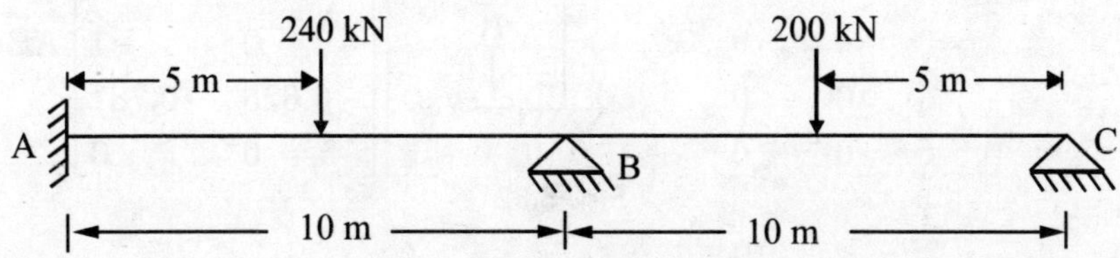

Fig. 5.79

[**Ans.** $M_{AB} = -355.8$ kNm ; $M_{BA} = 428.5$ kNm]

3. Find the end moments of the beam shown in Fig. 5.80 by stiffness matrix method and sketch the bending moment diagram.

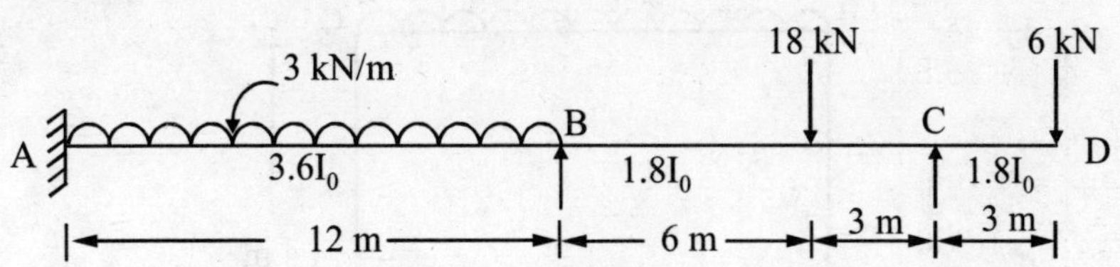

Fig. 5.80

[**Ans.** $M_{AB} = -43.005$ kNm ; $M_{BA} = +21.990$ kNm ;
$M_{BC} = -22.005$ kNm ; $M_{CB} = +18.000$ kNm]

4. Formulate the stiffness matrix for the bent shown in Fig. 5.81 treating rotations at A and D as unknowns and neglecting axial deformations.

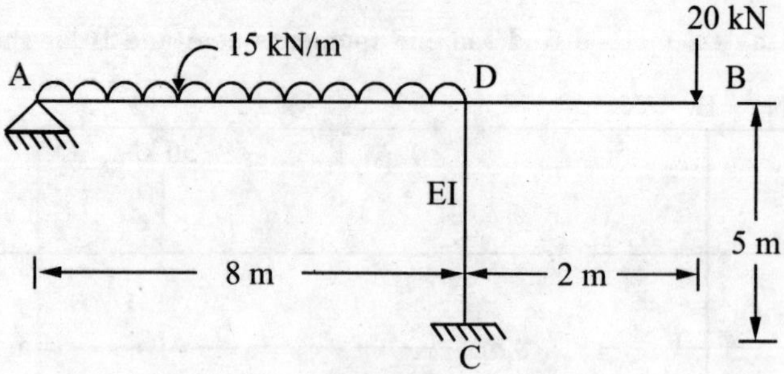

Fig. 5.81

[**Ans.** $M_{AB} = 0$; $M_{BA} = 81.3$ kNm ; $M_{BC} = -41.3$ kNm ; $M_{CD} = -20.6$ kNm]

5. Find the reactions at joint A and D for the plane frame shown in Fig. 5.82, considering only the flexural deformations. Assume that all members have the same flexural rigidity EI and that $l = 1.5$ H.

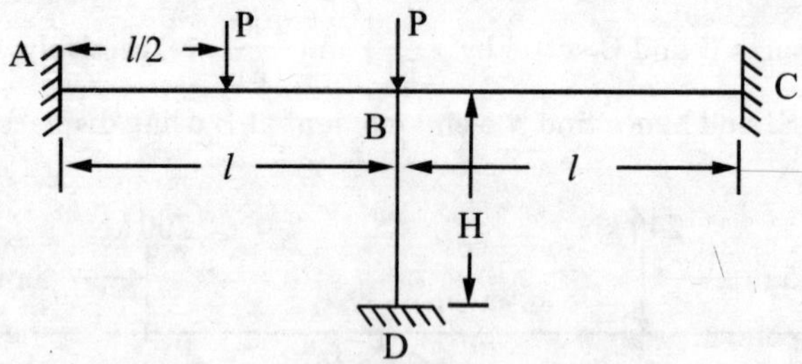

Fig. 5.82

$$\left[\textbf{Ans.}\ H_A = \frac{27\,P}{448}\ ;\ V_A = \frac{31\,P}{56}\ ;\ M_A = \frac{Pl}{7}\ ;\ H_D = \frac{-27\,P}{224}\ ;\ V_D = \frac{3\,P}{2}\ ;\ M_D = \frac{3\,Pl}{112}\right]$$

6. Analyse the portal frame with fixed base shown in Fig. 5.83 using the matrix stiffness method, given.

$$I_{AB} = 3I_0\ ;\ I_{BC} = 2I_0\ ;\ I_{CD} = 3I_0$$

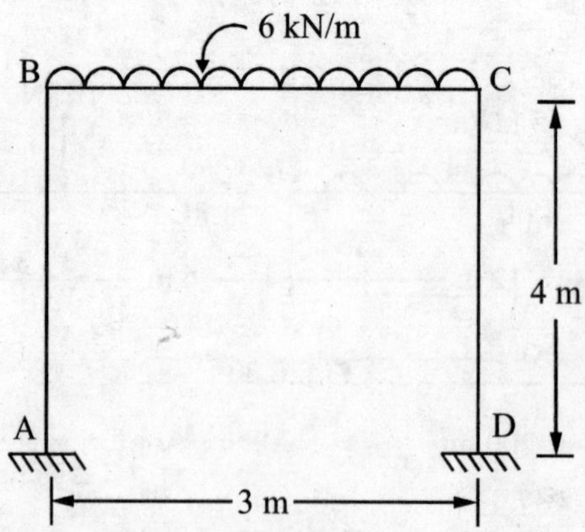

Fig. 5.83

[**Ans.** $M_A = -1.54$ kNm ; $M_B = 3.10$ kNm ; $M_C = 3.10$ kNm ; $M_D = 1.54$ kNm]

7. Analyse the frame shown in Fig. 5.84 using the matrix stiffness method and draw the bending moment diagram.

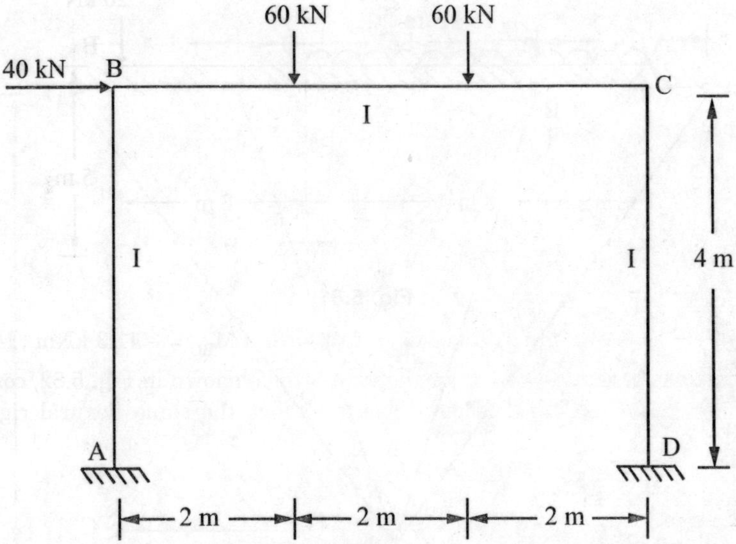

Fig. 5.84

[**Ans.** M_{AB} = 18 kNm ; M_{BA} = – 28 kNm ; M_{BC} = 28 kNm ;
M_{CB} = – 92 kNm ; M_{CD} = 92 kNm ; M_{DC} = 78 kNm]

8. A beam AB is suspended from a wire CB as shown in Fig. 5.85. The beam carries a central concentrated load P. It may be assumed that $E = 7 \times 10^{10}$ N/m², I = 800 mm⁴, A = 1.2 mm² and P = 300 N. Determine using the stiffness matrix approach, the deflection of point B. What would be this deflection if the load P were to be applied upwards, instead of downwards? The axial deformation of the beam AB may be ignored.

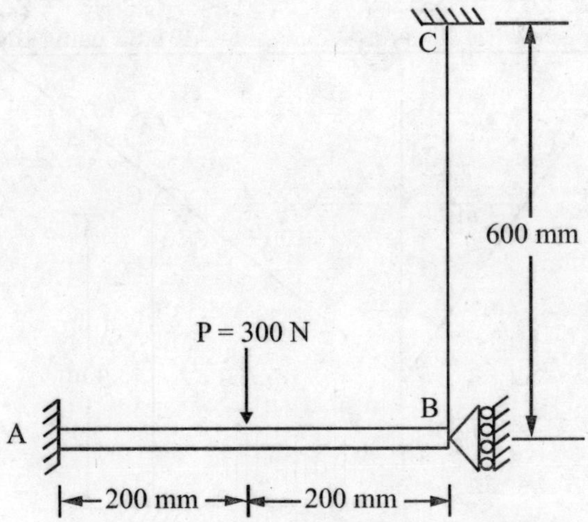

Fig. 5.85

[**Ans.** Deflection of point B = 0.64 mm (Downwards)]

9. Four steel wires OA, OB, OC and OD jointly support a load as shown in Fig. 5.86. Area of wires OB and OC is twice that of wires OA and OD. Formulate the stiffness matrix for the system and solve for the forces in the wires.

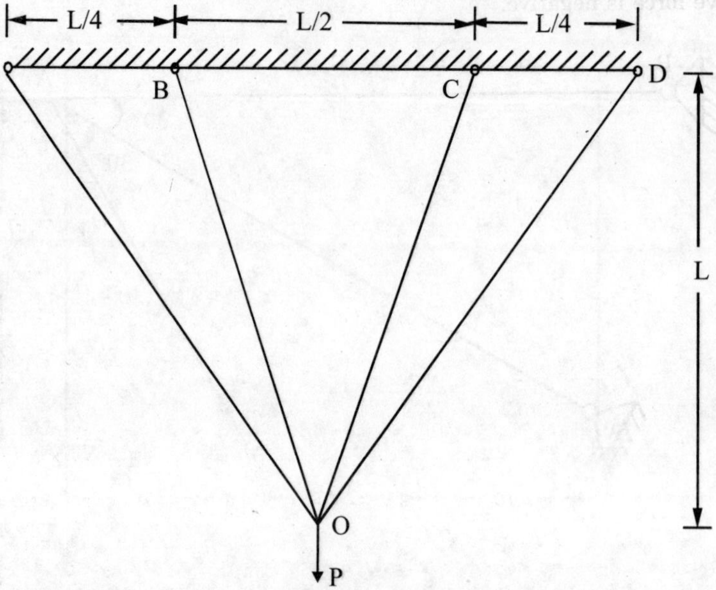

Fig. 5.86

[**Ans.** $F_{OA} = 0.1576$ P ; $F_{OB} = 0.3704$ P ; $F_{OC} = 0.3704$ P ; $F_{OD} = 0.1576$ P]

10. A pin jointed frame consists of three members connected as shown in Fig. 5.87. Compute the internal forces in members using matrix displacement method. Take E = 10^5 N/mm² for all members. Area of members are equal, A = 600 mm².

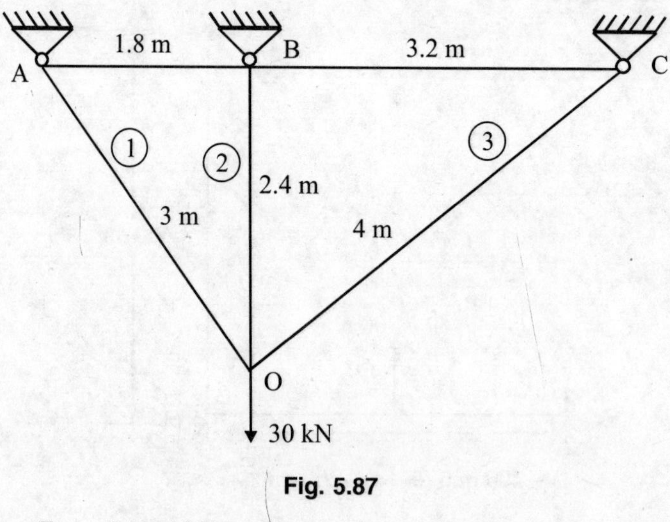

Fig. 5.87

[**Ans.** $F_{OA} = 9.9934$ kN (tension) ; $F_{OB} = 17.49$ kN (tension) ; $F_{OC} = 7.5$ kN (tension)]

11. Find the axial forces in all members of the truss shown in Fig. 5.88. The truss is subjected to a horizontal force P at joint A. Neglect the weights of the members from the analysis. Each member has a length L and axial rigidity EA. Assume that tensile force in a member is positive and compressive force is negative.

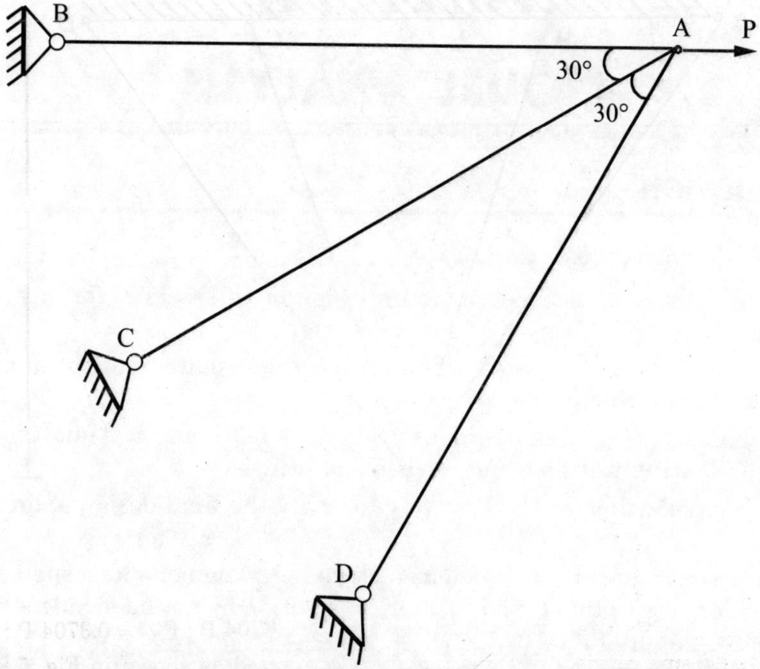

Fig. 5.88

[**Ans.** $F_{AB} = \frac{P}{2}$; $F_{AC} = 0.433P$; $F_{AD} = \frac{P}{4}$]

CHAPTER 6

MODEL ANALYSIS

6.1 INTRODUCTION

6.1.1 Need for Model Testing

Model analysis may be needed either in design or in research for one or more of the following reasons :

(*i*) A model analysis is indicated when an alternate mathematical analysis is either impossible or impracticable.

(*ii*) Model analysis is required in order to verify the mathematical analysis of unusual or complicated structures, such as suspension bridges.

(*iii*) Model analysis will offer the best way to verify the final design of an indeterminate structure.

(*iv*) In the case of an important problem such as the design of a long span bridge, if there is a need for verification of the theoretical solution by an independent method, model analysis will be useful.

(*v*) Model analysis will be useful for the confirmation of a newly developed theoretical method of analysis to verify its validity and accuracy.

6.1.2 Limitations of Model Analysis

(*i*) Design and loading of a model to exactly simulate the behaviour of the prototype will not be possible in all cases. For instance, the structural details such as welding or riveting do not scale up or down exactly. Hence, model analysis in such cases will not yield the correct stresses.

(*ii*) Different conditions that must be satisfied to achieve complete similarity between the model and the prototype may be incompatible. This is often the difficulty with dynamic loads. Model analysis in such cases is either impossible or doubtful.

(*iii*) Self weight and gravity loads are difficult to be simulated with the densities of model materials available.

6.1.3 Objectives in a Model Analysis of Structures

(*i*) Determination of stress values (may be bending moment, shear force or deflection) at various points in the model.

(*ii*) Determination of stress distribution, stress pattern or stress concentration in structures.

(*iii*) Determination of critical or buckling loads.

(*iv*) Analysis of characteristics of normal modes of vibration.

6.1.4 Types of Model Analysis

Model analysis is of two types :

(*i*) Direct model analysis

(*ii*) Indirect model analysis

In direct model analysis, the model is usually loaded in exactly the same manner as in the prototype. The response of the prototype is determined from measurements directly made on the model.

In the indirect model analysis, the response of the prototype or actual structure is studied indirectly. Influence lines are first obtained from the model. Using the influence lines, required stress components due to the given loading on the prototype are computed.

The fundamental difference between the two types of model analysis are compared and presented below.

S.No.	*Direct Method of Analysis*	*Indirect Method of Analysis*
1.	Model is loaded usually in exactly the same manner as the prototype.	The loading of the model is not directly related to the loading of the prototype.
2.	Strain measuring devices are mounted on the model and the recorded strains are related to the prototype.	Strain measurements are not made on the model.
3.	Strains and deflections are observed on the model and corresponding stresses and deflections of the prototype are determined using the principles of similitude.	Known displacement or distortion is impressed at chosen points of action and in the direction of the desired external reaction or internal stress components and the resulting displacements of all load points are measured.
4.	Load scaling is the distinguishing feature of direct model analysis.	The object here is to obtain the influence lines (by using Muller Breslau Principle) for reactions or stress components.

6.2 SIMILITUDE

Similitude means similarity between two objects that exist with regard to their physical characteristics.

6.2.1 Principles of Similitude

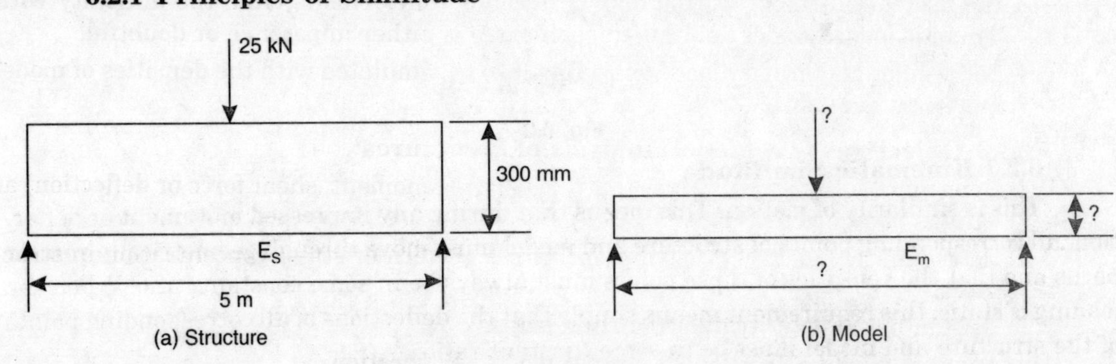

Fig. 6.1

Experience has shown that as far as practicable experiments performed on real structures yield results more readily and are of a higher reliability. This is obvious, because, the use of models add some additional sources of error and uncertainties due to change of size, materials and loading system. However, there are a variety of occasions where the use of reduced scale models is necessary either due to reasons of economy or practicability (*e.g.*, Models of large dams, the test requiring high capacity loadings). While using these models, scaling is always to be done, but sometimes it is more convenient to use different materials as well as different loading systems. Thus the theory of model to structure correlation must consider geometric scaling, load scaling and scaling of material properties. The principles which govern the scaling down so that the response and behaviour of the model would be exactly the same as that of the prototype are known as the Principles of Similitude.

These principles will ensure the condition for similarity of behaviour and response between the model and the structure and will also give the relationships (size, loading, etc) between the model and the structure. Thus, the result obtained from model testing can be used to extrapolate and predict the structure response.

In general, there are three kinds of similitude:

1. Geometric Similitude,
2. Kinematic Similitude, and
3. Dynamic (and/or Mechanical) Similitude.

6.2.2 Geometric Similitude

This is similarity of form. This means that all homologous (corresponding) dimensions of the structure and model must be in the some constant ratio.

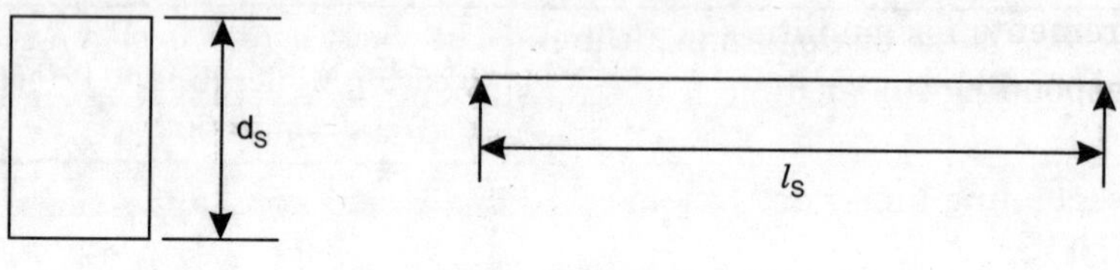

(a) Structure

(b) Model

Fig. 6.2

6.2.3 Kinematic Similitude

This is similarity of motion. This means that during any impressed movement or deflection, all corresponding points of structure and model must move through geometrically similar paths and that the velocities of these points must always be in some constant ratio. When the loading is static, this requirement means simply that the deflections of all corresponding points of the structure and model must be in some constant ratio.

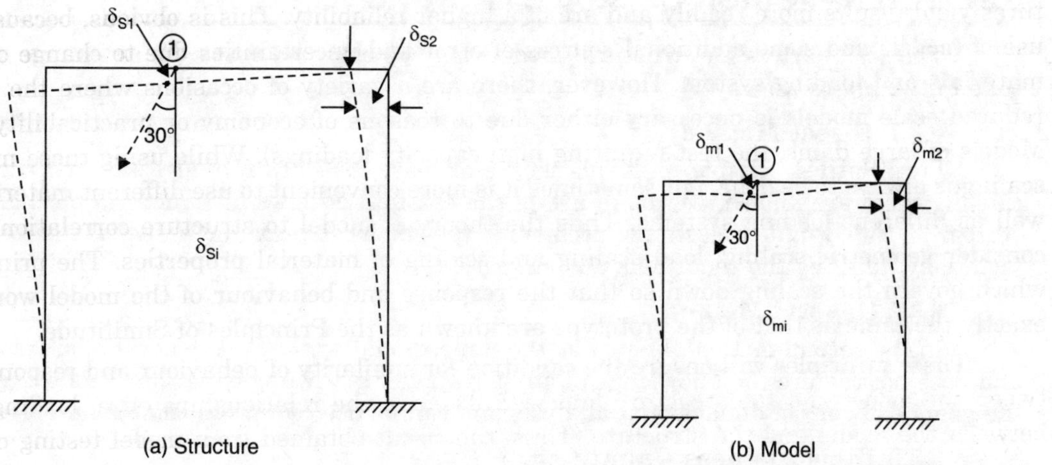

Fig. 6.3

6.2.4 Dynamic (and/or Mechanical) Similitude

This means similarity of masses and/or forces. This means that Kinematic similitude must, first of all, exist. In addition, the masses of all corresponding parts of the structure and model, and all corresponding forces which affect the motions of the structure and model must be in some constant ratio. In the case of static loadings, mechanical rather than dynamic similitude is required.

To summarize the above as they apply to statically loaded structural prototypes and models, the requirements for similitude are:

(*a*) All corresponding linear dimensions must be in some constant ratio *i.e.*, $L_S/L_m = k$,

(*b*) All corresponding linear deflections must be in some constant ratio *i.e.*, $\frac{\delta_S}{\delta_m} = c$, and

(*c*) All corresponding external loads, internal elastic forces and dead loads must be in some constant ratio *i.e.*, $F_S/F_m = j$.

6.2.5 Primary Scale Factors

These constant ratios are known as Primary Scale Factors. When these scale factors are determined and applied to the several dimensions, properties and loads of the prototype structure the corresponding dimensions and loads for the model will result.

Any two of the three primary scale factors may be assigned arbitrary values, the third must be computed from these two.

6.2.6 Secondary Scale Factors

These define ratios between prototype and model of such measurable quantities as cross sectional areas, masses, moments of inertia or moments of forces. When a model is designed for direct analysis, the procedure is to assign a value for the linear scale factor k, force scale factor j or deflection scale factor c. Secondary scale factors are determined from the primary scale factors.

6.3 DETERMINATION OF SCALE FACTORS

The scale factors for structural models may be obtained using the conditions of similitude in two alternate ways.

6.3.1 Direct Method

For relatively simple problems where at least the nature of stresses and the factors causing them are known, we make use of the established laws of structural mechanics and start from the formula of a simple problem of the same category and arrive at the correlation factors by fulfilling the conditions of similarity.

6.3.2 General Method

In the second method, known as the dimensional analysis, a general approach is used and does not require the knowledge of any formula or governing equation. This makes use of the general theory of dimensional analysis and can be used for more complex problems.

6.3.3 Dimensionless Quantities

In model analysis, this is the most important aspect. Whereas linear and force dimensions can get scaled down or up, all quantities that have no dimensions will have no scaling down (or up). The scale will be unity.

Breadth to depth ratio is a dimensionless quantity. Angles are dimensionless. Strains have no dimension.

Hence, linear ratios, angles and strains will be the same between the model and the structure it represents.

Very often analysts take great pains to coin dimensionless parameters for phenomena, structural actions etc. These parameters would define the phenomena whatever the linear or force scales. The following are some examples of dimensionless parameters.

(*i*) $p = \frac{A_S}{bd} \times 100$ ... Percentage of steel in R.C.

(*ii*) $r = \frac{A_S f_y}{bd f_{ck}}$ ... Ratio of forces on steel and concrete

(*iii*) $R = \frac{M}{bd^2 f_{ck}}$ ... Moment ratio in R.C.

The advantage of using parameters p, r, R above is that they are valid for all values of b, d, f_{ck} etc. [Needless to add that A_S is area in L^2 units, f_y, f_{ck} are in WL^{-2} units, b and d are in L units and M is in LW unit]

6.4 DERIVATION OF VARIOUS SCALE FACTORS

Suppose we have a structure and a model thereof constructed to the following scales. Linear scale – k ; Force scale – j. We must evolve methods of deriving secondary scale factors for areas, forces, moments etc.

6.4.1 Direct Method

(*i*) **Area** $\frac{A_S}{A_m} = \frac{{L_S}^2}{{L_m}^2} = k^2$

Hence, $A_m = \frac{A_S}{k^2}$

In actual laboratory practice we would find it inappropriate to use the same scale k for length dimensions and cross-sectional dimensions. We can then use a slicing factor z_1 so that

$$\frac{A_S}{A_m} = k^2 z_1 \qquad ...(6.1)$$

This means that we have a scale k for linear (centre line) dimensions and another scale for member cross-sections.

(*ii*) **Forces**

$$\frac{W_S}{W_m} = \frac{F_S}{F_m} = j, \text{ W being loads, F being forces.}$$

Since $F = \sigma A = A.E\varepsilon$,

$$j = \frac{A_S}{A_m}\frac{E_S}{E_m}\frac{\varepsilon_S}{\varepsilon_m} = (k^2\ z_1)\frac{E_S}{E_m}.1$$

We can define z such that,

$$z = z_1\frac{E_S}{E_m} \qquad ...(6.2)$$

then

$$j = k^2 z \qquad ...(6.3)$$

This is the force scale.

In this case we cannot use the force scale j arbitrarily. Once the model materials are chosen, then j scale gets fixed.

(*iii*) **Moments, bending moments, torsion**

$$\frac{M_S}{M_m} = \frac{F_S\ L_S}{F_m\ L_m} = k^2\ z.k = k^3\ z \qquad ...(6.4)$$

(*iv*) **Moments of inertia I**

For flexural members $\quad \frac{M}{I} = \frac{E}{R}; \quad I = \frac{MR}{E}$

Hence, $\quad \frac{I_S}{I_m} = \frac{M_S\ R_S}{E_S}\frac{E_m}{M_m\ R_m} = \frac{M_S}{M_m}.\frac{R_S}{R_m}.\frac{E_m}{E_S}$

$$\frac{I_S}{I_m} = k^3.z.k.\frac{E_m}{E_S} = k^4\ z\left(\frac{E_m}{E_S}\right) \qquad ...(6.5)$$

(*v*) **Material density γ**

$$W = \gamma\ A\ L; \quad \gamma = W/AL$$

$$\frac{\gamma_S}{\gamma_m} = \frac{W_S}{W_m}.\frac{A_m}{A_S}.\frac{L_m}{L_S} = k^2\ z.\frac{1}{k^2 z_1}\frac{1}{k}$$

Since $\quad z = \frac{E_S}{E_m}z_1,$

$$\frac{\gamma_S}{\gamma_m} = \frac{1}{k}\left(\frac{E_S}{E_m}\right) \qquad ...(6.6)$$

This will decide the kind of material we must choose. The inference drawn as given by equation (6.6) is of no practical value because we cannot choose materials of any chosen density. We have to choose our model materials based on other criteria such as machinability, low modulus of elasticity, fairly linear elastic behaviour etc. Then we cannot, in addition say, "it must have a density of 11.34 g/cc".

We must realize that in static testing of structures, density is important in simulating self weight. We must realize that this cannot be done. So in model tests self weight forces have to be substituted by surface forces only.

(*vi*) **Deflections**

Generally, $$\Delta = \int \frac{M_m\, dx}{EI}$$

Hence, $$\frac{\Delta_S}{\Delta_m} = \frac{M_S}{M_m} \cdot \frac{m_S}{m_m} \cdot \frac{L_S}{L_m} \frac{E_m}{E_S} \cdot \frac{I_m}{I_S}$$

$$\frac{\Delta_S}{\Delta_m} = k^3\, z\,.\,k\,.\,k\left(\frac{E_m}{E_S}\right)\frac{1}{k^4 z} \cdot \frac{E_S}{E_m} = k \qquad \text{...(6.7)}$$

We see that in spite of choosing a slicing factor, we get our kinematic scale to be the same as our linear scale.

Example 6.1 shows how a truss can be modelled.

Examples 6.2 and 6.3 model I girders.

Example 6.4 shows how a column is modelled.

Example 6.1 *Design a model of the prototype truss shown in Fig. 6.4 for direct analysis. The cross-sectional areas of the members shown against each member are in sq mm. The prototype is of steel with E = 2.1 × 10⁵ N/mm². The model material is aluminium with E = 0.7 × 10⁵ N/mm².*

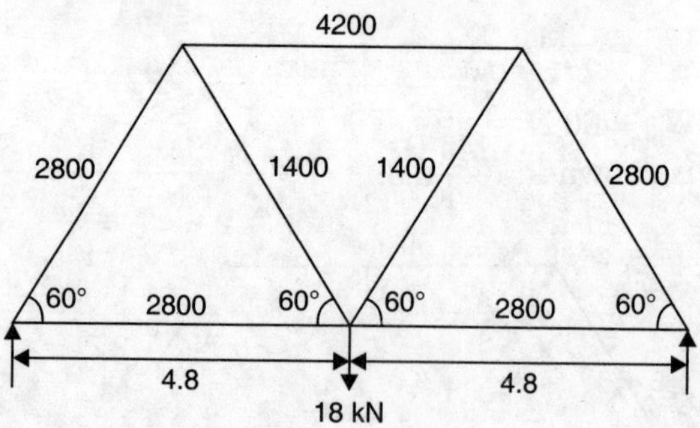

Fig. 6.4. Prototype.

Solution **Span :**

Let us assume that the span of the model is 0.96 m

Therefore, scale for length, $k = \dfrac{L_S}{L_m} = \dfrac{9.6}{0.96} = 10\,;\quad k = 10$

Areas :

We know $$\frac{A_S}{A_m} = k^2 z_1$$

$$z = z_1 \cdot \frac{E_S}{E_m}; \qquad z_1 = z \cdot \frac{E_m}{E_S}$$

$$\frac{A_S}{A_m} = k^2 z \frac{E_m}{E_S}$$

$$A_S = k^2 z \cdot \frac{E_m}{E_S} \cdot A_m$$

$$A_m = \frac{E_S}{E_m} \cdot \frac{1}{k^2 z} A_S$$

$$A_m = \frac{2.1 \times 10^5}{0.7 \times 10^5} \times \frac{1}{10^2 z} \cdot A_S$$

$$A_m = \frac{3A_S}{100z}$$

Adopting $z = 3$

$$A_m = \frac{3A_S}{100z} = \frac{3A_S}{100 \times 3} = \frac{A_S}{100}$$

Therefore, $$A_m = \frac{A_S}{100}$$

Forces :

$$\frac{F_S}{F_m} = \frac{W_S}{W_m} = k^2 z$$

$$W_m = \frac{W_S}{k^2 z} = \frac{18}{10^2 \times 3} = \frac{6}{100} \text{ kN} = \frac{6000}{100} \text{ N}$$

$$W_m = 60 \text{ N}$$

The designed model is shown in Fig. 6.5.

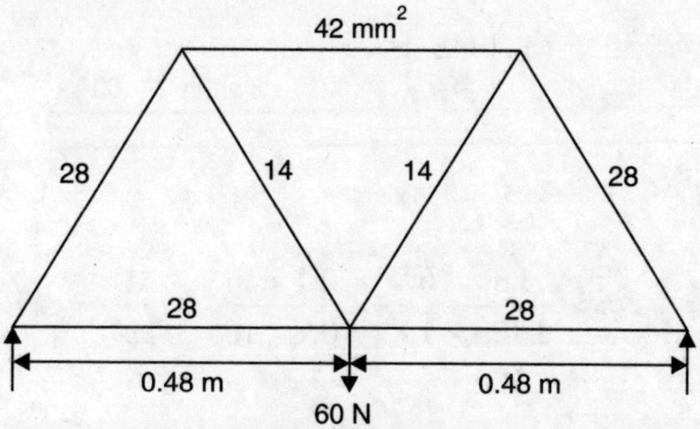

Fig. 6.5. Model.

Example 6.2 *A steel beam is simply supported over a span of 7.5 m. The cross-section of the beam is an I section with a depth of 550 mm. The moment of inertia with respect to the major axis is 1.8×10^9 mm^4 and the modulus of elasticity, E = 2.1 × 10^5 N/mm^2. The beam is subjected to a central concentrated load of 90 kN. Design a model for direct analysis using aluminium plates which are available in 3 mm thickness. E for aluminium = 0.7×10^5 N/mm^2.*

Solution

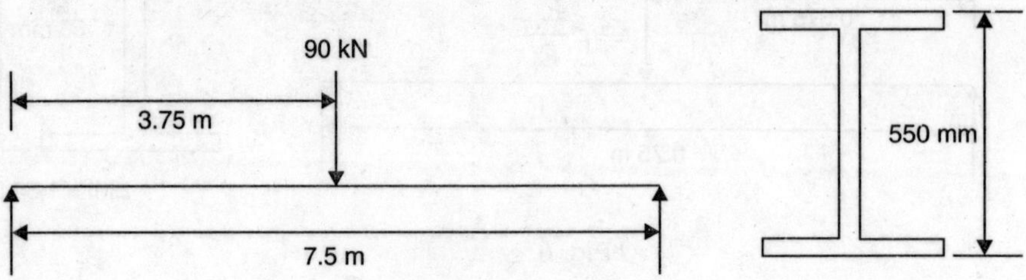

Fig. 6.6. Prototype structure.

Structure : $l_S = 7.5$ m ; $d_S = 550$ mm ; $I_S = 1.8 \times 10^9$ mm^4

$E_S = 2.1 \times 10^5$ N/mm^2 ; $W_S = 90$ kN

Model : $t_m = 3$ mm ; $E_m = 0.7 \times 10^5$ N/mm^2

Adopt a span of 0.75 m for the model

$$k = \frac{L_S}{L_m} = \frac{7.5}{0.75} = 10 \qquad k = 10$$

Depth of the model, $d_m = \dfrac{d_S}{k} = \dfrac{550}{10} = 55$ mm

Assuming an arbitrary width of 25 mm for the flange plates,

Moment of inertia of the model

$$I_m = \left[\frac{25 \times 3^3}{12} + 3 \times 25 \times \left(\frac{55}{2} - \frac{3}{2}\right)^2\right] \times 2 + \frac{3 \times 49^3}{12}$$

$$= [56.25 + 50{,}700] \times 2 + 29{,}412 = 1{,}30{,}925 \text{ mm}^4$$

We know, $$\frac{I_S}{I_m} = k^4 z\left(\frac{E_m}{E_S}\right) \qquad \text{...[Eqn. (6.5)]}$$

$$\frac{I_S}{I_m} \cdot \frac{E_S}{E_m} \cdot \frac{1}{k^4} = z$$

$$z = \frac{1.8 \times 10^9}{1.309 \times 10^5} \times \frac{2.1 \times 10^5}{0.7 \times 10^5} \times \frac{1}{10^4} = 4.125$$

$$z = 4.125$$

Load :

Load to be applied, $F_m = \dfrac{F_S}{k^2 z}$

$$F_m = \frac{90 \times 1000}{10^2 \times 4.125} = 218.18 \text{ N}$$

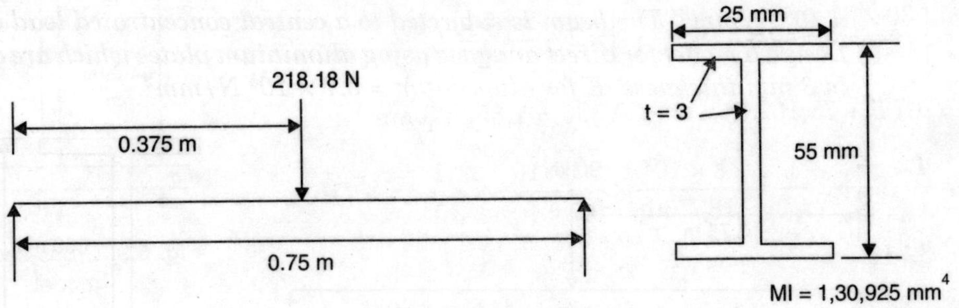

Fig. 6.7. Aluminium model.

Example 6.3 *A steel cantilever beam is shown in Fig. 6.8. The Young's modulus of the material (steel) of the beam is 2.1 × 10^5 N/mm^2 and the moment of inertia I_{xx} = 2.3 × 10^9 mm^4. Design a model for direct analysis. Use aluminium plates of 3 mm thickness with a modulus of elasticity of 0.7 × 10^5 N/mm^2.*

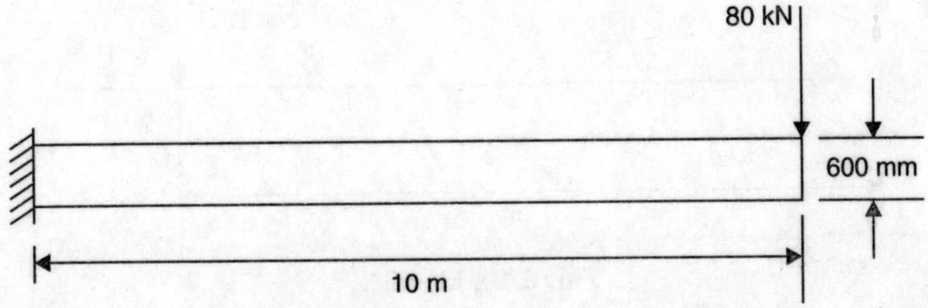

Fig. 6.8. Prototype structure.

Solution *Prototype Structure* :

Steel I beam

$$l_S = 10 \text{ m}\ ;\quad d_S = 600 \text{ mm}$$
$$E_S = 2.1 \times 10^5 \text{ N/mm}^2\ ;\quad I_S = 2.3 \times 10^9 \text{ mm}^4$$

Model *(Aluminium)*

$$E_m = 0.7 \times 10^5 \text{ N/mm}^2.$$

Let us choose a length of 1 m for the model

$$k = \frac{l_S}{l_m} = \frac{10}{1} = 10$$

For geometric similitude,

$$d_m = \frac{d_S}{k} = \frac{600}{10} = 60 \text{ mm} \qquad \left(\text{Since } \frac{d_S}{d_m} = k\right)$$

We will assume the flange width of the model as 30 mm.

Thickness of plates to be used = 3 mm.

The cross-section of the model is shown in Fig. 6.9.

Moment of Inertia of model,

$$I_m = \left[\frac{30\times3^3}{12} + 30\times3\times\left(\frac{57}{2}\right)^2\right]\times 2 + \frac{3\times54^3}{12}$$

$$I_m = [67.5 + 73,102.5] \times 2 + 39,366 = 1,85,706 \text{ mm}^4$$

$$z = \frac{I_P}{I_m}\cdot\frac{E_P}{E_m}\cdot\frac{1}{k^4} = \frac{2.3\times10^9}{185706}\times\frac{2.1\times10^5}{0.7\times10^5}\times\frac{1}{10^4} = 3.716$$

$$z = 3.716$$

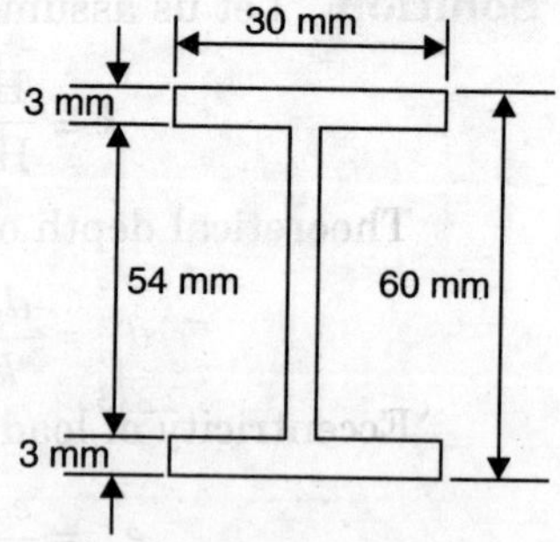

Fig. 6.9. Cross-section of model.

Load :

$$\frac{W_P}{W_m} = k^2 z$$

$$W_m = \frac{W_P}{k^2 z} = \frac{80\times1000}{10^2\times3.716} = 215.3 \text{ N}$$

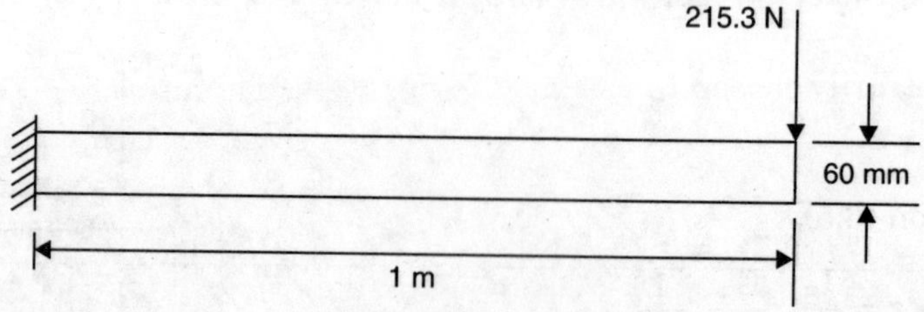

Fig. 6.10. Model.

Example 6.4 *An eccentrically loaded column made of a steel I section, is shown in Fig. 6.11. The overall depth of cross-section of the column is 450 mm. Design a model for direct analysis, using aluminium plates 2 mm thick. Take E for steel as 2.1×10^5 N/mm² and E for aluminium as 0.7×10^5 N/mm². I for the prototype may be taken as 9.2×10^8 mm⁴.*

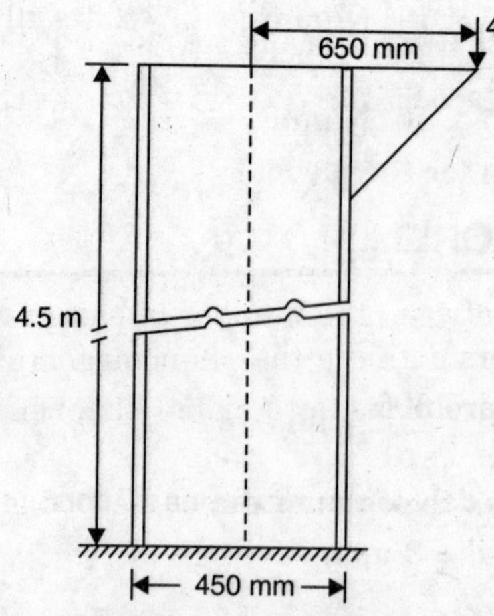

Fig. 6.11. Prototype structure.

Solution Let us assume a model of 450 mm height for the column.

$$k = \frac{H_S}{H_m} = \frac{4.5}{0.45} = 10 \quad k = 10$$

Theoretical depth of the model,

$$d_m = \frac{d_S}{k} = \frac{450}{10} = 45 \text{ mm}$$

Eccentricity of load in the model,

$$e_m = \frac{e_S}{k} = \frac{650}{10} = 65 \text{ mm}$$

Assume the width of aluminium flange plates as 20 mm.

Moment of inertia of the model,

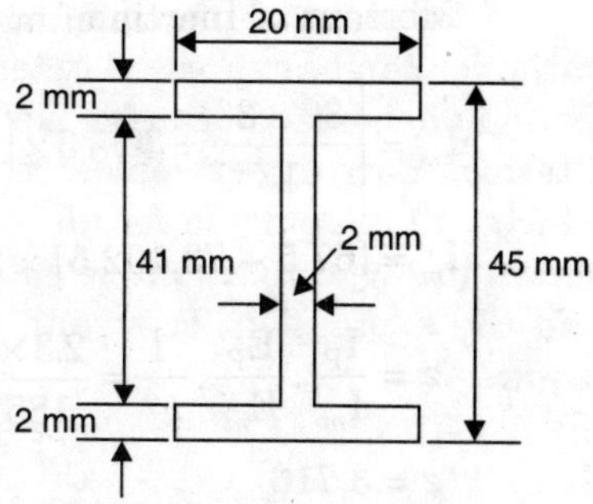

Fig. 6.11. (*a*) Model.

$$(I_{xx})_m = I_m = 2\left[\frac{20 \times 2^3}{12} + 20 \times 2 \times (21.5)^2\right] + \frac{2 \times 41^3}{12}$$

$$= 2\,[13.33 + 18{,}490] + 11486.83 = 48{,}493 \text{ mm}^4$$

$$I_m = 48{,}493 \text{ mm}^4$$

We have arbitrarily chosen I_m as above. From the actual ratio of $\frac{I_S}{I_m}$ we can deduce the factor z.

From equation (6.5),

$$z = \frac{I_S}{I_m} \cdot \frac{E_S}{E_m} \cdot \frac{1}{k^4}$$

$$= \frac{9.2 \times 10^8}{48793} \times \frac{2.1 \times 10^5}{0.7 \times 10^5} \times \frac{1}{10^4}$$

$$= 5.692$$

$$z = 5.692$$

Load to be applied to the model,

$$W_m = \frac{W_S}{k^2 z} = \frac{40 \times 1000}{(10)^2 \times 5.692} = 70.27 \text{ N}$$

$$W_m = 70.27 \text{ N}$$

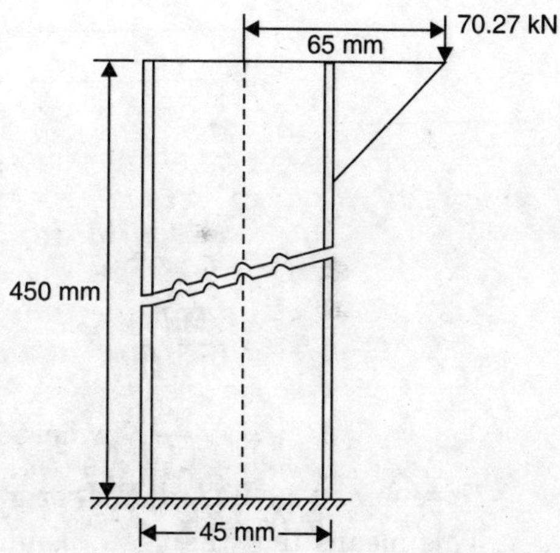

Fig. 6.12. Model.

6.5 BUCKINGHAM'S π THEOREM

This theorizes the relation between the number of basic dimensions in a phenomenon, number of dimensionless parameters defining the phenomenon and the number of variables.

By "phenomenon" we mean problems (or actions) like, bending, buckling, truss action, arch action etc.

Use of this theorem helps us develop an indirect method of deriving scale factors.

Buckingham's π Theorem reads as follows :

If a physical phenomenon can be defined in terms of **n** variables, and if each of these *n* variables can be expressed in terms of no more than **m** dimensions, then the general equations for the phenomenon can be expressed as functions of **(n-m)** dimensionless parameters (called π terms). Each dimensionless π term will be composed of **not greater than n** variables, two of which will be common for all the π terms. For example, let us consider a flexural member as shown in Fig. 6.13. Let it be subjected to some kind of loading. Applying dimensional analysis and the expression for strain, the π terms necessary to define similitude will be found as below.

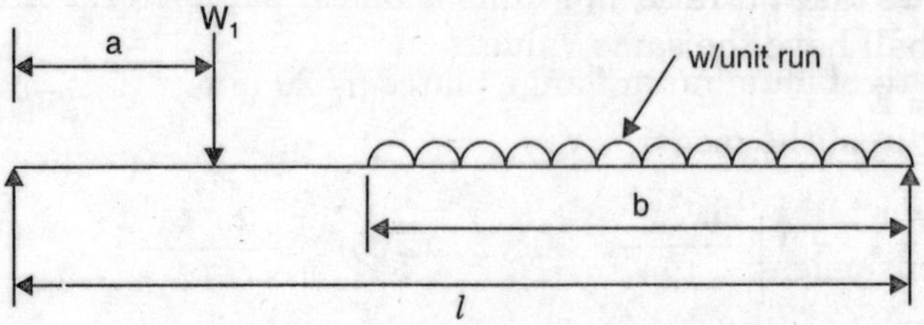

Fig. 6.13

For flexural members,

$$\text{Strain} = \frac{\text{Stress}}{\text{Young's Modulus}}$$

$$\varepsilon = \frac{\sigma}{E}$$

$$\frac{\sigma}{y} = \frac{M}{I}$$

$$\sigma = \frac{M}{I}\cdot y = \frac{M}{I}\frac{d}{2} \quad \text{(for sections symmetrical about both axes)}$$

$$\varepsilon = \frac{Md}{2I \times E}$$

$$\varepsilon = \frac{K\,Wld}{EI}, \text{ where K is a constant based on the type and position of loads.}$$

In general $\varepsilon = f(W, l, d, E, I)$ or $g(\varepsilon, W, l, d, E, I) = 0$

This means that the beam bending phenomenon is expressible in terms of the above 6 variables. $n = 6$.

These involve only 2 dimensions L and F. Now we must choose our dimensionless π terms. According to the π theorem, two of the variables may be common to all the π terms. These common variables are so chosen that the dimension of the force must be used at least once and the dimension of the length must be used at least once in the expressions for these common variables. Since $n = 6$ and $m = 2$, the number of π terms is $n - m = 4$. Choosing W and l as the common variables, the general expression for π terms is,

$$\pi_i = W^a\, l^b\, Q_i^{-1}$$

$$= F^a\, L^b\, Q_i^{-1}$$

Here π_i represents π_1, π_2, π_3 and π_4, and Q_i stands for Q_1, Q_2, Q_3 and Q_4. Out of the 6 variables, we will try to find π terms for ε, *d*, E and I, since W and *l* are to be common variables.

For π_1, $Q_1 = d$

$$\pi_1 = W^a\, l^b\, d^{-1}$$
$$= F^a\, L^b\, L^{-1}$$

Since π_1 is dimensionless, the powers of F and L shall be zero. So,

$$a = 0\ ;\quad b - 1 = 0 \quad \text{or } b = 1$$

Therefore, $\pi_1 = W^0\, l^1\, d^{-1}$

$$\pi_1 = \frac{l}{d}$$

In other words, we take l/d ratio as a dimensionless parameter. For the given beam and any of its models l/d shall have the same value.

For π_2, $Q_2 = E$

$$\pi_2 = W^a\, l^b\, E^{-1}$$
$$= F^a\, L^b\, F^{-1}\, L^{+2} = F^{a-1}\, L^{b+2}$$

Since π_2 is dimensionless,

$$a - 1 = 0\ ; \qquad a = 1$$

Since π_2 is dimensionless,

$$b + 2 = 0\ ; \qquad b = -2$$

Therefore, $\pi_2 = W^1\, l^{-2}\, E^{-1}$

$$\pi_2 = \frac{W}{l^2 E}$$

The inference is that for a perfect model, we must so choose W_m, that $\frac{W_S}{l_S^{\,2}\, E_S} = \frac{W_m}{l_m^{\,2}\, E_m}$, so that π_2 is same for structure and model.

For π_3, $Q_3 = I$

$$\pi_3 = W^a\, l^b\, I^{-1}$$
$$= F^a\, L^b\, L^{-4} = F^a\, L^{(b-4)}$$
$$a = 0\ ;\quad b - 4 = 0,\ b = 4$$

Therefore, $\pi_3 = W^0\, l^4\, I^{-1}$

$$\pi_3 = \frac{l^4}{I}\ , \quad \text{So } \frac{I_S}{I_m} = \left(\frac{l_S}{l_m}\right)^4$$

For π_4, $Q_4 = \varepsilon$

$$\pi_4 = W^a\, l^b\, \varepsilon^{-1}$$
$$= F^a\, l^b$$
$$a = 0\ ;\quad b = 0$$

Therefore, $\pi_4 = W^0\, l^0\, \varepsilon^{-1}$

$$\pi_4 = \frac{1}{\varepsilon}$$

Strain, being a non-dimensional quantity will be the same for the structure and its models.

For complete similitude, the four π terms obtained must be equal for the model and prototype structure.

Supposing we first fix the linear scale k.

From π_1, $\quad \dfrac{l_S}{d_S} = \dfrac{l_m}{d_m} \quad$ then $\quad \dfrac{d_S}{d_m} = k$

From π_2, $\quad \dfrac{W_S}{l_S^2 E_S} = \dfrac{W_m}{l_m^2 E_m} \quad$ or $\quad \dfrac{W_S}{W_m} = \dfrac{l_S^2 W_S}{l_m^2 E_m} = k^2 \dfrac{E_S}{E_m}$

From π_3, $\quad \dfrac{l_S^4}{I_S} = \dfrac{l_m^4}{I_m} \quad$ or $\quad \dfrac{I_S}{I_m} = \dfrac{l_S^4}{l_m^4} = k^4$

From π_4, $\quad \dfrac{1}{\varepsilon_S} = \dfrac{1}{\varepsilon_m} \quad$ or $\quad \dfrac{\varepsilon_S}{\varepsilon_m} = 1$

It is also possible to take all the loads and their positions (vide Fig. 6.13) as variables. Then the phenomenon would be $g(\varepsilon, W_1, w, a, b, l, d, E, I) = 0$ and there would be 7 π terms. The student is advised to write down π_1 to π_7 in that case.

Example 6.5 *A 150 mm × 250 mm wooden beam is simply supported over a span of 4 m. It carries a central concentrated load of 25 kN. Design a true model for predicting deflections, the model having a length ratio of 10. Use (i) Wood, (ii) Steel as model material. Write the prediction equation.*

Solution

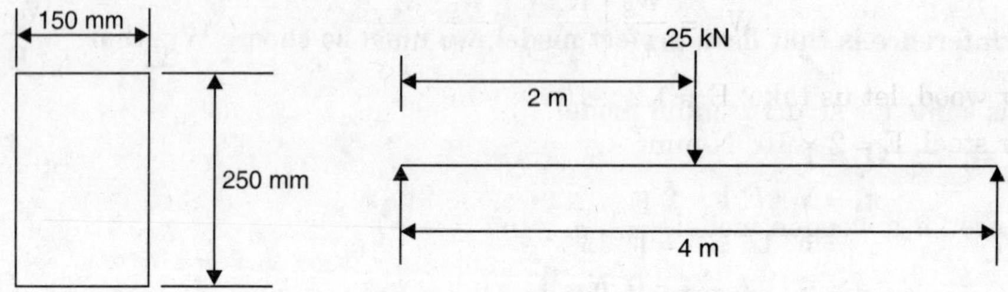

Fig. 6.14. Prototype structure.

Length ratio, $k = 10$

The beam bending phenomenon can be written as,

$$\delta = K \frac{Wl^3}{EI}$$

$$g(\delta, W, l, E, I) = 0$$

Number of variables $n = 5$.

Number of dimensions $m = 2$, (F and L)

We can choose W and l as common variables to define our dimensionless π terms.

Number of π terms $= n - m = 5 - 2 = 3$.

$$\pi_i = W^a \, l^b \, Q_i^{-1}$$

For π_1, $Q_1 = \delta$

$$\pi_1 = W^a l^b \delta^{-1}$$
$$= F^a L^{b-1}$$

Since π_1 is dimensionless, $a = 0$, $b = 1$.

$$\pi_1 = \frac{l}{\delta}$$

$$\frac{l_S}{\delta_S} = \frac{l_m}{\delta_m}\ ; \quad \delta_m = \frac{l_m}{l_S} . \delta_S = \frac{\delta_S}{k} = \frac{\delta_S}{10} .$$

The deflections in the model will be $\frac{1}{10}$th the deflections in the structure.

For π_2, $Q_2 = E$

$$\pi_2 = W^a l^b E^{-1}$$
$$= F^a L^b F^{-1} L^2$$
$$a - 1 = 0\ ; \quad a = 1$$
$$b + 2 = 0\ ; \quad b = -2$$

$\pi_2 = \frac{W}{l^2} E^{-1}$ is dimensionless.

$$\frac{W_S}{l_S{}^2 E_S} = \frac{W_m}{l_m{}^2} . \frac{1}{E_m}\ ; \quad W_m = W_S . \left(\frac{l_m}{l_S}\right)^2 . \frac{E_m}{E_S}$$

$$W_m = \frac{W_S}{k^2}\left(\frac{E_m}{E_S}\right) = \frac{W_S}{100} . \frac{E_m}{E_S}$$

For wood, let us take, $E = 1.2 \times 10^4$ N/mm^2

For steel, $E = 2 \times 10^5$ N/mm^2

Hence for a wooden model, $\frac{E_m}{E_S} = 1$ and $W_m = \frac{W_S}{100}$

For a steel model, $\frac{E_m}{E_S} = \frac{20}{1.2} = 16.67$

$$W_m = \frac{W_S}{100} \times 16.67 = 0.1667\ W_S$$

[This is unusual. We normally choose a model material with a low Young's modulus so that the deflections and strains are easily measurable].

For π_3, $Q_3 = I$

$$\pi_3 = W^a l^b I^{-1}$$
$$= F^a L^b L^{-4}$$

Since π_3 is dimensionless, $a = 0$, $b = 4$.

$$\pi_3 = \frac{l_m{}^4}{I}$$

$$\frac{l_S^4}{I_S} = \frac{l_m^4}{I_m} \; ; \qquad I_m = \left(\frac{l_m}{l_S}\right)^4 I_S = \frac{I_S}{k^4} = \frac{I_S}{10^4}.$$

This is straight away achieved by scaling down b and d by k (= 10)

$$b_m = \frac{150}{10} = 15 \text{ mm} \; ; \qquad d_m = \frac{250}{10} = 25 \text{ mm} \; ;$$

$$I_m = \frac{1}{12} \,.\, 15 \times 25^3 \left(= \frac{1}{10^4} \cdot \frac{1}{12} 150 \times 250^3 \right) = \frac{1}{12} \, 15 \times 25^3$$

$$I_m = 19{,}531 \text{ mm}^4.$$

PROBLEMS

1. Design the model of an RC beam using moulded resin as the model material. Span = 22 m. The beam has a depth of 600 mm, breadth 250 mm and equivalent moment of Inertia of 4720×10^6 mm^4. $E_C = 2 \times 10^4$ N/mm^2. E for resin = 0.4×10^4 N/mm^2. The RC beam is to be loaded with 8000 N/m over the whole span. Compute the scaled down load and the maximum deflection in the model for this load.
2. An Aluminium I section has to simulate a plated girder of span 27 m, overall depth, 660 mm and moment of inertia 2.32×10^9 mm^4. Design a model using aluminium strips of thickness 3 mm. [$E_S = 2 \times 10^5$ N/mm^2; $E_A = 0.8 \times 10^5$]. Load on the structure : 2 Nos. of 88 kN each at one third points.

CHAPTER 7
STRAIN ENERGY METHOD

7.1. INTRODUCTION

A structural system such as a beam, continuous beam, frames, pinjointed trusses, arches carries loads. They deflect a little due to the loads. So the loads lose some potential energy. The members in the structure develop stresses and suffer strains. Hence they store energy, called strain energy.

How much is the displacement caused by each load (or force) on the structure? The answer is provided by Castigliano's I theorem. Each load undergoes an amount of displacement in its direction so that the strain energy induced in the structure minus the P.E. lost by the load is a minimum.

If U is the strain energy and P is the force in question, then $E = U - P\delta$ is the total energy.

When we minimise this, $\frac{\partial E}{\partial P} = 0$. Hence $\frac{\partial U}{\partial P} = \delta$. This actually is Castigliano's I theorem.

If P is not an impressed force but a passive force (such as a reaction at unyielding support) we know that $\delta = 0$. Hence $\frac{\partial U}{\partial P} = 0$. This sometimes is called Castigliano's II theorem. This is applicable both to determinate and to indeterminate structures. But its use is mainly for solving indeterminate structures.

Minimisation of the strain energy is the tool invariably used in the solution of flexural frames, pinjointed frames, arches and other complex structures.

The method works even when we take into account strain energy due to tension, compression, shear, bending moment or / and torsional moments.

7.2. FLEXURAL MEMBERS

Let M_x be the bending moment at a variable position xx in a propped cantilever beam (Fig. 7.1). M_x can be expressed in terms of R_A, the unknown prop reaction.

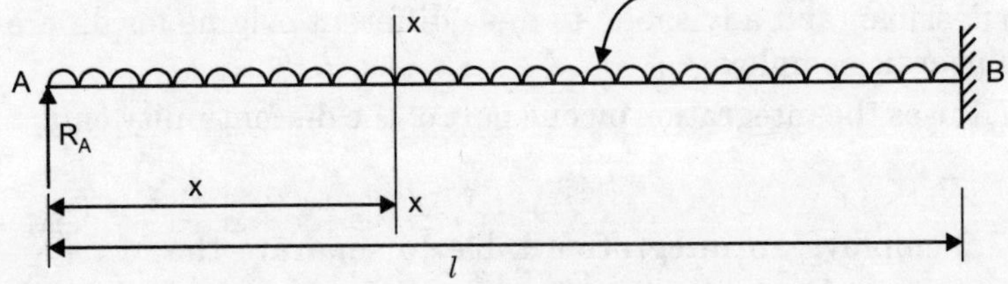

Fig. 7.1

$$M = M_x = R_A \cdot x - \frac{wx^2}{2} \qquad \text{...(7.1)}$$

Here, the strain energy in the beam AB is

$$U = \int_0^l \frac{M^2\, dx}{2EI}$$

Suppose we want to find out R_A by minimising U,

$$\frac{\partial U}{\partial R_A} = 0\,; \quad \int_0^l \frac{2M \dfrac{\partial M}{\partial R_A} dx}{2EI} = 0$$

If EI is constant, we end up with

$$\int_0^l M \frac{\partial M}{\partial R_A} dx = 0 \qquad \text{...(7.2)}$$

Taking M from equation (7.1),

$$\frac{\partial M}{\partial R_A} = x;$$ So, equation (7.2) can be rewritten as

$$\int_0^l \left(R_A\, x - \frac{wx^2}{2} \right) x\, dx = 0 = \int_0^l \left(R_A\, x^2 - \frac{wx^3}{2} \right) dx = 0\,;$$

$$\left[R_A \cdot \frac{x^3}{3} - \frac{w}{2} \cdot \frac{x^4}{4} \right]_0^l = 0$$

$$R_A \cdot \frac{l^3}{3} - \frac{wl^4}{8} = 0\,;\; R_A = \frac{3wl}{8}$$

This is known result, now verified by the strain energy method.

In the expression for bending moment [equation (7.1)] we have to keep in mind that

1. We would need differed expressions for different regions with mathematical discontinuity.

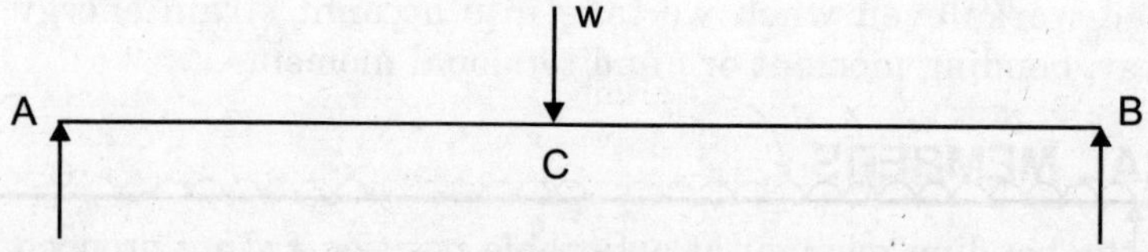

Fig. 7.2

In Fig. 7.2, M_x will have different expression for region AC and region CB.

2. It is permissible (and advisable) to have different origins for different regions. That will reduce the algebraic complexity.

Example 7.1 uses the integration on one part of the discontinuity only, taking advantage of the symmetry.

Example 7.2 employs an integration table to separate the M_x, $\dfrac{\partial M_x}{\partial R}$ values for the 3 discontinuous regions in the problem.

Example 7.3 deals with a bent having 2 redundant forces. Each redundant, say R_1 and R_2 shall minimise the strain energy U. Hence, we would have 2 energy equations like

$$\frac{\partial U}{\partial R_1} = 0 \text{ and } \frac{\partial U}{\partial R_2} = 0, \text{ from which we can isolate } R_1 \text{ and } R_2.$$

Example 7.1 *Analyse the continuous beam loaded as shown in Fig. 7.3 by the strain energy method. EI is constant.*

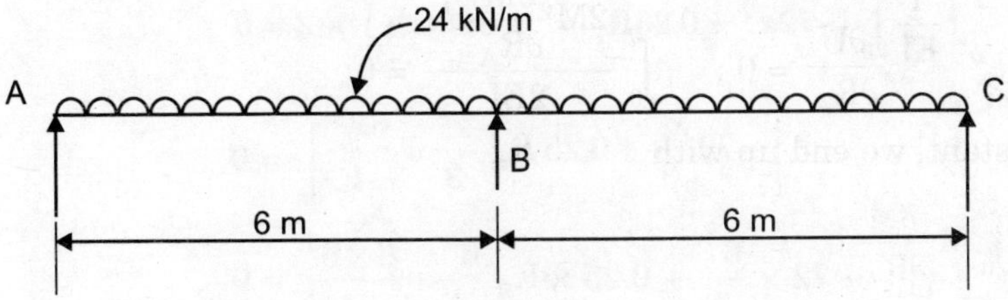

Fig. 7.3

Solution For beams with vertical loads only,

Static indeterminacy, I = number of vertical reactions + number of end moments – 2

$$= 3 + 0 - 2 = 1$$

Hence, the beam is statically indeterminate to the first degree. Let us treat R_B as the redundant. Since the beam and loading are symmetrical,

$$R_A = R_C = \frac{\text{Total load} - R_B}{2}$$

$$R_A = R_C = \frac{(24 \times 12 - R_B)}{2} = (144 - 0.5\ R_B)$$

The partial derivative of the total strain energy U in the beam AC with respect to R_B is zero, since $\delta_B = 0$

$$\frac{\partial U_{AC}}{\partial R_B} = 0 \quad i.e., \quad \frac{1}{EI}\int M \frac{\partial M}{\partial R_B} dx = 0$$

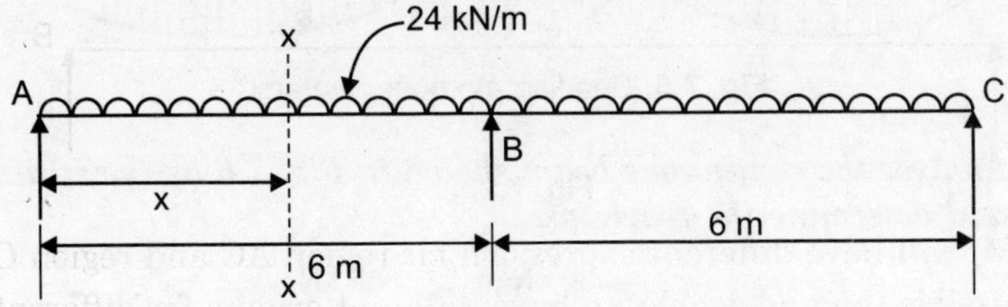

Fig. 7.4

Considering a section *xx* at a distance *x* from A in AB,

$$M_x = R_A \,.\, x - wx\,\frac{x}{2}$$

$$= (144 - 0.5\ R_B)\,x - 24\,\frac{x^2}{2}$$

$$M_x = 144x - 0.5\, R_B.x - 12\, x^2 \quad ...(1)$$

$$\frac{\partial M_x}{\partial R_B} = -0.5x$$

The integration limits are 0 and 6 m.

$$\frac{\partial U}{\partial R_B} = \left\{\frac{1}{EI}\int_0^6 [144x - 0.5R_B\, x - 12x^2][-0.5x]\, dx\right\} \times 2 = 0$$

$$\left\{\frac{1}{EI}\int_0^6 [-72x^2 + 0.25R_B\, x^2 + 6x^3]\, dx\right\} \times 2 = 0$$

$$\left[-72\frac{x^3}{3} + 0.25\, R_B\, \frac{x^3}{3} + \frac{6x^4}{4}\right]_0^6 = 0$$

$$-72 \times \frac{6^3}{3} + 0.25 \times R_B\, \frac{6^3}{3} + \frac{6 \times 6^4}{4} = 0$$

$$-5184 + 18\, R_B + 1944 = 0$$

$$18R_B = 5184 - 1944$$

$$R_B = \frac{5184 - 1944}{18}$$

$$R_B = 180 \text{ kN} = R_A$$

Substituting this in eqn. (1)

$$M_x = 144x - 0.5 \times 180x - 12x^2$$

When $x = 0$, $\quad M_A = 0$

When $x = 6$ m, $\quad M_B = 144 \times 6 - 0.5 \times 180 \times 6 - 12 \times 6^2 = -108$ kNm

$M_C = 0$

Simply supported beam moment in AB and BC $= \dfrac{wl^2}{8} = \dfrac{24 \times 6^2}{8} = 108$ kNm

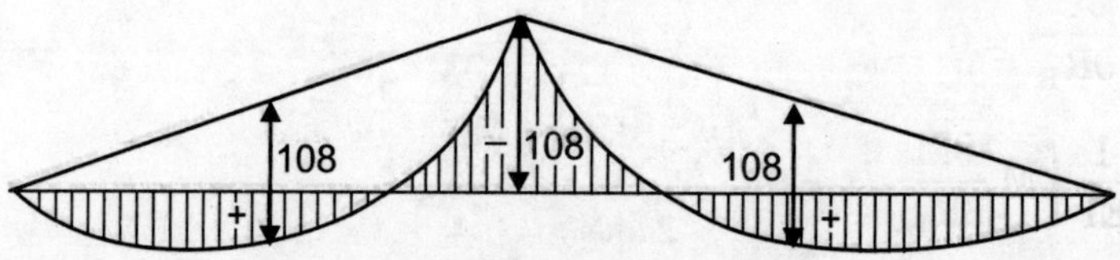

Fig. 7.5. Bending moment diagram.

Example 7.2 *Analyse the continuous beam shown in Fig. 7.6 by the strain energy method, and determine the reactions.*

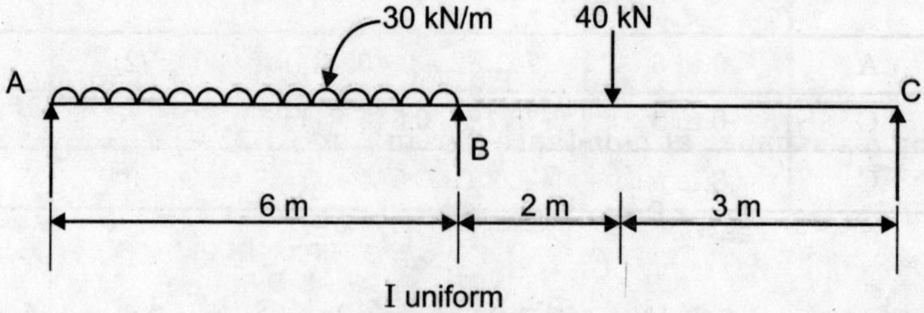

Fig. 7.6

Solution Degree of static indeterminacy of the given structure,

$$I = V.R. + E.M. - 2 \text{ (for beams with vertical loads only)}$$
$$= 3 + 0 - 2 = 1$$

Treating R_B as redundant and taking of moments about C,

$$R_A \times 11 + R_B \times 5 - 30 \times 6 \times (5 + 6/2) - 40 \times 3 = 0$$
$$11\,R_A + 5\,R_B = 1440 + 120 = 1560$$
$$R_A = \frac{1560 - 5R_B}{11}$$
$$R_A = (141.82 - 0.455\,R_B)$$
$$R_C = \text{Total load} - R_A - R_B$$
$$= 30 \times 6 + 40 - (141.82 - 0.455\,R_B) - R_B$$
$$= (78.18 - 0.545\,R_B)$$

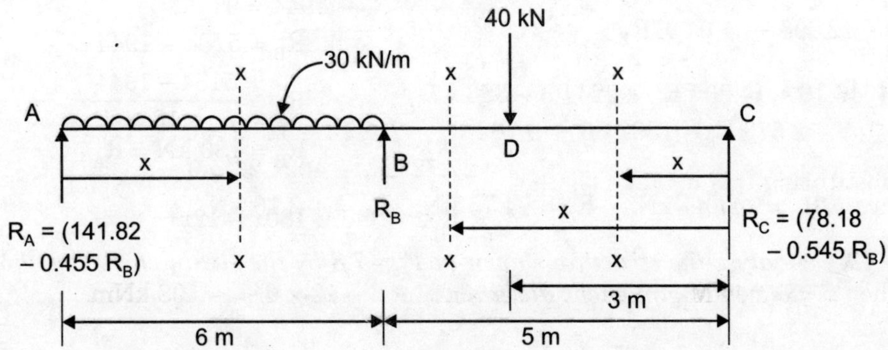

Fig. 7.7

$$\frac{\partial U}{\partial R_B} = 0 \qquad \text{(Since } \delta_B = 0\text{)}$$

$$\frac{1}{EI}\int M \frac{\partial M}{\partial R_B}\, dx = 0 \qquad \ldots(1)$$

EI is constant and can be removed

The integration will be done separately for the 3 zones, AB, CD and DB

Portion	*Origin*	*Limits (m)*	M_x	$\dfrac{\partial M_x}{\partial R_B}$
AB	A	0 to 6	$(141.82 - 0.455\,R_B)\,x - 30\,(x^2/2)$	$-0.455\,x$
CD	C	0 to 3	$(78.18 - 0.545\,R_B)\,x$	$-0.545\,x$
DB	C	3 to 5	$(78.18 - 0.545\,R_B)\,x - 40\,(x-3)$	$-0.545\,x$

Substituting the values from the tables in equation (1) above

$$\int_0^6 (141.82x - 0.455R_B\,x - 15x^2)(-0.455x)\,dx$$

$$+\int_0^3 (78.18x - 0.545\,R_B\,x)(-0.545x)$$

$$+\int_3^5 (78.18x - 0.545R_B\,x - 40x + 120)(-0.545x)\,dx = 0$$

$$\int_0^6 (-64.53x^2 + 0.207R_B\,x^2 + 6.825x^3)\,dx + \int_0^3 (-42.608x^2 + 0.297\,R_B\,x^2)\,dx$$

$$+\int_3^5 (-42.608x^2 + 0.297\,R_B\,x^2 + 21.8x^2 - 65.4x)\,dx = 0$$

$$\left(-64.53\frac{x^3}{3} + 0.207R_B\frac{x^3}{3} + 6.825\frac{x^4}{4}\right)_0^6 + \left(-42.608\frac{x^3}{3} + 0.297R_B\frac{x^3}{3}\right)_0^3$$

$$+\left(-42.608\frac{x^3}{3} + 0.297R_B\frac{x^3}{3} + 21.8\frac{x^3}{3} - 65.4\frac{x^2}{2}\right)_3^5 = 0$$

$$-4646.16 + 14.904\,R_B + 2211.3 - 383.47$$

$$+ 2.673\,R_B - 1391.86 + 9.702\,R_B + 712.13 - 523.2 = 0$$

$$27.729\,R_B - 4021.26 = 0$$

Hence, $R_B = 147.41$ kN; $R_A = 74.75$ kN; $R_C = -2.16$ kN

Example 7.3 *Analyse the structure shown in Fig. 7.8 by the strain energy method. Sketch the bending moment diagram.*

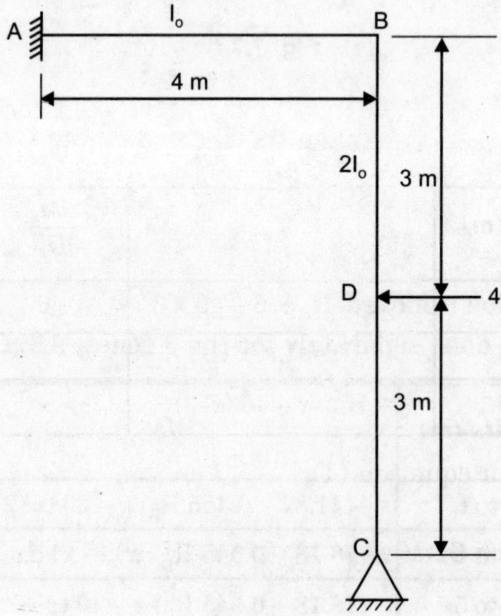

Fig. 7.8

Solution (*a*) **Finding redundants :**

The given structure is statically indeterminate to 2 degrees. The horizontal and vertical reactions at C (H_C, V_C) will be treated as the two redundants

Since C is hinged, $\dfrac{\partial U}{\partial H_C} = 0$ *i.e.,* $\displaystyle\int \frac{M}{EI}\frac{\partial M}{\partial H_C}\,dx = 0$ ...(1)

and $\dfrac{\partial U}{\partial V_C} = 0$ *i.e.,* $\displaystyle\int \frac{M}{EI}\frac{\partial M}{\partial V_C}\,dx = 0$ ...(2)

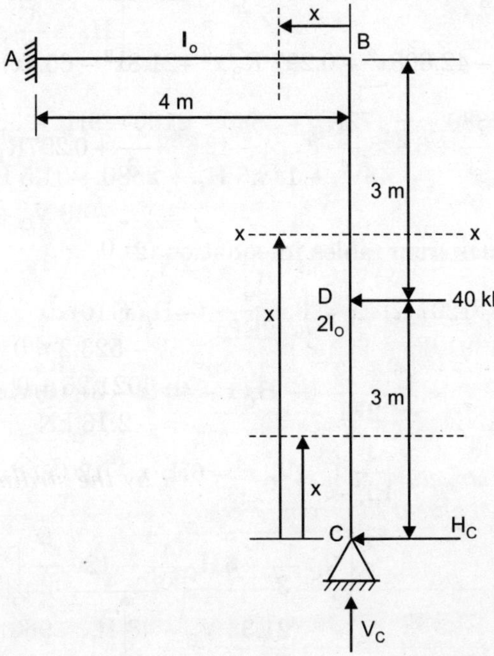

Fig. 7.9

Portion	*Origin*	*Limits (m)*	M_x	$\dfrac{\partial M}{\partial H_C}$	$\dfrac{\partial M}{\partial V_C}$	*I*
BA	B	0 to 4	$V_C\,x - H_C \times 6 - 40 \times 3$	-6	x	I_o
CD	C	0 to 3	$-H_C\,.\,x$	$-x$	0	$2I_o$
DB	C	3 to 6	$-H_C\,.\,x - 40\,(x-3)$	$-x$	0	$2I_o$

Substituting the values in equation (1),

$$\int_0^4 \frac{1}{EI_o}(V_C\,x - 6H_C - 120)\,(-6)\,dx + \int_0^3 \frac{1}{E\times 2I_o}\,(-H_C\,x)\,(-x)\,dx$$

$$+ \int_3^6 \frac{1}{E\times 2I_o}\,(-H_C\,x - 40x + 120)\,(-x)\,dx = 0$$

$$\left\{\int_0^4 (-6V_C\,x + 36H_C + 720)\,dx + \int_0^3 \frac{1}{2}H_C\,x^2\,dx + \int_3^6 \frac{1}{2}(H_C\,x^2 + 40x^2 - 120x)\,dx\right\} = 0$$

$$\left[-6V_C\frac{x^2}{2} + 36H_C x + 720x\right]_0^4 + \left[\frac{1}{2}H_C\frac{x^3}{3}\right]_0^3 + \left[\frac{1}{2}\left(H_C\frac{x^3}{3} + 40\frac{x^3}{3} - 120\frac{x^2}{2}\right)\right]_3^6 = 0$$

$$-48V_C + 144H_C + 2880 + 4.5\,H_C + \frac{1}{2}\left[\left(H_C \times \frac{6^3}{3} + 40 \times \frac{6^3}{3} - 120 \times \frac{6^2}{2}\right)\right.$$

$$\left. - \left(H_C \times \frac{3^3}{3} + 40 \times \frac{3^3}{3} - 120 \times \frac{3^2}{2}\right)\right] = 0$$

$$-48\,V_C + 148.5\,H_C + 2880 + \frac{1}{2}(72H_C + 2880 - 2160 - 9H_C - 360 + 540)$$

$$-48V_C + 148.5\,H_C + 2880 + 31.5\,H_C + 450 = 0$$

$$-48\,V_C + 180\,H_C = -3330 \qquad \ldots(3)$$

Substituting the values from tables in equation (2)

$$\int_0^4 \frac{1}{EI_o}(V_C\,x - 6H_C - 120)\,(x)\,dx + \int_0^3 \frac{1}{2EI_o}(-H_C\,x)\,(0)\,dx$$

$$+ \int_3^6 \frac{1}{2EI_o}((-H_C x - 40\,(x-3))\,(0)\,dx = 0$$

$$\frac{1}{EI_o}\int_0^4 (V_C x^2 - 6H_C x - 120\,x)\,dx = 0$$

$$\left(V_C\frac{x^3}{3} - 6H_C\frac{x^2}{2} - 120\frac{x^2}{2}\right)_0^4 = 0$$

$$(21.33\,V_C - 48\,H_C - 960) = 0$$

$$21.33\,V_C - 48\,H_C = 960 \qquad \ldots(4)$$

Solving (3) and (4)

$$V_C = 8.44\text{ kN};\quad H_C = -16.25\text{ kN}$$

(*b*) **Determining bending moments :**

AB:

$$M_x = V_C\,x - H_C \times 6 - 40 \times 3$$
$$= 8.44\,x - (-16.25) \times 6 - 40 \times 3$$
$$M_x = 8.44\,x - 22.5$$

At $x = 0$; $M_B = -22.5$ kNm

At $x = 4$ m; $M_A = 8.44 \times 4 - 22.5 = 11.26$ kNm

CD:

$$M_x = -H_C\,x = -(-16.25)\,x$$
$$M_x = 16.25\,x$$

At $x = 0$; $M_C = 0$

At $x = 3$ m; $M_D = 16.25 \times 3 = 48.75$ kNm

DB:

$$M_x = -H_C\,x - 40(x-3) = -(-16.25\,x) - 40\,(x-3)$$

$$M_x = -23.75\,x + 120$$

At $x = 3$ m; $\quad M_D = -23.75 \times 3 + 120 = 48.75$ kNm

At $x = 6$ m; $\quad M_B = -23.75 \times 6 + 120 = -22.5$ kNm

Figure 7.10 shows the bending moment diagram.

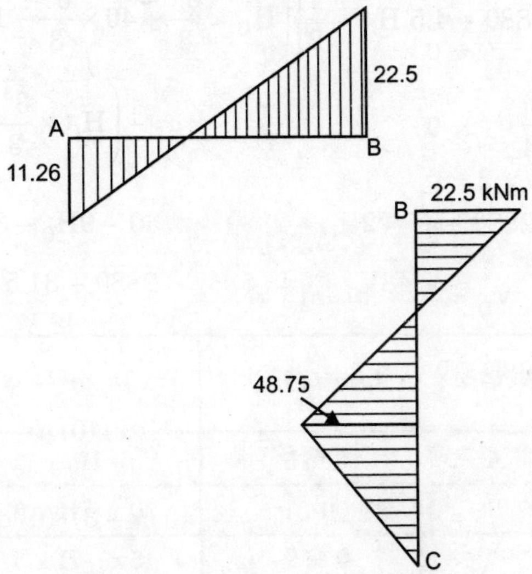

Fig. 7.10. Bending moment diagram.

Examples 7.4 and 7.5 analyse portal frames with non central loads. Moment distribution and slope deflection methods would have conditioned us to think that this is a sway problem (and hence tedious to solve). However, strain energy method is a flexibility based method (or compatibility method) and is concerned only with indeterminate forces and not displacements / rotations. So, sway will not bother us.

Example 7.4 *The simple portal frame shown in Fig. 7.11 is asymmetrically loaded. EI is constant. Analyse the frame by the strain energy method. Sketch the bending moment diagram.*

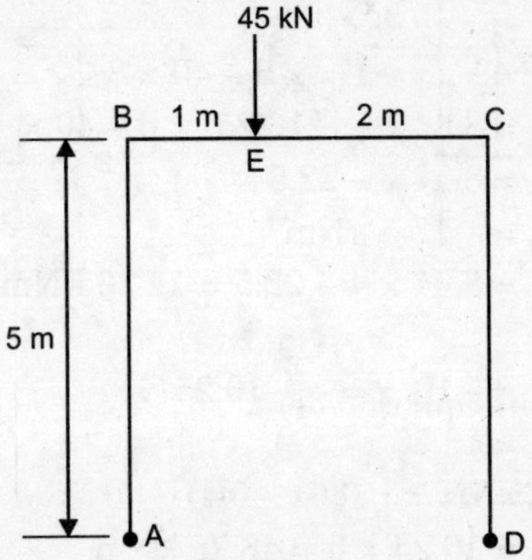

Fig. 7.11

Solution (*a*) **Finding the redundant force :**

Degree of static indeterminacy = 1 × 3 – 2 = 1

Let us treat the horizontal reaction at D as redundant. Since there is no other horizontal force,

$$H_A = -H_D = H$$

Since D is hinged, $\delta_D = 0$

$$\frac{\partial U}{\partial H} = 0$$

$$\frac{1}{EI}\int M\frac{\partial M}{\partial H}dx = 0 \qquad \text{...(1)}$$

$$V_A \times 3 - 45 \times 2 = 0$$

$$V_A = \frac{45 \times 2}{3} = 30 \text{ kN}$$

$$V_D = 45 - 30 = 15 \text{ kN}$$

Portion	*Origin*	*Limits (m)*	M_x *or M*	$\frac{\partial M}{\partial H}$
AB	A	0 to 5	$-H.x$	$-x$
BE	B	0 to 1	$30x - H \times 5$	-5
CE	C	0 to 2	$15x - H \times 5$	-5
DC	D	0 to 5	$-Hx$	$-x$

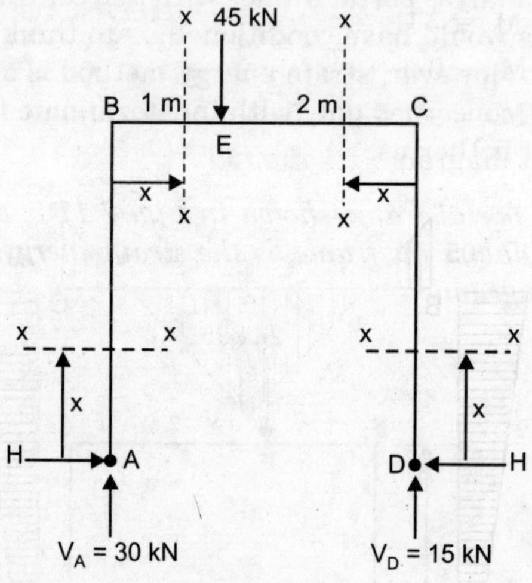

Fig. 7.12

Substituting the values in equation (1)

$$\frac{1}{EI}\left\{\int_0^5(-Hx)(-x)\,dx + \int_0^1(30x - 5H)(-5)\,dx + \int_0^2(15x - 5H)(-5)\,dx + \int_0^5(-Hx)(-x)\,dx\right\} = 0$$

$$2\int_0^5 Hx^2\,dx + \int_0^1 (-150x + 25H)\,dx + \int_0^2 (-75x + 25H)\,dx = 0$$

$$2\left[H\frac{x^3}{3}\right]_0^5 + \left[-\frac{150x^2}{2} + 25Hx\right]_0^1 + \left[-75\frac{x^2}{2} + 25Hx\right]_0^2 = 0$$

$$83.33\,H - 75 + 25\,H - 150 + 50H = 0$$

$$158.33\,H = 225$$

$$H = 1.421 \text{ kN}$$

(*b*) **Determining the bending moments :**

AB:

$M_x = -H\,x = -1.421\,x$

When $x = 0$, $M_A = 0$

$x = 5$ m, $M_B = -7.105$ kNm

BE:

$M_x = 30x - 5\,H = 30\,x - 5 \times 1.421 = 30x - 7.105$

At $x = 0$, $M_B = -7.105$ kNm

At $x = 1$ m, $M_E = 30 \times 1 - 7.105 = 22.895$ kNm

CE:

$M_x = 15x - 5H = 15x - 5 \times 1.421 = 15x - 7.105$

At $x = 0$, $M_C = -7.105$

At $x = 2$ m, $M_E = 15 \times 2 - 7.105 = 22.895$ kNm

DC:

$M_x = -Hx = -1.421\,x$

At $x = 0$; $M_D = 0$

At $x = 5$ m, $M_C = -1.421 \times 5 = -7.105$ kNm

The bending moment diagram is plotted in Fig. 7.13.

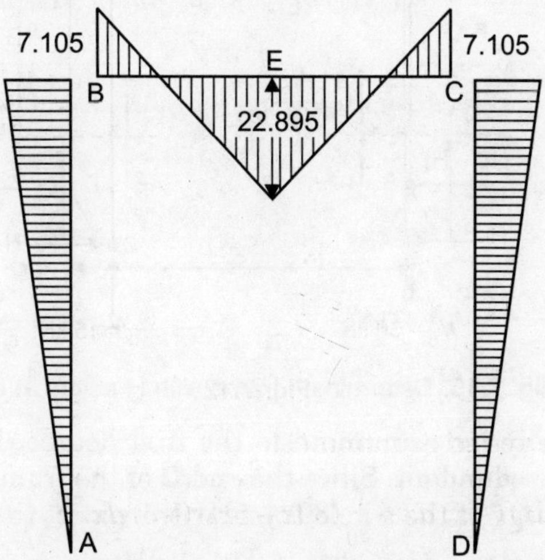

Fig. 7.13. Bending moment diagram.

We must remember that in earlier chapters, we were determining **moments** at the ends of members. They were distinguished as clockwise or anti-clockwise. Here in energy method what we find as M_A, M_B, M_C etc. are bending moments and are distinguished as sagging or hogging.

Example 7.5 *Analyse the portal frame with hinged base, loaded as shown in Fig. 7.14 by the strain energy method. Sketch the bending moment diagram.*

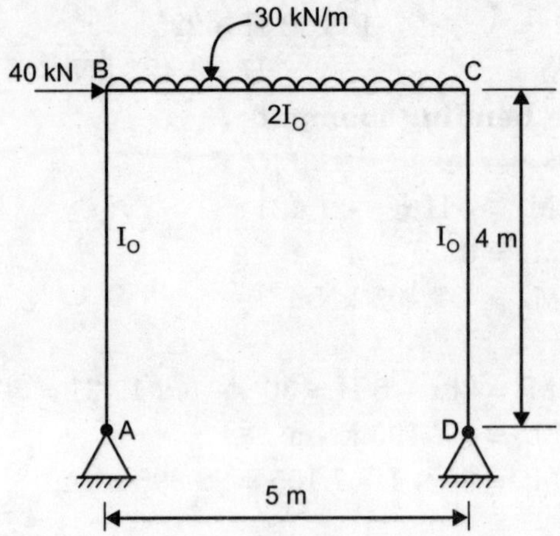

Fig. 7.14

Solution (*a*) **Computation of redundant :**

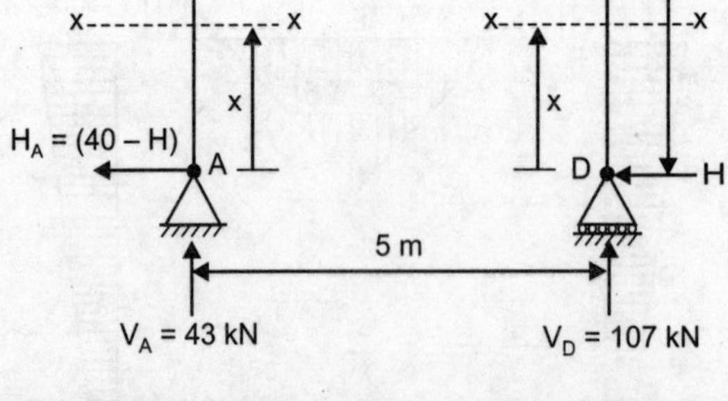

Fig. 7.15. Determinate frame with H acting at D.

The frame is statically indeterminate to the first degree. Let us treat the horizontal reaction at H_D at D, as the redundant. Since the end D of the frame does not yield, the partial derivative of the strain energy of the whole frame with respect to H_D or H will be zero.

$$\frac{\partial U}{\partial H} = 0\;;\quad i.e.,\quad \frac{\partial U}{\partial H} = \frac{1}{EI}\int M\,\frac{\partial M}{\partial H}\,dx = 0$$

Reactions:

$$\Sigma H = 0 \text{ gives}$$
$$H_A = (40 - H)$$

Taking moments about D,

$$V_A \times 5 + 40 \times 4 - 30 \times 5 \times \frac{5}{2} = 0$$

$$V_A = \frac{(-40 \times 4 + 30 \times 5 \times 2.5)}{5}$$

$$V_A = 43 \text{ kN}$$

$$V_D = 30 \times 5 - 43 = 107 \text{ kN}$$

Considering sections *xx* for various members of the frame as shown, the moments and their partial derivatives are tabulated under.

S.No.	*Portion*	*Origin*	*Limits (m)*	M_x *or M*	$\frac{\partial M}{\partial H}$	*I*
1.	AB	A	0 to 4	$(40 - H)\,x = 40x - H\,x$	$-x$	I_o
2.	BC	B	0 to 5	$43x + (40 - H) \times 4 - 30 \times (x^2/2)$	-4	$2I_o$
3.	DC	D	0 to 4	$H\,x$	$-x$	I_o

$$\frac{\partial U}{\partial H} = \frac{1}{EI}\int M \frac{\partial M}{\partial H}\,dx = 0$$

$$\left(\frac{\partial U}{\partial H}\right)_{AB} + \left(\frac{\partial U}{\partial H}\right)_{BC} + \left(\frac{\partial U}{\partial H}\right)_{DC} = 0$$

$$\frac{1}{EI_0}\int_0^4 (40x - Hx)(-x)\,dx + \frac{1}{E \times 2I_0}\int_0^5 \left[43x + (40 - H)\,4 - 15x^2\right](-4)\,dx$$

$$+ \frac{1}{EI_o}\int_0^4 (-Hx)(-x)\,dx = 0$$

$$\int_0^4 (-40x^2 + Hx^2)\,dx + \frac{1}{2}\int_0^5 (-172x - 640 + 16H + 60x^2)\,dx + \int_0^4 Hx^2\,dx = 0$$

$$\left[-40\frac{x^3}{3} + H\frac{x^3}{3}\right]_0^4 + \frac{1}{2}\left[-172\frac{x^2}{2} - 640x + 16Hx + 60\frac{x^3}{3}\right]_0^5 + \left[H\frac{x^3}{3}\right]_0^4 = 0$$

$$-853.33 + 21.33\,H - 1075 - 1600 + 40\,H + 1250 + 21.33\,H = 0$$

$$82.66\,H - 2258.33 = 0$$

$$H = \frac{2258.33}{82.66} = 27.32 \text{ kN}$$

$$H_A = 40 - H = 40 - 27.32 = 12.68 \text{ kN}$$

(*b*) **Final moments :**

AB:

$$M_x = 40x - Hx$$
$$= 40x - 27.32\,x = 12.68\,x$$

At $x = 0$, $M_A = 0$

At $x = 4$, $M_B = 50.72$ kNm

BC:

$$M_x = 43x + (40 - H) \times 4 - 15x^2$$
$$= 43x + (40 - 27.32) \times 4 - 15\,x^2$$
$$= 43x + 50.72 - 15x^2$$

At $x = 0$, $M_B = 50.72$ kNm

At $x = 5$ m, $M_C = 43 \times 5 + 50.72 - 15 \times 5^2$
$$= -\,109.38 \text{ kNm}$$

DC:

$$M_x = -\,Hx$$
$$= -\,27.32\,x$$

At $x = 0$, $M_D = 0$

At $x = 4$ m, $M_C = -\,27.32 \times 4 = -\,109.28$ kNm

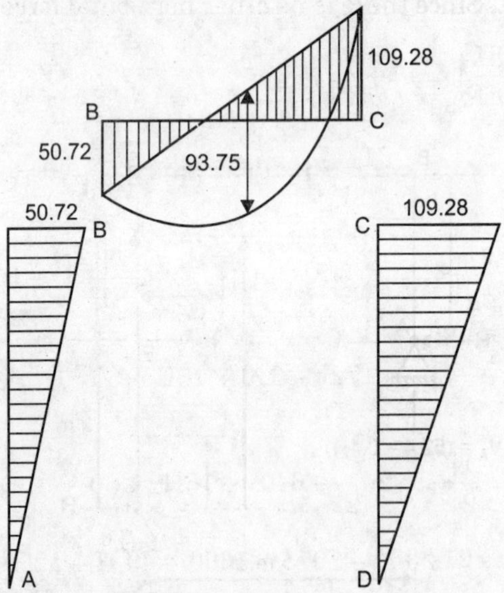

Fig. 7.16. Bending moment diagram.

Example 7.6 deals with frames with applied moments. This normally happens in workshop sheds housing a gantry. The gantry girder normally transfers loads to columns (of portals or gable frames) through rigid brackets projecting from columns.

Example 7.6 *Analyse the frame shown in Fig. 7.17 by the strain energy method. Sketch the bending moment diagram. All the members of the frame have uniform flexural rigidity throughout.*

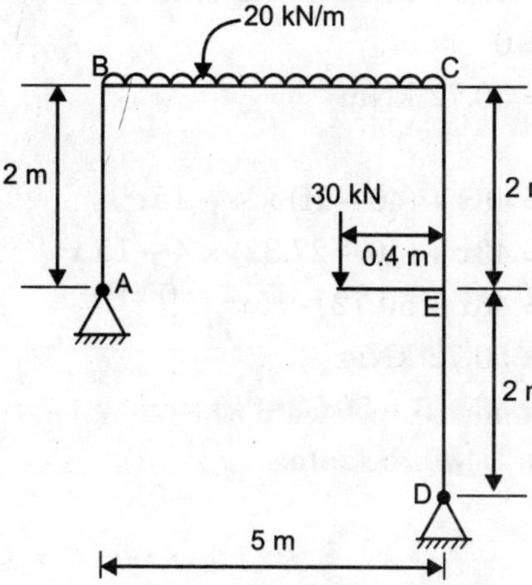

Fig. 7.17

Solution (*a*) **Finding redundant forces :**

The frame is statically indeterminate to the first degree. Let us treat the horizontal reaction H at A as redundant. Since there is no other horizontal force, H at D will be equal and opposite to H at A.

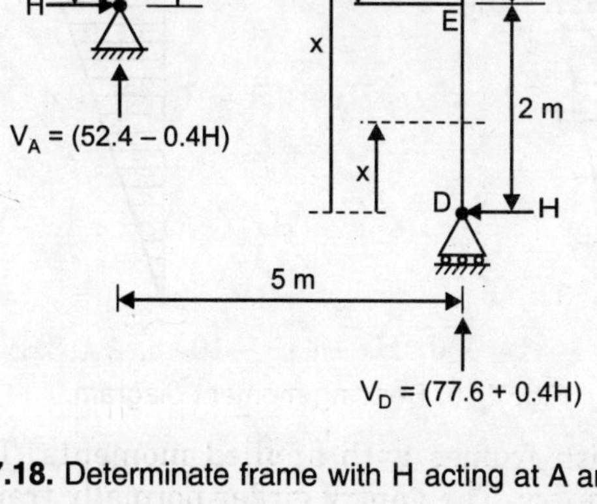

Fig. 7.18. Determinate frame with H acting at A and D.

Reactions:

Taking moments about A,

$$V_D \times 5 - 30 \times (5 - 0.4) - 20 \times 5 \times \left(\frac{5}{2}\right) - H \times 2 = 0$$

$$5\,V_D = 30 \times 4.6 + 10 \times 25 + 2H$$

$$V_D = \frac{1}{5}\,(388 + 2H)$$

$$V_D = (77.6 + 0.4H)$$

$$V_A = 20 \times 5 + 30 - V_D = 130 - (77.6 + 0.4H)$$

$$V_A = (52.4 - 0.4H)$$

Since the end A of the frame does not yield, the partial derivative of the strain energy of the whole frame with respect to H will be zero.

$$\frac{\partial U}{\partial H} = 0$$

$$\frac{\partial U}{\partial H} = \frac{1}{EI}\int M \frac{\partial M}{\partial H}\,dx = 0$$

S.No.	*Portion*	*Origin*	*Limits (m)*	*M*	$\frac{\partial M}{\partial H}$
1.	AB	A	0 to 2	$-Hx$	$-x$
2.	BC	B	0 to 5	$(52.4 - 0.4H)\,x - H \times 2 - 20\,(x^2/2)$	$(-0.4\,x - 2)$
3.	DE	D	0 to 2	$-Hx$	$-x$
4.	EC	D	2 to 4	$-Hx + 30 \times 0.4$	$-x$

$$\left(\frac{\partial U}{\partial H}\right)_{AB} + \left(\frac{\partial U}{\partial H}\right)_{BC} + \left(\frac{\partial U}{\partial H}\right)_{DE} + \left(\frac{\partial U}{\partial H}\right)_{EC} = 0$$

$$\frac{1}{EI}\left\{\int_0^2 (-Hx)(-x)\,dx + \int_0^5 [52.4x - 0.4Hx - 2H - 10x^2](-0.4x - 2)dx\right.$$

$$\left. + \int_0^2 (-Hx)(-x)\,dx + \int_2^4 (-Hx + 12)(-x)\,dx\right\} = 0$$

$$\int_0^2 Hx^2\,dx + \int_0^5 (-20.96x^2 + 0.16Hx^2 + 0.8Hx + 4x^3 - 104.8x + 0.8Hx + 4H + 20x^2)\,dx$$

$$+ \int_0^2 (-Hx^2)\,dx + \int_2^4 (-Hx^2 - 12x)\,dx = 0$$

$$\left[\frac{Hx^3}{3}\right]_0^2 + \left[-20.96\frac{x^3}{3} + 0.16H\frac{x^3}{3} + 0.8H\frac{x^2}{2} + 4\frac{x^4}{4} - 104.8\frac{x^2}{2}\right.$$

$$\left. + 0.8H\frac{x^2}{2} + 4Hx + 20\frac{x^3}{3}\right]_0^5 + \left[H\frac{x^3}{3}\right]_0^2 + \left[H\frac{x^3}{3} - 12\frac{x^2}{2}\right]_2^4 = 0$$

$$2.667H - 873.33 + 6.667\ H + 10\ H + 625 - 1310 + 10H + 20\ H + 833.33$$

$$+ 2.667\ H + 18.667\ H - 72 = 0$$

$$70.668\ H - 797 = 0$$

$$H = 11.278\ \text{kN}$$

(*b*) **Final moments :**

AB:

$M_x = -Hx = -11.278\,x$

when $x = 0$, $M_A = 0$

$x = 2$, $M_B = -22.556$ kNm

BC:

$$M_x = (52.4 - 0.4H)\,x - 2H - 10x^2$$

$$= (52.4 - 0.4 \times 11.278)\,x - 2 \times 11.278 - 10\,x^2$$

$$= 47.889x - 22.556 - 10x^2$$

At $x = 0$, $M_B = -22.556$ kNm

At $x = 5$ m, $M_C = 47.889 \times 5 - 22.556 - 10 \times 5^2 = -33.111$ kNm

DE:

$$M_x = -Hx = -11.278x$$

At $x = 0$, $M_D = 0$

At $x = 2$, $M_E = -11.278 \times 2$

$= -22.556$ kNm

EC:

$$M_x = -Hx + 12$$

$$= -11.278\,x + 12$$

At $x = 2$ m, $M_E = -10.556$ kNm

At $x = 4$ m, $M_C = -11.278 \times 4 + 12$

$= -33.112$ kNm

Simply supported beam moment in

$$BC = \frac{20 \times 5^2}{8} = 62.5\ \text{kNm}$$

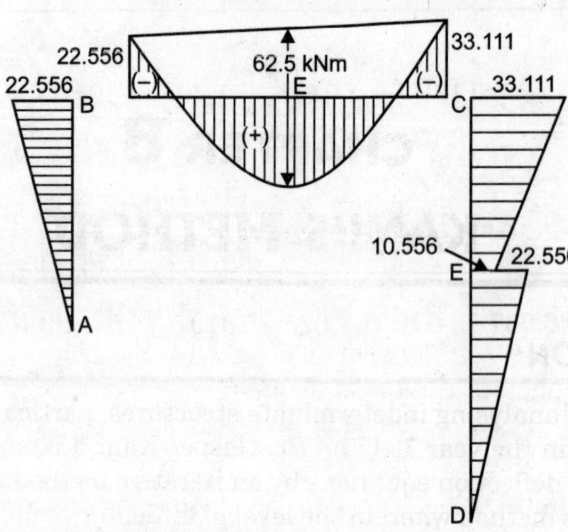

Fig. 7.19. Bending moment diagram plotted on tension side.

PROBLEMS

1. Solve problem 3 of chapter 2 by strain energy method.
2. Solve example 3.6 by strain energy method.
3. Analyse example 4.17 by strain energy method.
4. Solve the portal frame problem 6 of chapter 3 by strain energy method.
5. Analyse problem 7 of chapter 2 by strain energy method.
6. Analyse problem 11 of chapter 2 by strain energy method.

CHAPTER 8

KANI'S METHOD

8.1 INTRODUCTION

Kani's method of analysing indeterminate structures, particularly, building frames was developed in Germany in the year 1947 by Dr. Gasper Kani. Like moment distribution it is a method of solving slope deflection equations by an iterative method. Hence, this will fall under the category of stiffness method wherein the level of difficulty would be decided by the degrees of freedom (and not the degree of redundancy).

Actually Kani's method uses the slope deflection equations much more directly than moment distribution method does.

We may recall that in moment distribution method because of the erroneous initial assumption that rotation of every joint is zero (that all members are fixed) we get an unbalanced moment at every joint which has to be nullified by introducing an equal and opposite balancing moment. We then apportion this balancing moment between members in the joint. Thus the iteration process is carried out on the imbalance.

In Kani's method, the entire joint moment (and not just the imbalance) is taken for iterative correction. We treat the iterative process as completed when we find that 2 successive values of moments are nearly equal.

8.2 TERMINOLOGY

In Kani's method we have to get familiar with some terms. We will do well to start with a frame in Fig. 8.1.

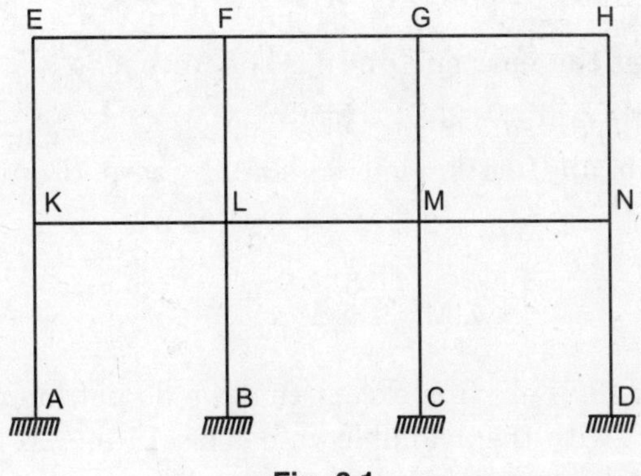

Fig. 8.1

Let us consider a typical member FG and a typical joint F. After loading, the members would undergo deformations and joints would undergo rotations and displacements. All members at a joint will undergo the same rotation. For example at joint L, θ_L would be the ***joint rotation***. θ_L is the rotation of LK, LB, LM and LF at L.

Since this is essentially a method of analysing moments, we would consider only rotational deformations and ignore axial deformations.

Hence, there will be no vertical displacements of joints (unless there is sinking of any of the supports). In addition, there can be only one horizontal displacement at any given level, say level EFGH. In other words, joint displacements

$$\Delta_E = \Delta_F = \Delta_G = \Delta_H.$$

Net support moment on any beam at a joint would then be made up of

1. Fixed end moments (due to given loading)
2. Rotation contributions
   (*a*) Moment due to rotation at near end, say end *i*.
   (*b*) Moment due to rotation at far end, say end *j*.
3. Displacement contributions

These are moments developed in vertical members due to relative horizontal displacements of floors or moments in horizontal members due to sinking of one or more columns.

8.3 FUNDAMENTAL EQUATIONS

Refering to Fig. 8.1, the slope deflection equation for M_{FG} can be written as

$$M_{LM} = M_{FLM} + \frac{4EI}{l}\theta_L + \frac{2EI}{l}\theta_M + \frac{6EI\delta}{l^2}, \quad \text{...(8.1)}$$

where θ_L and θ_M are rotations at L and M and δ is the horizontal displacement at level KLMN.

Equation (8.1) can be rewritten as (taking $\delta = 0$)

$$M_{LM} = M_{FLM} + 2M'_{LM} + M'_{ML} \quad \text{...(8.2)}$$

where $M'_{LM} = \frac{2EI}{l}\theta_L$ and $M'_{ML} = \frac{2EI}{l}\theta_M$

More generally if *i* is the near end and *j* is the far end

$$M_{ij} = M_{Fij} + 2M'_{ij} + M'_{ji} \quad \text{...(8.3)}$$

At joint *i*, the sum of all (final) moments shall be zero. (Equilibrium condition)

Hence, $$\Sigma M_{ij} = \Sigma M_{Fij} + 2\,\Sigma M'_{ij} + \Sigma M_{ji} = 0 \quad \text{...(8.4)}$$

$$\Sigma M'_{ij} = -\frac{1}{2}(\Sigma M_{Fij} + \Sigma M'_{ji}) \quad \text{...(8.5)}$$

Equations (8.4) and (8.5) are fine except that we do not know θ_L and θ_M. So we do not know M'_{ij}, M'_{ji}. So we start with the available values of M'_{ij} and M'_{ji} namely 0,0. Then we get $\Sigma M'_{ij}$ from equation (8.5) for ***every joint*** that rotates. These will cover M'_{ji} s also. Hence, we improve the value of $\Sigma M'_{ij}$ from equation (8.5) again and so on. This iterative procedure is continued cycle after cycle till we find that 2 successive values of M_{ij} are sensibly equal.

8.3.1 Rotation Factor (u)

Rotation factor in Kani's method is akin to distribution factor in moment distribution method.

Actually $u = -0.5$ DF

u values help us find each M_{ij} from $\Sigma M'_{ij}$.

Kani's procedure adopts a pictorial scheme of marking the various quantities at each joint for doing the iterative process as shown in Fig. 8.2. The resultant restraint moment ΣM_{Fij} is marked in the inner cell. The outer cells are meant for rotation factors u_{ij}.

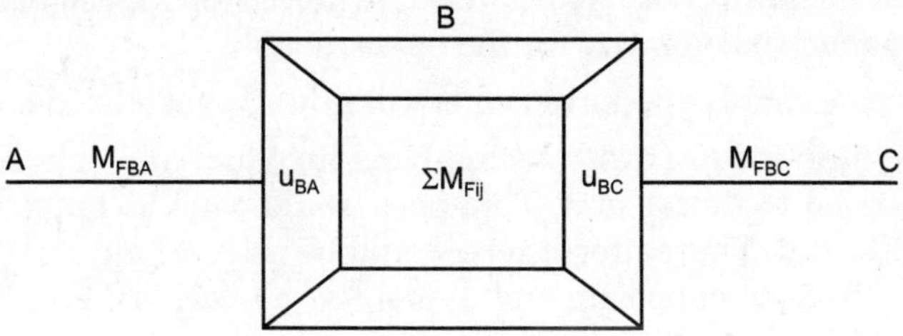

Fig. 8.2. Kani's scheme for a beam ABC.

Example 8.1 *For a beam ABC (SS at A and C and continuous at B), AB = 5 m; BC = 4 m. $I_{BA} = I_{BC} = I_o$. The fixed end moments are $M_{FBA} = 25$ kNm, $M_{FBC} = -40$ kNm. Fill in the Kani's scheme for joint B.*

Solution

$$u_{BA} = -0.5 \times DF_{BA}$$

$$k_{BA} = \frac{I_o}{5}\ ;\ k_{BC} = \frac{I_O}{4}\ \ \Sigma k = I_o\ \frac{9}{20}$$

$$DF_{BA} = \frac{k_{BA}}{\Sigma k} = \frac{1}{5} \times \frac{20}{9} = \frac{4}{9}$$

$$DF_{BC} = \frac{k_{BC}}{\Sigma k} = \frac{1}{4} \times \frac{20}{9} = \frac{5}{9}$$

$$u_{BA} = -\frac{1}{2} \times \frac{4}{9} = -\frac{2}{9}\ ;\ u_{BC} = -\frac{5}{18}$$

These values are marked in the appropriate outer boxes in Fig. 8.3.

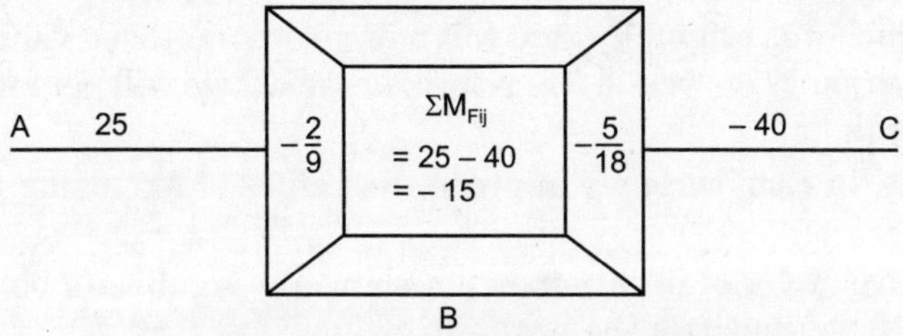

Fig. 8.3. Kani's scheme for joint B.

8.4 KANI'S ITERATION PROCEDURE FOR BEAMS

Now we are ready to launch the iterative procedure to successively refine the values of rotation contributions such as M_{ij}. That would lead us to the final moments. Example 8.2 will help us grasp the procedure. In the solution elaborate explanatory steps are included to help comprehend the process. In later examples most of these explanatory part can be dropped.

While solving structures by this method the following steps may be kept in mind :

Step 1. Compute all fixed end moments.

Step 2. Compute and tabulate all rotation factors for all joints that would have rotation.

Step 3. Fixed ends will not have rotation factors. Nor rotation contributions either to the same (fixed end) or to the opposite end.

Step 4. Extreme simply supported ends will initially get a fixed end moment (in step 1).

This would be nullified forthwith by applying an equal and opposite moment. Half this moment would be carried to its far end. This part would thus increase the 'initial' fixed end moment of the opposite end. Thereafter, there would be no rotation contribution to the SS end from the same end or from the opposite end. [vide Fig. 8.4 (*a*), (*b*)]

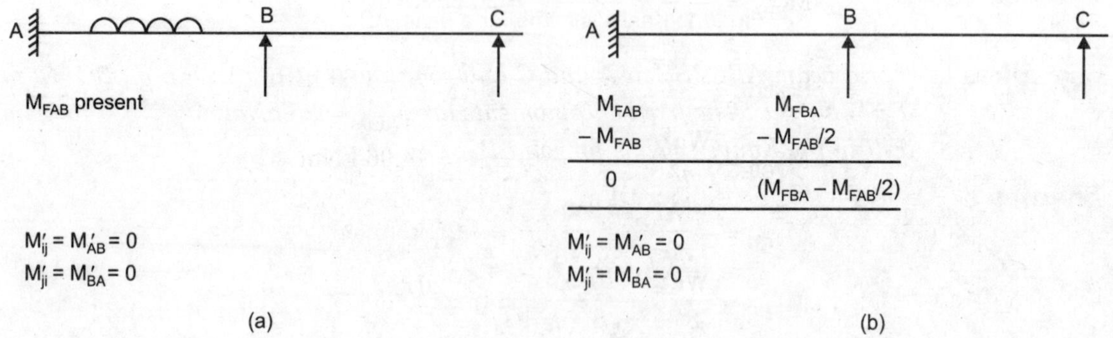

Fig. 8.4

Step 5. In the I cycle of iteration.

In the iterative process, proceed joint by joint to realise the condition.

$$M'_{ij} = u_{ij} \left(\Sigma M_{Fi} + \Sigma M'_{ji} \right) \quad \text{...(8.6), for joint } i.$$

If there are 2 members at joint *i*, the Σ *s* will contain 2 terms. To start with, in cycle 1, all M'_{ji} values would be taken as zero (since we say all M'_{ji}, this will include all M'_{ji} values too). When we first use equation (8.6), we therefore will have to take $\Sigma M_{ji}'$ on the right side as zero. M'_{ji} on the L.H.S. which was originally zero will now get an improved value. This improvement is thus given to all the joints in cycle 1. Therefore, in cycle 2 we will get improved values of M'_{ji} (and M'_{ji}) for all joints.

In other words, in each cycle we improve the values of M'_{ij} using the latest value(s) of M'_{ji} .

When successive cycles of iteration do not show any significant changes in the rotation contributions ('M'_{ji} 's) we conclude the iteration.

Step 6. Now that we have all M'_{ji} values we can get the final joint moments using equation (8.3). Now we will go to example 8.2.

Example 8.2 *Analyse the continuous beam loaded as shown in Fig. 8.5, by Kani's method. Sketch the bending moment diagram.*

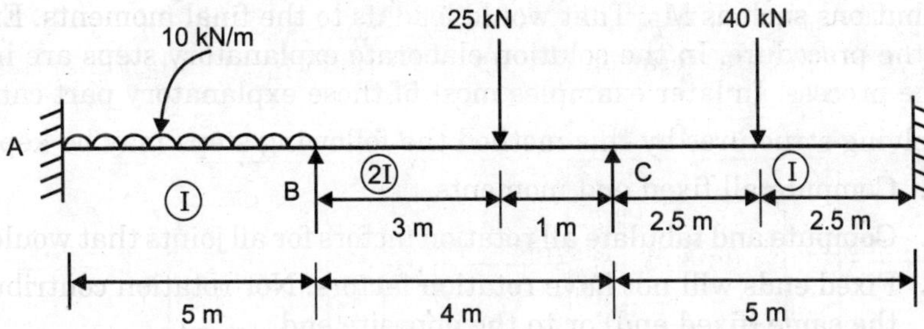

Fig. 8.5

Solution (*a*) **Fixed end moments :**

$$M_{FAB} = -\frac{wl^2}{12} = -\frac{10 \times 5^2}{12} = -20.83 \text{ kNm}$$

$$M_{FBA} = \frac{wl^2}{12} = \frac{10 \times 5^2}{12} = 20.83 \text{ kNm}$$

$$M_{FBC} = -\frac{Wab^2}{l^2} = -\frac{25 \times 3 \times 1^2}{4^2} = -4.69 \text{ kNm}$$

$$M_{FCB} = \frac{Wa^2b}{l^2} = \frac{25 \times 3^2 \times 1}{4^2} = 14.06 \text{ kNm}$$

$$M_{FCD} = -\frac{Wl}{8} = -\frac{40 \times 5}{8} = -25 \text{ kNm}$$

$$M_{FDC} = \frac{Wl}{8} = \frac{40 \times 5}{8} = 25 \text{ kNm}$$

(*b*) **Rotation factors :**

Joint	*Member*	*Relative stiffness*	*Sum of stiffness*	*Distribution factor*	*Rotation factor = – 0.5 × Dist. factor*
B	BA	$\frac{I}{5}$	$\frac{I}{5} + \frac{I}{2} = \frac{7I}{10}$	$\frac{I/5}{7I/10} = 0.286$	– 0.143
	BC	$\frac{2I}{4} = \frac{I}{2}$		$\frac{I/2}{7I/10} = 0.714$	– 0.357
C	CB	$\frac{2I}{4} = \frac{I}{2}$	$\frac{I}{2} + \frac{I}{5} = \frac{7I}{10}$	$\frac{I/2}{7I/10} = 0.714$	– 0.357
	CD	$\frac{I}{5}$		$\frac{I/5}{7I/10} = 0.286$	– 0.143

(*c*) **Resultant restraint moments :**

Joints A and D are fixed (ends). Hence, there are only two joints *i.e.*, B and C at which iterations are to be carried out.

Sum of the fixed end moments at $B = M_{FBA} + M_{FBC}$
$= 20.83 - 4.69 = 16.14$ kNm

Sum of the fixed end moments at $C = M_{FCB} + M_{FCD} = 14.06 - 25 = -10.94$ kNm

The fixed end moments, rotation factors and sum of fixed end moments are entered in appropriate places in Fig. 8.6.

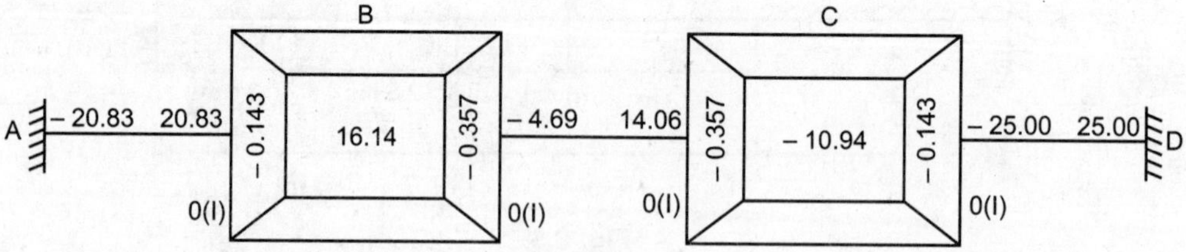

Fig. 8.6

Rotation contributions at fixed ends A and D are zero. Rotation contributions at joints B and C are initially assumed as zero arbitrarily. These values are entered below the line with I in parenthesis denoting initial values.

The iterations are carried out in Table 8.1 and the values are also written in Fig. 8.7.

(*d*) **Iteration cycles :**

Table 8.1

Joint	***B***		***C***	
Rotation contribution	M'_{BA}	M'_{BC}	M'_{CB}	M'_{CD}
Rotation factor, u	– 0.143	– 0.357	– 0.357	– 0.143
ΣM_F	16.14		– 10.94	
Formula for calculation ($M'_{AB} = 0$; $M'_{DC} = 0$)	$u_{BA}(\Sigma M_{FB} + M'_{AB} + M'_{CB})$ $= u_{BA}(\Sigma M_{FB} + M'_{CB})$	$u_{BC}(\Sigma M_{FB} + M'_{CB} + M'_{AB})$ $= u_{BC}(\Sigma M_{FB} + M'_{CB})$	$u_{CB}(\Sigma M_{FC} + M'_{BC} + M'_{DC})$ $= u_{CB}(\Sigma M_{FC} + M'_{BC})$	$u_{CD}(\Sigma M_{FC} + M'_{DC} + M'_{BC})$ $= u_{CD}(\Sigma M_{FC} + M'_{BC})$
Iteration 1*	– 0.143 (16.14 + 0) = – 2.31	– 0.357 (16.14 + 0) = – 5.76	– 0.357 (–10.94 – 5.76) = 5.96	– 0.143 (– 10.94 – 5.76) = 2.39
Iteration 2**	– 0.143 (16.14 + 5.96) = – 3.16	– 0.357 (16.14 + 5.96) = – 7.89	– 0.357 (–10.94 – 7.89) = 6.72	– 0.143 (– 10.94 – 7.89) = 2.69
Iteration 3	– 0.143 (16.14 + 6.72) = – 3.27	– 0.357 (16.14 + 6.72) = – 8.16	– 0.357 (– 10.94 – 8.16) = 6.82	– 0.143 (– 10.94 – 8.16) = 2.73
Iteration 4	– 0.143 (16.14 + 6.82) = – 3.28	– 0.357 (16.14 + 6.82) = – 8.20	– 0.357 (– 10.94 – 8.20) = 6.83	– 0.143 (– 10.94 – 8.20) = 2.74
Iteration 5	– 0.143 (16.14 + 6.83) ≐ – 3.28	– 0.357 (16.14 + 6.83) = – 8.20	6.83	2.74

* In iteration 1, M'_{CB} is assumed to be zero *i.e.,* θ_C is assumed zero and M'_{BA} and M'_{BC} will be obtained. The value of M'_{BC} will be used in the calculation of M'_{CB} and M'_{CD}.

** In iteration 2, M'_{CB} obtained in iteration 1 will be used in the calculation of M'_{BA} and M'_{BC}. This value of M'_{BC} will be used in calculating M'_{CB} and M'_{CD}.

Similarly, iterations will be continued till two successive cycles of iterations yield practically the same values.

(*e*) **Final moments :**

$$M_{AB} = M_{FAB} + 2M'_{AB} + M'_{BA}$$
$$= -20.83 + 0 - 3.28 = -24.11 \text{ kNm}$$
$$M_{BA} = M_{FBA} + 2M'_{BA} + M'_{AB}$$
$$= 20.83 - 2 \times 3.28 + 0 = 14.27 \text{ kNm}$$
$$M_{BC} = M_{FBC} + 2M'_{BC} + M'_{CB}$$
$$= -4.69 - 2 \times 8.20 + 6.83 = -14.26 \text{ kNm}$$
$$M_{CB} = M_{FCB} + 2M'_{CB} + M'_{BC}$$
$$= 14.06 + 2 \times 6.83 - 8.20 = 19.52 \text{ kNm}$$
$$M_{CD} = M_{FCD} + 2M'_{CD} + M'_{DC}$$
$$= -25.00 + 2 \times 2.74 + 0 = -19.52 \text{ kNm}$$
$$M_{DC} = M_{FDC} + 2M'_{DC} + M'_{CD}$$
$$= 25.00 + 2 \times 0 + 2.74 = 27.24 \text{ kNm}$$

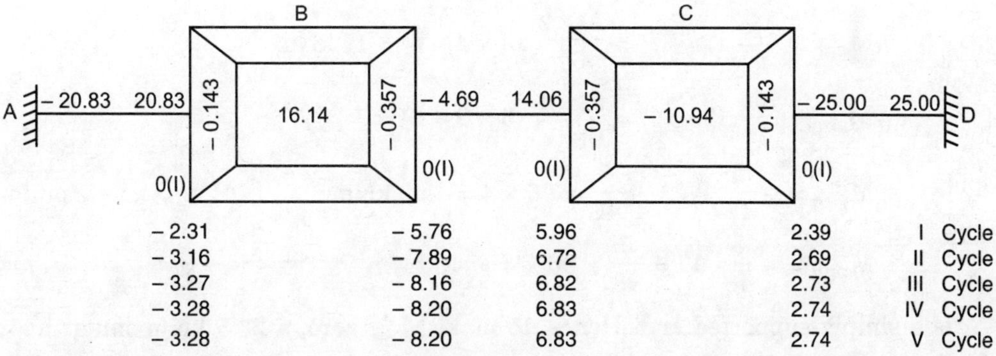

Fig. 8.7

(*f*) **Simply supported beam moments :**

$$\text{in AB} = \frac{wl^2}{8} = \frac{10 \times 5^2}{8} = 31.25 \text{ kNm}$$

$$\text{in BC} = \frac{Wab}{l} = \frac{25 \times 3 \times 1}{4} = 18.75 \text{ kNm}$$

$$\text{in CD} = \frac{Wl}{4} = \frac{40 \times 5}{4} = 50 \text{ kNm}$$

(*g*) **The bending moment diagram** (drawn on the tension side) is shown in Fig. 8.8 below.

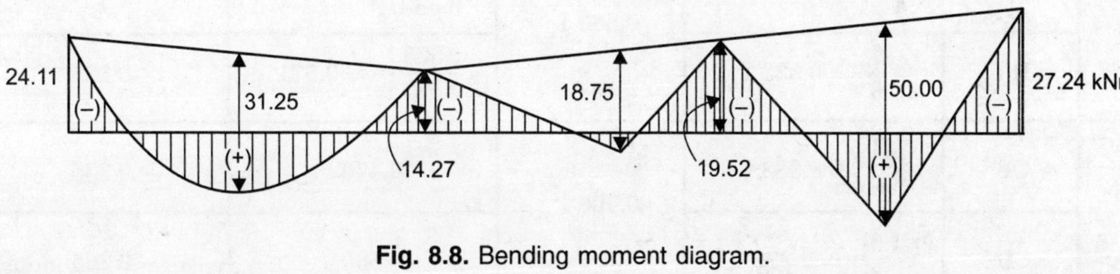

Fig. 8.8. Bending moment diagram.

Example 8.3 *Analyse the continuous beam loaded as shown in Fig. 8.9, by Kani's Method.*

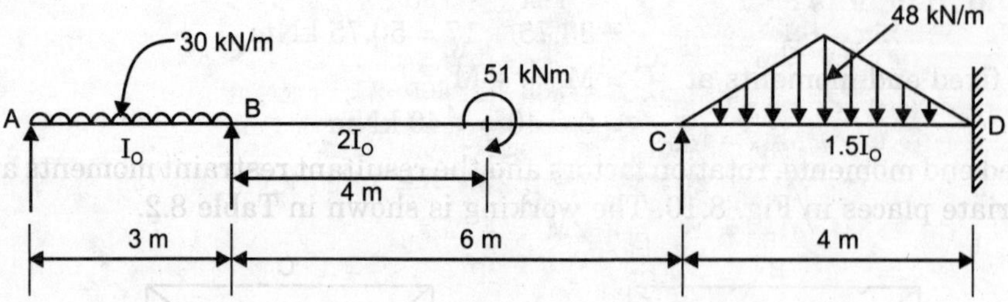

Fig. 8.9

Solution (*a*) **Fixed end moments :**

$$M_{FAB} = -\frac{30 \times 3^2}{12} = -22.5 \text{ kNm}$$

$$M_{FBA} = \frac{30 \times 3^2}{12} = 22.5 \text{ kNm}$$

$$M_{FBC} = \frac{Mb}{l^2}(3a - l) = \frac{51 \times 2}{6^2}(3 \times 4 - 6) = 17 \text{ kNm}$$

$$M_{FCB} = \frac{Ma}{l^2}(3b - l) = \frac{51 \times 4}{6^2}(3 \times 2 - 6) = 0$$

$$M_{FCD} = -\frac{5}{48} Wl = -\frac{5}{48} \times 96 \times 4 = -40 \text{ kNm} \qquad \left(W = \frac{1}{2} \times 4 \times 48 = 96 \text{ kN}\right)$$

$$M_{FDC} = +\frac{5}{48} Wl = \frac{5}{48} \times 96 \times 4 = 40 \text{ kNm}$$

A is a simply supported end. Hence, to make M_{AB} zero, a 22.5 kN moment has to be applied at A. Due to this the carry over B *i.e.*, M_{FBA} is 22.5/2 = 11.25 kNm. Hence, the fixed end moments get modified as,

$M_{FAB} = 0$ $\qquad$ $M_{FBA} = 22.5 + 11.25 = 33.75$ kNm

$M_{FBC} = 17$ kNm $\qquad$ $M_{FCB} = 0$

$M_{FCD} = -40$ kNm $\qquad$ $M_{FDC} = 40$ kNm

(*b*) **Rotation factors :**

Joint	*Member*	*Relative stiffness*	*Sum of stiffness* (D_f)	*Distribution factor*	*Rotation factor,* $u = -0.5\ D_f$
B	BA	$\frac{3}{4} \times \frac{I_o}{3} = 0.25\ I_o$	$0.583\ I_o$	$\frac{0.25 I_o}{0.583 I_o} = 0.429$	– 0.215
	BC	$\frac{2I_o}{6} = 0.333\ I_o$		$\frac{0.333\ I_o}{0.583\ I_o} = 0.571$	– 0.285
C	CB	$\frac{2I_o}{6} = 0.333\ I_o$	$0.708\ I_o$	0.470	– 0.235
	CD	$\frac{1.5 I_o}{4} = 0.375\ I_o$		0.530	– 0.265

(*c*) **Resultant restraint moments :**

Sum of fixed end moments at $B = M_{FBA} + M_{FBC}$

$= 33.75 + 17 = 50.75$ kNm

Sum of fixed end moments at $C = M_{FCB} + M_{FCD}$

$= 0 - 40 = -40$ kNm

The fixed end moments, rotation factors and the resultant restraint moments are entered in the appropriate places in Fig. 8.10. The working is shown in Table 8.2.

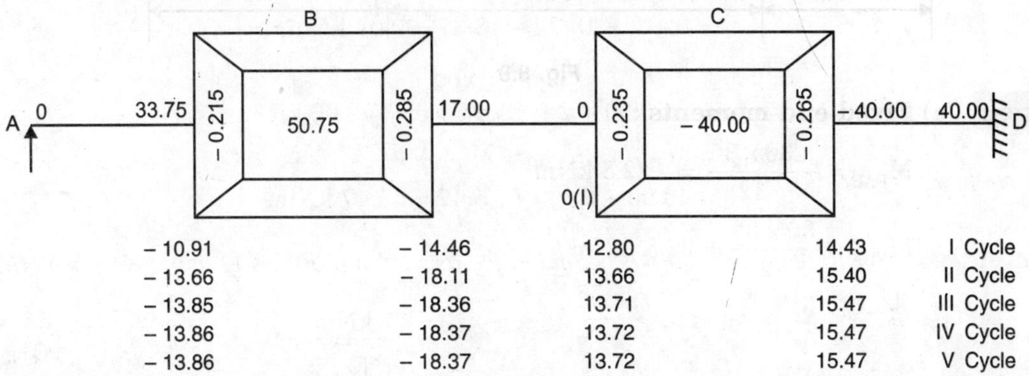

Fig. 8.10

(*d*) **Iteration process :**

Table 8.2. Iteration process.

Joint	***B***		***C***	
Rotation contribution	M'_{BA}	M'_{BC}	M'_{CB}	M'_{CD}
Rotation factors, u	– 0.215	– 0.285	– 0.235	– 0.265
ΣM_F	50.75		– 40.00	
Formula for calculation ($M'_{AB} = 0$; $M'_{DC} = 0$)	$u_{BA}(\Sigma M_{FB} + M'_{AB} + M'_{CB}) = u_{BA}(\Sigma M_{FB} + M'_{CB})$	$u_{BC}(\Sigma M_{FB} + M'_{CB} + M'_{AB}) = u_{BC}(\Sigma M_{FB} + M'_{CB})$	$u_{CB}(\Sigma M_{FC} + M'_{BC} + M'_{DC}) = u_{CB}(\Sigma M_{FC} + M'_{BC})$	$u_{CD}(\Sigma M_{FC} + M'_{DC} + M'_{BC}) = u_{CD}(\Sigma M_{FC} + M'_{BC})$
Iteration 1 (started at B with $M'_{CB} = 0$)	– 0.215 (50.75 + 0) = – 10.91	– 0.285 (50.75 + 0) = – 14.46	– 0.235 (– 40 – 14.46) = 12.80	– 0.265 (– 40 – 14.46) = 14.43
Iteration 2	– 0.215 (50.75 + 12.80) = – 13.66	– 0.285 (50.75 + 12.80) = – 18.11	– 0.235 (– 40 – 18.11) = 13.66	– 0.265 (– 40 – 18.11) = 15.40
Iteration 3	– 0.215 (50.75 + 13.66) = – 13.85	– 0.285 (50.75 + 13.66) = – 18.36	– 0.235 (– 40 – 18.11) = 13.71	– 0.265 (– 40 – 18.36) = 15.47
Iteration 4	– 0.215 (50.75 + 13.71) = – 13.86	– 0.285 (50.75 + 13.71) = – 18.37	– 0.235 (– 40 – 18.37) = 13.72	– 0.265 (– 40 – 18.37) = 15.47
Iteration 5	– 0.215 (50.75 + 13.72) = – 13.86	– 0.285 (50.75 + 13.72) = – 18.37	13.72	15.47

(*e*) **Final moments :**

$$M_{AB} = 0$$

$$M_{BA} = M_{FBA} + 2M'_{BA} + M'_{AB}$$
$$= 33.75 - 2 \times 13.86 + 0 = 6.03 \text{ kNm.}$$

$$M_{BC} = M_{FBC} + 2M'_{BC} + M'_{CB}$$
$$= 17.00 - 2 \times 18.37 + 13.72 = -6.02 \text{ kNm.}$$

$$M_{CB} = M_{FCB} + 2M'_{CB} + M'_{BC}$$
$$= 0 + 2 \times 13.72 - 18.37 = 9.07 \text{ kNm.}$$

$$M_{CD} = M_{FCD} + 2M'_{CD} + M'_{DC}$$
$$= -40.00 + 2 \times 15.47 + 0 = -9.06 \text{ kNm.}$$

$$M_{DC} = M_{FDC} + 2M'_{DC} + M'_{CD}$$
$$= 40.00 + 2 \times 0 + 15.47 = 55.47 \text{ kNm.}$$

Example 8.4 *Analyse the continuous beam loaded as shown in Fig. 8.11 using Kani's method.*

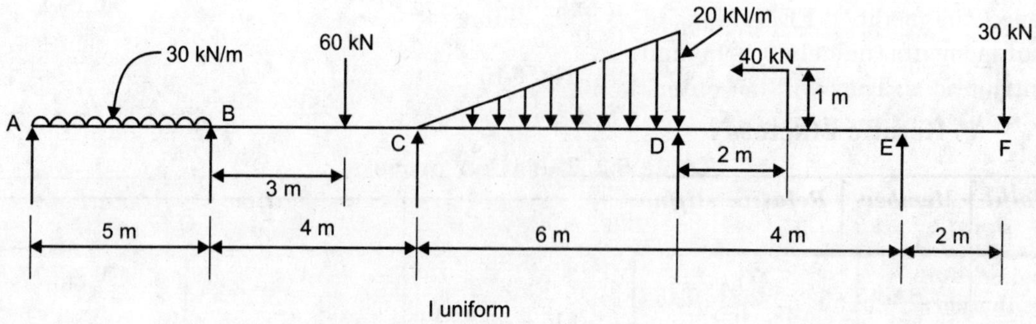

Fig. 8.11

Solution (*a*) **Fixed end moments :**

$$M_{FAB} = -\frac{30 \times 5^2}{12} = -62.5 \text{ kNm}$$

$$M_{FBA} = \frac{30 \times 5^2}{12} = 62.5 \text{ kNm}$$

A is a simply supported end. Hence to make M_{AB} zero a 62.5 kNm moment has to be applied at A, due to which the carry over to B *i.e.,* to $M_{FBA} = \frac{62.5}{2} = 31.25$ kNm.

Hence, $M_{FAB} = 0;$

$M_{FBA} = 62.5 + 31.25 = 93.75$ kNm.

$$M_{FBC} = -\frac{60 \times 3 \times 1^2}{4^2} = -11.25 \text{ kNm}$$

$$M_{FCB} = \frac{60 \times 3^2 \times 1}{4^2} = 33.75 \text{ kNm}$$

$$M_{FCD} = -\frac{20 \times 6^2}{30} = -24 \text{ kNm}$$

$$M_{FDC} = \frac{20 \times 6^2}{20} = 36 \text{ kNm}$$

$$M_{FDE} = -\frac{40 \times 1}{4} = -10 \text{ kNm}$$

$$M_{FED} = -\frac{40 \times 1}{4} = -10 \text{ kNm}$$

$$M_{EF} = -30 \times 2 = -60 \text{ kNm}$$

Since $M_{EF} = -60$ kNm, M_{ED} shall be 60 kNm. To realise this a moment of 70 kNm will be added to M_{FED}. This will make M_{FDE}, including the carry over, $= -10 + \frac{70}{2} = 25$ kNm. Hence, the modified FEM. M_{FDE} is 25 kNm. Supports A and E have now been assigned their final moments (namely 0, + 60 and – 60). They will not therefore receive or give any rotation contributions. They will not enter the iteration table either.

(*b*) **Rotation factors :**

Joint	*Member*	*Relative stiffness*	*Sum of stiffness*	*Distribution factor (DF)*	*Rotation factor, u = – 0.5 × (DF)*
B	BA	$\frac{3}{4} \times \frac{I}{5} = 0.15\ I$	0.15 I + 0.25 I = 0.40 I	$\frac{0.15I}{0.40I} = 0.375$	– 0.188
	BC	$\frac{I}{4} = 0.25\ I$		$\frac{0.25I}{0.40I} = 0.625$	– 0.312
C	CB	$\frac{I}{4} = 0.25\ I$	0.42 I	0.595	– 0.298
	CD	$\frac{I}{6} = 0.17\ I$		0.405	– 0.202
D	DC	$\frac{I}{6} = 0.17\ I$	0.36 I	0.472	– 0.236
	DE	$\frac{3}{4} \times \frac{I}{4} = 0.19\ I$		0.528	– 0.264

(*c*) **Resultant restraint moments :**

Sum of the fixed end moments at B, $\Sigma M_{FB} = M_{FBA} + M_{FBC}$
$= 93.75 - 11.25$
$= 82.5$ kNm.

Sum of the fixed end moments at C, $\Sigma M_{FC} = M_{FCB} + M_{FCD}$

$$= 33.75 - 24$$
$$= 9.75 \text{ kNm.}$$

Sum of the fixed end moments at D, $\Sigma M_{FD} = M_{FDC} + M_{FDE}$

$$= 36 + 25$$
$$= 61 \text{ kNm.}$$

The fixed end moments, rotation factors and resultant restraint moments are entered in appropriate places in Fig. 8.12.

(*d*) **Iteration process:**

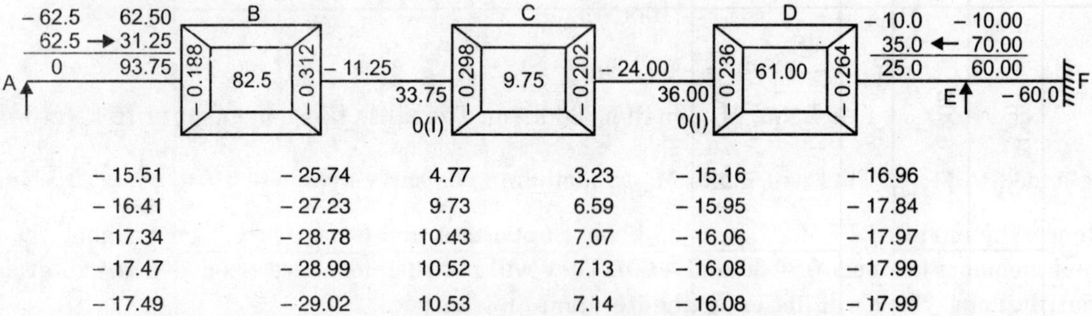

Fig. 8.12. Iteration process (to be read with Table 8.3).

(*e*) **Final moments :**

$$M_{AB} = 0$$

$$M_{BA} = M_{FBA} + 2M'_{BA} + M'_{AB}$$
$$= 93.75 - 2 \times 17.49 + 0$$
$$= 58.77 \text{ kNm}$$

$$M_{BC} = M_{FBC} + 2M'_{BC} + M'_{CB}$$
$$= -11.25 - 2 \times 29.02 + 10.53$$
$$= -58.76 \text{ kNm}$$

$$M_{CB} = M_{FCB} + 2M'_{CB} + M'_{BC}$$
$$= 33.75 + 2 \times 10.53 - 29.02$$
$$= 25.79 \text{ kNm.}$$

$$M_{CD} = M_{FCD} + 2M'_{CD} + M'_{DC}$$
$$= -24.00 + 2 \times 7.14 - 16.08$$
$$= -25.80 \text{ kNm.}$$

$$M_{DC} = M_{FDC} + 2M'_{DC} + M'_{CD}$$
$$= 36.00 - 2 \times 16.08 + 7.14 = 10.98 \text{ kNm.}$$

$$M_{DE} = M_{FDE} + 2M'_{DE} + M'_{ED}$$
$$= 25.00 - 2 \times 17.99 + 0 = -10.98 \text{ kNm}$$

$$M_{ED} = 60.00 \text{ kNm}$$
$$M_{EF} = -60.00 \text{ kNm.}$$

(*f*) **Iteration process :**

Table 8.3

Joint	***B***		***C***		***D***	
Rotation contribution	M'_{BA}	M'_{BC}	M'_{CB}	M'_{CD}	M'_{DC}	M'_{DE}
Rotation factors, u	– 0.188	– 0.312	– 0.298	– 0.202	– 0.236	– 0.264
ΣM_F	82.50		9.75		61.00	
Formula for calculation	$u_{BA}(\Sigma M_{FB} + M'_{CB})$ – 0.188 (82.50 + M'_{CB})	$u_{BC}(\Sigma M_{FB} + M'_{CB})$ – 0.312 (82.50 + M'_{CB})	$u_{CB}(\Sigma M_{FC} + M'_{BC} + M'_{DC})$ – 0.298 (9.75 + M'_{BC} + M'_{DC})	$u_{CD}(\Sigma M_{FC} + M'_{DC} + M'_{BC})$ – 0.202 (9.75 + M'_{DC} + M'_{BC})	$u_{DC}(\Sigma M_{FD} + M'_{DC})$ – 0.236 (61.00 + M'_{CD})	$u_{DE}(\Sigma M_{FD} + M'_{CD})$ – 0.264 (61.00 + M'_{CD})
Iteration 1 (started at B with M'_{CB} = 0)	– 0.188 (82.50 + 0) = – 15.51	– 0.312 (82.5 + 0) = – 25.74	– 0.298 (9.75 – 25.74 + 0) = 4.77	– 0.202 (9.75 + 0 – 25.74) = 3.23	– 0.236 (61.00 + 3.23) = – 15.16	– 0.264 (61.00 + 3.23) = – 16.96
Iteration 2	– 0.188 (82.50 + 4.77) = – 16.41	– 0.312 (82.5 + 4.77) = – 27.23	– 0.298(9.75 – 27.23 – 15.16) = 9.73	– 0.202(9.75 – 15.16 – 27.23) = 6.59	– 0.236(61.00 + 6.59) = – 15.95	– 0.264(61.00 + 6.59) = – 17.84
Iteration 3	– 0.188(82.50 + 9.73) = – 17.34	– 0.312 (82.50 + 9.73) = – 28.78	– 0.298 (9.75 – 28.78 – 15.95) = 10.42	– 0.202 (9.75 – 15.95 – 28.78) = 7.07	– 0.236 (61.00 + 7.07) = – 16.06	– 0.264(61.00 + 7.07) = – 17.97
Iteration 4	– 0.188 (52.50 + 10.42) = – 17.47	– 0.312 (82.50 + 10.42) = – 28.99	– 0.298 (9.75 – 28.99 – 16.06) = 10.52	– 0.202 (9.75 – 16.06 – 28.99) = 7.13	– 0.236(61.00 + 7.13) = – 16.08	– 0.264(61.00 + 7.13) = – 17.99
Iteration 5	– 0.188 (82.50 + 10.52) = – 17.49	– 0.312 (82.50 + 10.52) = – 29.02	– 0.298 (9.75 – 29.02 – 16.08) = 10.53	– 0.202 (9.75 – 16.08 – 29.02) = 7.14	– 0.236 (61.00 + 7.14) = – 16.08	– 0.264 (61.00 + 7.14) = – 17.99

Example 8.5 *Analyse the continuous beam loaded as shown in Fig. 8.13 by Kani's method. Given that*

$$2I_{AB} = I_{BC} = 2I_{CD} = 2I_o$$

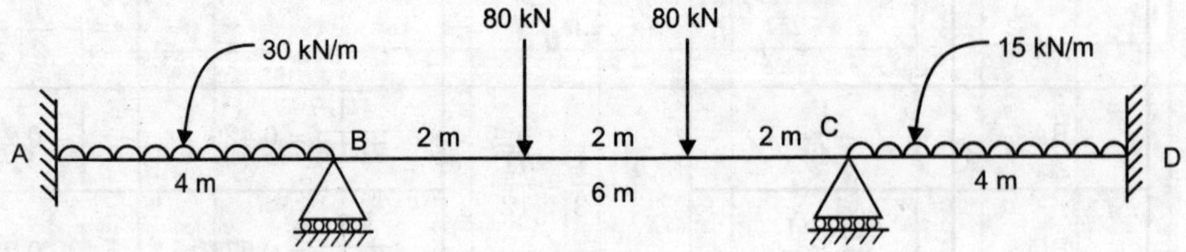

Fig. 8.13

Solution $I_{AB} = I_o$; $I_{BC} = 2I_o$; $I_{CD} = I_o$

$\theta_A = \theta_D = 0$ ($\because$ A and D are fixed ends)

unknowns: θ_B, θ_C.

(*a*) **Fixed end moments :**

AB :

$$M_{FAB} = -\frac{wl^2}{12} = -\frac{20 \times 4^2}{12} = -26.67 \text{ kNm}$$

$$M_{FBA} = \frac{wl^2}{12} = \frac{20 \times 4^2}{12} = 26.67 \text{ kNm}$$

BC :

$$M_{FBC} = -\left[\frac{Wa_1b_1^{\,2}}{l^2} + \frac{Wa_2b_2^{\,2}}{l^2}\right]$$

$$= -\left[\frac{80 \times 2 \times 4^2}{6^2} + \frac{80 \times 4 \times 2^2}{6^2}\right] = -106.67 \text{ kNm}$$

$$M_{FCB} = \left[\frac{Wa_1^{\,2}b_1}{l^2} + \frac{Wa_2^{\,2}b_2}{l^2}\right]$$

$$= \left[\frac{80 \times 2^2 \times 4}{6^2} + \frac{80 \times 4^2 \times 2}{6^2}\right] = 106.67 \text{ kNm}$$

CD :

$$M_{FCD} = -\frac{wl^2}{12} = \frac{-15 \times 4^2}{12} = -20 \text{ kNm}$$

$$M_{FDC} = \frac{wl^2}{12} = \frac{15 \times 4^2}{12} = 20 \text{ kNm}$$

(*b*) **Rotation factors :**

Joint	*Member (ij)*	*Relative stiffness* (k_{ij})	*Sum of stiffness* Σk_{ij}	*Distribution factor* $\frac{k_{ij}}{\Sigma k_{ij}}$	*Rotation factor* $= -\frac{1}{2}\left(\frac{k_{ij}}{\Sigma k_{ij}}\right)$
B	BA	$\frac{I}{4}$	$\frac{I}{4}+\frac{I}{3}=\frac{7I}{12}$	$\frac{I/4}{7I/12}=0.429$	– 0.214
	BC	$\frac{2I}{6}=\frac{I}{3}$		$\frac{I/3}{7I/12}=0.571$	– 0.286
C	CB	$\frac{2I}{6}=\frac{I}{3}$	$\frac{I}{3}+\frac{I}{4}=\frac{7I}{12}$	$\frac{I/3}{7I/12}=0.571$	– 0.286
	CD	$\frac{I}{4}$		$\frac{I/4}{7I/12}=0.429$	– 0.214

(*c*) **Resultant restraint moments :**

$\Sigma M_{FB} = -80$ $\Sigma M_{FC} = 86.67$

(*d*) **Iteration process :**

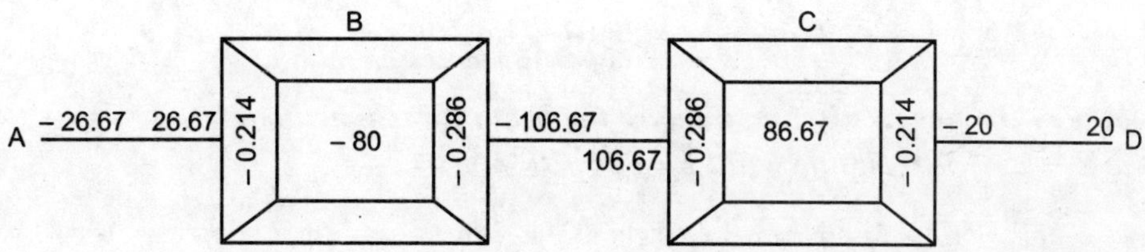

M'_{AB}	M'_{BA}	M'_{BC}	M'_{CB}	M'_{CD}	M'_{DC}
0	22.425	29.969	– 24.788	– 18.547	0
	24.259	32.421	– 33.359	– 24.961	
	24.409	32.621	– 34.060	– 25.485	
	24.421	32.637	– 34.117	– 25.528	
	24.422	32.639	– 34.122	– 25.532	
0	24.422	32.639	– 34.122	– 25.532	0

Fig. 8.14

(*e*) **Final moments :**

Member 'ij'	*Fixed end moment M_{Fij}*	*$2M'_{ij}$*	*M'_{ji}*	*Final moment (= M_{Fij}+ $2M'_{ij}$ + M'_{ji})*
AB	– 26.67	0	24.422	– 2.248
BA	+ 26.67	48.844	0	75.514
BC	– 106.67	65.278	– 34.122	– 75.514
CB	+ 106.67	– 68.244	32.639	71.065
CD	– 20	– 51.064	0	– 71.064
DC	+ 20	0	– 25.532	– 5.532

Example 8.6 *A continuous beam ABCD, 20 m long is simply supported at its ends and is propped at the same level at B and C as shown. It is loaded as shown in Fig. 8.15.*

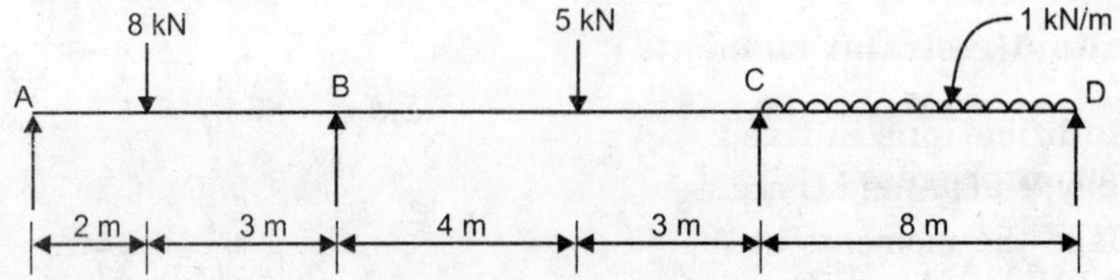

Fig. 8.15

If support B sinks by 10 mm, analyse the beam Kani's method. Take E = 2.1 × 10⁵ N/mm² and I = 85 × 10⁵ mm⁴.

Solution $E = 2.1 \times 10^5$ N/mm² $= 2.1 \times 10^8$ kN/m² ;

$EI = 85 \times 10^5$ mm⁴ $= 85 \times 10^{-7}$ m⁴ ;

$\delta = 10$ mm $= 0.01$ m

(*a*) **Fixed end moments :**

$$M_{FAB} = \frac{Wab^2}{l^2} - \frac{6EI\delta}{l^2}$$

$$= -\frac{8 \times 2 \times 3^2}{5^2} - \frac{6 \times 2.1 \times 10^8 \times 85 \times 10^{-7} \times 0.01}{5^2}$$

$$= -5.76 - 4.284 = -10.044 \text{ kNm}$$

$$M_{FBA} = \frac{Wa^2b}{l^2} - \frac{6EI\delta}{l^2}$$

$$= \frac{8 \times 2^2 \times 3}{5^2} - \frac{6 \times 2.1 \times 10^8 \times 85 \times 10^{-7} \times 0.01}{5^2}$$

$$= 3.84 - 4.284 = -0.444 \text{ kNm}$$

$$M_{FBC} = \frac{-Wab^2}{l^2} + \frac{6EI\delta}{l^2}$$

$$= \frac{-5 \times 4 \times 3^2}{7^2} + \frac{6 \times 2.1 \times 10^8 \times 85 \times 10^{-7} \times 0.01}{7^2}$$

$$= -3.673 + 2.186 = -1.487 \text{ kNm.}$$

$$M_{FCB} = \frac{Wa^2b}{l^2} + \frac{6EI\delta}{l^2}$$

$$= \frac{5 \times 4^2 \times 3}{7^2} + \frac{6 \times 2.1 \times 10^8 \times 85 \times 10^{-7} \times 0.01}{7^2}$$

$$= 4.898 + 2.186 = 7.084 \text{ kNm}$$

$$M_{FCD} = -\frac{wl^2}{12} = -\frac{1 \times 8^2}{12} = -5.333 \text{ kNm}$$

$$M_{FDC} = \frac{wl^2}{12} = -\frac{1 \times 8^2}{12} = 5.333 \text{ kNm}$$

(*b*) **Modifications in fixed end moment :**

Actually, ends A and D are free and moments at those ends must be equal to zero.

To make the moments at A and D as zero, applying the moments of + 10.044 kNm and – 5.333 kNm at A and D respectively and performing the corresponding carrying over,

$M_{FAB} = -10.044 + 10.044 = 0$

$M_{FBA} = -0.444 + 0.5 \times 10.044 = 4.578$ kNm

$M_{FDC} = 5.333 - 5.333 = 0$

$M_{FCD} = -5.333 - 0.5 \times 5.333 = -7.999$ kNm

Now, A and D will not enter the iteration process.

(*c*) **Calculation of rotation factors :**

Joint	*Member (ij)*	*Relative stiffness (k_{ij})*	*Sum of stiffness Σk_{ij}*	*Distribution factor $\frac{k_{ij}}{\Sigma k_{ij}}$*	*Rotation factor $= -\frac{1}{2}\left(\frac{k_{ij}}{\Sigma k_{ij}}\right)$*
B	BA	$\frac{3}{4}\frac{I}{l} = \frac{3}{4} \times \frac{I}{5} = \frac{3I}{20}$	$\frac{3I}{20} + \frac{I}{7} = \frac{21I + 20I}{140} = \frac{41I}{140}$	$\frac{3I/20}{41I/140} = 0.512$	– 0.256
	BC	$\frac{I}{l} = \frac{I}{7}$		$\frac{I/7}{41I/140} = 0.488$	– 0.244
C	CB	$\frac{I}{l} = \frac{I}{7}$	$\frac{I}{7} + \frac{3I}{32} = \frac{32I + 21I}{224} = \frac{53I}{224}$	$\frac{I/7}{53I/224} = 0.604$	– 0.302
	CD	$\frac{3I}{4l} = \frac{3}{4} \times \frac{I}{8} = \frac{3I}{32}$		$\frac{3I/32}{53I/224} = 0.396$	– 0.198

(*d*) **Net restraint moments at joints :**

$$M_{FB} = 3.091 \qquad M_{FC} = -0.915$$

(*e*) **Iteration process :**

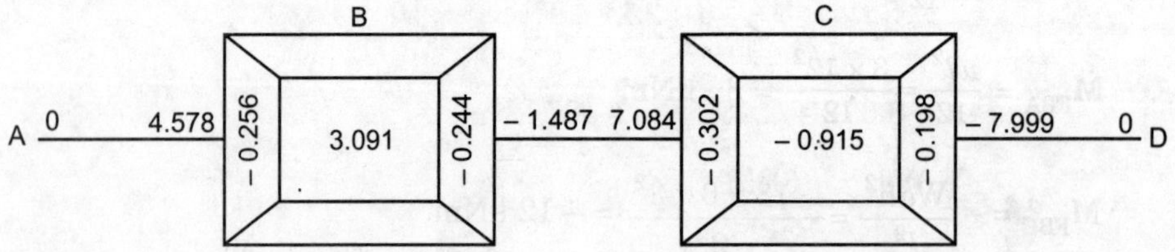

M'_{AB}	M'_{BA}	M'_{BC}	M'_{CB}	M'_{CD}	M'_{DC}
0	– 0.862	– 0.822	0.276	0.181	0
	– 0.926	– 0.882	0.525	0.344	
	– 0.930	– 0.887	0.543	0.356	
	– 0.930	– 0.887	0.544	0.357	

Fig. 8.16

(*f*) **Final moments :**

Member 'ij'	*Fixed end moment* M_{Fij}	$2M'_{ij}$	M'_{ji}	*Final moment* (= $M_{Fij} + 2M'_{ij} + M'_{ji}$)
AB	0	0	0	0
BA	4.578	– 1.860	0	+ 2.718
BC	– 1.487	– 1.774	0.544	– 2.717
CB	7.084	1.088	– 0.887	+ 7.285
CD	– 7.999	0.714	0	– 7.285
DC	–	–	–	0

Example 8.7 *Find the end moments in the beam shown in Fig. 8.17 by Kani's method and sketch the bending moment diagram.*

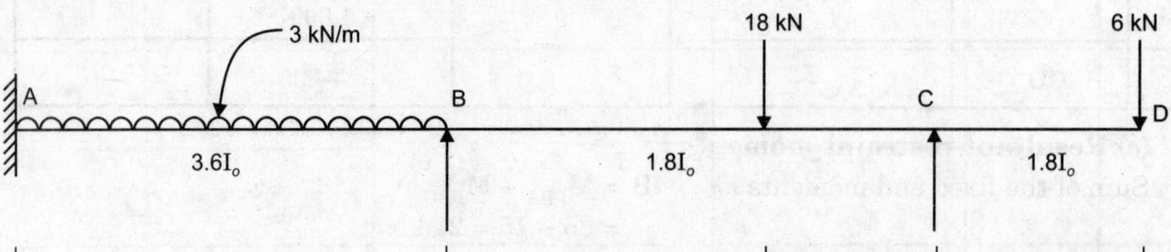

Fig. 8.17

Solution (*a*) **Fixed end moments :**

$$M_{FAB} = -\frac{wl^2}{12} = -\frac{3 \times 12^2}{12} = -36 \text{ kNm}$$

$$M_{FBA} = \frac{wl^2}{12} = \frac{3 \times 12^2}{12} = 36 \text{ kNm}$$

$$M_{FBC} = -\frac{Wab^2}{l^2} = -\frac{18 \times 6 \times 3^2}{9^2} = -12 \text{ kNm}$$

$$M_{FCB} = \frac{Wa^2b}{l^2} = -\frac{18 \times 6^2 \times 3}{9^2} = 24 \text{ kNm}$$

$$M_{CD} = -6 \times 3 = -18 \text{ kNm}$$

Since $M_{CD} = -18$ kNm, $M_{CB} = 18$ kNm. For this – 6 kNm will have to be added to M_{FCB} which will make $M_{FBC} = -12 - (1/2) \times 6 = -15$ kNm. (including the carry over).

Hence, the fixed end moments get modified as:

$M_{FAB} = -36$ kNm $\qquad M_{FBA} = 36$ kNm

$M_{FBC} = -15$ kNm $\qquad M_{CB} = 18$ kNm

$M_{CD} = -18$ kNm

(*b*) **Rotation factors :**

Joint	*Member (ij)*	*Relative stiffness* (k_{ij})	*Sum of stiffness* Σk_{ij}	*Distribution factor (DF)*	*Rotation factor = – 0.5 × DF*
B	BA	$\frac{3.6I_o}{12} = 0.3\ I_o$	$0.45\ I_o$	$\frac{0.3I_o}{0.45I_o} = 0.667$	– 0.333
	BC	$\frac{3}{4} \times \frac{1.8I_o}{9} = 0.15\ I_o$		$\frac{0.15I_o}{0.45I_o} = 0.333$	– 0.167
C	CB	$\frac{1.8I_o}{9}$	$\frac{1.8I_o}{9}$	$\frac{1.8I_o}{9} \cdot \frac{9}{1.8I_o} = 1.000$	– 0.500
	CD	—		—	—

(*c*) **Resultant restraint moments :**

Sum of the fixed end moments at $B = M_{FBA} + M_{FBC}$

$= 36 - 15 = 21$ kNm

The fixed end moments, rotation factors end the resultant restraint moment are entered in the appropriate places in Fig. 8.18.

(*d*) **Iteration process :**

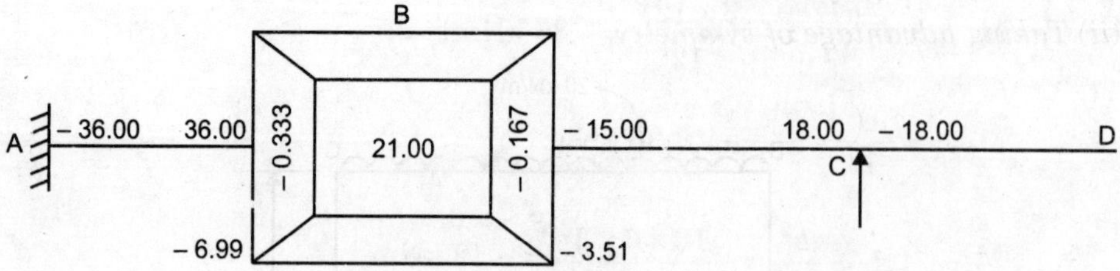

Fig. 8.18

Rotation contribution, $M'_{BA} = u_{BA}(\Sigma M_{FB} + M'_{AB} + M'_{CB})$

$M'_{CB} = 0$ since C is simply supported

$M'_{BA} = -0.333\,(21.00 + 0 + 0) = -6.99$

Rotation contribution, $M'_{BC} = u_{BC}\,(\Sigma M_{FB} + M'_{CB} + M'_{AB})$

$M'_{AB} = 0$ since A is fixed

$M'_{BC} = -0.167\,(21.00 + 0 + 0) = -3.51$

Since only B is the joint needing rotation correction, the iteration process will stop here.

(*e*) **Final moments :**

$M_{AB} = M_{FAB} + 2M'_{AB} + M'_{BA} = -36.00 + 2 \times 0 - 6.99 = -42.99$ kNm

$M_{BA} = M_{FBA} + 2M'_{BA} + M'_{AB} = 36.00 - 2 \times 6.99 + 0 = 22.02$ kNm

$M_{BC} = M_{FBC} + 2M'_{BC} + M'_{CB} = -15.00 - 2 \times 3.51 + 0 = -22.02$ kNm

$M_{CB} = 18$ kNm

$M_{CD} = -18$ kNm

(*f*) **Bending moment diagram :**

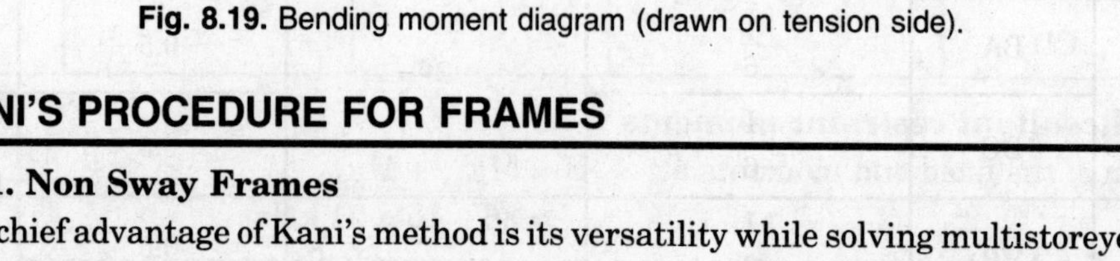

Fig. 8.19. Bending moment diagram (drawn on tension side).

8.5. KANI'S PROCEDURE FOR FRAMES

8.5.1. Non Sway Frames

The chief advantage of Kani's method is its versatility while solving multistoreyed frames.

In the case of non sway frames—frames that are symmetric and loaded symmetrically—there is hardly any change of procedure from what we used for beams. This is in evidence in the examples 8.8 and 8.9 that follows.

Example 8.8 *Analyse the frame shown in Fig. 8.20 by Kani's method,*

(i) Without using symmetry

(ii) Taking advantage of symmetry.

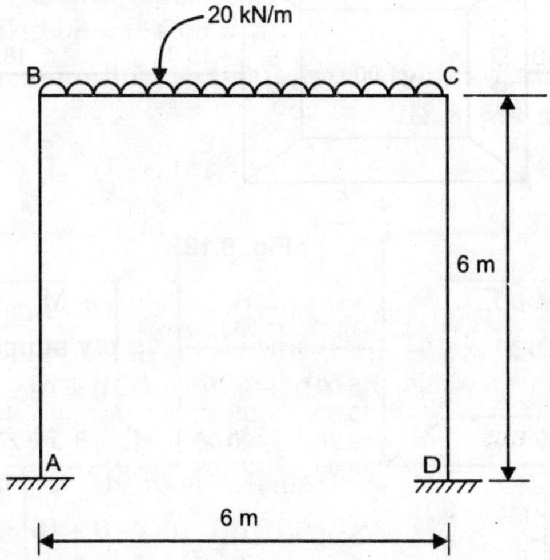

Fig. 8.20

Solution *(i)* **Without using symmetry :**

(a) Fixed end moments:

$$M_{FAB} = M_{FBA} = 0$$

$$M_{FBC} = -\frac{20 \times 6^2}{12} = -60 \text{ kNm}$$

$$M_{FCB} = +\frac{20 \times 6^2}{12} = 60 \text{ kNm}$$

$$M_{FCD} = M_{FDC} = 0$$

(b) Rotation factors :

Joint	*Member*	*Relative stiffness*	*Sum of stiffness*	*Distribution factor (DF)*	*Rotation factor u = – 0.5 DF*
B	BA	$\frac{I}{6}$	$\frac{2I}{6}$	0.5	– 0.25
	BC	$\frac{I}{6}$		0.5	– 0.25
C	CB	$\frac{I}{6}$	$\frac{2I}{6}$	0.5	– 0.25
	CD	$\frac{I}{6}$		0.5	– 0.25

(*c*) *Resultant restraint moments :*

Sum of fixed end moments at B, $\Sigma M_{FB} = M_{FBA} + M_{FBC}$
$= 0 - 60 = -60$ kNm

Sum of fixed end moments at C, $\Sigma M_{FC} = M_{FCB} + M_{FCD}$
$= 60 + 0 = 60$ kNm

The fixed end moments, rotation factors and the resultant restraint moments are entered in the appropriate places in Fig. 8.21.

(*d*) *Iteration process :*

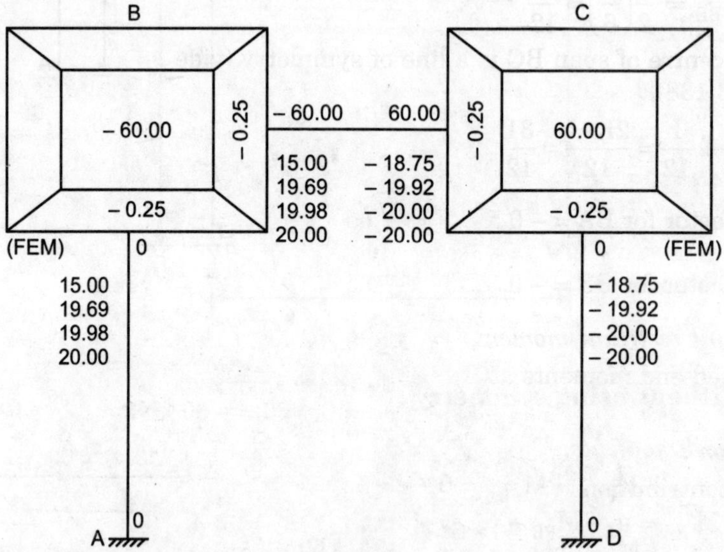

Fig. 8.21

Rotation contribution

$M'_{BA} = u_{BA}(\Sigma M_{FB} + M'_{CB})$ ...(1) $(\because\ M'_{AB} = 0)$

$M'_{BC} = u_{BC}(\Sigma M_{FB} + M'_{CB})$ ...(2) $(\because\ M'_{AB} = 0)$

$M'_{CB} = u_{CB}(\Sigma M_{FC} + M'_{BC})$ ...(3) $(\because\ M'_{DC} = 0)$

$M'_{CD} = u_{CD}(\Sigma M_{FC} + M'_{BC})$ ...(4) $(\because\ M'_{DC} = 0)$

The iteration will be started from joint B. Initially M'_{CB} will be assumed to be zero. The iterations are directly carried out in Fig. 8.21.

(*e*) *Final moments :*

$$M_{AB} = M_{FAB} + 2M'_{AB} + M'_{BA} = 0 + 0 + 20.00 = 20 \text{ kNm}$$
$$M_{BA} = M_{FBA} + 2M'_{BA} + M'_{AB} = 0 + 2 \times 20 + 0 = 40 \text{ kNm}$$
$$M_{BC} = M_{FBC} + 2M'_{BC} + M'_{CB} = -60 + 2 \times 20 - 20 = -40 \text{ kNm}$$
$$M_{CB} = M_{FCB} + 2M'_{CB} + M'_{BC} = 60 - 2 \times 20 + 20 = 40 \text{ kNm}$$
$$M_{CD} = M_{FCD} + 2M'_{CD} + M'_{DC} = 0 - 2 \times 20 + 0 = -40 \text{ kNm}$$
$$M_{DC} = M_{FDC} + 2M'_{DC} + M'_{CD} = 0 + 0 - 20 = -20 \text{ kNm}$$

Simply supported beam moment in BC $= \dfrac{wl^2}{8} = \dfrac{20 \times 6^2}{8} = 90$ kNm (*ii*) **Taking advantage of symmetry :**

(*a*) *Fixed end moments :*

Same as in solution (*i*)

(*b*) *Rotation factors :*

Joint B :

Relative stiffness:

$$k_{BA} = \frac{I}{6}$$

$$k_{BE} = \frac{1}{2}\left(\frac{I}{6}\right) = \frac{I}{12}$$

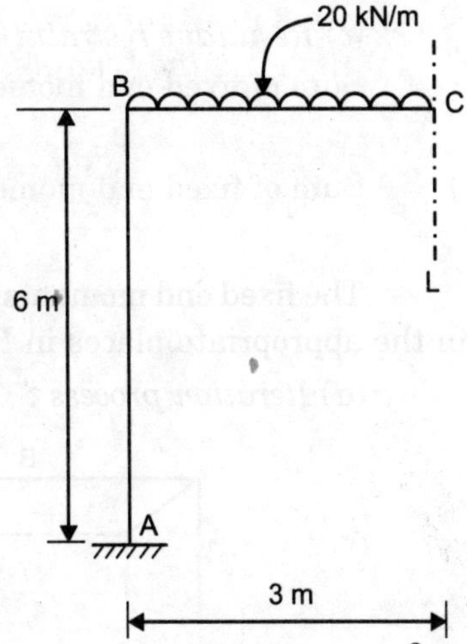

Fig. 8.22

Since the centre of span BC is a line of symmetry (vide article 3.8, Fig. 3.138)

$$\text{Sum} = \frac{I}{6} + \frac{I}{12} = \frac{2I + I}{12} = \frac{3I}{12}$$

$$\text{Rotation factor for BA} = -0.5 \text{ D.F} = -0.5\ \frac{I}{6} \times \frac{12}{3I} = -\frac{1}{3}$$

$$\text{Rotation factor for BE} = -0.5 \text{ D.F} = -0.5\ \frac{I}{12} \times \frac{12}{3I} = -\frac{1}{6}$$

(*c*) *Resultant restraint moments :*

Sum of fixed end moments at B, $\Sigma M_B = M_{FBA} + M_{FBC}$

$$= 0 - 60 = -60 \text{ kNm}$$

(*d*) *Iteration :*

Rotation contributions :

$$M'_{BA} = u_{BA}(M_{FB} + M'_{AB} + M'_{EB})$$

$$= -\frac{1}{3}(-60 + 0 + 0) = 20.0 \text{ kNm}$$

$$M'_{BC} = u_{BC}(\Sigma M_{FB} + M'_{EB} + M'_{AB})$$

$$= -\frac{1}{6}(-60 + 0 + 0) = 10.0 \text{ kNm}$$

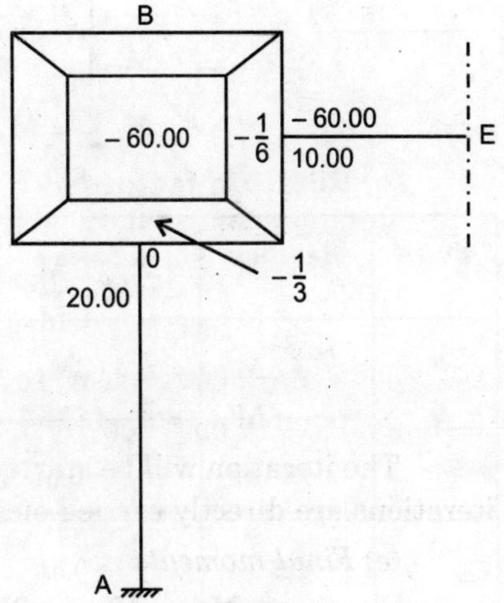

Fig. 8.23

Since there is only one interior joint, the iteration stops here.

(*e*) *Final moments :*

$$M_{AB} = M_{FAB} + 2M'_{AB} + M'_{BA}$$

$$= 0 + 0 + 20.0 = 20.0 \text{ kNm}$$

$$M_{BA} = M_{FBA} + 2M'_{BA} + M'_{AB}$$

$$= 0 + 2 \times 20.0 + 0 = 40.0 \text{ kNm}$$

$$M_{BE} = M_{FBE} + 2M'_{BE} + M'_{EB}$$

$$= -60 + 2 \times 10.0 + 0 = -40.0 \text{ kNm}$$

Due to symmetry in loading and geometry in structure

$$M_{DC} = -M_{AB} = -20.0 \text{ kNm}$$

$$M_{CD} = -M_{BA} = -40.0 \text{ kNm}$$

$$M_{CB} = -M_{BC} = 40.0 \text{ kNm}$$

Example 8.9 *Analyse the frame loaded as shown in Fig. 8.24 by Kani's method and sketch the bending moment diagram.*

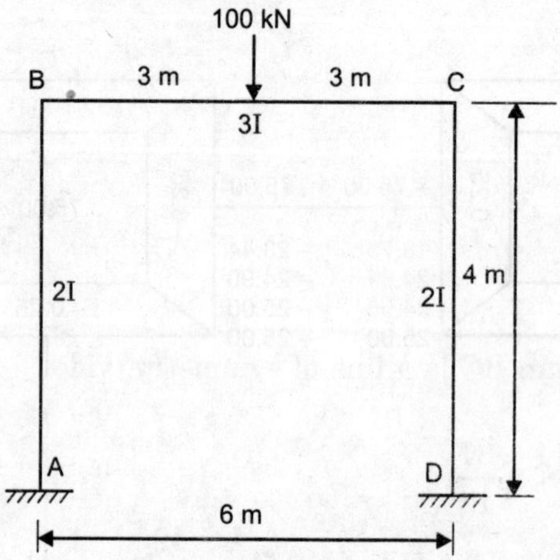

Fig. 8.24

Solution (*a*) **Fixed end moments :**

$$M_{FAB} = M_{FBA} = 0$$

$$M_{FBC} = -\frac{100 \times 6}{8} = -75 \text{ kNm}$$

$$M_{FCB} = \frac{100 \times 6}{8} = 75 \text{ kNm}$$

$$M_{FCD} = M_{FDC} = 0$$

(*b*) **Rotation factors :**

Joint	*Member*	*Relative stiffness*	*Sum of stiffness*	*Distribution factor (DF)*	*Rotation factor, u = – 0.5 DF*
B	BA	$\frac{2I}{4} = 0.5I$	$0.5I + 0.5I = I$	$\frac{0.51}{I} = 0.5$	$-0.5 \times 0.5 = -0.25$
	BC	$\frac{3I}{6} = 0.5I$		$\frac{0.5I}{I} = 0.5$	$-0.5 \times 0.5 = -0.25$
C	CB	$\frac{3I}{6} = 0.5I$	I	0.5	– 0.25
	CD	$\frac{2I}{4} = 0.5I$		0.5	– 0.25

(*c*) **Resultant restraint moments :**

Sum of fixed end moments at $B = M_{FBA} + M_{FBC}$
$= 0 - 75 = -75$ kNm

Sum of fixed end moments at $C = M_{FCB} + M_{FCD}$
$= 75 + 0 = 75$ kNm

The fixed end moments, rotation factors and the resultant restraint moments are entered in the appropriate places in Fig. 8.25.

(*d*) **Iteration cycles :**

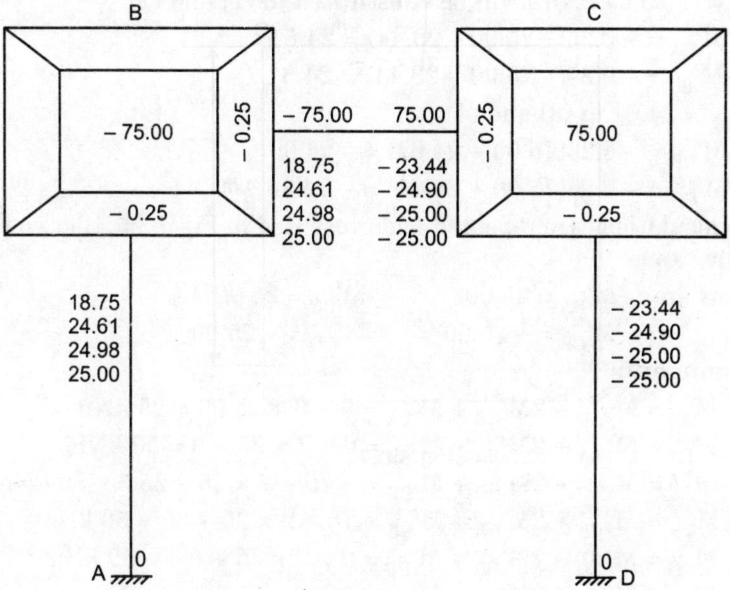

Fig. 8.25

Rotation contribution :

$M'_{AB} = 0$ since A is a fixed end

$$M'_{BA} = u_{BA}(\Sigma M_{FB} + M'_{AB} + M'_{CB})$$
$$= u_{BA}(\Sigma M_{FB} + M'_{CB}) \quad ...(1)$$
$$M'_{BC} = u_{BC}(\Sigma M_{FB} + M'_{CB} + M'_{AB})$$
$$= u_{BC}(\Sigma M_{FB} + M'_{CB}) \quad ...(2)$$
$$M'_{CB} = u_{CB}(\Sigma M_{FC} + M'_{BC} + M'_{DG})$$
$$= u_{CB}(\Sigma M_{FC} + M'_{BC}) \quad ...(3)$$

$M'_{DC} = 0$ since D is a fixed end

$$M_{CD} = u_{CD}\ (\Sigma M_{FC} + M'_{DC} + M'_{BC})$$
$$= u_{CD}\ (\Sigma M_{FC} + M'_{BC}) \quad ...(4)$$

I Cycle :

To start with, let us start the iteration at joint B, assuming $M'_{CB} = 0$.

Using (1) and (2)

$$M'_{BA} = -\ 0.25(-\ 75.00 + 0) = 18.75$$
$$M'_{BC} = -\ 0.25(75.00 + 0) = 18.75$$

Using (3) and (4) and substituting the value of $M'_{BC} = 18.75$ kNm

$$M'_{CB} = u_{CB}\ (\Sigma M_{FC} + M'_{BC})$$
$$= -\ 0.25\ (75.00 + 18.75) = -\ 23.44$$

$$M'_{CD} = u_{CD}\,(\Sigma M_{FC} + M'_{BC})$$
$$= -\,0.25\,(75.00 + 18.75) = -\,23.44$$

II Cycle :

Now M'_{CB} = (– 23.44 kNm) will be substituted in (1) and (2)

$$M'_{BA} = -\,0.25(-\,75.00 - 23.44) = 24.61$$
$$M'_{BC} = -\,0.25(-\,75.00 - 23.44) = 24.61$$

Using M'_{BC} = 24.61 in (3) and (4)

$$M'_{CB} = -\,0.25(75.00 + 24.61) = -\,24.90$$
$$M'_{CD} = -\,0.25(75.00 + 24.61) = -\,24.90$$

Similarly calculations are repeated and presented in Fig. 8.25, till two successive values are practically the same.

Final values are : $M'_{BA} = 25.00$ $\quad M'_{BC} = 25.00$

$M'_{CB} = -\,25.00$ $\quad M'_{CD} = -\,25.00$

(*e*) **Final moments :**

$$M_{AB} = M_{FAB} + 2M'_{AB} + M'_{BA} = 0 + 0 + 25.00 = 25 \text{ kNm}$$
$$M_{BA} = M_{FBA} + 2M'_{BA} + M'_{AB} = 0 + 2 \times 25 + 0 = 50 \text{ kNm}$$
$$M_{BC} = M_{FBC} + 2M'_{BC} + M'_{CB} = -\,75 + 2 \times 25 - 25 = -\,50 \text{ kNm}$$
$$M_{CB} = M_{FCB} + 2M'_{CB} + M'_{BC} = 75 - 2 \times 25 + 25 = 50 \text{ kNm}$$
$$M_{CD} = M_{FCD} + 2M'_{CD} + M'_{DC} = 0 - 2 \times 25 + 0 = -\,50 \text{ kNm}$$
$$M_{DC} = M_{FDC} + 2M'_{DC} + M'_{CD} = 0 + 0 - 25 = -\,25 \text{ kNm}$$

Simply supported beam moment :

$$\text{in BC} = \frac{wl}{4} = \frac{100 \times 6}{4} = 150 \text{ kNm}$$

(*f*) **Bending moment diagram :**

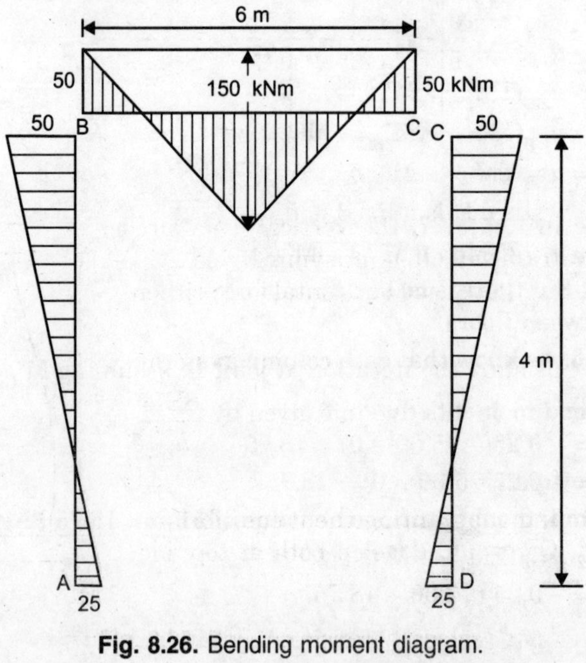

Fig. 8.26. Bending moment diagram.

8.5.2. Kani's Procedure for Frames that Sway

Non sway frames are ideal (non existent) cases. Sway frames are the reality. Now we have to move from the comfort of non sway cases to the reality of frames that sway. Since all building frames are likely to

(*i*) encounter wind or other horizontal forces

(*ii*) carry unsymmetrical vertical loads

(*iii*) be supported by columns of unequal heights and

(*iv*) be designed with unsymmetrical geometry they invariably suffer side sway.

Sway is the horizontal displacement at floor levels and causes bending moments in columns and consequently in beams too. This part of change in bending moment (from the original fixed end moments) is called rotation contribution.

Punmia (ref. 1) has dealt with 3 cases of displacement contributions due to sway.

1. Frames with equal column heights ; no horizontal forces.
2. Frames carrying horizontal forces ; equal column heights.
3. Frames carrying horizontal forces ; unequal column heights.

Obviously frames with unequal column heights without horizontal forces can be classed under case 2 above by equating horizontal forces to zero.

Case 1: Columns of same height, No horizontal forces.

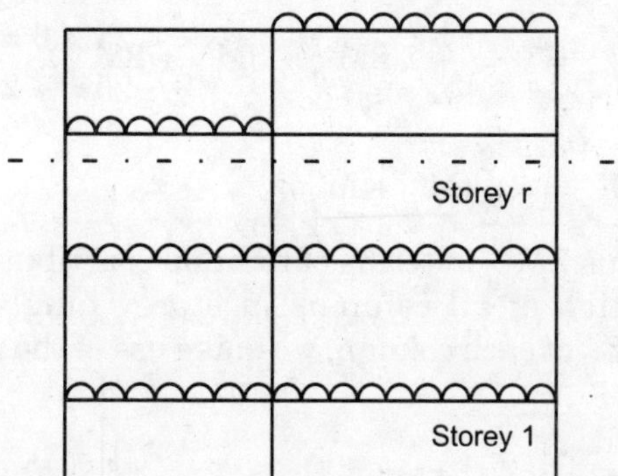

Fig. 8.27

For the frame in Fig. 8.27, the vertical loads are unsymmetrical. The frame itself is unsymmetrical. So there would be sway. But there is no horizontal force either at floor levels or between floors.

If the sway is δ, we know that each column at storey r will develop fixed end moments due to δ given by $\frac{6EI\delta}{h_r^2}$ both at top and at bottom.

For each column, to maintain moment equilibrium, shear forces Q_{r1}, Q_{r2}, Q_{r3} would develop both at top and bottom such that

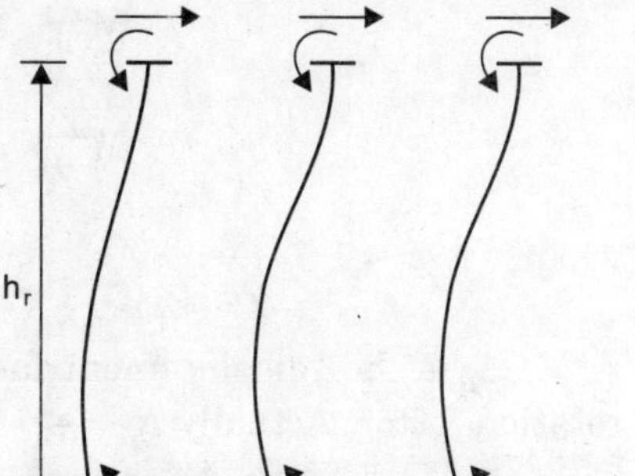

Fig. 8.28

$$Q_{rij}\,.h_r = M_{ij} + M_{ji} \quad \text{...(8.7)}$$

where M_{ij} and M_{ji} are the final moments in each column.

Since there are no external horizontal forces acting on the frame

ΣQ_{rij} shall be zero. This property was used in case 2. Therefore $\sum_r (M_{ij} + M_{ji}) = 0$ for the columns in storey number *r*. When we expand the final moments based on equation (8.3), we must take care to include a term M''_{ij} also as the displacement contribution, resulting in

$$M_{ij} = M_{Fij} + 2M'_{ij} + M'_{ji} + M''_{ij}$$

$$M_{ji} = M_{Fji} + 2M'_{ji} + M'_{ij} + M''_{ij} \quad \text{...(8.8)}$$

[This is also obtained by modifying the slope deflection equations (2.1) and (2.2)]. In equation (8.7),

(*i*) Since there are no loads on the columns, M_{Fij} s are zero.

(*ii*) $$M''_{ij} = -\frac{6EI\,\delta_{ij}}{h_r^{\,2}} = -6E\,k_{ij}\,\frac{\delta_{ij}}{h_r} \quad \text{...(8.9)}$$

Combining equations (8.7), (8.8) and (8.9), we get

$$Q_{rij} = \frac{1}{h_r}(3M'_{ij} + 3M'_{ji} + 2M''_{ij})$$

$$\Sigma Q_{rij} = 0 = \frac{1}{h_r}\Sigma(3M'_{ij} + 3M'_{ji} + 2M''_{ij})$$

$$\therefore \quad \Sigma M''_{ij} = -\frac{3}{2}(M'_{ij} + M'_{ji}) \quad \text{...(8.10)}$$

Equation (8.10) gives us $\Sigma M''_{ij}$ in terms of rotation contributions. $\Sigma M''_{ij}$ is the total (sway or) displacement contribution of all columns in storey number *r*. To obtain individual displacement contribution M''_{ij} of each column, we make use of the property implied in equation (8.9)— that $M''_{ij} \propto k_{ij}$

So, $$M''_{ij} = \frac{k}{\Sigma k_{ij}}\,\Sigma M''_{ij} \quad \text{...(8.11)}$$

Combining this with equation (8.10), we get

$$M''_{ij} = -1.5\,\frac{k_{ij}}{\Sigma k_{ij}}\,\Sigma(M'_{ij} + M'_{ji}) \quad \text{...(8.12)}$$

$$M''_{ij} = \Delta_{ij}\,\Sigma\,(M'_{ij} + M'_{ji}) \quad \text{...(8.13)}$$

where $\Delta_{ij} = -1.5\,\dfrac{k_{ij}}{\Sigma k_{ij}}$

Δ_{ij} is the "displacement factor" for each column, similar to u_{ij} we adopted earlier for rotation factor. Actually $\Delta_{ij} = -1.5$ DF and is applicable to columns only.

Example 8.10 shows us how to make use of the above ideas in solving a frame with sway. We observe that the iterative process would involve successively improving the values of M'_{ij} and M''_{ij} for all joints cyclically.

Example 8.10 *Analyse the portal frame shown in Fig. 8.29 by Kani's method. Sketch the bending moment diagram.*

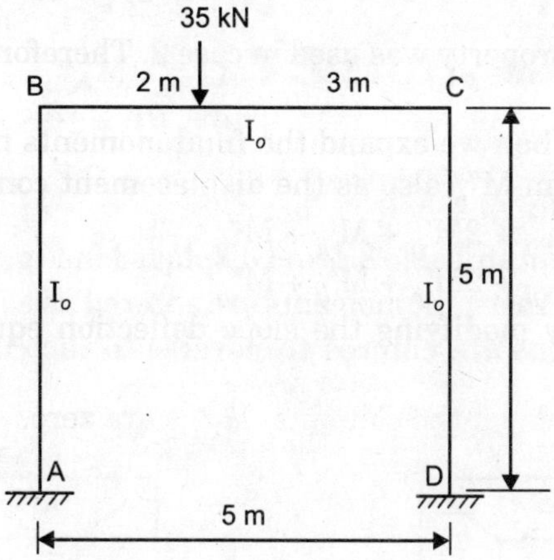

Fig. 8.29

Solution The given portal frame is geometrically symmetrical. Loading is asymmetrical. Hence, there will be side sway. Columns are of equal height and fixed at base. The frame is subjected to vertical load only.

(*a*) **Fixed end moments :**

$M_{FEB} = M_{FBA} = 0$, since there are no loads on the columns.

$$M_{FBC} = -\frac{35 \times 2 \times 3^2}{5^2} = -25.2 \text{ kNm}$$

$$M_{FCB} = +\frac{35 \times 2^2 \times 3}{5^2} = 16.8 \text{ kNm}$$

$$M_{FCD} = M_{FDC} = 0$$

(*b*) **Rotation factors and displacement factors :**

Joint	***Member***	***Relative stiffness***	***Sum of stiffness***	***Distribution factor (DF)***	***Rotation u = (– 0.5) DF***	***Displacement factors, (*) Δ = (– 1. 5) DF***
B	BA	$\frac{I_o}{5}$	$\frac{2I_o}{5}$	0.5	– 0.25	– 0.75
	BC	$\frac{I_o}{5}$		0.5	– 0.25	—
C	CB	$\frac{I_o}{5}$	$\frac{2I_o}{5}$	0.5	– 0.25	—
	CD	$\frac{I_o}{5}$		0.5	– 0.25	– 0.75

(*) Displacement factors are for columns only.

(*c*) **Resultant restraint moments :**

Sum of fixed end moments at B, $\Sigma M_B = M_{FBA} + M_{FBC}$

$= 0 - 25.2 = -25.2$ kNm

Sum of fixed end moments at C, $\Sigma M_C = M_{FCB} + M_{FCD}$

$= 16.8 + 0 = 16.8$ kNm

(*d*) **Iteration process :**

The process is carried out in Table 8.4 and displayed in Fig. 8.30. The fixed end moments, rotation factors and resultant restraint moments are entered as usual. The displacement factors and displacement contributions are entered transverse to each column.

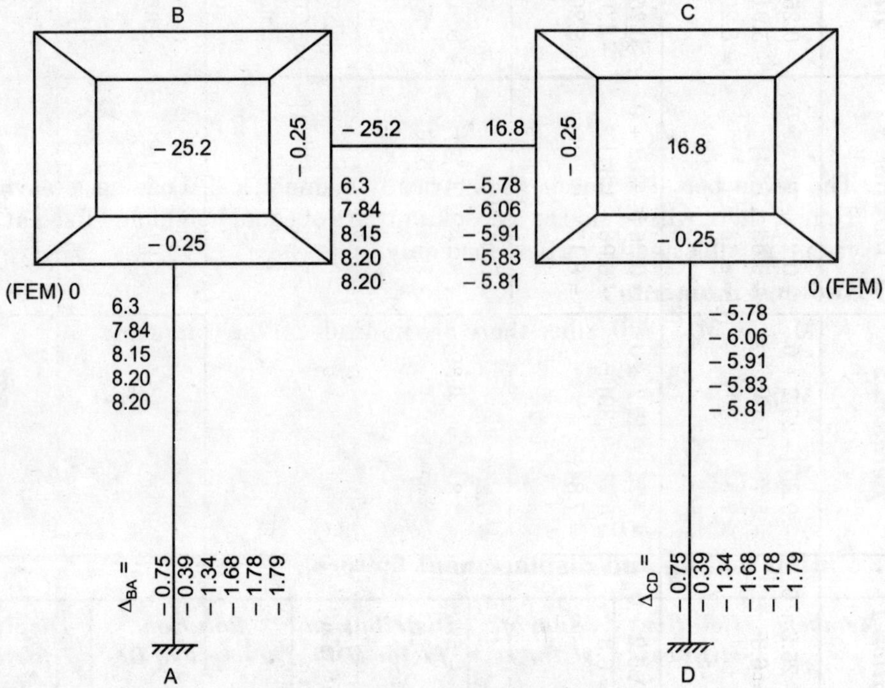

Fig. 8.30. Iteration for sway frame.

(*e*) **Calculation of rotation contributions and displacement contributions :**

Table 8.4

Formula for calculation	*Iteration 1*	*Iteration 2*	*Iteration 3*	*Iteration 4*	*Iteration 5*
$M'_{BA} = u_{BA} (\Sigma M_{FB} + M'_{AB} + M'_{CB} + M''_{BA}) = -0.25[-25.2 + 0 + M'_{CB} + M''_{BA}]$	$-0.25[-25.2 + 0 + 0 + 0] = 6.3$	$-0.25[-25.2 + 0 - 5.78 - 0.39] = 7.84$	$-0.25[-25.2 + 0 - 6.06 - 1.34] = 8.15$	$-0.25[-25.2 + 0 - 5.91 - 1.68] = 8.20$	$-0.25[-25.2 + 0 - 5.83 - 1.78] = 8.20$
$M'_{BC} = u_{BC} (\Sigma M_{FB} + M'_{AB} + M'_{CB} + M''_{BA}) = -0.25[-25.2 + 0 + M'_{CB} + M''_{BA}]$	$-0.25[-25.2 + 0 + 0 + 0] = 6.3$	$-0.25[-25.2 + 0 - 5.78 - 0.39] = 7.84$	$-0.25[-25.2 + 0 - 6.06 - 1.34] = 8.15$	$-0.25[-25.2 + 0 - 5.91 - 1.68] = 8.20$	$-0.25[-25.2 + 0 - 5.83 - 1.78] = 8.20$
$M'_{CB} = u_{CB} (\Sigma M_{FC} + M'_{BC} + M'_{DC} + M''_{CD}) = -0.25[16.8 + 0 + M'_{BC} + M''_{CD}]$	$-0.25[16.8 + 6.3 + 0 + 0] = -5.78$	$-0.25[16.8 + 7.84 + 0 - 0.39] = -6.06$	$-0.25[16.8 + 8.15 + 0 - 1.34] = -5.91$	$-0.25[16.8 + 8.20 + 0 - 1.68] = -5.83$	$-0.25[16.8 + 8.20 + 0 - 1.78] = -5.81$
$M'_{CD} = u_{CD} (\Sigma M_{FC} + M'_{DC} + M'_{BC} + M''_{CD}) = -0.25[16.8 + 0 + M'_{BC} + M''_{CD}]$	$-0.25[16.8 + 0 + 6.3 + 0] = -5.78$	$-0.25[16.8 + 0 + 7.84 - 0.39] = -6.06$	$-0.25[16.8 + 8.15 + 0 - 1.34] = -5.91$	$-0.25[16.8 + 0 + 8.20 - 1.68] = -5.83$	$-0.25[16.8 + 8.20 + 0 - 1.78] = -5.81$
$M''_{BA} = \Delta_{BA} (M_{BA} + M'_{AB} + M'_{CD} + M'_{DC}) = -0.75[M'_{BA} + 0 + M'_{CD} + 0]$	$-0.75[6.3 - 5.78] = -0.39$	$-0.75[7.84 - 6.06] = -1.34$	$-0.75[8.15 - 5.91] = -1.68$	$-0.75[8.20 - 5.83] = -1.78$	$-0.75[8.20 - 5.81] = -1.79$
$M''_{CD} = \Delta_{CD} (M'_{BA} + M'_{AB} + M'_{CD} + M'_{DC}) = -0.75[M'_{BA} + 0 + M'_{CD} + 0]$	$-0.75[6.3 - 5.78] = -0.39$	$-0.75[7.84 - 6.06] = -1.34$	$-0.75[8.15 - 5.91] = -1.68$	$-0.75[8.20 - 5.83] = -1.78$	$-0.75[8.20 - 5.81] = -1.79$

(*f*) **Final moments :**

$M_{AB} = M_{FAB} + 2M'_{AB} + M'_{BA} + M''_{AB} = 0 + 0 + 8.20 - 1.79 = 6.41$ kNm

$M_{BA} = M_{FBA} + 2M'_{BA} + M'_{AB} + M''_{BA} = 0 + 2 \times 8.20 + 0 - 1.79 = 14.61$ kNm

$M_{BC} = M_{FBC} + 2M'_{BC} + M'_{CB} + M''_{BC} = -25.2 + 2 \times 8.20 - 5.81 + 0 = -14.61$ kNm

$M_{CB} = M_{FCB} + 2M'_{CB} + M'_{BC} + M''_{CB} = 16.8 - 2 \times 5.81 + 8.20 + 0 = 13.38$ kNm

$M_{CD} = M_{FCD} + 2M'_{CD} + M'_{DC} + M''_{CD} = 0 - 2 \times 5.81 + 0 - 1.79 = -13.41$ kNm

$M_{DC} = M_{FDC} + 2M'_{DC} + M'_{CD} + M''_{DC} = 0 + 0 - 5.81 - 1.79 = -7.60$ kNm

(*g*) **Remarks :**

It can be observed that Kani's method employs one single analysis for frames with sway (unlike the moment distribution method, which uses two analysis, namely, non sway and sway).

The steps adopted in Table 8.4 will make it clear that we are improving the (initially erroneous) values of M'_{ij}, M'_{ji} and M''_{ij} successively. Using the conditions that at each joint, the total moment is zero and at each storey level, the net horizontal force is zero.

Case 2: Frames with horizontal forces.

When horizontal forces act at the floor levels, the net shear at a storey level will not be zero but will be the sum of the horizontal forces above that level.

While deriving equation (8.9), we used the condition $\sum_r Q_{ij} = 0$. This is not valid now.

We shall use $\sum_r Q_{ij} = -Q_r$, the shear at storey level *r*.

Hence, $$-Q_r = \frac{1}{h_r}\Sigma(3M'_{ij} + 3M'_{ji} + 2M''_{ij})$$

$$-2\sum_r M''_{ij} = Q_r h_r + \Sigma(3M'_{ij} + 3M'_{ji})$$

$$\sum_r M''_{ij} = -\frac{3}{2}\left[\frac{Q_r h_r}{3} + \sum_r (M'_{ij} + M'_{ji})\right] \quad \text{...(8.14)}$$

$$M''_{ij} = -\frac{3}{2}\frac{k_{ij}}{\Sigma k_{ij}}\left[\frac{Q_r h_r}{3} + \sum\left(M'_{ij} + M'_{ji}\right)\right]$$

$$M''_{ij} = \Delta_{ij}\,[M_r + \Sigma(M'_{ij} + M'_{ji})] \quad \text{...(8.15)}$$

where, $\Delta_{ij} = -\frac{3}{2}\frac{k_{ij}}{\Sigma k_{ij}}$ as before and $M_r = \frac{Q_r h_r}{3}$ called the **storey moment** for the *r*th storey.

Other than the inclusion of the external horizontal force in equation (8.15), the procedure is identical with case 1.

Case 3: Frames with unequal legs, carrying horizontal forces.

Case 3 can have some sub divisions taking into account the end conditions at the columns bases.

These conditions will primarily alter the displacement factor as under:

(*i*) If some columns are hinged at the base, their *k* values will get modified as $\frac{3}{4}\frac{I}{l}$. In addition, there are a few more changes in the expression for Δ_{ij}, as shown in this section.

(*ii*) If the column heights are different, we will choose some columns (with, let us say, the minimum height) as standard columns.

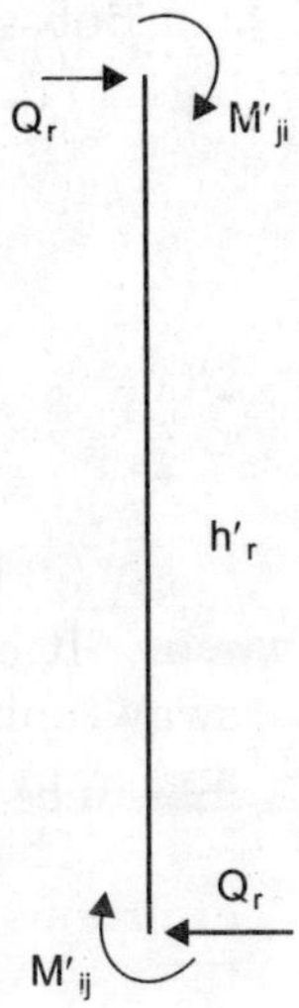

Fig. 8.31

Columns balance column moments as under (Fig. 8.31)

$$Q_r h'_r + M'_{ij} + M'_{ji} = 0,$$

where $Q_r = \sum_r Q_{ij}$, the shear at storey level *r*.

For the entire storey, the shear equation becomes

$$Q_r \,.\, h'_r + \sum_r \frac{h'_r}{h_{ij}}\left(M_{ij} + M_{ji}\right) = 0$$

where $Q_r = \sum_r Q_{ij}$, the shear at storey level *r*.

$$Q_r \,.\, h'_r + \sum_r C_{ij}\left(M_{ij} + M_{ji}\right) = 0 \qquad ...(8.16)$$

where $C_{ij} = \frac{h'_r}{h_{ij}}$, the column reduction factor.

This is justified because for every column, the shear force balancing the end moments (vide Fig. 8.31) varies inversely as the column height.

So we derive a new set of equations for the columns in this case.

$$M_{ij} = M^*_{Fij} + 2M'_{ij} + M'_{ji} + M''_{ij}$$
$$M_{ji} = M^*_{Fji} + 2M'_{ji} + M'_{ij} + M''_{ij} \qquad ...(8.17)$$

* Zero for columns

$$M_{ij} + M_{ji} = 3(M'_{ij} + M'_{ji}) + 2M''_{ij}$$

Using this in equation (8.16)

$$Q_r \,.\, h'_r + \sum_r C_{ij}\,(3M'_{ij} + 3M'_{ji} + 2M''_{ij}) = 0$$

$$\sum_r C_{ij}\, M''_{ij} = -\frac{3}{2}\left[\frac{Q_r\, h'_r}{3} + \sum_r (M'_{ij} + M'_{ji})\, C_{ij}\right]$$

$$\sum_r C_{ij}\, M''_{ij} = -\frac{3}{2}\left[M_r + \sum_r (M'_{ij} + M'_{ji})\, C_{ij}\right] \qquad ...(8.18)$$

where $M_r = \frac{Q_r h'_r}{3}$, the storey moment, as before.

Since, $M''_{ij} = -\frac{6EI\delta}{h^2} = -6Ek_{ij}\frac{\delta_{ij}}{h_{ij}}$, the moment shared by a column is proportional to k_{ij} and inversely proportional to h_{ij}.

Hence, $M''_{ij} \propto \frac{k_{ij}}{h_{ij}} \propto C_{ij}\, k_{ij}$. $\left(\text{Since } C_{ij} \propto \frac{1}{h_{ij}}\right)$

we get,

$$\frac{M''_{ij}}{C_{ij}\, k_{ij}} = \frac{\Sigma M''_{ij}}{\Sigma C_{ij}\, k_{ij}} = \frac{\Sigma C_{ij}\, M''_{ij}}{\Sigma C^2_{ij}\, k_{ij}}$$

Thus,

$$M''_{ij} = \frac{C_{ij}\, k_{ij}}{\Sigma C^2_{ij}\, k_{ij}} \,.\, \Sigma\, C_{ij}\, M''_{ij}$$

Substituting from eqn. (8.16),

$$M''_{ij} = -\frac{3}{2}\frac{C_{ij}\,k_{ij}}{\Sigma C^2_{ij}\,k_{ij}}\left[M_r + \sum_r (M'_{ij} + M'_{ji})\,C_{ij}\right]$$

$$M''_{ij} = \Delta_{ij}\left[M_r + \sum_r (M'_{ij} + M'_{ji})\,C_{ij}\right] \qquad ...(8.19)$$

where $\Delta_{ij} = -\dfrac{3}{2}\dfrac{C_{ij}\,k_{ij}}{\Sigma C_{ij}^{\;2} k_{ij}}$

Δ_{ij} has to be modified if some of the columns are hinged (and the others fixed). In that case

$$C_{ij} = \frac{h_r}{n\,h_{ij}}$$

Here, $n = 1.5$ for columns hinged at base and $n = 1$ for fixed columns.

In addition, $\Delta_{ij} = -\dfrac{3}{2}\dfrac{m\,C_{ij}\,k_{ij}}{\Sigma m^2\,C_{ij}^{\;2}\,k_{ij}}$

Here m accounts for the fact that in hinged columns the displacement contribution is only half that for fixed columns.

Therefore, $m = 0.75$ for hinged columns and 1.0 for fixed columns.

Example 8.11 demonstrates the use of the above concepts in solving a single bay single storey portal. The procedure is definitely in efficient when applied to simple frames. The real use of Kani's method lies in its application to multistorey multibay frames. This does not make gross approximations like in the case of portal method, cantilever method etc.

Example 8.11 *Analyse the portal frame loaded as shown in Fig. 8.32 by Kani's method.*

Solution The given portal frame is geometrically asymmetrical. So, there will be sideway. The columns are of unequal height. One column has a fixed base and other is with hinged base. The frame is subjected to vertical loading only.

(*a*) **Fixed end moments :**

$$M_{FAB} = M_{FBA} = 0$$

$$M_{FBC} = -\frac{50 \times 4}{8} = -25 \text{ kNm}$$

$$M_{FCB} = +\frac{50 \times 4}{8} = 25 \text{ kNm}$$

$$M_{FCD} = M_{FDC} = 0$$

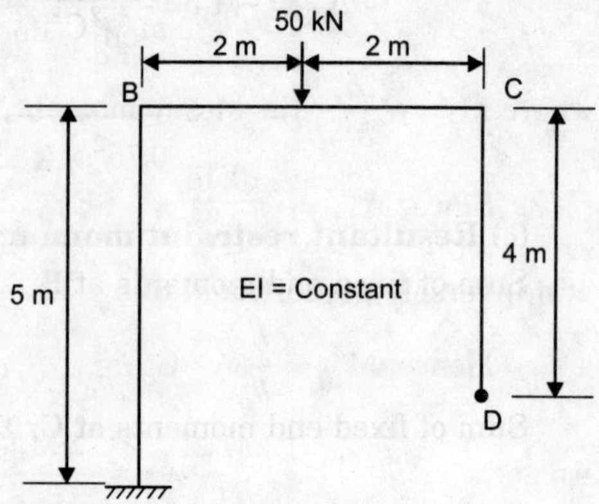

Fig. 8.32

(*b*) **Rotation factors and displacement factors :**

Joint	Member	Relative stiffness	Sum of stiffness	Distribution factor (DF)	Rotation factor, u = (– 0.5) DF	Displacement factors, Δ_{ij}
B	BA	$\frac{I}{5}$	$\frac{I}{5}+\frac{I}{4}=\frac{9I}{20}$	$\frac{I}{5}\times\frac{20}{9I}=0.444$	– 0.222	– 0.612
	BC	$\frac{I}{4}$		$\frac{I}{4}\times\frac{20}{9I}=0.556$	– 0.278	–
C	CB	$\frac{I}{4}$	$\frac{I}{4}+\frac{3I}{16}=\frac{7I}{16}$	$\frac{I}{4}\times\frac{16}{7I}=0.571$	– 0.286	–
	CD	$\frac{3}{4}\times\frac{I}{4}$		$\frac{3I}{16}\times\frac{16}{7I}=0.429$	– 0.214	– 0.957

Displacement factor, $\Delta_{ij} = -1.5\ \dfrac{m_{ij}\ C_{ij}\ k_{ij}}{\Sigma m_{ij}^2\ C_{ij}^2\ k_{ij}}$

Taking BA as the reference column, $h_r = 5$ m

$$C_{BA} = \frac{h_r}{n\,.\,h_{BA}} = \frac{5}{1\times5} = 1\ ;\quad C_{CD} = \frac{h_r}{n\,.\,h_{CD}} = \frac{0.5}{1.5\times4} = 0.833$$

$m_{BA} = 1$; $\quad m_{CD} = 0.75$

$C_{BA} = 1$; $\quad C_{CD} = 0.833$

$k_{BA} = \dfrac{I}{5}$; $\quad k_{CD} = \dfrac{3I}{16}$

Now, $\quad \Delta_{BA} = -1.5\ \dfrac{m_{BA}\ C_{BA}\ k_{BA}}{m_{BA}{}^2C_{BA}{}^2k_{BA} + m_{CD}{}^2C^2{}_{CD}k_{CD}}$

$$= -1.5\ \frac{1\times1\times I/5}{1^2\times1^2\times I/5 + 0.75^2\times0.833\times\dfrac{3I}{16}} = -1.042$$

$$\Delta_{CD} = -1.5\ \frac{m_{CD}\ C_{CD}\ k_{CD}}{m_{CD}{}^2C^2{}_{CD}k_{CD} + m^2{}_{BA}C^2{}_{BA}\ k_{BA}}$$

$$= -1.5\ \frac{0.75\times0.833\times3I/16}{0.75^2\times0.833\times3I/16 + 1^2\times1^2\times\dfrac{I}{5}} = -0.610$$

(*c*) **Resultant restraint moments :**

Sum of fixed end moments at B, $\Sigma M_B = M_{FBA} + M_{FBC}$

$= 0 - 25$

$= -25$ kNm

Sum of fixed end moments at C, $\Sigma M_C = M_{FCB} + M_{FCD}$

$= 25 + 0$

$= 25$ kNm

(*d*) **Calculation of rotation contributions and displacement contributions :**

Table 8.5

Formula for calculation	*Iteration 1*	*Iteration 2*	*Iteration 3*	*Iteration 4*	*Iteration 5*	*Iteration 6*	*Iteration 7*
$M'_{BA} = u_{BA} (\Sigma M_{FB} + M'_{AB} + M'_{CB} + M''_{BA})$ $= -0.222[-25 + 0 + M'_{CB} + M''_{BA}]$	– 0.222 [– 25 + 0 + 0 + 0] = 5.55	– 0.222 [– 25 + 0 – 9.14 + 0.154] = 7.545	– 0.222 [– 25 + 0 – 9.878 – 1.447] = 8.064	– 0.222 [– 25 + 0 – 9.796 – 2.04] = 8.178	– 0.222 [– 25 + 0 – 9.737 – 2.197] = 8.199	– 0.222 [– 25 + 0 – 9.719 – 2.231] = 8.203	– 0.222 [– 25 + 0 – 9.714 – 2.238] = 8.203
$M'_{BC} = u_{BC} (\Sigma M_{FB} + M'_{AB} + M'_{CB} + M''_{BA})$ $= -0.278[-25 + 0 + M'_{CB} + M''_{BA}]$	– 0.278 [– 25 + 0 + 0 + 0] = 6.95	– 0.278 [– 25 + 0 – 9.14 + 0.154] = 9.448	– 0.278 [– 25 + 0 – 9.878 – 1.447] = 10.098	– 0.278 [– 25 + 0 – 9.796 – 2.04] = 10.24	– 0.278 [– 25 + 0 – 9.737 – 2.197] = 10.268	– 0.278 [– 25 + 0 – 9.719 – 2.231] = 10.272	– 0.278 [– 25 + 0 – 9.714 – 2.238] = 10.272
$M'_{CB} = u_{CB} (\Sigma M_{FC} + M'_{DC} + M'_{BC} + M''_{CD})$ $= -0.286[25 + 0 + M'_{BC} + M''_{CD}]$	– 0.286 [25 + 0 + 6.95 + 0] = – 9.14	– 0.286 [25 + 0 + 9.448 + 0.09] = – 9.878	– 0.286 [25 + 0 + 10.098 – 0.847] = – 9.796	– 0.286 [25 + 0 + 10.24 – 1.194] = – 9.737	– 0.286 [25 + 0 + 10.268 – 1.286] = – 9.719	– 0.286[25 + 0 + 10.272 – 1.306] = – 9.714	– 0.286[25 + 0 + 10.272 – 1.310] = – 9.713
$M'_{CD} = u_{CD} (\Sigma M_{FC} + M'_{DC} + M'_{BC} + M''_{CD})$ $= -0.214[25 + 0 + M'_{BC} + M''_{CD}]$	– 0.214 [25 + 0 + 6.95 + 0] = – 6.84	– 0.214 [25 + 0 + 9.448 + 0.09] = – 7.391	– 0.214 [25 + 0 + 10.098 – 0.847] = – 7.330	– 0.214 [25 + 0 + 10.24 – 1.194] = – 7.286	– 0.214 [25 + 0 + 10.268 – 1.286] = – 7.272	– 0.214 [25 + 0 + 10.272 – 1.306] = – 7.269	– 0.214 [25 + 0 + 10.272 – 1.310] = – 7.268
$M''_{BA} = \Delta_{BA} (C_{BA} \times M'_{BA} + C_{CD} \times M'_{CD})$ $= -1.042[1 \times M'_{BA} + 0.833 \times M'_{CD}]$	– 1.042 [5.55 – 0.833 × 6.84] = 0.154	– 1.042 [7.545 – 0.833 × 7.391] = – 1.447	– 1.042 [8.064 – 0.833 × 7.330] = – 2.04	– 1.042 [8.178 – 0.833 × 7.286] = – 2.197	– 1.042 [8.199 – 0.833 × 7.272] = – 2.231	– 1.042 [8.203 – 0.833 × 7.269] = – 2.238	– 1.042[8.203 – 0.833 × 7.268] = – 2.239
$M''_{CD} = \Delta_{CD} (C_{BA} \times M'_{BA} + C_{CD} \times M'_{CD})$ $= -0.610[1 \times M'_{BA} + 0.833 \times M'_{CD}]$	– 0.610 [5.55 – 0.833 × 6.84 = 0.090	– 0.610 [7.545 – 0.833 × 7.391] = – 0.847	– 0.610 [8.064 – 0.833 × 7.330] = – 1.194	– 0.610 [8.178 – 0.833 × 7.286] = – 1.286	– 0.610 [8.199 – 0.833 × 7.272] = – 1.306	– 0.610 [8.203 – 0.833 × 7.269] = – 1.310	– 0.610[8.203 – 0.833 × 7.268] = – 1.310

(*e*) **Rotation contributions—Iteration :**

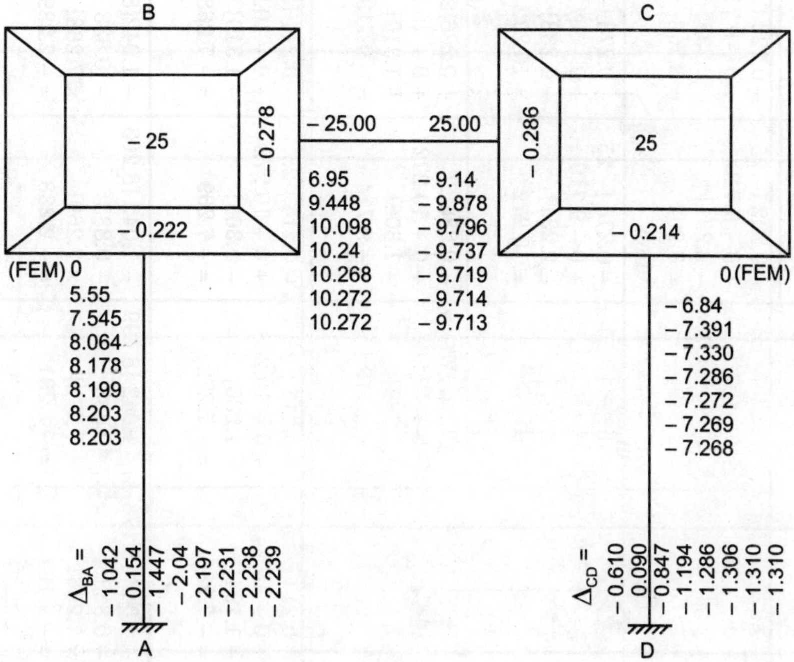

Fig. 8.33

(*f*) **Final moments :**

$M_{AB} = M_{FAB} + 2M'_{AB} + M'_{BA} + M''_{AB} = 0 + 0 + 8.203 - 2.239 = 5.964$ kNm

$M_{BA} = M_{FBA} + 2M'_{BA} + M'_{AB} + M''_{BA} = 0 + 2 \times 8.203 + 0 - 2.239 = 14.167$ kNm

$M_{BC} = M_{FBC} + 2M'_{BC} + M'_{CB} + M''_{BC} = -25 + 2 \times 10.272 - 9.713 + 0 = -14.169$ kNm

$M_{CB} = M_{FCB} + 2M'_{CB} + M'_{BC} + M''_{CB} = 25 - 2 \times 9.713 + 10.272 = 15.846$ kNm

$M_{CD} = M_{FCD} + 2M'_{CD} + M'_{DC} + M''_{CD} = 0 - 2 \times 7.268 + 0 - 1.310 = -15.846$ kNm

$M_{DC} = 0$

PROBLEMS

1. Solve the frame in example 3.16 by Kani's method (and compare the results there in).
2. Analyse the non-sway frame shown in example 3.18 by Kani's method.
3. Analyse the frame in example 2.14 by Kani's method.
4. Analyse the frame in problem 10 under chapter 2, (slope deflection method) by Kani's method.

Table 1. Flexibility matrices—Standard cases.

S.No.	*Co-ordinates*	[α]
1.	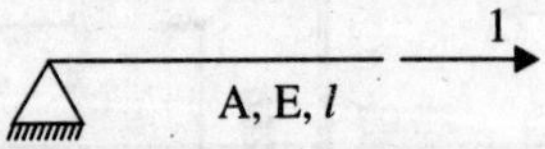	$\dfrac{l}{AE}$
2.	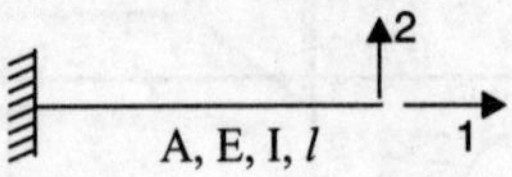	$\begin{bmatrix} \dfrac{l^3}{3EI} & 0 \\ 0 & \dfrac{l}{AE} \end{bmatrix}$
3.	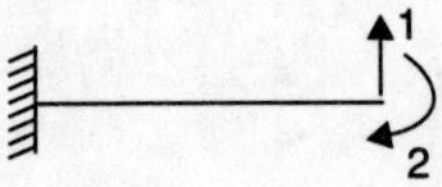	$\begin{bmatrix} \dfrac{l^3}{3EI} & \dfrac{l^2}{2EI} \\ \dfrac{l^2}{2EI} & \dfrac{l}{EI} \end{bmatrix}$
4.	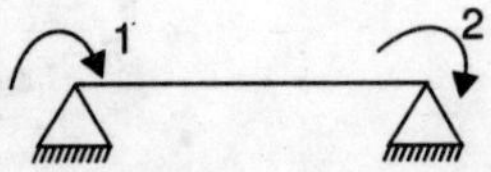	$\dfrac{l}{6EI}\begin{bmatrix} 2 & -1 \\ -1 & 2 \end{bmatrix}$
5.	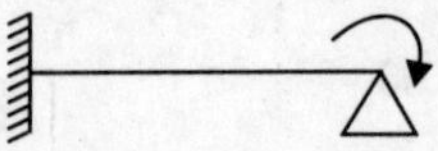	$\dfrac{l}{4EI}$

Table 2. Stiffness matrices—Standard cases.

S.No.	*Co-ordinates*	**[*k*]**
1.	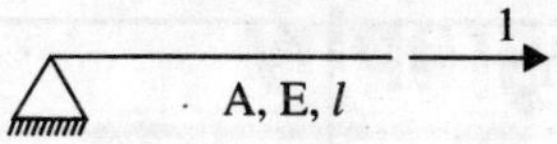	$\frac{AE}{l}$
2.	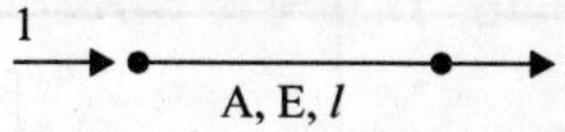	$\frac{AE}{l}\begin{bmatrix} 1 & -1 \\ -1 & 1 \end{bmatrix}$
3.	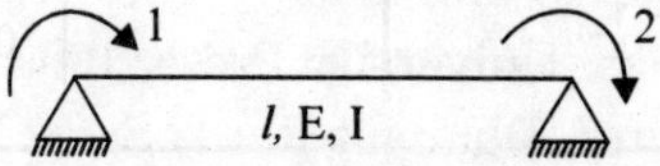	$\frac{EI}{l}\begin{bmatrix} 4 & 2 \\ 2 & 4 \end{bmatrix}$
4.	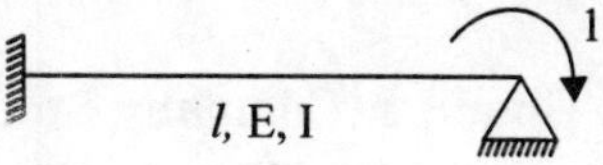	$\frac{4EI}{l}$
5.	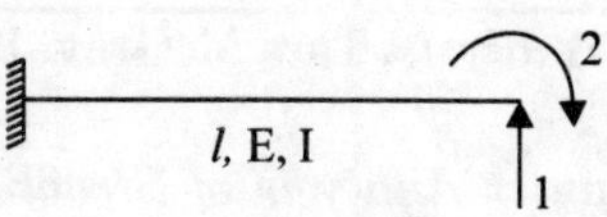	$\frac{EI}{l}\begin{bmatrix} \frac{12}{l^2} & \frac{6}{l} \\ \frac{6}{l} & 4 \end{bmatrix}$
6.	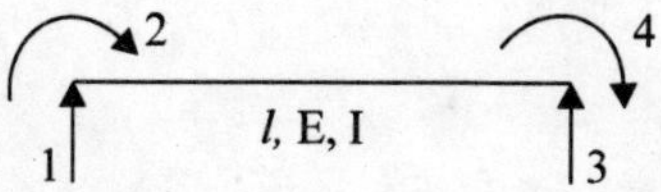	$\frac{EI}{l}\begin{bmatrix} \frac{12}{l^2} & \frac{-6}{l} & \frac{-12}{l^2} & \frac{-6}{l} \\ \frac{-6}{l} & 4 & \frac{6}{l} & 2 \\ \frac{-12}{l} & \frac{6}{l} & \frac{12}{l^2} & \frac{6}{l} \\ \frac{-6}{l} & 2 & \frac{6}{l} & 4 \end{bmatrix}$

Bibliography

1. Punmia, B.C., Ashok Jain and Arun Jain, *Theory of Structures*, Laxmi Publications, New Delhi, 1992.
2. Gupta, S.P., Pandit, G.S., Gupta, R., *Theory of Structures*, Vol. I & II, Tata McGraw-Hill Publishing Company Limited, New Delhi, 1999.
3. Prakash Rao, D.S., *Structural Analysis,* University Press, 1996.
4. Ramamrutham, S., *Theory of Structures*, Dhanpat Rai & Sons, Delhi, 1986.
5. Reddy, C.S., *Basic Structural Analysis*, Tata McGraw-Hill Publishing Company Limited, New Delhi, 1997.
6. Khurmi, R.S., *Theory of Structures*, S. Chand & Company Ltd., New Delhi, 1998.
7. Manickaselvam, V.K., *Elements of Matrix and Stability Analysis of Structures*, Khanna Publishers, Delhi, 1999.
8. Negi, L.S. and Jangid, R.S., *Structural Analysis*, Tata McGraw-Hill Publishing Company Limited, New Delhi, 1997.
9. Rubinstein, Mokshe, F., *Matrix Computer Analysis of Structures*, Prentice-Hall Inc., Englewood Cliffs, N.J., 1966.
10. Vazirani, V.N., Ratwani, M.M., *Analysis of Structures*, Khanna Publishers, Delhi, 1976.
11. Lecture Notes of the Authors.
12. Question papers from various universities.

INDEX